Advances in Intelligent Systems and Computing

Volume 920

Series editor

Janusz Kacprzyk, Systems Research Institute, Polish Academy of Sciences,
Warsaw, Poland
e-mail: kacprzyk@ibspan.waw.pl

The series "Advances in Intelligent Systems and Computing" contains publications on theory, applications, and design methods of Intelligent Systems and Intelligent Computing. Virtually all disciplines such as engineering, natural sciences, computer and information science, ICT, economics, business, e-commerce, environment, healthcare, life science are covered. The list of topics spans all the areas of modern intelligent systems and computing such as: computational intelligence, soft computing including neural networks, fuzzy systems, evolutionary computing and the fusion of these paradigms, social intelligence, ambient intelligence, computational neuroscience, artificial life, virtual worlds and society, cognitive science and systems, Perception and Vision, DNA and immune based systems, self-organizing and adaptive systems, e-Learning and teaching, human-centered and human-centric computing, recommender systems, intelligent control, robotics and mechatronics including human-machine teaming, knowledge-based paradigms, learning paradigms, machine ethics, intelligent data analysis, knowledge management, intelligent agents, intelligent decision making and support, intelligent network security, trust management, interactive entertainment, Web intelligence and multimedia.

The publications within "Advances in Intelligent Systems and Computing" are primarily proceedings of important conferences, symposia and congresses. They cover significant recent developments in the field, both of a foundational and applicable character. An important characteristic feature of the series is the short publication time and world-wide distribution. This permits a rapid and broad dissemination of research results.

More information about this series at http://www.springer.com/series/11156

Roman Szewczyk · Cezary Zieliński ·
Małgorzata Kaliczyńska

Editors

Automation 2019

Progress in Automation, Robotics and Measurement Techniques

 Springer

Editors
Roman Szewczyk
Industrial Research Institute for Automation
and Measurements PIAP Al
Warsaw, Poland

Cezary Zieliński
Industrial Research Institute for Automation
and Measurements PIAP Al
Warsaw, Poland

Małgorzata Kaliczyńska
Industrial Research Institute for Automation
and Measurements PIAP Al
Warsaw, Poland

ISSN 2194-5357   ISSN 2194-5365   (electronic)
Advances in Intelligent Systems and Computing
ISBN 978-3-030-13272-9   ISBN 978-3-030-13273-6   (eBook)
https://doi.org/10.1007/978-3-030-13273-6

Library of Congress Control Number: 2019931951

This Springer imprint is published by the registered company Springer Nature Switzerland AG
The registered company address is: Gewerbestrasse 11, 6330 Cham, Switzerland

Foreword

Automation, robotics and measurement techniques are the research fields underlying the design and implementation of cyber-physical systems. The main requirement imposed on such systems is their productivity, i.e. producing more using fewer resources. Currently, extra requirements are being imposed on cyber-physical systems, especially regarding their communication capabilities. There is an expectation that they will become components of the Internet of things and fulfil the principles of Industry 4.0. As a result, we can expect that the future cyber-physical systems will be the main drivers of the global industrial civilization. However, the only way to achieve this goal is synergetic optimization of all industrial system elements and to utilize artificial intelligence and machine learning. That is why engineers are aspiring to create self-optimized, adaptive cyber-physical systems. To that end further intensive research of component technologies is a must.

To follow radical changes, both scientists and engineers have to face the challenge of interdisciplinary approach directed at the development of cyber-physical systems. This approach encompasses interdisciplinary theoretical knowledge, numerical modelling and simulation as well as application of artificial intelligence techniques. Both software and physical devices are composed into systems that will increase production efficiency and resource savings.

This book contains papers dealing with three major aspects of creating cyber-physical systems: control, being the foundation of automation, robots, being both such systems by themselves and components of such systems, as well as permeating measurement techniques underlying perception. We strongly believe that the solutions and guidelines presented in this book will be useful to researchers and practitioners investigating, designing and implementing cyber-physical systems.

December 2018

Roman Szewczyk
Cezary Zieliński
Małgorzata Kaliczyńska

Contents

Robotics

About the Editors

Prof. Roman Szewczyk received both his PhD and DSc in the field of mechatronics. He is specializing in the modelling of properties of magnetic materials as well as in sensors and sensor interfacing, in particular magnetic sensors for security applications. He is leading the development of a sensing unit for a mobile robot developed for the Polish Police Central Forensic Laboratory and of methods of non-destructive testing based on the magnetoelastic effect. He was involved in over 10 European Union-funded research projects within the FP6 and FP7 as well as projects financed by the European Defence Organization. Moreover, he was leading two regional and national scale technological foresight projects and was active in the organization and implementation of technological transfer between companies and research institutes. He is Secretary for Scientific Affairs in the Industrial Research Institute for Automation and Measurements PIAP. He is also Associate Professor at the Faculty of Mechatronics, Warsaw University of Technology, and a Vice-Chairman of the Academy of Young Researchers of the Polish Academy of Sciences.

Prof. Cezary Zieliński received his M.Sc./Eng. in control in 1982, Ph.D. in control and robotics in 1988, the D.Sc. (habilitation) in control and robotics in 1996, all from the Faculty of Electronics and Information Technology, Warsaw University of Technology, Warsaw, Poland, and Full Professorship in 2012. Currently, he is Full Professor both in the Industrial Research Institute for Automation and Measurement PIAP and the Warsaw University of Technology, where he is Director of the Institute of Control and Computation Engineering. Since 2007, he has been Member of the Committee for Automatic Control and Robotics, the Polish Academy of Sciences. He is Head of the Robotics Group in the Institute of Control and Computation Engineering working on robot control and programming methods. His research interests focus on robotics in general and in particular include: robot programming methods, formal approach to the specification of architectures of multi-effector and multi-receptor systems, robot control, and design of digital circuits. He is the author/co-author of over 200 conference and journal papers as well as books concerned with the above-mentioned research subjects.

Małgorzata Kaliczyńska Ph.D. received her M.Sc. Eng. in cybernetics from the Faculty of Electronics, Wrocław University of Technology, and her Ph.D. in the field of fluid mechanics from the Faculty of Mechanical and Power Engineering in this same university. Now she is Assistant Professor in the Industrial Research Institute for Automation and Measurement PIAP and Editor of the scientific and technological magazine "Measurements Automation Robotics". Her areas of research interest include distributed control systems, Internet of things, Industry 4.0, information retrieval and Webometrics.

Control and Automation

Automatic Control and Feedback Loops in Biology and Medicine

Jaroslaw Smieja$^{(\boxtimes)}$

Institute of Automatic Control, Silesian University of Technology,
Akademicka 16, 44-100 Gliwice, Poland
Jaroslaw.Smieja@polsl.pl

Abstract. Biological systems, from individual cells to cells populations, to tissue and organs to whole organisms, are equipped in multiple positive and negative feedback control mechanisms. Knowledge of these mechanisms is crucial if we want to modify intracellular biochemical control systems (in genetic engineering), affect them during therapies, control populations metabolism or behavior (e.g. in bioreactors), or design artificial organs or devices supporting physiological processes. This work focuses on three aspects of interdisciplinary research in automatic control, biology and medicine: (i) modeling of physiological processes on a whole body level, aimed at supporting artificial organs development; (ii) using optimal control theory for designing anticancer therapy protocols and (iii) using simulation and analysis techniques for identification of complex intracellular regulatory mechanisms, aimed at expanding knowledge in this field.

Keywords: Biomedical systems · Modeling · Optimization · Systems biology

1 Introduction

Feedback control is characteristic not only in man-created technical systems but also in nature-created biological ones. Damage of the mechanisms involved in the latter may lead to ecological catastrophes (e.g., in the case of population dynamics this may be due to human actions) or disease or organ failure in a complex organism. Therefore, understanding these mechanisms is crucial for successful treatment of various diseases or design of devices supporting physiological functions.

It seems that in the field of automatic control the need for application of control theory methods in development of artificial organs or life-supporting equipment is widely recognized (see, e.g., [1–6]). In contrast, research that employs control theory or optimization theory to analyze and design treatment protocols or supports experimental work in molecular biology, seems to be underappreciated by researchers working in the field of automation, control or optimization

© Springer Nature Switzerland AG 2020
R. Szewczyk et al. (Eds.): AUTOMATION 2019, AISC 920, pp. 3–12, 2020.
https://doi.org/10.1007/978-3-030-13273-6_1

theory. However, taking into account complexity of the systems involved, without the use of advanced mathematical methods that have their origins in these fields, new developments in molecular biology or significant progress in efficacy of treatment in clinics are hardly possible.

2 Feedback Control of Physiological Systems

Of various physiological systems, the insulin-glucose system is arguably one of the most frequent subject of investigations conducted by researchers with automatic control background. This is due to both its clear description from the medical point of view and the rapidly increasing number of diabetic patients – it is estimated that one for every 11 persons in the world suffers from diabetes [7]. This section briefly describes the system and state-of-the art solutions that have been proposed and tested clinically in recent years.

The physiological control system is based on two feedback loops, with pancreas and liver acting together as a sensor, a controller and an actuator. If blood glucose level is too high, to increase its uptake by tissue cells (and to facilitate the conversion of glucose to glycogen in the liver) insulin is released by the pancreas. When the blood glucose level is too low, glucagon is released by the pancreas alpha cells to increase the rate of breakdown of glycogen into glucose in the liver, and subsequent release of glucose into blood. If insulin is not produced by the pancreas cells or the sensitivity of cells to insulin is reduced, then the cells in an organism cannot uptake the glucose and they are deprived of the main energy source. Moreover, the blood sugar is kept at a high level for prolonged time, which leads to damage of various organs.

The most common procedure for patients requiring insulin administration involves multiple daily insulin injections (MDI). Recent advances in technology changed this, with closed-loop control of blood glucose becoming a realistic solution, referred to as an artificial pancreas (Fig. 1). Existing sensors take the measurement every minute, providing information about the average every 5 min and the insulin pumps can be activated accordingly, which makes it possible to design almost continuous-time control subsystems (though, in fact, they are sampled data systems, they are widely referred to as continuous ones). There are reports about successful implementations (e.g. [8–10]) and more are under development [11], though these solutions have not been made widely available yet. A variety of algorithms and control structures closed loop control of BG have been developed, including PID controllers [12], predictive (MPC) (e.g. [13,14]), adaptive [15], systems with feedback and feedforward loops [16], run-to-run regulation [17], fuzzy-logic [18] or neural network controllers [19]. Much effort was put into identifying parameters and validating models (e.g. [20]).

Despite the existence of many different types of controllers mentioned above and proposed as the basis of the automatic glucose control system, actually only two have found their way into clinical trials (e.g., [21–23]). They are the PID [12,24,25] and MPC [14,26,27] controllers. They provide an adequate level of safety and resistance to changes in physiological parameters.

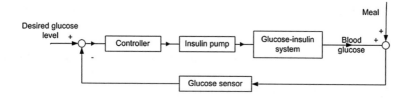

Fig. 1. A simplified block diagram of an glucose closed-loop control system

Another innovation that has been proposed recently is addition of a second control signal, associated with the administration of glucagon, aimed at elimination of hypoglycaemia events [28,29]. At the moment, however, these studies are in the initial phase and there are no such devices on the market.

It is widely known that physiological parameters may vary in a wide range between patients (e.g. [30]) and one patient (e.g. [31]), depending on body temperature or physical condition (exercise, illness, stress etc.). Taking into account increasing popularity of devices like smartwatches or smartphones with software tracking vital signals, data obtained from them could be used to improve quality of blood glucose management through automatic adjustment of controller parameters, following analysis of data obtained from those devices. The concept has already gained interest of some researchers [32,33]. However, due to the constraints imposed on control variable and delay caused by pharmacokinetics, it seems that in the case of physical exercise feedforward loop is necessary [34].

One of the most important impediments in successful mass-scale introduction of closed-loop glucose control systems are the problems with glucose measurement accuracy. Sensors used in these systems measure subcutaneous glucose, and these results are much less reliable than direct blood glucose measurements. In Europe, the acceptable quality of blood glucose monitoring systems is defined in EN ISO 15197. It takes into account the current state-of-the-art in CGM sensors, i.e., an acceptable error of 15% of the measured value for 95% of results (the required accuracy has been increased from 20% to 15% in relation to the previous version of this standard). Despite the visible progress in improving the accuracy of measurements, it is still unsatisfactory and any improvement is highly desirable. At the same time, the standard allows 5% of results to have a larger error. Comparative studies of commercially available CGM systems (see, e.g. [35]) indicate that the number of unreliable measurement data (which can be treated as so-called gross errors) from CGM systems can reach 7%–30%.

It should be also noted that patients with diabetes and their families create social networks, providing concepts for technical solutions without waiting for the launch of ready-made company solutions (www.openaps.org, www.nighscouts.pl).

3   Open Loop Control in Anticancer Treatment

Application of optimal control theory and methods in anticancer therapy planning has been the subject of research for many years (see, e.g., [36,37] and references therein). Practical applications of theoretical results obtained in this area, however, are scarce. Nevertheless, recent progress in linking various biomolecular markers to the stage of the disease as well as biomedical imaging has provided methods of monitoring cancer growth, studying cancer cell populations and that creates a new opening for mathematical modeling in this field and refining control theory methods so that they might gain acknowledgment in clinics. Nevertheless, taking into account the sampling period of measurements, one cannot attempt to create closed-loop system that could be used to control any anticancer therapy. Therefore, all models that are referred to further in this section assume an open-loop control structure and fixed control horizon (Fig. 2).

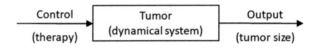

Fig. 2. Anticancer treatment form a control theoretic perspective.

There are numerous approaches to model cancer growth, depending on phenomena that the model is to capture. They include, among other, ordinary differential equations, partial differential equations and agent-based methods. Since this work is aimed at presentation of a compact view on control-related problems in biology and medicine, only the first group is discussed here. However, all models, regardless of their form, are aimed at one or more of the following goals:

– finding what conditions must be satisfied (in terms of model parameters) for the cancer population to be unstable;
– to evaluate effects of different possible therapies on cancer cells;
– to develop mathematically based therapy protocols that are the best possible, with respect to a chosen performance index.

In general, all models that describe dynamics of cell population are characterized by a positive feedback that results from cell division and negative feedback, associated with the death of cells. They can be represented by a generic state equation

$$\dot{N} = f(N, u), \tag{1}$$

where N and u denote the state vector, corresponding to patient's state and therapeutic actions, respectively. The state vector may consist of various variables representing, e.g., tumor size/volume, the number of cells of different types (both healthy and cancerous), vasculature volume, etc. Control variable u may be a scalar, in the case of a monotherapy, or a vector, in the case of combined therapy. The particular form of the function $f(N, u)$ and assumptions about the functional space, to which u belongs, may vary and depend, among other, on the type and stage of cancer and the type of therapies considered (see, e.g. [36, 38, 40] for a discussion of intricacies associated with the choice of a particular model).

In clinical terms, the therapy is aimed at maximization of tumor cure probability. This usually leads to minimization of the following functional:

$$J = \sum_i r_i N_i(T_k), \tag{2}$$

where $N_i(T_k)$ denote the number of cells of i–th type (in the i–th compartment), r_i are weight factors associated with them and T_k denotes the fixed therapy length, subject to constraints

$$0 \leq u_i \leq u_{i\_max} \leq 1 \tag{3}$$

and

$$\int_0^{T_k} u_i dt \leq \Xi_i. \tag{4}$$

The latter constraint may represent either cumulative effect of therapy on healthy tissue/cells or financial cost incurred by the therapy. Therefore, the optimization problem can be rewritten as minimization of

$$J = \sum_i r_i N_i(T_k) + \sum_i \lambda_i \int_0^{T_k} u_i(t) dt, \tag{5}$$

subject to constraints (3).

The simplest possible model assumes exponential growth of the population, as in the following example, in which $N(t)$ denotes the average number of cells in a population:

$$\dot{N}(t) = \lambda N(t), \quad N(0) = N_0 > 0, \tag{6}$$

where λ is the parameter equal to the inverse of the cell lifetime.

The control variable u, representing the effect of a single killing drug on the population, whose growth is described by (6), can be introduced in the following way:

$$\dot{N}(t) = (1 - 2u(t))\lambda N(t), \quad N(0) = N_0 > 0. \tag{7}$$

The control is bounded by $0 \leq u \leq u_{max} \leq 1$, $u = 0$ representing no drug or a completely ineffective drug and $u = u_{max}$ maximum efficiency of a killing agent, providing destruction of each cell undergoing division for $u_{max} = 1$. Such optimization problem can be solved using standard optimal control theory methods. Application of the Pontryagin's maximum principle leads to a hamiltonian that is linear with respect to the control variable. As a result, the optimal control is bang-bang, which represent alternative switchings between administering maximum tolerable dose of a killing drug and rest periods. Following application of a Legendre-Clebsch condition, one can reject singular controls as a part of optimal control trajectory.

Similar analysis can be performed for numerous types of models that describe cell-cycle specificity of the anticancer drugs, drug resistance [39], combined therapies that include antiangiogenic, chemo-, immuno- or radiotherapy [40]. Due to the particular form of the functional to be minimized and constraints imposed on control variables, optimal solutions in general are formed by bang-bang and singular arcs. However, practical application of singular arcs would require continuous measurement of state variables, which is infeasible in the systems considered. Therefore, such solutions should be used as benchmarks only, for suboptimal bang-bang solutions.

It should be noted, however, that population dynamics as well as individual cell behavior is governed by intracellular biochemical processes that take place inside each living cell. Therefore, it is important to gain insight into these processes in order to control processes at higher level of organization (populations, tissues, organs, whole body) better.

4   Feedback Loops in Intracellular Regulatory Networks

Cellular responses to external stimuli or internal events, such as DNA damage or cell division are governed by complex regulatory networks, built of cascades of biochemical processes, called signaling pathways. They involve creation or degradation of protein complexes, activation of enzymes and usually lead to activation or repression of transcription of genes specific for a given pathway. This results in production of new proteins (or their disappearance, if the genes are repressed) which may affect earlier stages of the cascade, thus creating positive or negative feedback loops. From control theory perspective, the cell can be presented as a closed-loop system, as in Fig. 3.

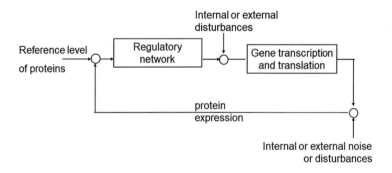

Fig. 3. A simplified block diagram of an intracellular regulatory network

The regulatory networks are created from basic building blocks that are similar in their actions to basic technical elements, such as, e.g., sensors, switches, repair elements, etc. [41]. Therefore, engineering perspective may provide a deep insight into possible dynamical behavior of these systems, their robustness, or sensitivity to external disturbances. Mathematical modeling of these systems may be used to indirectly verify biological hypotheses about their structure that are difficult to check with experimental methods [42], as well as to draw conclusions about how to influence cellular behavior to achieve therapeutic goals [43]. However, the models developed are often of a large order, with many parameters to be estimated. Therefore, sensitivity methods become increasingly popular as a tool supporting development of these models.

5   Sensitivity of Models of Biomedical Systems

Sensitivity analysis is an important tool used to determine how the change of parameters influence system behavior and originates in the field of technical system [44]. It helps to identify those parameters that have the greatest impact on the system output both in steady and transient states. Both global and local sensitivity methods have found their way into the research of biological systems. However, interpretation of their results should take into account particular features of biological systems, such as lack of measurements in absolute units [45, 46] and small number of measurements taken at discrete time points [47]. Therefore, special methods have been developed in this area, focused on analysis of system responses in frequency domain [48].

At the population level modeling, parametric sensitivity analysis is performed to find parameters that may dramatically change therapy outcome or to answer the question about which protocols are the most sensitive to parameter changes. Moreover, it may be a supporting tool in model evaluation [49] or choosing the appropriate model from several available that is the most suitable for a particular analysis [40].

6   Final Remarks

It should be noted that, though at the general level, knowledge about regulation of biochemical processes has been rapidly expanding in recent years, their exact nature and the feedback-associated phenomena are not always known. Even if they are known, it is rigorous mathematical analysis that can provide additional information on their influence on system behavior, their robustness or control laws that can be used to obtain desired system responses. Research conducted in various bio-areas (biomathematics, biomedicine, biocybernetics, bioinformatics, etc.) should aim at application of theoretical and engineering methods, first to unveil regulatory mechanisms built-in into live organisms and then to make practical use of that knowledge in biology, biotechnology and medicine.

References

1. Pietribiasi, M., et al.: Modelling transcapillary transport of fluid and proteins in hemodialysis patients. PLoS One **11**(8), e0159748 (2016). https://doi.org/10.1371/journal.pone.0159748
2. Waniewski, J., et al.: Changes of peritoneal transport parameters with time on dialysis: assessment with sequential peritoneal equilibration test. Int. J. Artif. Organs **40**(11), 595–601 (2017). https://doi.org/10.5301/ijao.5000622
3. Rodbard, D.: Continuous glucose monitoring: a review of recent studies demonstrating improved glycemic outcomes. Diabetes Technol. Ther. **19**(S3), S25–S37 (2017). https://doi.org/10.1089/dia.2017.0035
4. Aleppo, G., Webb, K.: Continuous glucose monitoring integration in clinical practice: a stepped guide to data review and interpretation. J. Diabetes Sci. Technol. **19** (2018). https://doi.org/10.1177/1932296818813581
5. Kim, J.H., Cowger, J.A., Shah, P.: The evolution of mechanical circulatory support. Cardiol. Clin. **36**(4), 443–449 (2018). https://doi.org/10.1016/j.ccl.2018.06.011
6. Branson, R.D.: Automation of mechanical ventilation. Crit. Care Clin. **34**(3), 383–394 (2018). https://doi.org/10.1016/j.ccc.2018.03.012
7. Ogurtsova, K., da Rocha Fernandes, J.D., Huang, Y., Linnenkamp, U., Guariguata, L., Cho, N.H., Cavan, D., Shaw, J.E., Makaroff, L.E.: IDF diabetes atlas: global estimates for the prevalence of diabetes for 2015 and 2040. Diabetes Res. Clin. Pract. **128**, 40–50 (2017). https://doi.org/10.1016/j.diabres.2017.03.024
8. Anderson, S.M., et al.: Multinational home use of closed-loop control is safe and effective. Diabetes Care **39**(7), 1143–1150 (2016). https://doi.org/10.2337/dc15-2468
9. Buckingham, B., Ly, T.: Closed-loop control in type 1 diabetes. Lancet Diabetes Endocrinol. **4**(3), 191–193 (2016). https://doi.org/10.1016/S2213-8587(16)00015-2
10. DeJournett, L., DeJournett, J.: In silico testing of an artificial-intelligence-based artificial pancreas designed for use in the intensive care unit setting. J. Diabetes Sci. Technol. **10**(6), 1360–1371 (2016)
11. Trevitt, S., Simpson, S., Wood, A.: Artificial pancreas device systems for the closed-loop control of type 1 diabetes: what systems are in development? J. Diabetes Sci. Technol. **10**(3), 714–23 (2016). https://doi.org/10.1177/1932296815617968
12. Ly, T.T., et al.: Automated overnight closed-loop control using a proportional-integral-derivative algorithm with insulin feedback in children and adolescents with type 1 diabetes at diabetes camp. Diabetes Technol. Ther. **18**(6), 377–384 (2016). https://doi.org/10.1089/dia.2015.0431
13. Hovorka, R., et al.: Nonlinear model predictive control of glucose concentration in subjects with type 1 diabetes. Physiol. Meas. **25**, 905–920 (2004)
14. Wang, Y., Xie, H., Jiang, X., Liu, B.: Intelligent closed-loop insulin delivery systems for ICU patients. IEEE J. Biomed. Health Inform. **18**(1), 290–299 (2014). https://doi.org/10.1109/JBHI.2013.2269699
15. Eren-Oruklu, M., Cinar, A., Quinnb, L., Smith, D.: Adaptive control strategy for regulation of blood glucose levels in patients with type 1 diabetes. J. Proc. Control **19**, 1333–1346 (2009)
16. Marchetti, G., Barolo, M., Jovanovic, L., Zisser, H., Seborg, D.E.: A feedforward-feedback glucose control strategy for type 1 diabetes mellitus. J. Proc. Control **18**(2), 149–162 (2008)
17. Palerm, C.C., Zisser, H., Jovanovic, L., Doyle, F.J.: A run-to-run control strategy to adjust basal insulin infusion rates in type 1 diabetes. J. Proc. Control **18**(3–4), 258–265 (2008)

18. Fereydouneyan, F., Zare, A., Mehrshad, N.: Using a fuzzy controller optimized by a genetic algorithm to regulate blood glucose level in type 1 diabetes. J. Med. Eng. Technol. **35**(5), 224–230 (2011). https://doi.org/10.3109/03091902.2011.569050

19. Fernandez de Canete, J., Gonzalez-Perez, S., Ramos-Diaz, J.C.: Artificial neural networks for closed loop control of in silico and ad hoc type 1 diabetes. Comput. Methods Programs Biomed. **106**(1), 55–66 (2012). https://doi.org/10.1016/j.cmpb.2011.11.006

20. Rahaghi, F.N., Gough, D.A.: Blood glucose dynamics. Diabetes Technol. Ther. **10**(2), 81–94 (2008). https://doi.org/10.1089/dia.2007.0256

21. Cameron, F.M., et al.: Closed-loop control without meal announcement in type 1 diabetes. Diabetes Technol. Ther. **19**(9), 527–532 (2017). https://doi.org/10.1089/dia.2017.0078

22. Bally, L., et al.: Day-and-night glycaemic control with closed-loop insulin delivery versus conventional insulin pump therapy in free-living adults with well controlled type 1 diabetes: an open-label, randomised, crossover study. Lancet Diabetes Endocrinol. **5**(4), 261–270 (2017). https://doi.org/10.1016/S2213-8587(17)30001-3

23. Thabit, H., et al.: Closed-loop insulin delivery in inpatients with type 2 diabetes: a randomised, parallel-group trial. Lancet Diabetes Endocrinol. **5**(2), 117–124 (2017). https://doi.org/10.1016/S2213-8587(16)30280-7

24. Ly, T.T., et al.: Automated hybrid closed-loop control with a proportional-integral-derivative based system in adolescents and adults with type 1 diabetes: individualizing settings for optimal performance. Pediatr. Diabetes **18**(5), 348–355 (2017). https://doi.org/10.1111/pedi.12399

25. Zavitsanou, S., Mantalaris, A., Georgiadis, M.C., Pistikopoulos, E.N.: In silico closed-loop control validation studies for optimal insulin delivery in type 1 diabetes. IEEE Trans. Biomed. Eng. **62**(10), 2369–2378 (2015). https://doi.org/10.1109/TBME.2015.2427991

26. Peyser, T., Dassau, E., Breton, M., Skyler, J.S.: The artificial pancreas: current status and future prospects in the management of diabetes. Ann. N. Y. Acad. Sci. **1311**, 102–123 (2014). https://doi.org/10.1111/nyas.12431

27. Pinsker, J.E., et al.: Randomized crossover comparison of personalized MPC and PID control algorithms for the artificial pancreas. Diabetes Care **39**(7), 1135–1142 (2016). https://doi.org/10.2337/dc15-2344

28. Christiansen, S.C., et al.: A review of the current challenges associated with the development of an artificial pancreas by a double subcutaneous approach. Diabetes Ther. **8**(3), 489–506 (2017). https://doi.org/10.1007/s13300-017-0263-6

29. Blauw, H., Keith-Hynes, P., Koops, R., DeVries, J.H.: A review of safety and design requirements of the artificial pancreas. Ann. Biomed. Eng. **44**(11), 3158–3172 (2016)

30. Akhlaghi, F., Matson, K.L., Mohammadpour, A.H., Kelly, M., Karimani, A.: Clinical pharmacokinetics and pharmacodynamics of antihyperglycemic medications in children and adolescents with type 2 diabetes mellitus. Clin Pharmacokinet. **56**(6), 561–571 (2017). https://doi.org/10.1007/s40262-016-0472-6

31. Zarkovic, M., et al.: Variability of HOMA and QUICKI insulin sensitivity indices. Scand. J. Clin. Lab. Investig. **77**(4), 295–297 (2017). https://doi.org/10.1080/00365513.2017.1306878

32. Dadiani, V., et al.: Physical activity capture technology with potential for incorporation into closed-loop control for type 1 diabetes. J. Diabetes Sci. Technol. **9**(6), 1208–1216 (2015). https://doi.org/10.1177/1932296815609949

33. Ben Brahim, N., Place, J., Renard, E., Breton, M.D.: Identification of main factors explaining glucose dynamics during and immediately after moderate exercise in patients with type 1 diabetes. J. Diabetes Sci. Technol. **9**(6), 1185–1191 (2015). https://doi.org/10.1177/1932296815607864
34. Smieja, J., Galuszka, A.: Rule-based PID control of blood glucose level. In: Automatyzacja Procesów Dyskretnych. Teoria i zastosowania. t.II, pp. 223–232 (2018)
35. Kovatchev, B., Anderson, S., Heinemann, L., Clarke, W.: Comparison of the numerical and clinical accuracy of four continuous glucose monitors. Diabetes Care **31**(6), 1160–1164 (2008). https://doi.org/10.2337/dc07-2401
36. Swierniak, A., Kimmel, M., Smieja, J., Puszynski, K., Psiuk-Maksymowicz, K.: System Engineering Approach to Planning Anticancer Therapies. Springer (2016). https://doi.org/10.1007/978-3-319-28095-0
37. Schaettler, H., Ledzewicz, U.: Optimal Control for Mathematical Models of Cancer Therapies. An Application of Geometric Methods. Springer, New York (2015). https://doi.org/10.1007/978-1-4939-2972-6
38. Swierniak, A., Kimmel, M., Smieja, J.: Mathematical modeling as a tool for planning anticancer therapy. Eur. J. Pharmacol. **625**(1–3), 108–121 (2009). https://doi.org/10.1016/j.ejphar.2009.08.041
39. Swierniak, A., Polanski, A., Smieja, J.: Modelling growth of drug resistant cancer populations as the system with positive feedback. Math. Comput. Model. **37**(11), 1245–1252 (2003)
40. Dolbniak, M., Smieja, J., Swierniak, A.: Structural sensitivity of control models arising in combined chemo-radiotherapy. In: Proceedings of the MMAR Conference, pp. 339–344 (2018)
41. Tyson, J.J., Novak, B.: Functional motifs in biochemical reaction networks. Ann. Rev. Phys. Chem. **61**, 219–240 (2010)
42. Smieja, J.: Model based analysis of signaling pathways. Int. J. Appl. Math. Comp. Sci. **18**(2), 139–145 (2008)
43. Kardynska, M., et al.: Quantitative analysis reveals crosstalk mechanisms of heat shock-induced attenuation of NF-κB signaling at the single cell level. Plos Comp. Biol. **14**(4), e1006130 (2018). https://doi.org/10.1371/journal.pcbi.1006130
44. Cruz, J.J.: Feedback Systems. McGraw-Hill, New York (1972)
45. Puszynski, K., Lachor, P., Kardynska, M., Smieja, J.: Sensitivity analysis of deterministic signaling pathways models. Bull. Pol. Acad. Sci. **60**(3), 471–479 (2012)
46. Smieja, J., Kardynska, M., Jamroz, A.: The meaning of sensitivity functions in signaling pathways analysis. Discret. Contin. Dyn. Syst. Ser. B **10**(8), 2697–2707 (2014). https://doi.org/10.3934/dcdsb.2014.19.2697
47. Kardyńska, M., Smieja, J.: Sensitivity analysis of signaling pathway models based on discrete-time measurements. Arch. Control Sci. **27**(2), 239–250 (2017). https://doi.org/10.1515/acsc-2017-0015
48. Kardynska, M., Smieja, J.: Sensitivity analysis of signaling pathways in the frequency domain. In: Advances in Intelligent Systems and Computing, vol. 472, pp. 2194–5357 (2016). https://doi.org/10.1007/978-3-319-39904-1_25
49. Dolbniak, M., Kardynska, M., Smieja, J.: Sensitivity of combined chemo-and antiangiogenic therapy results in different models describing cancer growth. Discret. Contin. Dyn. Syst. Ser. B **23**, 145–160 (2018). https://doi.org/10.3934/dcdsb.2018009

Normal Fractional Positive Linear Systems and Electrical Circuits

Tadeusz Kaczorek$^{(\boxtimes)}$

Białystok University of Technology, Wiejska 45D, 15-351 Białystok, Poland
kaczorek@ee.pw.edu.pl

Abstract. The notion of normal fractional positive electrical circuits is introduced and some their specific properties are investigated. New state matrices of fractional positive linear systems and electrical circuits are proposed and their properties are analyzed. The zeros and poles cancellation in the transfer functions of the fractional positive systems is discussed. It is shown that the fractional positive electrical circuits with diagonal state matrices are normal for all values of resistances, inductances and capacitances.

Keywords: Normal · Fractional · Positive · Linear · System · Electrical circuit

1 Introduction

A dynamical system is called positive if its trajectory starting from any nonnegative initial state remains forever in the positive orthant for all nonnegative inputs. An overview of state of the art in positive systems theory is given in the monographs [3, 13]. Variety of models having positive behavior can be found in engineering, economics, social sciences, biology and medicine, etc.

Mathematical fundamentals of the fractional calculus are given in the monographs [30–33]. Some selected problems of fractional systems theory are analyzed in [17, 19–21, 34–36] and of the fractional positive electrical circuits in [15, 18, 26].

The notions of controllability and observability have been introduced by Kalman in [28] and they are the basic concepts of the modern control theory [1, 2, 4, 5, 7, 8, 10, 11, 27, 37]. The controllability, reachability and observability of linear systems and electrical circuits have been investigated in [9, 14, 18, 19, 29].

The specific duality and stability of positive electrical circuits have been analyzed in [22] and positive systems and electrical circuits with inverse state matrices in [16]. The reduction of linear electrical circuits with complex eigenvalues to linear electrical circuits with real eigenvalues has been considered in [25].

Standard and positive electrical circuits with zero transfer matrices have been investigated in [23]. The asymptotic stability of positive standard and fractional linear systems has been addressed in [6, 13, 26].

The notion of normal positive linear electrical circuits has been introduced in [12].

In this paper the normal fractional positive linear systems and electrical circuits are investigated.

© Springer Nature Switzerland AG 2020
R. Szewczyk et al. (Eds.): AUTOMATION 2019, AISC 920, pp. 13–26, 2020.
https://doi.org/10.1007/978-3-030-13273-6_2

The paper is organized as follows. In Sect. 2 some preliminaries concerning the fractional and positive linear continuous-time systems are recalled. The transfer matrix of fractional positive systems with some specific forms of the state matrix is addressed in Sect. 3. The zeros and poles cancellation in the transfer functions is discussed in Sect. 4. The normal fractional positive linear systems are investigated in Sect. 5 and the normal fractional positive linear electrical circuits in Sect. 6. Concluding remarks are given in Sect. 7.

The following notation will be used: $\Re$ – the set of real numbers, $\Re^{n\times m}$ – the set of $n \times m$ real matrices, $\Re^{n\times m}_+$ – the set of $n \times m$ real matrices with nonnegative entries and $\Re^n_+ = \Re^{n\times 1}_+$, M_n – the set of $n \times n$ Metzler matrices (real matrices with nonnegative off-diagonal entries), I_n – the $n \times n$ identity matrix.

2 Preliminaries

Consider the fractional linear system

$$\frac{d^\alpha x(t)}{dt^\alpha} = Ax(t) + Bu(t), 0 < \alpha < 1 \tag{2.1a}$$

$$y(t) = Cx(t), \tag{2.1b}$$

where

$$_0D^\alpha_t f(t) = \frac{d^\alpha f(t)}{dt^\alpha} = \frac{1}{\Gamma(1-\alpha)} \int_0^t \frac{\dot{f}(\tau)}{(t-\tau)^\alpha} d\tau \tag{2.1c}$$

is Caputo definition of the fractional derivative of α order, $\dot{f}(\tau) = \frac{df(\tau)}{d\tau}$, $\Gamma(x) = \int_0^\infty t^{x-1} e^{-t} dt$, $\mathrm{Re}(x) > 0$ is the Euler gamma function, $x(t) \in \Re^n$, $u(t) \in \Re^m$, $y(t) \in \Re^p$ are the state, input and output vectors and $A \in \Re^{n\times n}$, $B \in \Re^{n\times m}$, $C \in \Re^{p\times n}$.

Definition 2.1. [21] The system (2.1a, 2.1b and 2.1c) is called (internally) positive if $x(t) \in \Re^n_+$ and $y(t) \in \Re^p_+$, $t \geq 0$ for any initial conditions $x_0 \in \Re^n_+$ and all inputs $u(t) \in \Re^m_+$, $t \geq 0$.

Theorem 2.1. [21] The system (2.1a, 2.1b and 2.1c) is positive if and only if

$$A \in M_n, B \in \Re^{n\times m}_+, C \in \Re^{p\times n}_+. \tag{2.2}$$

Definition 2.2. [21] The positive system (2.1a, 2.1b and 2.1c) for $u(t) = 0$ is called asymptotically stable (Hurwitz) if

$$\lim_{t\to\infty} x(t) = 0 \text{ for all } x_0 \in \Re^n_+. \tag{2.3}$$

Theorem 2.2. [21] The positive linear system (2.1a, 2.1b and 2.1c) for $u(t) = 0$ is asymptotically stable (Hurwitz) if and only if all coefficients of the characteristic polynomial

$$p_n(\lambda) = \det[I_n \lambda - A] = \lambda^n + a_{n-1}\lambda^{n-1} + \ldots + a_1\lambda + a_0, \quad \lambda = s^a \qquad (2.4)$$

are positive, i.e. $a_k > 0$ for $k = 0, 1, \ldots, n - 1$.

Consider the positive fractional system (2.1a, 2.1b and 2.1c) with the matrix A of the form

$$A_1 = \begin{bmatrix} \lambda_1 & a_1 & 0 & \cdots & 0 & 0 \\ 0 & \lambda_2 & a_2 & \cdots & 0 & 0 \\ \vdots & \vdots & \vdots & \ddots & \vdots & \vdots \\ 0 & 0 & 0 & \cdots & \lambda_{n-1} & a_{n-1} \\ 0 & 0 & 0 & \cdots & 0 & \lambda_n \end{bmatrix} \quad \text{or} \quad A_2 = \begin{bmatrix} \lambda_1 & 0 & 0 & \cdots & 0 & 0 \\ a_1 & \lambda_2 & 0 & \cdots & 0 & 0 \\ 0 & a_2 & \lambda_3 & \ddots & 0 & 0 \\ \vdots & \vdots & \vdots & \vdots & \vdots & \vdots \\ 0 & 0 & 0 & \cdots & a_{n-1} & \lambda_n \end{bmatrix}, \qquad (2.5)$$

$$a_k > 0, \, k = 1, \ldots, n - 1.$$

Theorem 2.3. [21] The positive system with (2.5) is asymptotically stable if and only if

$$\mathrm{Re}\lambda_k < 0 \, \text{for} \, k = 1, \ldots, n. \qquad (2.6)$$

3   Transfer Matrix of Fractional Positive Linear Systems

Using the Laplace transform $(\mathcal{L})$ to (2.1a, 2.1b and 2.1c) and taking into account that

$$\mathcal{L}\left[\frac{d^\alpha x(t)}{dt^\alpha}\right] = s^\alpha X(s), \mathcal{L}[x(t)] = \int_0^t x(t)e^{-st}dt, \, x(0) = 0 \qquad (3.1)$$

it is easy to show that the transfer matrix of the system has the form

$$T(\lambda) = C[I_n \lambda - A]^{-1}B \in \Re^{p \times m}(\lambda), \, \lambda = s^\alpha, \qquad (3.2)$$

where $\Re^{p \times m}(\lambda)$ is the set of $p \times m$ rational matrices in λ.

Theorem 3.1. If the matrices (2.5) are asymptotically stable (Hurwitz) and $B \in \Re_+^{n \times m}$, $C \in \Re_+^{p \times n}$ then all coefficients of the transfer matrices

$$T_1(\lambda) = C[I_n\lambda - A_1]^{-1}B, \, T_2(\lambda) = C[I_n\lambda - A_2]^{-1}B \qquad (3.3)$$

are nonnegative.

Proof. If the matrix A_1 is Hurwitz and $a_k > 0$, $k = 1, \ldots, n-1$ then the entries of the inverse matrix

$$[I_n\lambda - A_1]^{-1} = \begin{bmatrix} \lambda+\lambda_1 & -a_1 & 0 & \cdots & 0 & 0 \\ 0 & \lambda+\lambda_2 & -a_2 & \cdots & 0 & 0 \\ \vdots & \vdots & \vdots & \ddots & \vdots & \vdots \\ 0 & 0 & 0 & \cdots & \lambda+\lambda_{n-1} & -a_{n-1} \\ 0 & 0 & 0 & \cdots & 0 & \lambda+\lambda_n \end{bmatrix}^{-1}$$

$$= \frac{1}{(\lambda+\lambda_1)(\lambda+\lambda_2)\ldots(\lambda+\lambda_n)} \begin{bmatrix} \alpha_{11} & \alpha_{12} & \alpha_{13} & \cdots & \alpha_{1,n-1} & \alpha_{1,n} \\ 0 & \alpha_{22} & \alpha_{23} & \cdots & \alpha_{2,n-1} & \alpha_{2,n} \\ \vdots & \vdots & \vdots & \ddots & \vdots & \vdots \\ 0 & 0 & 0 & \cdots & \alpha_{n-1,n-1} & \alpha_{n-1,n} \\ 0 & 0 & 0 & \cdots & 0 & \alpha_{n,n} \end{bmatrix},$$

$\alpha_{11} = (\lambda+\lambda_2)\ldots(\lambda+\lambda_n)$, $\alpha_{12} = a_1(\lambda+\lambda_3)\ldots(\lambda+\lambda_n)$, $\alpha_{13} = a_1a_2(\lambda+\lambda_4)\ldots(\lambda+\lambda_n)$,
$\alpha_{1,n-1} = a_1\ldots a_{n-2}(\lambda+\lambda_n)$, $\alpha_{1,n} = a_1a_2\ldots a_{n-1}$,
$\alpha_{22} = (\lambda+\lambda_1)(\lambda+\lambda_3)\ldots(\lambda+\lambda_n)$, $\alpha_{23} = a_2(\lambda+\lambda_1)(\lambda+\lambda_4)\ldots(\lambda+\lambda_n)$,
$\alpha_{2,n-1} = a_2\ldots a_{n-2}(\lambda+\lambda_1)(\lambda+\lambda_n)$, $\alpha_{2,n} = a_2\ldots a_{n-1}(\lambda+\lambda_1)(\lambda+\lambda_n)$,
$\alpha_{n-1,n-1} = (\lambda+\lambda_1)\ldots(\lambda+\lambda_{n-2})(\lambda+\lambda_n)$, $\alpha_{n-1,n} = a_{n-1}(\lambda+\lambda_1)\ldots(\lambda+\lambda_{n-2})$,
$\alpha_{n,n} = (\lambda+\lambda_1)(\lambda+\lambda_2)\ldots(\lambda+\lambda_{n-1})$

$$(3.4)$$

are rational functions of λ with nonnegative coefficients.

Therefore, if $B \in \Re_+^{n \times m}$ and $C \in \Re_+^{p \times n}$ then all coefficients of the transfer matrix $T_1(\lambda)$ are nonnegative.

The proof for $T_2(\lambda)$ is dual. □

Example 3.1. Find the transfer function of the fractional positive system (2.1a, 2.1b and 2.1c) with the matrices

$$A = A_1 = \begin{bmatrix} -1 & 2 & 0 \\ 0 & -1 & 3 \\ 0 & 0 & -2 \end{bmatrix}, B = \begin{bmatrix} 0 \\ 0 \\ 1 \end{bmatrix}, C = [1 \quad 2 \quad 3]. \tag{3.5}$$

In this case using (3.3) and (3.5) we obtain

$$T_1(\lambda) = C[I_3\lambda - A_1]^{-1}B = \begin{bmatrix} 1 & 2 & 3 \end{bmatrix} \begin{bmatrix} \lambda+1 & -2 & 0 \\ 0 & \lambda+1 & -3 \\ 0 & 0 & \lambda+2 \end{bmatrix}^{-1} \begin{bmatrix} 0 \\ 0 \\ 1 \end{bmatrix}$$

$$= \frac{1}{(\lambda+1)^2(\lambda+2)} \begin{bmatrix} 1 & 2 & 3 \end{bmatrix} \begin{bmatrix} (\lambda+1)(\lambda+2) & 2(\lambda+2) & 6 \\ 0 & (\lambda+1)(\lambda+2) & 3(\lambda+1) \\ 0 & 0 & (\lambda+1)^2 \end{bmatrix} \begin{bmatrix} 0 \\ 0 \\ 1 \end{bmatrix}$$

$$= \frac{2\lambda^2 + 7\lambda + 15}{\lambda^3 + 4\lambda^2 + 5\lambda + 2}.$$

$$(3.6)$$

Note that the matrices (3.5) satisfy the assumptions of Theorem 3.1 and all coefficients of (3.6) are nonnegative.

4   The Zeros and Poles Cancellation

Consider the SISO (single-input ($m = 1$) single-output ($p = 1$)) fractional positive linear system with A_1 given by (2.5) and

$$B_1 = \begin{bmatrix} 0 \\ \vdots \\ 0 \\ 1 \end{bmatrix} \in \mathfrak{R}^n_+, \; C_1 \in \mathfrak{R}^{1 \times n}_+. \tag{4.1}$$

It is easy to check that if $a_k > 0$ for $k = 1, \ldots, n-1$ then

$$\text{rank}\begin{bmatrix} B_1 & A_1 B_1 & \cdots & A_1^{n-1} B_1 \end{bmatrix} = n. \tag{4.2}$$

Let $z_1, z_2, \ldots, z_{n-1}$ be the zeros (the roots of $n(s) = 0$) and $p_1, p_2, \ldots, p_n$ the poles (the roots of $d(s) = 0$) of the transfer function

$$T_1(\lambda) = C_1[I_n\lambda - A_1]^{-1}B_1 = \frac{n(\lambda)}{d(\lambda)}. \tag{4.3}$$

Theorem 4.1. At least one zero of (4.3) is equal to its poles if

$$\text{rank} \begin{bmatrix} C_1 \\ C_1 A_1 \\ \vdots \\ C_1 A_1^{n-1} \end{bmatrix} < n, \tag{4.4}$$

i.e. the pair (A_1, C_1) is unobservable in classical sense.

Proof. It is well-known [10] that if (4.4) holds then in (4.3) the zeros and poles cancellation occurs and this happens only if at least one zero of (4.3) is equal to its poles. $\qquad\square$

Example 4.1. Consider the fractional positive system (2.1a, 2.1b and 2.1c) with the matrices

$$A_1 = \begin{bmatrix} -1 & 2 & 0 \\ 0 & -1 & 1 \\ 0 & 0 & -2 \end{bmatrix}, \ B_1 = \begin{bmatrix} 0 \\ 0 \\ 1 \end{bmatrix}, \ C_1 = [0 \ \ 0 \ \ 2]. \tag{4.5}$$

The transfer function for (4.5) has the form

$$T_1(\lambda) = C_1[I_3\lambda - A_1]^{-1}B_1 = [0 \ \ 0 \ \ 2] \begin{bmatrix} \lambda+1 & -2 & 0 \\ 0 & \lambda+1 & -1 \\ 0 & 0 & \lambda+2 \end{bmatrix}^{-1} \begin{bmatrix} 0 \\ 0 \\ 1 \end{bmatrix} \tag{4.6}$$

$$= \frac{2(\lambda+1)^2}{(\lambda+1)^2(\lambda+2)} = \frac{2}{(\lambda+2)}.$$

In this case we have

$$\text{rank} \begin{bmatrix} C_1 \\ C_1A_1 \\ C_1A_1^2 \end{bmatrix} = \text{rank} \begin{bmatrix} 0 & 0 & 2 \\ 0 & 0 & -4 \\ 0 & 0 & 8 \end{bmatrix} = 1 < n = 3 \tag{4.7}$$

and the dual zero $z_1 = z_2 = -1$ has been cancelled with the dual pole $p_1 = p_2 = -1$. Now let us consider the SISO fractional positive system with A_2 given by (2.5) and

$$B_2 \in \Re_+^n, \ C_2 = [0 \ \ \cdots \ \ 0 \ \ 1] \in \Re_+^{1 \times n}. \tag{4.8}$$

It is easy to check that if $a_k > 0$ for $k = 1, \ldots, n-1$ then

$$\text{rank} \begin{bmatrix} C_2 \\ C_2A_2 \\ \vdots \\ C_2A_2^{n-1} \end{bmatrix} = n. \tag{4.9}$$

Let $\bar{z}_1, \bar{z}_2, \ldots, \bar{z}_{n-1}$ be the zeros and $\bar{p}_1, \bar{p}_2, \ldots, \bar{p}_n$ the poles of the transfer function

$$T_2(\lambda) = C_2[I_n\lambda - A_2]^{-1}B_2. \tag{4.10}$$

Theorem 4.2. At least one zero of (4.10) is equal to its poles if

$$\text{rank}[\,B_2 \quad A_2 B_2 \quad \cdots \quad A_2^{n-1} B_2\,] < n. \tag{4.11}$$

Proof. The proof is dual to the proof of Theorem 4.1.

The considerations can be easily extended to MIMO (multi-input multi-output) fractional positive linear systems.

5 Normal Fractional Positive Linear Systems

Consider the transfer matrix of fractional positive linear system of the form

$$T(\lambda) = \frac{N(\lambda)}{d(\lambda)} \in \Re^{p \times m}(\lambda), \tag{5.1a}$$

where $N(\lambda) \in \Re^{p \times m}[\lambda]$ is the polynomial matrix and $d(\lambda)$ is the least common denominator of the form

$$d(\lambda) = \lambda^n + a_{n-1}\lambda^{n-1} + \ldots + a_1\lambda + a_0. \tag{5.1b}$$

Definition 5.1. The positive linear system with (5.1a and 5.1b) is called normal if every nonzero second order minor of $N(\lambda)$ is divisible (with zero remainder) by the polynomial $d(\lambda)$.

The normal systems are insensitive to the change of their parameters [12].

Definition 5.2. The state matrix A of the linear system (2.1a, 2.1b and 2.1c) is called cyclic if its minimal polynomial $\Psi(\lambda)$ is equal to its characteristic polynomial

$$\varphi(\lambda) = \det[I_n\lambda - A]. \tag{5.2}$$

The minimal polynomial $\psi(\lambda)$ is related to its characteristic polynomial $\varphi(\lambda)$ by [12]

$$\psi(\lambda) = \frac{\varphi(\lambda)}{D_{n-1}(\lambda)}, \tag{5.3}$$

where $D_{n-1}(\lambda)$ is the greatest common divisor of all $n - 1$ order minors of the matrix $[I_n\lambda - A]$.

Therefore, $\psi(\lambda) = \varphi(\lambda)$ if and only if $D_{n-1}(\lambda) = 1$.

Theorem 5.1. The matrices A_1 and A_2 defined by (2.5) are cyclic.

Proof. By Definition 5.2 the matrices A_1 and A_2 are cyclic if and only if the greatest common divisor of all $n - 1$ order minors of the matrices

$$[I_n\lambda - A_1] = \begin{bmatrix} \lambda+\lambda_1 & -a_1 & 0 & \cdots & 0 & 0 \\ 0 & \lambda+\lambda_2 & -a_2 & \cdots & 0 & 0 \\ \vdots & \vdots & \vdots & \ddots & \vdots & \vdots \\ 0 & 0 & 0 & \cdots & \lambda+\lambda_{n-1} & -a_{n-1} \\ 0 & 0 & 0 & \cdots & 0 & \lambda+\lambda_n \end{bmatrix},$$

$$[I_n\lambda - A_2] = \begin{bmatrix} \lambda+\lambda_1 & 0 & 0 & \cdots & 0 & 0 \\ -a_1 & \lambda+\lambda_2 & 0 & \cdots & 0 & 0 \\ \vdots & \vdots & \vdots & \ddots & \vdots & \vdots \\ 0 & 0 & 0 & \cdots & \lambda+\lambda_{n-1} & 0 \\ 0 & 0 & 0 & \cdots & -a_{n-1} & \lambda+\lambda_n \end{bmatrix}$$

(5.4)

are $D_{n-1}(\lambda) = 1$. It is easy to see that the minors corresponding to the first column and the n-th row of the matrix $[I_n\lambda - A_1]$ and to the first row and the n-th column of the matrix $[I_n\lambda - A_2]$ are equal to $a_1a_2\ldots a_{n-1}$. Therefore, $D_{n-1}(\lambda) = 1$ and the matrices A_1 and A_2 are cyclic. $\square$

Theorem 5.2. The fractional positive linear system with the matrices A_1 and A_2 defined by (2.5) is normal for any $B \in \Re_+^{n\times m}$ and $C \in \Re_+^{p\times n}$.

Proof. By Definition 5.1 the positive linear system with A_1 (A_2) defined by (2.5) and any $B \in \Re_+^{n\times m}$, $C \in \Re_+^{p\times n}$ is normal if every nonzero second order minor of the matrix $N(\lambda) = C[I_n\lambda - A_1]_{ad}B$ is divisible by the polynomial $\det[I_n\lambda - A_1]$.

Let $Z_{j_1j_2\ldots j_q}^{i_1i_2\ldots i_q}$ be the minor of the matrix Z with its $i_1, i_2, \ldots, i_q$ rows and $j_1,j_2, \ldots, j_q$ its columns. Then it is well-known [24] that the q-minor of the matrix $Z = PQ$ is given by

$$Z_{j_1j_2\ldots j_q}^{i_1i_2\ldots i_q} = \sum_{1 < k_1 < \ldots < k_q} P_{k_1k_2\ldots k_q}^{i_1i_2\ldots i_q} Q_{j_1j_2\ldots j_q}^{k_1k_2\ldots k_q}. \tag{5.5}$$

Note that the minors of the matrices B and C are independent of λ. Using (5.5) for the matrix $C[I_n\lambda - A_1]_{ad}B$ it is easy to see that its every nonzero second order minor is divisible by $\det[I_n\lambda - A_1]$ since by Theorem 4.1 the matrix A_1 (A_2) is cyclic. Therefore, the positive linear system with A_1 (A_2) and any $B \in \Re_+^{n\times m}$, $C \in \Re_+^{p\times n}$ is normal. $\square$

Example 5.1. Consider the positive linear system with the matrices

$$A_1 = \begin{bmatrix} -2 & a_1 & 0 \\ 0 & -2 & a_2 \\ 0 & 0 & -3 \end{bmatrix}, \ a_k > 0 \text{ for } k = 1,2, \ B = \begin{bmatrix} b_{11} & b_{12} \\ b_{21} & b_{22} \\ b_{31} & b_{32} \end{bmatrix} \in \Re_+^{3\times2},$$

$$C = \begin{bmatrix} c_{11} & c_{12} & c_{13} \\ c_{21} & c_{22} & c_{23} \end{bmatrix} \in \Re_+^{2\times3}. \tag{5.6}$$

Taking into account that

$$d(\lambda) = \det[I_3\lambda - A_1] = \begin{vmatrix} \lambda+2 & -a_1 & 0 \\ 0 & \lambda+2 & -a_2 \\ 0 & 0 & \lambda+3 \end{vmatrix} \tag{5.7}$$
$$= (\lambda+2)^2(\lambda+3) = \lambda^3 + 7\lambda^2 + 16\lambda + 12$$

and

$$[I_3\lambda - A_1]_{ad} = \begin{bmatrix} (\lambda+2)(\lambda+3) & a_1(\lambda+3) & a_1a_2 \\ 0 & (\lambda+2)(\lambda+3) & a_2(\lambda+2) \\ 0 & 0 & (\lambda+2)^2 \end{bmatrix} \tag{5.8}$$

we obtain

$$N(\lambda) = C[I_3\lambda - A_1]_{ad}B$$
$$= \begin{bmatrix} c_{11} & c_{12} & c_{13} \\ c_{21} & c_{22} & c_{23} \end{bmatrix} \begin{bmatrix} (\lambda+2)(\lambda+3) & a_1(\lambda+3) & a_1a_2 \\ 0 & (\lambda+2)(\lambda+3) & a_2(\lambda+2) \\ 0 & 0 & (\lambda+2)^2 \end{bmatrix} \begin{bmatrix} b_{11} & b_{12} \\ b_{21} & b_{22} \\ b_{31} & b_{32} \end{bmatrix}$$
$$= \begin{bmatrix} n_{11}(\lambda) & n_{12}(\lambda) \\ n_{21}(\lambda) & n_{22}(\lambda) \end{bmatrix},$$

$$\tag{5.9a}$$

where

$$n_{11}(\lambda) = b_{21}b_{31}a_2c_{12}(\lambda+2) + b_{21}a_1c_{11}(\lambda+3) + b_{21}b_{31}c_{13}(\lambda+2)^2 + b_{21}b_{11}c_{11}(\lambda+2)(\lambda+3)$$
$$+ b_{12}c_{12}(\lambda+2)(\lambda+3) + b_{21}b_{31}a_1a_2c_{11},$$
$$n_{12}(\lambda) = b_{22}b_{32}a_2c_{12}(\lambda+2) + b_{22}a_1c_{11}(\lambda+3) + b_{22}b_{32}c_{13}(\lambda+2)^2 + b_{22}b_{12}c_{11}(\lambda+2)(\lambda+3)$$
$$+ b_{22}c_{12}(\lambda+2)(\lambda+3) + b_{22}b_{32}a_1a_2c_{11},$$
$$n_{21}(\lambda) = b_{21}b_{31}a_2c_{22}(\lambda+2) + b_{21}a_1c_{21}(\lambda+3) + b_{21}b_{31}c_{23}(\lambda+2)^2 + b_{21}b_{11}c_{21}(\lambda+2)(\lambda+3)$$
$$+ b_{21}c_{22}(\lambda+2)(\lambda+3) + b_{21}b_{31}a_1a_2c_{21},$$
$$n_{22}(\lambda) = b_{22}b_{32}a_2c_{22}(\lambda+2) + b_{22}a_1c_{21}(\lambda+3) + b_{22}b_{32}c_{23}(s+2)^2 + b_{22}b_{12}c_{21}(\lambda+2)(\lambda+3)$$
$$+ b_{22}c_{22}(\lambda+2)(\lambda+3) + b_{22}b_{32}a_1a_2c_{21}.$$

$$\tag{5.9b}$$

Therefore, the fractional positive linear system with (5.6) is normal.

Note that the matrices (2.5) for $a_k = 0$, $k = 1, \ldots, n-1$ are equal and have the diagonal form

$$A_d = \text{diag}[-\lambda_1 \quad -\lambda_2 \quad \cdots \quad -\lambda_n]. \tag{5.10}$$

In this particular case Theorem 5.2 has the following form.

Theorem 5.3. The fractional positive linear system with (5.10) and any $B \in \mathfrak{R}_+^{n \times m}$, $C \in \mathfrak{R}_+^{p \times n}$ is normal.

6   Normal Fractional Positive Linear Electrical Circuits

Consider fractional linear electrical circuits composed of resistors, capacitors, coils and voltage (current) sources. As the state variables (the components of the state vector $x(t)$) we choose the voltages on the capacitors and the currents in the coils. Using Kirchhoff's laws we may describe the fractional linear circuits in transient states by the state equations

$$\frac{d^\alpha x(t)}{dt^\alpha} = Ax(t) + Bu(t),\ 0 < \alpha < 1 \tag{6.1a}$$

$$y(t) = Cx(t), \tag{6.1b}$$

where $x(t) \in \mathfrak{R}^n$, $u(t) \in \mathfrak{R}^m$, $y(t) \in \mathfrak{R}^p$ are the state, input and output vectors and $A \in \mathfrak{R}^{n \times n}$, $B \in \mathfrak{R}^{n \times m}$, $C \in \mathfrak{R}^{p \times n}$.

Definition 6.1. [26] The fractional linear electrical circuit (6.1a and 6.1b) is called (internally) positive if the state vector $x(t) \in \mathfrak{R}_+^n$ and output vector $y(t) \in \mathfrak{R}_+^p$, $t \geq 0$ for any initial conditions $x_0 \in \mathfrak{R}_+^n$ and all inputs $u(t) \in \mathfrak{R}_+^m$, $t \geq 0$.

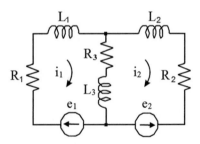

Fig. 1. Fractional electrical circuit

Theorem 6.1. [26] The fractional linear electrical circuit (6.1a and 6.1b) is positive if and only if

$$A \in M_n,\ B \in \mathfrak{R}_+^{n \times m},\ C \in \mathfrak{R}_+^{p \times n}. \tag{6.2}$$

The transfer matrix of the linear electrical circuit described by (6.1a and 6.1b) can be always written in the form (5.1a).

Definition 6.2. The fractional positive linear electrical circuit with (5.1a) is called normal if every nonzero second order minor of $N(\lambda)$ is divisible by $d(\lambda)$.

Example 6.1. Consider the fractional linear electrical circuit shown on Fig. 1 with given resistances R_k, inductances L_k, $k = 1, 2, 3$ and source voltages e_1, e_2.

Using the mesh method for the electrical circuit we obtain

$$\begin{bmatrix} L_{11} & -L_{12} \\ -L_{21} & L_{22} \end{bmatrix} \frac{d^\alpha}{dt^\alpha} \begin{bmatrix} i_1 \\ i_2 \end{bmatrix} = \begin{bmatrix} -R_{11} & R_{12} \\ R_{21} & -R_{22} \end{bmatrix} \begin{bmatrix} i_1 \\ i_2 \end{bmatrix} + \begin{bmatrix} e_1 \\ e_2 \end{bmatrix}, \quad 0 < \alpha < 1 \tag{6.3a}$$

where $i_1 = i_1(t)$, $i_2 = i_2(t)$ are the mesh currents and

$$R_{11} = R_1 + R_3, \ R_{12} = R_{21} = R_3, \ R_{22} = R_2 + R_3, \ L_{11} = L_1 + L_3,$$
$$L_{12} = L_{21} = L_3, \ L_{22} = L_2 + L_3. \tag{6.3b}$$

The inverse matrix

$$L^{-1} = \begin{bmatrix} L_{11} & -L_{12} \\ -L_{21} & L_{22} \end{bmatrix}^{-1} = \frac{1}{L_1(L_2 + L_3) + L_2 L_3} \begin{bmatrix} L_{22} & L_{12} \\ L_{21} & L_{11} \end{bmatrix} \tag{6.4}$$

has all positive entries.

From (6.3a) we obtain

$$\frac{d^\alpha}{dt^\alpha} \begin{bmatrix} i_1 \\ i_2 \end{bmatrix} = A \begin{bmatrix} i_1 \\ i_2 \end{bmatrix} + B \begin{bmatrix} e_1 \\ e_2 \end{bmatrix}, \tag{6.5a}$$

where

$$A = L^{-1} \begin{bmatrix} -R_{11} & R_{12} \\ R_{21} & -R_{22} \end{bmatrix}$$
$$= \frac{1}{L_1(L_2 + L_3) + L_2 L_3} \begin{bmatrix} -L_2(R_1 + R_3) - L_3 R_1 & L_2 R_3 - L_3 R_2 \\ L_1 R_3 - L_3 R_1 & -L_1(R_2 + R_3) - L_3 R_2 \end{bmatrix}, \tag{6.5b}$$
$$B = L^{-1} \in \Re_+^{2\times2}.$$

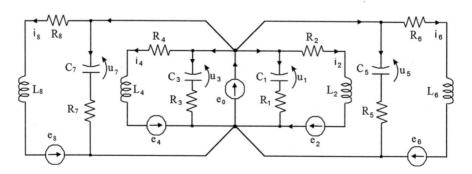

Fig. 2. Fractional positive electrical circuit

Note that if

$$L_1 R_3 = L_3 R_1 \text{ and } L_2 R_3 - L_3 R_2 > 0 \tag{6.6a}$$

then the matrix A has the form of the matrix A_1 defined by (2.5) and for

$$L_2 R_3 = L_3 R_2 \tag{6.6b}$$

the form of the matrix A_2. In both cases the fractional electrical circuit is positive.

These considerations can be easily extended to n-mesh fractional linear electrical circuits.

Following [26] let us consider the fractional linear electrical circuit shown in Fig. 2 with given resistances $R_k, k = 1, \ldots, 8$, inductances L_2, L_4, L_6, L_8, capacitances C_1, C_3, C_5, C_7 and source voltages e_0, e_2, e_4, e_6, e_8.

Using Kirchhoff's laws we may write the equations

$$e_0 = u_k + R_k C_k \frac{d^\alpha u_k}{dt^\alpha}, \, k = 1, 3, 5, 7, \tag{6.7a}$$

$$e_0 + e_j = R_j i_j + L_j \frac{d^\alpha i_j}{dt^\alpha}, \, j = 2, 4, 6, 8. \tag{6.7b}$$

The equations can be written in the form

$$\frac{d^\alpha}{dt^\alpha} \begin{bmatrix} u \\ i \end{bmatrix} = A \begin{bmatrix} u \\ i \end{bmatrix} + Be, \, 0 < \alpha < 1 \tag{6.8a}$$

where

$$u = \begin{bmatrix} u_1 \\ u_3 \\ u_5 \\ u_7 \end{bmatrix}, i = \begin{bmatrix} i_2 \\ i_4 \\ i_6 \\ i_8 \end{bmatrix}, e = \begin{bmatrix} e_0 \\ e_2 \\ e_4 \\ e_6 \\ e_8 \end{bmatrix} \tag{6.8b}$$

and

$$A = diag \begin{bmatrix} -\frac{1}{R_1 C_1} & -\frac{1}{R_3 C_3} & -\frac{1}{R_5 C_5} & -\frac{1}{R_7 C_7} & -\frac{R_2}{L_2} & -\frac{R_4}{L_4} & -\frac{R_6}{L_6} & -\frac{R_8}{L_8} \end{bmatrix},$$

$$B = \begin{bmatrix} B_1 \\ B_2 \end{bmatrix}, B_1 = \begin{bmatrix} \frac{1}{R_1 C_1} & 0 & 0 & 0 & 0 \\ \frac{1}{R_3 C_3} & 0 & 0 & 0 & 0 \\ \frac{1}{R_5 C_5} & 0 & 0 & 0 & 0 \\ \frac{1}{R_7 C_7} & 0 & 0 & 0 & 0 \end{bmatrix}, B_2 = \begin{bmatrix} \frac{1}{L_2} & \frac{1}{L_2} & 0 & 0 & 0 \\ \frac{1}{L_4} & 0 & \frac{1}{L_4} & 0 & 0 \\ \frac{1}{L_6} & 0 & 0 & \frac{1}{L_6} & 0 \\ \frac{1}{L_8} & 0 & 0 & 0 & \frac{1}{L_8} \end{bmatrix}. \tag{6.8c}$$

The matrix $A \in M_8$ is diagonal and asymptotically stable and $B \in \Re_+^{8 \times 5}$. Therefore, the electrical circuit is positive for any values of the resistances, inductances and capacitances and from Theorem 5.3 we have the following important theorem.

Theorem 6.2. Positive linear electrical circuit with diagonal matrix $A \in M_n$ and $B \in \Re_+^{n \times m}$, $C \in \Re_+^{p \times n}$ is normal for any values of the resistances, inductances and capacitances.

7 Concluding Remarks

The notion of normal positive electrical circuit has been introduced and some specific properties of this class have been investigated. New state matrices of the positive linear systems and electrical circuits have been introduced and their properties have been analyzed. Specific properties of the transfer matrices of fractional positive linear systems and electrical circuits have been discussed (Theorems 3.1 and 3.2). The zeros and poles cancellation of this class of fractional systems (Theorems 4.1 and 4.2) and cyclicity of the state matrices (Theorems 5.1 and 5.2) have been analyzed. It has been shown that the positive electrical circuits with diagonal state matrices are normal for all values of their resistances, inductances and capacitances (Theorem 6.2). The considerations have been illustrated by numerical examples.

The considerations can be extended to fractional linear systems and electrical circuits with different orders and with delays.

Acknowledgment. This work was supported by National Science Centre in Poland under work No. 2017/27/B/ST7/02443.

References

1. Antsaklis, E., Michel, A.: Linear Systems. Birkhauser, Boston (2006)
2. Borawski, K.: Characteristic equations for descriptor linear electrical circuits. In: Proceedings of the 22nd International Conference on Methods and Models in Automation and Robotics, Międzyzdroje (2017)
3. Farina, L., Rinaldi, S.: Positive Linear Systems: Theory and Applications. Wiley, New York (2000)
4. Gantmacher, F.R.: The Theory of Matrices. Chelsea Pub. Comp., London (1959)
5. Kaczorek, T.: A class of positive and stable time-varying electrical circuits. Electr. Rev. **91**(5), 121–124 (2015)
6. Kaczorek, T.: Asymptotic stability of positive fractional 2D linear systems. Bull. Pol. Acad. Sci. Tech. **57**(3), 289–292 (2009)
7. Kaczorek, T.: Characteristic equations of the standard and descriptor linear electrical circuits. Poznan Univ. Technol. Acad. J.: Electr. Eng. **89**, 11–23 (2017)
8. Kaczorek, T.: Characteristic polynomials of positive and minimal-phase electrical circuits. Electr. Rev. **92**(6), 79–85 (2016)
9. Kaczorek, T.: Controllability and observability of linear electrical circuits. Electr. Rev. **87**(9a), 248–254 (2011)
10. Kaczorek, T.: Linear Control Systems: Analysis of Multivariable Systems. Wiley, New York (1992)
11. Kaczorek, T.: Minimal-phase positive electrical circuits. Electr. Rev. **92**(3), 182–189 (2016)
12. Kaczorek, T.: Normal positive electrical circuits. IET Circ. Theory Appl. **9**(5), 691–699 (2015)

13. Kaczorek, T.: Positive 1D and 2D Systems. Springer, London (2002)
14. Kaczorek, T.: Positive electrical circuits and their reachability. Arch. Electr. Eng. **60**(3), 283–301 (2011)
15. Kaczorek, T.: Positive fractional linear electrical circuits. In: Proceedings of SPIE, vol. 8903, Bellingham WA, Art. No. 3903-35
16. Kaczorek, T.: Positive linear systems and electrical circuits with inverse state matrices. Electr. Rev. **93**(11), 119–124 (2017)
17. Kaczorek, T.: Positive systems consisting of n subsystems with different fractional orders. IEEE Trans. Circ. Syst. **58**(6), 1203–1210 (2011). Regular paper
18. Kaczorek, T.: Positivity and reachability of fractional electrical circuits. Acta Mechanica et Automatica **5**(2), 42–51 (2011)
19. Kaczorek, T.: Reachability and controllability to zero tests for standard and positive fractional discrete-time systems. J. Européen des Systemes Automatisés, JESA **42**(6–8), 769–787 (2008)
20. Kaczorek, T.: Responses of standard and fractional linear systems and electrical circuits with derivatives of their inputs. Electr. Rev. **93**(6), 132–136 (2017)
21. Kaczorek, T.: Selected Problems of Fractional Systems Theory. Springer, Berlin (2011)
22. Kaczorek, T.: Specific duality and stability of positive electrical circuits. Arch. Electr. Eng. **66**(4), 663–679 (2017)
23. Kaczorek, T.: Standard and positive electrical circuits with zero transfer matrices. Poznan Univ. Technol. Acad. J.: Electr. Eng. **85**, 11–28 (2016)
24. Kaczorek, T.: Vectors and Matrices in Automation and Electrotechnics. WNT, Warsaw, (1998). (in Polish)
25. Kaczorek, T., Borawski, K.: Reduction of linear electrical circuits with complex eigenvalues to linear electrical circuits with real eigenvalues. Measur. Autom. Monit. **61**(4), 115–117 (2015)
26. Kaczorek, T., Rogowski, K.: Fractional Linear Systems and Electrical Circuits. Studies in Systems, Decision and Control, vol. 13. Springer (2015)
27. Kailath, T.: Linear Systems. Prentice Hall, Englewood Cliffs, New York (1980)
28. Kalman, R.E.: On the general theory of control systems. In: Proceedings of the 1st International Congress on Automatic Control, London, pp. 481–493 (1960)
29. Klamka, J.: Controllability of Dynamical Systems. Kluwer Academic Press, Dordrecht (1991)
30. Kilbas, A.A., Srivastava, H.M., Trujilo, J.J.: Theory and Applications of Fractional Differential Equations. North-Holland Mathematics Studies, Amsterdam (2006)
31. Ostalczyk, P.: Epitome of the Fractional Calculus: Theory and its Applications in Automatics. Publishing Department of Technical University of Łódź, Łódź (2008). (in Polish)
32. Oldham, K.B., Spanier, J.: The Fractional Calculus. Academic Press, New York (1974)
33. Podlubny, I.: Fractional Differential Equations. Academic Press, New York (1999)
34. Radwan, A.G., Soliman, A.M., Elwakil, A.S., Sedeek, A.: On the stability of linear systems with fractional-order elements. Chaos, Solitons Fractals **40**(5), 2317–2328 (2009)
35. Tenreiro Machado, J.A., Ramiro Barbosa, S.: Functional dynamics in genetic algorithms. Workshop Fract. Differ. Appl. **1**, 439–444 (2006)
36. Vinagre, B.M., Monje, C.A., Calderon, A.J.: Fractional order systems and fractional order control actions. In: Lecture 3 IEEE CDC 2002 TW#2: Fractional Calculus Applications in Automatic Control and Robotics (2002)
37. Wolovich, W.: Linear Multivariable Systems. Springer, New York (1974)

Discrete, Fractional Order, Cancellation Controller. Part I: Idea and Simulations

Krzysztof Oprzędkiewicz[(✉)], Łukasz Więckowski, and Maciej Podsiadło

Department of Automatics and Robotics, Faculty of Electrotechnics, Automatics,
Informatics and Biomedical Engineering, AGH University of Science and Technology,
al. A Mickiewicza 30, 30-059 Krakow, Poland
kop@agh.edu.pl

Abstract. In the paper the proposition of discrete, fractional order cancellation controller dedicated to control a high order inertial plant is presented. The controller uses the hybrid transfer function model of the plant. Results of simulations show that the proposed controller assures the better control performance than PID controller tuned with the use of known methods.

Keywords: Fractional order systems ·
Fractional order transfer function · CFE approximation ·
Cancellation controller

1 Introduction

Fractional order (FO) models can properly and accurate describe a number of real physical phenomena. This problem is presented by many Authors, for example in: [6,8,9,16,17,19,20]. The usefulness of FO models is determined by the possibility of their approximation. Integer order, time-continuous approximation of elementary operator s^α is proposed by Oustaloup in [22], the approximation of elementary inertial transfer function $\frac{1}{(Ts+1)^\alpha}$ is proposed by Charef in [4]. The both approximations are also presented in [7].

An idea of cancellation control has been presented by many Authors, for example in books: [2,10,26]. Cancellation control methods has been also analysed in [13–15].

This paper is intented to present the new, proposed by authors, discrete cancellation controller. It uses hybrid transfer function model of the control plant. The hybrid transfer function can precisely describe a wide class of high order, aperiodic processes, for example heat transfer processes and it is more accurate than known Strejc model (see [18]).

The paper is organized as follows: at the beginning some preliminaries are recalled. An idea of cancellation control is given also. Next the proposed by authors discrete cancellation controller is proposed and illustrated by simulations.

© Springer Nature Switzerland AG 2020
R. Szewczyk et al. (Eds.): AUTOMATION 2019, AISC 920, pp. 27–38, 2020.
https://doi.org/10.1007/978-3-030-13273-6_3

2 Preliminaries

2.1 Elementary Ideas

The presentation of elementary ideas starts with definition of a non integer order, integro-differential operator. It is expressed as follows (see for example [11, 12]):

Definition 1. *The non integer order integro-differential operator*

$$
{}_0D_t^\alpha f(t) = \begin{cases} \frac{d^\alpha f(t)}{dt^\alpha} & \alpha > 0 \\ f(t) & \alpha = 0 \\ \int\limits_0^t f(\tau)(d\tau)^{-\alpha} & \alpha < 0 \end{cases} . \tag{1}
$$

where t denotes time limit to operator calculating, $\alpha \in \mathbb{R}$ denotes the non integer order of the operation.

The fractional-order, integro-differential operator (1) can be described by different definitions, given by Grünvald and Letnikov (GL definition), Riemann and Liouville (RL definition) and Caputo (C definition). The most natural way to discrete implementation is given by the GL definition:

Definition 2. *The Grünvald-Letnikov definition of the FO operator [3, 21].*

$$
{}_0^{GL}D_t^\alpha f(t) = \lim_{h \to 0} h^{-\alpha} \sum_{j=0}^{[\frac{t}{h}]} (-1)^j \binom{\alpha}{j} f(t - jh). \tag{2}
$$

In (2) $\binom{\alpha}{j}$ is a generalization of Newton symbol into real numbers:

$$
\binom{\alpha}{j} = \begin{cases} 1, & j = 0 \\ \frac{\alpha(\alpha-1)\dots(\alpha-j+1)}{j!}, & j > 0 \end{cases} \tag{3}
$$

2.2 The CFE Approximation

An implementation of operator s^α at each digital platform (PLC, microcontroller) requires to apply its integer-order, finite-length, discrete-time approximator. The well known approximators are based on Power Series Expansion (PSE) and Continuous Fraction Expansion (CFE). They allow to approximate a noninteger-order element with the use of digital FIR or IIR filters.

The PSE approximator is based directly on discrete-time version of the GL definition (2) and it has the form of an FIR filter containing only zeros. However its digital, high quality implementation requires to apply a long memory buffer (high order of the filter). The CFE approximator has the form of an IIR filter containing both poles and zeros. It is faster convergent and easier to implement because its order is relatively low, typically not higher that 5.

The discretization of fractional order element s^α can be done with the use of the so called generating function $s \approx \omega(z^{-1})$. The new operator raised to the power α has the following form (see for example [5], [23] p. 119):

$$
\begin{aligned}
\left(\omega(z^{-1})\right)^\alpha &= \left(\tfrac{1+a}{h}\right)^\alpha CFE\left\{\left(\tfrac{1-z^{-1}}{1+az^{-1}}\right)^\alpha\right\}_{M,M} \\
&= \frac{P_{\alpha M}(z^{-1})}{Q_{\alpha M}(z^{-1})} = \left(\tfrac{1+a}{h}\right)^\alpha \frac{CFE_N(z^{-1},\alpha)}{CFE_D(z^{-1},\alpha)} = \frac{\sum\limits_{m=0}^{M} w_m z^{-m}}{\sum\limits_{m=0}^{M} v_m z^{-m}}.
\end{aligned}
\tag{4}
$$

where h denotes the sampling time and M is the order of approximation. Numerical values of coefficients w_m and v_m and various values of the parameter a can be calculated for example with the use of the MATLAB function given in [24].

In Eq. (4), a is the coefficient depending on an approximation type. For $a = 0$ and $a = 1$ we obtain the Euler and Tustin approximations respectively. For $a \in (0,1)$ we arrive at the Al-Alaoui-based approximation, which is a linear combination of the Euler and Tustin approaches. Note that in this case the parameter a in Eq. (4) is equal to $a = \frac{1-\beta}{1+\beta}$, with β being the Al-Alaoui weighting coefficient (see [1, 25]). If the Tustin approximation is considered ($a = 1$) then $CFE_D(z^{-1},\alpha) = CFE_N(z^{-1}, -\alpha)$ and the polynomial $CFE_D(z^{-1},\alpha)$ can be given in the direct form (see [5]). Examples for the polynomial $CFE_D(z^{-1},\alpha)$ for $M = 1, 3, 5$ are given in Table 1. The approximator using the Muir recursion is presented for example in [27]. The detailed analysis of various forms of the CFE approximators has been given in [25].

Table 1. Coefficients of CFE polynomials $CFE_{N,D}(z^{-1},\alpha)$ for Tustin approximation

Order M	w_m	v_m
$M = 1$	$w_1 = -\alpha$	$v_1 = \alpha$
	$w_0 = 1$	$v_0 = 1$
$M = 3$	$w_3 = -\frac{\alpha}{3}$	$v_3 = \frac{\alpha}{3}$
	$w_2 = \frac{\alpha^2}{3}$	$v_2 = \frac{\alpha^2}{3}$
	$w_1 = -\alpha$	$v_1 = \alpha$
	$w_0 = 1$	$v_0 = 1$
$M = 5$	$w_5 = -\frac{\alpha}{5}$	$v_5 = \frac{\alpha}{5}$
	$w_4 = \frac{\alpha^2}{5}$	$v_4 = \frac{\alpha^2}{5}$
	$w_3 = -\left(\frac{\alpha}{5} + \frac{2\alpha^3}{35}\right)$	$v_3 = -\left(\frac{-\alpha}{5} + \frac{-2\alpha^3}{35}\right)$
	$w_2 = \frac{2\alpha^2}{5}$	$v_2 = \frac{2\alpha^2}{5}$
	$w_1 = -\alpha$	$v_1 = \alpha$
	$w_0 = 1$	$v_0 = 1$

2.3   Idea of Cancellation Control

Consider the closed loop control system shown in Fig. 1. It contains controller described by transfer function $G_c(s)$ and plant described by the transfer function $G(s)$. Denote the transfer function of the whole closed-loop control system by $G_{cl}(s)$. It is equal: $G_{cl}(s) = \frac{Y(s)}{R(s)} = \frac{G_c(s)G(s)}{1+G_c(s)G(s)}$. The idea of cancellation control consists in construction such a controller $G_c(s)$, which is able to assure a predefined form of closed-loop transfer function $G_{cl}(s) = G_m(s)$. Typically the transfer function $G_m(s)$ is expected to replace the transfer function of the plant $G(s)$.

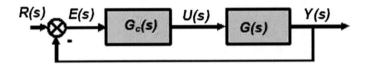

Fig. 1. Closed-loop control system.

Using the above idea the transfer function $G_c(s)$ of the cancellation controller is given as follows (see [26], pp. 205–207):

$$G_c(s) = \frac{1}{G(s)} \frac{G_m(s)}{1 - G_m(s)}. \tag{5}$$

The main problem during construction the controller in the form (5) is possibility of its realization - the order of numerator in (5) cannot be higher than order of denominator. The next problem during construction cancellation control system is caused by the fact that the "full cancellation" plant by controller is possible only if we know the exact model of plant. Such a situation is ideal and impossible to obtain in reality. Fortunately, experiments show that the proposed cancellation controller is able to assure good control performance without use of the exact model of the plant.

2.4   The Model of the Plant

The high order plant can be described by a number of transfer function models. In this paper the so called hybrid transfer function containing both integer order (IO) and fractional order (FO) parts will be employed. This model has been proposed and analysed in papers [18] and [20]. Generally it is able to more precisely describe a number of real plants than analogical IO transfer function. The hybrid transfer function employed in this paper has the following form:

$$G(s) = \frac{k}{(T_\alpha s^\alpha + 1)(T_n s + 1)}. \tag{6}$$

where k is the steady-state gain of the plant, $0 < \alpha < 2$ is the order of non integer order part, T_α and T_n are time constants for FO and IO parts respectively. This transfer function will be used to tune the proposed controller.

However in reality we do not have an exact knowledge about plant parameters. This makes us to express the real parameters of the model (6) as intervals, containing the exactly known, nominal parameters α, T_α and T_n. This can be described as follows:

$$G_i(s) = \frac{k}{(T_{\alpha_u} s^{\alpha_u} + 1)(T_u s + 1)}. \tag{7}$$

where:

$$
\begin{aligned}
k_u &= \left[\underline{k_u}; \overline{k_u}\right] \ni k. \\
\alpha_u &= (0.0; 2.0) \supset \left[\underline{\alpha_u}; \overline{\alpha_u}\right] \ni \alpha. \\
T_{\alpha_u} &= \left[\underline{T_{\alpha_u}}; \overline{T_{\alpha_u}}\right] \ni T_\alpha. \\
T_u &= \left[\underline{T_u}; \overline{T_u}\right] \ni T_n.
\end{aligned}
\tag{8}
$$

The above uncertain parameters build the vector q:

$$q = [k_u; \alpha_u; T_{\alpha_u}; T_u]. \tag{9}$$

All vectors q build the set of uncertain parameters Q:

$$q \in Q \subset I\left(\mathbb{R}^4\right). \tag{10}$$

The set Q can be interpreted in the 4-D plane as a hipercube with vertices described by border values of k_u, α_u, T_{α_u} and T_u. Properties of the control system containing plant (7)–(10) are determined by properties of its vertex subsystems, defined by vertices of hipercube (10).

3   Main Results

3.1   The Proposed FO Cancellation Controller

To construct the proposed FO cancellation controller assume that the control plant is described by the model in the form of hybrid transfer function (6). Of course, parameters of this transfer function should be well known. Next, let us assume that the predefined, "target" transfer function of the whole closed loop control system $G_m(s)$ is equal:

$$G_m(s) = \frac{1}{\delta T_n s + 1}. \tag{11}$$

where $0.0 < \delta < 1.0$ denotes the coefficient of dynamics reduction. The other words the controller is expected to eliminate the whole FO part of plant and decrease the time constant of the IO part with coefficient δ. The form of (11) with respect to (5) implies the following form of the FO cancellation controller:

$$G_c(s) = \frac{(T_\alpha s^\alpha + 1)(T_n s + 1)}{k\delta T_n s}. \tag{12}$$

After some elementary calculations the controller (12) takes the form close to the known PID IND formula:

$$G_c(s) = k_P + k_D s^\alpha + k_{ID} s^{\alpha-1} + k_I s^{-1}. \tag{13}$$

where:

$$
\begin{aligned}
k_P &= \frac{1}{k\delta}. \\
k_D &= \frac{T_\alpha}{k\delta}. \\
k_{ID} &= \frac{T_\alpha}{k\delta T_n}. \\
k_I &= \frac{1}{k\delta T_n}.
\end{aligned}
\tag{14}
$$

If we introduce the following symbols:

$$
\begin{aligned}
G_\alpha(s) &= s^\alpha \\
G_I(s) &= s^{-1}.
\end{aligned}
\tag{15}
$$

then the controller (13) takes the following form:

$$G_c(s) = k_P + k_D G_\alpha(s) + k_{ID} G_\alpha(s) G_I(s) + k_I G_I(s). \tag{16}$$

It can be proven that the use of cancellation controller in the form (12)–(16) with plant (6) allows to directly obtain the target transfer function (11).

3.2　Discrete Realization of Controller

Discrete realization of the controller (12)–(14) requires to express it as the discrete transfer function. With respect to (16) it takes the following form:

$$G_c^+(z^{-1}) = k_P + k_D G_\alpha^+(z^{-1}) + k_{ID} G_\alpha^+(z^{-1}) G_I^+(z^{-1}) + k_I G_I^+(z^{-1}). \tag{17}$$

If the Euler transformation is applied and order of CFE approximation is equal $M = 5$ then the transfer functions $G_I^+(z^{-1})$ and $G_\alpha^+(z^{-1})$ take the following form:

$$G_I^+(z^{-1}) = \frac{h}{1 - z^{-1}}. \tag{18}$$

$$G_\alpha^+(z^{-1}) = \frac{\sum\limits_{m=0}^{5} w_m z^{-m}}{\sum\limits_{m=0}^{5} v_m z^{-m}}. \tag{19}$$

Coefficients v_m and w_m can be calculated using MATLAB function *dfod1*, available at [24], h denotes the sampling time. The controller is ready to implement after conversion the discrete transfer function $G_c^+(z^{-1})$ to difference equation. Its tuning contains two steps. Firstly the identification of plant parameters: k, α, T_n and T_α must be done. Next, for identified plant the parameter δ assuring the optimal control performance is required to find. This can be done with the use of simulations.

4   Simulations

Simulations have been done using the SIMULINK diagrams shown in Fig. 2. The saturation block is necessary because the proposed controller will be implemented at real system with control signal limited to standard range 0–100%. The model with PID controller is employed to compare the work of the proposed controller with typical solution. The exactly known, nominal transfer function of the plant, applied to tuning the controller with respect to (14) is as follows:

$$G(s) = \frac{6.99}{\left(109s^{1.54} + 1\right)\left(49.9217s + 1\right)}. \tag{20}$$

The discrete cancellation controller has been tuned with the use of nominal transfer function (20) and the parameter $\delta = 0.7$. The s^α element is approximated using CFE approximation of 5 order. The PID controller has been tuned using standard *tune* function available in MATLAB with respect to obtain shortest settling time with minimal overshoot. Its parameters were equal: $P = 0.317$, $I = 0.00868$, $D = 2.54$.

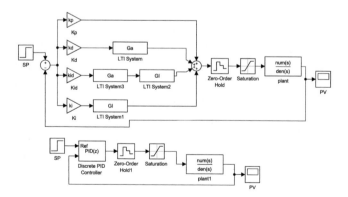

Fig. 2. SIMULINK model employed to tests

The control performance and robustness of the proposed controller were tested with assumption that the parameters of real, controlled plant are different from nominal, given by (20). All the parameters are described by intervals (8) equal:

$$k_u = [0.9k; 1.1k]$$
$$\alpha_u = [0.9\alpha; 1.1\alpha]$$
$$T_{\alpha_u} = [0.9T_\alpha; 1.1T_\alpha] \tag{21}$$
$$T_u = [0.9T_n; 1.1T_n]$$

The tests were executed for all vertices of cube (10), defined by (21). During tests the overshoot γ in % of set point and settling time T_s were examined. The servo control problem has been investigated. Results are given in the Table 2. In this table "C" denotes the cancellation controller, "PID" denotes the PID controller.

Table 2. Results of tests

Test No	Parameters	C γ [%]	PID γ [%]	C T_s [s]	PID T_s [s]
0	Nominal	0.82	5.41	97.01	113.93
1	$1.1k$, 1.1α, $1.1T_\alpha$, $1.1T_n$	2.15	5	92.61	129.61
2	$1.1k$, 1.1α, $1.1T_\alpha$, $0.9T_n$	0.44	4.78	89.03	130.58
3	$1.1k$, 1.1α, $0.9T_\alpha$, $1.1T_n$	1.96	4	99.21	116.97
4	$1.1k$, 1.1α, $0.9T_\alpha$, $0.9T_n$	0.32	3.06	97.64	118.83
5	$1.1k$, 0.9α, $1.1T_\alpha$, $1.1T_n$	7.98	20.19	197.64	241.41
6	$1.1k$, 0.9α, $1.1T_\alpha$, $0.9T_n$	5.85	15.66	160.98	217.7
7	$1.1k$, 0.9α, $0.9T_\alpha$, $1.1T_n$	6.08	14.99	177.01	157.92
8	$1.1k$, 0.9α, $0.9T_\alpha$, $0.9T_n$	4.04	10.67	94.27	140.77
9	$0.9k$, 1.1α, $1.1T_\alpha$, $1.1T_n$	0.81	4.01	125.55	89.99
10	$0.9k$, 1.1α, $1.1T_\alpha$, $0.9T_n$	−0.18	3.49	132.08	134.02
11	$0.9k$, 1.1α, $0.9T_\alpha$, $1.1T_n$	0.75	3.55	127.53	116.75
12	$0.9k$, 1.1α, $0.9T_\alpha$, $0.9T_n$	−0.2	2.28	132.99	123.08
13	$0.9k$, 0.9α, $1.1T_\alpha$, $1.1T_n$	3.88	19.01	118.7	256.84
14	$0.9k$, 0.9α, $1.1T_\alpha$, $0.9T_n$	1.61	14.69	114.64	225.22
15	$0.9k$, 0.9α, $0.9T_\alpha$, $1.1T_n$	2.47	14.25	116.98	174.11
16	$0.9k$, 0.9α, $0.9T_\alpha$, $0.9T_n$	0.2	10.1	113.52	154.37

Step responses for test no 0 (the nominal plant) and the two most far vertices of set Q (tests no 1 and no 16) are shown in Figs. 3, 4 and 5.

From Table 2 and Figs. 3, 4 and 5 it can be concluded that the use of the proposed cancellation controller allows to obtain the smaller overshoot in the each vertex of hipercube Q. The settling time assured by the proposed controller is shorter in the big part of tests.

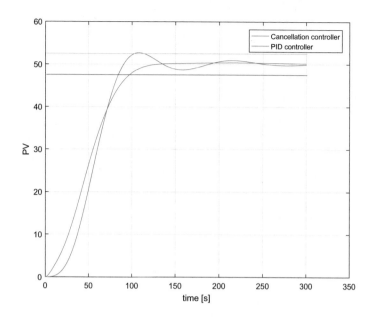

Fig. 3. Step response of the control system, nominal plant (test No 0)

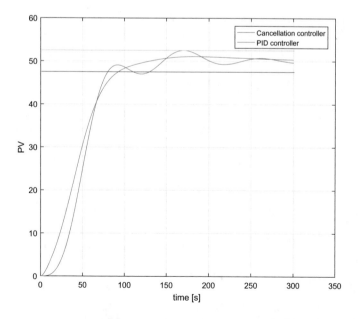

Fig. 4. Step response of control system, disturbed plant, test No 1

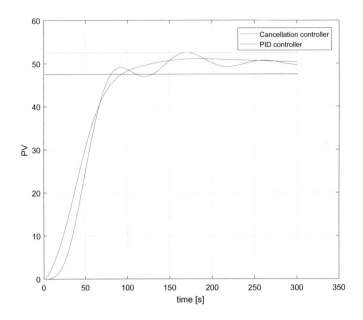

Fig. 5. Step response of control system, disturbed plant, test No 16

5   Final Conclusions

The main conclusion from this paper is that the proposed FO cancellation controller is able to assure the better control performance than typical PID. It is also easy to tune, additionally it is more robust to plant uncertainty than typical PID.

The practical PLC implementation of the proposed controller is shown in the second part of this paper.

The further investigations of the proposed controller cover its theoretical explanation and analysis, particularly the robust stability needs to be analysed.

Acknowledgements. This paper was sponsored partially by AGH UST grant no 11.11.120.815.

References

1. Al-Alaoui, M.A.: Novel digital integrator and differentiator. Electron. Lett. **29**(4), 376–378 (1993)
2. Astrom, K.J., Hagglund, T.: PID Controllers: Theory, Design and Tuning, ISA (1995)
3. Caponetto, R., Dongola, G., Fortuna, L., Petras, I.: Fractional Order Systems: Modeling and Control Applications. World Scientific Series on Nonlinear Science, Series A, vol. 72. World Scientific Publishing, Hackensack (2010)

4. Charef, A., Sun, H.H., Tsao, Y.Y., Onaral, B.: Fractional system as represented by singularity function. IEEE Trans. Aut. Control **37**(9), 1465–1470 (1992)
5. Chen, Y.Q., Moore, K.L.: Discretization schemes for fractional-order differentiators and integrators. IEEE Trans. Circ. Syst. I Fundam. Theory Appl. **49**(3), 363–367 March 2002
6. Das, S.: Functional Fractional Calculus for System Identification and Controls. Springer, Heidelberg (2008)
7. Das, S., Pan, I.: Fractional Order Signal Processing. Springer Briefs in Applied Sciences and Technology (2012). https://doi.org/10.1007/978-3-642-23117-9-2
8. Dlugosz, M., Skruch, P.: The application of fractional-order models for thermal process modelling inside buildings. J. Build. Phys. **I-13**, 1–13 (2015)
9. Dzielinski, A., Sierociuk, D., Sarwas, G.: Some applications of fractional order calculus. Bull. Pol. Acad. Sci. Tech. Sci. **58**(4), 583–592 (2010)
10. Ishihara, T., Hai-Jiao Guo, H.-J.: Design of optimal disturbance cancellation controllers via modified loop transfer recovery. Syst. Sci. Control Eng. **3**(1), 332–339 (2015). https://doi.org/10.1080/21642583.2015.1023470
11. Kaczorek, T.: Selected Problems in Fractional Systems Theory. Springer, Heidelberg (2011)
12. Kaczorek, T., Rogowski, K.: Fractional Linear Systems and Electrical Circuits. Bialystok University of Technology, Bialystok (2014)
13. Merrikh-Bayat, F.: Rules for selecting the parameters of Oustaloup recursive approximation for the simulation of linear feedback systems containing $PI^\lambda D^\mu$ controller. Commun. Nonlinear Sci. Numer. Simulat. **17**, 1852–1861 (2012)
14. Merrikh-Bayat, F.: Fractional-order unstable pole-zero cancellation in linear feedback systems. J. Process Control **23**(6), 817–825 (2013)
15. Merrikh-Bayat, F., Salimi, A.: Performance enhancement of non-minimum phase feedback systems by fractional-order cancellation of non-minimum phase zero on the Riemann surface: New theoretical and experimental results, Preprint submitted to Elsevier (2016)
16. Mitkowski, W., Skruch, P.: Fractional-order models of the supercapacitors in the form of RC ladder networks. Bull. Pol. Acad. Sci. Tech. Sci. **61**(3), 581–587 (2013)
17. Obraczka, A., Mitkowski, W.: The comparison of parameter identification methods for fractional partial differential equation. Solid State Phenom. **210**, 265–270 (2014)
18. Oprzedkiewicz, K., Mitkowski, W., Gawin, E.: Application of fractional order transfer functions to modeling of high order systems. In: MMAR 2015: 20th International Conference on Methods and Models in Automation and Robotics: 24–27 August 2015, Midzyzdroje, Poland: program, abstracts, proceedings (CD). Szczecin: ZAPOL Sobczyk Sp.j., [2015] + CD. Dod (2015). ISBN: 978-1-4799-8701-6, 978-1-4799-8700-9. ISBN: 978-83-7518-756-4
19. Oprzedkiewicz, K., Mitkowski, W., Gawin, E.: Parameter identification for non integer order, state space models of heat plant. In: MMAR 2016: 21st International Conference on Methods and Models in Automation and Robotics: 29 August–01 September 2016, Miedzyzdroje, Poland, pp. 184–188 (2016). ISBN: 978-1-5090-1866-6, ISBN: 978-837518-791-5
20. Oprzedkiewicz, K., Kolacz, T.: A non integer order model of frequency speed control in AC motor. In: Szewczyk, R., Zielinski, C. (eds.) Advances in Intelligent Systems and Computing, vol. 440, pp. 287–298. Springer, Switzerland (2016)
21. Ostalczyk, P.: Equivalent descriptions of a discrete-time fractional-order linear system and its stability domains. Int. J. Appl. Math. Comput. Sci. **22**(3), 533–538 (2012)

22. Oustaloup, A., Levron, F., Mathieu, B., Nanot, F.M.: Frequency-band complex nonin-teger differentiator: characterization and synthesis. IEEE Trans. Circ. Syst. I Fundam. Theory Appl. I **47**(1), 25–39 (2000)
23. Petras, I.: Fractional order feedback control of a DC motor. J. Electr. Eng. **60**(3), 117–128 (2009)
24. Petras I.: http://people.tuke.sk/igor.podlubny/USU/matlab/petras/dfod1.m
25. Stanislawski, R., Latawiec, K.J., Lukaniszyn, M.: A comparative analysis of laguerre-based approximators to the Grünwald-Letnikov fractional-order difference. Math. Prob. Eng. **2015**, 10 (2015). Article ID 512104https://doi.org/10.1155/2015/512104
26. Tzafestas, S.G. (ed.): Methods and Applications of Intelligent Control. Springer, New York (1997)
27. Vinagre, B.M., Chen, Y.Q., Petras, I.: Two direct Tustin discretization methods for fractional-order differentiator-integrator. J. Franklin Inst. **340**, 349–362 (2003)

Discrete, Fractional Order, Cancellation Controller. Part II: PLC Implementation

Krzysztof Oprzędkiewicz[(✉)], Łukasz Więckowski, and Maciej Podsiadło

Department of Automatics and Robotics,
Faculty of Electrotechnics, Automatics, Informatics and Biomedical Engineering,
AGH University of Science and Technology,
al. A Mickiewicza 30, 30-059 Krakow, Poland
kop@agh.edu.pl

Abstract. In the paper the PLC implementation of discrete, fractional order cancellation controller dedicated to control a high order inertial plant is given. The controller uses the hybrid transfer function model of the plant. It is implemented at PLC SIEMENS with respect to object-oriented approach recommended by IEC61131 standard. Results of tests agree with simulations. The proposed controller is easy to tune and assures the better control performance than typical PID.

Keywords: Fractional order systems ·
Fractional order transfer function · CFE approximation ·
Cancellation controller · PLC · 61131 standard

1 Introduction

Main areas of application the fractional order calculus in automation are: fractional order control and modeling of processes with dynamics hard to describe with the use of another approaches. Fractional order control covers mainly Fractional Order PID controllers (FO PID). FO PID controllers have been presented by many Authors and their usefulness has been proven (see for example: [2, 3, 12–14]).

An idea of cancellation control has been presented by many Authors, for example in books: [1, 4, 16]. Cancellation control methods have been also analysed in [5–7].

PLCs have been a workhorse of industrial automation for many years. Hardware and software of PLC systems are normalized (IEC standard 61131) and their programming platforms offer a powerful tool to implement each control algorithm. However, most implementations cover logic control, sequential control and PID control, although PLC platforms make possible to implement more complex tasks, for example model based control algorithms or model based fault detection systems. The PLC implementation of FO PID controller has been given for example in [13]. The PLC implementation of basic FO element at SIEMENS PLC is discussed in [9, 10].

© Springer Nature Switzerland AG 2020
R. Szewczyk et al. (Eds.): AUTOMATION 2019, AISC 920, pp. 39–46, 2020.
https://doi.org/10.1007/978-3-030-13273-6_4

This paper is intented to present the PLC implementation of the new, proposed by authors fractional order cancellation controller. It uses hybrid transfer function model of the control plant. To implement the CFE approximation was employed. This was caused by the fact that this method requires to use significantly shorter memory length, than Power Series Expansion (PSE) method.

The paper is organized as follows: at the beginning the experimental PLC system is presented. Next the discrete, fractional order (FO) cancellation controller is given and its implementation using standard software elements are proposed. Finally tests and their results are discussed.

2   The Experimental System

2.1   Construction of the System

The experimental system is shown in Fig. 1. It employes PLC SIEMENS S7 1500 with CPU 1516 and analog inputs and outputs modules. Detailed hardware configuration is shown in Fig. 2. The PLC system is connected to PC via PROFINET interface. All parameters to experiments were introduced using HMI panel SIEMENS KTP400. The 16-bit RTD analog module and 16-bit/0–20 mA analog output module are applied.

The control plant is the thin copper rod 260 [mm] long. For further considerations it will be assumed that the length of rod is equal 1.0. This allows to express the location and length of heater and RTD sensors with respect to 1.0. The rod is heated by the electric heater $\Delta x_0 = 0.14$ long, attached at the end. The output temperature is measured by RTD sensors Pt100 long Δx and localised in points: 0.29, 0.50 and 0.73 of rod length. The input signal of the system is the standard current 0.0–20.0 [mA]. It is amplified to the range 0.0–1.5 [A] and sent to the heater. Signals from RTDs are read directly in Celsius degrees by analog input module in the PLC. Data from PLC are collected by the SCADA. The time-spatial temperature distribution is shown in the Fig. 3.

2.2   Transfer Function Model of the Plant

The experimental heat plant can be described by the following transfer function, containing integer order (IO) and fractional order (FO) parts:

$$G(s) = \frac{k}{(T_\alpha s^\alpha + 1)(T_n s + 1)}. \tag{1}$$

where k is the steady-state gain of the plant, $0.0 < \alpha < 2.0$ is the order of non integer order part, T_α and T_n are time constants for FO and IO parts respectively. They can be identified via minimization the mean square error (MSE) cost function, describing the difference between step response of model (1) and step response of real plant, calculated at the same time grid (see: [8,11]). Parameters of the transfer function (1) for sensor no 3 employed to experiments are given in the Table 1. They are also used to tune the controller.

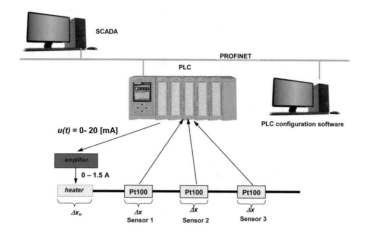

Fig. 1. The experimental system

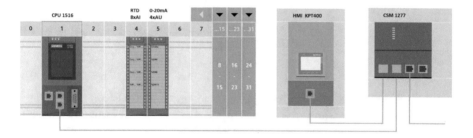

Fig. 2. The hardware configuration

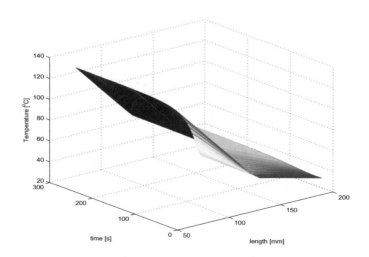

Fig. 3. The spatial-time temperature distribution in the plant

Table 1. Parameters of transfer function (1) for sensor no 3

k	α	T_α [s]	T_n [s]
6.99	1.54	109.00	49.92

3   Implementation of Controller

3.1   Difference Equations of Controller

The discrete transfer function of the controller has been given in the Part I of this paper. It has the following form:

$$G_c^+(z^{-1}) = k_P + k_D G_\alpha^+(z^{-1}) + k_{ID} G_\alpha^+(z^{-1}) G_I^+(z^{-1}) + k_I G_I^+(z^{-1}). \quad (2)$$

where:

$$
\begin{aligned}
k_P &= \frac{1}{\delta k}. \\
k_D &= \frac{T_\alpha}{\delta k}. \\
k_{ID} &= \frac{T_\alpha}{\delta T_n k}. \\
k_I &= \frac{1}{\delta T_n k}.
\end{aligned}
\quad (3)
$$

If the Euler transformation is applied and order of CFE approximation is equal $M = 5$ then the transfer functions $G_I^+(z^{-1})$ and $G_\alpha^+(z^{-1})$ take the following form:

$$G_I^+(z^{-1}) = \frac{h}{1 - z^{-1}}. \quad (4)$$

$$G_\alpha^+(z^{-1}) = \frac{\displaystyle\sum_{m=0}^{5} w_m z^{-m}}{\displaystyle\sum_{m=0}^{5} v_m z^{-m}}. \quad (5)$$

Coefficients v_m and w_m can be calculated using MATLAB function *dfod1*, available at [15], h denotes the sampling time.

The transfer functions G_α^+ and C_I^+ are realized as separated instances of Function Blocks (FB). This requires in turn to difference equations, analogically, as it has been done in [9]. Denote the input of transfer function G_α^+ by $u_a^+(k)$ and its output by $y_a^+(k)$ and analogically input and output of transfer function G_I^+ by $u_I^+(k)$ and $y_I^+(k)$. Then the difference equations equivalent to these functions are as follows:

$$y_I^+(k) = y_I^+(k-1) + h u_I^+(k). \quad (6)$$

$$y_a^+(k) = \frac{1}{v_0}\left[-\sum_{m=1}^{5} v_m y_a^+(k-m) + \sum_{m=0}^{5} w_m u_a^+(k-m)\right]. \quad (7)$$

Equations (6) and (7) are ready to digital implementation. They must be written as function blocks.

Table 2. Software components of controller

Formal name	Symbolic name	Description
Function (FC1)	FOCcoeffCalc	Function calculating the coefficients of CFE approximant with respect to (3)
Function Block (FB1)	FOCiPart	Function Block implementing the integral part of controller expressed by (6)
Function Block (FB2)	FOCdPart	Function Block implementing the differential component of controller expressed by (7)
Function Block (FB3)	FOCcontroller	Function Block calculating the response of cancellation controller with respect to (2). Both instances: **FB1** and **FB2** are called in this function block. The actual parameters and the static data of the above function block are stored to the instance data block (**DB2–DB5**)
Organization Block (OB1)	Main	Block contains the conditional call of function FOCcoeffCalc. The call is run via button F1 on HMI, results are written in retentive data block **DB1**
Organization Block (OB30)	Cyclic Interrupt	This block is activated by system clock interrupt with constant period, equal to sample time h. It calls the instance of **FB3**, calculating the output of cancellation controller
Data Block (DB1)	FOCcoeffDB	Data block saving the calculated coefficients of CFE approximation with respect to (3).
Data Blocks (DB2-DB5)	FOCparamDB	This data blocks accumulate recent parameters and the static data of the **FB1** and **FB2** instances used to perform correct calculations with respect to (2)

3.2 The PLC Software

The software has been prepared with the use of TIA PORTAL v.13 and STEP7-SCL language with respect to object-oriented approach recommended by 61131 standard and guidelines given in [9]. Controller coefficients are calculated in

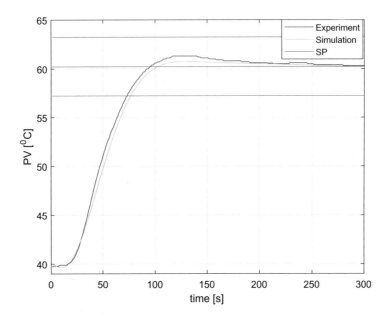

Fig. 4. Step responses of control system for experiment and simulation

function **FC1** with respect to (3) and stored to Data Block **DB1**. Both dynamic parts of controller expressed by (6) and (7) are written as two function blocks (FB). Instances of these blocks are called as local in the next bigger function block assembling the whole Eq. (2). The instance of this block is called by Organization Block OB30 with activating period equal h. Main components of the software are described in Table 2.

4 Experiments

Experiments were executed using the system presented in the previous section. The step response of set point SP between 40 and 60 ^oC has been tested (the servo control problem). The sampling time h was equal 1 [s], the number of all collected samples was equal 300. Results are shown in Fig. 4. The comparison of elementary performance indices: settling time and overshoot in % of set point SP are given in the Table 3.

Table 3. Comparison of performance indices for simulation and experiment

	Overshoot in % of SP	Settling time [s]
Experiment	2.19	72.72
Simulation	1.25	76.77

5   Final Conclusions

The main conclusion from this paper is that the proposed FO cancellation controller can be implemented at PLC with the use of standard software tools. Results of experiments are compliant to simulations given in the part I of this paper.

Further investigations of the presented controller cover its practical implementation at another digital platforms: microcontroller and Raspberry Pi with respect to requirements given by Industry 4.0.

Acknowledgements. This paper was sponsored partially by AGH UST grant no. 11.11.120.815.

References

1. Astrom, K.J., Hagglund, T.: PID Controllers: Theory, Design and Tuning. ISA (1995)
2. Caponetto, R., Dongola, G., Fortuna, l., Petras, I.: Fractional Order Systems: Modeling and Control Applications. World Scientific Series on Nonlinear Science, Series A, vol. 72. World Scientific Publishing (2010)
3. Das, S., Pan, I.: Intelligent Fractional Order Systems and Control: An Introduction. Springer (2013) https://doi.org/10.1007/s11633-010-0552-2
4. Ishihara, T., Hai-Jiao Guo, H.-J.: Design of optimal disturbance cancellation controllers via modified loop transfer recovery. Syst. Sci. Control Eng. **3**(1), 332–339 (2015). https://doi.org/10.1080/21642583.2015.1023470
5. Merrikh-Bayat, F.: Rules for selecting the parameters of Oustaloup recursive approximation for the simulation of linear feedback systems containing PID controller. Commun. Nonlinear Sci. Numer. Simulat. **17**, 1852–1861 (2012)
6. Merrikh-Bayat, F.: Fractional-order unstable pole-zero cancellation in linear feedback systems. J. Process Control **23**(6), 817–825 (2013)
7. Merrikh-Bayat, F., Salimi, A.: Performance enhancement of non-minimum phase feedback systems by fractional-order cancellation of non-minimum phase zero on the Riemann surface: new theoretical and experimental results. Preprint submitted to Elsevier (2016)
8. Oprzedkiewicz, K., Mitkowski, W., Gawin, E.: Parameter identification for non integer order, state space models of heat plant. In: MMAR 2016: 21th International Conference on Methods and Models in Automation and Robotics, 29 August 01 September 2016, Miedzyzdroje, Poland, pp. 184–188 (2016). ISBN 978-1-5090-1866-6, ISBN 978-837518-791-5
9. Oprzedkiewicz, K., Mitkowski, W., Gawin, E.: The PLC implementation of fractional-order operator using CFE approximation. In: Szewczyk, R., Zieliński, C., Kaliczyńska, M. (eds.) Automation 2017: Innovations in Automation, Robotics and Measurment Techniques, 15–17 March, Warsaw, Poland. Advances in Intelligent Systems and Computing, vol. 550, pp. 22–33. Springer, Cham (2017). ISSN 2194-5357; ISBN 978-3-31954041-2; e-ISBN 978-3-319-54042-9

10. Oprzedkiewicz, K., Gawin, E., Gawin, T.: Real-time PLC implementations of fractional order operator. In: Szewczyk, R., Zieliński, C., Kaliczyńska, M. (eds.) Automation 2018: Advances in Automation, Robotics and Measurement Techniques, Advances in Intelligent Systems and Computing, vol. 743. Springer, Cham (2018). ISSN 2194-5357, ISBN 978-3-319-77178-6; e-ISBN: 978-3-319-77179-3. https://goo.gl/RHBJfp
11. Oprzedkiewicz, K., Mitkowski, W.: Parameter identification for non integer order, discrete, state space model of heat transfer process using CFE approximation. In: 23rd International Conference on Methods and Models in Automation and Robotics, 27–30 August 2018, Miedzyzdroje, Poland, pp. 436–440 (2018). ISBN 978-1-5386-4324-2
12. Ostalczyk, P.: Discrete Fractional Calculus: Applications in Control and Image Processing. Series in Computer Vision, vol. 4. World Scientific Publishing (2016)
13. Petras, I.: Fractional order feedback control of a DC motor. J. Electr. Eng. **60**(3), 117–128 (2009)
14. Petras, I.: Tuning and implementation methods for fractional-order controllers. Fract. Calc. Appl. Anal. **15**(2), 2012 (2012)
15. Petras, I.: http://people.tuke.sk/igor.podlubny/USU/matlab/petras/dfod1.m
16. Tzafestas, S.G. (ed.): Methods and Applications of Intelligent Control. Springer (1997)

Identification of Fractional Order Transfer Function Model Using Biologically Inspired Algorithms

Klaudia Dziedzic$^{(\boxtimes)}$

AGH University of Science and Technology, Krakow, Poland
klaudia.dziedzic04@gmail.com

Abstract. This paper presents the identification of a non-integer order model for the heat transfer process using the particle swarm optimization algorithm (PSO), cockroach swarm optimization algorithm (CSO), gray wolf optimizer algorithm (GWO) and fminsearch function. In the beginning, fractional order systems have been discussed. Then an overview of individual optimization methods was prepared. Simulations have been carried out for all used the algorithms.

Keywords: Fractional order systems · Identification ·
PSO algorithm · fminsearch · CSO algorithm · GWO algorithm

1 Introduction

The fractional order differential equations exist almost as long as itself fractional calculus. The history of their creation is connected with the achievements of famous scholars such as Leibniz and Newton [1], which took place almost 300 years ago. The following works date back to the nineteenth century when Liouville together with Riemann created definitions of derivative-integrals of an fractional order. During all this time, numerous experiments were carried out, which proved that the fractional calculus can describe the real processes in a more accurate way. However, only at the end of the twentieth century, the appearance of digital computers offers an easy way to numerically simulate the calculation of the approximation of the non-integer order.

During the last two decades, can be observed the use of this mathematical tool in various fields. The specific system model is the first step in the design of the controller. That is why system identification is a very current topic. Several proposed methods of identification are available in the literature [2].

The identification process has a need to find transfer function parameters used to describe the object. Optimization methods can be used for this purpose. The task is to minimize the cost function, in this case, the mean square error. In such problems, it is necessary to use algorithms that are adapted to the number of unknown variables and the space of solutions. At the last decade one can observe the development of herd algorithms, which are inspired by the behavior

© Springer Nature Switzerland AG 2020
R. Szewczyk et al. (Eds.): AUTOMATION 2019, AISC 920, pp. 47–57, 2020.
https://doi.org/10.1007/978-3-030-13273-6_5

of animals [3]. The most popular algorithms are the algorithm to optimize the swarm of particles, or the cuckoo, firefly algorithms. This article discusses three algorithms in this field: the cockroach algorithm, particle swarm optimization algorithm and the gray wolf optimizer algorithm. These three algorithms are compared with the Matlab fminsearch function, which is based on the simplex method.

In this work, simulation experiments were carried out. They allow to verify the operation of individual methods in the process of identifying a fractional order model for the thermal conduction process. The step response of the object was obtained experimentally.

2   Fractional Order Systems

2.1   Definitions of Fractional Derivative-Integral Operator

The basic tool for non-integer systems is the differential-integer operator $_aD_t^\alpha$. This operator is defined as follows:

$$_aD_t^\alpha = \begin{cases} \frac{d^\alpha}{dt^\alpha} & \text{when } \alpha > 0 \\ 1 & \text{when } \alpha = 0 \\ \int_a^t (d\tau)^\alpha & \text{when } \alpha < 0 \end{cases} \tag{1}$$

There are several definitions of fractional derivative-integral operator, e.g. [4],[5]. Below, I will discuss two, most commonly used. The definition of Riemann-Liouville for $\alpha > 0$ is the function defined as follows

$$_aD_t^\alpha(t) = \frac{d^\alpha f(t)}{dt^\alpha} = \frac{1}{\Gamma(p-\alpha)} \frac{d^p}{dt^p} \int_a^t \frac{f(\tau)}{(t-\tau)^{\alpha+1-p}} d\tau, \tag{2}$$

where $\Gamma()$ is the gamma function of Euler (3), p is a positive integer satisfying limitation $p - 1 < \alpha < p$.

$$\Gamma(x) = \int_0^\infty t^{x-1} e^{-t} dt \tag{3}$$

The second important definition is the definition of Caputo, for $\alpha > 0$, which has the form

$$_aD_t^\alpha(t) = \frac{d^\alpha f(t)}{dt^\alpha} = \frac{1}{\Gamma(p-\alpha)} \int_0^t \frac{f^p(\tau)}{(t-\tau)^{\alpha+1-p}} d\tau, \tag{4}$$

2.2   Methods of Describing Fractional Order Systems

A fractional dynamic system with one input and one output is described by a non-integer differential equation of

$$\sum_{i=0}^n a_i D_t^{\alpha_i} y(t) = \sum_{k=0}^m b_k D_t^{\beta_i} u(t) \tag{5}$$

By applying the Laplace transform to both sides of the equation (5) with zero initial conditions, we get the transfer function [6]:

$$G(s) = \frac{Y(s)}{U(s)} = \frac{b_m s^{\beta_m} + b_{m-1} s^{\beta_{m-1}} + \dots + b_0 s^{\beta_0}}{a_m s^{\alpha_m} + a_{m-1} s^{\alpha_{m-1}} + \dots + a_0 s^{\alpha_0}} \qquad (6)$$

This article will look for parameters of individual coefficients and fractional order values to identify the model of a thermal object.

2.3   Methods of Approximation

Transfer function (6) is an irrational function of the variable s, therefore its approximations are realized with the help of rational functions, thanks to which it is possible determine a discrete, rational transfer function approximating a fractional order element. In order to obtain a discrete operator of the non-integer order of s^r a new operator is defined, this action can be implemented by various methods, Oustaloupa, CFE, CRONE, Carlsona.

The most popular method used in the Matlab program to calculate incomplete transmittance is the ORA - Oustaloup Recursive Approximation [7]. This method is based on the approximation of the continuous function s as follows:

$$H(s) = s^{\alpha}, \alpha \in R, \qquad (7)$$

for the frequency range ω using a rational function:

$$H(s) = C_o \Pi \frac{s + \omega'_k}{s + \omega k} \qquad (8)$$

where each coefficient have form:

$$\omega'_k = \omega_b \frac{\omega_h}{\omega_b}^{\frac{k+N+0.5(1-r)}{2N+1}}, \omega_k = \omega_b \frac{\omega_h}{\omega_b}^{\frac{k+N+0.5(1-r)}{2N+1}}, C_o = \frac{\omega_h}{\omega_b}^{\frac{-r}{2}} \Pi \frac{\omega_k}{s + \omega'_k} \qquad (9)$$

3   Optimization Methods

3.1   Fminsearch Function

The Matlab *fminsearch* function is a non-linear function for solving mathematical problems available on the Matlab platform [22]. It is used to optimize the solution by minimizing a given cost function. *Fminsearch* starts with the initial vector and tries to find the nearest local minimum. It is based on the Nelder-Mead simplex algorithm of direct search, which uses a simplex array for n + 1 points for n-dimensional x vectors [8].

3.2 PSO - Particle Swarm Optimization

Another method of optimization used in this work is Particle Swarm Optimization (PSO). It is a heuristic optimization method based on swarm intelligence. A single solution, called a particle, flies in space according to the positions and velocity of neighbors and their own parameters to find food, referred to as the optimal value. Performance is measured by the cost function associated with the specific problem [9].

   In the preliminary activities, set the initial values of the parameters and the limits of the values of each variable. The main loop of the algorithm is responsible for finding the best solution at every stage. At first, the cost function for each particle is calculated. The next step is to compare all calculated values to find the global minimum in this step. In addition, the speed and position of each particle are updated with respect to (10) and (11), and the calculations are repeated.

$$P_j(m + 1) = P_j(m) + V_j(m) \tag{10}$$

$$V_j(m + 1) = c_1 r_1 V_j(m) + c_2 r_2 (PBest - P_j(m)) + c_3 r_3 (GBest - P_j(m)) \tag{11}$$

where: $P_j(m)$ i $V_j(m)$ indicate the position and velocity of the jth particle in the mth iteration, $PBest$ is the best solution for a given particle, $GBest$ is the best global solution in the population at stage m, c1 and c2 are learning rates, r1 and r2 represent a random number between (0,1), $j = 1, ..., J$ - number of particles, $m = 1, ..., M$ - number of iterations.

3.3 Cockroach Swarm Optimization

Another optimization method that is biologically inspired is the cockroach algorithm CSO - Cockroach Swarm Optimization. It is modeled on three behaviors observed in these insects [10]. Each cockroach stores some solution to the problem that we are investigating at the moment. The algorithm is iterative, before starting it, we choose the initial values of the population of cockroaches. In each iteration, there are three phases that attempt to imitate insect behavior [11].

1. **Swarm**
   Every cockroach searches in population for a cockroach, that has a better solution than its own. If such a cockroach is achieved, the location in space changes towards an insect with a better solution (12). If not, the step is done in the direction of the current global solution. For this phase, visibility of the neighborhood is an important concept (13), defined by the *visual* parameter.

$$x_i = \begin{cases} x_i + step * rand * (pBest - x_i) \text{ gdy } x_i \neq p_i \\ x_i + step * rand * (gBest - x_i) \text{ gdy } x_i = p_i \end{cases} \tag{12}$$

$$x_i = Opt * (x_j, |x_i - x_j| \leq visual) \tag{13}$$

2. **Dispersing**

This phase introduces randomness to the algorithm. Each cockroach performs a random step (depending on the step parameter setting at the beginning) in the solution space (14). One additional component of hunger is added to this phase. A special *hungerthreshold* is set. Once it is reached, the cockroach migrates towards food (15).

$$x_i = x_i + rand(1, D) \tag{14}$$

$$x_i = x_i + (x_i - ct) + x_{food} \tag{15}$$

3. **Ruthless behavior**

This stage reflects the eating of cockroaches by each other. A stronger individual eats the weaker one. In the algorithm, this causes that the random cockroach takes on the current global optimum.

3.4 Grey Wolf Optimizer

Grey Wolf Optimizer (GWO) is a method of optimization based on the behavior of the apex predator (it is at the top of the food chain), a gray wolf living in the herd. Especially interesting is that they have a strictly defined hierarchy of individuals. The leader is an alpha wolf, who is mainly responsible for making decisions about hunting, sleeping, etc. The second level in hierachi is occupied by the beta wolf. They are subordinate wolves that help the alpha to make decisions or other activities related to the pack. It is the first candidate to become alpha at the moment when the alpha wolf dies. The beta wolf contacts the wolves from the herd in order to forward messages from alpha and gives feedback to the alpha. There is also a wolf delta, which is subordinate to alpha and beta, but dominates over the lowest rank omega. It is made up of wolves serving as scouts, guards, and hunters. Wolf omega plays the role of a scapegoat, they must give in to all other wolves. But often they play the role of taking care of the young members of the pack. The roles in the group are, for example, scouts who are responsible for observing the territory's boundaries and warning the herd in case of any threat. The guards protect and guarantee the security of the pack. Elders are experienced wolves who used to be alpha or beta. Hunters help Alpha and Beta when hunting for prey and providing food for the herd. Caretakers are responsible for looking after the weak, sick and wounded wolves in the herd. Group hunting is the most important and the most interesting social behavior of gray wolves [13].

The optimization algorithm is based on the social hierarchy and the gray wolf hunting phases.

1. **Tracking, racing and approaching the victim**

Gray wolves surround the prey during the hunt. The wolf tries to update his position towards the victim (16), (17). Mathematically, this is presented as follows:

$$D = |C * X_p(t) - X(t)| \tag{16}$$

$$X(t + 1) = X_p(t) - A * D \tag{17}$$

where t means current iteration, A and C are vector coefficients, Xp is the victim position vector, and X is the agent position, a is linearly reduced from 2 to 0 in relation to iteration, $r1$ and $r2$ represent a random number between (0.1). You can reach different places around the best agent by changing the value of a and c.

2. **Harassment, surrounding the victim**
   These individuals have the ability to recognize the victim's position and surround them. The hunt directs alpha, but beta and delta can help. In the case of an abstract search space, we have no idea about the location. To mathematically simulate this, we assume that alpha (the best solution), beta and delta have potentially better knowledge about the victim. That is why, we save and force other agents to search in given directions. This is done using (16), (17) for each individual in the herd. At the end, the location of the gray wolf is calculated according to

$$X(t + 1) = \frac{X_1 + X_2 + X_3}{3} \tag{18}$$

3. **Attack of prey**
   Gray wolves end up hunting by attacking the victim when he stops moving. To mathematically present this, we assume the value of a, which decreases as the iteration progresses.

4   Experimental System

We are considering the experimental heat plant shown in the Fig. 1. It has the form of a thin copper bar with a length of 260 [mm]. The rod is heated by an electric furnace with a length of Δ x0 = 0.14 located at one end of the rod. The output temperature is measured using Pt100 sensors located at points: 0.29, 0.50 and 0.73 of the rod length. The system input signal is a standard current signal in the range of 0–20 [mA]. It is amplified to the range 0–1.5 [A]. Next is the signal input for the radiator. Signals from Pt100 sensors are read directly by the analog input module in the PLC. The data from the PLC is read using the SCADA system. The entire system is connected via PROFINET [12].

5   Cost Function

As a cost function, MSE - Mean Squared Error was accepted for each method. The MSE cost function describes the difference between the step response of the model and the experimental heat plant, according to the formula:

$$MSE = \frac{1}{K_s} \sum_{k=1}^{K} (y_{model,j} - y_{real,j})^2. \tag{19}$$

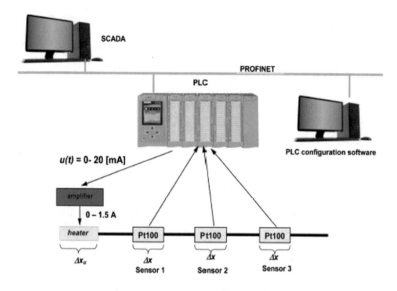

Fig. 1. The considered heat plant

In (19) $y_{model,j}$ means step response of model calculated using the appropriate optimization at j-th moment, $y_{real,j}$ is the step response of the real heating system at j-th moment. The considered transfer function is in the form:

$$G(s) = \frac{1}{a_2 s^{\alpha_1} + a_1 s^{\alpha_0} + a_0}$$ (20)

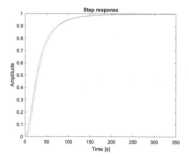

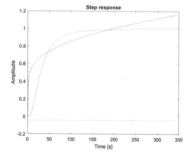

Fig. 2. Step response of the model with the fminsearch (blue) and plant (green) optimization. Transfer function (21).

Fig. 3. Step response of the model with the fminsearch (blue) and plant (green) optimization. Transfer function (22).

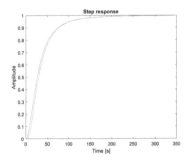
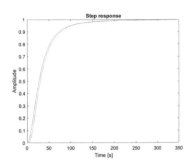

Fig. 4. Step response of the model with PSO optimization (blue) and plant (green). Transfer function (23).

Fig. 5. Step response of the model with PSO optimization (blue) and plant (green). Transfer function (24).

6   Numerical Tests

This section is divided into individual sub-chapters, which contain initial parameters, output transfer functions, and charts of the identification process. The Table 1 presents a comparison of all methods.

6.1   Fminsearch Function

The initial values of vector have the form $[0.5, 0.5, 1, 30, 30]$ (Figs. 2 and 3).

$$G(s) = \frac{1}{285.1s^{2.1237} + 31.638s^{0.98037} + 1} \tag{21}$$

The initial values of vector have the form $[0, 0, 1, 30, 30]$.

$$G(s) = \frac{1}{24.384s^{0.011007} + s^{0.007056} - 22.901} \tag{22}$$

In (22), it can be seen, how fminsearch can fail to locate the minimum of cost function at the time of choosing inappropriate values of the initial vector.

6.2   Particle Swarm Optimization

Initial values for swarm parameters are chosen from compartments (Figs. 4 and 5).

$$G(s) = \frac{1}{190.8s^{1.9425} + 27.646s^{0.95517} + 1} \tag{23}$$

$$G(s) = \frac{1}{248.66s^{2.0765} + 30.665s^{0.96878} + 1} \tag{24}$$

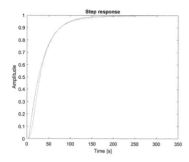

Fig. 6. Step response of the model with CSO optimization (blue) and plant (green). Transfer function (25).

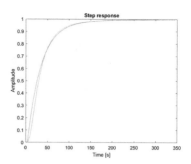

Fig. 7. Step response of the model with CSO optimization (blue) and plant (green). Transfer function (26).

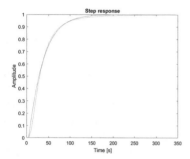

Fig. 8. Step response of the model with GWO optimization (blue) and plant (green). Transfer function (27).

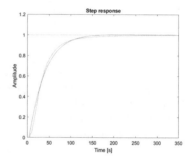

Fig. 9. Step response of the model with GWO optimization (blue) and plant (green). Transfer function (28).

6.3   Cockroach Swarm Optimization

The initial values of vector have the form $[1, 1, 1, 30, 30]$ (Figs. 6 and 7).

$$G(s) = \frac{1}{28.775s^{1.7775} + 29.069s^{0.98388} + 1} \tag{25}$$

The initial values of vector have the form $[0, 0, 1, 30, 30]$.

$$G(s) = \frac{1}{16.206s^{1.1659} + 16.998s^{0.91703} + 1} \tag{26}$$

6.4   Grey Wolf Optimizer

Initial values for swarm parameters are chosen from compartments (Figs. 8 and 9).

$$G(s) = \frac{1}{25.956s^{1.6828} + 31.756s^{0.9946} + 1} \tag{27}$$

$$G(s) = \frac{1}{0.2621s^{4.3291} + 38.956s^{1.0248} + 1} \tag{28}$$

Table 1. Summary of the results of the considered algorithms

Parameter	Optimization method							
	$fmins(20)$	$fmins(21)$	$PSO2(28)$	$PSO(23)$	$CSO(24)$	$CSO(25)$	$GWO(26)$	$GWO(27)$
Iteration	532	484	300	200	100	100	100	100
MSE	$6.89e-06$	$1.48e-02$	$2.32e-05$	$8.42e-06$	$5.56e-04$	$7.82e-04$	$3.72e-04$	$5.15e-04$
Time	$58.3sek$	$45.7sek$	$163.94sek$	$83.55sek$	$208.97sek$	$125.67sek$	$41.22sek$	$45.81sek$
Amount of swarm	4	4	10	10	5	5	10	10
Initial vector	TAK	TAK	–	–	TAK	TAK	–	–
Number of initial parameters	1	1	6	6	12	12	3	3

7   Conclusions

- Biologically inspired optimization methods used in this work are very well suited to identify models of complex dynamics systems with fractional order. Using a non-integral calculation allows the model to be approximated to the object's response.
- The table shows that the most effective algorithm is the gray wolf algorithm, because it not only has very good results and the shortest time, but also does not need a vector of initial variable values and does not set any additional initial parameters. Only scopes can be given, and in other cases, time should be spent to find optimization parameters.
- After analyzing the last row of the table, it is visible that the most important problem is setting the appropriate initial parameters, which complicates the whole process. This applies to CSO and PSO algorithms.

References

1. Weilbeer, M.: Efficient Numerical Methods for Fractional Differential Equations and their Analytical Background. Technischen Universitat Braunschweig (2005)
2. Malti, R., Aoun, M., Sabatier, J., Oustaloup, A.: Tutorial on system identification using fractional differentiation models. In: 14th IFAC Symposium on System Identification on International Federation of Automatic Control (IFAC), March 2006, Newcastle, Australia. pp. 606–611 (2006)
3. Kwiecien, J. Filipowicz, F.: Algorytmy stadne w problemach optymalizacji. Automatyka, Tom 11, Zeszyt 2 (2011)
4. Das, S.: Functional Fractional Calculus for System Identification and Controls. Springer, Heidelberg (2008)
5. Kaczorek, T.: Selected Problems of Fractional System Theory. Springer, Heidelberg (2011)
6. Meng, L., Wang, D., Han, P.: Identification of fractional order system using particle swarm optimization. In: 2012 International Conference on Machine Learning and Cybernetics, July 2012. https://doi.org/10.1109/ICMLC.2012.6359551
7. Petras I.: Fractional derivatives, fractional integrals and fraction differential equations. Technical University of Kosice (2012)
8. MathWorks. Documentation - fminsearch. https://www.mathworks.com/help/matlab/ref/fminsearch.html

9. Shi, Y., Eberhart, R.: A modified particle swarm optimizer. In: IEEE World Congress on Computational Intelligence, June 1998. https://doi.org/10.1109/ICEC.1998.699146
10. Kwiecien, J., Filipowicz, B.: Comparison of firefly and cockroach algorithms in selected discrete and combinatorial problems. Bull. Pol. Acad. Sci. Tech. **62**(4), 797–804 (2014)
11. Obagbuwa, I., Adewumi, A.: An improved cockroach swarm optimization. PubMed, May 2014. https://doi.org/10.1155/2014/375358
12. Oprzedkiewicz, K., Dziedzic, K.: New parameter identification method for the fractional order, state space model of heat transfer process. In: Automation 2018, pp. 401–417. Springer (2018). https://doi.org/10.1007/978-3-319-77179-3_38
13. Jangir, P., Bhesdadiya, R., Ladumor, D., Trivedi, I.: A multi-objective grey wolf optimization algorithm for economic/environmental dispatch. In: PNFE-2016, At St. Peters Engineering college, Hyderabad, India, October 2016. https://doi.org/10.13140/RG.2.2.21536.79364

Low Phase Shift and Least Squares Optimal FIR Filter

Mateusz Saków[✉]

Faculty of Mechanical Engineering and Mechatronics,
Department of Mechatronics, West Pomeranian University of Technology,
Szczecin, Poland
mateusz.sakow@zut.edu.pl

Abstract. The problem of a significant phase shift in a control loop is posing a lot of challenges to the control design. One of them is definitely the loss of performance with the increased phase shift when using a filter at the system's output. This paper contains a description of a new FIR weights determination method focused on low-pass filter design. The primary goal of this method is to minimize the phase shift caused by the filter. The filter theoretically fits a defined polynomial to an asymmetric data set. In this case, the nearest neighbour samples are only taken from the past side of the filtered vector of signal samples. This feature allows reducing the value of the phase shift, especially for a low-frequency spectrum. Therefore, the filter can be used directly in the closed-loop control and will minimize the loss of system performance.

Keywords: FIR filter design · Low phase shift filtering · LS optimal filtration

1 Introduction

The problem of a significant phase shift in the control loop is posing a lot of challenges to the control design. One of them is definitely the loss of performance with the increased phase shift when using a filter at the system's output. However, scientific literature provides multiple solutions that can tackle this problem. To this solutions, methods of Finite Impulse Response (FIR) filter design belong. One of them are the windowed approach that uses a Hamming [16] or a Kaiser [18] window to design low-pass FIR filters with a linear phase. To the group of FIR filters with a linear phase, we can also add a method that minimizes the weighted, integrated squared error between an ideal piecewise linear function and the magnitude response of the filter [18] or square-root raised cosine FIR filters with roll-off factor [35]. All three methods lead to comparable results. Unfortunately, they also lead to a delay problem in the system caused by the filter, and to improve the performance, these systems require methods of prediction e.g. by using the Smith Predictor [23, 24] or model-free prediction based on phase shifting [27, 29].

An alternative solution to the above was proposed in [10]. However, this approach of finding FIR filter coefficients leads to nonlinear phase diagram and finally to non-constant delay. Moreover, another alternative solution exists which was proposed by Kalman in 1960 [9], called later by the Kalman Filter. The filter was proposed to

© Springer Nature Switzerland AG 2020
R. Szewczyk et al. (Eds.): AUTOMATION 2019, AISC 920, pp. 58–66, 2020.
https://doi.org/10.1007/978-3-030-13273-6_6

overcome problems of random signal prediction and linear filtration, and it is used till this day as one of the advanced methods of filtration used to cope with modern engineering problems [4, 12, 19]. Nevertheless, the Kalman Filter is supported by a model of the system. Therefore, this feature leads to a problem of complex system modelling [11].

It is important to not forget about the linear and continues filters design like Bessel-Butterworth [6], Elliptic [1] or Chebyshev [10] filters which needs to be described by a FIR filter architecture design to be used in a discrete controller. In all cases, however, the phase diagrams remains significant and nonlinear. Furthermore, the continuous and linear filters could be adopted to fractional order theory. Nevertheless, they are still described by a significant phase shift [2, 20, 21, 36].

As another alternative solution to the problem, the adaptive filters were proposed [5, 7, 22]. Considered as simple, the Least Means Squares (LMS) filter [5] implements an adaptive FIR filter by using dedicated algorithms. The LMS algorithm estimates the filter weights needed to minimize the error, between the output signal and the desired signal. Moreover, a comparable approach is represented by the Recursive Least Squares (RLS) filter [7]. In this case, the filter recursively computes the least squares estimate of the FIR filter weights. It also estimates the filter weights that are needed to convert the input signal into the desired signal. Unfortunately, in both cases, adaptive filters require a reference signal. Therefore, their possible application is limited by random signals. Moreover, this algorithms are the most computationally intensively approaches from all of the proposed.

This paper contains a description of a new FIR weights determination method focused on low-pass filter design. The new method was based on the algorithm proposed in [26, 31]. The primary goal of this method is to minimize the phase shift caused by the filter for a low-frequency spectrum. The filter theoretically fits a defined polynomial to an asymmetric data set. In this case, the nearest neighbour samples are only taken from the past side of the filtered vector of signal samples. This feature allows reducing the value of the phase shift, especially for a low-frequency spectrum. Therefore, the filter can be used directly in the closed-loop control and will not become a cause of performance loss in the system. Moreover, the proposed solution results in local Least Squares (LS) optimal filtration of the signal but based on the filter's specific order and design.

In the case of relatively low frequency system responses, the proposed FIR filter has a potential of a widespread application. The proposed approach, can be used in teleoperation such as inverse model supported systems and their identification [25, 28–30, 32, 33] or in robotics and man machine control interfaces [13–15, 17, 19]. Thus, when the system response is characterized by a low frequency spectrum, the FIR filter can be simply implemented to multiply control schemes.

2 Determination of FIR Filter Weights

In this section, the mathematical description of the proposed method of determining weights of FIR filter is presented. Figure 1 presents the FIR filter scheme with the input signal $x(t)$ and the output signal $y(t)$.

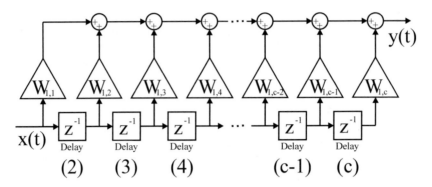

Fig. 1. FIR filter scheme.

The filter scheme in the Fig. 1 consists the delay line with a number of c unity delays z^{-1}; and weight gains $W_{1,c}$, where c is the c-th consecutive weight of the filter. In this case, the method depends on different parameters than parameters in the well-known approaches that are discussed in the introduction [18]. The damping or cut-off frequency can be obtained only experimentally – usually based on a signal's sample. The parameters that determine the filter coefficients are n – order of the polynomial; and c – the filter order (1 present sample and $c - 1$ consecutive samples from the past). Based on this two parameters X matrix is being obtained and described as (1):

$$X_{c,(n+1)} = \begin{bmatrix} 1 & 1 & 1 & \cdots & 1 \\ 1 & 2 & 4 & \cdots & 2^n \\ 1 & 3 & 9 & \cdots & 3^n \\ \vdots & \vdots & \vdots & \ddots & \vdots \\ 1 & c & c^2 & \cdots & c^n \end{bmatrix}. \tag{1}$$

The form of X matrix is very close to the matrix, that is being used to fit the n-th order polynomial in the LS sense. However, it consists only numerical sequences arranged in columns as presented above. When using a different description of the matrix X, the consecutive formulation might not be useful. After determination of the X matrix, it is required to calculate the Moore-Penrose pseudo-inverse matrix X^+ from the matrix X, e.g. (2):

$$X^+_{(n+1),c} = (X'X)^{-1}X', \tag{2}$$

where X' is the transpose of X. The pseudo-inverse matrix could be obtained in a couple of different ways, but Eq. 2 has linearly independent columns, and there is a possibility to use a basic Moore-Penrose inverse approach to calculate the inverse X^+ matrix from the X.

Then, while having the pseudo-inverse matrix X^+, a vector of weights W is calculated by summing elements in columns of pseudo-inverse matrix X^+, as (3):

$$W_{1,c} = I_{1,(n+1)}X^+_{(n+1),c},$$ (3)

where $I_{1x(n+1)}$ is a unity row vector with $n+1$ elements. Equation (3) is only true for the X matrix build as (1). This finds its confirmation in the mathematical analysis below.

If we are going to have a column vector Y (that was obtained from the standard delay line in FIR filter architecture) of samples of the signal collected from the present sample and consecutive samples from the past, fitted polynomial coefficients will be described as (4):

$$B_{(n+1),1} = X^+_{(n+1),c}Y_{c,1}.$$ (4)

However, due to the fact that the first row of the matrix X has only unity elements, it possible to just calculate the sum of actual polynomial coefficients by using the unity vector I:

$$I_{1,(n+1)}B_{(n+1),1} = \left(I_{1,(n+1)}X^+_{(n+1),c}\right)Y_{c,1}.$$ (5)

this, however, leads to a conclusion that it is possible to obtain filter weights directly from the pseudo inverse matrix X^+ by summing all elements in each column – Eq. (3).

This approach is relatively familiar to the interval fitting of polynomials [3] or smoothing methods [34]. However, it works in real-time and it is focused on minimization of the phase shift caused by the filter. Even relatively high order filter is characterized by a very low phase shift according to the group of regression type filters [8]. The phase shift caused by the filter is strongly minimized because the filter adopts the shape of the fitted polynomial. This sequence is repeated in each time it calculates the weighted sum of all signals from the present one to the all limited by order of the filter consecutive past samples. Moreover, this approach results in a local LS optimal filtration of the signal but based on the filter's specific order and the filter design.

3 Frequency Analysis of FIR Filters

In this section, a comparison in the frequency domain of FIR filters obtained by the well-known methods of FIR filter weights determination and the proposed method is carried out. Figure 2 presents a comparison of the proposed method with a variable n parameter (order of the polynomial), Fig. 3 presents a comparison with the variable c (order of the filter) and Fig. 4 presents a comparison between 3 different methods of FIR filter design and the proposed method.

The change of the *n*-th order of the polynomial fitting changes shapes of Magnitude (Fig. 2a) and Phase (Fig. 2b) diagrams. The increased parameter increases the size of the pass frequency band of the filter, however, it strongly reduces the damping for the cut-off frequency spectrum – Fig. 2a. Moreover, the phase shift is linear and closer to zero for a higher spectrum of frequency during the increase of the *n* – Fig. 2b. Unfortunately, the higher order of the fitting polynomial is, it increases the gain between the pass and the stop frequency band of the filter. The same thing could be noticed for Complex and nonlinear-phase equiripple FIR filter design [10].

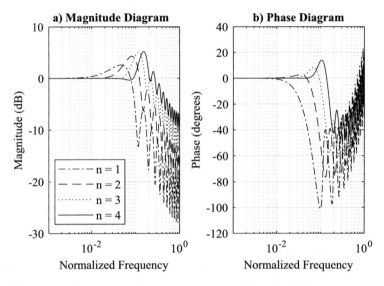

Fig. 2. Comparison of the proposed method with a variable order of the polynomial (c = 25) (a) Magnitude Diagram, (b) Phase Diagram.

The change of the filter order (c) does not have a significant influence on shapes of Magnitude (Fig. 3a) and Phase (Fig. 3b) diagrams. The increased order decreases size of the pass frequency band of the filter, however, it strongly increases the dumping for the cut-off frequency spectrum – Fig. 3a. The same thing could be noticed for the linear and close to zero part of the Phase diagram – Fig. 3b.

The comparison of the 4 types of FIR filter design leads to a conclusion that the proposed method due to other types of FIR filters design is characterized by an unnoticeable phase shift, especially for the pass frequency band – Fig. 4b. However, the method is characterized by a lower damping for the cut-off frequencies, then Window-based and Least-squares linear-phase FIR filter design methods – Fig. 4a. However, the proposed filter has a higher damping then obtained by the Complex and nonlinear-phase equiripple FIR filter design – Fig. 4a.

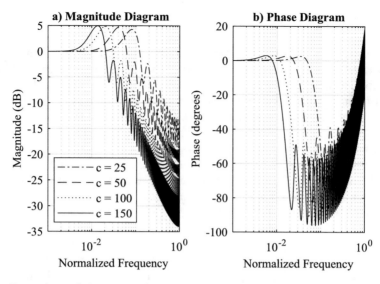

Fig. 3. Comparison of the proposed method with a variable order of the filter c (n = 2) (a) Magnitude Diagram, (b) Phase Diagram.

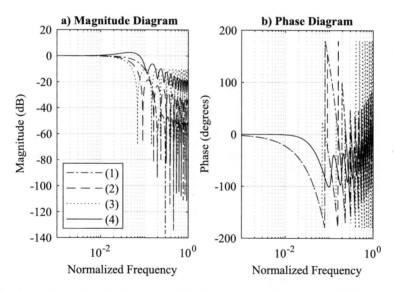

Fig. 4. Comparison: (1) Window-based; (2) Least-squares linear-phase; (3) Complex and nonlinear-phase equiripple FIR filter design; with (4) the proposed method (c = 25, n = 2), normalized cut-off frequency 0.05 and 25 samples (a) Magnitude Diagram, (b) Phase Diagram.

4   Experimental Valuation

For the experimental valuation of the proposed method, a hydraulic test stand was used. The measured control signal was a position of the 1 Degree of Freedom (DoF) hydraulic manipulator. The control unit was implemented in the dSpace 1104 programming platform and was working with the frequency of 10 kHz. In the position control loop, the proposed filter was placed as an output of the displacement sensors and was filtering the measured signal. Two different filters were compared. However, the proposed solution worked online while the results from the Window-based filter were computed offline due to the same data. The results are compared in the Fig. 5.

As the comparison signal, the Chirp was used. The signal was changing its frequency from 1 Hz to 20 Hz with 0.5 mm amplitude – Fig. 5a. The normalized cut-off frequency of the Window-based FIR filter was set to 0.05. Runs from Fig. 5a are zoomed in Fig. 5b (time period 1.8 s to 1.9 s), and Fig. 5c (time period 9.915 s to 9.925 s).

Experimental results confirm that the proposed method is characterized by a very low phase shift due to other types of FIR filter design methods. The difference is noticeable at lower frequency (1 Hz) of system operation – Fig. 5b, but also at higher frequencies (15 Hz) – Fig. 5c.

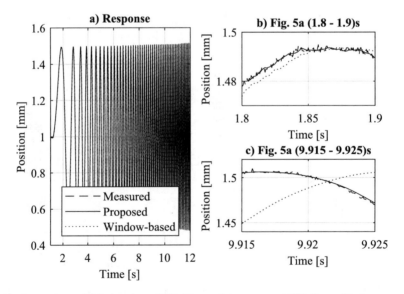

Fig. 5. Comparison of Window-based FIR filter and the proposed FIR filter with the same order of 200 samples at 10 kHz.

5 Conclusions

A new FIR weights determination method that results in low-pass filter design was presented. The primary goal of this method was to minimize the phase shift caused by the filter for a low-frequency spectrum. This was confirmed by analysis carried out in the frequency domain. The filter fitted a defined polynomial to an asymmetric data set. This feature allowed to reduce the value of the phase shift, especially for a low-frequency spectrum. It was also proven during experiments that the filter could be used directly in the closed-loop control and will not become a cause of the system's performance loss.

References

1. Ali, F., Jain, R., Gupta, D., et al.: Design and analysis of low pass elliptic filter. In: 2016 Second International Conference on Computational Intelligence & Communication Technology (CICT). Ghaziabad, India, pp. 449–451 (2016)
2. Baranowski, J., Piątek, P., Bauer, W., et al.: Bi-fractional filters, part 2: right half-plane case. In: 2014 19th International Conference on Methods and Models in Automation and Robotics (MMAR), Międzyzdroje, Poland, pp. 369–373. IEEE (2014)
3. Cleveland, W.S., Devlin, S.J.: Locally weighted regression: an approach to regression analysis by local fitting. J. Am. Stat. Assoc. **83**, 596–610 (1988)
4. Fatehi, A., Huang, B.: Kalman filtering approach to multi-rate information fusion in the presence of irregular sampling rate and variable measurement delay. J. Process Control **53**, 15–25 (2017)
5. Ferrara, E.: Fast implementations of LMS adaptive filters. IEEE Trans. Acoust. Speech Signal Process. **28**, 474–475 (1980)
6. Filanovsky, I.M.: Bessel-Butterworth transitional filters. In: 2014 IEEE International Symposium on Circuits and Systems (ISCAS), Melbourne, VIC, Australia, pp. 2105–2108 (2014)
7. Hayes, M.H.: Statistical Digital Signal Processing and Modeling. Wiley, New York (2009)
8. Janecki, D., Cedro, L.: Determining of signal derivatives with the use of regressive differential filters. Przegląd Elektrotechniczny **87**, 253–259 (2011). (in Polish)
9. Kalman, R.E.: A new approach to linear filtering and prediction problems. J. Basic Eng. **82**, 35–45 (1960)
10. Karam, L.J., Mcclellan, J.H.: Complex Chebyshev approximation for FIR filter design. IEEE Trans. Circuits Syst. II Analog. Digit. Signal Process. **42**, 207–216 (1995)
11. Kaya, I.: Obtaining controller parameters for a new PI-PD Smith predictor using autotuning. J. Process Control **13**, 465–472 (2003)
12. Krämer, D., King, R.: A hybrid approach for bioprocess state estimation using NIR spectroscopy and a sigma-point Kalman filter. J. Process Control (2017, in Press)
13. Miadlicki, K., Pajor, M., Sakow, M.: Loader crane working area monitoring system based on LIDAR scanner. In: Advances in Manufacturing, pp. 465–474. Springer (2017)
14. Miądlicki, K., Pajor, M., Saków, M.: Ground plane estimation from sparse LIDAR data for loader crane sensor fusion system. In: 2017 22nd International Conference on Methods and Models in Automation and Robotics (MMAR), Międzyzdroje, Poland, pp. 717–722. IEEE (2017)
15. Miądlicki, K., Pajor, M., Saków, M.: Real-time ground filtration method for a loader crane environment monitoring system using sparse LIDAR data. In: 2017 IEEE International Conference on INnovations in Intelligent SysTems and Applications (INISTA), pp. 207–212. IEEE (2017)

16. Mitra, S.K., Kuo, Y.: Digital Signal Processing: A Computer-Based Approach. McGraw-Hill Higher Education, New York (2006)
17. Okulski, M., Ławryńczuk, M.: A cascade PD controller for heavy self-balancing robot. In: Conference on Automation, pp. 183–192. Springer (2018)
18. Oppenheim, A.V.: Discrete-Time Signal Processing. Pearson Education India (1999)
19. Owczarek, P., Goslinski, J., Rybarczyk, D., et al.: Modeling and 3D simulation of an electro-hydraulic manipulator controlled by vision system with Kalman Filter. In: Advances in Manufacturing, pp. 375–384. Springer (2018)
20. Piątek, P., Baranowski, J., Zagórowska, M., et al.: Bi-fractional filters, part 1: left half-plane case. In: Advances in Modelling and Control of Non-integer-Order Systems, pp. 81–90. Springer (2015)
21. Psychalinos, C., Tsirimokou, G., Elwakil, A.S.: Switched-capacitor fractional-step butterworth filter design. Circuits Syst. Signal Process. **35**, 1377–1393 (2016)
22. Ra, W.S., Whang, I.H.: Recursive weighted robust least squares filter for frequency estimation. In: 2006 SICE-ICASE International Joint Conference, Busan, South Korea, pp. 774–778 (2006)
23. Raja, G.L., Ali, A.: Smith predictor based parallel cascade control strategy for unstable and integrating processes with large time delay. J. Process Control **52**, 57–65 (2017)
24. Rodríguez, C., Normey-Rico, J., Guzmán, J., et al.: On the filtered Smith predictor with feedforward compensation. J. Process Control **41**, 35–46 (2016)
25. Rybarczyk, D., Owczarek, P., Myszkowski, A.: Development of force feedback controller for the loader crane. In: Advances in Manufacturing, pp. 345–354. Springer (2018)
26. Saków, M.: Real-Time and low phase shift noisy signal differential estimation dedicated to teleoperation systems, pp. 132–141. Springer International Publishing, Cham (2018)
27. Saków, M., Marchelek, K.: Model-free and time-constant prediction for closed-loop systems with time delay. Control Eng. Pract. **81**, 1–8 (2018)
28. Saków, M., Marchelek, K., Parus, A., et al.: Signal prediction in bilateral teleoperation with force-feedback, pp. 311–323. Springer International Publishing, Cham (2018)
29. Saków, M., Miądlicki, K.: Transport delay and first order inertia time signal prediction dedicated to teleoperation, pp. 142–151. Springer International Publishing, Cham (2018)
30. Saków, M., Miądlicki, K., Parus, A.: Self-sensing teleoperation system based on 1-dof pneumatic manipulator. J. Autom. Mob. Robot. Intell. Syst. **11**, 64–76 (2017)
31. Saków, M., Parus, A., Miądlicki, K.: LS filter and its implementation into the control unit of the master-slave system with force-feedback (in Polish). In: Modelowanie inżynierskie, vol. 34 (2017)
32. Sakow, M., Parus, A., Pajor, M., et al.: Unilateral hydraulic telemanipulation system for operation in machining work area. In: Advances in Manufacturing, pp. 415–425. Springer (2018)
33. Saków, M., Parus, A., Pajor, M., et al.: Nonlinear inverse modeling with signal prediction in bilateral teleoperation with force-feedback. In: 2017 22nd International Conference on Methods and Models in Automation and Robotics (MMAR), Międzyzdroje, Poland, pp. 141–146. IEEE (2017)
34. Schoenberg, I.J.: Spline functions and the problem of graduation. Proc. Natl. Acad. Sci. **52**, 947–950 (1964)
35. Tranter, W.H., Rappaport, T.S., Kosbar, K.L., et al.: Principles of Communication Systems Simulation with Wireless Applications. Prentice Hall, Upper Saddle River (2004)
36. Tsirimokou, G., Psychalinos, C., Elwakil, A.S.: Digitally programmed fractional-order Chebyshev filters realizations using current-mirrors. In: 2015 IEEE International Symposium on Circuits and Systems (ISCAS), pp. 2337–2340 (2015)

Low Phase Shift Differential FIR Filter Design

Mateusz Saków[✉]

Faculty of Mechanical Engineering and Mechatronics,
Department of Mechatronics, West Pomeranian University of Technology,
Szczecin, Poland
mateusz.sakow@zut.edu.pl

Abstract. A low-phase shift and real-time differential estimation is a common problem in the control design. When linear phase shift filters are used at the system's output, a strong performance loss appears with the increased phase shift. This paper presents a new FIR weights determination method focused on low-pass and differential filter design. The primary goal of this method is to minimize the phase shift caused by the differential FIR filter inside the control unit. In this case, the filter fits a defined polynomial to an asymmetric data set, then calculates a specified weighted sum. This feature allows reducing the value of the phase shift, even for differential estimation, and especially for the low-frequency spectrum. Moreover, the proposed solution results in LS optimal local differentiation based on the filter design.

Keywords: FIR filter design · Low phase shift filtering · Differentiation

1 Introduction

A low-phase shift and real-time differential estimation is a common problem in the control design. When linear phase shift filters are used at the system's output, a strong performance loss appears with the increased phase shift. The scientific literature provides multiple solutions that can tackle this problem. To this solutions, methods of FIR filter design [21] belong. However, this leads to a problem of a delay in the system caused by the filter, and to improve the performance, these systems require methods of prediction e.g. by using the Smith Predictor [26, 27] or model-free prediction based on phase shifting [30, 32].

Many alternative solutions exists, and one of them was proposed by Kalman in 1960 [11], called later by the Kalman Filter. The filter was proposed to overcome problems of random signal prediction and linear filtration. It is used till this day as one of the advanced methods of differentiation solving modern engineering problems [5, 16, 22]. Nevertheless, the Kalman Filter is supported by a model of the system. Therefore, this feature leads to a problem of complex system modelling [13].

Scientific literature provides many alternative solutions, however, most of them are dedicated to offline operation. In 2014, Knowles and Renka divided methods for numerical differentiation of noisy data into six groups [14]: (i) least squares polynomial approximation [3]; (ii) Tichonov regularization [4]; (iii) smoothing splines [37]; (iv) convolution with a Friedrichs mollifier [8]; (v) Knowles and Wallace variational

R. Szewczyk et al. (Eds.): AUTOMATION 2019, AISC 920, pp. 67–76, 2020.
https://doi.org/10.1007/978-3-030-13273-6_7

methods; and (vi) total variation regularization [14, 15]. E.g., considered in literature as a simple method, the least squares polynomial approximation consisting of fitting the data points in the LS sense with a sequence of low-degree polynomials is dedicated mostly to offline estimation [3]. However, adopting these methods to online use leads to a constant time shift equal to half of the neighbours' samples used – see the Janecki and Cedro regression filter [10]. The Tichonov regularization is dedicated to offline data smoothing because of its computationally intensive algorithm [4]. However, as it turned out, this method gives the best mean estimation error result according to any other method presented in the literature [14].

Computational intensiveness of the Tichonov algorithm is incomparable with computational intensiveness of adaptive filters proposed in [6, 9, 25] that can be used for differential estimation. Considered as simple, the Least Means Squares (LMS) filter [6] estimates the filter weights needed to minimize the error, between the output signal and the desired signal. A comparable approach is represented by The Recursive Least Squares (RLS) filter [9] which also estimates the filter weights needed to convert the input signal into the desired signal. Unfortunately, in both cases, adaptive filters require a reference signal. Therefore, their possible application is limited by random signals. This algorithms are the most computationally intensively approaches from all of the proposed.

There is another possibility created by using linear and continues filters design filled with an additional differential component like Bessel-Butterworth [7], Elliptic [1] or Chebyshev [12] filters. In all cases, however, the phase diagrams remains significant and nonlinear. Furthermore, the continuous and linear filters could be adopted to fractional order theory. Nevertheless, they are still described by a significant phase shift [2, 23, 24, 38].

This paper presents a new Finite Impulse Response (FIR) weights determination method focused on low-pass and differential filter design. The method was based on the algorithm proposed in [29, 34]. The filter fits a defined polynomial to an asymmetric data set and then calculates the weighted sum from the fitted coefficients. In this case, the nearest neighbour samples are only taken from the past side of the signal samples that are collected in the vector. This feature allows reducing the value of the phase shift, especially for a low-frequency spectrum. For this reason, the differential filter can be used directly in the closed-loop control and will not become a cause of a loss of system performance. Moreover, the proposed solution results in the Least Squares (LS) optimal differentiation of the filtered signal but based on the differential filter's specific order and design.

In the case of relatively low frequency system responses, the proposed differential FIR filter has a potential of a widespread application. The proposed approach, can be used in teleoperation such as inverse model supported systems and their identification [28, 31–33, 35, 36] or in robotics and man machine control interfaces [17–20, 22]. Nevertheless, when the system response is characterized by a low frequency spectrum, the differential FIR filter can be simply implemented to multiply control schemes for e.g. velocity or acceleration estimation.

2 Determination of Differential FIR Filter Weights

In this section, the mathematical description of the proposed differential FIR filter weights determination method is presented. Figure 1 presents a standard FIR filter scheme with the input signal $x(t)$ and the output signal $y(t)$.

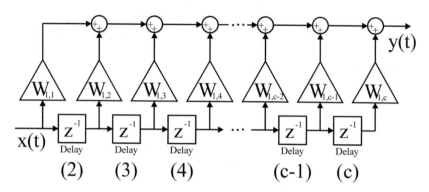

Fig. 1. FIR filter scheme.

The filter scheme that is presented in the Fig. 1 consists the delay line with a number of c unity delays z^{-1}; and weight gains $W_{1,c}$, where c is the c-th consecutive weight of the filter. In this case, the method depends on different parameters, then parameters in the well-known methods that are used to estimate weights for FIR filter [21]. The damping or cut-off frequency can be obtained only experimentally – usually based on a signal's sample. The parameters that determines the differential filter coefficients are: (i) n – order of the polynomial; (ii) c – filter order (1 present sample and $c-1$ consecutive samples from the past); (iii) p differential order; and (iv) f – controller frequency in [Hz]. Based on n and c parameters, X matrix is being obtained and described as (1):

$$X_{c,(n+1)} = \begin{bmatrix} 1 & 1 & 1 & \cdots & 1 \\ 1 & 2 & 4 & \cdots & 2^n \\ 1 & 3 & 9 & \cdots & 3^n \\ \vdots & \vdots & \vdots & \ddots & \vdots \\ 1 & c & c^2 & \cdots & c^n \end{bmatrix}. \tag{1}$$

Matrix X is very close to the matrix, that is used to fit the n-th order polynomial in the least squares sense. However, it consists only numerical sequences arranged in columns as (1). When a different description will be used, the consecutive description will not be useful.

After determination of the X matrix, it is required to calculate the Moore-Penrose pseudo-inverse matrix X^+ from the matrix X, as (2):

$$X^+_{(n+1),c} = (X'X)^{-1}X',\tag{2}$$

where X' is the transpose of X. The pseudo-inverse matrix could be obtained in couple of different ways, but (1) has linearly independent columns, and it is possible to use the More-Penrose inverse approach to estimate the X^+ inverse matrix from (1). When pseudo-inverse matrix X^+ was obtained, the vector of weights W is calculated by a weighted sum of elements in columns of (2), and then all weights are increased by the additional weight $-(-f)^p$ as (3):

$$W_{1,c} = \left(L_{1,(n+1)}X^+_{(n+1),c}\right)(-f)^p,\tag{3}$$

where $L_{1x(n+1)}$ is a row vector with $n+1$ elements that comes from the theoretical derivative of n-th order polynomial (example for the first differential and $n = 2$, $L = [0, 1, 2]$, or for the second differential and $n = 3$, $L = [0, 0, 2, 6]$). (3) is only true for the X matrix obtained as (1). Moreover, (3) is confirmed by mathematical analysis below.

If we have a column vector Y (that was obtained from the standard delay line in FIR filter architecture) of samples of the signal collected from the present sample and consecutive samples from the past, fitted polynomial coefficients will be described as (4):

$$B_{(n+1),1} = X^+_{(n+1),c}Y_{c,1}.\tag{4}$$

However, due to the fact that the first row of X is unitary, there is a possibility to calculate the weighted sum of polynomial coefficients by using the row vector L, and increase it by the $(-f)^p$ as (5):

$$L_{1,(n+1)}B_{(n+1),1}(-f)^p = \left(L_{1,(n+1)}X^+_{(n+1),c}\right)(-f)^pY_{c,1},\tag{5}$$

however, this leads to a conclusion that differential filter weights can be obtained directly from the pseudo inverse matrix X^+ by calculating a weighted sum of all elements in each column as (3).

This approach is familiar to the polynomials interval fitting [3] or smoothing approaches [37]. However, the presented differential FIR filter design works in real-time and is focused on minimization of the phase shift caused by the differential filter. Even relatively high order differential filter is characterized by a very linear and close to constant phase shift according to the group of regression type filters [10]. Moreover, the proposed approach results in local LS optimal differentiation of the filtered signal but based on the filter's specific order and the filter design.

3 Frequency Analysis of Differential FIR Filters

In this section, a comparison of differential FIR filters obtained by known methods of filter design and the proposed method is carried out. Figure 2 presents a comparison of the proposed method with a variable n (order of the polynomial), Fig. 3 presents a comparison with the variable c (order of the differential filter) and Fig. 4 presents a comparison between a regression differential FIR filter and the proposed differential FIR filter.

The change of a Magnitude (Fig. 2a) and Phase (Fig. 2b) diagrams shape is caused by setting the order n of the fitted polynomial. While n is increased, it increases the size of the pass frequency band of the filter, however, it strongly reduces the dumping for the cut-off frequency spectrum – Fig. 2a. Furthermore, the phase shift is linear and closer to constant for a higher frequency spectrum while n will be set with higher values – Fig. 2b. Unfortunately, by using a higher order of the polynomial will cause a deformation of the gain between the pass and the stop frequency bands of the differential FIR filter - Fig. 2a.

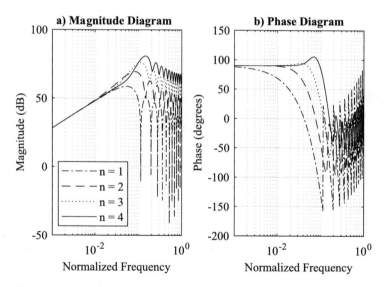

Fig. 2. Comparison of the proposed method with a variable order of the polynomial ($c = 25$) (a) Magnitude Diagram, (b) Phase Diagram.

Most importantly, the change of the differential filter order (c) do not have a significant impact on shape of Magnitude (Fig. 3a) and Phase (Fig. 3b) diagrams. However, the increased order (c) decreases the size of the differential filter linear pass frequency band, while instead, it increases the dumping for the cut-off frequency spectrum – Fig. 3a. The same remark is noticed for the linear and close to 90° part of the Phase diagram – Fig. 3b.

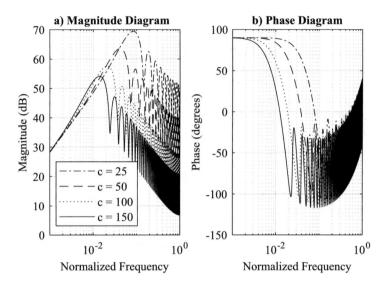

Fig. 3. Comparison of the proposed method with a variable order of the differential filter c (n = 2) (a) Magnitude Diagram, (b) Phase Diagram.

Two types of differential FIR filter design comparison leads to a conclusion that the proposed method is characterized by an unnoticeable phase shift due to other type of differential FIR filter design. This remark is visible for the pass frequency band – Fig. 4b. But the cost of the method is a lower damping for the cut-off frequency spectrum, then the same order Janecki/Cedro regression filter – Fig. 4a.

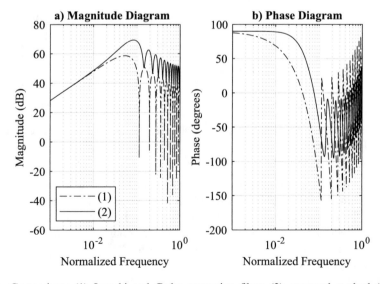

Fig. 4. Comparison: (1) Janecki and Cedro regression filter; (2) proposed method (c = 25, n = 2), (a) Magnitude Diagram, (b) Phase Diagram.

4   Experimental Valuation

For the experimental valuation of the proposed method, a hydraulic test stand was prepared. The measured control signal was a position of the 1 Degree of Freedom (1-DoF) hydraulic manipulator. The test stand was equipped with a current controlled 760 s Moog servo vale. The control unit was implemented in the DSpace 1104 programming platform and was working with a frequency of 10 kHz. In the position control loop, the proposed differential filter was placed as an output of the displacement sensor and was estimating the differential of the measured signal. Three different differential methods were compared. However, the proposed solution worked online while the results from the Janecki/Cedro regression filter, and Tichonov regularization approach were computed offline due to the same data. The results are compared in the Fig. 5.

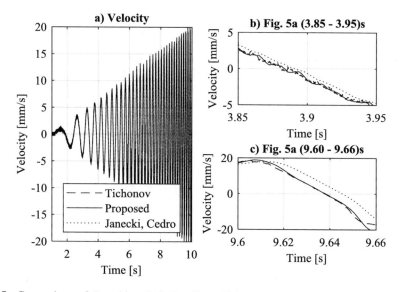

Fig. 5. Comparison of Janecki and Cedro filter, Tichonov regularization and the proposed differential FIR filter (order = 201 at 10 kHz).

As the comparison signal, the Chirp was used. The reference position signal was changing its frequency from 1 Hz to 20 Hz – Fig. 5a. The Janecki/Cedro regression filter was based on 1 present and 200 samples from the past – exactly the same order as the proposed method. The Tichonov regularization was used as a reference velocity signal and to valued the proposed differential FIR filter. Diagrams from Fig. 5a are zoomed in Fig. 5b (time from 3.85 s to 3.95 s), Fig. 5c (time from 9.60 s to 9.66 s). Experimental results clearly confirms that the proposed method caused a lower phase shift due to other types of differential FIR filter design methods, e.g. Janecki and Cedro differential regression FIR filter. The strong difference is noticeable at low frequency (1 Hz) of system operation – Fig. 5b, but also at higher frequencies (15 Hz) – Fig. 5c.

5 Conclusions

A new differential FIR filter weights determination method that results in low-pass filter design was proposed. The method minimizes the phase shift caused by the differential FIR filter for a low frequency spectrum. This remark is confirmed by analysis carried out in the frequency domain and experiments. The filter fits a polynomial to an asymmetric data set and calculates the weighted sum from the fitted coefficients. This feature allows to reduce the value of the phase shift. It was also proven that the differential filter could be used directly in the closed-loop control to estimate a velocity from an analogue position signal and will not cause a system's performance loss.

References

1. Ali, F., Jain, R., Gupta, D., et al.: Design and analysis of low pass elliptic filter. In: 2016 Second International Conference on Computational Intelligence & Communication Technology (CICT), Ghaziabad, India, pp. 449–451 (2016)
2. Baranowski, J., Piątek, P., Bauer, W., et al.: Bi-fractional filters, part 2: right half-plane case. In: 2014 19th International Conference on Methods and Models in Automation and Robotics (MMAR), Międzyzdroje, Poland, pp. 369–373. IEEE (2014)
3. Cleveland, W.S., Devlin, S.J.: Locally weighted regression: an approach to regression analysis by local fitting. J. Am. Stat. Assoc. **83**, 596–610 (1988)
4. Cullum, J.: Numerical differentiation and regularization. SIAM J. Numer. Anal. **8**, 254–265 (1971)
5. Fatehi, A., Huang, B.: Kalman filtering approach to multi-rate information fusion in the presence of irregular sampling rate and variable measurement delay. J. Process Control **53**, 15–25 (2017)
6. Ferrara, E.: Fast implementations of LMS adaptive filters. IEEE Trans. Acoust. Speech Signal Process. **28**, 474–475 (1980)
7. Filanovsky, I.M.: Bessel-Butterworth transitional filters. In: 2014 IEEE International Symposium on Circuits and Systems (ISCAS), Melbourne, VIC, Australia, pp. 2105–2108 (2014)
8. Friedrichs, K.O.: The identity of weak and strong extensions of differential operators. Trans. Am. Math. Soc. **55**, 132–151 (1944)
9. Hayes, M.H.: Statistical Digital Signal Processing and Modeling. Wiley, New York (2009)
10. Janecki, D., Cedro, L.: Determining of signal derivatives with the use of regressive differential filters. Przegląd Elektrotechniczny **87**, 253–259 (2011). (in Polish)
11. Kalman, R.E.: A new approach to linear filtering and prediction problems. J. Basic Eng. **82**, 35–45 (1960)
12. Karam, L.J., Mcclellan, J.H.: Complex Chebyshev approximation for FIR filter design. IEEE Trans. Circuits Syst. II Analog Digital Signal Process. **42**, 207–216 (1995)
13. Kaya, I.: Obtaining controller parameters for a new PI-PD Smith predictor using autotuning. J. Process Control **13**, 465–472 (2003)
14. Knowles, I., Renka, R.J.: Methods for numerical differentiation of noisy data. Electron. J. Differ. Equ. **21**, 235–246 (2014)
15. Knowles, I., Wallace, R.: A variational method for numerical differentiation. Numer. Math. **70**, 91–110 (1995)

16. Krämer, D., King, R.: A hybrid approach for bioprocess state estimation using NIR spectroscopy and a sigma-point Kalman filter. J. Process Control (2017)
17. Miadlicki, K., Pajor, M., Sakow, M.: Loader crane working area monitoring system based on LIDAR scanner. In: Advances in Manufacturing, pp. 465–474. Springer (2018)
18. Miądlicki, K., Pajor, M., Saków, M.: Ground plane estimation from sparse LIDAR data for loader crane sensor fusion system. In: 2017 22nd International Conference on Methods and Models in Automation and Robotics (MMAR), Międzyzdroje, Poland, pp. 717–722. IEEE (2017)
19. Miądlicki, K., Pajor, M., Saków, M.: Real-time ground filtration method for a loader crane environment monitoring system using sparse LIDAR data. In: 2017 IEEE International Conference on INnovations in Intelligent SysTems and Applications (INISTA), pp. 207–212. IEEE (2017)
20. Okulski, M., Ławryńczuk, M.: A cascade PD controller for heavy self-balancing robot. In: Conference on Automation, pp. 183–192. Springer (2018)
21. Oppenheim, A.V.: Discrete-Time Signal Processing. Pearson Education India, Bangalore (1999)
22. Owczarek, P., Goslinski, J., Rybarczyk, D., et al.: Modeling and 3D simulation of an electro-hydraulic manipulator controlled by vision system with Kalman filter. In: Advances in Manufacturing, pp. 375–384. Springer (2018)
23. Piątek, P., Baranowski, J., Zagórowska, M., et al.: Bi-fractional filters, part 1: left half-plane case. In: Advances in Modelling and Control of Non-Integer-Order Systems, pp. 81–90. Springer (2015)
24. Psychalinos, C., Tsirimokou, G., Elwakil, A.S.: Switched-capacitor fractional-step butterworth filter design. Circ. Syst. Sig. Process. 35, 1377–1393 (2016)
25. Ra, W.S., Whang, I.H.: Recursive weighted robust least squares filter for frequency estimation. In: 2006 SICE-ICASE International Joint Conference, Busan, South Korea, pp. 774–778 (2006)
26. Raja, G.L., Ali, A.: Smith predictor based parallel cascade control strategy for unstable and integrating processes with large time delay. J. Process Control 52, 57–65 (2017)
27. Rodríguez, C., Normey-Rico, J., Guzmán, J., et al.: On the filtered Smith predictor with feedforward compensation. J. Process Control 41, 35–46 (2016)
28. Rybarczyk, D., Owczarek, P., Myszkowski, A.: Development of force feedback controller for the loader crane. In: Advances in Manufacturing, pp. 345–354. Springer (2018)
29. Saków, M.: Real-time and low phase shift noisy signal differential estimation dedicated to teleoperation systems, pp. 132–141. Springer International Publishing, Cham (2018)
30. Saków, M., Marchelek, K.: Model-free and time-constant prediction for closed-loop systems with time delay. Control Eng. Pract. 81, 1–8 (2018)
31. Saków, M., Marchelek, K., Parus, A., et al.: Signal Prediction in Bilateral Teleoperation with Force-Feedback, pp. 311–323. Springer International Publishing, Cham (2018)
32. Saków, M., Miądlicki, K.: Transport delay and first order inertia time signal prediction dedicated to teleoperation, pp. 142–151. Springer International Publishing, Cham (2018)
33. Saków, M., Miądlicki, K., Parus, A.: Self-sensing teleoperation system based on 1-dof pneumatic manipulator. J. Autom. Mob. Robot. Intell. Syst. 11, 64–76 (2017)
34. Saków, M., Parus, A., Miądlicki, K.: LS filter and its implementation into the control unit of the master-slave system with force-feedback. Modelowanie inżynierskie 34, 107–117 (2017). (in Polish)
35. Sakow, M., Parus, A., Pajor, M., et al.: Unilateral hydraulic telemanipulation system for operation in machining work area. In: Advances in Manufacturing, pp. 415–425. Springer (2018)

36. Saków, M., Parus, A., Pajor, M., et al.: Nonlinear inverse modeling with signal prediction in bilateral teleoperation with force-feedback. In: 2017 22nd International Conference on Methods and Models in Automation and Robotics (MMAR), Międzyzdroje, Poland, pp. 141–146. IEEE (2017)
37. Schoenberg, I.J.: Spline functions and the problem of graduation. Proc. Natl. Acad. Sci. **52**, 947–950 (1964)
38. Tsirimokou, G., Psychalinos, C., Elwakil, A.S.: Digitally programmed fractional-order Chebyshev filters realizations using current-mirrors. In: 2015 IEEE International Symposium on Circuits and Systems (ISCAS), pp. 2337–2340 (2015)

The Problem of the Optimal Strategy of Minimax Control by Objects with Distributed Parameters

Igor Korobiichuk[1(✉)], Alexey Lobok[2], Boris Goncharenko[2],
Natalya Savitska[2], Marina Sych[3], and Larisa Vihrova[4]

[1] Industrial Research Institute for Automation and Measurements PIAP,
Warsaw, Poland
ikorobiichuk@piap.pl
[2] National University of Food Technologies, Kiev, Ukraine
{apl_apl, savitskanm}@ukr.net, goncharenkobn@i.ua
[3] National University of Life and Environmental Sciences of Ukraine,
Kiev, Ukraine
marina.sych@ukr.net
[4] Central Ukrainian National Technical University, Kropyvnytskyi, Ukraine
vihrovalg@ukr.net

Abstract. The problem of minimax control synthesis for objects that are described by a two-dimensional heat conduction equation of parabolic type is solved. It is assumed that the control object functions under uncertainty conditions, and the perturbations acting on the object belong to some given hyperelipsoid. The problem of constructing a regulator in the state of an object for cases of point and mobile limit control is considered in accordance with the integral-quadratic quality criterion. In the work, for the first time, a minimax approach was used to control the objects described by the two-dimensional parabolic type thermal conductivity equation; the theoretical positions of synthesis of minimax regulators for cases of lumped boundary (point) and moving regulators are considered; algorithmic software is developed that allows to simulate the dynamics of the constructed minimax-regulators and to investigate the corresponding transients.

Keywords: Minimax control · Regulators · Distributed parameter systems · Optimization · Gradient projection method · Point and mobile limit controls

1 Introduction

In connection with the widespread adoption of new advanced technologies related to the use of electronic, ion, laser and other radiation, in recent years intensive study of the possibilities of optimal control of distributed source systems by changing the location of point sources of radiation and laws of displacement.

The determination of the problems of point and motion control, some methods of their solution are given in the works [1–3]. One of the most important and complex tasks is the choice of an optimal point and move control strategy for systems that operate under

© Springer Nature Switzerland AG 2020
R. Szewczyk et al. (Eds.): AUTOMATION 2019, AISC 920, pp. 77–85, 2020.
https://doi.org/10.1007/978-3-030-13273-6_8

uncertainty [4–6]. This problem to which the article is devoted, which outlines ways of choosing the optimal location of point regulators and finding the optimal law of motion (moving) of a moving source at the boundary of a rectangular region for the process of heat transfer occurring under incomplete information [7, 8]. The theory of control moves towards the complexity of the phenomena studied, processes and the reduction of information about the control system, the object, its features, properties, characteristics, operating conditions, external influences. Taking into account all the above-mentioned, the chosen direction of research is perspective and has a high level of relevance.

2 Materials and Methods of Research

Let the process of heat transfer in a homogeneous thin rectangular plate be described by a function $\varphi(x,t)$, which is in the area $Q_T = \Omega \times (0,T)$, where $\Omega = \{(x_1, x_2) : 0 < x_1 < l_1, 0 < x_2 < l_2\}$, $l_1, l_2 > 0$, $T < \infty$, satisfies the equation

$$\frac{\partial \varphi(x,t)}{\partial t} = a\Delta_x \varphi(x,t) + f_1(x,t), \quad (x,t) \in Q_T, \tag{1}$$

but on the border Q_T − additional conditions

$$\varphi(x,0) = f_0(x), x \in \Omega; \quad \varphi(x,t) = \sum_{i=1}^{N} \delta(x - v_i(t))u_i(t), \quad (x,t) \in \Gamma \times (0,T). \tag{2}$$

where $\Delta_x = \frac{\partial^2}{\partial x_1^2} + \frac{\partial^2}{\partial x_2^2}$ − two-dimensional Laplace operator; $a > 0$ − coefficient of temperature conductivity; Γ − border of rectangular area Ω; $\delta(x - y)$ − Dirac's delta function; $t \to v_i(t) \in \Gamma$ − dimensional functions that determine the motion of boundary sources; $u_i(t) \in L_2(0,T)$ − control functions; $f_0(x) \in L_2(\Omega)$, $f_1(x,t) \in L_2(Q_T)$ − unknown functions belonging to the area

$$S_t = \{(f_0, f_1) : \ G(f_0; f_1(\tau), \ 0 < \tau < t) \le 1\}, t \in (0,T], \tag{3}$$

where $G(f_0; f_1(\tau), \ 0 < \tau < t) = F_0 \int_{\Omega} f_0^2(x)dx + F_1 \int_0^t \int_{\Omega} f_1^2(x,\tau)dxd\tau$,

and F_0, F_1 − positive constant values reflecting the contribution of noise f_0 and $f_1(t)$ in the final perturbation, acting on the system (1), (2).

Under the solution of the boundary value problem (1), (2) we will understand such a function $\varphi(x,t) \in L_2(Q_T)$, which satisfies the following integral identity

$$-\int_0^T \int_{\Omega} \varphi(x,t)\left(\frac{\partial \eta(x,t)}{\partial t} + a\Delta_x \eta(x,t)\right)dxdt = \int_{\Omega} f_0(x)\eta(x,0)dx + \int_0^T \int_{\Omega} f_1(x,t)\eta(x,t)dxdt$$

$$-a\sum_{i=1}^{N} \int_0^T u_i(t)\left.\frac{\partial \eta(x,t)}{\partial n}\right|_{x=v_i(t)} dt \forall \eta(x,t) \in \Phi, \tag{4}$$

where $\partial/\partial n$ − derivative of the external normal $\vec{n}$ to the border Γ of the area Ω,

$$\Phi = \{\eta(x,t) : \eta(x,t) \in H^{3,1}(Q_T), \eta(x,T) = 0, x \in \Omega; \eta(x,t) = 0, (x,t) \in \Gamma \times (0,T)\}$$

$H^{3,1}(Q_T)$ − Sobolevsky space [9].

It can be shown [10] that the solution of Eq. (4) with given controls $u_i(t) \in L_2(0,T)$ exists and is unique in space $L_2(Q_T)$. The task of choosing the optimal strategy for minimax control [11–13] will be to find vector functions $v^*(t) = [v_1^*(t), v_2^*(t), \ldots, v_N^*(t)]^T$ and $u^*(t) = [u_1^*(t), u_2^*(t), \ldots, u_N^*(t)]^T$ under conditions

$$I(u^*, v^*) = \inf_{v} \inf_{u} I(u, v), \tag{5}$$

where

$$I(u, v) = \sup_{S_T} \left[\int_{\Omega} S(x)\varphi(x,T)dx \right]^2 + \int_0^T \sum_{i=1}^N d_i \sup_{S_t} u_i^2(t)dt, \tag{6}$$

$S(x) \in L_2(\Omega)$, $d_i = const > 0$, $i = 1, 2, \ldots, N$, with a given control structure $u_i(t)$ in the form of a linear feedback

$$u_i(t) = \int_{\Omega} R_i(x,t)\varphi(x,t)dx. \tag{7}$$

3 The Results of Research

The solution of the formulated problem will be carried out in two stages: first we solve the problem of determining the optimal control $u^*(t)$ under condition

$$I(u^*, v) = \inf_{u} I(u, v) \tag{8}$$

for a fixed vector-function $v(t)$, and then find it $v^*(t)$, at which

$$I(u^*, v^*) = \inf_{v} I(u^*, v). \tag{9}$$

According to the results of [12, 13], the following theorem is proved: optimal control $u^*(t)$ of the optimization problem (1), (2), (6), (7) satisfying the necessary optimality conditions has the form

$$u_i^*(t) = \int_{\Omega} R_i^*(x,t)\varphi(x,t)dx, \quad R_i^*(x,t) = ad_i^{-1}\alpha^{-1}(t)g(x,t)h(v_i(t),t), \tag{10}$$

where

$$\alpha(t) = 1 + a^2 \sum_{k=1}^{N} d_k^{-1} \int_t^T h^2(v_k(\tau), \tau) d\tau,$$

$$\begin{bmatrix} g(x,t) \\ h(x,t) \end{bmatrix} = \sum_{i=1}^{\infty} s_i e^{\lambda_i(t-T)} \begin{bmatrix} \omega_i(x) \\ r_i(x) \end{bmatrix}, \quad s_i = \int_{\Omega} S(x) \omega_i(x) dx. \tag{11}$$

In the ratio (11) $i = (i_1, i_2)$ − multiindex

$$r_i(x) = r_{i_1, i_2}(x_1, x_2) = \frac{2\pi}{\sqrt{l_1 l_2}} \begin{cases} (-1)^{i_1} \frac{i_1}{l_1} \sin \frac{\pi i_2 x_2}{l_2}, & x_1 = l_1, \\ -\frac{i_1}{l_1} \sin \frac{\pi i_2 x_2}{l_2}, & x_1 = 0, \\ (-1)^{i_2} \frac{i_2}{l_2} \sin \frac{\pi i_1 x_1}{l_1}, & x_2 = l_2, \\ -\frac{i_2}{l_2} \sin \frac{\pi i_1 x_1}{l_1}, & x_2 = 0, \end{cases} \quad x = (x_1, x_2) \in \Gamma,$$

where λ_i, $\omega_i(x)$ − eigenvalues and corresponding orthonormalities in space $L_2(\Omega)$ own functions of the boundary value problem (1), (2) having the form

$$\lambda_i = \lambda_{i_1, i_2} = a\pi^2 \left[(i_1/l_1)^2 + (i_2/l_2)^2 \right],$$

$$\omega_i(x) = \omega_{i_1, i_2}(x_1, x_2) = \frac{2}{\sqrt{l_1 l_2}} \sin \frac{\pi i_1 x_1}{l_1} \sin \frac{\pi i_2 x_2}{l_2}.$$

The value of the functional (6) for optimal control (10) is determined by the formula

$$I(u^*, v) = \frac{W(0)}{F_0 \alpha(0)} + \frac{1}{F_1} \int_0^T \frac{W(t)}{\alpha(t)} dt, \tag{12}$$

where

$$W(t) = \sum_{i=1}^{\infty} s_i e^{2\lambda_i(t-T)}. \tag{13}$$

Let us now turn to the solution of the optimization problem (9), (12). Let's consider first a simpler case when $v_i(t) \equiv z_i \in \Gamma$, $i = 1, 2, \ldots, N$, that is, we solve the problem of optimal location of point boundary controls (10). Let's introduce the designation $z = [z_1, z_2, \ldots, z_N]^T$, $J(z) = I(u^*, z) \equiv I(u^*, v)$. Then the task under consideration will be to find the vector $z = [z_1^*, z_2^*, \ldots, z_N^*]^T$, at which

$$J(z^*) = \inf_{z \in \Omega_z} J(z) \tag{14}$$

where $\quad \Omega_z = \{z : z = [z_1, z_2, \ldots, z_N]^T, \ z_i = (z_{1i}, z_{2i}) \in \Gamma, \quad i = 1, 2, \ldots, N; \ z_i \neq z_j, \ i \neq j\}$.

Given that function $J(z)$ is a continuously differentiated function of its arguments, to solve the optimization problem (14) we use the gradient projection method [14]

$$z^{k+1} = \Pr_{\Omega_z}[z^k - \rho_k \nabla_z J(z^k)], \quad k = 0, 1, 2, \ldots, \tag{15}$$

where $\Pr_{\Omega_z}[z] = [y_1, y_2, \ldots, y_N]^T$, $y_i = \Pr_\Gamma[z_i]$ – projection of point z_i on the border Γ of a rectangular area Ω; $z^k = [z_1^k, z_2^k, \ldots, z_N^k]^T$ – approximate solution obtained on k-th iteration; z^0 – initial approximation; ρ_k – step of descent, which is chosen from the condition of the monotonous decline of the function of purpose $J(z)$ [14]; gradient $\nabla_z J(z)$ is determined by the formula

$$\nabla_z J(z) = [\nabla_{z_1} J(z), \nabla_{z_2} J(z), \ldots, \nabla_{z_N} J(z)]^T,$$

$$z = [z_1, z_2, \ldots, z_N]^T, \quad z_i = (z_{1i}, z_{2i}) \in \Gamma, \ i = 1, 2, \ldots, N,$$

$$\nabla_{z_n} J(z) = -2a^2 \left[\frac{W(0)\theta_n(0)}{F_0 \alpha^2(0)} + \frac{1}{F_1} \int_0^T \frac{W(t)\theta_n(t)}{\alpha^2(t)} dt \right],$$

$$\theta_n(t) = d_n^{-1} \sum_{i=1}^{\infty} \sum_{j=1}^{\infty} \frac{1 - e^{(\lambda_i + \lambda_j)(t-T)}}{\lambda_i + \lambda_j} s_i s_j r_i(z_n) P_j(z_n),$$

where $P_j(z_n) = \left[P_j^1(z_n), P_j^2(z_n) \right]$,

$$P_j^k(z_n) = P_{j_1 j_2}^k(z_{1n}, z_{2n}) = \frac{2\pi^2 j_1 j_2}{(l_1 l_2)^{3/2}} \cos \frac{\pi j_k z_{kn}}{l_k} \begin{cases} 0, & z_{kn} = 0, z_{kn} = l_k, \\ -1, & z_{3-k,n} = 0, \\ (-1)^{j_{3-k}}, & z_{3-k,n} = l_{3-k}, \end{cases}$$

$$k = 1, 2.$$

The condition of the stop was taken in the form $\left| J(z^{k+1}) - J(z^k) \right| < \varepsilon$, where $\varepsilon > 0$ – the accuracy of the solution is given.

This algorithm was programmed in the algorithmic language Fortran 90 with the following initial data: $l_1 = 2.0, l_2 = 1.0, T = 2.0, F_0 = 0.25, F_1 = 2.0, d_i = 1.0, i = 1, 2, \ldots, N, S(x) = 1.0, \varepsilon = 0.001, \rho_0 = 0.8$, number of regulators $N = 5$, for the value of the coefficient of thermal conductivity $a = 0.4$ was taken, which corresponds to the coefficient of thermal conductivity of the copper plate. The dimension of all quantities is given in the system [meter, time, deg. °C, kcal.]. The infinite series (11), (13) were broken off by the finite sum of the three first members. For numerical simulation of optimal controls $u_i^*(t)$ it was assumed that perturbation $f_0(x)$ and $f_1(x, t)$ is equal

$$f_0(x_1, x_2) = 2 \sin \frac{\pi x_1}{l_1} \sin \frac{\pi x_2}{l_2}, \quad f_1(x_1, x_2, t) = t \sin \frac{\pi x_1}{l_1} \cos \frac{\pi x_2}{l_2}.$$

We note that these perturbations are permissible, because $G(f_0, f_1(\tau), 0 < \tau < t) = F_0 l_1 l_2 + \frac{1}{12} F_1 l_1 l_2 t^3 = 0.5 + \frac{1}{3} t^3 < 1$, $\forall t \in (0, 0.2)$ and as a result, $f_0(x)$ and $f_1(x, t)$ belong to the area (3).

Table 1 gives the initial location $z^0 = [z_1^0, z_2^0, \ldots, z_N^0]^T$ point boundary regulators. Function value $J(z)$ at such an arrangement of controls equals $J(z^0) = 0.975632$. Optimal arrangement of regulators $z^* = [z_1^*, z_2^*, \ldots, z_N^*]^T$, obtained by the algorithm (15), is given in the Table 2 and $J(z^*) = 0.571874$.

<table>
<tr><td colspan="3">Table 1. Initial arrangement</td><td colspan="3">Table 2. Optimal arrangement</td></tr>
<tr><td>k</td><td>z_{1k}^0</td><td>z_{2k}^0</td><td>k</td><td>z_{1k}^*</td><td>z_{2k}^*</td></tr>
<tr><td>1</td><td>2.0</td><td>0.0</td><td>1</td><td>1.349</td><td>0.0</td></tr>
<tr><td>2</td><td>1.0</td><td>0.0</td><td>2</td><td>1.0</td><td>0.0</td></tr>
<tr><td>3</td><td>0.0</td><td>0.0</td><td>3</td><td>0.651</td><td>0.0</td></tr>
<tr><td>4</td><td>0.0</td><td>0.5</td><td>4</td><td>0.0</td><td>0.5</td></tr>
<tr><td>5</td><td>0.0</td><td>1.0</td><td>5</td><td>0.651</td><td>1.0</td></tr>
</table>

Figure 1 shows graphs of optimal point controls (10) that are optimally located on the boundary Γ of the area Ω in points $z_i^* \in \Gamma$, $i = 1, 2, \ldots, N$.

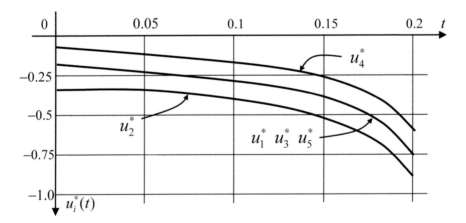

Fig. 1. Schedule of optimal boundary point controls

We now turn to the optimization problem (9), (12). Let's lick for simplicity one ($N = 1$) a moving source and let the perturbation $f_1(x, t)$ in the right side of Eq. (1) will be absent. Let's denote $u(t) = u_1(t)$, $v(t) = v_1(t)$, $d = d_1$, $J(v) = I(u^*, v)$. Then the task of minimizing the functional

$$J(v) = W(0)(F_0\alpha(0))^{-1} \tag{16}$$

is equivalent to the next optimization task

$$L(v) = \int_0^T h(v(\tau), \tau)d\tau \longrightarrow \sup_{\substack{t \to v(t) \in \Gamma \\ t \in (0,T)}},$$

where $\alpha(t)$, $h(x,t)$, $W(t)$ — functions determined by the formulas (11), (13).

To solve the last problem, the projection method of the gradient of the species was also used

$$v^{k+1}(t) = \Pr_\Gamma \left[v^k(t) + \rho_k \delta[L(v^k); t] \right], t \in (0, T)k = 0, 1, 2, \ldots, \tag{17}$$

where $v^0(t)$ — initial approximation; $v^k(t)$ — Approximate solution obtained at k-th step; ρ_k — step of descent to the minimum point; $\delta[L(v); t]$ — graceful Frechet functional $L(v)$ which is calculated by the formula

$$\delta[L(v); t] = 2h(v(t), t)\rho(v(t), t), \quad \rho(x, t) = \sum_{i=1}^{\infty} s_i e^{\lambda_i(t-T)} P_i(x).$$

The algorithm stops when is fulfilled the condition $\left| L(v^{k+1}) - L(v^k) \right| < \varepsilon$, where $\varepsilon > 0$ — the accuracy of the solution is given.

Numerical implementation of the algorithm (17) was carried out with the above earlier data. Below are the results of computational calculations. In Table 3 the initial law of motion is given $v^0(t) = \left(v_1^0(t), v_2^0(t) \right)$ for a moving boundary source. Optimal motion law $v^*(t) = \left(v_1^*(t), v_2^*(t) \right)$ of the moving controller (10), obtained by the algorithm (17), is given in Table 4.

<table>
<tr><td colspan="3">Table 3. Initial law</td><td colspan="3">Table 4. Optimal law</td></tr>
<tr><td>t</td><td>$v_1^0(t)$</td><td>$v_2^0(t)$</td><td>t</td><td>$v_1^*(t)$</td><td>$v_2^*(t)$</td></tr>
<tr><td>0.0</td><td>0.0</td><td>0.0</td><td>0.0</td><td>0.010</td><td>0.0</td></tr>
<tr><td>0.02</td><td>0.667</td><td>0.0</td><td>0.02</td><td>0.765</td><td>0.0</td></tr>
<tr><td>0.04</td><td>1.333</td><td>0.0</td><td>0.04</td><td>1.230</td><td>0.0</td></tr>
<tr><td>0.06</td><td>2.0</td><td>0.0</td><td>0.06</td><td>1.990</td><td>0.0</td></tr>
<tr><td>0.08</td><td>2.0</td><td>0.5</td><td>0.08</td><td>2.0</td><td>0.5</td></tr>
<tr><td>0.10</td><td>2.0</td><td>1.0</td><td>0.10</td><td>1.990</td><td>1.0</td></tr>
<tr><td>0.12</td><td>1.333</td><td>1.0</td><td>0.12</td><td>1.285</td><td>1.0</td></tr>
<tr><td>0.14</td><td>0.667</td><td>1.0</td><td>0.14</td><td>0.664</td><td>1.0</td></tr>
<tr><td>0.16</td><td>0.0</td><td>1.0</td><td>0.16</td><td>0.010</td><td>1.0</td></tr>
<tr><td>0.18</td><td>0.0</td><td>0.5</td><td>0.18</td><td>0.0</td><td>0.5</td></tr>
</table>

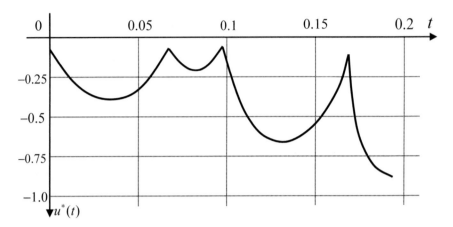

Fig. 2. Schedule of optimum limit moving control

The value of the functional (16) thus decreased from $J(z^0) = 0.639538$ to $J(z^*) = 0.438419$. In Fig. 2 shows an optimal control (10), the movement of which is carried out in the optimal trajectory, shown in the Table 4. The optimal trajectory consists of four parts, each of which resembles a parabola and defines (describes) the motion of the regulator along the corresponding boundary of the rectangular area.

The computational experiments also showed that the efficiency of point and moving boundary controls increases with a decrease in the coefficient of temperature conductivity, that is, with the decrease of this coefficient the value of the functional (8) after determining the optimal control strategy decreased by a larger value compared with the value of the same functional with a given initial control strategy.

4   Conclusions

In the work, for the first time, a minimax approach was used to control the objects described by the two-dimensional parabolic type thermal conductivity equation; the theoretical positions of synthesis of minimax regulators for cases of lumped boundary (point) and moving regulators are considered; algorithmic software is developed that allows to simulate the dynamics of the constructed minimax-regulators and to investigate the corresponding transients. The solution of the problem of finding the optimal placement strategy for point boundary regulators and the problem of determining the optimal trajectory of moving a regulator along the boundary of the region in which the distributed control object functions is achieved. The problem is solved in a minimal-scale setting, that is, an optimal controller is found for the state of the object, which functions in conditions of uncertainty, and the perturbation of the object belongs to a given bounded domain. The results of computational experiments are presented, which illustrate the efficiency of constructed lumped boundary point and moving regulators. The obtained results indicate that the control outputs are actually optimal and provide a minimum of errors (deviations from the given state) of the system's operation and energy costs for the control of the given conditions and the absence of any information on external influences, in addition to the area of permissible perturbations. Satisfactory performance indicators are observed even in the event of disturbance beyond the boundaries of a given area.

References

1. Butkovsky, A.G.: Metodi control systems with distributed parameters, 568 p. Moskva, Nauka (1975). (in Russian)
2. Butkovsky, A.G., Darinsky, Yu.V., Pustylnikov, L.M.: Mobile control systems with distributed parameters. Autom. Remote Control **2**, 15–25 (1976). (in Russian)
3. Butkovsky, A.G., Pustylnikov, L.M.: The theory of mobile control systems with distributed parameters, 397 p. Moskva, Nauka (1980, in Russian)
4. Korobiichuk, I., Siumachenko, D., Smityuh, Y., Shumyhai, D.: Research on automatic controllers for plants with significant delay. Advances in Intelligent Systems and Computing, vol. 519, pp. 449–457 (2017). https://doi.org/10.1007/978-3-319-46490-9_60
5. Korobiichuk, I., Ladanyuk, A., Shumyhai, D., Boyko, R., Reshetiuk, V., Kamiński, M.: How to increase efficiency of automatic control of complex plants by development and implementation of coordination control system. In: Recent Advances in Systems, Control and Information Technology. Advances in Intelligent Systems and Computing, vol. 543, pp. 189–195 (2017). https://doi.org/10.1007/978-3-319-48923-0_23
6. Korobiichuk, I., Lutskaya, N., Ladanyuk, A., Naku, S., Kachniarz, M., Nowicki, M., Szewczyk, R.: Synthesis of optimal robust regulator for food processing facilities. Advances in Intelligent Systems and Computing, vol. 550, pp. 58–66 (2017). https://doi.org/10.1007/978-3-319-54042-9_5
7. Baloev, A.A.: Synthesis of suboptimal control with incomplete measurement for systems with distributed parameters. Soviet Aeronautics (English translation of Izvestiya VUZ, Aviatsionnaya Tekhnika), vol. 27, no. 1, pp. 66–71 (1984)
8. Khamis, A., Subbaram Naidu, D.: Real-time algorithm for nonlinear systems with incomplete state information using finite-horizon optimal control technique. In: 7th International Symposium on Resilient Control Systems, ISRCS 2014, 6900094 (2014)
9. Lions, J.L.: Optimal control of systems described by partial differential equations, 414 p. Moskva, Mir (1972). (in Russian)
10. Ladyzhenskaya, O.A., Uraltseva, N.N., Solonnikov, V.A.: Linear and quasilinear equations of parabolic type, 736 p. Moskva, Nauka (1967). (in Russian)
11. Lobok, O.P., Goncharenko, B.M., Savitska, N.M.: Minimax control in linear systems of dynamic systems with distributed parameters. J. Naukovi pratsi NUHT **21**(6), 16–26 (2015). (in Ukrainian)
12. Kirichenko, N.F.: Minimax control and estimation in dynamic systems. Autom. Remote Control **1**, 32–39 (1982). (in Russian)
13. Lobok, O.P.: Minimax regulators in systems with distributed parameters. In: Modeling and Optimization of Complex Systems, no. 2, pp. 62–67 (1983). (in Russian)
14. Vasiliev, F.P.: Methods for solving extremal problems, 400 p. Moskva, Nauka (1981). (in Russian)

Transfer Matrices with Positive Coefficients of Positive Descriptor Continuous-Time Linear Systems

Tadeusz Kaczorek and Łukasz Sajewski

Białystok University of Technology, Wiejska 45D, 15-351 Białystok, Poland
kaczorek@ee.pw.edu.pl, l.sajewski@pb.edu.pl

Abstract. Descriptor positive continuous-time linear system with transfer matrices having only positive coefficients are analyzed. It is shown that if the positive descriptor system is asymptotically stable then its transfer matrix has only positive coefficients. The realization problem for the class of descriptor positive systems is formulated and solved. Considerations are illustrated by numerical examples.

Keywords: Positive · Descriptor · Linear · Continuous-time · System · Transfer matrix

1 Introduction

Descriptor (singular) linear systems have been considered in many papers and books [3–5, 7, 25, 26].

A dynamical system is called positive if its state variables take nonnegative values for all nonnegative inputs and nonnegative initial conditions. The positive linear systems have been investigated in [2, 6, 15, 17] and positive nonlinear systems in [10, 13, 20].

Examples of positive systems are industrial processes involving chemical reactors, heat exchangers and distillation columns, storage systems, compartmental systems, water and atmospheric pollution models. A variety of models having positive linear behavior can be found in engineering, management science, economics, social sciences, biology and medicine, etc.

The determination of the matrices A, B, C, D of the state equations of linear systems for given transfer matrices is called the realization problem. A tutorial on the positive realization problem has been given in the paper [1] and in the books [6, 15]. The positive realization problem for linear systems with delays has been analyzed in [8, 9, 16, 21], for cone systems in [12] and positive stable realizations in [11, 18, 19, 23]. The existence and determination of the set of Metzler matrices for given stable polynomials have been considered in [14].

In this paper the transfer matrices with positive coefficients of positive descriptor continuous-time linear systems will be investigated.

The paper is organized as follows. In Sect. 2 some basic definitions and theorems concerning descriptor linear systems are recalled. Conditions for positivity of the coefficients of transfer matrices of descriptor systems are established in Sect. 3.

© Springer Nature Switzerland AG 2020
R. Szewczyk et al. (Eds.): AUTOMATION 2019, AISC 920, pp. 86–94, 2020.
https://doi.org/10.1007/978-3-030-13273-6_9

Necessary conditions and the procedure for finding the realization is given in Sect. 4. Concluding remarks are given in Sect. 5.

The following notations will be used: $\Re$ – the set of real numbers, $\Re^{n\times m}$ – the set of $n \times m$ real matrices, $\Re_+^{n\times m}$ – the set of $n \times m$ real matrices with nonnegative entries and $\Re_+^n = \Re_+^{n\times 1}$, M_n – the set of $n \times n$ Metzler matrices (real matrices with nonnegative off-diagonal entries), I_n – the $n \times n$ identity matrix.

2 Preliminaries

Consider the descriptor continuous-time linear system

$$E\dot{x} = Ax + Bu,$$
$$y = Cx, \tag{1}$$

where $x = x(t) \in \Re^n$, $u = u(t) \in \Re^m$, $y = y(t) \in \Re^p$ are the state, input and output vectors and $E, A \in \Re^{n\times n}$, $B \in \Re^{n\times m}$, $C \in \Re^{p\times n}$.

It is assumed that $\det E = 0$ and

$$\det[Es - A] \neq 0 \text{ for some } s \in \mathbf{C} \text{ (the field of complex numbers)}. \tag{2}$$

In this case the system (1) has unique solution for admissible initial conditions $x(0) \in X \in \Re^n$ and admissible inputs $u(t) \in U \in \Re^m$.

It is well-known [22] that if (2) holds then there exists a pair of nonsingular matrices $P, Q \in \Re^{n\times n}$ such that

$$P[Es - A]Q = \begin{bmatrix} I_{n_1}s - A_1 & 0 \\ 0 & Ns - I_{n_2} \end{bmatrix}, A_1 \in \Re^{n_1\times n_1}, N \in \Re^{n_2\times n_2} \tag{3}$$

where $n_1 = \deg\{\det[Es - A]\}$ and N is the nilpotent matrix, i.e. $N^\mu = 0$, $N^{\mu-1} \neq 0$ (μ is the nilpotency index).

To simplify the considerations it is assumed that the matrix N has only one block. The nonsingular matrices P and Q can be found for example by the use of elementary row and column operations [22].

Definition 1. The descriptor system (1) is called (internally) positive if $x(t) \in \Re_+^n$, $t \geq 0$ for all admissible nonnegative initial conditions $x(0) \in \Re_+^n$.

Theorem 1. The descriptor system (1) is positive if and only if the matrix E has only linearly independent columns and the matrices $A_1 \in M_{n_1}$, $B_1 \in \Re_+^{n_1\times m}$, $-B_2 \in \Re_+^{n_2\times m}$, $C_1 \in \Re_+^{p\times n_1}$, $C_2 \in \Re_+^{p\times n_2}$, where $PB = \begin{bmatrix} B_1 \\ -B_2 \end{bmatrix}$, $CQ = [C_1 \quad C_2]$ and $Q \in \Re_+^{n\times n}$ is permutation matrix.

From (3) it follows that the system (1) for $B = 0$ has been decomposed into two independent subsystems

$$\dot{x}_1 = A_1 x_1, \; x_1 \in \mathfrak{R}^{n_1} \tag{4}$$

and

$$N\dot{x}_2 = x_2, \; x_2 \in \mathfrak{R}^{n_2}, \tag{5}$$

where

$$Q^{-1}x = \begin{bmatrix} x_1 \\ x_2 \end{bmatrix} \tag{6}$$

and Q and Q^{-1} are permutation matrices.

Definition 2. [6, 15] The positive system (4) is called asymptotically stable if

$$\lim_{t \to \infty} x_1(t) = 0 \text{ for all admissible } x_1(0) \in \mathfrak{R}_+^{n_1}. \tag{7}$$

Theorem 2. [5, 15] The positive system (4) is asymptotically stable if and only if one of the equivalent conditions is satisfied:

(1)　All coefficients of the polynomial

$$\det[I_{n_1} s - A_1] = s^{n_1} + a_{n_1-1} s^{n_1-1} + \ldots + a_1 s + a_0 \tag{8}$$

　　　are positive, i.e. $a_k > 0$ for $k = 0, 1, \ldots, n_1 - 1$.
(2)　All principal minors $\bar{M}_i$, $i = 1, \ldots, n_1$ of the matrix $-A_1$ are positive, i.e.

$$\bar{M}_1 = |-a_{11}| > 0, \; \bar{M}_2 = \begin{vmatrix} -a_{11} & -a_{12} \\ -a_{21} & -a_{22} \end{vmatrix} > 0, \ldots, \bar{M}_{n_1} = \det[-A_1] > 0. \tag{9}$$

(3)　There exists a strictly positive vector $\lambda = [\lambda_1 \quad \cdots \quad \lambda_{n_1}]^T$, $\lambda_k > 0$, $k = 1, \ldots, n_1$
　　　such that

$$A_1 \lambda < 0 \text{ or } A_1^T \lambda < 0. \tag{10}$$

If $\det A \neq 0$ then we may choose $\lambda = -A_1^{-1} c$, where $c \in \mathfrak{R}^{n_1}$ is any strictly positive vector.

Example 1. Consider the descriptor system (1) with the matrices

$$E = \begin{bmatrix} 0 & 0 & 0 & 2 \\ 0 & 1 & 0 & -2 \\ 1 & -2 & 0 & 0 \\ 0 & 0 & 0 & -2 \end{bmatrix}, \quad A = \begin{bmatrix} 0 & 1 & 0 & -4 \\ 1 & -4 & 0 & 4 \\ 0 & 6 & 1 & 0 \\ 1 & -1 & 0 & 4 \end{bmatrix}, \quad B = \begin{bmatrix} 1 \\ -3 \\ -1 \\ -3 \end{bmatrix},$$

$$C = \begin{bmatrix} 1 & 0 & 1 & 2 \end{bmatrix}. \tag{11}$$

The condition (2) for (11) is satisfied since

$$\det[Es - A] = \begin{vmatrix} 0 & -1 & 0 & 2s+4 \\ -1 & s+4 & 0 & -2s-4 \\ s & -2s-6 & -1 & 0 \\ -1 & 1 & 0 & -2s-4 \end{vmatrix} = -2s^2 - 10s - 12 \tag{12}$$

and $n_1 = 2$. In this case rank $E = 3$ and $\mu = \text{rank } E - n_1 + 1 = 2$.
   The matrices P and Q transforming the matrix $Es - A$ with (11) to the desired form

$$\begin{bmatrix} I_2 s - A_1 & 0 \\ 0 & Ns - I_2 \end{bmatrix} \text{ with } A_1 = \begin{bmatrix} -2 & 1 \\ 0 & -3 \end{bmatrix}, \ N = \begin{bmatrix} 0 & 1 \\ 0 & 0 \end{bmatrix} \tag{13}$$

have the form

$$Q = \begin{bmatrix} 0 & 0 & 0 & 1 \\ 0 & 1 & 0 & 0 \\ 0 & 0 & 1 & 0 \\ \frac{1}{2} & 0 & 0 & 0 \end{bmatrix}, \quad P = \begin{bmatrix} 1 & 0 & 0 & 0 \\ 0 & 1 & 0 & -1 \\ 0 & 2 & 1 & -2 \\ 1 & 0 & 0 & 1 \end{bmatrix}. \tag{14}$$

Note that the matrix A_1 defined by (13) is the stable Metzler matrix and the descriptor system with (11) is positive and asymptotically stable. In this case we also have

$$PB = \begin{bmatrix} B_1 \\ -B_2 \end{bmatrix} = \begin{bmatrix} 1 \\ 0 \\ -1 \\ -2 \end{bmatrix}, \ CQ = \begin{bmatrix} C_1 & C_2 \end{bmatrix} = \begin{bmatrix} 1 & 0 & 1 & 1 \end{bmatrix}. \tag{15}$$

Therefore by Theorem 1 the descriptor system (1) with (11) is positive.
   The transfer matrix of the system (1) has the form

$$T(s) = C[Es - A]^{-1}B. \tag{16}$$

Using (16) and (3) we obtain

$$T(s) = CQQ^{-1}[Es-A]^{-1}P^{-1}PB = CQ[P[Es-A]Q]^{-1}PB$$

$$= [C_1 \quad C_2] \begin{bmatrix} I_{n_1}s - A_1 & 0 \\ 0 & Ns - I_{n_2} \end{bmatrix} \begin{bmatrix} B_1 \\ -B_2 \end{bmatrix} \tag{17}$$

$$= C_1[I_{n_1}s - A_1]^{-1}B_1 + C_2[I_{n_2} - Ns]^{-1}B_2$$

$$= T_1(s) + P(s),$$

where

$$T_1(s) = C_1[I_{n_1}s - A_1]^{-1}B_1, \tag{18a}$$

$$P(s) = \sum_{i=0}^{\mu-1} C_2 N^i B_2 s^i \tag{18b}$$

since

$$[I_{n_2} - Ns]^{-1} = \sum_{i=0}^{\mu-1} N^i s^i. \tag{19}$$

Therefore, the improper transfer matrix of the descriptor system is the sum of strictly proper transfer matrix $T_1(s)$ defined by (18a) and the polynomial matrix $P(s)$ defined by (18b).

3 Transfer Matrices with Positive Coefficients of Positive Descriptor Systems

In the paper [24] the following theorem has been proved.

Theorem 3. If the matrix $A_1 \in M_n$ is Hurwitz and $B_1 \in \mathfrak{R}_+^{n \times m}$, $C_1 \in \mathfrak{R}_+^{p \times n}$ then all coefficients of the strictly proper transfer matrix $T_1(s)$ defined by (18a) are positive.

Note that the coefficients of the polynomial matrix $P(s)$ defined by (18b) are also positive. From (18a) it follows that all coefficients of the matrix $T(s) = T_1(s) + P(s)$ are positive.

Theorem 4. If the positive descriptor system (1) is asymptotically stable then all coefficients of its transfer matrix $T(s)$ are positive.

Example 2. (Continuation of Example 1). The descriptor system with the matrices (11) is positive and asymptotically stable and its improper transfer matrix $T(s)$ has the form

$$T(s) = C_1[I_{n_1}s - A_1]^{-1}B_1 + C_2[Ns - I_{n_2}]^{-1}(-B_2)$$

$$= [1 \quad 0]\begin{bmatrix} s+2 & -1 \\ 0 & s+3 \end{bmatrix}^{-1}\begin{bmatrix} 1 \\ 0 \end{bmatrix} + [1 \quad 1]\begin{bmatrix} -1 & s \\ 0 & -1 \end{bmatrix}^{-1}\begin{bmatrix} -1 \\ -2 \end{bmatrix} \tag{20}$$

$$= \frac{2s^2 + 7s + 7}{s+2}.$$

The transfer function (20) of the positive and asymptotically stable descriptor system with the matrices (11) has all positive coefficients.

4   Realization Problem for Positive Descriptor Linear Systems

Now let us consider the inverse problem called the realization problem which can be stated as follows.

Given the improper transfer matrix of positive asymptotically stable descriptor linear system $T(s) \in \Re^{p \times m}(s)$, find its matrices $E \in \Re^{n \times n}$, $A \in \Re^{n \times n}$, $B \in \Re^{n \times m}$, $C \in \Re^{p \times n}$.

The realization problem can be solved by the use of the following procedure.

Procedure 1.

Step 1. Decompose the improper transfer matrix $T(s)$ into the strictly proper matrix $T_1(s)$ and the polynomial matrix $P(s)$, i.e.

$$T(s) = T_1(s) + P(s). \tag{21}$$

Step 2. Find the matrices A_1, B_1, C_1 such that

$$C_1[I_{n_1}s - A_1]^{-1}B_1 = T_1(s) \tag{22}$$

and the matrices N, B_2, C_2 such that

$$C_2[I_{n_2} - Ns]^{-1}B_2 = P(s). \tag{23}$$

Step 3. Choose a permutation matrix $Q \in \Re^{n \times n}$, a nonsingular matrix $P \in \Re^{n \times n}$ and using the formula

$$E = P^{-1}\begin{bmatrix} I_{n_1} & 0 \\ 0 & N \end{bmatrix}Q^{-1}, \quad A = P^{-1}\begin{bmatrix} A_1 & 0 \\ 0 & I_{n_2} \end{bmatrix}Q^{-1},$$

$$B = P^{-1}\begin{bmatrix} B_1 \\ -B_2 \end{bmatrix}, \quad C = [C_1 \quad C_2]Q^{-1} \tag{24}$$

compute the desired matrices in the desired form. Note that the formula (24) follows from (3).

Theorem 5. If the coefficients of the strictly proper matrix $T_1(s)$ and of the polynomial matrix $P(s)$ are positive then there exists a positive asymptotically stable realization given by (24) of the improper transfer matrix $T(s)$.

Proof. If matrix $T_1(s)$ has positive coefficients then there exist Hurwitz Metzler matrix $A_1 \in M_{n_1}$ and $B_1 \in \mathfrak{R}_+^{n_1 \times m}$, $C_1 \in \mathfrak{R}_+^{p \times n_1}$ satisfying (22). Taking into account that (19) it is easy show that there exist matrices $B_2 \in \mathfrak{R}_+^{n_2 \times m}$ and $C_2 \in \mathfrak{R}_+^{p \times n_2}$ satisfying (23).

For different permutation matrix $Q \in \mathfrak{R}^{n \times n}$ and nonsingular matrix $P \in \mathfrak{R}^{n \times n}$ we obtain different positive asymptotically stable realizations of the matrix $T(s)$. $\square$

Example 3. Given the improper transfer matrix of the descriptor system

$$T(s) = \frac{2s^3 + 12s^2 + 23s + 17}{s^2 + 4s + 3} \tag{25}$$

with positive coefficients. Compute its positive asymptotically stable realization.

Using Procedure 1 we obtain the following.

Step 1. The improper transfer function (25) can be decomposed into the strictly proper part

$$T_1(s) = \frac{s + 5}{s^2 + 4s + 3} \tag{26}$$

and the polynomial part

$$P(s) = 2s + 4. \tag{27}$$

Step 2. The positive asymptotically stable realization of (26) has the form

$$A_1 = \begin{bmatrix} -1 & 2 \\ 0 & -3 \end{bmatrix}, \ B_1 = \begin{bmatrix} 1 \\ 1 \end{bmatrix}, \ C_1 = \begin{bmatrix} 1 & 0 \end{bmatrix} \tag{28}$$

and the matrices N, B_2, C_2 of (27) are

$$N = \begin{bmatrix} 0 & 1 \\ 0 & 0 \end{bmatrix}, \ B_2 = \begin{bmatrix} 2 \\ 1 \end{bmatrix}, \ C_2 = \begin{bmatrix} 2 & 0 \end{bmatrix} \tag{29}$$

since

$$C_1[I_2 s - A_1]^{-1} B_1 = \begin{bmatrix} 1 & 0 \end{bmatrix} \begin{bmatrix} s+1 & -2 \\ 0 & s+3 \end{bmatrix}^{-1} \begin{bmatrix} 1 \\ 1 \end{bmatrix} = \frac{s+5}{s^2 + 4s + 3}, \tag{30a}$$

$$C_2[I_2 - Ns]^{-1} B_2 = \begin{bmatrix} 2 & 0 \end{bmatrix} \begin{bmatrix} 1 & -s \\ 0 & 1 \end{bmatrix}^{-1} \begin{bmatrix} 2 \\ 1 \end{bmatrix} = 2(s+2) \tag{30b}$$

and the first possible realization has the form

$$\tilde{E} = \begin{bmatrix} I_{n_1} & 0 \\ 0 & N \end{bmatrix}, \tilde{A} = \begin{bmatrix} A_1 & 0 \\ 0 & I_{n_2} \end{bmatrix}, \tilde{B} = \begin{bmatrix} B_1 \\ -B_2 \end{bmatrix}, \tilde{C} = [\, C_1 \quad C_2 \,]. \tag{31}$$

Step 3. Choosing matrices P and Q in the form (14) we obtain

$$E = P^{-1}\tilde{E}Q^{-1} = \begin{bmatrix} 0 & 0 & 0 & 0 \\ -1 & 1 & 0 & 1 \\ -1 & 0 & 0 & 1 \\ 1 & 0 & 0 & 0 \end{bmatrix}, A = P^{-1}\tilde{A}Q^{-1}$$

$$= \begin{bmatrix} -2 & 0 & 0 & 2 \\ 3 & -3 & 0 & -3 \\ 0 & -2 & 1 & 0 \\ -3 & 2 & 0 & 2 \end{bmatrix},$$

$$B = P^{-1}\tilde{B} = \begin{bmatrix} -2 \\ 1 \\ -2 \\ 1 \end{bmatrix}, C = \tilde{C}Q^{-1} = [\, 1 \quad -4 \quad 2 \quad 0 \,]. \tag{32}$$

5 Concluding Remarks

Descriptor positive continuous-time linear system with transfer matrices having only positive coefficients have been analyzed. It has been shown that if the positive descriptor system is asymptotically stable then its transfer matrix has only positive coefficients. The realization problem for the class of descriptor positive systems has been formulated and solved. Considerations have been illustrated by numerical examples.

Acknowledgment. This work was supported by National Science Centre in Poland under work No. 2017/27/B/ST7/02443.

References

1. Benvenuti, L., Farina, L.: A tutorial on the positive realization problem. IEEE Trans. Autom. Control **49**(5), 651–664 (2004)
2. Berman, A., Plemmons, R.J.: Nonnegative Matrices in the Mathematical Sciences. SIAM (1994)
3. Bru, R., Coll, C., Romero-Vivo, S., Sanchez, E.: Some problems about structural properties of positive descriptor systems. Lecture Notes in Control and Information Sciences, vol. 294, pp. 233–240. Springer, Heidelberg (2003)
4. Dai, L.: Singular Control Systems. Lectures Notes in Control and Information Sciences. Springer, Heidelberg (1989)

5. Dodig, M., Stosic, M.: Singular systems state feedbacks problems. Linear Algebra Appl. **431**(8), 1267–1292 (2009)
6. Farina, L., Rinaldi, S.: Positive Linear Systems; Theory and Applications. Wiley, New York (2000)
7. Guang-Ren, D.: Analysis and Design of Descriptor Linear Systems. Springer, New York (2010)
8. Kaczorek, T.: A modified state variable diagram method for determination of positive realizations of linear continuous-time systems with delays. Int. J. Appl. Math. Comput. Sci. **22**(4), 897–905 (2012)
9. Kaczorek, T.: A realization problem for positive continuous-time linear systems with reduced numbers of delays. Int. J. Appl. Math. Comput. Sci. **16**(3), 325–331 (2006)
10. Kaczorek, T.: Analysis of positivity and stability of discrete-time and continuous-time nonlinear systems. Comput. Probl. Electr. Eng. **5**(1), 11–16 (2015)
11. Kaczorek, T.: Computation of positive stable realizations for linear continuous-time systems. Bull. Pol. Acad. Tech. Sci. **59**(3), 273–281 (2011)
12. Kaczorek, T.: Computation of realizations of discrete-time cone systems. Bull. Pol. Acad. Sci. Tech. Sci. **54**(3), 347–350 (2006)
13. Kaczorek, T.: Descriptor positive discrete-time and continuous-time nonlinear systems. In: Proceedings of SPIE, vol. 9290 (2014)
14. Kaczorek, T.: Existence and determination of the set of Metzler matrices for given stable polynomials. Int. J. Appl. Math. Comput. Sci. **22**(2), 389–399 (2012)
15. Kaczorek, T.: Positive 1D and 2D Systems. Springer, London (2002)
16. Kaczorek, T.: Positive minimal realizations for singular discrete-time systems with delays in state and delays in control. Bull. Pol. Acad. Sci. Tech. **53**(3), 293–298 (2005)
17. Kaczorek, T.: Positive singular discrete-time linear systems. Bull. Pol. Acad. Sci. Tech. **45**(4), 619–631 (1997)
18. Kaczorek, T.: Positive stable realizations for fractional descriptor continuous-time linear systems. Arch. Control Sci. **22**(3), 255–265 (2012)
19. Kaczorek, T.: Positive stable realizations with system Metzler matrices. Arch. Control Sci. **21**(2), 167–188 (2011)
20. Kaczorek, T.: Positivity and stability of discrete-time nonlinear systems. In: IEEE 2nd International Conference on Cybernetics, pp. 156–159 (2015)
21. Kaczorek, T.: Realization problem for positive discrete-time systems with delays. Syst. Sci. **30**(4), 117–130 (2004)
22. Kaczorek, T.: Theory of Control and Systems. PWN, Warsaw (1993). (in Polish)
23. Kaczorek, T., Sajewski, Ł.: The Realization Problem for Positive and Fractional Systems. Springer (2014)
24. Kaczorek, T., Sajewski, Ł.: Transfer matrices with positive coefficients for standard and fractional positive linear systems. In: 23rd International Conference on Methods and Models in Automation and Robotics, Miedzyzdroje, Poland, 27–30th August (2018)
25. Varga, A.: On stabilization methods of descriptor systems. Syst. Control Lett. **24**, 133–138 (1995)
26. Virnik, E.: Stability analysis of positive descriptor systems. Linear Algebra Appl. **429**, 2640–2659 (2008)

Study on Wave Simulator and Hydraulic Active Heave Compensation Structure

Arkadiusz Jakubowski[✉] and Arkadiusz Kubacki

Institute of Mechanical Technology, Poznan University of Technology,
Skłodowska-Curie Square 5, 60-965 Poznań, Poland
arkadiusz.z.jakubowski@doctorate.put.poznan.pl,
arkadiusz.kubacki@put.poznan.pl

Abstract. In the paper development of design wave simulator and hydraulic active heave compensation structure is presented. The system is consists of motion platform with two degrees of freedom and hydraulic motor which is used for compensation sea waves. The motion platform is used to simulate the wave effects, while the hydraulic motor is controlled by controller. Authors used an AHRS sensor to measure actual position of the steel load which is mounted at the end of the steel cable. On the AHRS sensor accelerometer, gyroscope and magnetometer are implemented. In addition, the Authors used two laser displacement sensors to measure the steel load position in two axes. Performed test included step response of hydraulic motor for different type of signals coming from motion platform and measured position of the steel load by laser displacement sensors and AHRS sensor.

Keywords: Active heave compensation · Heave compensation · Motion platform

1 Introduction

In most cases, an electrohydraulic servo-drive are used in maritime industry. Hydraulic actuators give the highest power to weight ratio of any other actuator. The hydraulic systems are well known in the maritime industry and are still used. The hydraulic drives are an important part of that industry, thanks to them it is possible to perform difficult operations on rough sea. More information about the electrohydraulic drives are in [1, 2].

Heave compensation systems are developed all the time. The use of these systems in maritime industry allows to perform tasks previously impossible to perform at sea during large waves. Two types of the heave compensation systems are currently used. The first is passive heave compensation (PHC). In this case, only mechanical components such as a damper and a spring are used. No any drives are used in PHC. On the other hand, active heave compensation (AHC) systems are used. One of the big advantage of this solution compared to the passive system is that we can control that system. In each AHC systems the drives are applied. In most cases that is the hydraulic drive. The most popular solution in load handling is the winch system, huge winch is put on the deck of the ship and hydraulic or electric motor drives the winch. In the

© Springer Nature Switzerland AG 2020
R. Szewczyk et al. (Eds.): AUTOMATION 2019, AISC 920, pp. 95–103, 2020.
https://doi.org/10.1007/978-3-030-13273-6_10

active heave compensation systems in order to control the drives it is necessary to measure the position of the ship which is not required in passive systems [3–9].

The paper is organized as follows. After the introduction, a review of the state of the art in the area of passive and active system are given in Sect. 2, in which examples of these systems are described. Wave simulator and active heave compensation structure, control system and test stand are given in Sect. 3. In Sect. 4 the Authors focus on Experimental tests. Conclusion and future work are shown in Sect. 5.

In this paper the Authors described study on wave simulator and hydraulic heave compensation system. The Authors presented 2 DoF wave simulator structure, control system and the test stand which combines the wave simulator and hydraulic active heave compensation system. The Authors of this paper performed basic experimental investigations using several sensors such as laser displacement sensor, AHRS sensor.

2 State of the Art

Examples of research on wave simulator include the work by Sanafilippo et al. [10]. Author presented motion platform which is used to simulate the sea wave's effects. The motion platform is drive by three electrical motors. On the top plate of the motion platform the industrial robot is placed. The robot is controlled by the user with joystick. Actual position of the top plate of the wave simulator was measured by means of an accelerometer. The wave simulator was built for testing different control algorithm. The Author presented simulation and experimental results which were carried out to validate the correct operation of the framework (Fig. 1).

Fig. 1. The physical motion platform [10].

Another work about simulation compensation system is by Kjelland and Hansen [11] who proposed offshore simulated loading and unloading of load. In this work a Stewart platform was used to simulate the motion of the vessel. They used a hydraulic vehicle loader crane to perform the payload transfer. The Authors measured the wire force by load cell. In this paper the simulation and experimental test are presented and the model of the hydro-mechanical winch system is experimentally verified.

In the paper [12] a review of the current state of the art in the range of heave compensation system is presented. The authors described in detail three types of heave compensation systems. In this paper the Authors shown their dedicated test stand for active heave compensation systems. The test stand is presented in Fig. 2. The main components of that test stand are hydraulic cylinder which was responsible for simulate sea waves and hydraulic motor that was used to provide a steady position of the steel load. A motion of the steel load which was fix at the end of the steel cable was measured via AHRS sensor. Roll, Pitch and Yaw were measured during movement of the hydraulic cylinder. This sensor used acceleration, gyroscope and magnetic field to estimate a roll, pitch and yaw angle. Simulation and experimental investigation results are presented.

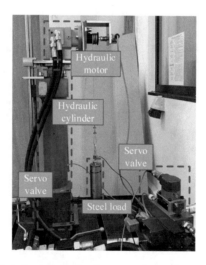

Fig. 2. Test stand for active heave compensation system [12].

3 Wave Simulator

3.1 Wave Simulator Structure

In this section Authors presented the wave simulator structure. The Author build the wave simulator and hydraulic active heave compensation system to give researchers the possibility of testing different control algorithm for heave compensation systems. The wave simulator has two DoFs. Sea waves are simulated by the performed wave simulator. The build wave simulator consists of two arms with steel connected to universal joints at the top plate. Each joint is set in motion by small electric servo motor for controlling the corresponding corner of the top plate. The Authors applied two two-stage planetary gear to increase the output torque. The output torque is equal to 40 Nm. The frame structure is made of aluminium profiles. The wave simulator and hydraulic active heave compensation system is presented in the Fig. 2.

The hydraulic motor is responsible for the movement compensation of the wave platform, for this reason it is necessary to convert the position of the top plate to the properly movement of the hydraulic motor. In the Figs. 3 and 4 basic dimensions of the test stand are presented. The equations below presents the conversion of the top plate position to the position of the hydraulic motor.

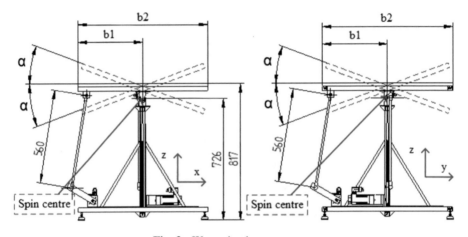

Fig. 3. Wave simulator structure.

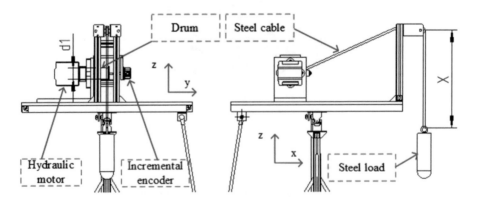

Fig. 4. Wave simulator structure.

$$y = \cos \alpha * b1 \tag{1}$$

$$z = \frac{R}{y} \tag{2}$$

$$R = 2\pi r \tag{3}$$

$$e = \frac{e_p}{z} \tag{4}$$

where:

$y-$ change in the height of the end point of the top plate,
b_1- the distance from the spin point to the end of the top plate,
$R-$ circumference of a circle,
d_1- diameter of the drum = 50 mm,
$r = \frac{d_1}{2}$,
$z-$ the number of the drum revolutions,
e_p- number of encoder pulses = 1000/revolution,
$e-$ desired signal sent to the hydraulic motor.

3.2 Control System

Wave simulator and heave compensation control system was based on PLC Power Panel 500 B&R Company. A PC was connected to the PLC via the Ethernet interface. The PLC controller has been equipped with additional modules that allowed to measure the displacement of the steel load, measure actual position of the shaft of the hydraulic motor via incremental encoder and dedicated module to control the electrical amplifier for servo valve. Two servo motors were controlled by PLC via servo drive ACOPOS P3. Scheme of the control system is shown in the Fig. 5.

3.3 Test Stand

In this section Authors presented the test stand which consist of wave simulator, hydraulic active heave compensation system and control system. The authors focused primarily on the main components of the wave simulator. The base of the wave simulator is made of aluminum profiles, thanks to which we can change its dimensions and add additional profiles, e.g. for screwing in sensors. The top plate is also made of aluminum profiles to which a wooden plate is screwed. A hydraulic motor through a special plate is screwed to the top plate. In order to ensure movement of the top plate in two axes at the same time, a cardan joint was used, which is mounted between the top plate and the base. On the shaft of the hydraulic motor the plastic drum is placed. The steel cable is wound on that plastic drum passes through the metal wheel with the bearing. At the end of the steel cable the steel load is mounted. The servo motors controller and all of dedicated modules for PLC are placed on the base of the test stand (Fig. 6).

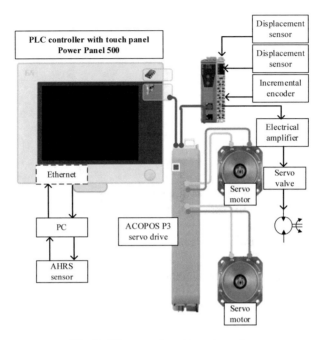

Fig. 5. The control system schematic.

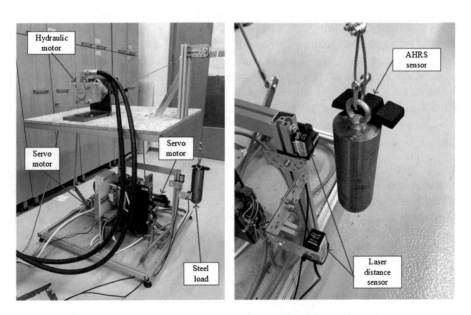

Fig. 6. The test stand for testing active heave compensation systems.

4 Experimental Tests

The study aimed to verify the position of the steel load changes during operation wave simulator. The change of the steel load position was checked when the simulator was operated in two axes. The supply pressure of the hydraulic motor was equal to 8 MPa. The hydraulic motor was controlled by the electrohydraulic servo valve. Authors used hydraulic power supply with parameters: maximum flow rate = 100 dm^3/min, maximum pressure p 0 = 40 MPa, electric motor power = 37 kW, filtration at 6 μ. As a drive-in heave compensation system, a Bosch Rexroth GMS 80 hydraulic motor was used. The maximum output power of the hydraulic motor is equal to 16 kW. The hydraulic motor can operate at a maximum pressure of 21 MPa. The nominal motor rotary speed is 810 rpm. In order to rotate the top plate of the wave simulator, the Authors used two 8LSA25 B&R electric servo motors. The basic parameters of the servo motors are as follows: nominal speed is 6000 rpm, nominal torque is 052 Nm, and nominal power is 327 W. The increase in torque was obtained through the use of 8GP40-060 planetary gears whose transmission gear ration is 25. During tests position of the steel load was measured by micro laser distance sensor Panasonic HG-C1050-P and Panasonic HG-C1100-P. Detection range is equal 50 mm and 100 mm respectively. The repeatability is 200 um and 300 um. In addition, the authors measured the steel load movement using an AHRS sensor manufactured by AISENS Company. This sensor is equipped in accelerometer, gyroscope and magnetometer. A resolution of this sensor for roll and pitch angle is under 1°. All data are streamed real time with frequency up to 200 Hz.

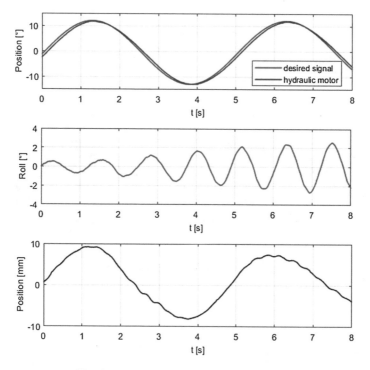

Fig. 7. Rotation of the steel load for X-axes.

The first experimental test was to check the steel load movement during the wave simulator operation in the X axis. The top plate of the wave simulator moved in X-axis in the rage of ±12°. The desired signal was sent to one of the wave simulator drives. The signal set to the wave simulator drive was converted to the rotation of the hydraulic motor (according to Sect. 3.1). The wave simulator movement caused the rotation of the hydraulic motor which caused the steel cable to be rolled up and unroll. The upper waveform in the Fig. 7 shows the desired signal and the response of the hydraulic motor. In the middle the signal from the AHRS sensor is shown. The measurement from the laser distance sensor is on the bottom chart. The obtained results show that the steel load moves in the range of 10 mm.

The next step was also to check the steel load movement, this time for the Y-axis. The top plate of the wave simulator moved in Y-axis in the rage of ±12°. In this case the steel load movement was almost twice as large as for the X-axis and was equal to 18 mm. The upper waveform in the Fig. 8 shows the desired signal and the response of the hydraulic motor. In the middle the signal from the AHRS sensor is shown. The measurement from the laser distance sensor is on the bottom chart.

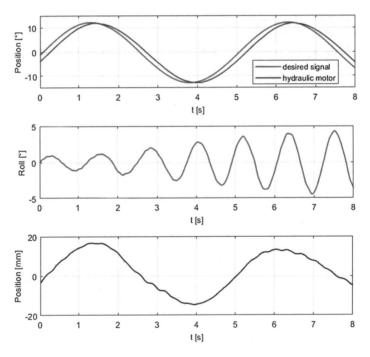

Fig. 8. Rotation of the steel load for Y-axes.

5 Conclusion

This paper presents study on wave simulator and hydraulic active heave compensation in a laboratory setup. Authors described wave simulator and active heave compensation system. The main components of the test stand are: the wave simulator, hydraulic motor and control system. The hydraulic motor was controlled by electrohydraulic servo valve. The control system was based on PLC type Power Panel 500. The proposed wave simulator gives possibility to testing active heave compensation system in laboratory conditions. In section experimental test the Authors accurately describe the equipment used and its parameters. In this section few of the experimental results made by the Authors are also showed. Presented results confirmed that the wave simulator can successfully simulated the sea waves.

The authors will continue research of the hydraulic active heave compensation systems. The next step will be to apply the second AHRS sensor to measure position of the top plate of the wave simulator. There is also important to check the force which is indicated between steel load and steel cable.

Acknowledgments. The work described in this paper was funded from 02/22/DSMK/1458.

References

1. Cundiff, J.S.: Fluid Power Circuits and Controls: Fundamentals and Applications. CRC Press, Boca Raton (2001)
2. Chapple, P.: Principles of Hydraulic Systems Design. Momentum Press (2014)
3. Jakubowski, A., Kubacki, A.: Modelling and simulation of a hydraulic active heave compensation system. In: ITM Web of Conferences, vol. 15. EDP Sciences (2017)
4. Jakubowski, A., Kubacki, A., Rybarczyk, D.: Design of the test stand for hydraulic active heave compensation system. Arch. Mech. Technol. Mater. **37**(1), 76–78 (2017)
5. Woodacre, J.K., Bauer, R.J., Irani, R.A.: A review of vertical motion heave compensation systems. Ocean Eng. **104**, 140–154 (2015)
6. Walid, A.A., Gu, P., Hovland, G., Hansen, M.R., Iskandarani, G.Y.: Modeling and simulation of an active heave compensated draw-works. In: Proceedings, pp. 291–296 (2011)
7. Ni, J., Liu, S., Wang, M., Hu, X., Dai, Y.: The simulation research on passive heave compensation system for deep sea mining. In: Mechatronics and Automation (2009)
8. El-Hawary, F.: Compensation for source heave by use of a Kalman filter. IEEE J. Oceanic Eng. **7**(2), 89–96 (1982)
9. Huang, L., Zhang, Y., Zhang, L., Meiying, L.I.U.: Semi-active drilling draw works heave compensation system. Petrol. Explor. Dev. **40**(5), 665–670 (2013)
10. Sanfilippo, F., Hatledal, L.I., Zhang, H., Rekdalsbakken, W., Pettersen, K.Y.: A wave simulator and active heave compensation framework for demanding offshore crane operations. In: Electrical and Computer Engineering (CCECE) (2015)
11. Kjelland, M.B., Hansen, M.R.: Offshore wind payload transfer using flexible mobile crane (2015)
12. Jakubowski, A., Milecki, A.: The investigations of hydraulic heave compensation system. In: Conference on Automation, pp. 380–391. Springer, Cham (2018)

UAVs Fleet Mission Planning Subject to Weather Fore-Cast and Energy Consumption Constraints

Amila Thibbotuwawa[1], Peter Nielsen[1], Grzegorz Bocewicz[2(✉)], and Zbigniew Banaszak[2]

[1] Department of Materials and Production, Aalborg University, Aalborg, Denmark
{amila, peter}@mp.aau.dk
[2] Faculty of Electronics and Computer Science, Koszalin University of Technology, Koszalin, Poland
bocewicz@ie.tu.koszalin.pl,
Zbigniew.Banaszak@tu.koszalin.pl

Abstract. The problems of mission planning for UAVs fleets are subject of intensive research. Their roots go back to the well-known extensions of VRP addressing the routing and scheduling of UAVs to deliver goods from a depot to customer locations. Rising expectations following the new outdoor applications besides seamless flow routing constraints forces to consider other aspects such as the weather forecast and energy consumption. In that context, this research concerns a declarative framework enabling to state a model aimed at the analysis of the relationships between the structure of a given UAVs driven supply network and its behavior resulting in a sequence of submissions following a required delivery. Because of the Diophantine character of the considered model the main question concerns its solvability. The provided illustrative example shows an approach leading to sufficient conditions guaranteeing solutions existence and as a consequence providing requirements for a solvable class of UAV driven mission planning problems.

Keywords: UAV fleet mission planning · UAV routing and scheduling · Delivery service

1 Introduction

Materials and goods distribution in outdoor transportation systems can be carried out using various modes of transportation means, such as ground vehicles: cars, trains, trucks, automated guided vehicles (AGVs); watercrafts: ships, ferryboats [1, 2] etc. In addition to the aforementioned means, unmanned aerial vehicles (UAVs) are increasingly considered for materials movements between distributor and customers, as UAVs significantly reduce the cost and time required to deliver materials as they have an increased flexibility than other transportation modes.

UAV technologies, on one hand, give the opportunity for more flexible transfer of materials and goods between customers, but, on the other hand, they generate new

R. Szewczyk et al. (Eds.): AUTOMATION 2019, AISC 920, pp. 104–114, 2020.
https://doi.org/10.1007/978-3-030-13273-6_11

problems related to the organization and maintenance of the planned routes and schedules. Typical limitations on the implementation of UAVs in transportation systems include the limited distance (dependent on: the battery capacity, payload of UAV, weather conditions), overlapping air corridors designated for UAV movement (leading to collisions and deadlocks) as well as selected technical parameters (speed, maximum pay-load). The present research addresses the problems of routing and scheduling of a UAV fleet, taking into account the changing weather (wind speed and directions) conditions. The focus of the study are solutions that allow to find admissible (collision-free, non-empty battery) plan of UAVs flights which guarantee the satisfaction of all given customers' orders.

In that context, the research presents a declarative framework enabling stating a reference model aimed at the analysis of the relationships between the structure of a given UAV driven supply network and its potential behavior resulting in a sequence of submissions following a required delivery. Because of the Diophantine character of the considered model the main question concerns its solvability. The provided illustrative example shows the way leading to sufficient conditions guaranteeing existence of solutions and as a consequence providing requirements for a solvable class of UAVs driven mission planning problems subject to weather forecast and energy consumption constraints. The results fall within the scope of research, reported in previous papers [1, 3].

The remainder of the paper is structured as follows. Section 2 provides an overview of literature. An example of a UAV Fleet Routing and Scheduling problem is discussed in Sect. 3. Section 4 presents approach before conclusions are formulated and main directions of future research are suggested in Sect. 5.

2 Literature Review

In general, consumer goods delivery boils down to routing problems. The basis of routing literature comes from the vehicle routing problem (VRP), which is a well-studied field and still very much applicable for the advancement of new technology [4]. In its simplest form, the VRP addresses the routing of a fleet of homogeneous vehicles to deliver identical packages from a depot to customer locations while minimizing the total travel cost [4]. VRPs have been applied to solve delivery problems [5] which could appear similar to the UAV routing as a VRP attempts to find the optimal routes for one or more vehicles to deliver commodities to a set of locations [6].

UAV routing problems involve a huge amount of stochastic information in contrast to vehicle routing problems in general, as UAVs should be able to change, adapt, modify, and optimize their routes. Finding feasible trajectories for UAVs in a routing problem is a complex task, which is why very few studies have attempted to simultaneously solve the routing and the trajectory optimization problems for a fleet of UAVs [3].

According to literature, UAV routing problems are normally modeled as traditional vehicle routing problems where side constraints will be added to reflect natures of applications. In early years, most research on UAV routing focuses on developing heuristics and simulations to find a quick solution with an acceptable quality. In contrast to general routing problems, several individual objective functions can be used in UAV routing such as reducing individual UAV costs, enhancing its profit, increasing safety in operations, reducing lead time, and increasing the load capacity of the entire system [7, 8]. Moreover, UAVs' nature is a 3-D environment [9], whereas land and maritime based transportation are 2-D [10].

Limited contributions have been presented [11, 12] in the area of UAV routing in 3-D environments. What has been accomplished in the field has focused on UAV routing for transporting materials and surveillance [13] without considering the stochastic conditions in weather and non-linear energy consumption models [14]. However, as the influence of weather is very important in terms of energy consumption for UAVs, the non-linier energy models proposed in our earlier studies are used for calculating the energy consumption of UAVs [1, 15]

In UAV routing problems where a fleet of UAVs has to visit a set of waypoints assuming generic kinematics and dynamics constraints, the solutions have to satisfy constraints imposed by wind conditions and those related to collision avoidance between UAVs [3]. Collision avoidance and deadlock avoidance is seldom considered in UAV routing literature. Most recent studies regarding routing problems in communication networks do not explicitly avoid collisions between UAVs during the routing because of recent advances in collision avoidance technology as most small UAVs can sense the air traffic and alter flight altitudes or turn in order to avoid collision [16]. AGV routing studies have considered deadlock avoidance methods with respect to flexible manufacturing systems [17]. In our previous work, methods are proposed to avoid collisions by traveling at different heights and by using mutual exclusion constraints on UAVs in collision zones [3].

In conclusion, existing literature have seldom considered the effects of weather conditions changing and their relation to the non-linier behavior of energy consumption of UAVs. Moreover, a common feature of the all research conducted in the considered domain accept a priori presumption assuming the existence of at least one admissible solution.

3   Description of the Problem

In this section a new version of UAV routing problems considering the above presented factors is given.

Let us consider the transportation system from Fig. 1(b). It is represented as a network composed of one depot (node 0) and 5 customers (nodes 1–5). Each customer demands (the number of packages which should be delivered) are given. To transport packages a fleet composed of two UAVs is used. The depot acts like a charging station where each UAV may return to the depot multiple times to collect more packages and replace its batteries. Each UAV route can be seen as a path that starts and ends at the depot. For example in Fig. 1(b) route of the first UAV (blue color line) is: (0, 1, 3, 0) and for second UAV (orange color line) is: (0, 2, 4, 5, 0). The weather forecast shown in Fig. 1(a) distinguishes the three weather time windows (determining different wind speed and direction). The weather forecast determine the time horizon in which the missions of UAVs will be planed. In that context the main research question can be stated as follows: is it possible to deliver all the customer demands using a fleet of UAV ensuring collision avoidance under given weather forecast before the end of time horizon?

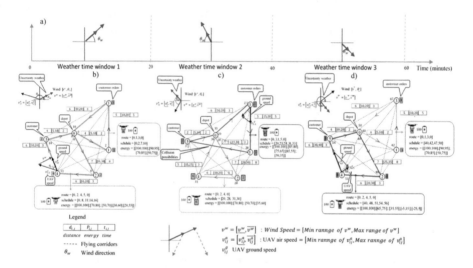

Fig. 1. Illustrative example: description of example submissions (routes and schedules) considering changing weather conditions (a) for first weather time window (b) second weather time window (c) and third weather time window (d).

It is important to note, that this type of problem is not always solvable (i.e. a solution exists guaranteeing collision avoidance and non-empty battery). An example of this case is shown on the Fig. 1(b) (c) (d). In the considered example it is not possible to conduct the entire mission in one weather time window due to the changing weather conditions and the limited energy capacity of the UAVs.

Hence, the mission is divided in two submissions (Fig. 1(b) (c), and (d)) and each submission is conducted separately in different weather time windows. In the first weather time window (Fig. 1b) two UAVs simultaneously fly on two routes and there is no possibility of collisions. In the second weather time window (Fig. 1c), two UAVs fly two different routes where there is a possibility of collisions. In the third weather time window (Fig. 1d) two UAVs fly simultaneously executing two different routes. However, for the second UAV it is not possible to complete the mission, as the energy is not sufficient to complete the route. The main goal is to propose a model that allows one to search for admissible solutions of the fleet mission planning problem subject to weather forecast and energy consumption constraints.

3.1 Transportation System

This study considers routes flown by a fleet of UAVs in order to deliver packages to a set of targets N. Let $G = (N, E)$ be a graph with $N = \{0, 1, ..., n\}$ as the set of nodes and $E = \{\{i, j\} \mid i, j \in N, i \neq j\}$ as the set of edges defined between each pair of nodes. Node 0 is the depot. The distance d_{ij} between each node are given. All UAVs have the same loading capacity Q and a maximum fuel capacity of P_{max}.

Connectivity between the nodes in the network is provided in a connectivity matrix. A blocking matrix is derived by finding the combinations of arcs which cross each other in close proximity. When a given edge is in use by a UAV that arc and the corresponding crossing arcs are blocked for other UAVs. These matrixes are made in advance and used in the experiments as input data. Each location $i \in N_0$, where the set $N_0 = N \backslash \{0\}$, has a demand D_i which represents the weight of the package that will be delivered to location i. The demands could be delivered to the customers at any time during the time horizon. This study considers the UAVs with VTOL (Vertical Take-Off and Landing).

3.2 Assumptions

UAV Characteristics

- All the UAVs in the UAV fleet have similar characteristics, i.e. flying ground speed is constant (20 m/s), a maximum carrying payload is (20 kg) per UAV and maximum energy capacity is (40,000 kJ).

Weather

- Weather forecast is known in advance with sufficient accuracy to specific so called weather time windows, in which constant weather conditions exists, such as speed and direction of wind.
- Weather time windows can be subdivided into flying time windows of the same length (size of the time used in flying of UAV considering the maximum fuel limit and maximum carrying payload).
- The minimum and maximum ranges of wind speed for each weather time window is known in advance. Wind direction is the same inside a given weather time window.

Transportation/Delivery Network

- Only one UAV can occupy a corridor in a given time step. Loitering is not allowed.
- When a UAV is travelling in a given corridor in a given time step, other UAVs cannot occupy the corresponding crossing corridors.
- Every travelled route of a UAV starts and finishes within a given weather window.

Customer Demands

- A similar kind of material is delivered to customers in different amounts [Kgs].
- End of time horizon is one day (24 h)
- Customers can accept deliveries at any time during the time horizon.

3.3   Questions?

- Is it possible to deliver all the customer demands using a fleet of UAVs ensuring collision avoidance under given weather forecast before the end of the time horizon?
- How to divide the whole mission into several submissions?
    - How many weather time windows will guarantee a feasible solution?
    - How many times should each UAV fly to deliver all the customer demands?
- How to deliver all the customer demands using a fleet of UAVs ensuring collision avoidance minimizing the total travel time under uncertainty in weather conditions?

4   Approach

Following previously mentioned assumptions, a weather time window is equal to multiplications of the flying time windows (the time used in flying of UAV considering for example criteria such as: the maximum fuel limit and maximum carrying payload). For each flying time window a set of arbitrarily selected clusters (subsets) of customers is considered. In turn, for each cluster a set of feasible UAVs fleet routings and accompanying schedules taking into account the weather conditions imposing energy consumption constraints is calculated. For instance, according to the Fig. 2, in the first two flying time windows, customers 1, 3, 4 and 5 are included in the alternative clusters of A and B. In the third and fourth flying time windows, customers 1, 2, 4 and 5 are inside the selected alternative clusters of B and C. In the fifth time flying time windows customers, 1, 2, 3 and 5 are inside the selected alternative clusters of E and F. Routes of 0-3-0 and 0-1-0 are shown and fall under cluster A.

Figure 3 illustrates the set of routes and schedules (submissions S1–S4) that is possible under alternative cluster A and B. For each flying time window, there are various possible routes and schedules (submissions). In other words, each flying time

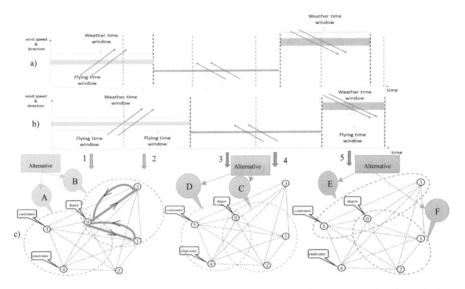

Fig. 2. Alternative clustering of customers (c) following assumption that a weather time window is equal to multiplications of the flying time windows (b) following weather forecast time windows vs. energy limited flying time windows (a).

window consists of submissions, which is a choice of customers to be served depending on the choice from alternative clusters. Consequently, each submission have alternative routes and schedules to serve the selected customers in each alternative cluster. From the possible combinations of submissions, the challenge is to find the admissible solutions.

Tables 1 and 2 presents two possible sequences of submissions. According to Table 1, customer 1, 2 and 4 have not received 100% of the demand at the end of time horizon. Table 2 provides a solution, i.e. sequence of submissions S3, S6, and S12 (see Fig. 4) guaranteeing assumed demands to all the customers.

Figure 4 presents a methodology sketching an approach providing a set of possible sequences of submissions which has to be searched for admissible ones guaranteeing all customers demand fulfillment. At stage 4 aimed at sub-scenarios prototyping while implementing a declarative model of possible solutions a set of sufficient conditions guaranteeing admissibility, i.e. collision and deadlock-free, routings of UAV fleet, have to be employed [1, 3, 18].

Besides of above mentioned conditions the other ones, for example linking a fleet size with a given time horizon and weather conditions, have to be determined as to guarantee admissible sequence of submissions.

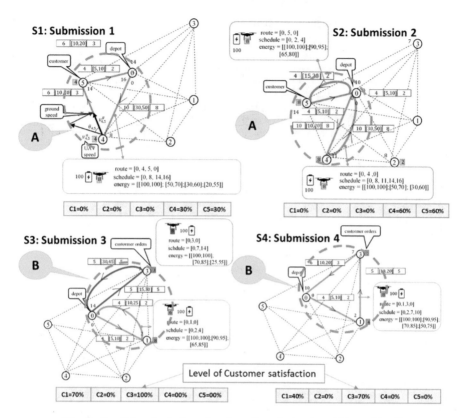

Fig. 3. Possible submissions for alternative cluster A and B, see Fig. 2.

Table 1. Final level of customers demand fulfillment following sequences of submissions S2, S7 and S10 (see Fig. 4)

						UAVs	Customer number				
							1	2	3	4	5
lying tim window	1	S2	0-4-0	A	0-5-0	A	2				60%	60%
	2	S7	0-1-0	C	0-2-0	C	2	80 %	70%			
	3	S10	0-3-0	E	0-5-0	E	2			100%		40%
Final level of customer demand fulfillment							80%	70%	100%	60%	100%

Table 2. Final level of customers demand fulfillment following sequences of submissions S3, S6, and S12 (see Fig. 4).

								Customer number				
Submission			Routes and selected Cluster				UAVs	1	2	3	4	5
Flying time window	1	S3	0-3-0	B	0-1-0	B	2	70%		100%		
	2	S6	0-4-0	D	0-5-0	D	2				100%	100%
	3	S12	0-1-0	F	0-2-0	F	2	30%	100%			
Final level of customer demand fulfillment								100%	100%	100%	100%	100%

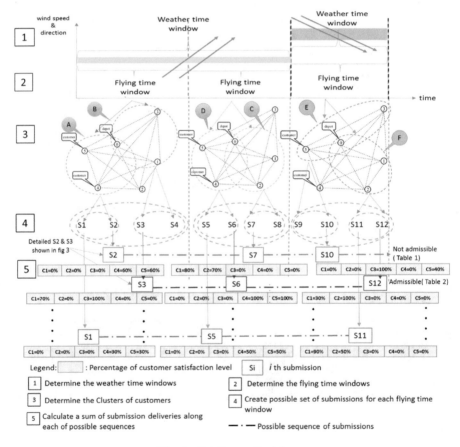

Fig. 4. Solution methodology.

5   Conclusions

A reference model that allows one to search for admissible solutions of UAVs fleet mission planning problem subject to weather forecast and energy consumption constraints is proposed. In that context, the presented study provides the ground for future research aimed at finding the sufficient conditions guaranteeing the existence of at least one feasible solution, and consequently to develop algorithms dedicated to UAV routing and scheduling problems subject to collision avoidance, weather changes and energy consumption constraints. Therefore, in a future research the proposed solution methodology will be used in order to find sufficient conditions and following them algorithms supporting decision making within domain of UAVs fleet mission planning problems.

References

1. Thibbotuwawa, A., Peter, N., Zbigniew, B., Bocewicz, G.: Energy consumption in unmanned ariel vehicles: a review of energy consumption models and their relation to the UAV routing. In: Information Systems Architecture and Technology: Proceedings of 39th International Conference on Information Systems Architecture and Technology (2018)
2. Khosiawan, Y., Khalfay, A., Nielsen, I.: Scheduling unmanned aerial vehicle and automated guided vehicle operations in an indoor manufacturing environment using differential evolution-fused particle swarm optimization. Int. J. Adv. Robot. Syst. **15**, 1–15 (2018). https://doi.org/10.1177/1729881417754145
3. Bocewicz, G., Nielsen, P., Banaszak, Z., Thibbotuwawa, A.: Routing and scheduling of unmanned aerial vehicles subject to cyclic production flow constraints. In: Proceedings of 15th International Conference on Distributed Computing and Artificial Intelligence. Advances in Intelligent Systems and Computing, vol. 801, pp. 75–86 (2019). https://doi.org/10.1007/978-3-319-99608-0_9
4. Golden, B.L., Raghavan, S., Wasil, E.A.: The Vehicle Routing Problem: Latest Advances and New Challenges. Springer Science + Business Media, New York (2010)
5. Eksioglu, B., Vural, A.V., Reisman, A.: The vehicle routing problem: a taxonomic review. Comput. Ind. Eng. **57**, 1472–1483 (2009). https://doi.org/10.1016/j.cie.2009.05.009
6. Dorling, K., Heinrichs, J., Messier, G.G., Magierowski, S.: Vehicle routing problems for drone delivery. IEEE Trans. Syst. Man Cybern. Syst. **47**, 70–85 (2017)
7. Coelho, B.N., Coelho, V.N., Coelho, I.M., et al.: A multi-objective green UAV routing problem. Comput. Oper. Res. **88**: 1–10. https://doi.org/10.1016/j.cor.2017.04.011
8. Enright, J.J., Frazzoli, E., Pavone, M., Savla, K.: Handbook of Unmanned Aerial Vehicles (2015). https://doi.org/10.1007/978-90-481-9707-1
9. Goerzen, C., Kong, Z., Mettler, B.: A survey of motion planning algorithms from the perspective of autonomous UAV guidance. J. Intell. Robot. Syst. Theory Appl. (2010). https://doi.org/10.1007/s10846-009-9383-1
10. Karpenko, S., Konovalenko, I., Miller, A., et al.: UAV control on the basis of 3D landmark bearing-only observations. Sensors (Switzerland) **15**, 29802–29820 (2015). https://doi.org/10.3390/s151229768
11. Guerriero, F., Surace, R., Loscrí, V., Natalizio, E.: A multi-objective approach for unmanned aerial vehicle routing problem with soft time windows constraints. Appl. Math. Model. **38**, 839–852 (2014). https://doi.org/10.1016/j.apm.2013.07.002

12. Khosiawan, Y., Park, Y., Moon, I., et al.: Task scheduling system for UAV operations in indoor environment. Neural Comput. Appl. (2018). https://doi.org/10.1007/s00521-018-3373-9
13. Dorling, K., Heinrichs, J., Messier, G.G., Magierowski, S.: Vehicle routing problems for drone delivery. IEEE Trans. Syst. Man. Cybern. Syst. 1–16 (2016). https://doi.org/10.1109/tsmc.2016.2582745
14. Wang, X., Poikonen, S., Golden, B.: The Vehicle Routing Problem with Drones : A Worst-Case Analysis Outline Introduction to VRP Introduction to VRPD, pp. 1–22 (2016)
15. Thibbotuwawa, A., Nielsen, P., Zbigniew, B., Bocewicz, G.: Factors affecting energy consumption of unmanned aerial vehicles: an analysis of how energy consumption changes in relation to UAV routing. In: Advances in Intelligent Systems and Computing, pp. 228–238. Springer International Publishing (2018)
16. Shetty, V.K., Sudit, M., Nagi, R.: Priority-based assignment and routing of a fleet of unmanned combat aerial vehicles. Comput. Oper. Res. 35, 1813–1828 (2008). https://doi.org/10.1016/j.cor.2006.09.013
17. Bocewicz, G., Nielsen, I.E., Banaszak, Z.A.: Production flows scheduling subject to fuzzy processing time constraints. Int. J. Comput. Integr. Manuf. 29, 1105–1127 (2016). https://doi.org/10.1080/0951192X.2016.1145739
18. Krystek, J., Kozik, M.: Analysis of the job shop system with transport and setup times in deadlock-free operating conditions. Arch. Control Sci. 22, 417–425 (2012). https://doi.org/10.2478/v10170-011-0032-0

Nested NARIMA Model
of the Atmospheric Distillation Column

Michał Falkowski[1][(✉)] and Paweł D. Domański[2]

[1] Faculty of Automotive and Construction Machinery Engineering,
Warsaw University of Technology, Narbutta 84, 02-524 Warsaw, Poland
mfalko@stud.elka.pw.edu.pl
[2] Institute of Control and Computation Engineering,
Warsaw University of Technology, ul. Nowowiejska 15/19, 00-665 Warsaw, Poland
p.domanski@ia.pw.edu.pl

Abstract. Atmospheric distillation column plays a very important role
in the crude oil processing. It is multivariate, strongly nonlinear process.
Its operation and proper control impacts significantly the overall refinery
performance. One may find in literature several papers describing column
modeling with full spectrum of approaches staring from the first principle models up to regression *black-boxes*. Presented model composes of
subsequent sections associated with technology. Each section is modeled
as the nested NARIMA model. This methodology has been previously
tested in several other chemical applications. It is extended by decomposition and coordination of subsequent sub-models in the considered
case. Such a structure enables to obtain process nonlinear model with
clear technological meaning of all considered elements. Further, it may be
directly used in the process of control philosophy design. The procedure
is illustrated with real data originating from the industrial installation.

Keywords: Atmospheric distillation column · NARIMA ·
Nonlinear process · Black-box

1 Introduction

Presented paper addresses two important issues: nonlinear methodology of
nested NARIMA models and evaluation of atmospheric distillation column simulation model. In presented case the need for simulation model has appeared first.
The modeling using NARIMA approach was adopted as it fits the best technology requirements and identification constraints. The subject of atmospheric
column modeling exists in the research for several years. The first principle models are the most popular candidates [1–3], especially once unit simulation is an
ultimate goal. They give an insight into the process and can be implemented in
a simple way.

However, they are simplified and their calibration with industrial data might
be difficult or even impossible due to the limited information and data. Secondly, it may be hard or not reasonable to use such a model in real time control

© Springer Nature Switzerland AG 2020
R. Szewczyk et al. (Eds.): AUTOMATION 2019, AISC 920, pp. 115–124, 2020.
https://doi.org/10.1007/978-3-030-13273-6_12

or optimization applications. Empirical models deliver an alternative. One may choose between different classes [4], like linear or nonlinear regression, Wiener-Hammerstein, neural networks, fuzzy, hybrid models, etc. Hammerstein [5] approach was selected, although not in a standard form. It has given the inspiration, further developed into the novel methodology.

The next design decision is associated with the structure identification [6]. Appropriate model variables (model inputs and outputs) have to be selected. Output selection is straightforward. They directly originate from the modeling goals. Other limiting variables are also often taken into consideration. Selection of input signals [7] is not simple and frequently depends on data availability and quality, process knowledge, statistical analysis, applied benchmarking indexes, etc. Concluding, model structure identification is rather the art.

Once the structure identification phase is achieved, we may progress towards estimation of model parameters. We focus on Hammerstein structure in our case. Thus, our model consists of two elements: nonlinear static part and linear dynamic one. There are many various algorithms [8]. Parameters of the nonlinear part are determined by any non-linear optimization, while the linear part may be determined by least squares approach. We may find many papers considering various approaches, like blind approach [9], iterative [10], multi-variable methods [11] and aspects such as an identification in closed loop [12].

In course of the work, various approaches were tested. Linear dynamic element could be satisfactory estimated with least squares. Nonlinear part was designed as polynomials using the least squares algorithm as well.

The paper is organized as follows. In the starting Sect. 2 the modeling methodology is presented. Then Sect. 3 presents the process of the crude oil atmospheric column. Those two background sections are followed by presentation of the modeling results (Sect. 4). The paper concludes in Sect. 5 with the discussion of obtained results and presentation of open issue for future research.

2   Modeling Methodology

Modeling methodology for the complex installation consists of four main phases:

1. process decomposition into separable dynamic subprocesses that may be identified independently, ideally, with one common approach,
2. structure identification for selected subprocesses, i.e. proper selection of model inputs and outputs, its type, size, order, etc.
3. estimation of the model parameters and its validation,
4. consolidation and coordination of all sub-models into single model.

2.1   Process Decomposition

Process decomposition derives from the considered modeling problem. The main goal is to select such sub-processes that are clearly separable, being independent on the rest of the process. They are often associated with the process technological design. The main two approaches for the selection are serial or parallel structures, however hybrid configurations are frequently used as well [4]. In presented work selection is done according to the P&ID (Process & Instrumentation Diagram).

2.2   Model Structure Identification

Model structure identification is quite a tedious task, not well defined and very often customized to the problem. It frequently consists of several steps, while each of them gives fragmentary contribution to the overall methodology [13,14]. In the considered case the process is nonlinear and identification is based on the historical data. The results significantly depend on the data thus variables time series review and proper selection of data files constitutes an important step (time trends, histograms and statistical analysis). Methodology includes combination of various methods, like autocorrelation and cross-correlation backed with the basic modeling of different structures and their performance comparison.

2.3   Model Parameter Estimation

Estimation of the model parameters is strictly connected with the selection of the process model type. In the considered case the Nested ARIMA (NARIMA) model structure was selected (see Fig. 1) with modeled output in form of (1).

$$y_m(k+1) = y_m(k) + \Delta y_m(k+1) \tag{1}$$

It belongs to the class of nonlinear dynamical models called block-oriented ones. It consists of a nonlinear memoryless static element followed by linear dynamics and as such is very similar to the Hammerstein models. While the nonlinear part is interpreted in a similar way to Hammerstein models (with output denoted as $y_m(k)$), the approach to dynamics modeling differs. Dynamic part identifies model error of the nonlinear static element $\Delta y_m(k+1)$. Static nonlinear function and linear dynamic part can be generally of different origin and structure [8]. In our case we use static MISO polynomials and ARIMA regression models.

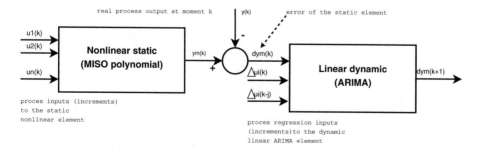

Fig. 1. Nested ARIMA (NARIMA) model scheme

Model selection determines estimation algorithms. Identification of the static MISO polynomials is done using standard least squares. ARIMA regression models are also derived with the least squares method implemented within Matlab System Identification Toolbox. In all cases data are divided into two date sets, one used for identification and the other one for validation purposes.

Identification of the structures is done separately for each of the sub-models. At first the size of the static polynomial is selected. In most cases it is first or the second order. Orders of the dynamic regression models are chosen in an experimental way, selecting those number which result in the best model fitting.

It has to be noted that all the models are multivariate. Selection of the appropriate input signals is done on the basis of the process technology consulted with site experts accompanied by correlation analysis.

3   Atmospheric Column for Crude Oil

The crude oil Atmospheric Distillation unit (column), is one of the most important plants in a refinery. It is crucial to have a proper simulation tools to study thoroughly such a multicomponent distillation process, because of its complexity and the fact that its products are the feeds to other site production units. The crude oil is distilled to produce distillate streams, which are the basic streams for the refinery product slate. These streams are either subject to further treating downstream or become feed stock for other conversions in the refinery.

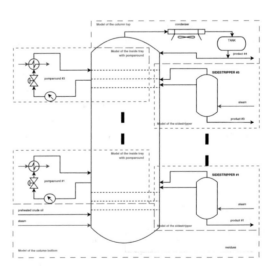

Fig. 2. Crude oil atmospheric column with the decomposition scheme to subprocesses

A schematic flow diagram of an atmospheric crude column is sketched in Fig. 2. The process is considered in literature for many years and thus its discussion may be found in several reports [15–17]. The process is complex, however it consists of several similar elements. For the sake of decomposition the following sub-process are defined and considered, like [18]:

- model of the column bottom,
- model of the inside tray with a pumparound,
- model of the sidestripper,
- model of the column top.

Although models of the top and bottom of the column appear only once there might be several models of the inside tray and sidestrippers. In our case there are three inside tray models and three sidestrippers. Modeled variables are interesting product flows and temperature of products and at each tray level.

The structure of the selected sub-models, i.e. inputs and outputs selection is the same for each model instance. Model of the column bottom has four inputs: inflow of the superheat steam, inflow and temperature of the crude oil and outflow of the atmospheric residues. Temperature at the top of the column section forms the output of the sub-model.

Inside tray with pomparound model has tray top temperature as the output. The inputs are as follows: tray bottom temperature (it is the tray top temperature of the lower section), pumparound flow and the tray product outflow.

In case of sidestrippers models each one has temperature on the output. There are four inputs: tray bottom product flow and temperature (within the same section), superheated steam and product flow which is further processed. All the data used originates from the full-scale operational anonymous industrial installation.

4   Modeling Process

The modeling described in the paper is performed according to the industry requirements constrained by technology and limits of Instrumentation & Control infrastructure. There are two main issues that should be addressed. Historical data access, collection and preprocessing embedded in plant data historian is the first one. Aspects of identification in the mixed mode with some process loops operating in MANUAL and unfortunately the most of them in AUTO (resulting in close loop identification).

4.1   Data Collection and Preprocessing

Working with the real object is associated with a number of complications and problems. Considerations of economy and security of the plant, which is modeled object does not allow for testing of identification. The only solution appears to work with historical data collected during normal process operation.

Target plant is working under control of the well-tuned system. Operation is additionally frequently altered by human interventions (MANUAL model operation). Process frequently is non-stationary, due to external weather conditions, varying parameters of substrates and volumes of consumption.

It is very important to notice that consultation process with plant personnel is crucial and may significantly help. However, before it will be possible to work with data, they must be properly selected, grouped, filtered. Data may be subject to various errors and modifications, like for instance due to the properties of data compression in SCADA systems, transmission failures, etc. The best practice shows that human data inspection is very important. Autonomous identification often leads toward unreliable results.

4.2 Close Loop Identification

Identification of the dynamics, which is inside of the close loop is a challenge. In industrial reality it is often impossible to open the loop, apply external identification input (step, PRBS, ...) and collect data. Apart from the cost of altering normal production and adding of external disturbance, i.e. identification signals, everybody have concerns about installation safety and associated risks. However, in closed loop case standard identification policies, like for instance least squares fail due to the convergence criteria. Thus formally there are two options:

- Indirect Process Identification: An overall model of the closed-loop is identified. The controller has to be known and than process model is evaluated.
- Direct Process Identification: Process is identified as it is.

In reality it is difficult to practically use knowledge about real controller. It is complex structure several other blocks, like nonlinear functions $f(x)$, feedforwards, override controls, interlocking, normalizations, interactions with other loops and human interventions (biases, setpoint changes, etc.). Thus we are left to the Direct Process Identification option despite all deficiencies.

4.3 Identification Results

The results of the modeling process of the atmospheric column are presented for the complete column model in the form of trends of the static and NARIMA models. In addition, a residual analysis was performed to determine the quality of these models. The analysis procedure follows the modern approach presented for the chemical engineering plants [19].

Figure 3 presents the results of identification of the static and NARIMA model of the temperature of the upper part of column. Sampling is equal to Ts = 1 [min].

Analysis of trends of received global models of atmospheric column for crude oil shows that static model despite following real signal trend achieves worse results than NARIMA model - Fig. 3. Both identified models do not fully accurately reflect real signal however the NARIMA model in all cases follows a real signal better and keeps dynamics of atmospheric column for crude oil. For more accurate comparison series of coefficients determining the quality of the obtained global models of crude distillation unit were placed in Table 1.

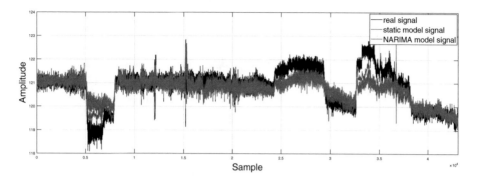

Fig. 3. Comparison of global static and NARIMA atmospheric column for crude oil models

When comparing coefficients listed in Table 1 for both global static and NARIMA models error, it can be noticed that static model has better quality compared to dynamic one. Value of the integral absolute error is in both cases comparable, however the value of mean square error indicates in favor of a static model. Analysing trends and coefficients, it can be concluded that, after all global NARIMA model of atmospheric column for crude oil has a higher quality compared to a global static model.

Table 1. Coefficients comparision of the atmospheric column for crude oil models error

Coefficient	Value	
	Static model	NARIMA model
Mean	−0.261	−0.177
Logistic location	−0.257	−0.172
Bias	0.009	0.098
Standard deviation	0.751	0.861
Scale (Cauchy)	0.407	0.356
Scale (Laplace)	0.568	0.599
Scale (Logistic)	0.663	0.587
IAE	0.601	0.613
MSE	0.633	0.722

Further Figs. 4 and 5 show global error histogram of a NARIMA model. First obtained histogram is characterized by two occurring maxima. This situation indicates that the measurement data set should be divided and some data rejected. Re-analysis of input and output signals trends allowed to reject a certain range of data and to re-determine the error histogram of the NARIMA model. Results

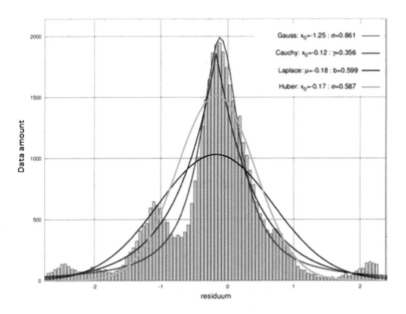

Fig. 4. Error histogram of NARIMA model of the atmospheric column for crude oil with distributions

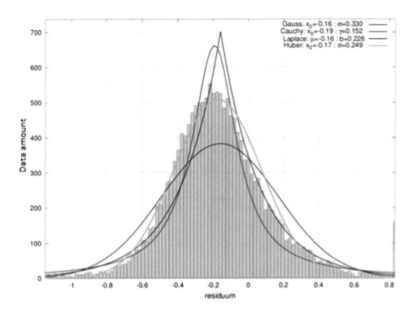

Fig. 5. Error histogram of NARIMA model of the atmospheric column for crude oil with distributions after data rejection

of such an operation are shown in Fig. 5. It is clearly visible that the undesirable maxima from Fig. 4 disappeared and Gauss, Cauchy, Laplace and Huber distributions are better matched [20].

5   Conclusions and Further Research

The paper includes results of the dynamic modeling of the crude oil atmospheric distillation unit based on the data originating from the real industrial site.

First step was to analyze received real data and actual object diagrams, which results in division the atmospheric column for crude oil into numerous local models. For the sake of decomposition the process are defined and considered like model column bottom, insidetray with a pumparound, sitestripper and column top.

Second step was to extract the correct input and output signals. Inputs like steam flow were considered as energy of this steam, referring to such an approach used in literature. Many methods have been used to verify the correctness of chosen input and output signals, e.g. by determining their correlation.

An important step is also selection of delays in increments of input signals. This element is used in implementation of NARIMA model dynamic block by identifying error model obtained from static block.

During identification process of the atmospheric column for crude oil model a number of problems were found. Operator's influence or various unknown external factors had negative impact on the quality of obtained static and NARIMA model. Situation presented in histogram (Fig. 4) indicates that part of the entire measurement data set should be rejected. This is also evidenced by the values of coefficients presented in Table 1.

In conclusion, it can be concluded that final result is satisfactory after a detailed analysis of the trends of both global static and NARIMA models and residual analysis. Quality of the developed models can be described as good with a slight advantage in favor of a NARIMA model.

References

1. Gani, R., Ruiz, C., Cameron, I.: A generalized model for distillation columns - I. Comput. Chem. Eng. **10**(3), 181–198 (1986)
2. Cameron, I.T., Ruiz, C.A., Gani, R.: A generalized model for distillation columns - II. Comput. Chem. Eng. **10**(3), 199–211 (1986)
3. Chaudhuri, U.R.: Fundamentals of Petroleum and Petrochemical Engineering. Chemical Industries. CRC Press, Boca Raton (2010)
4. Nelles, O.: Nonlinear System Identification: From Classical Approaches to Neural Networks and Fuzzy Models. Springer, Heidelberg (2001)
5. Hammerstein, A.: Nichtlineare Integralgleichungen nebst Anwendungen. Acta Math. **54**(1), 117–176 (1930)
6. Nelles, O.: Process Dynamics and Control: Modeling for Control and Prediction. Wiley, Hoboken (2007)

7. Sindelar, R., Babuska, R.: Input selection for nonlinear regression models. IEEE Trans. Fuzzy Syst. **12**(5), 688–696 (2004)
8. Isermann, R., Münchhof, M.: Identification of Dynamic Systems: An Introduction with Applications. Springer, Heidelberg (2011)
9. Bai, E.W.: A blind approach to Hammerstein model identification. IEEE Trans. Signal Process. **50**(7), 1610–1619 (2002)
10. Vörös, J.: An iterative method for Hammerstein-Wiener systems parameter identification. J. Electr. Eng. **55**(11), 328–331 (2004)
11. Wills, A., Schön, T.B., Ljung, L., Ninness, B.: Identification of Hammerstein-Wiener models. Automatica **49**(1), 70–81 (2013)
12. Han, Y., de Callafon, R.: Closed-loop identification of Hammerstein systems using iterative instrumental variables. In: Proceedings of the 18th IFAC World Congress, Miedzyzdroje, Poland, vol. 18, pp. 13930–13935 (2011)
13. Haber, R., Keviczky, L.: Nonlinear System Identification - Input-Output Modeling Approach. Volume 2: Nonlinear System Structure Identification. Kluwer Academic Publishers, Dordrecht, Boston, London (1999)
14. Sun, N.Z., Sun, A.: Model Calibration and Parameter Estimation: For Environmental and Water Resource Systems. Springer, New York (2015)
15. Watkins, R.N.: Petroleum Refinery Distillation, 2nd edn. Gulf Publishing Company, Houston (1979)
16. Hsie, W.H.L., McAvoy, T.J.: Modeling, simulation and control of crude towers. Chem Eng Comm. **98**, 1–29 (1990)
17. Green, D.W., Perry, R.H.: Perry's Chemical Engineers' Handbook, 8th edn. McGraw-Hill Professional, New York (2008)
18. Radulescu, G., Paraschiv, N., Marinoiu, V.: A model for the dynamic simulation of a crude oil unit. Control. Eng. Appl. Inform. **2**(1), 43–50 (2000)
19. Domański, P.D., Golonka, S., Jankowski, R., Kalbarczyk, P., Moszowski, B.: Control rehabilitation impact on production efficiency of ammonia synthesis installation. Ind. Eng. Chem. Res. **55**, 10366–10376 (2016). https://doi.org/10.1021/acs.iecr.6b02907
20. Domański, P.D., Ławryńczuk, M.: Assessment of predictive control performance using fractal measures. Nonlinear Dyn. **89**, 773–790 (2017)

A Diagnostic System for Remaining Useful Life of Ball Bearings

Bogdan Lipiec[(⊠)] and Marcin Witczak

Institute of Control and Computation Engineering, University of Zielona Gora,
65-246 Zielona Gora, Poland
{b.lipiec,m.witczak}@issi.uz.zgora.pl
https://www.issi.uz.zgora.pl/?en

Abstract. This document describes estimation and approximation process of bearings Remaining Useful Life from 2012 Data Challenge. Data were received with the help of PRONOSTIA Platform, constructed for the needs of IEEE Data Challenge. This paper shows different methods of data processing to approximate Remaining Useful Life, which is very important in industries and it is the main part of maintenance. Tests were made on learning set provided by Data Challenge. Raw Data were extracted, filtered and analyzed using algorithms implemented in Matlab. During tests there were used kurtosis, root mean square algorithms and moving average, which helped to process data to be useful for next tests. Remaining useful life was approximated using exponential fitting and different length of original data. Process of analysis boils down to determine at what point it is possible to correctly determinate the bearing remaining useful life.

Keywords: Remaining useful life · Kurtosis · Exponential fitting · Root mean square · Approximation

1 Introduction

Engineering development enables to build more efficient systems. System performance consists of many factors, such as: durability, reliability and many other. Successful fault diagnostics and health monitoring of structures and machines was for many years (and still is) the main target towards a safe operation of machines and structures [4]. Furthermore condition-based maintenance, which includes both diagnosis and prognosis of faults, is a topic of growing interest for improving the reliability of electrical drives [1,10] and many other systems. System designers can mitigate some of the risks posed by potential component failures through a more conservative specification of system redundancies, system operating policies, and system maintenance policies [2].

Remaining Useful Life (RUL) [8] is a part of durability. It makes it possible to estimate how long the system will work without maintenance. The maintenance of crafts, helicopters, ground vehicles, ships, as well as in complex engineering

© Springer Nature Switzerland AG 2020
R. Szewczyk et al. (Eds.): AUTOMATION 2019, AISC 920, pp. 125–134, 2020.
https://doi.org/10.1007/978-3-030-13273-6_13

systems in air civil structures has gained a lot of attention in the last 50 years in the framework of aging machinery and structures, and the need for prolonging their service life beyond the intent of the initial design [4]. Equally important is fault diagnosis and failure prognosis [2]. The latter makes it possible to create a resistant system to unplanned repairs. An example could be a maintenance monitoring system for wind power generation system [12, 13].

There are many research results for RUL of components. The research was carried out among other things over the life of batteries [5, 7] and ball bearings. Accurate RUL prediction is the key process for successful implementation of condition based maintenance program in any industry [3]. In practice, it is difficult for researchers to obtain data from industrial equipment. [9] Industry data sets are rarely shared and hardly ever published. In 2012, IEEE 2012 PHM Date Challenge competition was conducted to specify the effective method of determining the RUL of ball bearings [6]. The testing platform of bearings has been designed and realized at AS2M department of FEMTO-ST Institute. Platform provide real experimental data, which reflects degradation of components. Center for Advanced Life Circle Engineering (CALCE) from University of Maryland won the competition by using three different methods of approximation [11].

2   Organization of Data

PRONOSTIA platform enables to perform run-to failure experiments. Tests were stopped when the amplitude of the vibration signal over-passed 20 g. Data were provided from two acceleration sensors. Sensor were attached vertically and horizontally to the tested bearing, that is a mean of sensor measure vibrations in two axes [11]. All bearings used in experiment were constructed with seven balls.

2.1   Challenge Data Sets

To make the paper self-contained, this section provides a fundamental description of the data sets. Data represents 3 different loads:

- First conditions: 1800 rpm and 4000 N;
- Second conditions: 1650 rpm and 4200 N;
- Third conditions: 1500 rpm and 5000 N.

Life duration of all bearings was varying from 1 h to 7 h. Table 1 presents structure of learning set. Every learning set contains three bearings, which were tested with different loads.

Table 1. Data sets of IEEE 2012 PHM.

Data set	Condition 1	Conditions 2	Conditions 3
Learning set 1	Bearing1_1	Bearing2_1	Bearing3_1
Learning set 2	Bearing1_2	Bearing2_2	Bearing3_2

2.2 ASCII Files

Learning data sets were given in 7z compressed files. Each one contains vibration ASCII files named acc_xxxxx.csv Specification of vibration signals (Fig.1):

- – Sampling frequency: 25.6 kHz
- – Recording: 2560 samples (i.e. 1/10 s) are recorded each 10 s

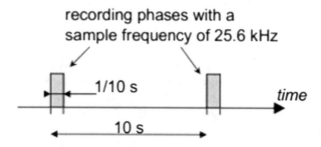

Fig. 1. Acquisition parameters for vibrations signal [6].

All data were arranged in order presented in Table 2. The most important are columns 5 and 6, which contain horizontal and vertical accelerations. During all test only this part of files were used.

Table 2. Arrangement of data

Signal	1	2	3	4	5	6
Vibrations	H	Min	S	ms	Horiz. Acc.	Vert. Acc.

2.3 Raw Data

Figure 2 shows graph of raw acceleration. Vibrations of bearing increases with time. Unfortunately the graph is not legible and it is impossible to determinate RUL of bearing.

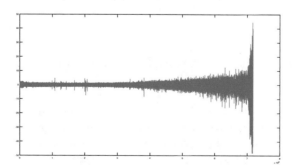

Fig. 2. Raw signal of horizontal acceleration of 1st bearing in first condition.

3   Data Analysis

For every data set of bearing was made a kurtosis. A moving average filter was applied to the time series kurtosis to identify trends over time. If a vibrations signal was similar to Gaussian distribution then this means a good condition of bearing. In this case kurtosis is close to 3.0. Value 2.8 was observed at the beginning of the experiment.

3.1   Kurtosis

Every raw signal of acceleration in two axes was passed through the kurtosis to determinate a probability distribution. Kurtosis window was set to 100. This strategy provides frequent value changes and less data are skipped during kurtosis. Equation (1) represents general formula of kurtosis.

$$\frac{1}{N} \sum_{i}^{N} \frac{(x_i - \overline{x})^4}{\sigma^4} \tag{1}$$

where N is number of samples, x_i represents value of sample, $\overline{x}$ is mean, and σ^4 stands for the standard deviation.

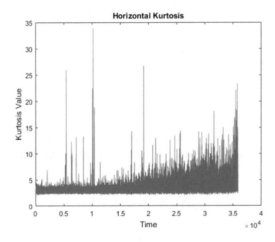

Fig. 3. Vertical kurtosis of Bearing 1 Condition 1 before moving average

Figure 3 presents data from first bearing after kurtosis. In this case maximum value of kurtosis reached 50. Signal contains noises and graph is still unclear. Data overlap and it is difficult to approximate the RUL. It is hard to use any estimation processes. This, it is needed to use moving average, which will help to filter the data from noises. A general form of moving average was presented

in Eq. (2). In this case, this formula is similar to the normal average, but it is
made on specific number of samples unless on all data.

$$\frac{p_0 + p_1 + \dots + p_n}{n} \tag{2}$$

where n stands for number of samples and $p_0 + p_1 + \dots + p_n$ are samples.

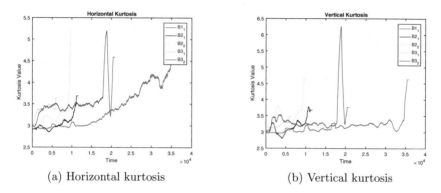

(a) Horizontal kurtosis   (b) Vertical kurtosis

Fig. 4. Comparison of bearings horizontal (a) and vertical (b) kurtosis

Figure 4a presents comparison of Horizontal kurtosis while Fig. 4b shows vertical kurtosis. Every line shows a life of single bearing in one axis. A correlation
coefficient was calculated for each of the vibration features: rms and kurtosis.
Based on the correlation analysis using the extracted features from the six training bearings, the kurtosis extracted from the band-pass filtered vibration signals
with a frequency range of 5.5 to 6.0 kHz was identified as a best feature for bearing prognostics. The window size of the moving average for the time-series of
kurtosis was 1000 data points. The best option in test would be window of the
size of 100 data points, but data contains still many noises and are hard to read.
Window at the size of 1000 allows to clearly work on this data.

When kurtosis of bearing reached value of 3.4, the bearing is at the end of
life. This means a significant degradation of the bearing's components. Bearings
from first condition lives the longest time. It is caused by a mild version of
condition and life of the bearings lasted approx. 8 h. The shortest life are from
bearings from third condition and lasted approx. 4 h. This is due to the hardest
condition.

3.2   Root Mean Square (RMS)

Root mean square from raw data used during test gave better results than kurtosis. Data created by RMS are better to analyze, because graph is smoother
with fewer fluctuations. Point of death of the bearing is better visible using this
method.

Figure 5a and b show comparison of bearings condition using RMS. Bearings in good conditions have RMS value around 0.5 and bearings reach point of dead in value 2 or more. All RMS value were calculated using Eq. (3). After RMS calculations, moving average filter was applied to the data and to separate unnecessary noises.

$$\sqrt{\left(\frac{1}{N} \sum_i^N x_2^i\right)} \tag{3}$$

where N is number of data points and x_2^i stands for data point value.

Contrarily to data filtered by kurtosis, graphs are better readable and easier to understand and determinate important parameters of RUL. Root mean square from vertical axes are more linear. Vertical data from $B2_2$ are the most unclear but despite this Point of Death near of 2 is clearly visible.

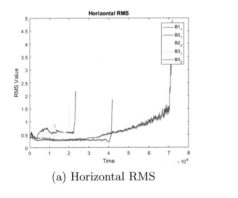
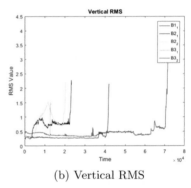

(a) Horizontal RMS   (b) Vertical RMS

Fig. 5. Comparison of bearings horizontal (a) and vertical (b) Root Mean Square

4   Exponential Fitting

The objective of this section is to provide the essential framework used for predicting RUL. The task is exponential fitting, which approximates the shape of kurtosis/rms and predict its development

$$f(x) = p_1 e^{p_2 x} + p_3 e^{p_4 x} \tag{4}$$

where p_i are parameters to be determined.

All parameters were estimated using Matlab function `fit` and provide the best fitting to original data. Function can make fitting along with 95% confidence region. Equation (4) shows general form of exponential fitting.

Figure 6a and b represents comparison of two methods of fitting of Bearing 1 in 1st condition. Red solid line represents fitted curve estimate using Matlab and red doted line show upper and lower bounds of the confidence interval.

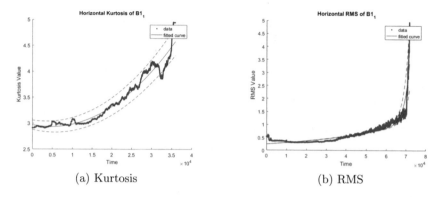

(a) Kurtosis   (b) RMS

Fig. 6. Exponential fitting of bearing's vertical kurtosis (a) and Root Mean Square (b)

Figure 6a shows exponential fit for kurtosis of Bearing 1 in 1st condition along with confidence interval. Figure 6b presents exponential fitting used on Root Mean Square data. Fitting is not so clear and is hard to estimate remaining useful life. Bounds are very small and do not cover 95% of data. Approximation using any length of data from RMS is not effective. Most part of data is almost linear and it caused linear character of approximation. Using RMS is not possible to determinate correctly Point of Death of bearings. Tests were made on all of bearings from learning set. Because of the unsuitability of data, they have not been used in this paper.

Analysis of data was made only on kurtosis because of better fitting with less errors. Fitting from 4/5 of life provide almost the same results as effect from 2/3.

Figure 7 shows an attempt to approximation of point of death using 2/3 data from kurtosis of Bearing 2 from 1 Condition. point of death is well estimated

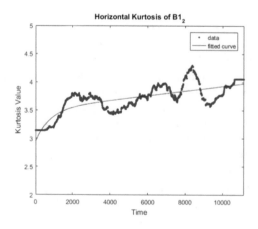

Fig. 7. Exponential fitting 2/3 of data of bearing 2 from condition 1 vertical Kurtosis

despite of noises and not clearly exponential. For this example, after 2/3 of life
is possible to define, when ball bearing is no longer usable.

From the obtained results, it is evident that using exponential fitting, it is
possible to estimate RUL only if data used to create fitting is longer than 2/3 of
whole life of bearing. In case when used data are shorten than this value fitting
function is not efficient and it is impossible to correctly estimate reaming useful
life. Figure 8a presents fitting test using 2/3 of original kurtosis data. Estimation
shows the life shorter of around 10% than the real one. Using 4/5 of life fitting
is nearly to original signal and Point of Death is properly estimated.

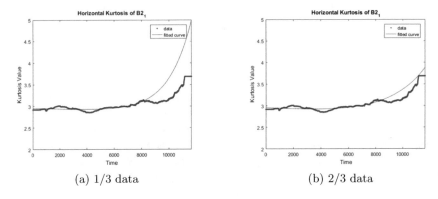

(a) 1/3 data   (b) 2/3 data

Fig. 8. Exponential fitting of bearing 1 condition 2 vertical Kurtosis using different
length of data: 1/3 (a) and 2/3(b)

5   Conclusions

Problem of remaining useful life is one of the most important factors in planning
maintenance of any system. Solving this problem will make the system more
resistant to unappealing effects and more efficient. More efficient system generates more profit and less costs. As it was shown, even the easiest method of
estimation results in improvement of efficient.

RUL is possible to estimate using Kurtosis and exponential fitting. After
analyzing of all estimation results, it is possible to conclude, that RUL can be
correctly approximated after 2/3 of bearings life. Error around 10% is acceptable.
If bearing will be replaced slightly faster, it is fine result. Using this estimation it
is possible to plan maintenance earlier and do not let to stop the entire system for
a long time because of wrong estimation or lack of estimation. The problem would
be when approximations will show life longer than real. In this case planned
maintenance would be to late and due to damage bearings system will not be
working for a longer time. Incorrectly selected point of death results in more
damages in system, because failure of one element can affect other parts of the
system. Using 4/5 of data provided the best approximation. However, in this

case, it can be to late for replacement. This results in loss of additional time to plan maintenance. Comparison of all exponential fitting from one bearing shows improvement of estimation. If results from the last couple curves are similar, it is possible to determinate closely point of death of bearing. Difference between 2/3 data and 4/5 is small enough, and hence 2/3 estimation can be use as a good approximation.

Acknowledgements. The work was supported by the National Science Centre, Poland under Grant: UMO-2017/27/B/ST7/00620.

References

1. Zhirabok, A., Shumsky, A.: Fault diagnosis in nonlinear hybrid systems. Int. J. Appl. Math. Comput. Sci. **28**(4), 635–648 (2008). https://doi.org/10.2478/amcs-2018-0049
2. Bole, M.B.: Load allocation for optimal risk management in systems with incipient failure modes. Georgia Institute of Technology, Zielona Góora (2013)
3. Kundu, P., Chopra, S., Lad, B.K.: Multiple failure behaviors identification and remaining useful life prediction of ball bearings. J. Intell. Manufact. (2017). ISSN: 1572-8145. https://doi.org/10.1007/s10845-017-1357-8
4. Loutas, H.T., Roulias, D., Geogoulos, G.: Remaining useful life estimation in rolling bearings utilizing data-driven probabilistic E-support vectors regression. IEEE Trans. Reliab. **62**(4), 821–832 (2013). https://doi.org/10.1109/TR.2013.2285318
5. Miao, Q., et al.: Remaining useful life prediction of lithium-ion battery with unscented particle filter technique. Microelectron. Reliab. **53**(6), 805–810 (2012). https://doi.org/10.1016/j.microrel.2012.12.004
6. Gouriveau Nectoux, P.R., et al.: PRONOSTIA: an experimental platform for bearings accelerated life test. In: IEEE International Conference on Prognostics and Health Management, Denver, CO, USA (2012)
7. Saha, B., et al.: Prognostics methods for battery health monitoring using a bayesian framework. IEEE Trans. Instrum. Meas. **58**(2), 291–296 (2009). https://doi.org/10.1109/TIM.2008.2005965
8. Si, X.-S., et al.: Remaining useful life estimation - a review on the statistical data driven approaches. Eur. J. Oper. Res. **213**(1), 1–14 (2011). https://doi.org/10.1016/j.ejor.2010.11.018
9. Sikorska, J., et al.: A collaborative data library for testing prognostic models. In: Third European Conference of the Prognostics and Health Management Society 2016 (2016)
10. Singleton, K.R., et al.: Extended Kalman filtering for remaining-useful-life estimation of bearings. IEEE Trans. Ind. Electron. **62**(3), 1781–1790 (2015). https://doi.org/10.1016/j.microrel.2012.12.004
11. Sutrisno, E., Oh, H., Vasan, A.S.S.: Estimation of remaining useful life of ball bearings using data driven methodologies. In: 2012 IEEE Conference on Prognostics and Health Management (PHM) (2012). https://doi.org/10.1109/ICPHM.2012.6299548

12. Tian, Z., et al.: Condition based maintenance optimization for wind power generation systems under continuous monitoring. Renewable Energy **36**(5), 1502–1509
(2011). https://doi.org/10.1016/j.renene.2010.10.028
13. Simani, S., Farsoni, S., Castaldi, P.: Data-driven techniques for the fault diagnosis
of a wind turbine benchmark. Int. J. Appl. Math. Comput. Sci. **28**(2), 247–268
(2008). https://doi.org/10.2478/amcs-2018-0018

A Data-Driven Approach to Constraint Optimization

Jarosław Wikarek and Paweł Sitek[✉]

Department of Information Systems,
Kielce University of Technology, Kielce, Poland
{j.wikarek, sitek}@tu.kielce.pl

Abstract. Many problems occurring in production, transport, supply chains and everyday life problems can be formulated in the form of constraint optimization problems (COPs). Most often these are issues related to planning and scheduling, distribution of resources, fleet selection, route and network optimization, configuration of machines and manufacturing systems, timetabling, etc. In the vast majority of cases, these are discrete problems of a combinatorial nature. Significant difficulties in modelling and solving COPs are usually the magnitude of real problems, which translates into a large number of variables and constraints as well as high computational complexity (usually NP-hard problems). The article proposes a data-driven approach, which allows a significant reduction in the magnitude of modelled problems and, consequently, the possibility of solving many real problems in an acceptable time.

Keywords: Data-driven · Constraint optimization · Mathematical programing · Problem modeling

1 Introduction

Constraint optimization problems (COPs) have been raising interest among scientists and practitioners in recent decades [1]. This is due to the fact that they are often encountered in modern processes, including manufacturing, distribution, transport and supply chains. They are also encountered in management, organizational and social processes. Practically, each of the above-mentioned processes is characterized by many constraints of different character and structure. The most important constraints include: resources, time, cost, environment, technology, organization, etc. COPs in production are usually planning and scheduling, resource allocation, control and automation of manufacturing processes, machine configuration, routing, etc. COPs in distribution and supply chains involve route planning, fleet selection, location of warehouses, etc. COPs in organization and management are timetabling processes, order scheduling, and selection of employees with specific competences, etc., to name only the most important ones. Constraint optimization problem usually adopts the form (1)..(3) [2]. In general, the COP solution consists in the optimization of one or multi-criteria objective function (1) under certain restrictions (2), (3), which must be met.

© Springer Nature Switzerland AG 2020
R. Szewczyk et al. (Eds.): AUTOMATION 2019, AISC 920, pp. 135–144, 2020.
https://doi.org/10.1007/978-3-030-13273-6_14

$$\min f(x) \tag{1}$$

$$\text{subject to } g_i(x) = c_i \text{ for } i = 1, \ldots, n \text{ Equality constraints} \tag{2}$$

$$h_i(x) \geq d_j \text{ for } j = 1, \ldots, m \text{ Inequality constraints} \tag{3}$$

The accurate methods to solve COPs include mathematical programming, dynamic programming etc. and approximate methods. These are usually stochastic and heuristic methods as well as the recently popular metaheuristics (GA-Genetic Algorithms, SA-Simulated Annealing, ACO-Ant Colony Optimization, etc.) [3]. The main difficulties that arise when solving COPs usually result from their combinatorial character and the magnitude of real problems. These are usually NP-hard problems. This results in the necessity of searching large spaces for potential solutions, extending the calculation time and the presence of the so-called combinatorial explosion. In order to somehow sort this difficulty, we propose a new COPs modelling method based on data (Sect. 2), which under certain conditions guarantees a significant reduction in the magnitude of the modelled problem, in particular, a reduction in the number of decision variables and constraints and simplification of other constraints. The main contribution of the presented paper is a data-driven approach to the modeling and solving of COPs. This approach is based on data instances and uses a unique procedure to reduce the solution space for the modeled problem. The proposed modelling method has been tested using two illustrative examples. Mathematical programming methods have been used for the solution, although in general, it can be any method. The experiences related to the development of a hybrid approach to modelling and solving problems [4] with constraints were behind the inspiration to develop the presented modelling method. The paper structure is as follows. Section 2 presents the general assumptions of the data-driven approach. Section 2.1 presents a procedure that reduces the space of solutions to the modeled problem (Proc_R). Sections 2.2 and 2.3 provide, respectively, a description and calculation experiments for illustrative examples. The final chapter includes the conclusions and future research.

2   A Data-Driven Approach to Modelling COPs

The general COPs modelling scheme using standard IT tools is shown in Fig. 1. In simple terms, modelling is performed in the following way. On the basis of a verbal description and/or a formal (mathematical) model using the API of a given tool or a universal modelling language, e.g. AMPL [5], every model constraint is mapped in the form of a set of equations/inequalities using the appropriate quantifiers (e.g. for each, sum, etc.). At this stage, data instance values are not used and are replaced with appropriate coefficients/parameters in modelled equations/inequalities. As a result of this process, a "high-level" model is obtained in the appropriate language or format of the given API. Then, based on this model, the "low-level" model is generated, usually in a matrix form and the MPS (Mathematical Programming System) format [6]. The low-level model is solved by a solver of a given tool. This modelling method has two major disadvantages. First of all, it requires proper data preparation and conversion to the matrix format, i.e. replacing and supplementing the missing parameter values with zeros, etc. Secondly, such a modelling method leads to a considerable redundancy of constraints and decision variables due to the occurrence of quantifiers for each, totals of quantifiers and taking into account all values of even zero coefficients (see Note_1).

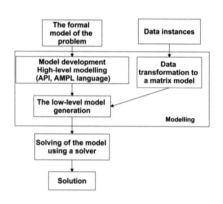

Fig. 1. A classic approach to modelling and solving COPs

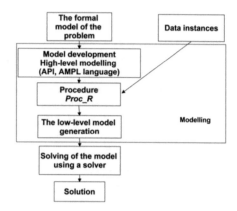

Fig. 2. A data-driven approach to modelling and solving COPs

Note_1: There are 100 customers and 200 different products. Suppose customers want to buy 1200 products. This modelling using quantifiers for each (100 customers and 200 products) will generate 20,000 (100 × 200) constraints. If only placed orders are considered, then there are 1,200 constraints.

Considering the above-mentioned drawbacks that result in the creation of high redundancy models, an alternative COPs modelling method was proposed, the general concept of which is shown in Fig. 2. The basic difference from the method shown in Fig. 1 is the introduction of an extra stage between the high-level and low-level models. As part of this stage, a reduction procedure *Proc_R* (Sect. 2.1) is launched based on the high-level model and the data instance values. As a result of its operation, a low-level model is generated automatically, which can be solved with any mathematical programming solver in this version. The conditions of *Proc_R* procedure applicability are presented in Sect. 2.1, while the examples illustrating the procedure operation and the comparison of both approaches (Figs. 1 and 2) are presented in Sect. 2.2.

2.1 Reducing Procedure *Proc_R*

The algorithm is programmed as part of the procedure *Proc_R* presented in the form of a pseudocode in Fig. 3. The following is needed to run the procedure: a high-level model and data instances written in the form of relations/facts [7]. The effectiveness of the procedure depends on the data instance value. In the relational/fact-based data form, inversely to the matrix form, the inadmissible values do not occur, and more specifically there is no tuple/fact instance for the inadmissible value. The more inadmissible values, the more efficient the procedure, because fewer tuples in a given relations or facts, and this translates into fewer variables and constraints. In the matrix model, all values must be supplemented – not allowed with zeros.

Step 1. For each constraint
    Step 2. For each relation describing the problem
        Step 3. If the relation attributes are a subset of the constraint attributes
            Step 4. If there is no tuple in the relation for the value of the constraint attributes
                Step 5. Delete this constraint
                Step 6. Go to the next constraint
        Step 7. Otherwise
            Step 8. For each variable occurring in the constraint
                Step 9. If the relation attributes are a subset of the variable attributes
                    Step 10. If there is no tuple in the relation for the value of the variable attributes
                        Step 11. Delete this variable from the constraint

Fig. 3. The low-level model generation procedure –*Proc_R*

2.2   Illustrative Examples – Description

In order to present the proposed approach, and in particular the method and effectiveness of the reducing procedure *Proc_R*, the implementation of two illustrative examples were proposed – *ex_01* and *ex_02*. The example *ex_01* illustrates how the procedure *Proc_R* works, while the example *ex_02* was used to carry out the effectiveness tests of the proposed data-driven approach (Fig. 2) in relation to the classic approach (Fig. 1). The example *ex_01* is a simplified version of the transport problem [8].

Example ex_01

Manufacturers $I \in F$ produce products $j \in P$ that are ordered by customers' $k \in O$. To build a model of this problem, the appropriate parameters and decision variables have been introduced. If a manufacturer i manufactures a product j, it is the parameter $a_{i,j} = 1$, otherwise $a_{i,j} = 0$. The value parameter $W_{i,j}$ determines production capacity of a manufacturer i in the field of product manufacturing j. Whereas, the parameter value $z_{j,k}$ determines the product order value j by a customer k. Decision variable $X_{i,j,k}$ determines the volume of product j manufactured by a manufacturer i for the needs of an ordering party k. An additional decision variable $Y_{j,k}$ was also introduced. It determines the amount of product j which could not be delivered to the ordering party k to fulfil its order. The introduction of this additional variable always provides a solution to the problem from the example *ex_01*. The mathematical model for this problem consists of a goal function (4), which minimizes the undelivered quantities of products and constraints on production capacity (5) and customer orders realization (6).

$$\min \sum_{j\in P} \sum_{k\in O} Y_{j,k} \tag{4}$$

$$\sum_{i\in F} a_{i,j} \cdot X_{i,j,k} \leq W_{i,j} \forall i \in F, j \in P \tag{5}$$

$$\sum_{i\in F} X_{i,j,k} = z_{j,k} - Y_{j,k} \forall j \in P, k \in O \tag{6}$$

The second of illustrative examples *ex_02* applies to cost optimization of supply chain with multi-modal transport, the formalization of which is presented in [9]. Based on this example, a comparison has been made between both modelling approaches (*Model1* – Fig. 1) and (*Model2* – Fig. 2) in the context of problem solution effectiveness. To solve the example *ex_02*, both mathematical programming (MP) and the hybrid method were used [4, 9].

2.3   Illustrative Examples – Modelling and Computational Experiments

Based on the example *ex_01* both COPs modelling methods are presented, according to the diagrams that are shown in Figs. 1 and 2. The data recorded in the form of the *R1..R5* relations (Appendix A) were used when developing the models. The representation using the relational model is common in databases, data warehouses and the facts. Using the approach from Fig. 1, a high-level model is built using the API LINGO [10] and data transformation is performed in the first step. The transformation completes the missing data with zeros, this way the modified relations *R4a* and *R5a* are formed (Appendix A), creating a matrix of parameters (i.e. the parameter value was determined for each combination of indexes *i, j, k*). In the next step, based on the data after the transformation and the high-level model, the low-level model (Model1) is automatically generated by the LINGO environment, which may already be sent to the MP solver. Fragments of the low-level model (Model1) are shown in Fig. 4. The model contains 41 constraints and 150 decision variables.

```
MODEL:
[_1]MIN=Y_1_1+Y_1_2+Y_1_3+Y_1_4+Y_1_5+Y_1_6+Y_2_1+Y_2_2+Y_2_3+Y_2_4+Y
   _2_5+Y_2_6+Y_3_1+Y_3_2+Y_3_3+Y_3_4+Y_3_5+Y_3_6+Y_4_1+
Y_4_2+Y_4_3+Y_4_4+Y_4_5+Y_4_6+Y_5_1+Y_5_2+Y_5_3+Y_5_4+Y_5_5+
Y_5_6;

[_2]X_1_1_1+X_1_1_2+X_1_1_3+X_1_1_4+X_1_1_5+X_1_1_6<=100;
[_3]X_1_2_1+X_1_2_2+X_1_2_3+X_1_2_4+X_1_2_5+X_1_2_6<=100;
[_4]X_1_3_1+X_1_3_2+X_1_3_3+X_1_3_4+X_1_3_5+X_1_3_6<=100;
[_5]X_2_3_1+X_2_3_2+X_2_3_3+X_2_3_4+X_2_3_5+X_2_3_6<=100;

  . . . .
[_38]X_1_5_3+X_2_5_3+X_3_5_3+X_4_5_3+Y_5_3=10;
[_39]X_1_5_4+X_2_5_4+X_3_5_4+X_4_5_4+Y_5_4=10;
[_40]X_1_5_5+X_2_5_5+X_3_5_5+X_4_5_5+Y_5_5=0;
[_41]X_1_5_6+X_2_5_6+X_3_5_6+X_4_5_6+Y_5_6=10;
END
```

Fig. 4. The low-level model (*Model1*) generated according to the diagram of Fig. 1

The second model was built based on a data-driven approach using a reducing procedure *Proc_R* according to the diagram of Fig. 2. The high-level model was written analogously to the previous approach using API LINGO and then reduced using the algorithm from Fig. 3. The algorithm uses data from the relation *R1..R5*. It was implemented in the form of the procedure *Proc_R*. In the next step, a low-level model (*Model2*) was created (Fig. 5), which can be sent to the MP solver. This time, the low-level model (*Model2*) includes 44 decision variables and 25 constraints, which makes it significantly smaller than the previous model (*Model1*). Additionally, many constraints have a simpler structure, i.e. have fewer decision variables. Figure 6 shows the constraint on factory production capacity in both models.

```
MODEL:
[_1]MIN=Y_1_1+Y_1_5+Y_1_6+Y_2_6+Y_3_1+Y_3_2+Y_3_3+Y_4_4+Y_4_5+
Y_4_6+Y_5_2+Y_5_3+Y_5_4+Y_5_6;
[_2]X_1_1_1+X_1_1_5+X_1_1_6<=100;
[_3]X_1_2_6<=100;
[_4]X_1_3_1+X_1_3_2+X_1_3_3<=100;
[_5]X_2_3_1+X_2_3_2+X_2_3_3<=100;
 . . . .
[_20]X_2_4_5+X_3_4_5+Y_4_5=10;
[_21]X_2_4_6+X_3_4_6+Y_4_6=10;
[_22]X_2_5_2+X_4_5_2+Y_5_2=10;
[_23]X_2_5_3+X_4_5_3+Y_5_3=10;
[_24]X_2_5_4+X_4_5_4+Y_5_4=10;
[_25]X_2_5_6+X_4_5_6+Y_5_6=10;
END
```

Fig. 5. The low-level model (Model2) generated according to the diagram of Fig. 2

```
Model1: [_2]X_1_1_1+X_1_1_2+X_1_1_3+X_1_1_4+X_1_1_5+X_1_1_6<=100;
Model2: [_2]X_1_1_1+X_1_1_5+X_1_1_6<=100;
```

Fig. 6. Constraints structure comparison for the Model1 and Model2 (production capacity of the first manufacturer for the first product)

The same modelling methods were used for the illustrative example *ex_02*. The example shows the optimization of costs in the supply chain, which consists of three layers: factories, distribution centers and customers. The layers are connected by means of a multi-modal transport. The formalization and description of this illustrative example is shown in [9]. At this stage, the research was meant to determine the modelling effectiveness and its impact on the time of solving the problem and the amount of the reduction of the modelled problem. Computational experiments were performed for both models (*Model1* and *Model2*) of the illustrative example *ex_02* in two environments. The first was the classic mathematical programming environment (LINGO), while the second was the author's environment using a hybrid approach that integrates Eclipse_CLP [11] and Lingo [10]. The results are shown in Table 1. The data instances presented in [9] were used for the experiments. Individual examples E(n) differ in the number of orders from consignees.

Table 1. Results of numerical experiments

E(No)	Mathematical programming							
	Model1				Model2			
	Fc	T	V	C	Fc	T	V	C
E1(10)	19 699	353	1541	1631	19 699	78	431	228
E2(15)	27 345*	500**	1541	1891	27 306	89	462	340
E3(20)	42 286*	500**	1541	2161	41 993	156	523	453
E4(25)	44 124*	500**	1541	2431	43 347	297	616	632
E5(30)	45 945*	500**	1541	2701	44 632	310	736	810
E(No)	Hybrid approach							
	Model1				Model2			
	Fc	T	V	C	Fc	T	V	C
E1(10)	19 699	34	186	174	19 699	22	133	113
E2(15)	27 306	82	246	174	27 306	47	189	118
E3(20)	41 993	123	280	174	41 993	67	226	123
E4(25)	43 347	235	340	174	43 347	93	282	127
E5(30)	44 632	258	378	174	44 632	134	325	130

No: The number of orders
Fc: The optimal value of the objective function
T: Calculation time in seconds
V/C: The number of integer variables/constraints
*: The feasible value of the objective function after the time T
**: Calculation was stopped after T = 400 s

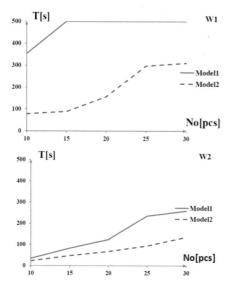

Fig. 7. Calculation times for *Model1* and *Model2* in MP (upper) and Hybrid (bottom) environments

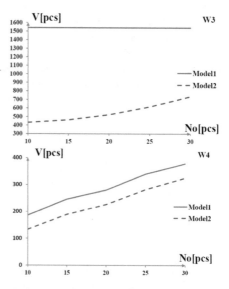

Fig. 8. The number of variables for *Model1* and *Model2* in MP (upper) and Hybrid (bottom) environments

The graph W1 presents the calculation times for both models in the MP environment and the W2 graph in the hybrid environment. Analogously, the graph W3 shows the number of the decision variables for both models in the MP environment and the W4 graph in the hybrid environment.

3 Conclusion

The proposed approach based on data for COPs significantly reduces the size of the modelled problems and, in consequence, shortens the optimization time. A reduction in the number of decision variables was obtained achieved up to 20 times for the example *ex_02* using the mathematical programming environment and up to 40% using the hybrid method [4, 9, 12]. Similarly, the number of constraints was reduced over 50 times using the mathematical programming and up to 50% using the hybrid method. The most interesting is the analysis of the obtained optimization times. In both environments, the optimization time was reduced. The classic modelling (*Model1*) using the mathematical approach for the larger examples made it impossible to find the optimum solution in an acceptable time. The detailed optimization times are shown in Figs. 7, 8 and Table 1. The proposed approach has one more practical advantage. Many solvers on the market have some licensing restrictions on the number of decision variables and the number of constraints. This results in the fact that it becomes impossible to implement a classic model in a solver environment for the real COPs and then to solve it due to exceeding the number of decision variables and the number of constraints. It can therefore be concluded that the proposed approach is a very universal way of presolving the modelled problem, which reduces its magnitude. The proposed approach is very versatile and can be applied to various practical problems. They do not have to be classic COPs. These can be both decision-making and optimization problems with logical, non-linear constraints, etc. This results from the representation of the problem data as facts and the universality of the procedure *Proc_R*. Further research will focus on the application of the proposed approach to the productions [13–15], transport networks [16], logistics [17], and IoT [18].

Appendix A

See Table 2.

Table 2. Data instances for *ex_01*

R1	R2	R3	R4				R5					
i	j	k	i	j	$a_{i,j}$	$w_{i,j}$	j	k	$z_{j,k}$	j	k	$z_{j,k}$
1	1	1	1	1	1	100	1	1	10	4	4	10
2	2	2	1	2	1	100	1	5	10	4	5	10
3	3	3	1	3	1	100	1	6	10	4	6	10
4	4	4	2	3	1	100	2	6	10	5	2	10
	5	5	2	4	1	100	3	1	10	5	3	10
		6	2	5	1	100	3	2	10	5	5	10
			3	3	1	100	3	3	10			
			3	4	1	100						
			4	1	1	100						
			4	5	1	100						

R4a

i	j	$a_{i,j}$	$w_{i,j}$	i	j	$a_{i,j}$	$w_{i,j}$
1	1	1	100	3	1	0	0
1	2	1	100	3	2	0	0
1	3	1	100	3	3	1	100
1	4	0	0	3	4	1	100
1	5	0	0	3	5	0	0
2	1	0	0	4	1	1	100
2	2	0	0	4	2	0	0
2	3	1	100	4	3	0	0
2	4	1	100	4	4	0	0
2	5	1	100	4	5	1	100

R5a

j	k	$z_{j,k}$	j	k	$z_{j,k}$
1	1	10	3	4	0
1	2	0	3	5	0
1	3	0	3	6	0
1	4	0	4	1	0
1	5	10	4	2	0
1	6	10	4	3	0
2	1	0	4	4	10
2	2	0	4	5	10
2	3	0	4	6	10
2	4	0	5	1	0
2	5	0	5	2	10
2	6	10	5	3	10
3	1	10	5	4	0
3	2	10	5	5	10
3	3	10	5	6	0

References

1. Tang, B., Zhu, Z., Luo, J.: A framework for constrained optimization problems based on a modified particle swarm optimization. Math. Probl. Eng. **2016**, Article no. 8627083, 19 pages. http://dx.doi.org/10.1155/2016/8627083
2. Antoniou, A., Lu, W.-S.: Practical Optimization Algorithms and Engineering Applications. Springer, New York (2007)

3. Dash, S., Tripathy, B.K., Rehman, A.: Handbook of Research on Modeling, Analysis, and Application of Nature-Inspired Metaheuristic Algorithms. IGI GLOBAL. ISBN 9781522528579

4. Sitek, P., Wikarek, J.: A hybrid programming framework for modeling and solving constraint satisfaction and optimization problems. Sci. Program. **2016**, Article no. 5102616 (2016). https://doi.org/10.1155/2016/5102616

5. Home – AMPL. https://ampl.com/. Accessed 19 Oct 2018

6. MIPLIB – Mixed Integer Problem. http://miplib.zib.de. Accessed 19 Oct 2018

7. Apt, K., Wallace, M.: Constraint Logic Programming using Eclipse. Cambridge University Press, New York (2006)

8. Díaz-Parra, O., Ruiz-Vanoye, J.A., Loranca, B.B., Fuentes-Penna, A., Barrera-Cámara, R. A.: A survey of transportation problems. Hindawi Publ. Corp. J. Appl. Math. **2014**, Article no. 848129, 17 pages. http://dx.doi.org/10.1155/2014/848129

9. Wikarek, J.: Implementation aspects of hybrid solution framework. In: Recent Advances in Automation, Robotics and Measuring Techniques, vol. 267, pp. 317–328 (2014). https://doi.org/10.1007/978-3-319-05353-0_31

10. Home LINDO. www.lindo.com. Accessed 19 Oct 2018

11. Eclipse - The Eclipse Foundation open source community. www.eclipse.org. Accessed 19 Oct 2018

12. Sitek, P., Wikarek, J.: A multi-level approach to ubiquitous modeling and solving constraints in combinatorial optimization problems in production and distribution. Appl. Intell. **48**, 1344–1364 (2018). 10.1007/s10489-017-1107-9

13. Nielsen, I., Dang, Q.-V., Nielsen, P., Pawlewski, P.: Scheduling of mobile robots with preemptive tasks. In: DCAI, Advances in Intelligent Systems and Computing, vol 290, Springer (2014). https://doi.org/10.1007/978-3-319-07593-8_3

14. Krystek, J., Kozik, M.: Analysis of the job shop system with transport and setup times in deadlock-free operating conditions. Arch. Control. Sci. **22**(4), 371–379 (2012)

15. Janardhanan, M.N., Li, Z., Bocewicz, G., Banaszak, Z., Nielsen, P.: Metaheuristic algorithms for balancing robotic assembly lines with sequence-dependent robot setup times. Appl. Math. Model. **65**, 256–270 (2019). https://doi.org/10.1016/j.apm.2018.08.016

16. Sitek, P., Wikarek, J., Nielsen, P.: A constraint-driven approach to food supply chain management. Ind. Manag. Data Syst. **117**(9): 2115–2138. https://doi.org/10.1108/IMDS-10-2016-0465

17. Grzybowska, K., Gajšek, B.: Regional logistics information platform as a support for coordination of supply chain. In: Highlights of Practical Applications of Scalable Multi-Agent Systems, The PAAMS Collection, pp. 61–72 (2016). https://doi.org/10.1007/978-3-319-39387-2_6

18. Deniziak, S., Michno, T., Pieta, P.: IoT-based smart monitoring system using automatic shape identification. In: Advances in Intelligent Systems and Computing book series (AISC), vol. 511, pp. 1–18 (2015). https://doi.org/10.1007/978-3-319-46535-7_1

Follow-Up Sequencing Algorithm for Car Sequencing Problem 4.0

Sara Bysko$^{(\boxtimes)}$ and Jolanta Krystek

Silesian University of Technology, Akademicka 16, 44-100 Gliwice, Poland
{sara.alszer, jolanta.krystek}@polsl.pl

Abstract. The problem of effective car sequencing in the paint shop results from the specifics of production process itself and from the structure of production line. Sequencing of cars intended to painting process is justified by economic reasons. The paper describes the new concept of this problem – Car Sequencing Problem 4.0, which takes into account the real structure of paint shop and buffers used there. The main goal of the research is to minimize the number of costly changeovers of painting guns, which results only from painting color changes. Each color change is related to the need to clean paint system, loss of paint and gun changeover, which increases production cost. The aim should be synchronization of changeovers, resulting from color changes, with periodic cleanings of painting guns, forced by technological requirements. Periodic cleanings are made to ensure good quality of painting process. In the paper the Follow-up Sequencing Algorithm is proposed to solve the considered sequencing problem. The purpose of the conducted research was to verify the influence of periodic cleaning interval on the effectiveness of the presented sequencing algorithm.

Keywords: Sequencing · Follow-up algorithm · Car production · Car sequencing problem

1 Introduction

For manufacturing enterprises, the flexibility of production is becoming more and more important. It enables faster response to market and specific customer needs, timeliness and reduction of production costs. Production flexibility is typical for multiversion, repetitive make to order production (MTO) and is obtained by sequencing (mixing) of product variants within the same process. As a result, the production in appropriate machine tooling can be realized in smaller batches, but it should be noted that machines changeover is a time-consuming and costly process. Therefore the batch size should be carefully selected.

Sequencing can be understood as a short-term decision-making process, which task is to determine the order in which variants of repetitive production are produced so that the demand for all products planned under Master Production Schedule (MPS) is fully met. Sequencing is used e.g. in multiversion production of cars, bicycles, for which different product variants are determined. The authors have focused their attention since 2015 on problems related to paint shops in the automotive industry. Their researches

© Springer Nature Switzerland AG 2020
R. Szewczyk et al. (Eds.): AUTOMATION 2019, AISC 920, pp. 145–154, 2020.
https://doi.org/10.1007/978-3-030-13273-6_15

are aimed at increasing automation level of painting lines and their efficiency. The paper presents the Follow-up Sequencing Algorithm (FuSA), which is based on the idea of the follow-up production control (FPC). The algorithm is applied to sequencing of cars (production variants) in a multiversion, repetitive process of car painting.

Effectively operating car factory products about 1000 vehicles in two shifts working day. Each car is produced in several successive stages – production begins in press shop and ends on assembly line. The transport system used to transfer cars between different production stages includes not only conveyor system but also increasingly used buffers. In these buffers vehicles may be temporarily stored, because some stages cannot be stopped, while others require periodic downtimes. Inter-departmental buffers are located between body shop, paint shop and assembly line and ensure continuity of production. In addition, within the paint shop it is often used a buffer, which precedes the process of painting cars with base paint. This buffer is also applied for car sequencing (Fig. 1).

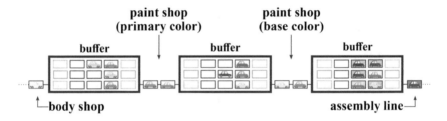

Fig. 1. The structure of the car production line.

Such a modification of the paint shop structure allows increasing the flexibility of painting process. The important question is therefore: how to use this buffer to create an optimal sequence from the point of view of the optimization criteria for the painting process.

The color in which a car is to be painted is specified by the MPS. If subsequent vehicles to be painted are not assigned the same color, then the paint in the guns must be replaced and the gun head is cleaned with a solvent. This process not only wastes time and money, but is also a significant environmental burden. That is why it is so important to paint cars in the longest batches characterized by a uniform color option. In connection with the above, taking into account the vehicle's feature, which is the color, it is necessary to set the car sequences in such a way that the number of painting gun changeovers are as small as possible. In car production only changeovers resulting from color changes are usually taken into account. The paper considers a more advanced concept of the sequencing problem – Car Sequencing Problem 4.0 (CSP 4.0). This concept includes the additional technological requirement: periodic cleaning of the paint system. Taking into account periodic cleaning is necessary due to the need to maintain good quality of painting process – if the guns are not cleaned systematically, the remaining paint will agglutinate, which results in poor quality of the paint coat [4]. From the point of view of optimization of the painting process, the aim should be synchronization of changeovers, resulting from color changes, with periodic cleanings of painting system. Previous work [1] confirmed that the primary concept of FuSA exceeds the simple priority algorithms. The purpose of this article was to verify whether the FuSA is independent to changing the technological parameter like interval of periodic cleaning. Investigations carried out by the authors have an additional and huge value resulting from the fact that developed solutions are implemented and tested on the real factory.

The article is organized as follows. Section 2 reviews related literature in order to present a similar sequencing problem and methods to solve it. Section 3 formulates a new sequencing problem that takes into account the current requirements of the automotive industry, and contains proposed quality indicators used to evaluate potential solutions. Section 3 describes the proposed sequencing algorithm in the context of follow-up control algorithm. Section 4 contains experimental research and discussion of obtained results. The final Section concludes the paper and presents further researches.

2   Literature Review

The car sequencing problem solved using a buffer is considered in the literature as Color Batching Problem (CBP). The goal is to find such a solution which minimizes the number of color changes or equivalently, maximizes the average size of color blocks.

Spieckermann et al. [7] introduced a formulation of the CBP as a Sequential Ordering Problem (SOP) and proposed a Branch and Bound (B&B) algorithm to find the optimal sequence. Moon et al. [6], based on simulation study, designed and implemented a color rescheduling storage (CRS), which was used in an automotive factory, and proposed some simple rules intended to operate the buffer. Hartmann and Runkler [3] presented two ant colony optimization (ACO) algorithms to enhance simple rules, which are based on color batching methods. Proposed algorithms were used for handling two stages of online resequencing, i.e. filling and releasing the buffer. Sun et al. [8] developed two heuristics named arraying and shuffling in order to achieve a quick and effective solution to the CBP. The arraying approach was used in the filling stage, while shuffling was applied to unloading cars from the buffer. Based on the conducted research, the authors found that the proposed heuristics can cooperate with each other in order to obtain competitive solutions to CBP. In addition the computational time was short. Ko et al. [5] analyzed the CBP in the context of application in M-to-1 conveyor systems. The motivation for such an approach was the resequencing problem occurring in one of the Korean car manufactures. The authors developed a mixed-integer linear programming model for the CBP and used dynamic programming for a special case of the considered problem, i.e. 2-to-1 conveyor system. They proposed two genetic algorithms to find near-optimal solutions for the general case.

The CBP is the close to the industrial car sequencing problem, but does not fully reflect the real problem. In most cases, decisions are made based on full access to information, e.g. the order of cars transported to the buffer is known. In addition, more importantly it is assumed that there are no periodic cleanings, due to which the purpose of optimization process does not correspond to the real requirements of paint shop. Although attempts to solve the CBP were based on the existence of a physical buffer, none of the presented approaches can be considered as fully comprehensive. This is due to the fact that the developed methods were either intended exclusively for sequencing on the buffer output (e.g. [7]) or did not take into account the need for correlation of heuristics used at the stage of loading and unloading the buffer (e.g. [8]). For the above mentioned reasons, it is necessary to define a new concept of sequencing problem that takes into account industry requirements and develop advanced sequencing algorithm.

3   Problem Formulation

If a buffer is used for sequencing, cars transported from the body shop to the paint shop create sequences in real time. A color of a car is not known until the car reaches the buffer input, where the Quick Response code (QR code), containing color information, is read. This means that the input sequence to the buffer is not known *a priori*. Only production plan determined for a certain time horizon is available. Each procedure for the sequencing problem should therefore be based on limited information (Fig. 2), i.e. only the following data is known at the time:

- The color of the car located on the loading shuttle (1).
- The color of the car located on the buffer input (2).
- The color of all cars located in the buffer (3).
- The colors of all cars that left the buffer (4).

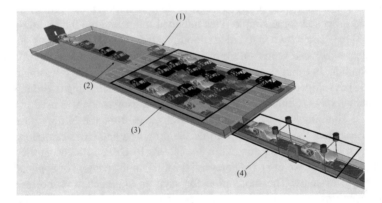

Fig. 2. Illustration of information available in the sequencing process.

Therefore, the considered sequencing problem can be included in the *online* problem [2]. It consists of making two decisions in real time. When a car appears on the buffer input, it is determined the line to which the car is directed. On the other hand, on the unloading side of the buffer, it is necessary to decide which car should be directed to the paint station.

In addition to limited access to information and real-time decision making requirements, the problem of car sequencing in the paint shop is subject to an additional restriction, which results from the applied paint system. This conditions the need to clean painting guns after a certain number of cars is painted. This limitation determines the optimal length of sequence composed of cars in the same color. This problem is called Car Sequencing Problem 4.0.

An instance of the considered problem is defined as tuple $(V, C, NoRowBuff, NoColBuff, TPerClean)$, based on production plan and technical parameters, where:

- $V = \{v_1, .., v_N\}$ – set of vehicle to be produced,
- $C = \{c_1, .., c_D\}$ – set of available colors and function c: $V \rightarrow C$, that associates color c_i to each vehicle v_i,
- *NoRowBuff, NoColBuff* – buffer size defined by number of buffer rows and number of buffer columns,
- *TPerClean* – periodic cleaning interval.

The solution of the CSP 4.0 is an order in which cars are delivered to the paint station. The sequence is generated in real time, therefore the solution is known only after the current production plan is completed. This means that it is difficult to predict the consequences of decision made on an ongoing basis.

The optimization problem is to find such *TPerClean*-element subsequences, for which the number of color changes requiring gun changeovers is minimal. Consequently, the number of changeovers for the whole sequence, which is a solution of the problem, is also minimal. For economic reasons, the most favorable situation is when the gun changeover coincides with periodic cleaning. In other words, the optimal solution is a sequence composed of single-color *TPerClean*-element subsequences (major optimization goal), whereby two consecutive subsequences should be in different color (minor optimization goal). The Fig. 3. presents an example of an optimal solution for *TPerClean* = 3.

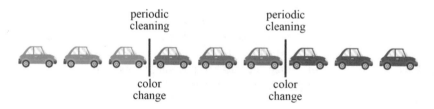

Fig. 3. An example optimal solution for *TPerClean* = 3

The authors propose two quality indicators allowing evaluating the solution of proposed sequencing problem:

- Number of Color Changes (*NCC*) – determines the total number of color changes.

$$NCC = \sum_{n=1}^{N-1} isCC_{n,n+1} \qquad (1)$$

$$\text{where } is\ CC_{n,n+1} = \begin{cases} 0, & c(v_n) = c(v_{n+1}) \\ 1, & else \end{cases};$$

- Effectiveness of Synchronization (*ES*) – determines the total number of color changes occurring between two subsequences in relation to the number of all color changes.

$$ES = \frac{\displaystyle\sum_{\substack{1 \leq n \leq N-1 \\ n \ mod TPerClean = 0}} isCC_{n,n+1}}{NCC} \cdot 100\% \qquad (2)$$

4   Follow-Up Sequencing Algorithm

The cars painting process carried out in batches, which are described with a specific color, can be treated as a repetitive process. Repetitive production is a type of production realized in work cells, where at a certain interval assortments of manufactured products, as well as machine tooling and their allocation to operations are repeated.

Because machine changeover, related to change of production assortment, is an expensive and time-consuming process, the aim is to extend the time between changeovers that is to increase the size of batches of manufactured products. On the other hand, it is known that the increase in the batch size results in an increase in the level of stocks, and this leads to an increase in storage costs. Equations are known for so-called the optimal batch size which minimizes the sum of changeover and storage costs [4] but in practice the real-time control of changeovers yields batch sizes significant different from the optimal ones. The purpose of this control is primarily to adjust the load of each machine or production line to the changing demand for products.

The repetitive process of car painting is characterized by a limited number of machine tooling variants (paint color) and the corresponding product range (version and color of the car – product variants), repeated in generally irregular intervals.

In the FPC method each order (car) for a production system (paint shop) appears at the beginning of a certain current operational planning period and contains the identifier of the process to be performed. Production orders for the paint shop are operational plans.

The FuSA makes the current decision about work or downtime of the paint shop. In the case of a decision to work, the algorithm specifies car sequence. In addition, in both cases, the expected moment of ending the start-up work period or downtime period. Because each time this moment is a forecast of the beginning of the next period of work or downtime, it is possible to simulate multiple operation of FuSA over period of any length. The justification for the name of the Follow-up Sequencing Algorithm is that at any time its past decisions have a schedule structure.

The FuSA is designed for real-time generation of executive plans for the paint shop in such a way that these plans follow the operational plans coming from the production control system. Decision of the FuSA are made based on the algorithm state, which consists of recently made decisions and state of backlogs in the implementation of operational plans by executive plans. The backlogs are cumulative difference between these plans in subsequent executive planning periods. Therefore, they can be considered as a quality measure of following up implementation plans for operational plans. Operational plans are introduced with a fixed period. The FuSA works in a feedback system, analogous to the follow-up control system, while operational plans play the role of set points, executive plans – follow-up values, and backlogs are integrals of adjustment errors, understood as differences between the sizes of these plans.

The FuSA concept is illustrated in Fig. 4. Due to complexity of the algorithm, presented block diagrams show a general concept. The loading (Fig. 4a) and unloading (Fig. 4b) methods work parallel and are coupled together – in order to optimally used empty line, if the *cIn* is the same as the *cOut* the car located on loading shuttle is directly transfer into this line. The algorithm follows: the state at the input/output of the buffer, buffer state, occurrences of periodic cleanings and production plan.

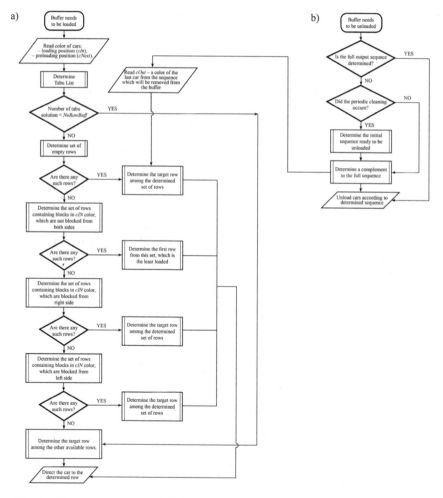

Fig. 4. The general concept of FuSA: (a) loading algorithm, (b) unloading algorithm.

5 Numerical Experiments

The only parameter of the instance whose value can be modified is the *TPerClean* parameter. The aim of the conducted research was to verify the influence of this parameter on the effectiveness of the follow-up sequencing algorithm. Four values of *TPerClean* were tested, namely: *TPerClean* ={3, 5, 7, 9}.

The case study assumed consideration of buffer presented in Fig. 5. The buffer consisted of 25 positions (5 × 5) intended for car buffering (conveyors), and of 2 shuttles used to transport the cars to the correct row. The car transport for each conveyor is possible only in one direction, while shuttles move in two directions, allowing the vehicles to be directed to individual lines.

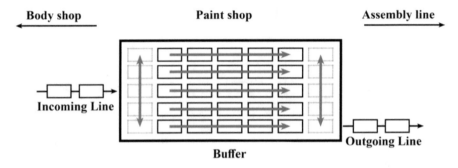

Fig. 5. The structure of the considered buffer.

For the purpose of the research, 5 sets of experimental data were used. Each set consisted of 1000 cars painted in 6 different colors. The color distribution in each set was the same and as follows: C1: 6%, C2: 38%, C3: 29%, C4: 14%, C5: 10%, C6: 3%.

All the obtained solutions were evaluated based on the quality indicators proposed in Sect. 3. The obtained values of Number of Changeovers (NC) and Effectiveness of Synchronization (ES) are presented in Table 1 and Table 2, respectively.

Table 1. Experimental results – NCC for tested *TPerClean* values

Data No.	*TPerClean* values			
	3	5	7	9
Data_01	312	249	249	258
Data_02	326	246	254	278
Data_03	320	257	260	272
Data_04	303	251	261	278
Data_05	310	243	256	283

Table 2. Experimental results – ES for tested *TPerClean* values

Data No.	*TPerClean* values			
	3	5	7	9
Data_01	89%	72%	45%	24%
Data_02	84%	70%	50%	35%
Data_03	83%	69%	50%	24%
Data_04	89%	69%	43%	34%
Data_05	86%	70%	41%	31%

Based on the results contained in Table 1 it can be deducted that the value of *TPerClean* parameter has a significant impact on the number of color changes. The follow-up sequencing algorithm gives the best solutions when the *TPerClean* parameter is equal to 5, what corresponds to the number of buffer columns *NoCollBuff*. If the parameter *TPerClean* is different from *NoCollBuff*, the NCC indicator achieves higher values. The number of color changes is higher if the *TPerClean* is smaller than *NoCollBuff* in comparison to the case when *TPerClean* is greater than *NoCollBuff*.

Minor optimization goal, i.e. ensuring that two consecutive subsequences are in different color, is more often met, if the *TPerClean* parameter reaches lower values. The values of ES indicator increase with the increase of the *TPerClean* parameter. This is due to the fact that creating a subsequences composed of fewer cars in the same color is easier than creating such subsequences, but composed of a larger number of cars.

6 Conclusions and Further Research

Over the years, researchers conducted a lot of discussion about the problem of car sequencing. In many cases several real industrial aspects, which call into questions the approach to these problem, were not taken into account. The paper introduced in details the new concept of the problem with regard to car sequencing in the paint shop. This concept uses buffers for car sequencing – a correct combination of conveyors, sensors, and actuators, connected and commanded by computing systems, allows carrying out automatic car flow through the buffer, avoiding routine tasks having to be performed by human operators.

Based on the carried out research, presented results and conducted discussion, it can be clearly stated that the presented algorithm FuSA meets the requirements of the industry – includes the use of buffers for the sequencing process and takes into account the need to synchronize changeovers resulting from color changes with necessary periodic cleanings. The painting process is the primary source for air emissions of regulated chemicals. From the perspective of economic indicators, it represents also a major cost of production. Consequently, optimization of this process in terms of the number of color changes results in factory's savings and better environmental protection. The previous researched presented in [1] confirmed the possibility of reducing the number of changeovers by 50% compared to commonly used simple algorithm. Research conducted in this paper allowed to estimate the optimal value of the *TPerClean* parameter for the tested case.

In the near future a number of new scenarios will be simulated by introducing stochastic variables replicating different potential uncertain events, including faults of machines, delayed arrivals of incoming cars, and lack of paint.

Acknowledgement. This work has been supported by Polish Ministry of Science and Higher Education under internal grants 02/010/BK_18/0102 and BKM-2018 for Institute of Automatic Control, Silesian University of Technology, Gliwice, Poland and by the company ProPoint Sp. z o.o. Sp. K.

References

1. Alszer, S., Krystek, J., Bysko, S.: Complex approach to the control of car body buffering in the paint shop. In: Szewczyk, R., Zieliński, C., Kaliczyńska, M. (eds.) Automation 2018. Advances in Automation, Robotics and Measurement Techniques, vol. 743 of AISC, pp. 152–161. Springer, Berlin (2018). https://doi.org/10.1007/978-3-319-77179-3_14
2. Fiat, A., Woeginger, G.: Online Algorithms—The State of the Art. Springer, Berlin (1998)
3. Hartmann, S.A., Runkler, T.A.: Online optimization of a color sorting assembly buffer using ant colony optimization. In: Kalcsics, J., Nickel, S. (eds.) Operations Research Proceedings, vol. 2007, pp. 415–420. Springer, Heidelberg (2008)
4. Krystek, J., Alszer, S.: Contemporary aspects of car sequencing problem in a paint shop. Mechanik **7**, 527–529 (2017). https://doi.org/10.17814/mechanik.2017.7.67
5. Ko, S.S., Han, Y.H., Choi, J.Y.: Paint batching problem on M-to-1 conveyor systems. Comput. Oper. Res. **74**, 118–126 (2016)
6. Moon, D.H., Kim, H.S., Song, C.: A simulation study for implementing color rescheduling storage in an automotive factory. Simulation **81**, 625–635 (2005)
7. Spieckermann, S., Gutenschwager, K., Voß, S.: A sequential ordering problem in automotive paint shops. Int. J. Prod. Res. **42**, 1865–1878 (2004)
8. Sun, H., Fan, S., Shao, X., Zhou, J.: A colour-batching problem using selectivity banks in automobile paint shops. Int. J. Prod. Res. **53**, 1124–1142 (2015)

Neural Network Control Systems for Objects of Periodic Action with Non-linear Time Programs

Victor Tregub[1], Igor Korobiichuk[2(✉)], Oleh Klymenko[1],
Alena Byrchenko[1], and Katarzyna Rzeplińska-Rykała[3]

[1] National University of Food Technologies, Kyiv, Ukraine
`tregubvg70@gmail.com`, {`frank._.s,alena3095`}`@ukr.net`
[2] Warsaw University of Technology, Institute of Automatic Control
and Robotics, Warsaw, Poland
`igor@mchtr.pw.edu.pl`
[3] Industrial Research Institute for Automation and Measurements PIAP,
Warsaw, Poland
`krykala@piap.pl`

Abstract. The research is devoted to the development and comparison of one-circuit and combined automatic control systems (ACS), as well as combined ACS with logic devices and ACS with a neural network controller for program control of periodic action apparatuses. The effectiveness of the listed systems was evaluated by the accuracy of the implementation of time programs with non-linear distributions in different variants of transition from one section to another.

Keywords: Neural network controller · Apparatus of periodic action ·
Non-linear time programs · Logic devices

1 Introduction

In the food industry, both continuous and periodic apparatuses are used. Non-stationary and non-linearity are inherent to the last. In most cases, these apparatuses are objects of program control with a program that must meet the requirements of the techno logical rules or be created by solving the variational problem of optimal control [1–3].

There is a large group of apparatus of periodic action (sterilizers, autoclaves, reactors, heat treatment furnaces, etc.), for which the program of temperature changes or other values is given by the technological regulations and consists of three sections: growth-endurance-decline. For such objects, research has been carried out on the selection of the most effective software regulation systems (programmers) in the case of programs with linear time sections [4–6] and programs with a nonlinear decline section [7]. It has been proved that the advantage in these cases has the combined system of programming control (CSPC) with logical devices that allow you to change the settings of CSPC when switching from one program section to another.

R. Szewczyk et al. (Eds.): AUTOMATION 2019, AISC 920, pp. 155–164, 2020.
https://doi.org/10.1007/978-3-030-13273-6_16

At the same time, to increase the productivity of such apparatus of periodic action (APA) by reducing the cycle of its work, programs with nonlinear-time sections of growth and decline and programs with a step transition to the decline section may be used. The task of choosing a programmer for these cases is not resolved.

2   Materials and Methods of Research

To compare possible variants of programmers, the following four programming control systems (PCS) have been selected: one-circuit, combined, combined with logic devices and a neural network (NN). Neural network control has a number of advantages, since artificial neural networks are capable of self-learning, and neurocontrol does not require a complex mathematical apparatus and a large amount of a priori information about an object and is one of the most effective types of intellectual controllers of a new generation, which are used along with typical linear controllers, and separately [8].

In the MATLAB system for implementation of the neural network control there is a special neural network extension package – the Neural Networks Toolbox, and Simulink is used to simulate dynamic systems. The Neural Networks Toolbox provides tools for designing, modeling, learning and using many of the paradigms of the device for artificial neural networks [9]. NN Predictive Controller is chosen for the study. This controller uses a controlled process model in the form of a neural network in order to predict future reactions of the process to random control signals. The optimization algorithm computes control signals that minimize the difference between the desired and actual signal changes at the output of the model and thus optimize the controlled process. Studying the NN takes place autonomously with the use of any training algorithms that are based on reverse error propagation.

An autoclave, in which sterilization programs are implemented, is selected as an object of control. The object is described [10] by the transfer functions of the autoclave $W_a(p)$ and the resistance thermometer $W_t(p)$:

$$W_{ob}(p) = W_a(p) \cdot W_t(p) = K_a K_t \exp(-p\tau_a)/(T_a^2 p^2 + 2\lambda T_a p + 1)(T_t p + 1) \qquad (1)$$

where K_a – is the transfer coefficient of the autoclave, T_a is the time constant of the inertial link of the second order, λ is the damping factor, τ_a is the time constant of the delay line, K_t is the transfer coefficient of the resistance thermometer, T_t is the time constant of the resistance thermometer. For further calculations, the following values of the parameters of functions (1) are taken: $K_a = 1$; $T_a = 250$ s; $\tau_a = 20$ s; $\lambda = 2$; $K_t = 1$; $T_t = 100$ s.

As programs of sterilization with "heating – shutter – cooling" areas with corresponding time intervals "20–10–20 min" and "20–30–20 min" were selected. The shutter speed was set at 120 °C for the first program and 100 °C for the second, and the cooling temperature was 45 °C and 35 °C respectively. Experiments were carried out for system of programming control (SPC) with a continuous transition and for SPC with a continuous transition between the heating and cooling areas and with a step transition to the cooling area. An integral quadratic criterion was used to assess the quality of SPC.

Implementation of one-circuit SPC with a continuous transition in MATLAB system is shown in Fig. 1.

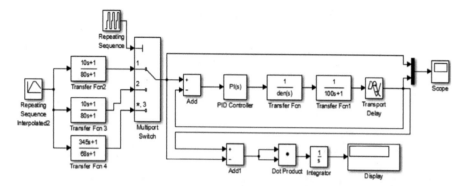

Fig. 1. One-circuit SPC with a continuous transition in MATLAB.

The object in Fig. 1 has two sequentially connected Transfer Fcn and Transfer Fcn1 blocks and a Transport Delay block. Transfer Fcn is a linear link definition due to assignment of its transfer function, and Transport Delay provides a signal delay for a given number of steps in the model time.

Scheme for calculating the quality criterion of SPC consists of the following blocks: Dot Product – calculates and outputs the value of the product of input signals; Integrator – integrates in the continuous time of the input; Display – is intended to display the numerical values of the input value.

To implement the device that defines the program, the Repetition Sequence Interpolated block was applied. Further, Transfer Fcn2, Transfer Fcn3, Transfer Fcn4 blocks, which characterize the nonlinear sections of the program, goes parallel to each other. After them, the Multiport Switch unit, which is a switch of large number of inputs, is located, a signal from the Repeating Sequence block is provided to the block input, which specifies the time and switch settings. Add Block combines signal from the Multiport Switch and a feedback signal, which covers the control object. The PI controller is used to control. It's implemented by the PID Controller unit, whose setting disables the D-component.

The optimal settings of the controller are found by using the Ziegler-Nichols method. To do this, first we bring the system to the limit of stability, and then we calculate the value of the settings in accordance with known formulas.

One-circuit SPC with a continuous transition for the heating and shutter sections and with a step transition to the cooling section (Fig. 2) is implemented similarly to the SPC with a continuous transition, but for the first two sections we put the unit Repeating Sequence Interpolated2, and for the third section – the Step block, in which the initial value and the set temperature are set (according to the sterilization program).

The task of determining the transfer function of the compensator appears for the implementation of a combined SPR for the already selected object. To avoid reduction of system stability, it should not be covered by feedback. The transfer function of the compensator is calculated by the following formula:

$$W_K(p) = \frac{1}{W_{ob}(p)} \tag{2}$$

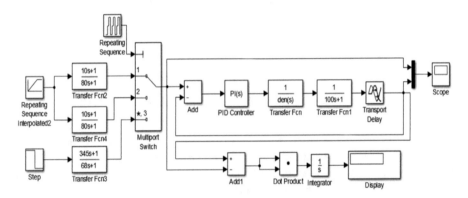

Fig. 2. One-circuit SPC with a continuous transition for the heating and shutter sections and with a step transition to the cooling section in MATLAB.

Physical realization of the transfer function of the compensator by formula (2) is impossible, since the degree of polynomial numerator is higher than the polynomial denominator; therefore, the transfer function of the compensator is replaced by an approximate, which satisfies the conditions of physical and program realization.

A scheme of a combined SPC with a continuous transition in MATLAB Simulink is shown in Fig. 3, where Transfer Fcn5 is a link that simulates the transfer function of the compensator. This scheme creates a task similar to the scheme in Fig. 1, but after Multiport Switch, the signal goes to the Transfer Fnc5 compensator and to the block Add, where the signal from the feedback is also received. The value of the discrepancy is given to the controller (PID Controller). The signal from the controller and the signal from the compensator are summed with the use of the Add2 block.

A scheme of a combined SPC with a continuous transition for the heating and shutter sections and with a step transition to the cooling section in MATLAB Simulink is shown in Fig. 4. Here, the task, as in the scheme in Fig. 2, is formed by using the Repeating Sequence Interpolated2 and Step blocks.

Logical devices were added to improve the quality of regulation of combined SPC. To do this, a separate PI controller was used for each section, with the switching between them using another Multiport Switch, which connected the output of the PI controller to the input of the Multiport Switch1 block, the number of which was fed to the first Multiport Switch input. In Repeating Sequence, Time values and input numbers (Output values) are specified.

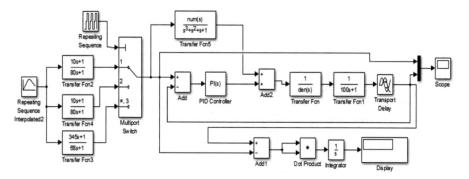

Fig. 3. Combined SPC with a continuous transition in MATLAB.

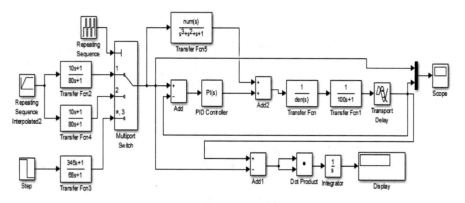

Fig. 4. Combined SPC with a step transition in MATLAB.

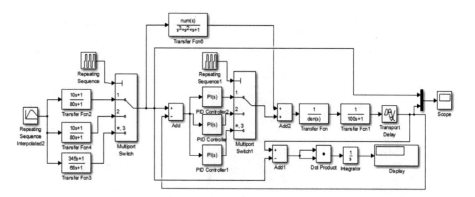

Fig. 5. Combined SPC with LD with continuous transition in MATLAB.

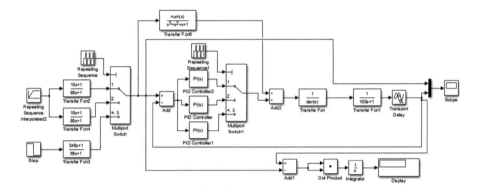

Fig. 6. Combined SPC with LD with step transition in MATLAB.

The settings for the controllers were initially set to the same as in the previous scheme, and then adjusted according to the optimal values for each section. The scheme of combined SPC with logic devices (LDs) and the continuous transition is shown in Fig. 5, and the scheme of combined SPC with logical devices and with a step transition – in Fig. 6.

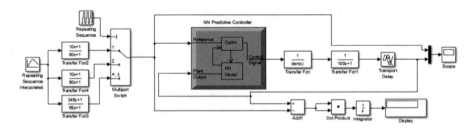

Fig. 7. SPC with NN Predictive Controller with continuous transition between program sections.

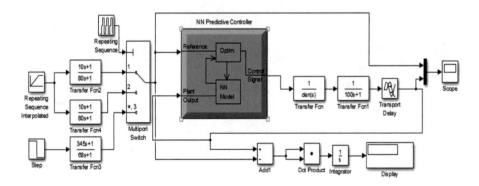

Fig. 8. SPC with NN Predictive Controller with step transition between program sections.

The simulation schemes neural networks with continuous and step transitions in the MATLAB are shown in Fig. 7 and Fig. 8, respectively. To configure the parameters of the predictive neural network controller, the following data is used: the values from the controller input at the previous schemes, as well as the behavior of the sterilizer model during the control of the corresponding schemes.

In the diagrams shown in Figs. 7 and 8, the given values are served at the Reference input of the controller, and the simulated temperature values are served at the Plant Output input. At the Control Signal output of the controller, a signal is generated to control the sterilization process. The NN Model block is a neural network block used to predict the future performance of a controlled process, and Optim is an optimization unit that optimizes this performance.

3   The Results of Research

The value of the integral quadratic criterion (IQC) for the optimality of the implementation of programs for all SPC with nonlinear sections and continuous transition between the sections is given in Table 1; step transition to the descending section is given in Table 2. The IQC was calculated by the formula:

$$I = \int_{\tau_0}^{\tau_k} \Delta t^2(\tau)\,d\tau \qquad (3)$$

where $\Delta t(\tau)$ is the value of the controlled variable in time.

Table 1. IQC of programs with nonlinear sections and continuous transition.

Type of program, min	Optimization criterion, °C²·s			
	One-circuit SPC	Combined SPC without LD	Combined SPC with LD	NN SPC
20-10-20	$2.6{\cdot}10^5$	$6.9{\cdot}10^4$	$2.6{\cdot}10^4$	$0.47{\cdot}10^4$
20-30-20	$1.8{\cdot}10^5$	$6.1{\cdot}10^4$	$1.5{\cdot}10^4$	$0.42{\cdot}10^4$

Table 2. IQC of programs with nonlinear sections and step transition to the descending section.

Type of program, min	Optimization criterion, °C²·s			
	One-circuit SPC	Combined SPC without LD	Combined SPC with LD	NN SPC
20-10-20	$33.2{\cdot}10^5$	$11.3{\cdot}10^5$	$9.6{\cdot}10^4$	$8.2{\cdot}10^4$
20-30-20	$25.9{\cdot}10^5$	$8.4{\cdot}10^5$	$7.9{\cdot}10^4$	$6.9{\cdot}10^4$

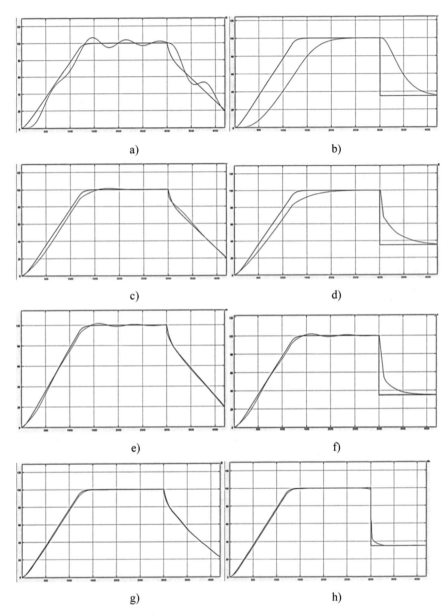

a)

b)

c)

d)

e)

f)

g)

h)

Fig. 9. Graphs of temperature changes during the implementation of the SPC.

Figure 9 presents the simulation results of comparable SPC:

(a) when implementing one-circuit SPC with a continuous transition in the coordinates "temperature (°C) – time (s)";
(b) when implementing one-circuit SPC with a continuous transition for the heating and shutter sections and with a step transition to the cooling section in the coordinates "temperature (°C) – time (s)";
(c) when implementing Combined SPC without LD with a continuous transition in the coordinates "temperature (°C) – time (s)";
(d) when implementing Combined SPC without LD with a continuous transition for the heating and shutter sections and with a step transition to the cooling section in the coordinates "temperature (°C) – time (s)";
(e) when implementing Combined SPC with LD with a continuous transition in the coordinates "temperature (°C) – time (s)";
(f) when implementing Combined SPC with LD with a continuous transition for the heating and shutter sections and with a step transition to the cooling section in the coordinates "temperature (°C) – time (s)";
(g) when implementing NN SPC with a continuous transition in the coordinates "temperature (°C) – time (s)";
(h) when implementing NN SPC with a continuous transition for the heating and shutter sections and with a step transition to the cooling section in the coordinates "temperature (°C) – time (s)".

4   Conclusions

The research and comparison of different systems of programming control of the periodic action apparatus on the example of a sterilizer was carried out. The three sections sterilization program "growth-endurance-decline" was used for the most common time intervals "20-10-20 min" and "20-30-20 min". At the same time, to reduce the cycle duration, both continuous and step transition between the last two sections was investigated.

In the case of an assessment of the quality of program implementation with the use of the Integral Quadratic Criterion (IQC), it was found that the worst results are characteristic for one-circuit SPC. When using combined SPC, the IQC rates have decreased significantly. In the case of the introduction of logical devices into combined SPC, it was possible to reproduce the program more precisely by setting the optimal system parameters for each section, which contributed to the further reduction of the IQC. The best results and simplifying the scheme of implementation of the system are obtained for the neural network programming control systems, especially in the absence of step transitions between program sections. At the same time, this system requires more complicated procedures and more time to configure it.

References

1. Tregub, V.: Automated control devices periodic action on food enterprises. Sci. Works Nat. Univ. Food Technol. **16**, 143–145 (2005). (in Ukranian)
2. Korobiichuk, I., Lutskaya, N., Ladanyuk, A., Naku, S., Kachniarz, M., Nowicki, M., Szewczyk, R.: Synthesis of optimal robust regulator for food processing facilities. In: Advances in Intelligent Systems and Computing, ICA 2017, Automation 2017, vol. 550, pp. 58–66 (2017). https://doi.org/10.1007/978-3-319-54042-9_5
3. Korobiichuk, I., Ladanyuk, A., Zaiets, N., Vlasenko, L.: Modern development technologies and investigation of food production technological complex automated systems. In: ACM International Conference Proceeding Series, pp. 52–57 (2018). https://doi.org/10.1145/3185066.3185075
4. Klymenko, O.: Automatic control of batch sterilizers using methods of development programmers: the author's abstract of the PhD dissertation: specialty 05.07.13 "Automation of control processes", NUFT, Kyiv, 25 p. (2015). (in Ukranian)
5. Korobiichuk, I., Ladanyuk, A., Shumyhai, D., Boyko, R., Reshetiuk, V., Kamiński, M.: How to increase efficiency of automatic control of complex plants by development and implementation of coordination control system. Adv. Intell. Syst. Comput. **543**, 189–195 (2017). https://doi.org/10.1007/978-3-319-48923-0_23
6. Korobiichuk, I., Lysenko, V., Reshetiuk, V., Lendiel, T., Kamiński, M.: Energy-efficient electrotechnical complex of greenhouses with regard to quality of vegetable production. Adv. Intell. Syst. Comput. **543**, 243–251 (2017). https://doi.org/10.1007/978-3-319-48923-0_30
7. Dovzhenko, Y.: Development of a programmer for realization of time programs with nonlinear sites: the author's abstract of the master's work, NUFT, Kyiv, 0 p. (2013). (in Ukranian)
8. Slany, Y., Tregub, V.: Automatic control system with neural network controllers. Food Ind. (3), 163–165 (2004). (in Ukranian)
9. MATLAB online documentation. https://www.mathworks.com/help/nnet/neural-network-control-systems.html
10. Klymenko, O., Tregub, V.: Research programmers to systems management apparatus batch. Sci. Works Nat. Univ. Food Technol. (42), 11–15 (2012). (in Ukranian)

Technological Monitoring in the Management of the Distillation-Rectification Plant

Vasii Kyshenko[1], Igor Korobiichuk[2(✉)], and Katarzyna Rzeplińska-Rykała[3]

[1] National University of Food Technologies, Kiev, Ukraine
`vdk.nuft@gmail.com`
[2] Warsaw University of Technology,
Institute of Automatic Control and Robotics, Warsaw, Poland
`igor@mchtr.pw.edu.pl`
[3] Industrial Research Institute for Automation and Measurements PIAP,
Warsaw, Poland
`krykala@piap.pl`

Abstract. The analysis of time series of the process variables was conducted by methods of nonlinear dynamics, which allowed to determine the randomness values that are based on the depth of an object prediction. The filtering of the time series was obtained experimentally, using wavelet analysis. Defined the fractal properties of chaotic information flow, correlation dimension and the Hurst parameter were defined. The intensity of the impact of a technological parameter value on the process using Kohonen maps was identified.

Keywords: Monitoring · Distillation-ractification facility · Data mining · Wavelet analysis · Neural networks · Uncertainty

1 Introduction

Modern industrial plants and large technical systems are complex processes that form the basis of production of the finished product. The difficulties of managing such complexes arise from a large number of alternatives, multicriteriality choice decisions and uncertainty of initial information [1–3]. In terms of analysis and synthesis of automated control systems, distilleries are complex object control, characterized by multidimensionality [4], multiconnection [5] and nonstationary and treated as objects of series-parallel structure. In critical situations, a person is not able to cope with the influx of information and often is mistaken in identifying situations that arise. Recognizing the situation at local level means automatic and even microprocessor technology is not possible because the necessary analysis of parameters of the whole or of its individual functional – related consolidated parts. This reduces the speed and quality of decision – making them largely dependent on the experience and personal characteristics of the human operator.

The systems approach to the development of alcohol production, along with the improvement of technology and equipment requires the use of automation of production. Modern structured management system must implement automation strategy that provides complete operation and management of distilleries. Making the transition to structured automation systems – an effective way of intensifying production [6, 7]. Currently, the majority of factories operate complex hierarchical automation system ethanol production based on the structures of 3 levels creating an automated process control system.

For the analysis of quantitative and qualitative characteristics of object behavior and training necessary to organize data management strategies and decision management solutions for SCADA must have a membership subsystem process monitoring. The traditional role of monitoring in enterprises was mainly to give an answer to the management of the quantitative and qualitative characteristics of the production processes. Problem analysis of operational data stream monitoring of complex technological systems has so far not given due attention. At the same time, in many other areas, such as telecommunications, masses of content in databases are widely used to detect hidden internal laws followed by the practical application of the knowledge gained.

In this regard, the urgent task is the development of methods and algorithms for vector data analysis system operational monitoring of complex technical objects with the use of modern methods of data mining in order to improve performance of existing monitoring systems. Effective analysis of monitoring data is possible while creating the means of data mining techniques using induction on statistical models that can detect the internal structure of the vector of parameters of operational monitoring. Creation of such "intelligent" monitoring tools and automated analysis of monitoring data will increase the efficiency of the monitoring system, enhancing the effectiveness of controlling the technological object.

Development of methods and algorithms to obtain knowledge of the dataset operational monitoring of complex technical objects with the use of modern methods of artificial intelligence and data mining, followed by the use of algorithms and programs for automation of complex objects to make theme development subsystem process monitoring relevant as in respect of the application, and in scientific and technical terms.

2   Materials and Methods of Research

Low quality (unreliable) accumulated real-time data entered manually or automatically collected on objects cannot directly use them to monitor and predict the state of the technological process. To solve these problems it is necessary to use the object model of the complex process of ethanol production and specialized data models that characterize their condition. Quality control of technological complex distillation distillery department directly affects the quality of the process of obtaining the highest concentration of ethanol.

To assess the effectiveness of management decisions appropriate to use targeted analysis of work situations.

If this target analysis works situations simultaneously solved three problems:

- Formation of classes of situations;
- Classification of the current situation to one of the existing classes;
- Directed search terms compensate deviations, given the current situation indicators.

Implementation of the first task on the basis of statistical data of the company for the last period, which should include information on the technical condition of the equipment, process conditions and deviations for performance indicators characterizing the production situation.

The main functions of technological monitoring subsystem and its structure were developed based on the analysis of technological monitoring problems. This subsystem provides a comprehensive treatment of the input – output information, identifying patterns, analysis of work situations and technological forecasting.

The technical result of the application of this structure is achieved particularly through the block abnormal discharge measurements, which include emissions erroneous measurements due to faulty instrumentation; recovery unit passes data such as laboratory measurements at regular intervals; block filtering to improve the signal/noise ratio. The system presents the information in qualitative linguistic values, which makes it possible to develop intelligent logical – linguistic model that is the basis of knowledge bases.

For efficient process control as part of computer – integrated systems proposed structure allows devices using structural and parametric identification of diverse mathematical models developed for the purposes of prediction and control.

To evaluate the decisions made by management and their implementation on-site management unit is used for evaluating the effectiveness of strategies control that allows together with the block forecasting system to estimate trends in results of operation of process control and establish preventive measures to correct deviations in the generated script control, defined in the block assessment and implementation scenarios.

3   The Results of Research

In the study of time series describing the technological facilities to automate the process of detection of events and states of objects, the signal is distorted influenced by a wide variety of sounds. These signals may use technology Data Mining [8]. The study of signals by Data Mining held in several stages.

Use of discharge characteristics of signals based on the use of segmentation methods and the subsequent unification of segments is a key step. After receiving the prepared material is necessary to organize it with allocation classes that correspond to the main groups of events.

Since there are no *a priori* segments of a given class, then they need to apply a clustering technique. Comparison of event clusters segments occur chronologically, based on the hypothesis that at one point in time there is one event.

The final step is the construction and training of a neural network capable of real-time process data time series process and classifies events and the state of technological department, and further testing and evaluation of performance of the model.

Pretreatment of the original data. The purpose of pre-treatment – to reduce background noise [9]:

– Removing the main trend and the transition to the unbiased estimate number;
– Normalization of baseline data;
– Conducting wavelet analysis.

Increasingly used wavelet analysis, which gives an idea of the signal in frequency and time domain.

The introduction of mechanisms for data processing method of wavelet analysis demonstrates their ability to complex approach to the task. Real data often contain drop-down area; to handle such signals developed adaptive wavelet methods. Implementation of all these attractive properties of wavelets sometimes constrained by significant amount of computation required, which leads to low processing speed. The result of the wavelet transform anticipated image as a vector of properties that corresponds to the current state of the process.

Methods of wavelet analysis may be applied to data of different nature. It may be, for example, one-dimensional or two-dimensional image function. Rough classification of wavelet algorithms can be done by highlighting the continuous (CWT – Continuous Wavelet Transform) and discrete (DWT – Discrete Wavelet Transform) of the wavelet transform.

Controller's "low (all-mode) tuning" is the main disadvantage of existing control systems, which reduces profits. Frequent changes of plant parameters are the prime cause of low quality tuning and reduced efficiency of control systems. These changes are caused by changeable mode of plants. The quality of work with such tuning is obviously worse [5].

Real data derived from the measurement of some variables are almost always subject to distortion – to a greater or lesser extent. In particular, in addition to the desired signal, measurements contain unwanted additional noise [10].

Calculating wavelet spectra was performed using the package Wavemenu Tool in MATLAB [11]. We evaluated the possibility of identifying the characteristics of wavelet spectra corresponding to different patterns.

Our results were obtained graphs mash temperature signal (Fig. 1). Consider the possibility of using Kohonen maps subsystem monitoring the production alcohol as part of industrial control systems, because information about the current state and trends of the process of change should be the basis for developing measures for decision-making and taken into account in predicting the development of object.

As was teaching set stands a database that contains information about the technological process of distillation: concentration of epurate, steam flow to epuration column, bottom pressure, and top temperature. Based on this data to build a model that shows the relationship of the above parameters and their influence on the process.

To solve the problem we use analytical package Deductor Studio Academic 4.1 [12].

The first phase of construction is loading data.

The next step – to start the wizard finishes and choice of treatment method list "Kohonen Map". Next, it is necessary to configure the destination column, for each column to select one of the appointments: input, output, and information is available.

The third step proposes to split the original set for training, test and validation. The fourth step starts the process of learning the network and observes the change in the quantity and percentage of errors detected examples in teaching and test sets. In this case, the teaching set recognized 93.98%, while the test – 82.69% examples. In the analysis of input cards is recommended to use several cards. Investigate the map fragment consisting of four cards of inputs that shown in Fig. 1.

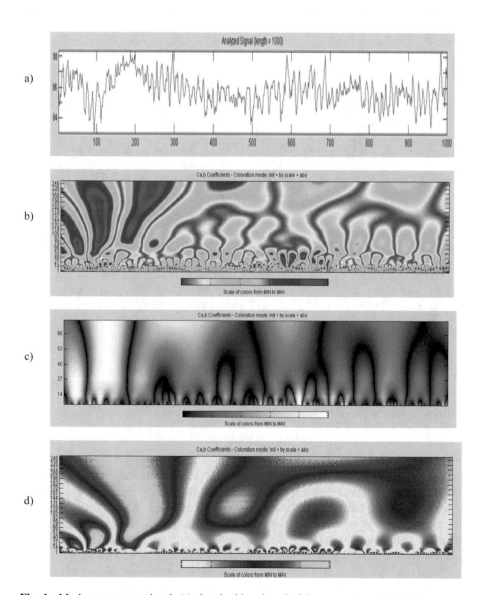

Fig. 1. Mash temperature signal: (a) signal with noise; (b) Meyer wavelet; (c) MHAT wavelet; (d) Gauss wavelet.

As an example, we can consider the individual values of time series that are beyond the permissible limits of the technological regime. Defining these objects on Kohonen maps, we can predict the behavior of the system over time.

Using Kohonen maps to process monitoring systems, particularly in the subsystem technology monitoring department for rectification distillery, makes it possible to determine the intensity of the impact of a technological parameter value to the process, to assess the relationship of the main technological parameters.

On one of the maps highlight the region with the largest value of the index. Next, you need to examine these same neurons to other cards (Fig. 2).

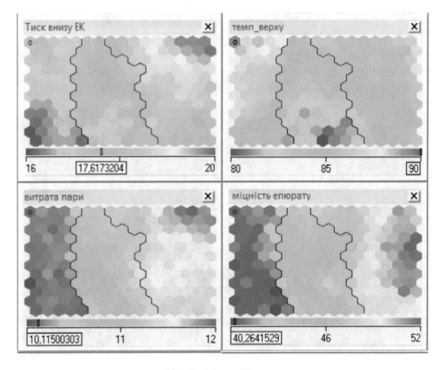

Fig. 2. Maps of inputs.

The first map of the most important are the facilities located in the higher left corner. Considering the four cards at the same time, we can say that these same objects are most important indicator, shown at three cards remaining. Also on the coloring of maps can be concluded that there is a relationship between these parameters. Here we see a few cards inputs (process indicators) and formed clusters, each of which has a separate color.

Thus, the intensity of exposure was determined by a particular value of the process parameter on the process of using Kohonen maps. Using maps, self-organizing, makes it possible to assess the relationship of the main technological parameters.

Equally important, especially for chaotic processes that are inherent in the technology of alcohol, acquire methods of fractal analysis. One method of fractal analysis method is based on the algorithm of R/S – time series analysis.

For the classification system can be used to calculation of correlation dimension and obtain Hurst parameter [13]. Correlation dimension Dc, based on the calculation of correlation integral, is an important quantitative characteristic of the attractor, which carries information about the complexity of the behavior of dynamic systems.

Correlation integral C(r) calculates the average difference between the points of the reconstructed phase space coordinates which are the values of the time series with increasing number of delays over time. If the time series is completely deterministic, its behavior is determined by some dependencies containing k variables. Then with increasing delay order ascending order correlation between the number of stabilizing integral k and k + 1, and it is taken for evaluating the fractal dimension of time series. If a series of chaotic, random, the ascending order of the correlation integral is growing about as fast as the dimension of the phase space.

The function C(r) for each r is normalized to the number of pairs of points considered plural (object), the distance between them does not exceed r

$$C(r) = \frac{1}{n^2} \sum_{\substack{i,j = 1 \\ i \neq j}}^{n} H\left(r - |y_i - y_j|\right) \tag{1}$$

where the Heaviside function $H(x) = 0$, if $x < 0$; $H(x) = 1$, if $x \geq 0$, for all pairs of values i and j, if $i \neq j$, $|y_i - y_j|$ – absolute value of the distance between the points of the set, $i, j = 1, 2, 3, \ldots,$ n, where n – number of points. For practical calculation of the dimension of the graph isolated region of linear dependence (scaling region) and the function is approximated by a straight line by least squares. Then the slope of the graph is the dimension Dc [13]. Figure 3 shows the definition of the correlation dimension of the time series epurate concentration.

To investigate the performance of time series by fractal dimension and the Hurst parameter. The main characteristic of a fractal object is the fractal dimension.

Of particular importance fractal time series analysis is that it takes into account not only the behavior of the system during the measurement, but also his background.

Fractal dimension curve is a measure of the complexity of time series.

In analyzing alternating sections with different fractal dimension, you can learn to predict the behavior of the system. And, most importantly, diagnose and predict the unstable state.

An important aspect of this approach is the availability of a critical value of the fractal dimension of the time curve, when approaching which the system loses its stability and becomes unstable condition, options or quickly increasing, or decreasing, depending on the current trends.

To investigate the performance of time series by fractal dimension and the Hurst parameter. The main characteristic of a fractal object is the fractal dimension.

Of particular importance fractal time series analysis is that it takes into account not only the behavior of the system during the measurement, but also his background.

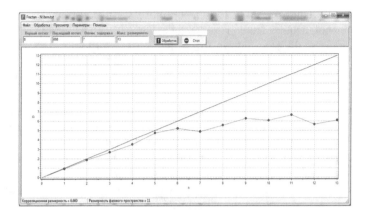

Fig. 3. Determination of the correlation dimension of the time series epurate concentration.

Fractal dimension curve is a measure of the complexity of time series.

In analyzing alternating sections with different fractal dimension, you can learn to predict the behavior of the system. And, most importantly, diagnose and predict the unstable state.

An important aspect of this approach is the availability of a critical value of the fractal dimension of the time curve, when approaching which the system loses its stability and becomes unstable condition, options or quickly increasing, or decreasing, depending on the current trends.

Today, with the development of the theory of stochastic coat metals, this should become a popular time series as an indicator of H. Hurst is known that it is associated with the traditional "cell" fractal dimension D simple equation:

$$D + H = 2 \qquad (2)$$

Hurst index is a measure of persistence – tendency to process trends [13]. Value $H > \frac{1}{2}$ means aimed at certain aspects of the process dynamics in the past is likely to result in the continuation of the movement in the same direction. If $H < \frac{1}{2}$, it is predicted that the process of changing direction. $H = \frac{1}{2}$ means uncertainty – Brownian motion.

Calculation of Hurst for output time series, which characterizes the change epurate concentration and is shown in Fig. 4.

Calculated values Hurst $H > 0.5$, indicating a characteristic persistence considered variable and possible deep enough predictability.

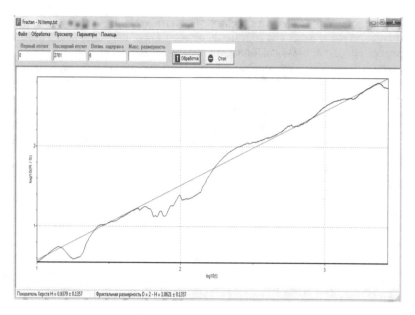

Fig. 4. Calculation of Hurst parameter of RC's control plate temperature.

4 Conclusions

Algorithms detection system variable in an object management, identify the type of object behavior by analyzing time series of process variables, analysis and classification of situations with self-organizing Kohonen maps. It is possible to increase the efficiency of decision-making on management and form-situational important area for control scenario distillation area of ethanol production plant.

References

1. Ladanyuk, A., Shkolna, O., Kyshenko, V.: Automation of evaporation plants using energy-saving technologies. Advances in Intelligent Systems and Computing, Warsaw, Poland, vol. 543, pp. 220–226 (2016)
2. Korobiichuk, I., Ladanyuk, A., Shumyhai, D., Boyko, R., Reshetiuk, V., Kamiński, M.: How to increase efficiency of automatic control of complex plants by development and implementation of coordination control system. Advances in Intelligent Systems and Computing, vol. 543, pp. 189–195 (2017). https://doi.org/10.1007/978-3-319-48923-0_23
3. Korobiichuk, I., Ladanyuk, A., Zaiets, N., Vlasenko, L.: Modern development technologies and investigation of food production technological complex automated systems. In: ACM International Conference Proceeding Series, pp. 52–57 (2018). https://doi.org/10.1145/3185066.3185075
4. Novakovska, N., Kyshenko, V.: Fractal analysis of distillation unit time series in prediction and control problems. Ukrainian J. Food Sci. **3**(2), 243–253 (2015)

5. Kishenko, V.: The problems of technological monitoring in the control of industrial enterprise. Pressing issues and priorities in development of the scientific and technological complex, pp. 62–68. B&M Publishing, San Francisco (2013)
6. Korobiichuk, I., Siumachenko, D., Smityuh, Y., Shumyhai, D.: Research on automatic controllers for plants with significant delay. Advances in Intelligent Systems and Computing, vol. 519, pp. 449–457 (2017). https://doi.org/10.1007/978-3-319-46490-9_60
7. Korobiichuk, I., Lutskaya, N., Ladanyuk, A., Naku, S., Kachniarz, M., Nowicki, M., Szewczyk, R.: Synthesis of optimal robust regulator for food processing facilities. In: Automation 2017, ICA 2017. Advances in Intelligent Systems and Computing, vol. 550, pp. 58–66 (2017). https://doi.org/10.1007/978-3-319-54042-9_5
8. Cios, K.J.: Data Mining: A Knowledge Discovery Approach, p. 123. Springer, Heidelberg (2007)
9. Ladanyuk, A., Kyshenko, V., Smityuh, Y.: The biotech complexes control in conditions of situational. Ann. Warsaw Univ. Life Sci. **60**, 149–154 (2012)
10. Mallat, S.: A wavelet tour of signal processing, 637 p. Academic Press (1999)
11. Grinsted, A., Moore, J.C., Jevrejeva, S.: Application of the cross wavelet transform and wavelet coherence to geophysical time series. Nonlinear Process. Geophys. **11**, 561–566 (2004). https://doi.org/10.5194/npg-11-561-2004
12. Kohonen, T.: Self-Organization and Associative Memory, p. 255. Springer, Heidelberg (1984)
13. Peters, E.E.: Fractal Market Analysis, Applying Chaos Theory to Investment and Economics, 336 p. Wiley, New York (1994)

Development of the Structure of an Automated Control System Using Tensor Techniques for a Diffusion Station

Viktor Sidletskyi[1], Igor Korobiichuk[2(✉)], Anatolii Ladaniuk[1],
Ihor Elperin[1], and Katarzyna Rzeplińska-Rykała[3]

[1] National University of Food Technologies, Kiev, Ukraine
vmsidletskiy@gmail.com, ladanyuk@ukr.net,
ivelperin@gmail.com
[2] Warsaw University of Technology, Institute of Automatic Control
and Robotics, Warsaw, Poland
igor@mchtr.pw.edu.pl
[3] Industrial Research Institute for Automation and Measurements PIAP,
Warsaw, Poland
krykala@piap.pl

Abstract. In modern automation systems, when creating of the regulating action the predicted values obtained from mathematical models are used and so, as a consequence, the efficiency of the enterprise will depend on the developed mathematical model adequacy. If the mathematical model is formulated in a tensorial form, then the prerequisite is created for the model to describe the process in an adequate manner. In this paper, we give an example of the development of a tensor model for a sugar house juice extraction complex and its use on the example of a structural diagram of an operating system. This structural diagram of the operating system consists of six main components. At the first stage the radius vectors (input and regulated values) are formed. At the second and third stage tensors are calculated. That will mathematically describe the connection between the input and output parameters of the operating system. At the fourth step, an unbalance signal in local coordinates and the controller coefficients is calculated. At the fifth and sixth steps, values of control signals in local coordinates are calculated. This approach allows us to calculate a regulating action to improve all the performance indicators of the technological site.

Keywords: Diffusion · Extraction · Automation · Operating system · Tensor analysis

1 Introduction

One of the important and relevant problems of the sugar producing process is the maintenance of high quality sugar indicators, and with a decrease in its cost. That is why the site of sugar extraction from beets is important, both in terms of the impact on the regularity of pace of the sugar house operation (which depends on the cost of the final product) and in terms of sugar quality indicators, which depend on the transition of

© Springer Nature Switzerland AG 2020
R. Szewczyk et al. (Eds.): AUTOMATION 2019, AISC 920, pp. 175–185, 2020.
https://doi.org/10.1007/978-3-030-13273-6_18

non-sugars from beet chips to diffusion juice in the extraction process, their type and quantity. At the same time, the diffusion on the sugar houses is not an exception. A large number of different types of equipment is used which is primarily due to constant technological improvement. This requires the need for an integrated and systematic approach to the technological process, which includes a comprehensive analysis of all the essential components of the diffusion operation [1], what requires high qualification, both from the operative personnel and from the automation system, which should not only handle all the management terms, but also conduct a deep analysis of the functioning of the diffusion and make recommendations when making decisions.

One of the most important ways to improve the work efficiency is reducing of the production cost by reduction of energy consumption in production, what can be achieved in the process of determining and maintaining optimal conditions for the development of heat-engineering equipment [2], which will also depend on both the productivity of the entire enterprise and the regularity of pace of production, as well as the quality and cost of production [3, 4].

As stated above, as perfect equipment so the introduction of additional heat engineering equipment, in order to reduce energy costs, everything requires an efficient automated operating system that can only be developed through the usage of modern approaches in the process of building of a technological process operating system for the food industry [5].

For modern diffusion operating systems, it is typical to use microprocessor technology and the presence of an automated workplace of the operator from which the operator can observe the process [6], review the historical trends, conduct an analysis and when necessary to interfere in the technological process [7], which requires the need to supplement traditional operating systems with additional add-ins with decision support subsystems.

When developing such systems, modern software packages MATLAB can be used, which allow to model and analyze the functioning of the operating system at the development stage. In general, the use of mathematical models is also necessary for the study and modeling of the equipment at the stage of its development, as well as for optimal automatic control of equipment during its operation [8], and the diffusion plant is one of the most important elements of the heat engineering complex of sugar beet production. Its model is considered as a component of the mathematical model of the heat engineering complex of a sugar house and is based on the modeling of sugar beet extraction [9, 10].

It should also be noted that the modern approaches to the construction of operating systems are their development in accordance with international (ISA-95) and European standards (IEC 62264), as a hierarchical system, where at the lower level there is an automated process operating system [11], and at the highest – a business process operating system [12]. That is, the general approach to the development of the automation system is that a universal approach is developed in which actual processes are replaced by the model, and the operating system itself is already being explored as an abstract system working with signals from the model. But the model may not always be able to objectively and adequately describe a complex system in which it is necessary to take into account the physical real-world structure that is not included in the model [13], which is why in this paper the tensor methodology is used to describe both technological processes and the formation of regulating actions [14]. Tensor techniques for describing tasks are progressive and are currently used in various fields, for example: tensors are used for data analysis [15]; in the field of artificial vision [16]; the methods of tensor analysis are used to manage mobile robots [17].

That is, the purpose of this work is to synthesize the structure of the operating system, a system that will form a regulating action aimed at achieving the specified performance indicators of the technological complex, which will take into account the current state of the technological process. To achieve this goal, the following tasks need to be solved: (1) to analyze the work of the sites of the technological complex to obtain the necessary parameters to be taken into account in the mathematical model; (2) develop a method for calculating a mathematical model in a tensor form; (3) to develop a methodology for calculating the regulating act according to the calculated tensor models; (4) develop a structural diagram of the operating system, which allows to find a regulating act for tensor models.

2   Materials and Methods of Research

Colonial, rotary and inclined extraction plant have become the most widespread in sugar houses of Ukraine. Advantages of the sloping extraction plant are: stability of work; the ability to regulate the selection of diffusion juice in a wide range, the ability to reduce the temperature of the diffusion juice at the outlet, which allows adjusting the use of energy resources, as well as reducing the loss of sucrose from thermal decomposition. According to [1–9] for the technological site of the diffusion, all technological parameters are controlled and managed. Direct control of technological processes is carried out by an automated operating system.

The device of the inclined extraction plant (Fig. 1) has a trough-shaped body, inside of which there are screw augers that are rotating and move the chips upwards. Feeding water is handed down, in counterflow with chips. Diffusion juice is taken from the bottom of the device and is applied for cleaning. The body of the device is equipped with steam heater cameras, which allows to maintain optimum inside temperature.

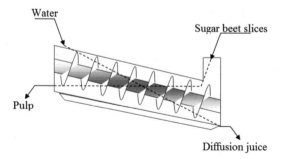

Fig. 1. Diffusion device for inclined extraction plant.

The site of juice extraction is characterized by the following physical and chemical processes: moving of chips by auger conveyor, heating the coot-flake mixture, denaturation of sugar beet cells, diffusion of sugar from the cell into hot water. The most influential parameters are: the temperature in the diffusion device and the time the chips spent in the machine, this is due to the specifics of the process. So at the beginning there is diffusion of sugar from the broken cells, then the penetration of water into the cellular juice begins and after heating the chips above 60 °C and denaturation of the protoplasm the main process of extracting sugar from the vacuoles of the cells of the beet cell to the diffusion juice begins. The attention paid to these parameters is also associated with the ambiguity present in the expected consequences when managing them. For example: the temperature increase contributes to the increase in the coefficient of diffusion of sucrose, so more sucrose will shift from chips to water. On the other hand, at temperatures above 75 °C, there is a rapid swelling of pectin substances of chips and decreases of its elasticity, therefore, the chips in the device will cling together and most of it will not be washed with water. And at temperatures below 70 °C, microorganisms are intensively developing, resulting in the destruction of chips. This also applies to the time during which chips stay in the device. If the chips will contact with water longer, then more sucrose is washed out of it, but the result of increasing the extraction time is a more complete transition from the chips to the solution of not only sucrose, but also pectin's and other non-sugars.

That is, changing these parameters can lead to various consequences, and changing one of the parameters (temperature, time) for one single quality indicator (sugar loss, non-sugar content), will to that another indicator will change at the same time.

At the technological site of the sugar house, the process of sugaring beet chips for the production of diffusion juice has a number of important quality indicators, namely the quality of diffusion juice and the loading of the device. Loading the device affects the performance and regularity of pace of the whole plant, and the quality of the juice influences the process further.

In this paper, each of these quality indicators is adopted as a radius vector in a rectangular coordinate system, where the components (coordinates) have parameter values along each axis. Thus, the selected parameters as axes of coordinates, made it possible to fully determine the vectors of quality indicators. So for the quality of diffusion juice they are: sugar content >15%; nitrogen content <45 mg/100 m; soluble ash content, 0.50–0.65%; alkali content of carbonaceous ash 0.21–0.25%; coefficient of stability >6; pulp, 4.0–4.4%; the content of reducing agents <0.07%. That is, the radius vector will have coordinates $\vec{a}_1 = \{15.00, 45.00, 0.50, 0.21, 6.00, 4.00, 0.07\}$. In turn, the loading of the device depends on such parameters as the uniformity of the supply of chips, the number of revolutions 0.6–0.7/min, the current consumed by motors 85 A, the length of chips 9–12 m. But the loading of the device affects not only the regularity of pace and productivity of the plant, but also the time of chips stay in the device, and therefore the quality of diffusion juice. Usually reducing the length of stay leads to a decrease in the sugar content of juice and a greater loss of sugar in the pulp. At the same time, an increase in the time of stay leads to an increase in the content of non-sugars in diffusion juice. That is, the radius vector for loading the device will have next coordinates $\vec{a}_2 = \{0.60, 85.00, 10.00\}$.

In developing the model, it was taken into account that the vector of loading affects the quality of the diffusion juice vector, that is, firstly, the space for the first vector was calculated, and then the plane on the edge of the vector was the basis for the next vector. In graphical form, for a mathematical model, the indicated sequence is shown on (Fig. 2). So, for the technological site of juice extraction, each process that affects the quality index is indicated as P_i ($i = 1, \ldots, l$) where l – the number of processes for the site (in this case it is accepted $i = 2$, load capacity of the device and the quality of the diffusion juice), and the corresponding parameter values for each process as $x_{P_i} = (x_1, x_2, \ldots, x_j)$ where j is a number of parameters in the process. The radius vector which is the basic one will be denoted as $r_{0,1}$, that is the radius vector which connects the first point with the origin of the coordinates. For a common denominator (for example, for a radius vector with parameter values P_{i+1} relative to the radius of the vector with the values of the parameters P_i) will be next $r_{P_i, P_{i+1}}$. Each qualitative index of the work of the site is connected with the process on the technological site for example f_i for the i-th process and the vector that will bind the process with the quality indicator through the function will be r_{P_i, f_i}.

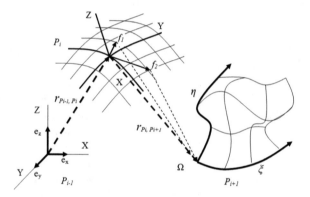

Fig. 2. Technological processes in local coordinates.

As can be seen from (Fig. 2), each vector has an initial and final surface site, and for the diffusion, the final surface of the "loading device" index will be the initial value for the "quality of diffusion juice". Therefore, it was accepted that the vector of the input parameters $\vec{x} = (x_1, x_2, \ldots, x_n)^T$ and a vector of output parameters $\vec{y} = (y_1, y_2, \ldots, y_m)^T$ are linked together in the form of a system of equations which is convenient to submit in a matrix form $\vec{y} = A\vec{x}$. The columns of the matrix A are the components of the vectors for each parameter. Thus, the model of change of quality indicators is presented as a sequence of transformations from the starting point (initial set of parameters) to the final one by using the transformation matrix. In addition, matrix A allows you to find not only the quality indicators by parameters, but also to calculate the inverse matrix A^{-1} to make a recalculation, that is, to find parameters in terms of quality $\vec{x} = A^{-1}\vec{y}$.

According to [18] this approach to the description of the problem area allows us to operate the properties of the linear space, that is, we accept that the model of conducting the technological process of juice extraction and the relationship of the dependence of the quality indicators to the technological parameters are linear spaces, that is, the quality indicators represent a set of radius vectors that are consistently directed from the first surface to the next, which is the origin of the coordinates for the following radius vector. That is the expression $A\overrightarrow{x} = \theta$ where θ is a zero vector. According to [10] Zero space $N(A)$ is obtained after transformation T_A. As shown in [10], $N(A)$ is the set of all points in the space by which the radius vector is transposed to the origin of coordinates $x \in N(A) \xrightarrow{T_A} \theta$. That is, $N(A)$ is a linear space and consists of one point – the origin of the coordinates.

Accordingly, using the transformation of T_A to the set of points x we obtain a set of points y $x \in R^3 \xrightarrow{T_A} y \in R(A)$. Where $R(A)$ is the set of points (or radius vectors) in which the points of space R3 transform T_A. This transformation can be done by orthogonalization of vectors [10]. For the possibility of using in subsequent calculations it is required that the vectors are pairwise orthogonal. To find orthonormal vectors, orthogonalization algorithms are used [10].

Then the components of the matrix A are tensors of the second rank. In this case, the axes X and Y (Fig. 2) form a coordinate plane, and Z is a perpendicular area of the coordinate axis. For such a representation vectors f_i can be represented as a projection vector on a coordinate plane and a coordinate axis, respectively. After modifications, the components of the tensor a_{ij} depend on the coordinate system and they change when the coordinate system changes with the rule of transformation. That is, when changing the coordinate system, in this case, when changing from the indicator "load in the device" to the "quality of diffusion juice", the tensor changes.

Transformation of the coordinate system may include the displacement of the origin of the coordinates without changing the direction of the axes (translation), changing the orientation of the axes without transferring the beginning, as well as both together. The first type of transformation is translation, is not of interest, because in this case the coordinates of all points of space change on the same value. Therefore, we will consider only the second type of conversion, by considering the origin of the coordinates unchanged. Changing the orientation of the axes of coordinates can occur either as a result of rotation around any axis, or when reflecting (inversion) in a certain plane.

For the technological process, the choice of the coordinate system, the base point, i.e. the origin of the coordinates in this paper is indicated by the upper indices, that is, if the origin of the coordinates will be at the point P_i for the radius vector $r_{P_i,P_{i+1}}$, then this radius vector will have a denotation $r_{P_i,P_{i+1}}^{P_i}$, and the radius vector relative to the quality index of the technological site will be as follows $r_{P_i,f_i}^{P_i}$. The transition from the point P_i to the point P_{i+1} is executed using the matrix of the transition $A_{P_{i+1}}^{P_i}$, which has the form of changing the technological parameters in the pixel array P_i and P_{i+1} with the help of the following equality $r_{P_{i+1}} = A_{P_{i+1}}^{P_i} r_{P_i}$ for the reverse path, this dependence will be as follows

$$r_{P_i} = \left(A^{P_i}_{P_{i+1}}\right)^{-1} r_{P_{i+1}} \tag{1}$$

For an array of all current quality indicators of the technological site $f_i = (1, 2, ..., n)$, there are regulated (given) indicators $\gamma_i = (1, 2, ..., n)$. That is, in the operating system of value, both calculated indicators and given quality indicators that are transferred to local coordinates on the plane Ω with coordinate of axes η and ξ. The transition of the quality indicator f_i from the current values, in the form of technological parameters to sites in local coordinates is performed in the form of a vector $\Phi^{\Omega}_{\Omega,f_i}$ and, accordingly, for regulated quality indicators $\Phi^{\Omega}_{\Omega,\gamma_i}$.

Transition matrix $A^{P_i}_{P_{i+1}}$ from one coordinate system (in this case from P_i to P_{i+1}) taken metric tensor that was found as the product of the tensors of the corresponding surfaces, and they in their turn, as a product of radius vectors coordinates, for example, for local coordinate axes η and ξ components of the tensor are an array of coefficients a_{ii} quadratic form

$$Q = a_{11}d\xi^2 + 2a_{12}d\xi d\eta + a_{22}d\eta^2 = (d\xi, d\eta)\begin{pmatrix} a_{11} & a_{12} \\ a_{12} & a_{22} \end{pmatrix}\begin{pmatrix} d\xi \\ d\eta \end{pmatrix}, \tag{2}$$

that is, in this case, the tensor T_1 is a model consisting of small sites that are tangent to the surface and which consists of four components $T1 = \begin{pmatrix} a_{11} & a_{12} \\ a_{12} & a_{22} \end{pmatrix}$.

Using these approaches, the following algorithm for finding the regulating action was formed.

Algorithm 1

Step 1. For the radius vectors that are formed from the measured and stored data of the technological site (these vectors set the initial and final position of the technological process), we find quadratic forms.

Step 2. By a quadratic form we find the tensor.

Step 3. Behind the deviations at the initial (final) state we find the transition tensor.

Step 4. After the transition tensor we calculate the necessary regulating action.

Step 5. After the transition tensor and the radius vector of the current (measured) values, we calculate the predicted state of the system.

Step 6. By the difference between the predicted state of the system and the measured one, we adjust the regulating action.

The use of the "Algorithm 1" allowed us to find and correct the regulating action as the difference between the regulated and measured input and output values of the technological process.

3 The Results of Research

For the system's operation are used: the sets of measured values of technological parameters and the sets of regulated values introduced by operational personnel, both technological parameters and performance indicators of the site operation are used. Structural diagram of the manager formation that has been developed, finds control signals based on the calculated tensor models of the technological process and transformation tensors. The developed structural diagram of the operating system is presented on (Fig. 3).

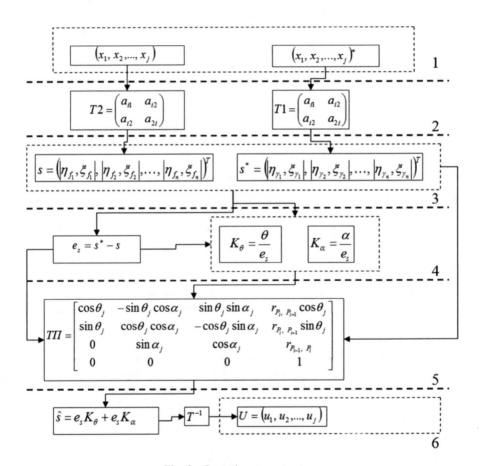

Fig. 3. Control system structure.

For the given system it is possible to allocate such basic six stages. At the first step, sets of parameters which subsequently are presented as radius vectors (input and regulated values) are formed. Further, at the second stage, by the first quadratic form these radius vectors are found in local coordinates and the calculated tensors are mathematical models constructed according to the input values achieved from the operating system. These models are elementary planes that are tangent to the surface and are designed for current and regulated values. In the structural diagram, this process is indicated at the third stage.

At the fourth stage of the structural diagram, the dissolution signal in the local coordinates is calculated, and the regulator coefficients are calculated too. The calculated data is used to calculate the projection of the vector of regulated values into a plane of current values, which is a control signal in local coordinates. At the sixth stage, the calculated control signals enter the operating system.

4   Conclusions

1. The analysis showed that the main component of the system that forms the regulating act is the mathematical model of the technological process and therefore the quality of the operating system will significantly dependent on the adequacy of the mathematical model. At the same time, the most progressive approach to the mathematical description of the problem area is the use of methods of tensor analysis, which involves the development of a tensor as the process itself and transformation tensors when describing the change of conditions of the process. At the same time, tensor models were not used for the development of hierarchical operating systems, and the possibilities are shown in this article.

2. The results of the analysis and synthesis make it possible to determine that the control corresponds to the optimal transition processes, uniquely associated with reducing of the loss of energy resources. That is why the solution of the problems by tensor analysis allows us to formulate a mathematical description in such a form that it does not depend on the specific values of the process parameters, and all the dependencies will be linear, and the parameter values will be used only when selecting the coordinate axes.

3. When building an operating system using tensor analysis methods, the components of the metric tensor were found. The tensor components are calculated from the values of the technological parameters that formed the radius vector. The tensor was found to be used as a model that translates the values of the process parameters into a vector in local coordinates. These tensors are used both for transition to local coordinates and for calculation of values in real values, for this purpose it is necessary to find the inverse tensor. When finding the regulating action, it was assumed that the vectors change both its magnitude and the angle of incline, these values are used to find the misalignment signal and the coefficients of the regulator.

4. For reasonable application of tensor methods it is necessary to take into account that in technological objects there are a few processes that run simultaneously, such as of heat and mass transfer, hydrodynamics and physical and chemical transformations of substances in one technological device. This leads to the need to take into account the non-stationary, nonlinearity and multidimensions, which are imposed as additional technological requirements. Thus, it is shown that on the basis of tensor analysis it is necessary to conduct further targeted research to determine and take into account all possible reactions in the behavior of the technological process.

References

1. Pushanko, M.M., Verkhola, L.A.: Ekstraktsiia tsukru z buriakiv: mozhlyvosti naiavnoho obladnannia. Tsukor Ukrainy, No. 11. pp. 33–41 (2011)
2. Korobiichuk, I., Lutskaya, N., Ladanyuk, A., Naku, S., Kachniarz, M., Nowicki, M., Szewczyk, R.: Synthesis of optimal robust regulator for food processing facilities. Adv. Intell. Syst. Comput. **550**, 58–66 (2017). https://doi.org/10.1007/978-3-319-54042-9_5
3. Boiko, V.O., Maslikov, M.O., Priadko, M.O.: Vyznachennia kontsentratsii i temperatury dyfuziinoho soku. Naukovi pratsi Natsionalnoho universytetu kharchovykh tekhnolohii. **18**, 86–89 (2006)
4. Korobiichuk, I., Ladanyuk, A., Shumyhai, D., Boyko, R., Reshetiuk, V., Kamiński, M.: How to increase efficiency of automatic control of complex plants by development and implementation of coordination control system. Adv. Intell. Syst. Comput. **543**, 189–195 (2017). https://doi.org/10.1007/978-3-319-48923-0_23
5. Sidletskyi, V.M.: Vdoskonalennia avtomatyzovanoi systemy keruvannia tekhnolohich-nym protsesom pidpryiemstva kharchovoi promyslovosti. Problemy ta perspektyvy rozvytku enerhetyky, elektrotekhnolohii ta avtomatyky v APK: IV Mizhnarodna naukovo-praktychna konferentsiia, 21–22 lystopada 2016 r, NUBIP, Kyiv, pp. 71–72 (2016)
6. Korobiichuk, I., Dobrzhansky, O., Kachniarz, M.: Remote control of nonlinear motion for mechatronic machine by means of CoDeSys compatible industrial controller. Tehnički vjesnik/Technical Gazette **24**(6), 1661–1667 (2017). https://doi.org/10.17559/TV-20151110164217
7. Petrushka, V.P., Buryy, N.A., Seregin, A.A., Lyulka, D.N.: Avtomatizatsiya i opyt ekspluatatsii diffuzionnoy ustanovki EKA-3. Sakhar, No. 1, p. 52 (2005)
8. Kharlamenko, V., Ruban, S., Korobiichuk, I., Petruk, O.: Adaptive control of dynamic load in blooming mill with online estimation of process parameters based on the modified Kaczmarz Algorithm. Adv. Intell. Syst. Comput. **543**, 227–233 (2017). https://doi.org/10.1007/978-3-319-48923-0_28
9. Trehub, V.H., Yu, D.Y.: Avtomatyzatsiia potoku dyfuziinoho soku. Avtomatyzatsiia vyrobnychykh protsesiv, no. 1(14), 66–68 (2002)
10. Ladaniuk, A.P., Zaiets, N.A., Vlasenko, L.O.: Suchasni tekhnolohii konstruiuvannia system avtomatyzatsii skladnykh obiektiv (merezhevi struktury, adaptatsiia, diahnostyka ta prohnozuvannia): monohrafiia, Nats. un-t kharch. tekhn, Lira-K, Kyiv, 312 p. (2016). ISBN 978-617-7320-34-9
11. Korobiichuk, I., Ladanyuk, A., Zaiets, N., Vlasenko, L.: Modern development technologies and investigation of food production technological complex automated systems. In: ACM International Conference Proceeding Series, pp. 52–57 (2018). https://doi.org/10.1145/3185066.3185075
12. Sidletskyi, V.M., Elperin, I.V., Polupan, V.V.: Analiz ne vymiriuvalnykh parametriv na rivni rozpodilenoho keruvannia dlia avtomatyzovanoi systemy, obiektiv i kompleksiv kharchovoi promyslovosti. Naukovi pratsi Natsionalnoho universytetu kharchovykh tekhnolohii **22**(3), 7–15 (2016)
13. Korobiichuk, I., Siumachenko, D., Smityuh, Y., Shumyhai, D.: Research on automatic controllers for plants with significant delay. Adv. Intell. Syst. Comput. **519**, 449–457 (2017). https://doi.org/10.1007/978-3-319-46490-9_60

14. Varychev, A.I., Sidletsky, V.M.: Avtomatyzovana systema upravlinnia dyfuziinoiu stantsi-ieiu, yak skladovoiu kompiuterno-intehrovanoi systemy tsukrovoho zavodu, z vykorystan-niam tezornoho analizu. Suchasni metody, informatsiine, prohramne ta tekhnichne zabezpechennia system keruvannia orhanizatsiino-tekhnichnymy ta tekhnolohichnymy kompleksamy: materialy IV mizhnarodnoi naukovo-tekhnichnoi Internet - konferentsii, 22 lystopada 2017 r., NUKhT, Kyiv, pp. 185–186 (2017)
15. Fletcher, P.T., Joshi, S.: Riemannian geometry for the statistical analysis of diffusion tensor data. Sig. Process. **87**(2), 250–262 (2007). https://doi.org/10.1016/j.sigpro.2005.12.018
16. Vasilescu, M.A., Terzopoulos, D.O.: Multilinear analysis of image ensembles: tensorfaces. In: Lecture Notes in Computer Science, pp. 447–460 (2002)
17. Chen, J., Jia, B., Zhang, K.: Trifocal tensor-based adaptive visual trajectory tracking control of mobile robots. IEEE Trans. Cybern. **47**(11), 3784–3798 (2017). https://doi.org/10.1109/TCYB.2016.2582210
18. Strang, G.: The fundamental theorem of linear algebra. Am. Math. Mon. **100**(9), 848–855 (1993)

Infrastructure of RFID-Based Smart City Traffic Control System

Bartosz Pawłowicz$^{(\boxtimes)}$ ⓘ, Mateusz Salach$^{(\boxtimes)}$ ⓘ,
and Bartosz Trybus$^{(\boxtimes)}$ ⓘ

Rzeszow University of Technology,
al. Powstancow Warszawy 12, 35-959 Rzeszow, Poland
{barpaw, m.salach, btrybus}@prz.edu.pl

Abstract. Taking as a basis the concept of urban development referred to as smart city 2.0, the article presents the possibilities offered by modern urban infrastructure in the field of intelligent traffic control. In particular, this applies to algorithms of optimal traffic distribution and rapid response to emergency situations. Various scenarios for improving this control using RFID (Radio-Frequency IDentification) technology are discussed. Thanks to RFID it is possible to control access to zones with traffic restrictions, including city centers, areas available for specific types of vehicles or for vehicles meeting certain environmental standards. An important novelty is the solution involving the installation of RFID readers in vehicles, and not only in road infrastructure. A hybrid solution that could be used during the transitional period is also presented.

Keywords: Smart City · Traffic management · RFID

1 Introduction

More and more advanced traffic control systems are being implemented in urban areas. Three processes can be distinguished related to traffic management:

- **Monitoring** includes acquisition of traffic data in an urban area and sending them to a data warehouse. Current road conditions are determined on the basis of data collected from specific time intervals [1]. The data is often collected in a computing cloud and can be analyzed to set appropriate vehicle routes or to react to events that may occur. Availability of parking spaces can be obtained from city cameras, parking sensors and data collected from the machines issuing tickets in paid parking lots.
- **Traffic flow control** covers vehicle traffic, integrated public transport and pedestrian behavior. It is done usually by means of traffic lights control to ensure the most fluent traffic. For this purpose, algorithms based on artificial intelligence can be used to perform the control in near real time. By using analysis methods, e.g. deep or machine learning, the system may be able to predict traffic load based on historical data. Thus, it is possible to precisely determine e.g. rush hours in which traffic is the

© Springer Nature Switzerland AG 2020
R. Szewczyk et al. (Eds.): AUTOMATION 2019, AISC 920, pp. 186–198, 2020.
https://doi.org/10.1007/978-3-030-13273-6_19

largest. Currently, intelligent traffic control systems as part of smart city solutions are still at the conceptual stage or are being under tests.

- **Supervised and controlled traffic** is possible via a city traffic management and control centre. Vehicles communicate with the centre, either to provide data useful for the system, or to accept control commands. The communications must allow responding in as close as possible to real time. In such conditions, personal vehicles should be able to achieve autonomy at least at the L2 level while public transport vehicles and other vehicles moving within the city area of the L4 level. Thanks to this, such a system will be able to dynamically determine the best routes for the drivers and dynamically regulate traffic permissions in designated zones of the city. The system will also be able to respond to unexpected events such as accidents or traffic jams. In case of an accident, the system will allow faster action of emergency units and automatic detours.

This paper takes supervised and controlled traffic as a reference solution and deals with appropriate infrastructure supported with control algorithms.

2   Traffic Solutions for Smart Cities

Recently popular solutions of urban development are based on the idea of Internet of Things [2, 3]. Smart benches or "intelligent" bus stops are examples. Along with the development of the Internet of Things concept, urban facilities and vehicles exchange data and establish communication with each other. In road traffic solutions, two types of communication are used – V2V (vehicle-to-vehicle) and V2I (vehicle-to-infrastructure) [4]. The first attempts to establish communication between vehicles began in the 1970s with the introduction of the EGRS (Electronic Route Guidance System, [5]) based on geographical data. In the 1980s, the Global Positioning System (GPS) appeared, which is still being used today.

Computer vision is used to analyze traffic using city cameras mounted on pedestrian crossings or roadside cameras mounted above the road, e.g. for the analysis of car license plates [4]. One of the known solutions is the Japanese Smartway system based on the VICS (Vehicle Information Communication System) introduced in 2003 [5]. It uses telecommunications masts and a sensor system located along the road to analyze the vehicle location. A project presented at the UCSI University in Kuala Lumpur uses GPS and GSM/GPRS techniques to identify objects in space [6].

Data obtained from vehicles must be stored in a secure location with various security measures implemented. A secure system must be well-developed, built and managed through integrated functions, sensors or electronic modules exchanging information only with a dedicated database [7]. Biometric solutions and smartphones must be properly secured [8].

Urban development and management techniques are expanding more and more to the newly introduced technologies included in the Smart City 2.0 concept [9, 10].

3 Urban Traffic Control System Infrastructure

Current urban traffic control systems use data obtained by various types of sensors located at specific points of the road infrastructure (Fig. 1). Information on the route load or current traffic intensity is usually provided by traffic monitoring cameras. In order to implement traffic control, it is necessary to provide a two-way communication of the supervised vehicle with the traffic control centre [11].

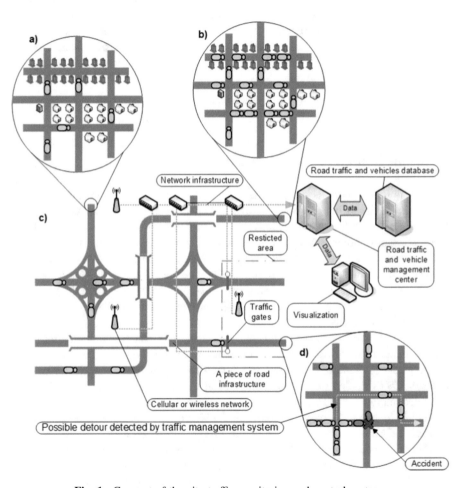

Fig. 1. Concept of the city traffic monitoring and control system.

One of the basic information that is necessary for the system to operate is the position of individual vehicles. The position can be obtained using GPS receivers installed in vehicles [12]. An alternative method is the use of a wireless Wi-Fi network (Fig. 1c). Vehicles communicate with the city infrastructure using embedded Wi-Fi modules and access points are located along the route [13]. The data collected from moving vehicles by the access points is then sent to a dedicated database in the cloud [14].

The operation of the traffic control system focuses mainly on the monitoring and management of vehicle traffic that can communicate with the system. It is possible to control the flow of traffic by means of traffic lights and to quickly inform traffic participants about accidents [15]. The system may suggest alternative routes to avoid traffic incidents and possibly to force an alternative road through appropriate signalling. It is also possible to analyze the load of a given road or street [11].

By acquiring information from many vehicles, the system can monitor a road and route vehicles to sideways in the event of a heavy load [16]. A simplified algorithm of traffic detection which redirects vehicles in case of heavy load is presented in Fig. 2. The analysis of the route and its choice for a given vehicle are determined not only by parameters of the vehicle or a road. Particularly, the system should predict the load on a particular route at a given time (Fig. 1a, b). This can be done by solutions based on machine learning or artificial intelligence [17]. The system anticipating a greater load may propose alternative routes leaving the main roads for vehicles with a higher priority (rescue services, public transport). Factors such as atmospheric conditions can also be taken into account. The information may come from automated meteorological stations.

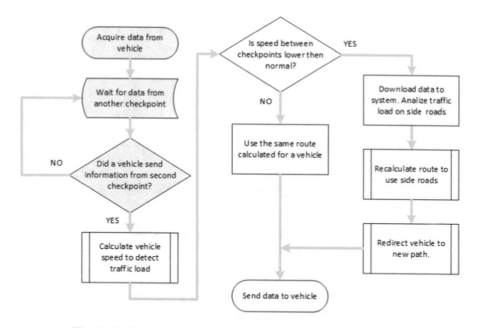

Fig. 2. Traffic detection algorithm using data acquired from a vehicle.

One of the most important functions of such systems is the reaction to sudden road events. This scenario is implemented in algorithm presented in Fig. 3, which notifies emergency services (police, ambulance, etc.). After reporting an accident, a new route is established in accordance with the assumptions presented in the Fig. 1d. There may some areas not available for entry for some vehicles, such as special roads around airports, in parks or city centres with restricted entry for vehicles with a particular type of fuel. The reserved areas may have an additional layer of security applied. By marking each vehicle with its own unique identification number and using traffic gates, the system may verify access to a given destination. A database is maintained with the identification numbers of vehicles that are allowed to enter the restricted areas. If the vehicle approaching the gate is present in the database, the system will raise the gate.

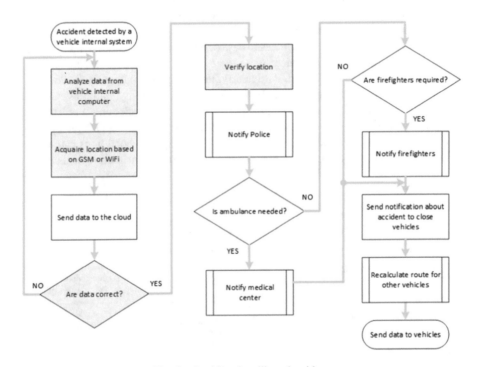

Fig. 3. Accident handling algorithm.

4   Improving Traffic Management with RFID Solutions

RFID (Radio Frequency Identification) technology can significantly improve traffic management. In particular, this applies to automated identification of vehicles. The contactless identification system can be used in two ways. In the common approach, RFID tags are installed in vehicles and reader devices are installed in road infras-tructure [9]. Thus, it is possible to read important information directly from the vehicle. The identifiers can be programmed at the production stage to include information on the vehicle identification number, model, engine type, type of fuel, and much more.

The infrastructure of the system resembles the one described in p. 3, but the introduction of the RFID technique is a fundamental novelty. Full identification of individual vehicles and their location takes place at RFID gates located at some points of the infrastructure, where vehicle RFID transponders are read.

In the case of areas with restrictions, it is possible to use dedicated road barriers coupled with RFID readers. Each vehicle is assigned a unique identification number stored in the RFID transponder. The approaching vehicle is identified and the information retrieved by the reader is compared with the database records to check if the vehicle is authorized to enter a given restricted area. If so, the barrier is raised, allowing the vehicle to enter (Fig. 4). The use of RFID increases the security of the system and its integrity. A unique approach not yet applied is the installation of RFID tags in the road surface or at the roadside and RFID readers installed in vehicles (Fig. 5). In this solution, traffic analysis is based on data read by vehicles e.g. at given time intervals and transferred to the traffic management centre. The information can be processed by the system to determine the speed of vehicles in the urban area. Using the data, the system can also determine the road load in real time [17]. The example on Fig. 5a shows a vehicle entering a road and moving over the first RFID transponder (A). The vehicle retrieves data from the transponder and sends it to the management centre. When the vehicle reaches the next identifier, it also reads the data and sends it to the system. At this point, the algorithm starts the route analysis based on the time interval

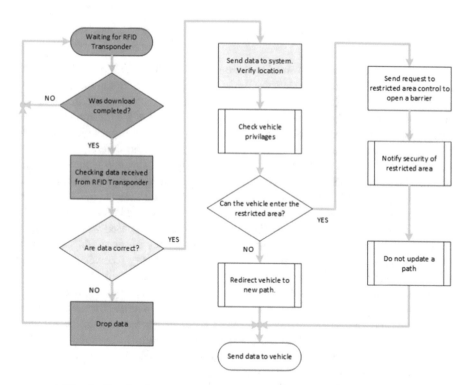

Fig. 4. Restricted access control algorithm for RFID-based identification

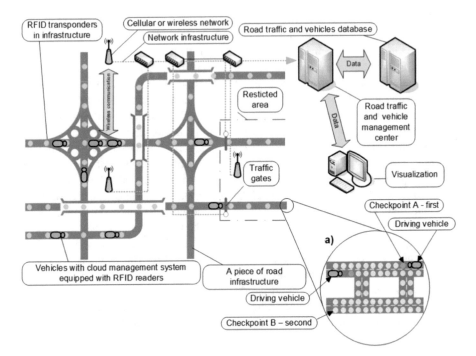

Fig. 5. City traffic infrastructure with RFID transponders placed in the road surface

between the two readings. Information is acquired from all vehicles on a given section of a road. This allows to accurately determine the load of the route in both directions.

Each vehicle navigating the route section reads the data stored in the RFID transponder. This information includes the exact coordinates of the transponder and the vehicle. By setting the appropriate arrangement of RFID tags, it is possible to use them in a lane assistant. The vehicle scanning RFID tags is able to determine whether it approaches the edge of the lane or crosses it in a non-safe way, e.g. if the driver falls asleep. This solution guarantees enhanced safety due to the fact that radio systems are more resistant to weather conditions. Current solutions using cameras may fail in such situations, especially in winter, when the road surface is covered with snow.

The information contained in RFID transponder can also be used when the vehicle detects a traffic accident. It will be able to call emergency services and provide the accurate location of the event (Fig. 6). The management center receives more precise information and sent them to proper services. No more data and verification is required.

However, one should consider a case when a city is entered by a vehicle that does not meet the required conditions, e.g. it lacks the RFID reader. The traffic control system must be adapted to many vehicles of different generations, hence a hybrid solution should be established with a cohesive urban infrastructure. Such solution is presented in Fig. 7 and assumes three possible types of vehicle identification: (a) a vehicle not equipped with an RFID system, (b) a vehicle with an RFID transponder, (c) a vehicle equipped with RFID readers and transponder. In any case, the traffic

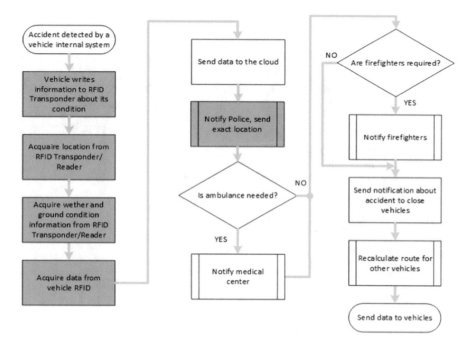

Fig. 6. RFID-based accident handling

monitoring system will exchange information with vehicles regardless of the solutions used. Vehicles that do not have RFID elements will be identified by cellular networks or wireless networks. A solution with a dedicated application installed in the driver's mobile device or in on-board navigation system is also possible. Vehicles equipped with RFID transponders will use RFID readers located in the urban road infrastructure (e.g. on masts or booms). The readers will send data directly to the system, indicating current position the vehicle. In case of vehicles equipped with both the RFID reader and the transponder, the traffic control system may select one of the identification solutions, for example when one of them fails.

Emission of electromagnetic radiation is one of the most important elements of the automatic identification with RFID. This mainly determines the shape of the interrogation zone [18], within which it is possible to exchange data between elements of the system of contactless identification, especially in the field of transport and traffic. From the point of view of the correctness of the design and operation of the RFID system and other radiocommunication devices, it is necessary to determine the acceptable radiation levels.

In the area of contactless identification systems, the limits of the intensity of the electromagnetic field have been included in four standards, EN 300 330, EN 300 440, EN 302 208 [19–21], based on the document CEPT/ERC Recommendation 70-03 [22] prepared by European Telecommunications Standards Institute (ETSI). The document covers standards in the area of widely understood telecommunications systems. Table 1 summarizes the existing emission standards for RFID systems applicable for transportation and traffic management.

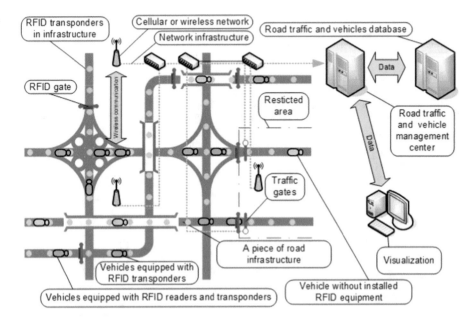

Fig. 7. Hybrid city traffic management infrastructure

Table 1. Limits of the intensity of the magnetic field or radiated power for RFID applications that can be used in the area of transport and road traffic.

Frequency f, MHz	Intensity of the magnetic field (10 m from the radiation source)/radiation power	Standard/comment
$0.119 \leq f < 0.135$	66 dBµA/m with the drop of 3 dB/oct above 0.119 MHz	EN 300 330 – passive RFID systems with inductive coupling
$13.553 \leq f < 13.567$	Typically 42 dBµA/m or 60 dBµA/m for f = 13.56 MHz	EN 300 330 – passive RFID systems with inductive coupling
$865 \leq f < 868$	2 W ERP (in selected channels): 865.7 MHz, 866.3 MHz, 866.9 MHz and 867.5 MHz	EN 302 208 – passive, microwave RFID systems
2446–2454 MHz	In built-up area for continuous operation: 500 mW EIRP without restrictions; or up to 4 W with restrictions (Annex G)	EN 300 440 – active, microwave RFID systems

Despite the existing standardization in the scope of RFID applications, the possibility of effective use of the RFID systems in transport and road traffic, in each case consideration should be given to the restrictions that apply in individual ETSI member states.

5   Test Traffic Infrastructure with a Cloud Service

In order to test the discussed traffic management scenarios in practice, a laboratory system was developed that uses embedded solutions and cloud computing. Real traffic data was also used.

Raspberry Pi 3 and Raspberry Pi 2 computers were used in the experiments as the on-board platform for vehicles. They were equipped with RFID readers based on MRF-522 chip. Several test boards with a grid of RFID transponders have been prepared for different scenarios. One of them, dedicated for testing parking management is shown in Fig. 8. It uses 8 × 6 two-dimensional matrix of RFID transponders. When a vehicle approaches the RFID transponder, the RFID data is read by MRF-522. A dedicated script in Python has been prepared for reading the required data.

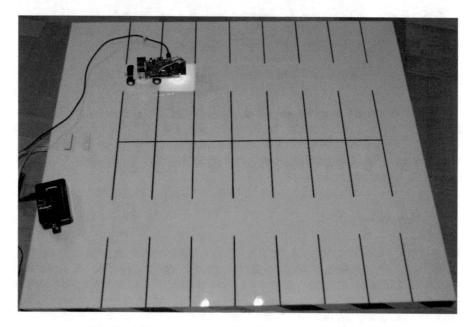

Fig. 8. A laboratory model with RPi vehicle and RFID reader

The vehicles are connected to the internet via a Wi-Fi module, either built-in (Pi 3) or using the USB port (Pi 2). All the information read from the RFID transponder is sent to the cloud. The Python script marks the data with the current time stamp. Messages that are exchanged during the communication between the vehicle and the cloud are shown in Fig. 9.

The Microsoft Azure cloud service with IoT Hub as the main control system has been used in the laboratory model as a computing service. The data is stored in the CosmosDB non-relational database. The Azure cloud's Location Based Services have been employed for vehicle location.

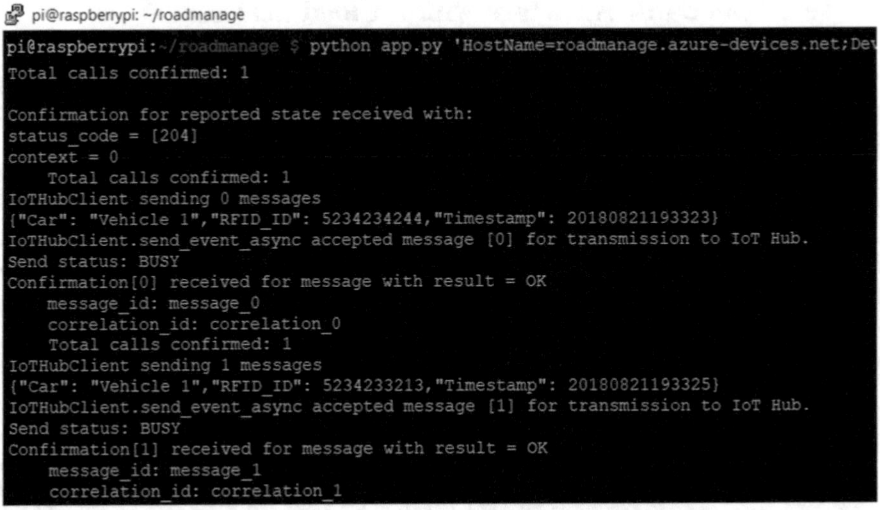

Fig. 9. Vehicle communication with the IoT cloud service

In some of the experiments, real traffic data was used. They were obtained from the Municipal Road Administration in Rzeszów and Public Transport Authority in Rzeszów. The data has been used for machine learning algorithms involved for traffic management [17]. The algorithms were tested under Machine Learning Studio, being a part of the Microsoft Azure cloud.

6   Conclusions

It seems that the RFID technique can significantly improve traffic control in Smart City solutions. Current systems based on GPS or traffic cameras to monitor traffic and vehicle identification are characterized by a fair degree of uncertainty. RFID devices operate in areas where existing solutions fail (covered parking lots, darkened streets, etc.). The increasing availability of RFID infrastructure allows for the consideration of a scenario where RWD devices are installed in vehicles. It should also be remembered that the data contained in RFID transponder can be modified during operation, what creates additional possibilities.

Acknowledgement. This work is financed by Polish Ministry of Science and Higher Education under the program "Regional Initiative of Excellence" in 2019–2022.

Project number 027/RID/2018/19, funding amount 11 999 900 PLN.

References

1. Wong, S.F., Mak, H.C., Ku, C.H., Ho, W.I.: Developing advanced traffic violation detection system with RFID technology for smart city. In: 2017 IEEE International Conference on Industrial Engineering and Engineering Management, IEEM, Singapore, pp. 334–338 (2017)
2. Rassia, S.T., Pardalos, P.M. (eds.): Smart City Networks. Through the Internet of Things. Springer, Cham (2017)
3. Tokoro, N.: The Smart City and the Co-creation of Value. Springer, Tokyo (2016)
4. McClellan, S., Jimenez, J.A., Koutitas, G. (eds.): Smart Cities. Applications, Technologies, Standards, and Driving Factors. Springer, Cham (2018)
5. An, S.-H., Lee, B.-H., Shin, D.-R.: A survey of intelligent transportation systems. In: 2011 Third International Conference on Computational Intelligence, Communication Systems and Networks. IEEE Xplore (2011)
6. Rathod, R., Khot S.T.: Smart assistance for public transport system. In: 2016 International Conference of Inventive Computation Technologies. IEEE Xplore (2017)
7. Hall, E.R.: The vision of a smart city. In: 2nd International Life Extension Technology Workshop, Paris, p. 2 (2000)
8. Mahmood, Z. (ed.): Smart Cities. Development and Governance Frameworks, pp. 49–54. Springer, Cham (2018)
9. Chowdhury, P., Bala, P., Addy, D., Giri, S., Chaudhuri, A.R.: RFID and Android based smart ticketing and destination announcement system. In: 2016 International Conference on Advances in Computing, Communications and Informatics, ICACCI, Jaipur, pp. 2587–2591 (2016)
10. Etezadzadeh, C.: Smart City – Future City? Smart City 2.0 as a Livable City and Future Market. Springer, Wiesbaden (2016)
11. Haroon, P.S.A.L., Eranna, U., Irudayaraj, I.R., Ulaganathan, J., Harish, R.: Paradoxical monitoring of urban areas & mailbags tracking system using RFID & GPS. In: 2017 International Conference on Electrical, Electronics, Communication, Computer, and Optimization Techniques, ICEECCOT, Mysuru, pp. 1–4 (2017)
12. Vakula, D., Raviteja, B.: Smart public transport for smart cities. In: 2017 International Conference on Intelligent Sustainable Systems, ICISS, Palladam, India, pp. 807–808 (2017)
13. Zhang, R., Liu, W., Jia, Y., Jiang, G., Xing, J., Jiang, H., Liu, J.: WiFi sensing-based real-time bus tracking and arrival time prediction in urban environments. IEEE Sens. J. 18(11), 4760–4765 (2018)
14. Jawhar, I., Mohamed, N., Al-Jaroodi, J.: Networking and communication for smart city systems. In: 2017 IEEE SmartWorld, Ubiquitous Intelligence & Computing, Advanced & Trusted Computed, Scalable Computing & Communications, Cloud & Big Data Computing, Internet of People and Smart City Innovation, San Francisco, CA, USA (2017)
15. Gupta, V., Kumar, R., Srikanth, K., Panigrahi, B.K.: Intelligent traffic light control for congestion management for smart city development. In: 2017 IEEE Region 10 Symposium, Cochin, India (2017)
16. Arbi, Z., Driss, O.B., Sbai, M.K.: A multi-agent system for monitoring and regulating road traffic in a smart city. In: 2017 International Conference on Smart, Monitored and Controlled Cities. IEEE Xplore (2017)
17. Pawłowicz, B., Salach, M., Trybus, B.: Smart city traffic monitoring system based on 5G cellular network, RFID and machine learning. In: KKIO 2018, pp. 5–9 (2018)
18. Finkenzeller, K.: RFID Handbook – Fundamentals and Applications in Contactless Smart Cards, Radio Frequency Identification and Near-Field Communication, 3rd edn. Wiley, Hoboken (2010)

19. ETSI EN 300 330 V2.1.1: Short Range Devices (SRD), Radio equipment in the frequency range 9 kHz to 25 MHz and inductive loop systems in the frequency range 9 kHz to 30 MHz; Harmonised Standard covering the essential requirements of article 3.2 of Directive 2014/53/EU, February 2017
20. ETSI EN 300 440 V2.2.1: Short Range Devices (SRD); Radio equipment to be used in the 1 GHz to 40 GHz frequency range; Harmonised standard for access to radio spectrum; harmonised European, July 2018
21. Draft ETSI EN 302 208 V3.2.0: Radio Frequency Identification Equipment operating in the band 865 MHz to 868 MHz with power levels up to 2 W and in the band 915 MHz to 921 MHz with power levels up to 4 W; Harmonised standard for access to radio spectrum, February 2018
22. ERC Rec. 70-03 - ERC recommendation 70-03 relating to the use of short range devices (SRD), ERO (2007)

Low Power Wireless Protocol for IoT Appliances Using CSMA/CA Mechanism

Tymoteusz Lindner$^{(\boxtimes)}$, Daniel Wyrwał$^{(\boxtimes)}$, and Arkadiusz Kubacki

Poznan University of Technology, Piotrowo Street 3, 60-965 Poznan, Poland
{tymoteusz.lindner,daniel.wyrwal}@put.poznan.pl

Abstract. Authors propose solution in communication for low power embedded appliances, for system consists of many low power transmitters and one receiver. Transmitter is supplied from battery and works in loop: run program, send data, go sleep. Communication is based on ISM band on frequency 868 MHz. Estimated worktime of transmitter on one battery is 12 years, with average current in sleep about 2 uA. Communication between transmitter and receiver is one-sided. The transmitter does not receive any acknowledgement if the receiver has received data packet, due to reduce energy consumption by transmitter. In case to avoid lost packets, authors applied CSMA/CA (Carrier-sense multiple access with collision avoidance) mechanism. CSMA/CA mechanism is designed to check if channel is free before sending data. Data are sent, only if channel is sensed to be free.

Keywords: IoT · Embedded · Low power · Wireless · CSMA

1 Introduction

In case of battery-powered devices, an important task, is to develop a device operation strategy, that will allow it to work for a very long time without changing the battery. In the era of IoT where devices communicate wirelessly, there are many factors that can disturb communication.

It is important, to check, whether the device received the sent data. A good way to check, if the receiver has received data, is after sending information, switching to listen state and waiting for receiving confirmation, from the device, to which the data was sent. This approach involves increasing the time in the active mode, in which the device consumes a much larger current, compared to the inactive state (sleep), and thus increases the energy consumption.

In situation when we want to consciously resign from bidirectional communication, it means, that only one device can send data. In some cases, it is uncertain, whether the sent data by the transmitter will be received by the receiver. In case, that two transmitters sent data to the receiver at the same time, no information will be received. The frames will overlap each other and will be misread. Data will be rejected by the receiver due to incorrect CRC or other control checking the frame's correctness. Information will be lost.

In order to minimize the possibility of data loss, the authors propose a protocol based on the CSMA/CA (Carrier-sense multiple access with collision avoidance)

© Springer Nature Switzerland AG 2020
R. Szewczyk et al. (Eds.): AUTOMATION 2019, AISC 920, pp. 199–207, 2020.
https://doi.org/10.1007/978-3-030-13273-6_20

mechanism. This approach provides greater efficiency, in the case of one directional communication of multiple transmitting devices with one receiving device (star topology) with relatively low energy consumption by transmitting device.

1.1 State of the Art

CSMA/CA mechanism is used in LoRa (Long Range), which is digital wireless data communication technology. LoRa is widely used in IoT appliances. In [4] authors presented investigating and experimenting using CSMA mechanism for LoRa technology. Authors explained the various approaches in adapting CSMA mechanism for LoRa networks. Authors showed experimental results using low cost IoT LoRa framework to implement innovative long-range image sensor nodes. In the paper [5] authors showed how a CSMA mechanism can be adapted for LoRa networks to decrease collisions with variety size of message.

CSMA/CA mechanism is used successfully in IEEE 802 – family standards dealing with local area networks (LAN) and metropolitan area networks (MAN) [1–3].

In the paper [6] authors presented the performance of the state-of-the-art wireless sensor networks medium access control (MAC) protocols such as IEEE 802.15.4 (ZigBee), IEEE 802.11 (WLAN) using CSMA mechanism. They present advantages and disadvantages of the protocol in harsh smart grid spectrum environments.

A hybrid MAC protocol, called Z-MAC has been introduced in [7]. Author present design, implementation and performance evaluation. Z-MAC is protocol for wireless sensor networks, that using advantages of CSMA and TDMA while decreasing their weaknesses, reduces collision among two-hop neighbors at a low cost.

Experimental results using nRF24 transceiver in transmitting data with different data rate has been presented in [8]. Authors used CSMA algorithm and analyzed the packet loss and lifetime estimation.

In the paper [9] authors proposed an adaptive MAC protocol based on IEEE 802.15.4, named Ada-MAC. It is protocol address to health-care services and medical cyber-physical systems. The proposed protocol combines schedule-based on time-triggered protocol and CSMA/CA mechanism.

Correct communication between wireless low power devices is important, because they are used in many areas such as agriculture [10] or home automation [11, 12].

2 Test Stand

2.1 Hardware

The proposed protocol can be used in the network consisting sensors and receivers. The authors have simulated this environment by creating a test stand consist of 10 transmitters and 1 receiver, which collects all data. As a transmitter authors used Nucleo board with microcontroller STM32L152 [13] and eval board STEVAL-FKI868V1 [14], what has been shown in Fig. 2. RF transceiver is S2-LP [15]. On the PCB there are jumpers used to measure current.

As receiver authors used USB dongle STEVAL-IDS001V4 [16] with STM32L151 and SPIRIT1 [17] as RF transceiver, what has been shown in Fig. 1.

Fig. 1. Receiver based on SPIRIT1.   **Fig. 2.** Transmitter based on S2-LP and NUCLEO board.

2.2 Connection

The receiver is connected to a PC via USB. Receiver receives all transmitted data from the transmitters via radio. An application was designed to receive data on the notebook,

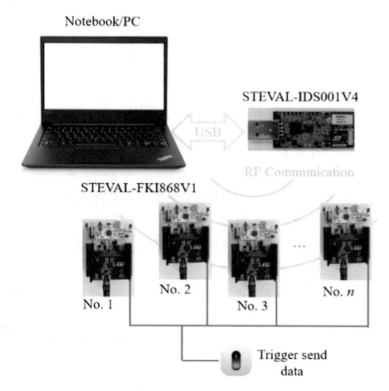

Fig. 3. Test stand connection.

which displays whether data from all transmitters have been received by the receiver. The connection has been shown in Fig. 3. Sending is triggered with a relay, after closing the relay contacts, transmitters sent a frame of data to the receiver at the same time. The distance between the transmitters and the receiver is about 2 m.

2.3 Low Power Strategy

A typical approach in the design of low power devices operation has been shown in Fig. 4 [18, 19]. The device starts work from the initialization of the entire system. Then the device goes into sleep mode - inactive, in which device consumes very little energy, compared to the run mode – active. To the active mode, the device is awoken by a signal (IRQ), it can be a push of a button or RTC interrupt. The time, when the device is in inactive mode is t_{INA}, while the time in which the device is in active mode is t_A. In order to minimize the energy consumption, the current consumption in the active and inactive mode and the time, when the device is in active mode are minimized.

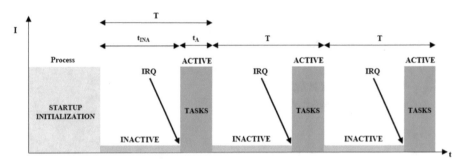

Fig. 4. Typical work timeline of low power embedded devices.

3   Wireless Protocol

Authors propose algorithm of communication for low power embedded devices, communicating in ISM band using CSMA/CA mechanism. All transmission between transmitter and receiver is crypted. The transmission time t_S of 20 Bytes of data is equal 5.7 ms (without CSMA mechanism). Encrypting time t_{CRYP} of the data is 10 ms. That gives total time in active state t_A 15.7 ms, where $t_A = t_S + t_{CRYP}$.

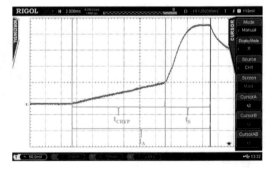

Fig. 5. Voltage proportional to current consumption in encrypting state (t_{CRYP}) and transmission state (t_S).

Figure 5 shows voltage proportional to current consumption in particular states, transmission and encrypting. Voltage is smooth, because of filtering capacitors. There is a big difference between these two states in current consumption. For this reason, less time in transmission state, gives less current consumption of whole devices.

3.1 Transmission Parameters

RF transmission parameters are:

- Frequency base: 868.15 MHz,
- Data rate: 100 ksps,
- Modulation: 2-(G)FSK BT = 0.5,
- Output power: 10 dBm,
- Channel filter: 200 kHz.

Packet parameters are:

- Preamble length: 8 Bytes,
- Sync word length: 4 Bytes,
- Data length: 20 Bytes,
- CRC Polynomial: 0x8005.

Current consumption of RF transceiver S2-LP in TX state is 13 mA, in sleep state 500 nA. Current consumption of microcontroller STM32L152 in sleep state is 1,4 uA and in active state is 5 mA. The measurements were made with a precision current adapter.

3.2 CSMA/CA Mechanism

The CSMA/CA mechanism is a channel access mechanism based on the rule of sensing the channel before transmitting. This avoids the simultaneous use of the channel by different transmitters and increases the probability of correct reception of data being transmitted. Mechanism works based on a comparison of the RSSI sensed with the threshold.

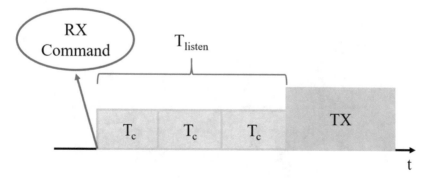

Fig. 6. CSMA mechanism if channel is free.

If the channel is busy, a back-off procedure is activated to repeat the process of listening a certain number of times, until the channel is found to be free. When the limit is reached, an interrupt notifies that the channel has been found busy. While in back-off procedure, the RF transceiver S2-LP stays in sleep state in order to reduce current consumption.

If the channel is free, the device must assert channel free for a certain number of T_C periods (T_{listen}) before transmitting. T_C is a time of one period of listening. Total time of listening before transmitting is T_{listen}. Timeline has been shown in Fig. 6.

If channel becomes free, transmitter senses whether the channel becomes free. After every period T_C of listening, transmitter stays in sleep. To avoid any wait synchronization between different channel contenders, which may cause successive failing listening in T_C, the back-off wait time (T_{BO1}, T_{BO2}, T_{BO3}) is calculated randomly. During this time, the S2-LP is kept in the sleep state. Timeline has been shown in Fig. 7.

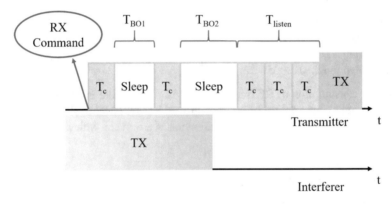

Fig. 7. CSMA mechanism if channel becomes free.

If channel is still busy after reached maximum number of transmitting attempts (*NBOM – NumberBackOffMax*) interrupt is generated. This case has been shown in Fig. 8.

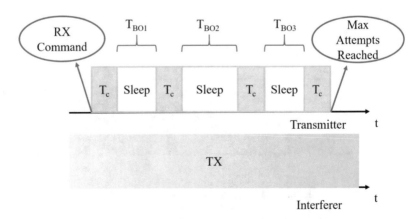

Fig. 8. CSMA mechanism if channel is busy.

4   Experimental Investigation

The test was carried out for 10 transmitters and 1 receiver. The transmitters remained in inactive mode and when trigger occurred, the transmitters were woken up and sent data. The relay was controlled from a separate microcontroller and relay contacts was closed every 2 s for 60 s. During the test, the current consumption by the transmitters was measured. The receiver sent data to a PC computer, where there was information about how many data frames should come, and how many came in reality. In time of 60 s every transmitter should send 30 frames, which gives 300 frames for 10 transmitters.

The CSMA parameters that were changed:

- T_C – the time of one period of listening whether the channel is busy,
- $NBOM$ – maximum number of back-off cycles,
- T_{listen} – total time of listening before transmitting when channel is sensed to be free.

RSSI threshold was set to 92 dB.

Table 1. Results of experiment $T_{listen} = T_C$, $NBOM = 2$.

T_C [ms]	$NBOM$	T_{listen} [ms]	Current [µA]	% received frames
0.8	2	0.8	90	28%
1.6	2	1.6	95	30%
3.2	2	3.2	106	32%
6.4	2	6.4	128	55%

Table 2. Results of experiment $T_{listen} = T_C$, $NBOM = 7$.

T_C [ms]	$NBOM$	T_{listen} [ms]	Current [µA]	% received frames
0.8	7	0.8	94	77%
1.6	7	1.6	100	80%
3.2	7	3.2	111	83%
6.4	7	6.4	135	91%

Table 3. Results of experiment $T_{listen} = 2 \cdot T_C$, $NBOM = 2$.

T_C [ms]	$NBOM$	T_{listen} [ms]	Current [µA]	% received frames
0.8	2	1.6	92	46%
1.6	2	3.2	100	48%
3.2	2	6.4	118	48%
6.4	2	12.8	151	57%

Table 4. Results of experiment $T_{listen} = 2 \cdot T_C$, $NBOM = 7$.

T_C [ms]	$NBOM$	T_{listen} [ms]	Current [µA]	% received frames
0.8	7	1.6	98	81%
1.6	7	3.2	104	82%
3.2	7	6.4	122	85%
6.4	7	12.8	156	89%

In Tables 1 and 3, the percentage of received frames by the receiver is not acceptable. For the number of sensors under experiment, single time of listening (T_C), the number of attempts (*NBOM*) and the time in which the device was in the sleep state were not sufficient, due to the time it takes for the transmitter to send a frame of data.

For *NBOM* = 7 (Tables 2 and 4) statistically 1 frame of all sent was not received. Increasing the *NBOM* causes increase of the current consumption of the device, however, the greatest impact on current consumption has T_{listen} – the time, when the device checks whether the channel is free before sending data.

A better approach is to increase *NBOM* instead of T_{listen}. The device listens for a certain period of time whether the channel is free, then goes into sleep mode for a random time. This time increases with each next attempt. Finally, the current consumption does not increase as much compared to the increase T_{listen}.

The average current is very high due to the frequency of data sent. If the data were sent every 30 s, not 2 s the average current would be around 14 uA with *NBOM* = 7, T_C = 6.4 ms and T_{listen} = 2·T_C. For battery 1400 mAh time life of the device will be about 12 years. It is theoretical divagation; self-discharge of the battery is not included. Most producer of the batteries gives 10 years warranty.

5   Summary

Investigation shows, that with increasing *NBOM* parameter increases efficiency of communication. Efficiency of communication increased by about 40%, comparing *NBOM* = 2 and *NBOM* = 7. For *NBOM* = 0, CSMA/CA is disabled, so with CSMA/CA mechanism, efficiency of communication is almost twice larger. The solution proposed by the authors increased the percentage of received frames almost 2 times with properly selected CSMA parameters with acceptable power consumption.

In the paper [8] authors performed research for 250 kbps, 1 Mbps and 2 Mbps. The number of devices varies from 10 to 80. Data length was 32 Bytes. Simulation was launched for 2 s, data was sent with 100 ms intervals. For similar CSMA parameters the authors obtained the result of over 90% of received frames. But devices begin to sense the first data packet at a random instant between the beginning of the simulation and 100 ms, so it cannot unambiguously compare the results with this paper.

Application of CSMA mechanism for low power embedded devices, increases probability of correct reception, in the case of one-way communication at the expense of a slight increase in the current consumption of the device. A compromise must be found, between current consumption and communication reliability.

Acknowledgement. The work described in this paper was funded from 02/22/DSPB/1434.

References

1. Lou, H., et al.: Multi-user parallel channel access for high efficiency carrier grade wireless LANs. In: IEEE International Conference on Communications (ICC), pp. 3868–3870. IEEE (2014)
2. Wu, S., Mao, W., Wang, X.: Performance study on a CSMA/CA-Based MAC protocol for multi-user MIMO wireless LANs. IEEE Trans. Wirel. Commun. **13**, 3153–3166 (2014)
3. Doost-Mohammady, R., Naderi, M.Y., Chowdhury, K.R.: Performance analysis of CSMA/CA based medium access in full duplex wireless communications. IEEE Trans. Mob. Comput. **15**, 1457–1470 (2016)
4. Pham, C.: Investigating and experimenting CSMA channel access mechanisms for LoRa IoT networks, pp. 1–6. IEEE (2018)
5. Pham, C.: Robust CSMA for long-range LoRa transmissions with image sensing devices, pp. 116–222. IEEE (2018)
6. Yigit, M., Yoney, E.A., Gungor, V.C.: Performance of MAC protocols for wireless sensor networks in harsh smart grid environment, pp. 50–53. IEEE (2013)
7. Rhee, I., Warrier, A., Aia, M., Min, J.: Z-MAC: a hybrid MAC for wireless sensor networks, pp. 90–101. IEEE (2008)
8. Zhu, N., O'Connor, I.: Performance evaluations of unslotted CSMA/CA algorithm at high data rate WSNs scenario, pp. 406–411. IEEE (2013)
9. Xia, F., Wang, L., Zhang, D., He, D., Kong, X.: An adaptive MAC protocol for real-time and reliable communications in medical cyber-physical systems. Telecommun. Syst. **58**, 125–138 (2015)
10. Aquino-Santos, R., González-Potes, A., Edwards-Block, A., Virgen-Ortiz, R.A.: Developing a new wireless sensor network platform and its application in precision agriculture. Sensors **11**, 1192–1211 (2011)
11. Tudose, D.S., et al.: Home automation design using 6LoWPAN wireless sensor networks, pp. 1–6. IEEE (2011)
12. Mukhopadhyay, S.C., Suryadevara, N.K.: Internet of Things: Challenges and Opportunities. Internet of Things, pp. 1–17 (2014)
13. NUCLEO-L152RE - STM32 Nucleo-64 development board with STM32L152RE MCU, supports Arduino and ST morpho connectivity – STMicroelectronics. https://www.st.com/en/evaluation-tools/nucleo-l152re.html. Accessed 13 July 2018
14. STEVAL-FKI868V1 - Sub-1 GHz transceiver development kit based on S2-LP – STMicroelectronics. https://www.st.com/content/st_com/en/products/evaluation-tools/solution-evaluation-tools/communication-and-connectivity-solution-eval-boards/steval-fki868v1.html. Accessed 13 July 2018
15. S2-LP - Ultra-low power, high performance, sub-1 GHz transceiver – STMicroelectronics. https://www.st.com/en/wireless-connectivity/s2-lp.html. Accessed 13 July 2018
16. STEVAL-IDS001V4 - SPIRIT1 - Low Data Rate Transceiver - 868 MHz - USB dongle – STMicroelectronics. https://www.st.com/en/evaluation-tools/steval-ids001v4.html. Accessed 13 July 2018
17. SPIRIT1 - Low data rate, low power Sub 1 GHz transceiver - STMicroelectronics'. https://www.st.com/en/wireless-connectivity/spirit1.html. Accessed 13 July 2018
18. Rhee, S., Seetharam, D., Liu, S.: Techniques for minimizing power consumption in low data-rate wireless sensor networks, pp. 1727–1731. IEEE (2004)
19. Heinzelman, W.R., Chandrakasan, A., Balakrishnan, H.: Energy-efficient communication protocol for wireless microsensor networks. In: Proceedings of the 33rd Annual Hawaii International Conference on System Sciences, vol. 2, 10 pp. (2000)

Minimum Energy Control and Reachability of Continuous-Time Linear Systems with Rectangular Inputs

Krzysztof Rogowski$^{(\boxtimes)}$

Faculty of Electrical Engineering, Bialystok University of Technology, Wiejska 45D, 15-351 Bialystok, Poland
k.rogowski@pb.edu.pl

Abstract. In the paper a reachability property of continuous-time linear systems with rectangular type inputs vector is addressed. Necessary and sufficient conditions for the existence of that type of input signals that steer the system from zero initial conditions to desired final state in assumed time are derived and proved. The computation method of the input signals vector set that are the solution to the problem is presented. Next, a minimum energy control problem is considered. Using an integral control performance index the minimum energy input vector is chosen from the set of consistent inputs. The considerations are illustrated by a numerical example of electrical circuit with pulse wave voltage sources, where the input signals that minimise assumed performance index are computed and applied to steer the electrical circuit to desired final state in given time.

Keywords: Reachability · Minimum energy control · Electrical circuit control · Rectangular inputs

1 Introduction

Reachability is a fundamental property of linear systems in the area of control systems theory [1–4]. This notion describes a structural property of dynamic systems, especially the ability to achieve a certain state of the system in a given time. It appears in many different areas of science and applications, such as control systems analysis, graph theory, Petri nets, game theory, decision procedures, electrical circuits design, etc.

The reachability problem has been solved for many different classes of dynamic systems [1–4]. In [5] the reachability of interval max-plus systems has been considered. Local positive reachability of positive nonlinear systems has been addressed in [6]. In [7] the reachability notion is used to analyze the impact of uncertain pulsed loads on shipboard power systems. Reachability of continuous-time systems and its application to electrical circuits with constant inputs have been addressed in [8] and with piecewise constant inputs in [9].

© Springer Nature Switzerland AG 2020
R. Szewczyk et al. (Eds.): AUTOMATION 2019, AISC 920, pp. 208–217, 2020.
https://doi.org/10.1007/978-3-030-13273-6_21

The optimal control of dynamic systems is another important problem in control theory [10]. For reachable dynamic system there exist an infinite number of inputs that steers this system to the desired final state in a given interval of time. The minimum energy control assumes that we are able to choose such an input vector that some performance index takes its minimal possible value. The most popular type of performance indexes is an integral of inputs and/or state variables of the dynamic system with some weighting coefficients. It takes only non-negative values, therefore in minimum energy control we are looking for such an input signals for which the value of this performance index is the smallest possible. Minimum energy control problem for different classes of dynamic systems has been investigated in [11–15]. In [14,16] the reachability and minimum energy control problem for continuous-time fractional order systems with different orders have been solved.

In this paper the reachability and minimum energy control of continuous-time linear dynamic systems with input signals in the form of rectangular (pulse wave) type functions will be considered and solved. To the best knowledge of the author such aspects of control systems in continuous-time domain have not been analysed yet.

The following notation will be used. $\mathbb{R}$ will be the set of real numbers, $\mathbb{R}^{n \times m}$ – the set of $n \times m$ matrices with real entries. The identity matrix of rank n will be denoted by $\mathbb{I}_n$ and by $0_{n \times m}$ – $n \times m$ matrix with all zero elements. Symbol T will denotes the transpose of a matrix or vector.

2   Problem Formulation

In this paper we will consider the standard continuous-time system described by the following state-space equation [2,4]

$$\frac{\mathrm{d}x(t)}{\mathrm{d}t} = Ax(t) + Bu(t) \tag{1}$$

where $x(t) \in \mathbb{R}^n$ is the state vector, $u(t) \in \mathbb{R}^m$ is the input vector of the dynamic system for $t \geq 0$ and the matrices with real elements $A \in \mathbb{R}^{n \times n}$, $B \in \mathbb{R}^{n \times m}$ depending on a structure and parameters of considered physical phenomena.

The solution to the state-space equation (1) with initial state $x_0 = x(0) \in \mathbb{R}^n$ for $t \geq 0$ and arbitrary inputs $u(t) \in \mathbb{R}^m$ is given by [2,4]

$$x(t) = e^{At}x_0 + \int_0^t e^{A\tau}Bu(t - \tau)\mathrm{d}\tau. \tag{2}$$

In the further part of this paper we will consider system (1) with rectangular type input signals (known also as pulse wave signals) of the form

$$u(t) = \begin{cases} U & \text{for } (i-1)T \geq t \geq (i-1)T + \Delta T \\ 0 & \text{for } (i-1)T + \Delta T > t > iT \end{cases} \tag{3}$$

where $i = 1, 2, \ldots$ and $U \in \mathbb{R}^m$ is the vector of m constant inputs magnitudes, T is the period of the periodic signal and ΔT is the pulse width.

In this research we will look for an input signals vector of the type defined by (3) that steers the dynamic system (1) in time $t_f = qT$ from zero initial state $x(0) = x_0 = 0$ to the desired arbitrary given final state $x_f = x(t_f) \in \mathbb{R}^n$ and minimises an integral performance index

$$I(u(t)) = \int_0^{t_f} u^T(\tau) Q u(\tau) \mathrm{d}\tau, \tag{4}$$

where $Q \in \mathbb{R}^{m \times m}$ is a symmetric positive definite matrix, i.e. $Q = Q^T$ and $v^T Q v > 0$ for every nonzero column vector $v \in \mathbb{R}^m$.

The main purpose of considerations conducted in this paper is to determine whether such vector of input signals (3) exists and minimises the performance index (4) for given matrices A, B of dynamical system and time period T, pulse width ΔT of the input signal and final time t_f.

3   Reachability of Continuous-Time Systems with Rectangular Inputs

To find an minimum energy control of dynamic system we shall check whether such input, which minimises performance index (4), exists. For this purpose we will define a notion of reachability of continuous-time systems with rectangular inputs.

Definition 1. *The continuous-time system (1) is called reachable for rectangular inputs if there exists such input vector (3) that steers the system from zero initial condition $x_0 = x(0)$ to arbitrary final state $x_f = x(t_f) \in \mathbb{R}^n$.*

To establish necessary and sufficient conditions of reachability we use solution (2) for zero initial condition and input signal (3) and we obtain

$$x_f = x(t_f) = \int_0^{t_f} e^{A\tau} B u(t_f - \tau) \mathrm{d}\tau$$

$$= \int_{T-\Delta T}^{T} e^{A\tau} B U \mathrm{d}\tau + \int_{2T-\Delta T}^{2T} e^{A\tau} B U \mathrm{d}\tau + \cdots + \int_{kT-\Delta T}^{kT} e^{A\tau} B U \mathrm{d}\tau. \tag{5}$$

Taking into account that

$$A \int_{t_1}^{t_2} e^{A\tau} \mathrm{d}\tau = A \sum_{k=0}^{\infty} \frac{A^k}{k!} \int_{t_1}^{t_2} \tau^k \mathrm{d}\tau = A \sum_{k=0}^{\infty} \frac{A^k \left(t_2^{k+1} - t_1^{k+1} \right)}{(k+1)!} = e^{At_2} - e^{At_1} \tag{6}$$

and premultiplying (5) by the matrix A we obtain

$$
\begin{aligned}
Ax_f &= A\left[\int_{T-\Delta T}^{T} e^{A\tau}d\tau + \int_{2T-\Delta T}^{2T} e^{A\tau}d\tau + \cdots + \int_{kT-\Delta T}^{kT} e^{A\tau}d\tau\right] BU \\
&= \left[e^{AT} - e^{A(T-\Delta T)} + e^{2AT} - e^{A(2T-\Delta T)} + \cdots + e^{AkT} - e^{A(kT-\Delta T)}\right] BU \\
&= \left[\mathbb{I}_n - e^{-A\Delta T}\right]\left[e^{AT} + e^{2AT} + \cdots + e^{AqT}\right] BU.
\end{aligned}
\tag{7}
$$

Premultiplying both sides of (7) by $\left[\mathbb{I}_n - e^{-A\Delta T}\right]^{-1}$ we have

$$
\begin{aligned}
\left[\mathbb{I}_n - e^{-A\Delta T}\right]^{-1} Ax_f &= \left[e^{AT} + \left(e^{AT}\right)^2 + \cdots + \left(e^{AT}\right)^q\right] BU \\
&= \left[\sum_{k=1}^{q} \left(e^{AT}\right)^k\right] BU.
\end{aligned}
\tag{8}
$$

Therefore

$$
\left[\sum_{k=1}^{q} \left(e^{AT}\right)^k\right]^{-1} \left[\mathbb{I}_n - e^{-A\Delta T}\right]^{-1} Ax_f = BU.
\tag{9}
$$

Using Kronecker-Capelli theorem to (9) and taking into account that the matrix exponential function is nonsingular for any matrix power we may formulate the following theorem.

Theorem 1. *The continuous-time linear system (1) is reachable for rectangular inputs (3) if and only if the matrices A and B of the system have full row rank, i.e.*

$$
\text{rank} A = \text{rank} B = n.
\tag{10}
$$

Remark 1. The reachability of the continuous-time systems is independent of the input signal pulse width ΔT, its period T and the number of pulses q (time t_f) and is dependent on the rank of the system matrices A and B.

The input that steers the system (1) from zero initial condition $x(0) = x_0 = 0$ to the desired final state x_f in time $t_f = qT$ can be computed by the following formula

$$
U = B^{+}\left[\sum_{k=1}^{q} \left(e^{AT}\right)^k\right]^{-1} \left[\mathbb{I}_n - e^{-A\Delta T}\right]^{-1} Ax_f
\tag{11}
$$

where $B^{+} \in \mathbb{R}^{m\times n}$ is the right pseudoinverse of the rectangular matrix $B \in \mathbb{R}^{n\times m}$, which can be computed using [17,18]

$$
B^{+} = B^{T}\left[BB^{T}\right]^{-1} + \left(\mathbb{I}_m - B^{T}\left[BB^{T}\right]^{-1} B\right) K_1 \quad \text{for arbitrary} \quad K_1 \in \mathbb{R}^{m\times n}
\tag{12a}
$$

or

$$
B^{+} = K_2\left[BK_2\right]^{-1} \quad \text{for arbitrary} \quad K_2 \in \mathbb{R}^{m\times n}, \quad \det\left[BK_2\right] \neq 0.
\tag{12b}
$$

The formula (11) with pseudoinverses (12) defines the set of inputs that steers the system to the desired final state x_f. This set is dependent on matrices K_1 and K_2, which can be chosen arbitrary to compute the pseudoinverse matrix B^+. The problem of minimum energy control of the system (1) requires to find an optimum input vector from the whole set, i.e. we should find the values of matrices K_1 or K_2 in (12) such that the performance index (4) takes the minimal value.

4   Minimum Energy Control for Rectangular Inputs

To find an optimum (minimum energy) values of input vector we shall analyse the performance index (4). In the case of rectangular type of input signals (3) the performance index (4) takes the form

$$I(u(t)) = q\Delta T U^T Q U. \tag{13}$$

Let us denote

$$M = \left[\sum_{k=1}^{q} \left(e^{AT}\right)^k\right]^{-1} \left[\mathbb{I}_n - e^{-A\Delta T}\right]^{-1} A x_f \in \mathbb{R}^{n \times 1}. \tag{14}$$

The performance index (13) for (11) with the right pseudoinverse (12a) with arbitrary chosen matrix K_1 is given by

$$
\begin{aligned}
I(u(t)) = q\Delta T U^T Q U &= q\Delta T M^T \left[B^+\right]^T Q B^+ M \\
&= q\Delta T M^T \left\{\left[BB^T\right]^{-1} B + K_1^T \left(\mathbb{I}_m - B^T \left[BB^T\right]^{-1} B\right)\right\} Q \times \\
&\quad \times \left\{B^T \left[BB^T\right]^{-1} + \left(\mathbb{I}_m - B^T \left[BB^T\right]^{-1} B\right) K_1\right\} M.
\end{aligned}
\tag{15}
$$

Note, that for every positive definite matrix $Q \in \mathbb{R}^{m \times m}$ we have $v^T Q v > 0$ for any nonzero vector $v \in \mathbb{R}^m$. Taking this into account, in this case from (15) it follows that the performance index (13) with positive definite matrix Q takes its minimal value for $K_1 = 0_{m \times n}$. The minimal value of the performance index has the form

$$I_{opt} = I(u(t)) = q\Delta T M^T \left\{\left[BB^T\right]^{-1} B\right\} Q \left\{B^T \left[BB^T\right]^{-1}\right\} M. \tag{16}$$

Therefore the optimal value of input can be computed using (11) with the right pseudoinverse B^+ given by formula (12a) with $K_1 = 0_{m \times n}$

$$U_{opt} = B^T \left[BB^T\right]^{-1} \left[\sum_{k=1}^{q} \left(e^{AT}\right)^k\right]^{-1} \left[\mathbb{I}_n - e^{-A\Delta T}\right]^{-1} A x_f = B^T \left[BB^T\right]^{-1} M. \tag{17}$$

5   Numerical Analysis of Electrical Circuit

In this section we will show that the notions of reachability and minimum energy control for dynamic systems with rectangular inputs can be applied to a control process of the currents and voltages in electrical circuits. We will find the minimum energy control inputs (values of source voltages magnitudes) that steers the system to desired final values of the currents and voltages in this circuit in assumed time t_f.

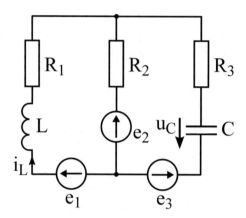

Fig. 1. Electrical circuit of Example 1.

Example 1. Let us consider the electrical circuit shown in Fig. 1 composed of resistances $R_1 = 0.1 \ \Omega$, $R_2 = 1 \ \Omega$, $R_3 = 2 \ \Omega$, inductance $L = 0.5$ H and capacitance $C = 0.1$ F and three voltage sources $e_1(t)$, $e_2(t)$, $e_3(t)$.

   Assuming that as the state variables we choose the current in the coil and the voltage across the capacitor, the state vector of electrical circuit shown in Fig. 1 will take the following form

$$x(t) = \begin{bmatrix} i_L(t) \\ u_C(t) \end{bmatrix} \tag{18}$$

and the voltages sources form the input vector

$$u(t) = \begin{bmatrix} e_1(t) \\ e_2(t) \\ e_3(t) \end{bmatrix} . \tag{19}$$

   Using Kirchhoff's laws we may formulate the state-space Eq. (1) describing the electrical circuit shown in Fig. 1

$$\frac{\mathrm{d}}{\mathrm{d}t} \begin{bmatrix} i_L(t) \\ u_C(t) \end{bmatrix} = A \begin{bmatrix} i_L(t) \\ u_C(t) \end{bmatrix} + B \begin{bmatrix} e_1(t) \\ e_2(t) \\ e_3(t) \end{bmatrix} , \tag{20}$$

where

$$A = \begin{bmatrix} -\dfrac{R_1}{L} - \dfrac{R_2 R_3}{L(R_2 + R_3)} & \dfrac{R_2}{L(R_2 + R_3)} \\ -\dfrac{R_2}{C(R_2 + R_3)} & -\dfrac{1}{C(R_2 + R_3)} \end{bmatrix} = \begin{bmatrix} -1.53 & 0.67 \\ -3.33 & -3.33 \end{bmatrix}, \quad (21a)$$

$$B = \begin{bmatrix} \dfrac{1}{L} - \dfrac{R_3}{L(R_2 + R_3)} & -\dfrac{R_2}{L(R_2 + R_3)} \\ 0 & -\dfrac{1}{C(R_2 + R_3)} & \dfrac{1}{C(R_2 + R_3)} \end{bmatrix} = \begin{bmatrix} 2 & -1.33 & -0.67 \\ 0 & -3.33 & 3.33 \end{bmatrix}. \quad (21b)$$

Let us assume that we want to steer our circuit from zero initial conditions $\begin{bmatrix} i_L(0) & u_C(0) \end{bmatrix}^T = \begin{bmatrix} 0 \text{ A} & 0 \text{ V} \end{bmatrix}^T$ in time $t_f = 1.25$ s to the desired final state

$$x_f = \begin{bmatrix} i_L(t_f) \\ u_C(t_f) \end{bmatrix} = \begin{bmatrix} 0.5 \text{ A} \\ 2 \text{ V} \end{bmatrix} \quad (22)$$

with rectangular input which pulse has the width equal $\Delta T = 0.15$ s and the period $T = 0.25$ s.

First we check whether electrical circuit under our consideration is reachable for rectangular input using the condition formulated in Theorem 1. We have

$$\text{rank} A = \text{rank} B = 2 = n \quad (23)$$

and the condition is satisfied. Therefore the problem of control from zero initial conditions $x_0 = 0$ to desired final state x_f in electrical circuit under consideration has a solution.

Now let us compute the optimal (minimum energy) input vector that steers the system to the desired final state x_f and minimise the performance index (13) with positive definite matrix $Q = \mathbb{I}_m$. Using (17) for matrices (21a, 21b) we obtain

$$U_{opt} = B^T \begin{bmatrix} BB^T \end{bmatrix}^{-1} M = \begin{bmatrix} -0.95 \\ -1.93 \\ 2.88 \end{bmatrix}. \quad (24)$$

The minimal value of the performance index for optimal rectangular input signal vector can be computed using (16)

$$I_{opt} = 9.69. \quad (25)$$

The waveforms of current in the coil i_L and voltage across the capacitor u_C of the electrical circuit shown in Fig. 1 for rectangular input with $\Delta T = 0.15$ s period $T = 0.25$ s and input voltages magnitudes vector (24) are shown in Fig. 2.

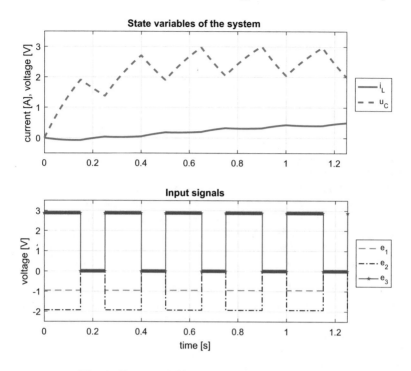

Fig. 2. State variables and inputs of Example 1.

6   Concluding Remarks

The reachability problem for continuous-time systems described by the state-space equations with rectangular type inputs has been considered. Necessary and sufficient conditions for the solution to this problem have been established and proved (Theorem 1). It has been shown that this type of reachability is independent of the number of pulses and the period (frequency) of the input signal (Remark 1).

Finally the minimum energy problem for such a system has been considered and solved. It has been shown that using the right pseudoinverse of the rectangular matrix the minimal value of the magnitudes of the input signal can be found. The reachability conditions of continuous-time systems with rectangular type inputs are much more restrictive than the conditions for reachability of systems with arbitrary shape of input signals (see [2,4]).

An open problem is a generalisation of considerations proposed in this paper for the case of systems controlled by inputs of the pulse-width modulation type. In such a problem we will have one more degree of freedom in choosing of the input signal shape and parameters. Such problem is more complex and will be the subject of subsequent research article of the author.

Acknowledgements. This work was supported by National Science Centre in Poland under work No. 2017/27/B/ST7/02443.

References

1. Antsaklis, P., Michel, A.N.: Linear Systems. Birkhäuser, Boston (2006)
2. Kaczorek, T.: Linear Control Systems: Analysis of Multivariable Systems. Research Studies Press, New York (1992)
3. Kaczorek, T.: Positive 1D and 2D systems. In: Communication and Control Engineering. Springer, London (2002)
4. Kailath, T.: Linear Systems. Prentice-Hall, New York (1980)
5. Wang, C., Tao, Y., Yang, P.: Reachability of interval max-plus linear systems. In: Proceedings of the 36th Chinese Control Conference, Dalian, China, 26–28 July 2017, pp. 2392–2396 (2017). https://doi.org/10.23919/ChiCC.2017.8027716
6. Bartosiewicz, Z.: Local positive reachability of nonlinear continuous-time systems. IEEE Trans. Autom. Control **12**, 4217–4221 (2016). https://doi.org/10.1109/TAC.2015.2511921
7. Villegas Pico H.N., Aliprantis D.C., Sudhoff S.D.: Reachability analysis of shipboard power systems with uncertain pulsed loads. In: IEEE Electric Ship Technologies Symposium (ESTS), Old Town Alexandria, 21–24 June 2015, pp. 395–402 (2015). https://doi.org/10.1109/ESTS.2015.7157925
8. Rogowski, K.: Reachability of standard and fractional continuous-time systems with constant inputs. Arch. Control Sci. **2**, 147–159 (2016). https://doi.org/10.1515/acsc-2016-0008
9. Rogowski, K.: Reachability of standard and fractional continuous-time systems with piecewise constant inputs. In: Szewczyk, R., Zielinski, C., Kaliczynska, M. (eds.) Challenges in Automation, Robotics and Measurement Techniques, Proceedings of Automation–2016, Warsaw, Poland, 2–4 March 2016, vol. 440, pp. 241–249. Springer, Cham (2016). https://doi.org/10.1007/978-3-319-29357-8_31
10. Lewis, F., Vrabie, D., Syrmos, V.: Optimal Control, 3rd edn. Wiley, New York (2012)
11. Kaczorek, T.: An extension of Klamka's method of minimum energy control to fractional positive discrete-time linear systems with bounded inputs. Bull. Pol. Acad. Sci. Tech **2**, 227–231 (2014). https://doi.org/10.2478/bpasts-2014-0022
12. Kang, J., Tang, W.: Minimum energy control of 2D singular systems with constrained controls. In: International Conference on Machine Learning and Cybernetics, Guangzhou, China, New York, 18–21 August 2005, pp. 1244–1248 (2005). https://doi.org/10.1109/ICMLC.2005.1527134
13. Klamka, J.: Controllability and minimum energy control problem of fractional discrete-time systems. In: New Trends in Nanotechnology and Fractional Calculus Applications, pp. 503–509. Springer (2010)
14. Sajewski, L.: Reachability, observability and minimum energy control of fractional positive continuous-time linear systems with two different fractional orders. Multidimension. Syst. Signal Process. **1** 27–41 (2016). https://doi.org/10.1007/s11045-014-0287-2
15. Schwartz, C.A., Maben, E.: A minimum energy approach to switching control for mechanical systems. In: Stephen Morse, A. (ed.) Control Using Logic-Based Switching. LNCIS. Springer, Germany (1997)

16. Sajewski, L.: Reachability of fractional positive continuous-time linear systems with two different fractional orders. In: Szewczyk, R., Zielinski, C., Kaliczynska, M. (ed.) Recent Advances in Automation, Robotics and Measuring Techniques: Proceedings of Automation–2014, Warsaw, Poland, 26–28 March 2014, vol. 267, pp. 239–249. Springer, Cham (2014). https://doi.org/10.1007/978-3-319-05353-0_24
17. Gantmacher, F.R.: The Theory of Matrices. vol. I and II (translated by Hirsch KA). Chelsea Publishing Co., New York (1959)
18. Kaczorek, T.: Vectors and Matrices in Automation and Electrotechnics. WNT, Warsaw (1998). (in Polish)

Battery Voltage Estimation Using NARX Recurrent Neural Network Model

Adrian Chmielewski[1]([✉]) [ID], Jakub Możaryn[2] [ID], Piotr Piórkowski[3] [ID], and Krzysztof Bogdziński[1]

[1] Institute of Vehicles, Warsaw University of Technology,
Narbutta 84 Street, Warsaw, Poland
{a.chmielewski,k.bogdzinski}@mechatronika.net.pl
[2] Institute of Automatic Control and Robotics, Warsaw University
of Technology, Sw. A. Boboli 8, Warsaw, Poland
J.Mozaryn@mchtr.pw.edu.pl
[3] Institute of Construction Machinery Engineering, Warsaw University
of Technology, Narbutta 84 Street, Warsaw, Poland
piotr.piorkowski@simr.pw.edu.pl

Abstract. This work presents a prediction of battery terminal voltage in subsequent charging/discharging cycles. To estimate chosen signals the NARX (AutoRegressive with eXogenous input) model based on Recurrent Neural Network has been employed. A training and testing data were gathered at the laboratory test stand with the Lithium Iron Phosphate (LiFePO$_4$) battery in different working conditions. Test stand research was conducted for 40 charging/discharging cycles. Furthermore, the paper presents the results of the identification of double RC model parameters for a specified state of charge level. As a result, the analysis of the proposed methodology has been discussed.

Keywords: Artificial neural network · LiFePO4 battery ·
Experimental research · Recurrent neural networks

1 Introduction

Currently, there are being prepared the European Union climate and energy policies for years 2030 and 2050 [1]. In 2030 perspective [1] new goals are set, including improvement in energetic efficiency by 27%, the increase of RES share to 27% of the energy used in the EU and decrease of greenhouse gasses emissions by nearly 40%. The proposed goals will have a great impact on technology development, particularly distributed energy generation devices [1], which will have an impact on the competitiveness of the European Union economy. One of the crucial strategies to improve energy efficiency is the advancement of energy storage technology. In scientific literature and in industry various methods of energy storage are distinguished [2] based on the type of energy stored, e.g.: in reversible 2nd type cells (electrochemical) [2–4], flywheels (mechanical), supercapacitors (electrical) [3, 4], magnetic superconductive coils (electrical) [2], pumped storage systems (mechanical) or in underground rock

© Springer Nature Switzerland AG 2020
R. Szewczyk et al. (Eds.): AUTOMATION 2019, AISC 920, pp. 218–231, 2020.
https://doi.org/10.1007/978-3-030-13273-6_22

cavities (mechanical), where stored air can be used again for powering machines, or in heat storages (heat energy) [2].

In future, simultaneously to the liberalization of the energy market and prosumer support [1], a rise in interest in electric vehicles is expected [3, 5–10], which will become an integral part of distributed energy generation devices [1, 11]. Currently, in Polish law for example, there are no regulations governing the massive introduction of such vehicles into use. Restrictions for massive introduction are related to vehicle construction and external factors. Specific barriers to overcome are limited generation and transmission capabilities of the Polish energy industry [1, 3]. Increased peak power use, caused by ungoverned charging of electric vehicles may lead to overloading of power plants and the power network and, as consequence, cause so-called "black-outs". These assumptions are supported by, for example, simulations done for the city of Los Angeles [6].

From this point of view, the battery control and the prediction of its parameters are necessary. To improve algorithms for prediction and control in Battery Management Systems (BMS) for distributed systems and electric vehicles, it is crucial to know the operating characteristics of typical electrochemical batteries [12–20], which can be supported by mathematical models [3, 12–37]. This approach plays the essential role in hybrid systems [11]. Research of a dynamic operation of an electrochemical battery in a specific load cycle supported by a mathematical model e.g. energy balance model, neural network model [5, 9, 11, 12, 15, 20], ARMA model [5, 11], fuzzy model [5, 11], can give answers to different questions regarding the operation of batteries [2, 3, 10, 12, 34, 36, 37] in technical facilities. It should be emphasized, that the energy storage modeling gets recently an increasing interest connected with new developments in the field of electric vehicles [3, 5–9] or polygeneration systems [2, 11].

Charging/discharging dynamics of an energy storage system are highly nonlinear, especially during a discharging phase. Therefore, to simplify the identification procedure the use of artificial neural networks (ANN) is proposed [5, 9, 12, 15, 20, 38, 39]. They have some important advantages over standard linear identification methods i.e. an approximation of multivariable nonlinear functions [38], an easy adaptation of model parameters and a rapid calculation of model equations [39, 40]. In contrary to energy–balance analytical models, designing ANN does not require an exact knowledge of model physical equations, and physical parameters that describe the model, but only values of model variables (cause-effect relationships).

In the energy storage systems usefulness of ANNs for identification were proposed mostly in the form of feed-forward models [15, 39–41] and recurrent models [12, 15, 21, 38, 42]. In particular, interesting for our research are Recurrent Artificial Neural Networks (R-ANNs), where connections between units form a directed cycle. Such networks exhibit dynamic temporal behavior that can be useful for the prediction of the chosen signals, and indirectly the State of Charge (SOC) of the non-stationary energy storage systems in subsequent charge/discharge cycles. Previous research [12] show, that with properly chosen and pre-processed data, system parameters and operating conditions, recurrent ANN model can be used in prediction to some extent an energy storage performance under a wide variety of operating conditions. Considered in this paper, NARX (Nonlinear AutoRegressive with eXogenous input) model based on R-

ANN plays a significant role because it can be used as a predictor, nonlinear filter and as a nonlinear model of the dynamical system.

The article is organized as follows. In Sect. 2 a test stand of the LiFePO4 battery is described and presented. Section 3 presents an analytical model and an identification results of the LiFePO4 battery for SOC = 0.25 based on the experimental research. In Sect. 4 NARX model of the LiFePO4 battery is presented. In Sect. 5 different NARX models are subsequently validated and compared to experimental research results. In Sect. 6 concluding remarks are given.

2 Laboratory Test Stand Description

The LiFePO4 battery was evaluated on the laboratory test stand, where the measurement chain consists of the following components: PC class computer with LabVIEW software, application for registration of selected experimental data by using the National Instruments hardware e.g.: NI9206 card – for registration of voltage at the battery terminal, NI 9213 card – for measurement of battery body, battery terminal and ambient temperatures with use of type K thermocouples; NI 9263 analog output card for controlling a regulated power load unit – external input allows for generation of specific load profile) and a NI 9401 digital TTL input/output card for relay control of charge and discharge – run load circuit (discharging processes). The maximum power range of a single load unit is around 1 kWe. The test stand features temperature and current safety monitoring (Fig. 1a, b) essential in long-time dynamic operational tests. The test stand also informs the user about any exceeding of values by an email and SMS over TCP/IP gate.

The test stand was designed for long-duration operational research on electrochemical batteries (i.e.: Li-ion, Li-poly, Lithium-titanate battery – LTO, nickel cadmium, ultracapacitors), as well as hybrid energy storages consisting of electrochemical batteries and ultracapacitors. The laboratory test bench allows conducting a wide arrange of operational test, to the point of complete exploitation and loss of properties of a researched energy storage (usable capacity at the end of test is in range from near zero to a few percents of nominal capacity).

In Fig. 1(a) a photo of the measurement chain and control unit for a single load unit is illustrated. Figure 1(b) presents the scheme of the test stand with the distinction of parameters measured in course of operational tests of an electrochemical Lithium iron phosphate battery – LiFePO4.

3 Analytical Model of the Battery SOC Estimation

In this section the battery state of charge (SOC) description is presented. In Fig. 2 a scheme of LiFePO4 electrical equivalent circuit model is illustrated.

The battery's SOC is the ratio of its effective capacity Q_{bat} to nominal capacity Q_n described as:

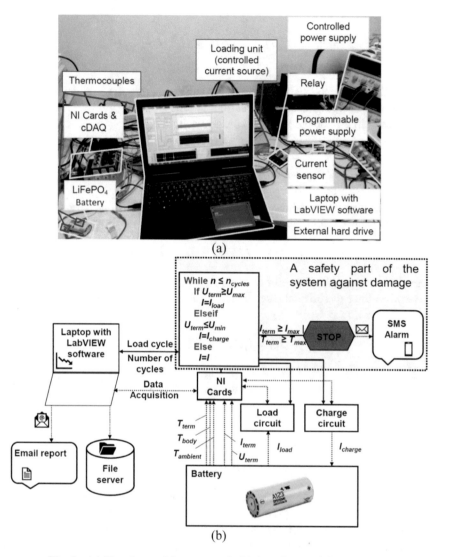

(a)

(b)

Fig. 1. (a) The photo of the test stand, (b) the scheme of the test stand.

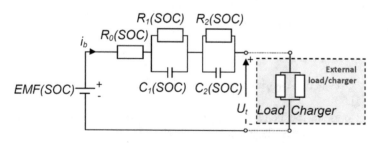

Fig. 2. A scheme of the LiFePO$_4$ electrical model.

$$SOC = Q_{bat}Q_n^{-1} = SOC_0 \pm Q_n^{-1} \int_0^t \eta_c \cdot i_b(t)dt \tag{1}$$

where: SOC_0 – the initial state of charge for time $t = t_0$, η_c – the coulombic efficiency, i_b – the battery charging/discharging current at time t.

The open circuit voltage can be expressed as follows:

$$U_{OCV} = EMF(SOC) = EMF_{min} + \Delta U_{term}SOC =$$
$$EMF_{min} + (EMF_{max} - EMF_{min})SOC \tag{2}$$

The EMFmin and EMFmax values in (2) can be determined on the basis of experiments from the following equation:

$$U_0 = EMF \pm i_b R_0 \Rightarrow EMF = U_0 \mp i_b R_0 \tag{3}$$

where: R_0 – internal resistance of the battery.

Based on the experimental research for $I = I_{ch} = I_{dch} = 1$ C the value of R_0 can be calculated as follows:

$$R_0 = (|U_{ch}| - |U_{dch}|)/2I \tag{4}$$

where: U_{ch} – the charging voltage, U_{dch} – the discharging voltage.

Based on analysis of the Fig. 2, the terminal voltage U_t can be expressed as follows:

$$\begin{cases} U_t = U_{OCV} - U_1 - U_2 - U_0 \\ \dot{U}_1 = -(R_1C_1)^{-1}U_1 + i_bC_1^{-1} \\ \dot{U}_2 = -(R_2C_2)^{-1}U_2 + i_bC_2^{-1} \end{cases} \tag{5}$$

where: C_1, C_2 – the effective capacitances which were used to present the transient response while transfer of power to and from the battery, and to describe the electrochemical polarization and the concentration polarization, respectively, R_1 – an effective resistance characterizing electrochemical polarization, R_2 – the effective resistance characterizing concentration polarization, U_1 and U_2 – the voltages across C_1 and C_2.

Usually, to identify the model parameters: C_1, C_2, R_0, R_1, R_2 it is necessary to solve the minimization problem [17, 22, 24–27]:

$$\|U_{measurement} - U_t(I_{measurement}, R_0, R_1, R_2, C_1, C_2)\| \rightarrow \min \tag{6}$$

by using e.g. Genetic Algorithm [35], Levenberg–Marquardt algorithm [20] or other methods [5].

Table 1 presents the identified parameters for the double RC model at SOC = 0.25, whose physical scheme is illustrated in Fig. 2. The parameters were identified using the method of least squares optimization procedure and the optimization procedure correction of model parameters in the response to obtain the best fit model to the measured data (6).

The identified parameters of double RC model (Table 1) were used in the simulation model in order to prepare the voltage curve which is presented in Fig. 3. The curves of simulated voltage and data from the experimental research are given for the current excitation shown in Fig. 5. The error between the simulated voltage and the voltage from the experimental research did not exceed 0.025 V.

In the literature [4], the presented method of identifying model parameters based on experimental data is well known as off-line method.

In order to identify the battery parameters in the entire range, i.e. from SOC = 0 (full discharge) to SOC = 1 (full charge), it is necessary to perform several iterations of the identification data obtained during the experimental research. Based on obtained waveforms, there can be calculated a set of points for each of parameters which could be described in the form of the polynomial.

Consequently, a set of identified data is obtained in the following form: $R_0 = f$ (SOC), $R_1 = f(SOC)$, $R_2 = f(SOC)$, $C_1 = f(SOC)$, $C_2 = f(SOC)$ and EMF $= f(SOC)$, accordingly. The identification of parameters using, among others, the iteration–approximation method, is presented in [10, 18, 19, 21, 24, 27, 28].

4 Design of the NARX–RANN Model

In alternative approach, to estimate the changes of the voltage at the battery terminals, the Nonlinear AutoRegressive with eXogenous input (NARX) model has been employed. The NARX model is the nonlinear extension of well-known linear AutoRegressive with eXogenous input (ARX) model, which is commonly used in time–series modeling [38].

At the laboratory test stand, a LiFePO4 battery was recharged in subsequent cycles. There are measured: the voltage at the terminals $U_{term}(t)$, a charging/discharging current load $I_{load}(t)$, an ambient temperature $T_{ambient}(t)$, a battery body temperature $T_{body}(t)$ and temperature at battery terminals $T_{term}(t)$. Therefore, the defining equation of the NARX model of the LiFePO4 battery is the following:

$$
\begin{aligned}
U_{term\,NN}(k) = \quad & f[U_{term}(k-1), I_{load}(k), \ldots, I_{load}(k-n), \\
& T_{ambient}(k), \ldots, T_{ambient}(k-n), \\
& T_{body}(k), \ldots, T_{body}(k-n), \\
& T_{term}(k), \ldots, T_{term}(k-n)]
\end{aligned}
\tag{7}
$$

where: n – the number of delay steps.

There are two NARX common architectures, that are used in practical applications: feed-forward architecture (Fig. 3a), and recurrent architecture (Fig. 3b). The first one is useful during a training phase, when a static back–propagation can be used [41]. Then in the testing phase, it can be easily transformed into the second one. Another approach is to learn recurrent neural network, that later doesn't need transformation. However, this significantly impedes the training phase.

Table 1. The LiFePO4 battery identified parameters for SOC = 0.25.

Parameter name	Parameter value
C1	2.6278e+05 F
C2	451.39 F
R1	3.4054e+05 Ω
R2	0.0033 Ω
R0	0.0594 Ω
U_{OCV}	3.2860 V

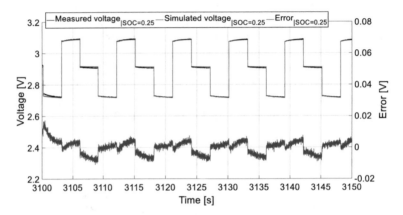

Fig. 3. The voltage signals from the experimental research and from simulation at SOC = 0.25 (for model parameters given in Table 1).

5   NARX Model Evaluation

The experimental research was conducted on the LiFePO4 which had the following nominal parameters: nominal capacity Q_n = 2.5 Ah, nominal voltage U_n = 3.3 V, nominal current I_n = 2.5 A. The LiFePO4 battery charge cycle was preset to I_{ch} = 3.1 A, and was performed, until a maximum voltage of 3.6 V, was reached, after which the battery was switched to load circuit. Load cycle selected for the experiment was a repeating sequence of 3 A, 6 A and 9 A – each load of 3 s duration (Fig. 4). The load cycle was performed until a minimum voltage of 0.7 V was achieved.

During an evaluation, there was investigated the efficiency of NARX models trained in a batch mode using Lavenberg-Marquardt (LM) and Conjugate Gradient Backpropagation with Fletcher-Reeves restarts (CGF) [40]. For each model, all weights and biases were initialized using Nguyen-Widrow initialization procedure [39].

There were trained and tested NARX models with 10 neurons in the hidden layer (71 weights) as our previous research showed the good properties of such models in energy storage identification. Preliminary analysis showed, that within the datasets concerned, variables are correlated to a very small extent and the lag time was significantly larger than the sampling of signals measured at the laboratory stand.

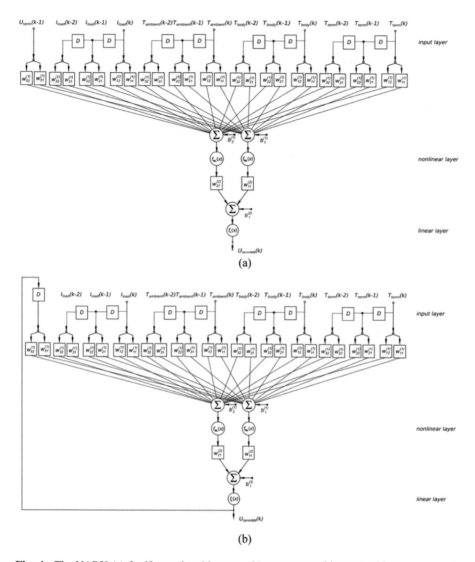

Fig. 4. The NARX (a) feedforward architecture, (b) recurrent architecture with two neurons in hidden layer, and one neuron in output layer (D – tapped delay line).

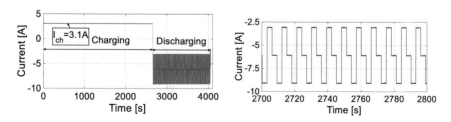

Fig. 5. The charging/discharging current value in specific cycle

Therefore there were selected every 30th measurement sample as a data for training of NARX model. It still ensured adequate coverage of the measurement range for both training data and testing data, and significantly reduced the NARX training time. This is in compliance with the required number of samples in the training set, that should be at least two times bigger than the estimated number of weights [40].

During NARX model identification there were compared two training/testing strategies, namely:

'r/r' – NARX model during testing and training had recurrent architecture,

'ff/r' – NARX model during training had feed-forward architecture, and during testing had recurrent architecture.

Let us denote error of the terminal voltage estimation, for the sample k, as follows:

$$e(k) = U_{termNN}(k) - U_{term}(k) \tag{8}$$

The performance function of ANN model was chosen as a *Mean Squared Error*:

$$MSE = \frac{1}{N} \sum_{k=1}^{N} e^2(k) \tag{9}$$

where: N – the number of data samples.

During R-ANN training the following stop conditions were chosen: epochs < 1000 or MSE = 0:001. The quality of a single NARX model of the battery was evaluated using (8) and the goodness of fit between the NARX data and the reference data was calculated as a Normalized Root Mean Square Error[1]:

$$NRMSE(k) = 1 - \frac{\|e(k)\|}{\|y(k) - \bar{y}\|} \tag{10}$$

where: $\bar{y}$ – mean value of the modeled signal.

There were calculated and compared results of NARX models, for 1st (approximation), 10th (prediction) and 40th (prediction) cycles, trained with different algorithms (LM or CGF) and using different strategies (Table 2, Fig. 6).

Results gathered in Table 1 show, that both training methods allowed to improve the approximation and prediction properties of evaluated NARX models, in comparison with untrained models. It can be observed that, when the ANN was recurrent during training and testing ('r/r' strategy), quality indices clearly decreased when the number of predicted cycle increased – that property can be useful in further usage and improvement of NARX models. Moreover, the CGF method gave more accurate results, than LM training method.

For the best model (10–CGF–r/r), the testing data was predicted with less accuracy than training data. As it can be seen in Table 1 the data for the 1st cycle (used for training) was approximated with significant MSE decrease (around 2800 times), and the very good fit measured by NRMSE index. Prediction properties worsen for the

---

[1] The NRSME index vary between $-\infty$ (no fit) to 1 (perfect fit).

Table 2. Quality indices for NARX models of the LiFePO4 battery (To simplify the description, network structure is represented as *'mnn–method–strategy'* where: *mnn* – the number of neurons in the nonlinear hidden layer, *method* – the training method acronym, *strategy* – training/testing strategy. **Bold fonts** were used to emphasize the worst, and *italic fonts* were used to emphasize best values for the evaluated models.)

Model structure	Data set	Training time	MSE	NRMSE
Untrained NARX	Train, cycle 1	–	0.5636	−8.5292
	Test, cycle 10		**0.9046**	**−8.8209**
	Test, cycle 40		0.7817	−7.8709
10–LM–ff/r	Train, cycle 1	∼60 s	**0.1311**	**−3.5962**
	Test, cycle 10		0.0227	−0.5570
	Test, cycle 40		0.0507	−1.2599
10–LM–r/r	Train, cycle 1	∼18 s	0.0082	−0.1496
	Test, cycle 10		0.0270	−0.6954
	Test, cycle 40		**0.0315**	**−0.7795**
10–CGF–ff/r	Train, cycle 1	∼2 s	0.0019	0.4429
	Test, cycle 10		0.0077	0.0953
	Test, cycle 40		**0.0142**	**−0.1955**
10–CGF–r/r	Train, cycle 1	∼15 s	0.0002	0.8437
	Test, cycle 10		0.0039	0.3588
	Test, cycle 40		**0.0051**	**0.2846**

testing data: 10th cycle and 40th cycle – NRMSE decreased 2 and 3 times respectively, compared with 1st cycle. However, the goodness of fit was visibly high comparing with other evaluated models.

The State of Charge (SoC) for each NARX model in the recharging phase can be predicted using the knowledge of voltage changes as a function of SoC, as it is presented in Fig. 6.

6   Summary and Conclusions

The article presents the possibility of estimating voltage and indirectly State of Charge of the LiFePO4 battery in subsequent charging/discharging cycles.

The double RC battery model presented in this paper gives good results for identification carried out at SOC = 0.25. The error for the voltage curve at the battery terminals (U_t) obtained from the experimental research and from the simulation for the identified parameters of the double RC model (for the identified parameters: R_0, R_1, R_2, C_1, C_2 and EMF, respectively) did not exceed 0.025 V. In double RC model based on iteration–approximation method it is necessary to perform several iterations in several points of the identification data obtained during experimental research: $R_0 = f(SOC)$, $R_1 = f(SOC)$, $R_2 = f(SOC)$, $C_1 = f(SOC)$, $C_2 = f(SOC)$ and EMF = $f(SOC)$, respectively, from SOC = 0 (full discharge) to SOC = 1 (full charge).

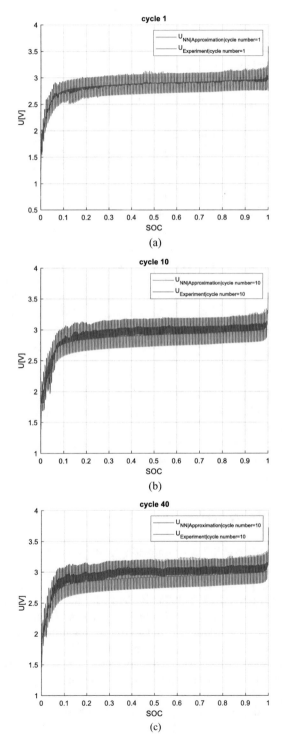

Fig. 6. Approximation of the voltage as a SoC function in (a) training (1st cycle), (b) testing data (10th cycle), (c) testing data (40th cycle) for the 10–CGF–r/r NARX model.

Three data sets from the laboratory test stand were selected for calculation of parameters and verification of analytical and NARX models. Comparing with previous research results [12, 42] one can observe significant improvement of the prediction accuracy. The main advantage of using the neural model in comparison to the iteration-approximation method and analytical model is that the identification data for the parameters measured at numerous points is not required, since neural model is non-linear and has good generalization properties.

In order to properly correct and improve its properties, it is possible to take into account the changes of the electrical energy storage system properties in subsequent cycles (charging/discharging). On the basis of studies of several cycles, by obtaining a set of cycle time points, an attempt can be made to determine the general form of the polynomial of such changes which can improve SOC prediction and allow to estimate the State of Health (SOH) [18, 34, 36, 37] of the energy storage system, especially for the batteries with non–stationary performance (e.g. LiFePO$_4$, VRLA AGM).

The presented modeling methodology can be useful in the practical applications e.g. BMS for distributed polygeneration systems and electric vehicles.

Currently, the authors are working on on-line an application that will use implemented Kalman filter [4, 13, 14, 32, 35], which will allow for estimation the voltage of the open battery circuit as well as the state of charge (SOC) and state of health (SOH) of the battery. In the future models will be implemented in a BMS research system, designed to work with the wide variety of batteries, e.g. lithium iron phosphate batteries.

References

1. Chmielewski, A., Gumiński, R., Mączak, J., Radkowski, S., Szulim, P.: Aspects of balanced development of RES and distributed micro cogeneration use in Poland: case study of a µCHP with stirling engine. Renew. Sustain. Energy Rev. **60**, 930–952 (2016)
2. Luo, X., Wang, J., Dooner, M., Clarke, J.: Overview of current development in electrical energy storage technologies and the application potential in power system operation. Appl. Energy **137**, 511–536 (2015)
3. Chmielewski, A., Piórkowski, P., Bogdziński, K., Szulim, P., Gumiński, R.: Test bench and model research of hybrid energy storage. J. Power Technol. **97**(5), 406–415 (2017)
4. Chmielewski, A., Piórkowski, P., Gumiński, R., Bogdziński, K., Możaryn, J.: Model-based research on ultracapacitors. In: Advances in Intelligent Systems and Computing, Automation 2018, vol. 743, pp. 254–264 (2018)
5. Hannan, M.A., Lipu, M.S.H., Hussain, A., Mohamed, A.: A review of lithium-ion battery state of charge estimation and management system in electric vehicle applications: challenges and recommendations. Renew. Sustain. Energy Rev. **78**, 834–854 (2017)
6. Kim, J.D., Rahimi, M.: Future energy loads for a large-scale adoption of electric vehicles in the city of Los Angeles: impacts on greenhouse gas (GHG) Emissions. Energy Policy **73**, 620–630 (2014)
7. Chmielewski, A., Szulim, P., Gregorczyk, M., Gumiński, R., Mydłowski, T., Mączak, J.: Model of an electric vehicle powered by a PV cell – a case study. In: Proceedings of the International Conference on Methods and Models in Automation and Robotics, MMAR 2017, pp. 1009–1014. IEEE (2017)

8. Xiong, R., Li, L., Li, Z., Yu, Q., Mu, H.: An electrochemical model based degradation state identification method of lithium-ion battery for all-climate electric vehicles application. Appl. Energy **219**, 264–275 (2018)
9. Yang, D., Wang, Y., Pana, R., Chenb, R., Chen, Z.: A neural network based state-of-health estimation of lithium-ion battery in electric vehicles. Energy Procedia **105**, 2059–2064 (2017)
10. Szumanowski, A., Chang, Y.: Battery management system based on battery nonlinear dynamics modeling. IEEE Trans. Veh. Technol. **57**(3), 1425–1432 (2008)
11. Kalogirou, S.A., Mellit, A.: Artificial intelligence techniques for photovoltaic applications: a review. Prog. Energy Combust. Sci. **34**, 574–632 (2008)
12. Możaryn, J., Chmielewski, A.: Selected parameters prediction of energy storage system using recurrent neural networks. In: Proceedings of the 10th IFAC Symposium on Fault Detection, Supervision and Safety for Technical Processes – SAFEPROCESS 2018, IFAC–PapersOnLine. (in Press)
13. He, H., Xiong, R., Zhang, X., Sun, F., Fan, J.X.: State-of-charge estimation of the lithium-ion, battery using an adaptive extended Kalman filter based on an improved Thevenin model. IEEE Trans. Veh. Technol. **60**(4), 1461–1469 (2011)
14. Mu, H., Xiong, R., Zheng, H., Chang, Y., Chen, Z.: A novel fractional order model based state-of-charge estimation method for lithium-ion battery. Appl. Energy **207**, 384–393 (2017)
15. Chmielewski, A., Możaryn, J., Piórkowski, P., Gumiński, R., Bogdziński, K.: Modelling of ultracapacitors using recurrent artificial neural network. In: Proceedings of the Advances in Intelligent Systems and Computing, Automation 2018, vol. 743, pp. 713–723 (2018)
16. Bottinger, M., Paulitschke, M., Bocklisch, T.: Systematic experimental pulse test investigation for parameter identification of an equivalent circuit based lithium-ion battery model. Energy Procedia **135**, 337–346 (2017)
17. He, H., Xiong, R., Fan, J.: Evaluation of lithium-ion battery equivalent circuit models for state of charge estimation by an experimental approach. Energies **4**, 582–598 (2011)
18. Sepasi, S., Ghorbani, R., Liaw, B.Y.: Inline state of health estimation of lithium-ion batteries using state of charge calculation. J. Power Sources **299**, 246–254 (2015)
19. Cheng, P., Zhou, Y., Song, Z., Ou, Y.: Modeling and SOC estimation of LiFePO4 battery. In: Proceedings of the IEEE International Conference on Robotics and Biomimetics, Qingdao, China, pp. 2140–2144 (2016)
20. Tong, S., Lacap, J.H., Park, J.W.: Battery state of charge estimation using a load-classifying neural network. J. Energy Storage **7**, 236–243 (2016)
21. Zhang, C., Allafi, W., Dinh, Q., Ascencio, P., Marco, J.: Online estimation of battery equivalent circuit model parameters and state of charge using decoupled least squares technique. Energy **142**, 678–688 (2018)
22. Lai, X., Zheng, Y., Sun, T.: A comparative study of different equivalent circuit models for estimating state-of-charge of lithium-ion batteries. Electrochim. Acta **259**, 566–577 (2018)
23. Gallien, T., Brasseur, G.: State of charge estimation of a LiFePO$_4$ battery: a dual estimation approach incorporating open circuit voltage hysteresis. In: Proceedings of the 2016 IEEE International Instrumentation and Measurement Technology Conference, pp. 1–6. IEEE (2016)
24. Feng, T., Yang, L., Zhao, X., Zhang, H., Qiang, J.: On-line identification of lithium-ion battery parameters based on an improved equivalent-circuit model and its implementation on battery state-of-power prediction. J. Power Sources **281**, 192–203 (2015)

25. Li, Z., Xiong, R., He, H.: An improved battery on-line parameter identification and state-of-charge determining method. Maldives, Energy Procedia **103**, 381–386 (2016). Applied Energy Symposium and Forum, REM2016: Renewable Energy Integration with Mini/Microgrid, 19–21 April 2016

26. Xiong, R., Yu, Q., Wang, L.Y.: Open-circuit-voltage and state-of-charge online estimation for lithium ion batteries. Energy Procedia **142**, 1902–1907 (2017). In: 9th International Conference on Applied Energy, ICAE2017, 21–24 August 2017, Cardiff, UK

27. Nikolian, A., Firouz, Y., Gopalakrishnan, R., Timmermans, J.M., Omar, N., van den Bossche, P., Mierlo, J.: Lithium ion batteries-development of advanced electrical equivalent circuit models for nickel manganese cobalt lithium-ion. Energies **360**(9), 1–23 (2016)

28. Ke, M.-Y., Chiu, Y.-H., Wu, C.-Y.: Battery modelling and SOC estimation of a LiFePO4 battery. In: 2016 International Symposium on Computer, Consumer and Control. IEEE, pp. 208–211 (2016). 978-1-5090-3071-2/16 © 2016. https://doi.org/10.1109/is3c.2016.63

29. Pattipati, B., Balasingam, B., Avvari, G.V., Pattipati, K.R., Bar-Shalom, Y.: Open circuit voltage characterization of lithium-ion batteries. J. Power Sources **269**, 317–333 (2014)

30. Wang, A., Jin, X., Li, Y., Li, N.: LiFePO4 battery modeling and SOC estimation algorithm. In: Proceedings of the 29th Chinese Control and Decision Conference (CCDC), pp. 7574–7578. IEEE (2017). 978-1-5090-4657-7/17/$31.00_c 2017

31. Xing, Y., He, W., Pecht, M., Tsui, K.L.: State-of-charge estimation of lithium-ion batteries using the open-circuit voltage at various ambient temperatures. Appl. Energy **113**, 106–115 (2014)

32. Afshar, S., Morris, K., Khajepour, A.: State of charge estimation via extended Kalman filter designed for electrochemical equations. IFAC PapersOnLine **50–1**, 2152–2157 (2017)

33. Xavier, M.A., Trimboli, M.S.: Lithium-ion battery cell-level control using constrained model predictive control and equivalent circuit models. J. Power Sources **285**, 374–384 (2015)

34. He, Q., Zha, Y., Sun, Q., Pan, Z., Liu, T.: Capacity fast prediction and residual useful life estimation of valve regulated lead acid battery. Math. Probl. Eng. **2017**, 1–9 (2017). Article ID 7835049

35. Ting, T.O., Man, K.L., Lim, E.G., Leach, M.: Tuning of Kalman filter parameters via genetic algorithm for state-of-charge estimation in battery management system. Math. Probl. Eng. **2014**, 1–11 (2014). Article ID 176052

36. Zhou, D., Yin, H., Fu, P., Song, X., Lu, W., Yuan, L., Fu, Z.: Prognostics for state of health of lithium-ion batteries based on Gaussian process regression. Math. Probl. Eng. **2018**, 1–11 (2018). Article ID 8358025

37. Kim, J., Cho, B.H.: State-of-charge estimation and state-of-health prediction of a li-ion degraded battery based on an EKF combined with a per-unit system. IEEE Trans. Veh. Technol. **60**(9), 4249–4260 (2011)

38. Box, G.E.P., Jenkins, G.M., Reinsel, G.C., Ljung, G.M.: Time Series Analysis: Forecasting and Control. Wiley Series in Probability and Statistics, 3rd edn. Wiley, Hoboken (2015)

39. Nguyen, D., Widrow, B.: Improving the learning speed of 2-layer neural networks by choosing initial values of the adaptive weights. In: Proceedings of the International Joint Conference on Neural Networks, vol. 3, pp. 21–26 (1990)

40. Hagan, M.T., Demuth, H.B., Beale, M.H.: Neural Network Design. PWS Publishing, Boston (1996)

41. Narendra, K.S., Parthasarathy, K.: Learning automata approach to hierarchical multiobjective analysis. IEEE Trans. Syst. Man Cybern. **20**(1), 263–272 (1991)

42. Chmielewski, A., Możaryn, J., Gumiński, R., Szulim, P., Bogdziński, K.: Experimental evaluation of mathematical and artificial neural network modeling of energy storage system. In: Springer Proceedings in Mathematics and Statistics, 14th International Conference Dynamical Systems Theory and Applications, DSTA'2017 (in Press)

Realization of the Descriptor Continuous-Time Fractional System Consist of Strictly Proper Part and Polynomial Part

Konrad Andrzej Markowski[✉]

Faculty of Electrical Engineering, Institute of Control and Industrial Electronics,
Warsaw University of Technology, Koszykowa 75, 00-662 Warsaw, Poland
Konrad.Markowski@ee.pw.edu.pl
http://markowski.edu.pl

Abstract. In this paper, method for computation of a realization of a given proper transfer function of descriptor fractional continuous-time one-dimensional linear systems has been presented. For the proposed method, an algorithm a based on digraphs theory was constructed. The proposed solution allows determine sets of realizations as a $(\mathbf{E}, \mathbf{A}, \mathbf{B}, \mathbf{C})$ matrices pairs. Method was discussed and illustrated with numerical examples.

1 Introduction

Descriptor (or singular) linear systems were considered in many monographs: [3,6,11,12] and research papers: [1,2,5,27]. The properties and the use matrix theory of the descriptor systems were established in [7,13,14]. There are many problems associated with the analysis and synthesis of the descriptor systems. One of the very important and very difficult problem is realization and minimal realization problem. In many papers and books we can find only one realization as so-called canonical form which satisfy the system described by the transfer function. In fact, there are many sets of matrices.

In recent years integral and differential calculus of a fractional order has become a subject of great interest in different areas of physics, biology, economics and other sciences. The fractional calculus found to be a very useful tool for modelling systems. Mathematical fundamentals of fractional calculus are given in the monographs: [4,22,23,25] and some other applications of fractional-order systems can be found in: [15,16,21,24,26].

The main purpose of this paper is to present a new method based on one-dimensional digraph theory for computation the set of minimal realization of a given proper transfer function of the one-dimensional continuous-time fractional descriptor system. In this paper digraph structure corresponding to descriptor linear fractional system has been considered. It should be noted, that in analysis of descriptor systems a digraph mask is very important. The first time concept of

© Springer Nature Switzerland AG 2020
R. Szewczyk et al. (Eds.): AUTOMATION 2019, AISC 920, pp. 232–244, 2020.
https://doi.org/10.1007/978-3-030-13273-6_23

digraph-mask was presented in paper [17]. As a result, we proposed the procedure for determining the set of descriptor realizations. The realization corresponding to determine digraph-structure in class $\mathcal{K}_1$ defined in [10]. This work is the next step in the research on the determination of the realization problem of the continuous-time fractional singular systems by using a digraph theory, which was started in the publication [18, 19] and [20].

2   Fractional Continuous-Time Descriptor System

Let be given the fractional descriptor continuous-time linear system

$$\mathbf{E}\, _0\mathfrak{D}_t^\alpha x(t) = \mathbf{A}x(t) + \mathbf{B}u(t), \tag{1}$$
$$y(t) = \mathbf{C}x(t),$$

where $x(t) \in \mathbb{R}^n$, $u(t) \in \mathbb{R}^m$ and $y(t) \in \mathbb{R}^p$ are the state, input and output vectors, respectively and

$$\mathbf{E} \in \mathbb{R}^{n \times n},\ \mathbf{A} \in \mathbb{R}^{n \times n},\ \mathbf{B} \in \mathbb{R}^{n \times m},\ \mathbf{C} \in \mathbb{R}^{p \times n}. \tag{2}$$

The following Caputo definition of the fractional derivative will be used:

$$_a^C\mathfrak{D}_t^\alpha = \frac{d^\alpha}{dt^\alpha} = \frac{1}{\Gamma(n-\alpha)} \int_a^t \frac{f^{(n)}(\tau)}{(t-\tau)^{\alpha+1-n}} d\tau, \tag{3}$$

where $\alpha \in \mathbb{R}$ is the order of a fractional derivative, $f^{(n)}(\tau) = \frac{d^n f(\tau)}{d\tau^n}$ and $\Gamma(x) = \int_0^\infty e^{-t} t^{x-1} dt$ is the gamma function.

The Laplace transform of the derivative-integral (3) has the form

$$\mathcal{L}\left[_0^C\mathfrak{D}_t^\alpha \right] = s^\alpha F(s) - \sum_{k=1}^n s^{\alpha-k} f^{(k-1)}(0^+).$$

After using the Laplace transform to (1) and taking into account

$$X(s) = \mathcal{L}\left[x(t)\right] = \int_0^\infty x(t)e^{-st} dt;\ \ \mathcal{L}\left[\mathfrak{D}^\alpha x(t)\right] = s^\alpha X(s) - s^{\alpha-1}x_0 \tag{4}$$

we obtain:

$$X(s) = \left[\mathbf{E}s^\alpha - \mathbf{A}\right]^{-1}\left[s^{\alpha-1}x_0 + \mathbf{B}U(s)\right],$$
$$Y(s) = \mathbf{C}X(s),\ \ \ U(s) = \mathcal{L}\left[u(t)\right]. \tag{5}$$

After using (5) we can determine the transfer matrix of the system in the following form:

$$\mathbf{T}(s) = \mathbf{C}\left[\mathbf{E}s^\alpha - \mathbf{A}\right]^{-1}\mathbf{B} \in \mathbb{R}^{p \times m}(s). \tag{6}$$

It is assumed that $\det \mathbf{E} = 0$ and the pencil of the system (1) is regular, that is

$$\det\left[\mathbf{E}s^\alpha - \mathbf{A}\right] \neq 0 \tag{7}$$

for some $z \in \mathbb{C}$ (where $\mathbb{C}$ is the field of the complex numbers).

Definition 1. *The system* (1) *is called singular system if and only if* $\det \mathbf{E} = 0$ (rank $\mathbf{E} = r < n$).

The matrices (2) are called a realisation of a given transfer matrix $\mathbf{T}(s)$ if they satisfy the equality (6). The realisation is called minimal if the dimension of the sate matrix $\mathbf{A}$ is minimal among all possible realisations of $\mathbf{T}(s)$.

3   Main Results

The solution of the realisation problem will be presented for one-dimensional single-input single-output (SISO) transfer function ($m = p = 1$). The proposed method will be based on the one-dimensional digraph theory, and it will determine solutions in class $\mathcal{K}_1$. The classes of the digraph structure ware considered in detail in [10].

Consider the irreducible transfer function:

$$T(s) = \frac{n(s)}{d(s)} = \frac{b_q s^{\alpha \cdot q} + b_{q-1} s^{\alpha(q-1)} + b_{q-2} s^{\alpha(q-2)} + \ldots + b_1 s^\alpha + b_0}{a_r s^{\alpha \cdot r} + a_{r-1} s^{\alpha(r-1)} + a_{r-2} s^{\alpha(r-2)} + \ldots + a_1 s^\alpha + a_0}, \quad (8)$$

where b_i, $i = 0, 1, \ldots, q$ and a_j, $j = 0, 1, \ldots, r-1$ are given real coefficients and $r < n$. The transfer function (8) can be considered as a pseudo-rational function of the variable $\lambda = s^\alpha$ in the form:

$$T(\lambda) = \frac{n(\lambda)}{d(\lambda)} = \frac{b_q \lambda^q + b_{q-1} \lambda^{q-1} + b_{q-2} \lambda^{q-2} + \ldots + b_1 \lambda + b_0}{a_r \lambda^r + a_{r-1} \lambda^{r-1} + a_{r-2} \lambda^{r-2} + \ldots + a_1 \lambda + a_0}, \quad (9)$$

Theorem 1. *The improper transfer matrix* (9) *can be always written as the sum of strictly proper part:*

$$T_{sp}(\lambda) = \widetilde{\mathbf{C}} \left[\widetilde{\mathbf{E}} \lambda - \widetilde{\mathbf{A}} \right] \mathbf{B} \quad (10)$$

and the polynomial part

$$P(\lambda) = d_p \lambda^p + d_{p-1} \lambda^{p-1} + d_{p-2} \lambda^{p-2} + \ldots + d_1 \lambda + d_0 \,_{|p \in \mathbb{N} = \{1, 2, \ldots\}} \quad (11)$$

Proof. Let be given transfer function in the form (9) and assume that $q > r$ and $q = r + p$, then:

$$T_{sp}(\lambda) = T(\lambda) - P(\lambda) = \quad (12)$$

$$= \frac{\begin{array}{l}(b_q - a_r d_p) \cdot \lambda^q + (b_{q-1} - a_r d_{p-1} - a_{r-1} d_p) \cdot \lambda^{q-1} + \\ (b_{q-2} - a_r d_{p-2} - a_{r-1} d_{p-1} - a_{r-2} d_p) \cdot \lambda^{q-2} + \\ (b_{q-3} - a_r d_{p-3} - a_{r-1} d_{p-2} - a_{r-2} d_{p-1} - a_{r-3} d_p) \cdot \lambda^{q-3} + \ldots + \\ (b_2 - a_2 d_0 - a_1 d_1 - a_0 d_2) \cdot \lambda^2 + (b_1 - a_1 d_0 - a_0 d_1) \cdot \lambda + (b_0 - a_0 d_0)\end{array}}{a_r \lambda^r + a_{r-1} \lambda^{r-1} + a_{r-2} \lambda^{r-2} + \ldots + a_1 \lambda + a_0} =$$

The strictly proper transfer function we can write in the form:

$$T_{sp}(\lambda) = \frac{\widetilde{b}_q \lambda^q + \widetilde{b}_{q-1} \lambda^{q-1} + \widetilde{b}_{q-2} \lambda^{q-2} + \widetilde{b}_{q-3} \lambda^{q-3} + \ldots + \widetilde{b}_2 \lambda^2 + \widetilde{b}_1 \lambda + \widetilde{b}_0}{a_r \lambda^r + a_{r-1} \lambda^{r-1} + a_{r-2} \lambda^{r-2} + \ldots + a_1 \lambda + a_0}$$

If the degree of the nominator is one grater than the degree of denominator, then to the (12) we substitute $\lambda^q = \lambda^{r+1}$ and we obtain the following conditions:

$$\tilde{b}_q \lambda^{r+1} = 0 \Rightarrow b_q - a_r d_p = 0, \tag{13}$$
$$\tilde{b}_{q-1} \lambda^r = 0 \Rightarrow b_{q-1} - a_r d_{p-1} - a_{r-1} d_p = 0,$$

that must be met. If the degree of the nominator is two grater than the degree of denominator, then to the (12) we substitute $\lambda^q = \lambda^{r+2}$ and we obtain the following conditions:

$$\tilde{b}_q \lambda^{r+2} = 0 \Rightarrow b_q - a_r d_p = 0,$$
$$\tilde{b}_{q-1} \lambda^{r+1} = 0 \Rightarrow b_{q-1} - a_r d_{p-1} - a_{r-1} d_p = 0, \tag{14}$$
$$\tilde{b}_{q-2} \lambda^r = 0 \Rightarrow b_{q-2} - a_r d_{p-2} - a_{r-1} d_{p-1} - a_{r-2} d_p = 0,$$

that must be met. Proceeding in the same way, we can determine the conditions for higher differences between nominator and denominator. The strictly proper transfer function (12) we can rewrite in general form:

$$T_{sp}(\lambda) = \frac{\displaystyle\sum_{k=0}^{q} \lambda^{q-k} \left[\overbrace{b_{q-k} - \left(\sum_{\substack{i=0 \\ i+j=k}}^{k} \sum_{\substack{j=k \\ }}^{0} a_{r-i} d_{q-r-j} \right)}^{\tilde{b}_{q-k}} \right]}{\displaystyle\sum_{l=0}^{r} a_{r-l} \lambda^{r-l}} \tag{15}$$

from which we can determine coefficients of the polynomial $P(\lambda)$ after fulfilling

$$\sum_{k=0}^{q-r} \left(\lambda^{q-k} \tilde{b}_{q-k} \right) = 0 \tag{16}$$

and determine coefficients of the strictly proper transfer function

$$\sum_{q-r+1}^{q} \lambda^{q-k} \tilde{b}_{q-k} \tag{17}$$

which depends on determined coefficients polynomial $P(\lambda)$.   ■

Now we can realization problem divide into two part. The first corresponding to determine realization of the strictly proper transfer function (10) and the second corresponding to determine realisation of the polynomial (11).

3.1   Realisation of the Function $T_{sp}(\lambda)$

Then using Theorem 1 presented in paper [8,9], we can create all digraph realizations of the characteristic polynomials. It should be noted that, each digraph corresponding to a characteristic polynomial must satisfy two conditions. The first condition (C1) relates to the existence in the common part of the digraph, the second condition (C2) relates to non-existence of additional cycles in the digraph. From the obtained digraph, we can write state matrix $\widetilde{\mathbf{A}}$. Then we can expanding the digraph created in the first step by connect source vertex s corresponding to matrix $\mathbf{B}$ with vertex $v_1, \ldots, v_n$ and output vertex y corresponding to matrix $\widetilde{\mathbf{C}}$ with vertex v_i belonging to a set of common parts of a digraph.

3.2   Realisation of the Polynomial $P(\lambda)$

The polynomial (11) can be presented in the following form:

$$P(\lambda) = \frac{n_p(\lambda)}{d_p(\lambda)} = \frac{d_p \lambda^p + d_{p-1} \lambda^{p-1} + \ldots + d_1 \lambda + d_0}{1} \tag{18}$$

In the first step, we must find matrix $\overline{\mathbf{A}}$. After multiplying the denominator of the (18) by $\lambda^{-(p+1)}$ we obtain the following polynomial:

$$d_p(\lambda) = \lambda^{-(p+1)} \tag{19}$$

for which we create digraphs representations (see Fig. 1). To the vertices belonging to the common part of the digraph we assigned weight equal to **1** (marked in dark gray) and for the other vertices – weight equal to **0** (marked in white).

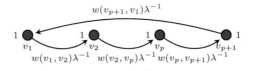

Fig. 1. Digraph corresponding to polynomial (19)

In the second step, we must find matrix $\overline{\mathbf{E}}$. The weight of the vertices in a digraph is associated with the digraph-mask. If the characteristic polynomial is in the form (19) then we can determine digraph-mask by the use the following theorem.

Theorem 2. *A digraph-mask $\mathfrak{M}(\mathcal{D})$ corresponding to a polynomial* (19) *consists of* $(p+1)$*–vertices and one vertex with the weight equal to* 0.

Proof. If digraph-mask $\mathfrak{M}(\mathcal{D})$ consists from all vertex with weight not equal to zero, this means that matrix $\overline{\mathbf{E}}$ is diagonal with all non-zero entries. In this case the condition $\det \overline{\mathbf{E}} = 0$ is not satisfied and the system is not descriptor.   ∎

All possible digraph-masks of the characteristic polynomial (19) presented in Fig. 2. It should be noted, that vertices with weight equal to 1 are marked in

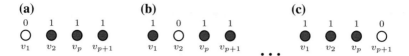

Fig. 2. (a)–(c) All possible digraph-masks of the (19).

dark gray and vertices with a weight equal to 0 are marked in white. Then, using composition relative to vertices and digraph-mask determined by the using Theorem 2, we can create all possible digraph realizations as combinations of the monomial representations and a digraph-mask:

$$\mathcal{D} = \mathfrak{M}_j(G) \circ \mathcal{G}_{d_p} \tag{20}$$

where: $\mathfrak{M}_j(G)$ is j-th possible digraph-mask; $\circ$ is an operation of the superposition digraph-mask and digraph realisations $\mathcal{G}_{d_p}$ of the polynomial (19). It should be noted that superposition operation $\circ$ on digraphs vertices corresponds to logical operation OR. We choose digraph-structure in which all vertices have weight equal to 1 – presented in Fig. 3. In details this kind of relation is presented in paper [20].

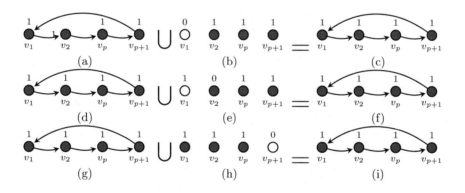

Fig. 3. All possible realizations of the characteristic polynomial (19).

It should be noted that other realization can be obtained from digraph presented in Fig. 1 by change of the direction of the arcs in the graph. This operation is similar to a transposition matrix used in the matrix theory.

Now from obtain digraph we can write the following matrices:

$$
\mathbf{A} \in \mathbb{R}^{(p+1)\times(p+1)}
$$

$$
\begin{array}{c|ccccc}
v_i\backslash^{v_j} & v_1 & v_2 & \cdots & v_p & v_{p+1} \\
\hline
v_1 & 0 & 0 & \cdots & 0 & w(v_{p+1},v_1) \\
v_2 & w(v_1,v_2) & 0 & \cdots & 0 & 0 \\
\vdots & \vdots & & \ddots\ddots & \vdots & \vdots \\
v_p & 0 & 0 & \ddots & 0 & 0 \\
v_{p+1} & 0 & 0 & \cdots & w(v_p,v_{p+1}) & 0
\end{array}
\tag{21}
$$

$$
\underbrace{
\begin{bmatrix}
0 & 0 & \cdots & 0 & 0 \\
0 & 1 & \cdots & 0 & 0 \\
\vdots & \vdots & \ddots & \vdots & \vdots \\
0 & 0 & \cdots & 1 & 0 \\
0 & 0 & \cdots & 0 & 1
\end{bmatrix}
}_{\mathbf{E}\in\mathbb{R}^{(p+1)\times(p+1)}},
\underbrace{
\begin{bmatrix}
1 & 0 & \cdots & 0 & 0 \\
0 & 0 & \cdots & 0 & 0 \\
\vdots & \vdots & \ddots & \vdots & \vdots \\
0 & 0 & \cdots & 1 & 0 \\
0 & 0 & \cdots & 0 & 1
\end{bmatrix}
}_{\mathbf{E}\in\mathbb{R}^{(p+1)\times(p+1)}},\ \cdots,\
\underbrace{
\begin{bmatrix}
1 & 0 & \cdots & 0 & 0 \\
0 & 1 & \cdots & 0 & 0 \\
\vdots & \vdots & \ddots & \vdots & \vdots \\
0 & 0 & \cdots & 1 & 0 \\
0 & 0 & \cdots & 0 & 0
\end{bmatrix}
}_{\mathbf{E}\in\mathbb{R}^{(p+1)\times(p+1)}}
\tag{22}
$$

(a) *(b)* *(c)*

In the last stage we must <u>find matrix $\overline{\mathbf{B}}$ and $\overline{\mathbf{C}}$</u>. After multiplying the nominator of the (18) by λ^{-p} we obtain the following polynomial:

$$
n_p(\lambda) = d_p + d_{p-1}\lambda^{-1} + \ldots + d_1\lambda^{1-p} + d_0\lambda^{-p}
\tag{23}
$$

For this purpose, we are expanding the digraph created in the first part of the algorithm. Let us consider realization presented in Fig. 3c. From vertex v_1 we determine a path of the maximum length equal to p. Now we can add to the digraph the source vertex s_1 corresponding to matrix $\overline{\mathbf{B}}$ and output vertex y_1 corresponding to matrix $\overline{\mathbf{C}}$, and we combine them. We obtain the digraph structure presented in Fig. 4.

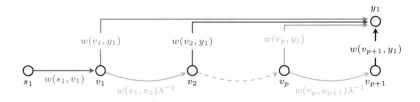

Fig. 4. Digraph corresponding to polynomial (23)

Using obtain digraph, we can write the set of the equations

$$
\begin{cases}
\lambda^{-1} & \begin{vmatrix} w(s_1,v_1)\cdot w(v_1,y_1) \end{vmatrix} & = d_p \\
& \begin{vmatrix} w(s_1,v_1)\cdot w(v_1,v_2)\cdot w(v_2,y_1) \end{vmatrix} & = d_{p-1} \\
\vdots & \vdots & \vdots \ \vdots \\
\lambda^{1-q} & \begin{vmatrix} w(s_1,v_1)\cdot w(v_1,v_2)\cdot \ldots \cdot w(v_p,y_1) \end{vmatrix} & = d_1 \\
\lambda^{-q} & \begin{vmatrix} w(s_1,v_1)\cdot w(v_1,v_2)\cdot \ldots \cdot w(v_p,v_{p+1}) \cdot w(v_{p+1},y_1) \end{vmatrix} & = d_0
\end{cases} \tag{24}
$$

and after solving them we can write $\overline{\mathbf{B}}$ and $\overline{\mathbf{C}}$ matrix in the following form:

$$
\underbrace{
\begin{array}{c}
v_i \backslash^s \\
v_1 \\
v_2 \\
\vdots \\
v_p \\
v_{p+1}
\end{array}
\begin{array}{c}
s_1 \\
\begin{bmatrix} w(s_1,v_1) \\ 0 \\ \vdots \\ 0 \\ 0 \end{bmatrix}
\end{array}
}_{\overline{\mathbf{B}} \in \mathbb{R}^{(p+1)\times 1}}
,\quad
\underbrace{
\begin{array}{cccccc}
y\backslash^{v_j} & v_1 & v_2 & \cdots & v_p & v_{p+1} \\
y_1 & \begin{bmatrix} w(v_1,y_1) & w(v_2,y_1) & \cdots & w(v_p,y_1) & w(v_{p+1},y_1) \end{bmatrix}
\end{array}
}_{\overline{\mathbf{C}} \in \mathbb{R}^{(1\times p+1)}}
\tag{25}
$$

where:

$$
w(v_1,y_1) = \frac{d_p}{w(s_1,v_1)},\ w(v_2,y_1) = \frac{d_{p-1}}{w(v_1,v_2)\cdot w(s_1,v_1)},
$$

$$
\ldots,\ w(v_p,y_1) = \frac{d_1}{w(s_1,v_1)\cdot w(v_1,v_2)\cdot \ldots}
$$

$$
w(v_{p+1},y_1) = \frac{d_0}{w(s_1,v_1)\cdot w(v_1,v_2)\cdot \ldots \cdot w(v_p,v_{p+1})}.
$$

The realization of the polynomial (18) is given by (21), (22) and (25) for $w(s_1,v_1) \in \mathbb{R}$.

3.3 Connection Realisations

If we have determine all possible realisation of the strictly proper transfer function $T_{sp}(\lambda)$ as a set of triple $(\widetilde{\mathbf{A}}, \widetilde{\mathbf{B}}, \widetilde{\mathbf{C}})$ and determine all possible realization of the polynomial $P(\lambda)$ as a set of fours $(\overline{\mathbf{E}}, \overline{\mathbf{A}}, \overline{\mathbf{B}}, \overline{\mathbf{C}})$ then we can connect them in the following form:

$$
\mathbf{E} = \begin{bmatrix} \mathbf{I}_r & 0 \\ \hline 0 & \overline{\mathbf{E}} \end{bmatrix},\
\mathbf{A} = \begin{bmatrix} \widetilde{\mathbf{A}} & 0 \\ \hline 0 & \overline{\mathbf{A}} \end{bmatrix} \in \mathbb{R}^{(r+p+1)\times(r+p+1)}, \tag{26}
$$

$$
\mathbf{B} = \begin{bmatrix} \widetilde{\mathbf{B}} \\ \hline \overline{\mathbf{B}} \end{bmatrix} \in \mathbb{R}^{(r+p+1)\times 1},\
\mathbf{C} = \begin{bmatrix} \widetilde{\mathbf{C}} & \overline{\mathbf{C}} \end{bmatrix} \in \mathbb{R}^{1\times(r+p+1)}.
$$

It should be noted that in proof of the Theorem 1 we assume that $q = r + p$. Therefore, the size of the system can be written in the form $\mathbf{E}, \mathbf{A} \in \mathbb{R}^{(q+1)\times(q+1)}$, $\mathbf{B} \in \mathbb{R}^{(q+1)\times 1}$ and $\mathbf{C} \in \mathbb{R}^{1\times(q+1)}$, and the size of the realization is smallest of the possible. Additionally, by the use of his method we can determine all possible realization which depends on combination of the realization strictly proper transfer function and realisation of polynomial.

4   Numerical Example

Compute realization of the transfer function

$$T(s) = \frac{s^4 + 0.3s^{3.2} + 1.2s^{2.4} + 2.82s^{1.6} + 0.92s^{0.8} + 2}{s^{2.4} - 0.7s^{1.6} - 0.1s^{0.8} - 0.08} \tag{27}$$

Solution: The transfer function (27) can be considered as a pseudo-rational function of the variable $\lambda = s^{0.8}$ in the form:

$$T(\lambda) = \frac{\lambda^5 + 0.3\lambda^4 + 1.2\lambda^3 + 2.82\lambda^2 + 0.92\lambda + 2}{\lambda^3 - 0.7\lambda^2 - 0.1\lambda - 0.08} \tag{28}$$

From relation $q = p + r$ we know that polynomial $P(\lambda)$ have form $P(\lambda) = d_2\lambda^2 + d_1\lambda + d_0$. In the first step using (16) we write set of the equations

$$k = 0 \Rightarrow \lambda^5 \widetilde{b}_5 = 0 \Rightarrow b_5 - a_3 d_2 = 0 \Rightarrow d_2 = 1;$$
$$k = 1 \Rightarrow \lambda^4 \widetilde{b}_4 = 0 \Rightarrow b_4 - a_3 d_1 - a_2 d_2 = 0 \Rightarrow d_1 = 1;$$
$$k = 2 \Rightarrow \lambda^3 \widetilde{b}_3 = 0 \Rightarrow b_3 - a_3 d_0 - a_2 d_1 - a_1 d_2 = 0 \Rightarrow d_0 = 2.$$

In the second step using (17) we determine nominator of the strictly proper transfer function

$$k = 3 \Rightarrow \lambda^2(b_2 - a_2 d_0 + a_1 d_1 + a_0 d_2) = 4.4\lambda^2;$$
$$k = 4 \Rightarrow \lambda(b_1 - a_1 d_0 - a_0 d_1 = 1.2\lambda;$$
$$k = 5 \Rightarrow \lambda^0(b_0 - a_0 d_0) = 2.16 \tag{29}$$

Finally we can write:

$$T_{sp}(\lambda) = \frac{4.4\lambda^2 + 1.2\lambda + 2.16}{\lambda^3 - 0.7\lambda^2 - 0.1\lambda - 0.08} \tag{30}$$

$$P(\lambda) = d_2\lambda^2 + d_1\lambda + d_0 = \lambda^2 + \lambda + 2 \tag{31}$$

In the next step we determine realization of the strictly proper transfer function. Using digraph based algorithm we obtain 9 possible digraph-structure which satisfy characteristic polynomial of (30). One of them presented in Fig. 5 (vertices: v_1, v_2, v_3 and black arcs). Now we can: add source vertex s_1 and output vertex y_1 and connect them with vertices: v_1, v_2, v_3 as presented in Fig. 5; write the set of the equation. After solving them we obtain triple

$$\left(\underbrace{\begin{bmatrix} 0.7 & 0.1 & 0.08 \\ 1 & 0 & 0 \\ 0 & 1 & 0 \end{bmatrix}}_{\widetilde{\mathbf{A}}}, \underbrace{\begin{bmatrix} w(s_1, v_1) \\ 0 \\ 0 \end{bmatrix}}_{\widetilde{\mathbf{B}}}, \frac{1}{w(s_1, v_1)} \underbrace{\begin{bmatrix} 4.4 & 1.2 & 2.16 \end{bmatrix}}_{\widetilde{\mathbf{C}}} \right) \tag{32}$$

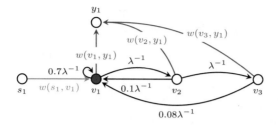

Fig. 5. Realisation of the (30)

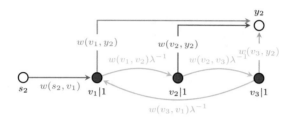

Fig. 6. Digraph corresponding to polynomial (31)

which is one of the possible realisation of the (30).

In the next step we must determine realisation of the polynomial $P(\lambda)$. Using method presented in Sect. 3.2 we can construct the digraph shown in Fig. 6. Now we can write matrices:

$$\overline{\mathbf{E}} = \begin{bmatrix} 0\,0\,0 \\ 0\,1\,0 \\ 0\,0\,1 \end{bmatrix}, \quad \overline{\mathbf{A}} = \begin{bmatrix} 0\,0\,1 \\ 1\,0\,0 \\ 0\,1\,0 \end{bmatrix} \tag{33}$$

and set of the equations:

$$\begin{cases} \lambda^{-1} \\ \lambda^{-2} \end{cases} \begin{Vmatrix} w(s_2, v_1) \cdot w(v_1, y_2) = 1 \\ w(s_2, v_1) \cdot w(v_2, y_2) = 1 \\ w(s_2, v_1) \cdot w(v_3, y_2) = 2 \end{Vmatrix} \tag{34}$$

After solving (34), we can write input and output matrix in the following form:

$$\overline{\mathbf{B}} = \begin{bmatrix} w(s_2, v_2) \\ 0 \\ 0 \end{bmatrix}, \overline{\mathbf{C}} = \begin{bmatrix} \dfrac{1}{w(s_2, v_1)} & \dfrac{1}{w(s_2, v_1)} & \dfrac{2}{w(s_2, v_1)} \end{bmatrix}. \tag{35}$$

Finally we must connection of the strictly proper realisation (30) and polynomial realisation (31) and we obtain the following matrices:

$$
\underbrace{\begin{bmatrix} 1 & 0 & 0 & 0 & 0 & 0 \\ 0 & 1 & 0 & 0 & 0 & 0 \\ 0 & 0 & 1 & 0 & 0 & 0 \\ \hline 0 & 0 & 0 & 0 & 0 & 0 \\ 0 & 0 & 0 & 0 & 1 & 0 \\ 0 & 0 & 0 & 0 & 0 & 1 \end{bmatrix}}_{\mathbf{E}}
\underbrace{\begin{bmatrix} 0.7 & 0.1 & 0.08 & 0 & 0 & 0 \\ 1 & 0 & 0 & 0 & 0 & 0 \\ 0 & 1 & 0 & 0 & 0 & 0 \\ \hline 0 & 0 & 0 & 0 & 0 & 1 \\ 0 & 0 & 0 & 1 & 0 & 0 \\ 0 & 0 & 0 & 0 & 1 & 0 \end{bmatrix}}_{\mathbf{A}}
\underbrace{\begin{bmatrix} w(s_1, v_1) \\ 0 \\ 0 \\ \hline w(s_2, v_1) \\ 0 \\ 0 \end{bmatrix}}_{\mathbf{B}},
\tag{36}
$$

$$
\underbrace{\left[\begin{array}{ccc|ccc} \dfrac{4.4}{w(s_1, v_1)} & \dfrac{1.2}{w(s_1, v_1)} & \dfrac{2.16}{w(s_1, v_1)} & \dfrac{1}{w(s_2, v_1)} & \dfrac{1}{w(s_2, v_1)} & \dfrac{2}{w(s_2, v_1)} \end{array} \right]}_{\mathbf{C}}
$$

The desired realisation of the (27) is given by (36) for $w(s_1, v_1) \in \mathbb{R}$ and $w(s_2, v_1) \in \mathbb{R}$.

5 Concluding Remarks

A method for computation of a minimal realisation of a given proper transfer function of fractional singular one-dimensional continuous-time linear systems has been proposed. Sufficient conditions for the existence of a minimal realisation of a given proper transfer function have been established. Proposed method based on one-dimensional digraph theory. The effectiveness of the algorithm has been illustrated with some numerical examples. It should be noted, that, the strength of the method is possibility of the division system into two simplest sub-systems. Finally, we receive a set of solutions which is equal to the product of the strictly proper transfer function realisation and polynomial realisation. Extension of those considerations for fractional singular discrete-time linear systems is possible.

References

1. Bru, R., Coll, C., Romero-Vivo, S., Sánchez, E.: Some problems about structural properties of positive descriptor systems, pp. 233–240. Springer, Heidelberg (2003). https://doi.org/10.1007/978-3-540-44928-7_32
2. Bru, R., Coll, C., Sánchez, E.: Structural properties of positive linear time-invariant difference-algebraic equations. Linear Algebra Appl. **349**(1), 1–10 (2002). https://doi.org/10.1016/S0024-3795(02)00277-X
3. Dai, L. (ed.): System analysis via transfer matrix, pp. 197–230. Springer, Heidelberg (1989). https://doi.org/10.1007/BFb0002482
4. Das, S.: Functional Fractional Calculus. Springer, Heidelberg (2011). https://doi.org/10.1007/978-3-642-20545-3
5. Dodig, M., Stošic, M.: Singular systems, state feedback problem. Linear Algebra Appl. **431**(8), 1267–1292 (2009). https://doi.org/10.1016/j.laa.2009.04.024

6. Farina, L., Rinaldi, S.: Positive Linear Systems: Theory and Applications. Series on Pure and Applied Mathematics. Wiley-Interscience, New York (2000)
7. Horn, R.A., Johnson, C.R.: Topics in Matrix Analysis. Cambridge University Press, Cambridge (1991)
8. Hryniów, K., Markowski, K.A.: Parallel digraphs-building algorithm for polynomial realisations. In: Proceedings of 2014 15th International Carpathian Control Conference (ICCC), pp. 174–179 (2014)
9. Hryniów, K., Markowski, K.A.: Optimisation of digraphs creation for parallel algorithm for finding a complete set of solutions of characteristic polynomial. In: Proceedings of 20th International Conference on Methods and Models in Automation and Robotics, MMAR 2015, Miedzyzdroje, Poland, 24–27 August, pp. 1139–1144 (2015). https://doi.org/10.1109/MMAR.2015.7284039
10. Hryniów, K., Markowski, K.A.: Classes of digraph structures corresponding to characteristic polynomials. In: Challenges in Automation, Robotics and Measurement Techniques: Proceedings of AUTOMATION-2016, March 2-4, 2016, Warsaw, Poland, pp. 329–339. Springer International Publishing (2016). https://doi.org/10.1007/978-3-319-29357-8_30
11. Kaczorek, T.: Positive 1D and 2D Systems. Springer, London (2001)
12. Kaczorek, T., Sajewski, L.: The Realization Problem for Positive and Fractional Systems. Springer, Berlin (2014). https://doi.org/10.1007/978-3-319-04834-5
13. Kublanovskaya, V.N.: Analysis of singular matrix pencils. J. Sov. Math. **23**(1), 1939–1950 (1983). https://doi.org/10.1007/BF01093276
14. Luenberger, D.G.: Positive linear systems. In: Introduction to Dynamic Systems: Theory, Models, and Applications. Wiley, New York (1979)
15. Machado, J., Lopes, A.M.: Fractional state space analysis of temperature time series. Fractional Calc. Appl. Anal. **18**(6), 1518–1536 (2015)
16. Machado, J., Mata, M.E., Lopes, A.M.: Fractional state space analysis of economic systems. Entropy **17**(8), 5402–5421 (2015)
17. Markowski, K.A.: Digraphs structures corresponding to minimal realisation of fractional continuous-time linear systems with all-pole and all-zero transfer function. In: 2016 IEEE International Conference on Automation, Quality and Testing, Robotics (AQTR), pp. 1–6 (2016). https://doi.org/10.1109/AQTR.2016.7501367
18. Markowski, K.A.: Realisation of continuous-time (fractional)descriptor linear systems. In: Szewczyk, R., Zieliński, C., Kaliczyńska, M. (eds.) Automation 2017, pp. 204–214. Springer, Cham (2017)
19. Markowski, K.A.: Realisation of linear continuous-time fractional singular systems using digraph-based method. First approach. J. Phys. Conf. Ser. **783**(1), 012,052 (2017). http://stacks.iop.org/1742-6596/783/i=1/a=012052
20. Markowski, K.A.: Minimal positive realizations of linear continuous-time fractional descriptor systems: two cases of an input-output digraph structure. Int. J. Appl. Math. Comput. Sci. **28**(1), 9–24 (2018)
21. Martynyuk, V., Ortigueira, M.: Fractional model of an electrochemical capacitor. Sig. Process. **107**, 355–360 (2015)
22. Nishimoto, K.: Fractional Calculus. Decartess Press, Koriama (1984)
23. Ortigueira, M.D.: Fractional Calculus for Scientists and Engineers. Academic Press, Springer, Netherlands (2011). https://doi.org/10.1007/978-94-007-0747-4
24. Ortigueira, M.D., Rivero, M., Trujillo, J.J.: Steady-state response of constant coefficient discrete-time differential systems. J. King Saud Univ. Sci. (2015)
25. Podlubny, I.: Fractional Differential Equations. Academic Press, San Diego (1999)

26. Podlubny, I., Skovranek, T., Datsko, B.: Recent advances in numerical methods for partial fractional differential equations. In: 2014 15th International Carpathian Control Conference (ICCC), pp. 454–457. IEEE (2014)
27. Virnik, E.: Stability analysis of positive descriptor systems. Linear Algebra Appl. **429**(10), 2640–2659 (2008). https://doi.org/10.1016/j.laa.2008.03.002

Digraphs Structures with Weights Corresponding to One-Dimensional Fractional Systems

Konrad Andrzej Markowski$^{(\boxtimes)}$

Faculty of Electrical Engineering, Institute of Control and Industrial Electronics,
Warsaw University of Technology, Koszykowa 75, 00-662 Warsaw, Poland
Konrad.Markowski@ee.pw.edu.pl
http://markowski.edu.pl

Abstract. In this paper, after extensive study and experimentation, the first classification of digraphs structures $\mathcal{D}$ corresponding to one-dimensional (1D) fractional continuous-time and discrete-time systems has been presented. It was found that digraph structures created can be divided into three classes $\mathcal{K}_1$, $\mathcal{K}_2$, $\mathcal{K}_3$ with a different feasibility for different polynomials. Additional two cases of possible input-output digraph structure $\mathcal{IO}_1$, $\mathcal{IO}_2$ was investigated and discussed. It should be noted, that the proposed digraph classes give the opportunity to easily determine the realization of the dynamic system as a set of matrices $(\mathbf{A}, \mathbf{B}, \mathbf{C})$. The proposed digraphs classification was illustrated with some numerical examples.

1 Introduction

In recent years, many researchers have been interested in fractional continuous-time and discrete-time linear systems. The first definition of the fractional derivative was introduced by Liouville and Riemann at the end of the 19th century [23]. Mathematical fundamentals of fractional calculus are given in the monographs: Nishimoto [23], Miller and Ross [20], Podlubny [27], Das [2], Ortigueira [24] or Kaczorek and Sajewski [8]. Some others applications of fractional-order systems can be found in Petras et al. [26], Podlubny et al. [28], Machado and Lopes [10], Machado et al. [11], Martynyuk and Ortigueira [19], Ortigueira et al. [25] or Muresan et al. [22].

In analysis of fractional systems there exists many difficult problems. One of them is the realization problem. In many books: Farina and Rinaldi [3]; Dai [1]; Luenberger [9]; Kaczorek and Sajewski [8] or Monje et al. [21] we can find constant matrices form, which satisfies the system described by transfer matrix. It should be noted that in general there exists a set of possible realizations.

In last three years, there was a presentation of the first version of the parallel algorithm for finding the characteristic polynomial realizations of a two-dimensional dynamic system which is based on the multi-dimensional digraphs theory [4]. It is an alternative method to determination of the realization in

© Springer Nature Switzerland AG 2020
R. Szewczyk et al. (Eds.): AUTOMATION 2019, AISC 920, pp. 245–257, 2020.
https://doi.org/10.1007/978-3-030-13273-6_24

the constant matrix forms, which satisfy the system described by the transfer function. In the next paper, the algorithm has been improved for finding all possible sets of matrices which fit into the system transfer function [5,7]. As a result, in paper [6] three classes of digraphs structures were proposed. It should be noted that, these classes are correct only for two-dimensional systems. In the next stage of the research, digraph-based method have been used to solve the realization problem of fractional systems. In paper [14,16,18] some modifications of the digraphs based method was proposed and the authors tried to use digraphs method in the area of the electrical circuit [13,15] and descriptor systems [16,17].

Still, one-dimensional digraphs used for the realizations problem are not fully defined and determined. It is known that some structures obtained will not satisfy the polynomial or there will be a need for solving a system of polynomial equations to determine the coefficients of state matrices from them. On the other hand, there exist some input-output digraphs structures which can be useful during determination realization (especially minimal realization).

The main propose of this paper is to present some digraph-structures which can be grouped into some different classes. The problem is related to linear one-dimensional continuous-time and discrete-time non-commensurate fractional systems. In this paper after extensive study and experimentation, for the first time digraphs have been divide into classes. The division was made from the point of view relationship between digraph-structures and state, input and output matrices-structures. Determination of digraphs structures in class $\mathcal{K}_1$, $\mathcal{K}_2$, $\mathcal{K}_3$ and connect them with input and output $\mathcal{IO}_1$, $\mathcal{IO}_2$ digraph structures, gives us the opportunity to find all possible proper digraph structures corresponding to the given transfer function. It is a very important especially in state-space an analysis of dynamical systems. It should be noted, that concepts of reachability and controllability of the dynamic system are related to the structure of the graphs.

2   Fractional Systems

2.1   Continuous-Time Fractional System

Let be given the continuous-time fractional linear system described by equations:

$$_0\mathfrak{D}_t^\alpha x(t) = \mathbf{A}x(t) + \mathbf{B}u(t), \quad 0 < \alpha \leqslant 1, \tag{1}$$
$$y(t) = \mathbf{C}x(t) + \mathbf{D}u(t),$$

where $x(t) \in \mathbb{R}^n$, $u(t) \in \mathbb{R}^m$, $y(t) \in \mathbb{R}^p$ are the state, input and output vectors, respectively and $\mathbf{A} \in \mathbb{R}^{n \times n}$, $\mathbf{B} \in \mathbb{R}^{n \times m}$, $\mathbf{C} \in \mathbb{R}^{p \times n}$ and $\mathbf{D} \in \mathbb{R}^{p \times m}$. The following Caputo definition of the fractional derivative will be used:

$$_a^C\mathfrak{D}_t^\alpha = \frac{d^\alpha}{dt^\alpha} = \frac{1}{\Gamma(n-\alpha)} \int_a^t \frac{f^{(n)}(\tau)}{(t-\tau)^{\alpha+1-n}} d\tau, \tag{2}$$

where $\alpha \in \mathbb{R}$ is the order of fractional derivative, $f^{(n)}(\tau) = \frac{d^n f(\tau)}{d\tau^n}$ and $\Gamma(x) = \int_0^\infty e^{-t} t^{x-1} dt$ is the gamma function. The Laplace transform of the derivative-integral (2) is given in [27]. After using the Laplace transform to (1) we can determine transfer matrix of the system in the following form:

Transfer matrix of the system (1) has the following form:

$$\mathbf{T}(s) = \mathbf{C} [\mathbf{I}_n s^\alpha - \mathbf{A}]^{-1} \mathbf{B} + \mathbf{D}. \tag{3}$$

2.2   Discrete-Time Fractional System

Let us consider the fractional discrete-time linear system, described by the equations:

$$\Delta^\alpha x_{i+1} = \mathbf{A} x_i + \mathbf{B} u_i \tag{4a}$$

$$y_k = \mathbf{C} x_i + \mathbf{D} u_i, \tag{4b}$$

where $x_i \in \mathbb{R}^n$, $u_i \in \mathbb{R}^m$, $y_i \in \mathbb{R}^p$ for $i \in \mathbb{Z}_+$ are the state, input and output vectors, respectively and $\mathbf{A} \in \mathbb{R}^{n \times n}$, $\mathbf{B} \in \mathbb{R}^{n \times m}$, $\mathbf{C} \in \mathbb{R}^{p \times n}$, $\mathbf{D} \in \mathbb{R}^{p \times m}$ and $\alpha \in \mathbb{R}$. The following fractional α order difference of the function x_i will be used:

$$\Delta^\alpha x_i = \sum_{j=0}^{k} (-1)^j \binom{\alpha}{j} x_{i-j} \tag{5}$$

where $0 < \alpha < 1$, $\alpha \in \mathbb{R}$ and

$$\binom{\alpha}{i} = \begin{cases} 1 & for\ i = 0 \\ \dfrac{\alpha(\alpha - 1) \dots (\alpha - i + 1)}{i!} & for\ k = 1, 2, \dots \end{cases}$$

The $\mathcal{Z}$–transform of the (5) is given in [8].

The transfer matrix of the system (4a)–(4b) has the following form:

$$\mathbf{T}(z) = \mathbf{C} [\mathbf{I}_n (z - c_\alpha) - \mathbf{A}]^{-1} \mathbf{B} + \mathbf{D} \tag{6}$$

where:

$$c_\alpha = c_\alpha(i, z) = \sum_{j=1}^{i+1} (-1)^{j-1} \binom{\alpha}{j} z^{1-j} \tag{7}$$

3   Problem Formulation

Further considerations will be presented for the single-input single-output (SISO) one-dimensional fractional continuous-time linear system described by the model (1). In that case, the transfer matrix becomes a transfer function which can be considered as a pseudo-rational function of the variable $\lambda = s^\alpha$ in the following form:

$$T(\lambda) = \mathbf{C} \left[\mathbf{I}_n \lambda - \mathbf{A} \right]^{-1} \mathbf{B} + \mathbf{D}. \tag{8}$$

The transfer function (8), can be rewritten in the following form:

$$T(\lambda) = \frac{n(\lambda)}{d(\lambda)} = \frac{b_n \lambda^n + b_{n-1} \lambda^{n-1} + \ldots + b_1 \lambda + b_0}{\lambda^n + a_{n-1} \lambda^{n-1} + \ldots + a_1 \lambda + a_0}. \tag{9}$$

Remark 1. For single-input single-output (SISO) one-dimensional fractional discrete-time linear system described by the model (4a)–(4b), the transfer matrix (6) becomes a transfer function of the operator $\lambda = z - c_\alpha$ and can be written in the form (8) or as a polynomial function (9).

Using the transfer function (8) we have $\mathbf{D} = \lim_{\lambda \to \infty} T(\lambda) = [b_n]$, since $\lim_{\lambda \to \infty} [\mathbf{I}\lambda - \mathbf{A}] = 0$. The strictly proper transfer function is given by the equation:

$$T_{sp}(\lambda) = \frac{n(\lambda)}{d(\lambda)} = \frac{\widetilde{b}_{n-1} \lambda^{n-1} + \ldots + \widetilde{b}_1 \lambda + \widetilde{b}_0}{\lambda^n + a_{n-1} \lambda^{n-1} + \ldots + a_1 \lambda + a_0}, \tag{10}$$

where $\widetilde{b}_{n-j} = b_{n-j} - b_n a_{n-j}$ for $j = 1, \ldots, n$ and

$$d(\lambda) = \lambda^n - a_{n-1} \lambda^{n-1} - \ldots - a_1 \lambda - a_0 \tag{11}$$

$$n(\lambda) = \lambda^n - a_{n-1} \lambda^{n-1} - \ldots - a_1 \lambda - a_0 \tag{12}$$

The algorithm presented in [4,7] is based on the one-dimensional digraph theory to allow the creation of a complete set of solutions of transfer function realization. After using this method, we are able to create a set of solutions in the form of four $(\mathbf{A}, \mathbf{B}, \mathbf{C}, \mathbf{D})$. As the algorithm is able to find all the possible structures there is the need of checking the validity digraph representation of the characteristic polynomial. For some of them, it is impossible to obtain state, input and output matrices that will satisfy the transfer function. Other digraphs structures generate solutions for which it is needed to get the coefficients of matrices by solving a system of polynomial equations that in some cases can be under-determined.

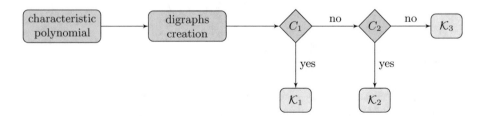

Fig. 1. Classes structure

4 Problem Solution

Digraphs structures can be considered in two spaces. The first is related to the characteristic polynomial (11) and the second with polynomial (12). In the first case we have three possible classes. In details this kine of class is consider in Sect. 4.1. In the second case we have two possible input-output digraphs structures. In details this kine of class is consider in Sect. 4.2.

4.1 Characteristic Polynomial Digraph-Structure

The digraph-structure can be divided into three classes. The first class denoted by $\mathcal{K}_1$ corresponding to a digraph consists of cycles corresponding to a binomial in a characteristic polynomial. This kind of digraph structure was investigated for example in papers [14], [12], [18]. There exists some digraph structures which contain additional arcs. In this case for determination of the realization of a characteristic polynomial, we should solve a set of linear equations to get wages of digraph arcs. This type of digraphs we denoted as class $\mathcal{K}_2$. Finally, there is a structure that cannot guarantee a proper solution for a given characteristic polynomial. This type of a digraph-structure we denoted by $\mathcal{K}_3$. Figure 1 illustrates relations between all digraphs structures divided into three classes $\mathcal{K}_1$, $\mathcal{K}_2$ and $\mathcal{K}_3$. It should be noted that for a multi-dimensional system the conditions which must be satisfied for digraphs structures are much more complicated. Classes and conditions for two-dimensional systems are presented in paper [6].

Let us consider the conditions C_1 and C_2 in details:

Condition C_1: There exist state matrices of the fractional linear continuous-time fractional system corresponding to the characteristic polynomial (11) if for digraph $\mathcal{D} = \mathcal{D}_1 \cup \mathcal{D}_2 \cup \ldots \cup \mathcal{D}_k$ all of the following conditions are met:

(C_{1a}): $\mathbb{V}_1(\mathcal{D}_1) \cap \mathbb{V}_2(\mathcal{D}_2) \cap \ldots \cap \mathbb{V}_k(\mathcal{D}_k) \neq \{\emptyset\}$, where $\mathbb{V}_k(\mathcal{D}_k)$ is a set of vertices of digraph $\mathcal{D}_k$ of k-th binomial;

(C_{1b}): in the obtained digraph does not appear additional cycles (one cycle in digraph corresponding to one binomial in characteristic polynomial).

Condition C_2: Digraph structures belonging to class $\mathcal{K}_2$ if for any term, existing in a characteristic polynomial (11), there exist any cycles corresponding to that term and we can determine the wages for all arcs in a digraph that satisfy a given characteristic polynomial.

Let us consider the classes $\mathcal{K}_1$, $\mathcal{K}_2$ and $\mathcal{K}_3$ in details:

Class $\mathcal{K}_1$: Digraph structures belonging to class $\mathcal{K}_1$ satisfy all characteristic polynomials for any coefficients of the characteristic polynomial (11). This type of digraph structures are the most examined in previous papers and they can be computed quickly in efficient way.

Class $\mathcal{K}_2$: Digraph structures belonging to this class satisfy a given characteristic polynomial with specific, but structures cannot be computed directly using the digraph-based method and we must solve a set of equations. This causes a significant slowdown in the state matrix determination algorithm.

Class $\mathcal{K}_3$: Digraph structures belonging to this class cannot satisfy the given characteristic polynomial or we are unable to determine the solution due to a problem with solving a system. These digraphs structures are considered as invalid for a given characteristic polynomial.

4.2   Input-Output Digraph-Structure

The input-output digraph-structure can be divide into two classes. The first class is denoted by $\mathcal{IO}_1$ corresponding to a digraph-structure received from a characteristic polynomial (see Sect. 4.1). We extend the digraph by adding two vertices: s corresponding to input matrix $\mathbf{B}$; y corresponding to output matrix $\mathbf{C}$. In this class, we connect: source vertex s with one vertex v_i $i \in \mathbb{Z}_+$ from the set of vertices $\mathbb{V}(\mathcal{D})$; all vertices v_i $i \in \mathbb{Z}_+$ from the set of vertices $\mathbb{V}(\mathcal{D})$ with output vertex y. In Fig. 2 presented digraph corresponding to class $\mathcal{IO}_1$. The second class is denoted by $\mathcal{IO}_2$ corresponding to a digraph-structure received from a characteristic polynomial (see Sect. 4.1). We extend the digraph by adding two vertices: s corresponding to input matrix $\mathbf{B}$; y corresponding to output matrix $\mathbf{C}$. In this class, we connect: source vertex s with all vertices v_i $i \in \mathbb{Z}_+$; one vertex v_i $i \in \mathbb{Z}_+$ from the set of vertices $\mathbb{V}(\mathcal{D})$ with output vertex y. In Fig. 3 presented digraph corresponding to class $\mathcal{IO}_1$.

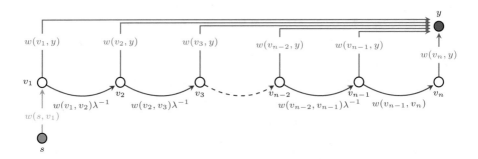

Fig. 2. Input-output $\mathcal{IO}_1$ digraph structure.

It should be noted that we have two possibilities of the connection source vertex s. The first $\mathcal{S}_1$ when s is connected with the vertex belonging to the common part of digraph vertices. Finally, we determine all paths from source s to output y of the length p equal to $2 \leqslant p \leqslant n + 1$. The second $\mathcal{S}_2$, when the source is connected with any vertex. Finally, we determine all paths from source s to output y of the length p equal to $2 \leqslant p \leqslant n + 1$, creation of sub-graphs

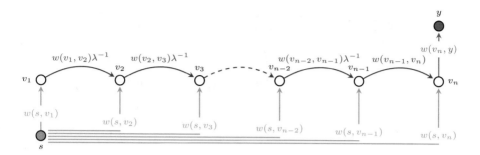

Fig. 3. Input-output $\mathcal{IO}_2$ digraph structure.

for each cycle by removal of certain vertices and multiplication of wages on sub-graphs by (-1). It should be noted that, the connection type $\mathcal{S}_2$ is much more complicated to analyse then the connection type $\mathcal{S}_1$.

5   Example

Let be given the following strictly proper transfer function of the continuous-time fractional system:

$$T_{sp}(s) = \frac{\widetilde{b}_3(s^\alpha)^3 + \widetilde{b}_2(s^\alpha)^2 + \widetilde{b}_1 s^\alpha + \widetilde{b}_0}{(s^\alpha)^4 - a_3(s^\alpha)^3 - a_2(s^\alpha)^2 - a_1 s^\alpha - a_0} \tag{13}$$

Determine state **A**, input **B** and output **C** matrices for $0 < \alpha < 1$.

Solution: The strictly proper transfer function (13) we can present as a pseudo-rational function of the variable $\lambda = s^\alpha$ in the form:

$$T_{sp}(\lambda) = \frac{n(\lambda)}{d(\lambda)} = \frac{\widetilde{b}_3 + \widetilde{b}_2 \lambda^{-1} + \widetilde{b}_1 \lambda^{-2} + \widetilde{b}_0 \lambda^{-3}}{1 - a_3 \lambda^{-1} - a_2 \lambda^{-2} - a_1 \lambda^{-3} - a_0 \lambda^{-4}} \tag{14}$$

The strictly characteristic polynomial $d(\lambda)$ we can write as a sum of binomials in the following form:

$$\underbrace{(1 - a_3 \lambda^{-1})}_{B_3} \cup \underbrace{(1 - a_2 \lambda^{-2})}_{B_2} \cup \underbrace{(1 - a_1 \lambda^{-3})}_{B_1} \cup \underbrace{(1 - a_0 \lambda^{-4})}_{B_0} \tag{15}$$

where $\cup$ is digraph operation on vertices called composition relative to vertices presented in details in paper [16]. Then, we must determine all possible digraphs realizations corresponding to binomials B_3, B_2, B_1 and B_0 (see Fig. 4), for $i, j, k, l \in \mathbb{Z}_+ = \{1, 2, 3, 4\}$ and $i \neq j \neq k \neq l$. It should be noted that we have: 4 possible realizations of the B_3; 6 possible realization of the B_2; 8 possible realizations of the B_1 and 2 possible realizations of the B_0. Finally we receive $4 \cdot 6 \cdot 8 \cdot 2 = 384$ variants of digraphs structures in class $\mathcal{K}_1$, $\mathcal{K}_2$ or $\mathcal{K}_3$.

Below we consider solutions for three possible classes.

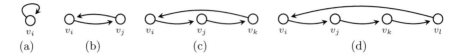

Fig. 4. Digraphs representations of the binomial (a) B_3; (b) B_2; (c) B_1; (d) B_0.

5.1   Class $\mathcal{K}_1$

By the use digraphs realizations we can obtain digraph-structure presented in Fig. 5. If condition C_1 is satisfy, then digraph-structure belonging to class $\mathcal{K}_1$. The first condition (C_{1a}) relates to the existence of the common part in the digraph (vertex v_1) is met. The second condition (C_{1b}) relates to non-existence of additional cycles in a digraph also is met. From the obtained digraphs presented in Fig. 5, we can write a state matrices in the form:

$$\mathbf{A} = \begin{bmatrix} w(v_1,v_1) & w(v_2,v_1) & w(v_3,v_1) & w(v_4,v_1) \\ w(v_1,v_2) & 0 & 0 & 0 \\ 0 & w(v_2,v_3) & 0 & 0 \\ 0 & 0 & w(v_3,v_4) & 0 \end{bmatrix} \qquad (16)$$

where:

$$a_3 = w(v_1,v_1); \; a_2 = w(v_1,v_2)w(v_2,v_1); \; a_1 = w(v_1,v_2)w(v_2,v_3)w(v_3,v_1)$$
$$a_0 = w(v_1,v_2)w(v_2,v_3)w(v_3,v_4)w(v_4,v_1) \qquad (17)$$

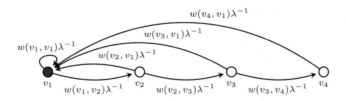

Fig. 5. Digraph in class $\mathcal{K}_1$

5.2   Class $\mathcal{K}_2$ and $\mathcal{K}_3$

By the use digraphs realizations we can obtain digraph-structure presented in Fig. 6. Investigated digraph-structure does not belong to the class C_1 because conditions (C_{1a}) and (C_{1b}) is not met. Therefore, investigated digraph does not belong to the class $\mathcal{K}_1$ as the conditions (C_{1a}) and (C_{1b}) are not met, and belong to the class $\mathcal{K}_2$ if it is possible determine coefficients a_3, a_2, a_1 and a_0 (see (19)) of the characteristic polynomial $d(\lambda)$ in (14). If determine coefficients (19) is not possible then digraph belong to class $\mathcal{K}_3$.

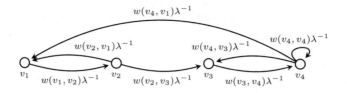

Fig. 6. Digraph in class $\mathcal{K}_2$ or $\mathcal{K}_3$

Using digraph structure presented in Fig. 6 we can write a state matrix in the form:

$$\mathbf{A} = \begin{bmatrix} 0 & w(v_2, v_1) & 0 & w(v_4, v_1) \\ w(v_1, v_2) & 0 & 0 & 0 \\ 0 & w(v_2, v_3) & 0 & w(v_4, v_3) \\ 0 & 0 & w(v_3, v_4) & w(v_4, v_4) \end{bmatrix} \tag{18}$$

where:

$$a_3 = -w(v_4, v_4); \quad a_2 = -w(v_1, v_2)w(v_2, v_1) - w(v_3, v_4)w(v_4, v_3); \tag{19}$$
$$a_1 = w(v_1, v_2)w(v_2, v_1)w(v_4, v_4);$$
$$a_0 = -w(v_1, v_2)w(v_2, v_3)w(v_3, v_4)w(v_4, v_1) + w(v_1, v_2)w(v_2, v_1)w(v_3, v_4)w(v_4, v_3)$$

5.3   Class $\mathcal{IO}_1$

Let us consider in details digraph-structure from class $\mathcal{K}_1$ presented in Fig. 5. Input-output digraph in class $\mathcal{IO}_1$ (corresponding to Fig. 5) presented on Fig. 2.

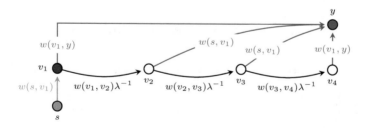

Fig. 7. Digraph in class $\mathcal{IO}_1$ corresponding to polynomial (14)

Then we can write the set of the equations in the following form (Fig. 7):

$$\begin{cases} & \|w(s, v_1)w(v_1, y)\| & = b_3 \\ \lambda^{-1} & \|w(s, v_1)w(v_1, v_2)w(v_2, y)\| & = b_2 \\ \lambda^{-2} & \|w(s, v_1)w(v_1, v_2)w(v_2, v_3)w(v_3, y)\| & = b_1 \\ \lambda^{-3} & \|w(s, v_1)w(v_1, v_2)w(v_2, v_3)w(v_3, v_4)w(v_4, y)\| & = b_0 \end{cases}$$

After solving them we can write **B** and **C** matrices in the following form:

$$\mathbf{B} = \begin{bmatrix} w(s,v_1) \\ 0 \\ 0 \\ 0 \end{bmatrix}, \quad \mathbf{C} = \begin{bmatrix} w(v_1,y) \ w(v_2,y) \ w(v_3,y) \ w(v_4,y) \end{bmatrix} \tag{20}$$

where:

$$w(v_1,y) = \frac{b_3}{w(s,v_1)}, \quad w(v_2,y) = \frac{b_2}{w(s,v_1)w(v_1,v_2)},$$

$$w(v_3,y) = \frac{b_1}{w(s,v_1)w(v_1,v_2)w(v_2,v_3)}$$

$$w(v_4,y) = \frac{b_0}{w(s,v_1)w(v_1,v_2)w(v_2,v_3)w(v_3,v_4)}$$

5.4 Class $\mathcal{IO}_2$

Let us consider in details digraph-structure from class $\mathcal{K}_1$ presented in Fig. 5. Input-output digraph in class $\mathcal{IO}_2$ (corresponding to Fig. 5) presented on Fig. 3.

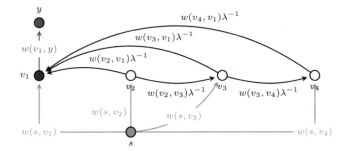

Fig. 8. Digraph in class $\mathcal{IO}_1$ corresponding to polynomial (14)

Then we can write the set of the equations in the following form (Fig. 8):

$$\begin{cases} \lambda^{-1} \left\| \begin{matrix} w(s,v_1)w(v_1,y) \\ w(s,v_2)w(v_2,v_1)w(v_1,y) + w(s,v_3)w(v_3,v_1)w(v_1,y) + \\ + w(s,v_4)w(v_4,v_1)w(v_1,y) \end{matrix} \right. & \begin{matrix} = \widetilde{b}_3 \\ \\ = \widetilde{b}_2 \end{matrix} \\ \lambda^{-2} \left\| \begin{matrix} w(s,v_2)w(v_2,v_3)w(v_3,v_1)w(v_1,y) + \\ w(s,v_3)w(v_3,v_4)w(v_4,v_1)w(v_1,y) \end{matrix} \right. & \begin{matrix} \\ = \widetilde{b}_1 \end{matrix} \\ \lambda^{-3} \left\| \begin{matrix} w(s,v_2)w(v_2,v_3)w(v_3,v_4)w(v_4,v_1)w(v_1,y) \end{matrix} \right. & = \widetilde{b}_0 \end{cases}$$

After solving them we can write **B** and **C** matrices in the following form:

$$\mathbf{B} = \begin{bmatrix} w(s,v_1) \\ w(s,v_2) \\ w(s,v_3) \\ w(s,v_4) \end{bmatrix}, \quad \mathbf{C} = \begin{bmatrix} w(v_1,y) \ 0 \ 0 \ 0 \end{bmatrix} \tag{21}$$

where:

$$w(s, v_1) = \frac{\widetilde{b}_3}{w(v_1, y)}; \quad w(s, v_2) = \frac{\widetilde{b}_0}{w(v_2, v_3)w(v_3, v_4)w(v_4, v_1)w(v_1, y)}; \quad (22)$$

$$w(s, v_3) = \frac{\widetilde{b}_1 w(v_3, v_4)w(v_4, v_1) - \widetilde{b}_0 w(v_3, v_1)}{w(v_3, v_4)^2 w(v_4, v_1)^2 w(v_1, y)};$$

$$w(s, v_4) = \frac{\begin{array}{l}\widetilde{b}_2 w(v_2, v_3)w(v_3, v_4)^2 w(v_4, v_1)^2 - \\ \widetilde{b}_1 w(v_3, v_4)w(v_4, v_1)w(v_3, v_1)w(v_2, v_3) - \\ \widetilde{b}_0 \left[w(v_2, v_1)w(v_3 v_4)w(v_4, v_1) - w(v_3, v_1)^2 w(v_2, v_3) \right]\end{array}}{w(v_1, v_2)w(v_2, v_3)w(v_3, v_4)w(v_1, y)}$$

Remark 2. We have two possible digraph-structures as: (a) connection class $\mathcal{K}_1$ and $\mathcal{IO}_1$; (b) connection class $\mathcal{K}_1$ and $\mathcal{IO}_2$. As a result, we receive the following minimal realizations as a set of the triples $(\mathbf{A}, \mathbf{B}, \mathbf{C})$: (16), (20) for $w(s, v_1) \in \mathbb{R}$ and (16), (21) for $w(v_1, y) \in \mathbb{R}$.

Remark 3. In this same way we can determine minimal realizations of the system for digraph-structures presented in Fig. 6. Then we get two additional digraph structures with classes $\mathcal{IO}_1$ and $\mathcal{IO}_2$.

6   Concluding Remarks

In this paper we have introduced the first classification of digraph structures that are used to solve a realization problem. Three classes $\mathcal{K}_1$, $\mathcal{K}_2$ and $\mathcal{K}_3$ of such structures are determined along with conditions how to classify digraph solutions. Thanks to these classes we can quickly determine state matrix $\mathbf{A}$ which corresponds to characteristic polynomial. To all of these classes, we can add two additional classes $\mathcal{S}_1$ and $\mathcal{S}_2$ by which we can determine input $\mathbf{B}$ and output $\mathbf{C}$ matrices. Through the use of such a division, we can easily use the digraph theory to determine the set of possible realization of the linear fractional dynamic systems.

Such basic classification is the first step for determination of properties of different digraph structures (for example how to check reachability and availability from digraph only, without need of system matrices) and introducing methods for finding best solutions fast.

References

1. Dai, L. (ed.): System Analysis Via Transfer Matrix, pp. 197–230. Springer, Heidelberg (1989). https://doi.org/10.1007/BFb0002482
2. Das, S.: Functional Fractional Calculus. Springer, Heidelberg (2011). https://doi.org/10.1007/978-3-642-20545-3
3. Farina, L., Rinaldi, S.: Positive Linear Systems: Theory and Applications. Series on Pure and Applied Mathematics. Wiley-Interscience, New York (2000)

4. Hryniów, K., Markowski, K.A.: Parallel digraphs-building algorithm for polynomial realisations. In: Proceedings of 2014 15th International Carpathian Control Conference (ICCC), pp. 174–179 (2014). http://dx.doi.org/10.1109/CarpathianCC.2014.6843592
5. Hryniów, K., Markowski, K.A.: Digraphs minimal realisations of state matrices for fractional positive systems. In: Szewczyk, R., Zielinski, C., Kaliczynska, M. (eds.) Progress in Automation, Robotics and Measuring Techniques. Advances in Intelligent Systems and Computing, vol. 350, pp. 63–72. Springer, Cham (2015). https://doi.org/10.1007/978-3-319-15796-2_7
6. Hryniów, K., Markowski, K.A.: Classes of digraph structures corresponding to characteristic polynomials. In: Challenges in Automation, Robotics and Measurement Techniques: Proceedings of AUTOMATION-2016, 2–4 March 2016, Warsaw, pp. 329–339. Springer (2016). https://doi.org/10.1007/978-3-319-29357-8_30
7. Hryniów, K., Markowski, K.A.: Parallel multi-dimensional digraphs-building algorithm for finding a complete set of positive characteristic polynomial realisations of dynamic system. In: Applied Mathematics and Computation (Submitted)
8. Kaczorek, T., Sajewski, L.: The Realization Problem for Positive and Fractional Systems. Springer, Berlin (2014). https://doi.org/10.1007/978-3-319-04834-5
9. Luenberger, D.G.: Positive linear systems. In: Introduction to Dynamic Systems: Theory, Models, and Applications. Wiley, New York (1979)
10. Machado, J., Lopes, A.M.: Fractional state space analysis of temperature time series. Fract. Calc. Appl. Anal. 18(6), 1518–1536 (2015)
11. Machado, J., Mata, M.E., Lopes, A.M.: Fractional state space analysis of economic systems. Entropy 17(8), 5402–5421 (2015)
12. Markowski, K.A.: Digraphs structures corresponding to minimal realisation of fractional continuous-time linear systems with all-pole and all-zero transfer function. In: IEEE International Conference on Automation, Quality and Testing, Robotics (AQTR), pp. 1–6 (2016). https://doi.org/10.1109/AQTR.2016.7501367
13. Markowski, K.A.: Digraphs structures corresponding to realisation of multi-order fractional electrical circuits. In: IEEE International Conference on Automation, Quality and Testing, Robotics (AQTR), pp. 1–6 (2016). https://doi.org/10.1109/AQTR.2016.7501368
14. Markowski, K.A.: Determination of minimal realisation of one-dimensional continuous-time fractional linear system. Int. J. Dyn. Control 5(1), 40–50 (2017). https://doi.org/10.1007/s40435-016-0232-3
15. Markowski, K.A.: Relations Between Digraphs Structure and Analogue Realisations with an Example of Electrical Circuit, pp. 215–226. Springer (2017). https://doi.org/10.1007/978-3-319-54042-9_20
16. Markowski, K.A.: Two cases of digraph structures corresponding to minimal positive realisation of fractional continuous-time linear systems of commensurate order. J. Appl. Nonlinear Dyn. 6(2), 265–282 (2017). https://doi.org/10.5890/JAND.2017.06.011
17. Markowski, K.A.: Minimal positive realisations of linear continuous-time fractional descriptor systems: two cases of input-output digraph-structure. Int. J. Appl. Math. Comput. Sci. 28(1), 9–24 (2018)
18. Markowski, K.A., Hryniów, K.: Finding a set of (A, B, C, D) realisations for fractional one-dimensional systems with digraph-based algorithm, vol. 407, pp. 357–368. Springer, Cham (2017). https://doi.org/10.1007/978-3-319-45474-0_32
19. Martynyuk, V., Ortigueira, M.: Fractional model of an electrochemical capacitor. Signal Process. 107, 355–360 (2015)

20. Miller, K., Ross, B.: An Introduction to the Fractional Calculus and Fractional Differenctial Equations. Wiley, New York (1993)
21. Monje, C.A., Chen, Y., Vinagre, B.M., Xue, D., Feliu, V.: Fractional-Order systems and Control: Fundamentals and Applications. Springer, London (2010). https://doi.org/10.1007/978-1-84996-335-0
22. Muresan, C.I., Dulf, E.H., Prodan, O.: A fractional order controller for seismic mitigation of structures equipped with viscoelastic mass dampers. J. Vibr. Control **22**(8), 1980–1992 (2016). https://doi.org/10.1177/1077546314557553
23. Nishimoto, K.: Fractional Calculus. Decartess Press, Koriama (1984)
24. Ortigueira, M.D.: Fractional Calculus for Scientists and Engineers. Springer, Dordrecht (2011). https://doi.org/10.1007/978-94-007-0747-4
25. Ortigueira, M.D., Rivero, M., Trujillo, J.J.: Steady-state response of constant coefficient discrete-time differential systems. J. King Saud Univ. Sci. **28**(1), 29–32 (2015)
26. Petras, I., Sierociuk, D., Podlubny, I.: Identification of parameters of a half-order system. IEEE Trans. Signal Process. **60**(10), 5561–5566 (2012)
27. Podlubny, I.: Fractional Differential Equations. Academic Press, San Diego (1999)
28. Podlubny, I., Skovranek, T., Datsko, B.: Recent advances in numerical methods for partial fractional differential equations. In: 15th International Carpathian Control Conference (ICCC), pp. 454–457. IEEE (2014)

The Issue of Adaptation of Diagnostic System to Protect Industrial Control Systems Against Cyber Threads

Paweł Wnuk$^{(\boxtimes)}$ ⓘ, Jan Maciej Kościelny ⓘ, Michał Syfert ⓘ,
and Piotr Ciepiela

Institute of Automatic Control and Robotics,
św. A. Boboli 8, 02-553 Warsaw, Poland
{p.wnuk,jmk,m.syfert}@mchtr.pw.edu.pl,
piotr.ciepiela@gmail.com

Abstract. The paper discusses issues related to the adaptation of diagnostic systems to the protection of industrial control systems (ICS) against cyber threats. Typical methods of attacking industrial systems are presented, along with a brief description of exemplary attacks. The potential consequences of attacks on both the operation of supervised industrial installations and the behaviour of operators were also discussed. The vulnerabilities of industrial control systems to attacks have been demonstrated, while the differences between them and typical IT systems were highlighted. Three main groups of ICS protection methods are discussed, their strengths and weaknesses were presented. An important part of the paper is the proposal to change the structure of the diagnostic system so that it would be able to better detect and distinguish between attacks and faults.

Keywords: Fault diagnostic · Security · Cyberattack ·
Industrial control systems

1 Introduction

In the sense of "safety" technical safeness is considered as a problem of preventing major industrial accidents caused by the unreliability of technological installation components (e.g. pipeline cracks or leaks), faults of control system components or human errors. A safeness, is the sense of "security", considered as the protection against intentional hostile external attacks (e.g. hacker attacks on control systems) or sabotage actions carried out from inside is another issue. Despite various reasons, the effects of dangerous faults and attacks can be the same, e.g. fire, explosion, environmental contamination, destruction of installation, stopping the process, etc. Therefore, the overall security strategy should include and integrate security issues both in the sense of "safety" as well as "security" (Kosmowski et al. 2006). The purpose of this paper is to indicate that this postulate should also apply to on-line diagnostic systems of industrial processes. It brings to the integration of the task of fault diagnosis and

R. Szewczyk et al. (Eds.): AUTOMATION 2019, AISC 920, pp. 258–267, 2020.
https://doi.org/10.1007/978-3-030-13273-6_25

cyberattacks recognition within common advanced diagnostic system. So far, such solutions have not been created.

2 Faults and Cyberattacks on ICS Systems

Cyberattack on ICS (*Industrial Control Systems*) – also called cyber-physical attack – is defined as an intentional disruption of the proper functioning of the control system made in a virtual space (usually via the Internet), the purpose of which is to take control of the control system. Fault (or failure) is understood as the loss of the functional unit's ability to fulfil the required function. It is usually assumed, in the scope of diagnostics, that the fault is any destructive event causing deterioration in the quality of operation of the process (or one of the process component) that should be detected in the process of diagnosing. Therefore, the basic difference between a cyberattack and fault concerns the origin of the threat. Cyberattack is carried out intentionally by a human in order to cause certain physical losses (cyber-physical). Whereas, the fault occurs in an abrupt or incipient (slowly increasing) way due to destructive physical processes taking place in the device (material aging, wear etc.) or incorrect (unintentionally) use of the device.

Faults arise on site in a controlled technological installation or in the control system itself. Therefore, the faults can appear in measurement devices, actuators, process installation components, control units, units enabling process monitoring and maintenance as well as in communication networks. Cyberattack can be derived from anywhere on the globe – therefore, the cybercriminals, especially groups responsible for the so-called APT (advanced persistent threats), can maintain a sense of impunity. Cyberattacks are directed at DCS (Distributed Control Systems), SCADA (Supervisory Control and Data Acquisition), PLC and PAC units (Programmable Logic Controllers, Programmable Automation Controllers), and HMI (Human Machine Interface) control panels. A critical control loop, as well as measuring and actuator devices in such a circuit may be their target.

Recent attacks on industrial facilities indicate, that the threat of critical infrastructure objects with cyberattacks having an ideological (terrorist) or political background is very real. It is especially true as the cost of preparing and conducting the cyberattack is disproportionately low compared to the cost of physical attack such as sabotage. At the same time, the propaganda effect may be large, while the political and criminal consequences for the perpetrators are negligible or even none, due to known difficulties in detecting the source of the attack and proving the blame (Kozak et al. 2016).

The ease of conducting the attack has been increased by the prevailing trend to ensure a high degree of openness of ICS systems (the use of standard operating systems and networks, open program platforms and connection to the IT network and the Internet) present in previous years (Kościelny et al. 2017).

Both faults and cyberattacks, if they got through the standard layers of protection, manifest in various changes in the operation of the automatic-control system and the course of the process deviating from its normal state. The arising changes are observed by the operator as a sequence of alarms informing about exceeding alarm limits by particular process variables. In the case of serious faults or cyberattacks, the number of alarms is very high. Overflow of alarms is a basic disadvantage of control systems, as it cause information overload. The EEMUA data shows that the average daily number of alarms in the petrochemical industry is about 1500, and in the energy industry 2000, while, according to the recommendations, it should not exceed 144. Large delays in detection and the effect of masking of symptoms by control loops are another disadvantages of automatic-control systems. These disadvantages result from the usage of a simple limits control for the detection of emergency states. The inference about the reasons of the alarms is not carried out in control systems. The process operators are responsible for the fulfilment of this task.

The interpretation of a large number of alarms generated in short-term is a serious problem for the operators. There appears the phenomenon of information overload that results with higher stress level. Under these conditions, the operators are unable to formulate a correct diagnosis, i.e. to recognize the existing hazards. This increases the likelihood of undertake improper protective actions, which consequences cumulate with the previously occurred faults causing serious failures. The mechanism of such unfavourable (positive) feedback was the cause of many serious failures in nuclear and conventional power plants as well as chemical plants (including explosions at the Texaco's Milford Haven refinery in 1994). In addition, if the cause of a faulty control system operation is a cyberattack, the interference in the control system may consist in modifying the operation of the alarm system in such a way as to hide the attack symptoms from the operator.

3   Characteristics of Attacks on ICS Systems

While the cyber security problem is known in the IT environment, it is still undervalued by an industry which is not free of cyber threats. The effects of cyberattacks on industrial facilities, widely known as Operation Technology (OT), can be catastrophic for property, health and life of employees and the environment. On the other hand, the ICS systems are less resistant to cyber threats than classic systems. A comparison of selected features of classic IT systems, ICS systems and controllers and field devices is given in Table 1.

Table 1. The specific features of ICS systems.

	IT	OT – ICS	OT – PLC
Computational power, resources	Large	Large	Small
Lifecyle	3–5 years	5–10 years	>10 years
Updates frequency	High	Medium	Small
Updates mode	Automatic	Manual	Manual
Work mode	Intermittent	Continuous	Continuous, real time
Security-related additional software	Yes	Yes/No	No
Schematic work	Low	High	High
Encrypted communication	Yes	Yes/no	No

Presented comparison shows that ICS systems, and in particular controllers, are more difficult to secure and protect. For example, for the security reasons it is recommended the IT systems should be updated regularly, often even more than once a day, but such approach is not possible for most OT systems. Their update usually involves stopping the operation of the device being updated, which is unacceptable in the case of controllers. It is acceptable only during planned stops, taking place relatively rarely, in a cycle that does not fit the requirements of the security update.

On the other hand, from the analysis of available data on the types of attacks on industrial control systems, it can be seen that the devices which can have the susceptibility to attacks similar to classical IT systems (from the group OT/ICS) have been much more often attacked so far (Fig. 1).

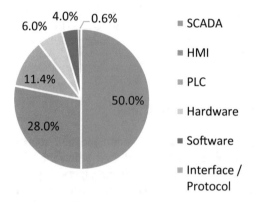

Fig. 1. Most often attacked ICS elements (Gritsay et al. 2012).

Nearly 80% of attacks, according to (Gritsai et al.), is on either the SCADA system or the operator's panel. Modern SCADA systems are created as applications that work under Windows operating system. It entails their sensitivity to any detected vulnerabilities in the operating system itself. Usually, modern operator's panels are also computers based on Intel or ARM architecture, and operate under the control of Windows, or less frequently, systems from the Unix/Linux or Android family. Analogously as in the case of SCADA systems, there is a problem of gaps in the operating system and their use to infect the ICS system. As a consequence, their operation is disturbed or even the destruction of the supervised process/device is possible.

The attacks on field devices like PLC controllers, sensors, intelligent actuators, or directly on industrial networks are still carried out much less often. There may be many reasons for this, but it seems that the basic two causes are:

1. Lack of deep knowledge of these techniques and devices among hackers. There are still not many commonly known vulnerabilities and gaps in the controllers protection systems, just like the controllers itself are not widely known. The market of PLCs, intelligent measuring devices and actuators is much more diversified than the market of standard PCs. It means less possibilities to use even already found vulnerabilities in protections. In the case of general-purpose computer systems, one has three

operating systems on the market which have to be considered in real-life, each implemented in millions of devices. In the case of PLCs, the number of manufacturers is much higher and often they have local coverage. At the same time, the amount of sold devices is much smaller. The limited hardware resources and programming possibilities of the devices from this group are also not without significance.

2. The separation of the OT network and the devices connected to it from the Internet. In most cases, such devices are not directly accessible, i.e. they do not have direct Internet connection. Such a way of connecting the devices implies the need for the hackers firstly to take control on the systems connecting controllers and field devices to the intranet, and then the Internet. Such systems are mostly SCADA systems. Thus, their penetration is necessary, in many cases, to carry out an effective attack on field devices. However, when the SCADA system has been taken over, there is no point in attacking devices dependent on it.

3. Despite the reasons above with the continuous growth of attacks on OT components there is a clear trend to identify and use vulnerabilities related to lower layers in particular PLCs.

4   Cyberattack Detection Methods

The evolution of ICS architecture and the communication technology makes them more and more susceptible to cyberattacks. Usually the vulnerabilities of modern ICS can be divided into three main categories: attack on supervisory control layer, attack on automatic control and communication layer, and attack on technical processes.

Commercially used security systems against cyberattacks, similarly like threats, originate from the IT environment. In this environment the methods for both the detection of intrusion and attack as well as prevention were developed. According to the work (Van Do 2015), the ways of protection of digital control systems can be divided as presented on Fig. 2.

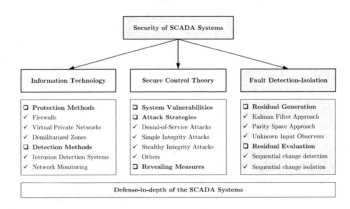

Fig. 2. Attack detection and isolation methods (Van Do 2015).

First group of methods are directly incorporated from IT systems. Those methods usually are focused on ensuring confidentiality and integrity of information (Bishop 2004). The confidentiality of data is usually performed by access control, authentication and data encryption methods. On the other hand, also the integrity of data (to ensure that there is no unauthorized modification of processed information) is also addressed by those methods.

There are also proposed modifications of IT methods (Krutz 2006; DOE 2002) in order to improve the security of SCADA/DCS systems against cyberattack, including designing specific firewalls between OT and corporate networks, using VPNs for data transmission over Internet, or developing dedicated Intrusion Detection Systems or network monitors.

It is believed that mentioned methods among with their modifications may improve security of the ICS as well as reduce consequences of successful cyberattack. However, known incidents (e.g. Stuxnet) have given evidence that these IT-based tools do not protect other components that SCADA/DCS in sufficient way. Moreover – ICS systems are in many aspects different form general purpose IT systems (as shown in Table 1). The requirement of continuous work prevents from installing updates to system itself as well as to additional security related software, like antivirus. Some of IT solutions (e.g. strong cryptography) require a lot of computational resources, unavailable on PLC controllers.

Some methods based on AI/learning of typical behaviour of ICS/typical network traffic are also hard to implement to ICS systems due to specific working conditions – typically very schematic (process is in steady state) with rare huge changes (start/stop of installation, changing a setpoints, product, etc.). Such tools do not have a possibility to learn typical, but extremely rare situations, what causes false alarms.

Second group of methods are techniques based on secure control approach. Those methods are focused mainly on analysing the security of networked control systems against cyberattack. In this approach two main types of attacks are distinguished: DoS (Denial of Service) and Integrity attacks. In first case an attacker tries to break into a communication channel between control centre and actuators/sensors, and thus changing their behaviour. Second case considers modification of data packets on control network, carried without knowledge of process model (simple attack case) or with knowledge of process and control model (stealth attack). Stealth attacks can bypass traditional anomaly detectors by simulating regular network traffic.

Third group of methods are Fault Detection and Isolation (*FDI*) techniques. It has been shown in (Kościelny 2017) that FDI systems can be used to detect cyberattacks as well as sabotage realized by employees (man in the middle attack). In FDI approach technical safety is considered as the problem of preventing major industrial failures caused by the unreliability of technological installation components, e.g. pipeline cracks, faults of control system components and human errors. On the other hand – the safety in the sense of security as a matter of protection against intentional hostile external attacks, e.g. hacker attacks on control systems, and sabotage activities conducted from the inside is another aspect. Despite the various causes, the effects of severe faults and cyberattacks can be the same, e.g. fire, explosion, environmental pollution, installation damage, process stoppage. The occurring symptoms of these hazards may be the same or very similar. Finally, as shown in (Kościelny 2017), cyberattacks can be detected in the control system by the same methods as faults. This all leads to the conclusion that the diagnostics of faults and cyberattacks should be implemented in a single diagnostic system (Fig. 3).

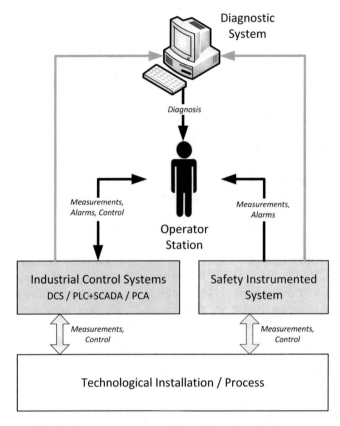

Fig. 3. Diagnostic system in the safety and control system.

The overall security strategy should cover and integrate safety issues in the sense of safety and security.

The above statement is confirmed by newest safety standards (e.g. IEC 61511) where security was added as one of the mandatory considerations for process safety.

The IT approach is dedicated to improving the security of SCADA systems by protection measures. The compatibility between the cyber layer and the physical infrastructure has not been considered. The secure control theory approach focuses mainly on investigating system vulnerabilities, designing stealthy/deception attacks, and proposing countermeasures for revealing undetectable attacks. The joint detection and isolation of attacks have not yet been considered in the comprehensive manner.

Above leads to conclusion that FDI methods, ready for dealing with abnormal situations occurring to stochastic-dynamical systems under the model uncertainties, disturbances and random noises are also good for detecting cyberattacks. All we need is to adapt them to the detection and isolation of cyberattacks.

5   Changes in FDI Systems Caused by New Area of Implementation

In order to adapt the diagnostic system to the detection of cyberattacks one should consider and analyse if several assumptions that have been tacitly assumed in the case of classic FDI systems are met.

The ability to read the same information, e.g. the value of a physical variable, from different sources is appreciate in the FDI systems. Such approach is used to confirm the value and/or to detect discrepancies and generate the appropriate diagnostic signal. However, it is not assumed, that redundancy is ensured by gathering the information from the same device (sensor) but with various communication paths, e.g. directly via a fieldbus protocol, such as Modbus, and downloaded from the DCS system. It is rather assumed, as a certainty, that the information ones entered into the IT system is no longer distorted, and even if it occurs happens, it is easily detected by methods typical for IT, such as checksum, retransmission of packets, etc.

In the case of systems designed to detect cyberattacks, the assumption about the possibility of changing the information during its processing by the computer system is the most legitimate. One of the system's goals will be to detect and inform about such changes. Therefore, the ability to read the same information from the same device should be treated as a desirable feature of such a system. In an optimal configuration, the FDI system should have an access to read analog signals directly from measuring devices. Nowadays, this postulate may be difficult to fulfil. However, it seems that providing an alternative channel of reading the information from the controller via fieldbus protocol (Modbus, HART, CAN, etc.), reading the same value from the controller, SCADA/DCS system, SIS systems or Historian systems makes sense. It seems especially reasonable considering the frequency of attacks on various elements of the control system, discussed in Sect. 3, in which the controller has not been attacked and was able to deliver correct measurements, while the information available (delivered) in the upper layers has been falsified.

The usage of a diagnostic systems for the detection of cyberattacks also requires the development of dedicated diagnostic tests. Examples of such tests are presented in (Kościelny et al. 2017, Cárdenas et al. 2011). The situation is similar when analysing the results of such tests. In addition to the typical assumption that each measuring device can be damaged, it should be also assumed that every measurement can be deceitful – the real physical value is measured correctly, correct control value affecting the process is generated based on this value, the only thing that is changed is the visible value passed to FDI/SCADA system or other observers.

The last important postulate concerns the passivity of the FDI system. Such system is also exposed to the same attacks as SCADA systems are because it is also an IT system operating on a classical operating system. The FDI system operation regime does not have to be as strong as in the case of control or supervisory systems. Temporary breaks in their operation and updates are allowed. Nevertheless, there is a risk of their taking over. The best way to minimize this risk is to protect the system from attackers e.g. by dedicated isolation.

In addition, one should consider the possibility of reading the information in a passive way where it is possible. Most fieldbus protocols do not provide/enable encryption of transmitted information. It means that it can be read by listening to communication between devices. In the case of data available in ICS of a higher-level, one should consider using read-only access in combination with hardware/software solutions that allow information flow only in one way – from the SCADA system to the FDI system.

6   Conclusions

FDI system can be treated as new layer of safety and security for ICS protection. It allows to detect, and isolate cyberattacks even in situation when other layers will fail. Despite various causes, the effects of serious damage and cyberattack can be similar (e.g. fire, explosion, environmental contamination, destruction of the installation). The symptoms of these risks may also be the same or similar.

Other described approaches to cyberattack detection has serious disadvantages. Security systems designed for typical IT systems does not considers sufficiently ICS specific. Secure control theory approach focuses mainly on investigating system vulnerabilities, and thus it's more a tool for proper system design than real on-line detection.

To fulfil security-related tasks, FDI system should be redesigned. Some remarks to the scope and type of these changes have been discussed.

References

Alonso-González, C., Rodriguez, J., Prieto, O., Pulido, B.: Ensemble methods and model based diagnosis using possible conflicts and system decomposition. In: Proceedings of the 23rd International Conference on Industrial Engineering and other Applications of Applied Intelligent Systems, IEA/AIE 10, pp. 116–125. Springer, Heidelberg (2010)

Kościelny, J.M., Syfert, M., Wnuk, P.: The idea of on-line diagnostics as a method of cyberattack. In: International Conference on Diagnostics of Processes and Systems, Advanced Solutions in Diagnostics and Fault Tolerant Control, DPS 2017, Sandomierz, 10–13 September 2017, pp. 449–457. Springer (2017). ISBN 978-3-319-64473-8 ISBN 978-3-319-64474-5 (eBook). https://doi.org/10.1007/978-3-319-64474-5_38

Kozak, A., Kościelny, J.M., Pacyna, P., Gołębiewski, D., Paturej, K., Swiątkowska, J.: Cybersecurity of industrial installations – the cornerstone of the "Industry 4.0" project and a chance for Poland. In: White paper on CYBERSEC 2016 (2016). (in Polish)

Van Do, L.: Sequential Detection and Isolation of Cyber-physical Attacks on SCADA Systems, Thèse de doctorat de l'UNIVERSITE DE TECHNOLOGIE DE TROYES (2015). 2015TROY0032

Bishop, M.: Introduction to Computer Security. Addison-Wesley Professional (2004)

DOE: 21 steps to improve cyber security of SCADA networks. Office of Energy Assurance, U.S. Department of Energy (2002)

Gritsai, G., Timorin, A., Goltsev, Y, Ilin, R., Gordeychik, S., Karpin, A.: SCADA safety in numbers, Positive Technologies (2012). https://www.ptsecurity.com/ww-en/

Krutz, R.L.: Securing SCADA Systems. Wiley Publication (2006)

Bajpai, S., Gupta, J.P.: Terror-proofing chemical process industries. Trans IChemE, Part B, Process Saf. Environ. Prot. **85**(B6), 559–565 (2007)

Kyoung-Dae, K., Kumar, P.R.: Cyber–physical systems: a perspective at the centennial. In: Proceedings of the IEEE, vol. 100, pp. 1287–1308, 13 May 2012

Kosmowski, K.T., Sliwinski, M., Barnert, T.: Functional safety and security assessment of the control and protection systems. In: Soares, G., Zio, E. (eds.) Safety and Reliability for Managing Risk. Taylor & Francis Group, London (2006). ISBN 0-415-41620-5

Pacyna, P., Rapacz, N., Chmielecki, T., Chołda, P., Potrawka, P., Stankiewicz, R., Wydrych, P., Pach, A.: OKIT. Metodyka ochrony teleinformacyjnych infrastruktur krytycznych. Wyd, PWN (2013)

Kościelny, J.M., Bartyś, M.: The requirements for a new layer in the industrial safety systems. In: 9th IFAC Symposium on Fault Detection, Supervision and Safety of Technical Processes, SafeProcess 2015, vol. 1333–1338, Paris, France, 2–4 September 2015. http://www.ifac-papersonline.net/

Cárdenas, A.A., Amin, S., Lin, Z.-S., Huang, Y.-L., Huang, C.-Y., Sastry, S.: Attacks against process control systems: risk assessment, detection, and response. In: Proceedings of the 6th ACM Symposium on Information, Computer and Communications Security, pp. 355–366. ACM (2011)

The Test Stand Research on HONDA NHX 110 Powered with Alternative Fuels: A Case Study

Adrian Chmielewski[(⊠)] [ID], Krzysztof Bogdziński,
Robert Gumiński [ID], Artur Małecki, Tomasz Mydłowski,
and Jacek Dybała [ID]

Institute of Vehicles, Warsaw University of Technology,
Narbutta 84 Street, Warsaw, Poland
{adrian.chmielewski,robert.guminski,
jacek.dybala}@pw.edu.pl, {k.bogdzinski,
t.mydlowski}@mechatronika.net.pl,
a.malecki85@gmail.com

Abstract. The work presents research on HONDA NHX 110 engine. The Internal Combustion engine (ICE engine) was fueled with gasoline RON95 and alternative fuels, including: methanol, ethanol and butanol. The experimental research was conducted for various ignition advance angle settings, at the authors' own dynamometer engine test stand. Tests were conducted for several hundred work cycles at maximum engine load with fully open throttle (the ICE engine works as distributed generation device in electricity generation configuration). In this work the effects of ignition advance angle change on indicated work, mechanical power, electrical power and torque figures produced by the engine has been shown. Moreover, the influence of ignition advance angle on the values of indicated and open pressure graphs for ICE engine have been analyzed and shown. Furthermore the influence of emission of nitrogen oxides – NOx and hydrocarbons – HC at different angles of the ignition advance have been presented.

Keywords: Alternative fuels · Measurement · Indicated work ·
Ignition advance angle · HC and NOx emissions

1 Introduction

Nowadays more and more attention is dedicated to distributed energy generation devices, especially devices that generate energy from renewable energy sources, and devices which utilize the waste heat energy to produce electricity in a single technological process (Combined Heat and Power devices – CHP). Development of such devices is essential in terms of energy efficiency improvement [1, 2] of European Union countries. In this context it is worth to focus particularly on small scale cogeneration devices, the so called microgeneration devices, in which the amount of electrical power produced does not exceed 50 kWe [2]. The Honda NHX 110 internal combustion engine considered in this article, using alternative fuels [3–14] or pure hydrogen as fuel [15, 16], could be categorized as such microgeneration device [17–19].

© Springer Nature Switzerland AG 2020
R. Szewczyk et al. (Eds.): AUTOMATION 2019, AISC 920, pp. 268–277, 2020.
https://doi.org/10.1007/978-3-030-13273-6_26

In such low power systems the information about long term, efficient operation [20–22] is extremely important. In order to determine the correct operational parameters (such as ignition advance angle, type of fuel, fuel dosage), not only the information about the operation conditions is required, but also a constant monitoring of toxic compound levels in exhaust gasses, such as: hydrocarbon emissions – HC, nitrous oxides – NOx, which are generated during the combustion process in such devices. The conducted quantitative and quality analysis allows for examination of influence of essential parameters (i.e. ignition advance angle, fuel type) on effective operation parameters of the engine, such as: electric power, mechanical power, torque, indicated work and indicated mean effective pressure in the cylinder.

In the following work, the research on low power Honda NHX internal combustion engine, fueled by RON95 gasoline, methanol, ethanol and butanol has been presented.

2   The Honda NHX 11 Engine Test Bench

The test bench consisted of the Honda NHX 110 internal combustion 4-stroke engine coupled with an electric machine in form of a brushless DC motor with permanent magnets, working as an electric generator. The combustion engine was equipped with a trifunctional catalytic converter and controlled by a programmable engine management unit ECU Master EMU [23]. The torque was transferred to the electric motor through a toothed belt transmission (ratio i = 1.42). The generated electric power was received by system consisting of a 3-phase bridge rectifier (rated for maximum voltage of 400 V and maximum current of 300 A), transistor module, resistor unit (0.05 Ω resistance). The parameters of the electric engine were: supply voltage range: 30–70 V, (with rotational speed increase of 150 rpm per 1 V, maximum of 10 500 rpm), current draw with no load: 13 A at 20 V. The transistor module was controlled by a custom made microprocessor controller, described in detail in [23, 24].

The measurement track of the test stand consisted of: Zemic L6N load cell for torque measurements (class C3 accuracy) and a SSI standard absolute encoder for combustion engine crankshaft angle acquisition (single turn encoder, 14 bit precission). The encoder – test stand communication allowed for measurement accuracy of 0.5° of crankshaft angle (CA) at 3800 rpm, and 1° CA at 7600 rpm respectively. The diagram of the test stand is presented in Fig. 1a. Figure 1b presents a photograph of the test stand.

A detailed description of the test stand has been presented in [23, 24].

In Sect. 3 the experimental research results for Honda NHX 110 fueled with: RON95 gasoline, methanol, ethanol and butanol, for various ignition advance angle values, are presented.

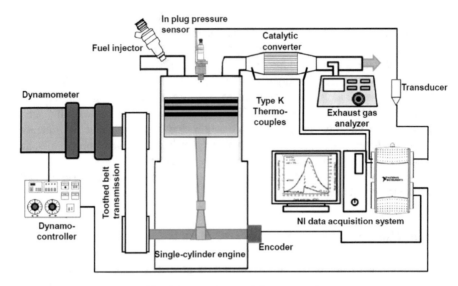

Fig. 1. The scheme of the test stand.

3 Experimental Research Results

The experimental research was conducted under steady operation conditions. The temperature of intake air was equal 298 ± 3 K, the atmospheric pressure was identical for the all measurements, at 1009 hPa. Experimental research was conducted for air-fuel equivalence ratio maintained at $\lambda = 1$. During measurements, the throttle was fully open and the load value set so the rotational speed of the engine was equal 5200 rpm.

3.1 Indicated Pressure Graphs and Mean Effective Pressure Graphs

In the following subsection presented are the open indicated pressure graphs acquired over several hundred work cycles for the following fuels: RON 95 gasoline (Fig. 2a), methanol (Fig. 2b), ethanol (Fig. 2c) and butanol (Fig. 2d). The measurements were conducted for ignition advance angle ranging from IAA = 10° to IAA = 60° with 5° increments for methanol and ethanol. RON 95 gasoline and butanol were measured for ignition advance angle ranging from IAA = 15° to IAA = 60° (the engine was unable to operate at ignition advance angle values lower than 15°).

It is worth pointing out that despite engine operation with very late ignition advance (IAA = 10°) was possible for methanol and ethanol, the pressure values for ethanol for several cycles (out of the nearly 350 cycles considered) were significantly lower in comparison to other fuels.

Figure 3 presents mean effective indicated pressure graphs (averaged between 0° CA and 720° CA) for various values of ignition advance angle values for the following fuels: RON 95 (Fig. 3a), Methanol (Fig. 3b), Ethanol (Fig. 3c) and Butanol (Fig. 4d).

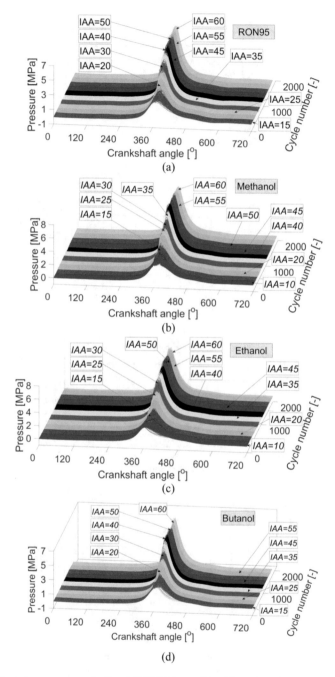

Fig. 2. Indicated pressure graphs for: (a) RON95 gasoline, (b) methanol, (c) ethanol, (d) butanol for several hundred consequitve cycles.

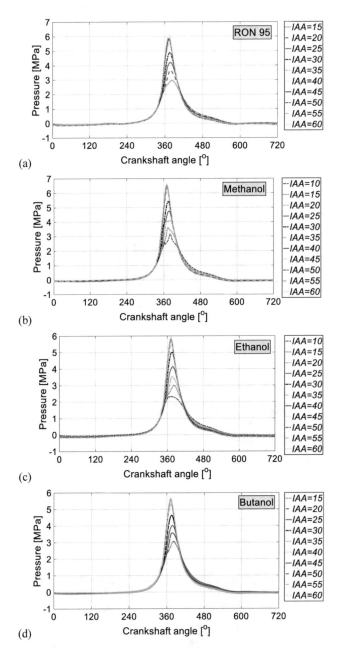

Fig. 3. The influence of the ignition advance angle on the average values of pressure in range from 0 to 720 CA for the ICE engine powered with: (a) gasoline RON95, (b) Methanol, (c) Ethanol, (d) Butanol.

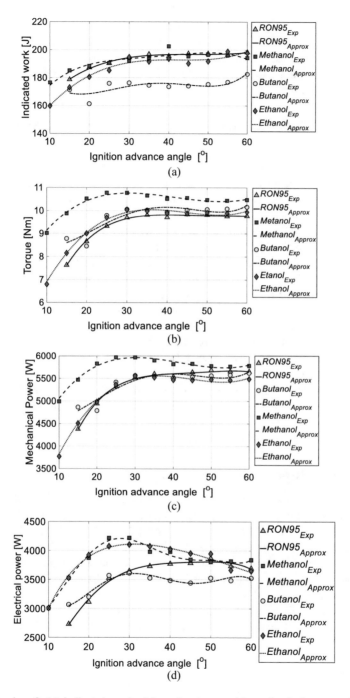

Fig. 4. Graphs of: (a) indicated work, (b) engine torque, (c) mechanical power, (d) electric power, for Honda NHX 110 for various ignition advance angle values (IAA).

From analysis of Figs. 3(a–d) it was concluded, that with increase of ignition advance angle the mean effective pressure value increases. Highest mean effective pressure value was observe for methanol at IAA = 60°, the value equal to 6.66 MPa (Fig. 3b). It is worth highlighting that this result was nearly 0.4 MPa greater than the mean effective pressure of RON95 at IAA = 60°.

3.2   Graphs of Indicated Work, Mechanical Power, Electric Power and Torque

In the following subsection presented are the graphs of indicated work (Fig. 4a), torque (Fig. 4b), mechanical power (Fig. 4c) and electrical power (Fig. 4d) of Honda NHX 110 fueled by: RON95, Methanol, Ethanol and Butanol, for various ignition advance angle values.

From analysis of Fig. 4(a) it can be concluded that the highest value of indicated work was achieved for IAA = 40° was equal to 202.5 J. Lower values of indicated work and electric power were measured for butanol, which has greater density and viscosity compared to other examined fuels. For butanol, the highest value of indicated work was observed for IAA = 60°, at 182.6 J, while the maximum electric power was observed for IAA = 30°, at 3.61 kWe. It is worth mentioning that for IAA = 60° for butanol, the emissions of hydrocarbons HC (Fig. 4a) and nitrous oxides NOx is highest, and was equal to respectively 163.5 ppm for HC and 2194 ppm for NOx.

With operational parameters taken into consideration, methanol is the best fuel, with highest values of torque measured (Fig. 4b) – T = 10.77 Nm for IAA = 25° and 30°, and highest mechanical power noted for IAA = 25°, at 5.97 kW (Fig. 4c) while the electric power equaled 4.22 kWe (Fig. 4d). It is worth mentioning that for methanol, the hydrocarbon emission (Fig. 5a) does not exceed 4 ppm, and is the lowest values out of all results for the fuels tested.

3.3   Graphs of Hydrocarbons and Nitrous Oxides Emissions

In the following subsection presented are graphs of hydrocarbon HC (Fig. 5a) emissions and nitrous oxides NOx (Fig. 5b) emissions for the engine fueled by: RON95 gasoline, methanol, ethanol and butanol. From analysis of Fig. 5 it can be concluded, that with increase of ignition advance angle, the emissions of both hydrocarbons and nitrous oxides increase. Highest hydrocarbon emission level was detected for RON95 and butanol at IAA = 60°. In case of RON95 the HC emission was 166 ppm and was the highest out of all tested fuels.

It is worth highlighting, that the increase of ignition advance angle lowers the temperature of exhaust gasses, which results in higher toxic compound emissions, such as HC and NOx.

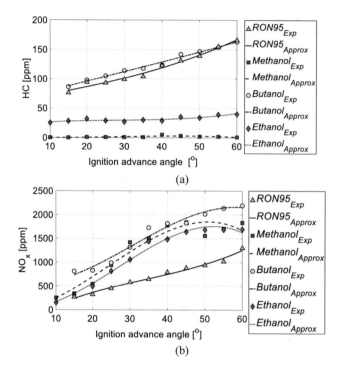

Fig. 5. Effect of ignition advance value change on: (a) hydrocarbon emissions, (b) nitrous oxides emission in exhaust gasses of the combustion engine fueled by RON95, methanol, ethanol and butanol.

4   Summary and Conclusions

Based on conducted research it was concluded that with increase of ignition advance angle up to IAA = 30°, for rotational speed of the shaft of ICE engine equal 5200 rpm, the values of: mean effective indicated pressure inside the cylinder, indicated work, torque, mechanical power and electrical power, increase for all tested fuels.

From the research conducted it was also concluded that methanol, ethanol and butanol can be used a substitute alternative fuels for RON95 gasoline. Particularly methanol, as increase the effective values of engine parameters, such as 11% increase in engine torque, mechanical and electrical power (compared to RON95) was observed. From an energy generation stand point, an internal combustion engine fueled by methanol could produce 11% more energy compared to RON95. It is worth highlighting that the emission of hydrocarbons in engine exhaust gasses for methanol was 95% lower in comparison to other tested fuels (RON95, ethanol and butanol).

Based on conducted research it was also concluded, that butanol can be used as an alternative fuel to RON95, however with worse results than methanol. The performance differences are caused mainly by the greater density and viscosity of butanol.

The disadvantages of using butanol as an alternative fuel to RON95 are: difficult engine start up and high emission of toxic compounds (e.g. 2 times greater nitrous oxides NOx emissions in comparison to RON95).

Currently the authors are proceeding with further research on the Honda NHX 110 engine, conducting tests for other alternative fuels, including hydrogen.

References

1. Directive 2012/27/EU of the European Parliament and of the Council of 25 October 2012 on energy efficiency, amending Directives 2009/125/EC and 2010/30/EU and repealing Directives 2004/8/EC and 2006/32/EC
2. Chmielewski, A., Gumiński, R., Mączak, J., Radkowski, S., Szulim, P.: Aspects of balanced development of RES and distributed micro cogeneration use in Poland: case study of a µCHP with Stirling engine. Elsevier Renew. Sustain. Energy Rev. **60**, 930–952 (2016)
3. Awad, O.I., Mamat, R., Ali, O.M., Sidik, N.A.C., Yusaf, T., Kadirgama, K., Kettner, M.: Alcohol and ether as alternative fuels in spark ignition engine: a review. Renew. Sustain. Energy Rev. **82**, 2586–2605 (2018)
4. Ilhak, M.I., Akansu, S.O., Kahraman, N., Ünalan, S.: Experimental study on an SI engine fuelled by gasoline/acetylene mixtures. Energy **151**, 707–714 (2018)
5. Chmielewski, A., Gumiński, R., Mydłowski, T., Małecki, A., Bogdziński, K.: Operation tests of an engine supplied with alternative fuels, working as a distributed generation device. J. Power Technol. **97**(5), 437–445 (2017)
6. Alrazena, H.A., Ahmad, K.A.: HCNG fueled spark-ignition (SI) engine with its effects on performance and emissions. Renew. Sustain. Energy Rev. **82**, 324–342 (2018)
7. Chmielewski, A., Gumiński, R., Małecki, A., Mydłowski, T., Bogdziński, K.: Test stand research on ICE engine powered by an alternative fuel. In: Proceedings of the Advances in Intelligent Systems and Computing, AUTOMATION 2018, vol. 743. Springer (2018)
8. Bae, C., Kim, J.: Alternative fuels for internal combustion engines. Proc. Combust. Inst. **36**, 3389–3413 (2017)
9. Chmielewski, A., Gumiński, R., Mydłowski, T., Małecki, A., Bogdziński, K.: Research on HONDA NHX 110 fueled with biogas, CNG and E85. In: International Conference on the Sustainable Energy and Environmental Development SEED 2017, Krakow, Poland, 14–17 November 2017, IOP Conference Series: Earth and Environmental Science (EES) (2018, in print)
10. Nadaleti, W.C., Przybyla, G.: Emissions and performance of a spark-ignition gas engine generator operating with hydrogen-rich syngas, methane and biogas blends for application in southern Brazilian rice industries. Energy **154**, 38–51 (2018)
11. Chmielewski, A., Gumiński, R., Mydłowski, T., Małecki, A., Bogdziński, K.: Model based research on ICE engine powered by alternative fuels. In: 14th International Conference Dynamical Systems Theory and Applications, Łódź, Poland, December 11–14, 2017. Springer Proceedings in Mathematics & Statistics (2018, in print)
12. Wanga, Z., Liua, H., Reitz, R.D.: Knocking combustion in spark-ignition engines. Prog. Energy Combust. Sci. **61**, 78–112 (2017)
13. Li, Y., Nithyanandan, K., Meng, X., Lee, T.H., Li, Y., Lee, C.F., Ning, Z.: Experimental study on combustion and emission performance of a spark-ignition engine fueled with water containing acetone-gasoline blends. Fuel **210**, 133–144 (2017)

14. Su, T., Ji, C., Wang, S., Cong, X., Shi, L., Yang, J.: Improving the lean performance of an n-butanol rotary engine by hydrogen enrichment. Energy Convers. Manage. **157**, 96–102 (2018)
15. Yan, F., Xu, L., Wang, Y.: Application of hydrogen enriched natural gas in spark ignition IC engines: from fundamental fuel properties to engine performances and emissions. Renew. Sustain. Energy Rev. **82**, 1457–1488 (2018)
16. Su, T., Ji, C., Wang, S., Shi, L., Yang, J., Cong, X.: Improving idle performance of a hydrogen-gasoline rotary engine at stoichiometric condition. Int. J. Hydrogen Energy **42**, 11893–11901 (2018)
17. Milewski, J., Szabłowski, Ł., Kuta, J.: Control strategy for an internal combustion engine fuelled by natural gas operating in distributed generation. Energy Procedia **14**, 1478–1483 (2012)
18. Szabłowski, Ł., Milewski, J., Kuta, J., Badyda, K.: Control strategy of a natural gas fuelled piston engine working in distributed generation system. Rynek Energii **3**, 33–40 (2011)
19. Milewski, J., Szabłowski, Ł., Kuta, J.: Optimal control strategy of NG piston engine as a DG unit obtained by an utilization of Artificial Neural Network. In: Power Engineering and Automation Conference (PEAM 2012), pp. 410–416. IEEE (2012)
20. Sendzikiene, E., Rimkus, A., Melaika, M., Makareviciene, V., Pukalskas, S.: Impact of biomethane gas on energy and emission characteristics of a spark ignition engine fuelled with a stoichiometric mixture at various ignition advance angles. Fuel **162**, 194–201 (2015)
21. Shi, L., Ji, C., Wang, S., Cong, X., Su, T., Wang, D.: Combustion and emissions characteristics of a S.I. engine fueled with gasoline-DME blends under different spark timings. Fuel **211**, 11–17 (2018)
22. Pham, P.X., Vo, D.Q., Jazar, R.N.: Development of fuel metering techniques for spark ignition engines. Fuel **206**, 701–715 (2017)
23. Małecki, A., Mydłowski, T., Dybała, J.: Stanowisko hamowniane do badań silników spalinowych o małych mocach (The engine dynamometer to test internal combustion engines with low power). Zeszyty Naukowe Instytutu Pojazdów **96**, 55–66 (2013). (in Polish)
24. Dybała, J., Mydłowski, T., Małecki, A., Bogdziński, K.: Dynamometer and test stand for low power internal combustion engine. Combust. Engines **162**, 996–1000 (2015)

Universal Data Acquisition Module PIAP-UDAM for INDUSTRY 4.0 Application in Agriculture

Roman Szewczyk[✉], Oleg Petruk, Marcin Kamiński, Rafał Kłoda,
Jan Piwiński, Wojciech Winiarski, Anna Stańczyk,
and Jakub Szałatkiewicz

Industrial Research Institute for Automation and Measurements PIAP,
Al. Jerozolimskie 202, 02-486 Warsaw, Poland
`rszewczyk@onet.pl`

Abstract. The paper presents the Internet of Things based concept of the universal system of agricultural production management and control accordingly to INDUSTRY 4.0 concept. Proposed system will be implemented in test sites in Poland and Israel and will cover four layers: decision support, data processing, data acquisition and transmission and sensors. The key for successful operation in the rural area is robust and efficient data acquisition layer. Solution for this layer – the Universal Data Acquisition Module PIAP-UDAM is presented in the paper together with example of application focused on the system for advanced fertilizer tanks monitoring.

Keywords: Agriculture · Internet of Things · Industry 4.0

1 Introduction

The INDUSTRY 4.0 ideas application in the agriculture (known as FARMING 4.0 or AGRICULTURE 4.0) was proposed soon after introduction of the idea of INDUSTRY 4.0 [1]. Similar to other areas, FARMING 4.0 is based on the "produce more with less" concept [2] It is estimated, that implementation of FARMING 4.0 in the agriculture will lead to the increase of efficiency of production up to 15% [3]. Considering, that agriculture (and accompanying services for agriculture) participation in gross domestic product (GDP) of Israel is about 2.4% whereas in Poland it is about 2.3% (accordingly to GUS, Eurostat and OECD), implementation of FARMING 4.0 in 35% of agriculture production process will lead to increase GDP of Israel and Poland by 0.37% and 0.42% respectively. It should be stressed, that this increase will happen without increase of intensity of exploitation of natural resources or energy consumption. For this reason, challenges of INDUSTRY 4.0 are in the focal point of interest of national governments, private companies offering solution for agricultural business as well as individual farmers and farming companies.

Recently developed solutions for FARMING 4.0 are based on the standardized industrial automation equipment, such as programmable logic controllers (PLC), industrial sensors or devices for specific methods data transmission (such as GMS or

© Springer Nature Switzerland AG 2020
R. Szewczyk et al. (Eds.): AUTOMATION 2019, AISC 920, pp. 278–285, 2020.
https://doi.org/10.1007/978-3-030-13273-6_27

low power radio) [4]. Such solutions are useful for research activities or for feasibility studies; however, its possibility of wide implementation to the agriculture is strongly limited [4]. This limitation is caused by the lack of energy consumption optimization, lack of information model of used sensors as well as limited knowledge about electromagnetic compatibility of applied hardware.

From the point of view of information technology, previously developed systems were oriented on demonstration of specific functionality of intelligent farming or predictive maintains in agriculture [2]. As a result, possibility of wide application of developed systems was limited. Moreover, due to the limited scale of implementation, the development of artificial intelligence-based optimization models was limited to specific cases due to lack of critical quantity of collected data, necessary for machine learning process.

Artificial intelligence oriented decision support systems were covering mainly different kinds of expert systems, implementing the specific, clearly determined farming knowledge. Such approach has significant limitations due to the changes of farming processes (e.g. consumption of materials in time, etc.) and necessity of automated, adaptive accommodation [5] of the control system to changes in operation environment.

The key barrier for machine learning and artificial intelligence based control in agriculture is efficient and reliable collection of large volumes, high quality of synchronised data [6]. To overcome this problem, Universal Data Acquisition Module PIAP-UDAM was developed. We hope that implementation of this module will give fresh impetus for development of agriculture-oriented machine learning systems for agriculture in Poland and Israel.

2 General System Structure

Proposed machine learning oriented system, accordingly to the conception of INDUSTRY 4.0 will cover four layers:

- decision support (fourth layer),
- data processing (third layer),
- data acquisition and transmission (second layer),
- sensors (first layer).

Sensor layer (the lowest layer) of the system will be not oriented on the specific sensors, but will enable plug-and-play connection of low level sensors to the developed system. For this purposes standard analogue industrial interfaces will be implemented (4–20 mA, 0–10 V, frequency) together with serial interfaces such as RS-232/485, SPI and Bluetooth. As a result commonly used industrial sensors (like liquid level sensor, temperature sensors, rotation and position as well as crystallization sensors) and more sophisticated modules (such as GPS positioning) will be efficiently connected to the system feeding it with real time, real world data.

Second layer of the system, data acquisition and data transmission layer will be developed on the base of real-time operating system (RTOS) implemented on the embedded ARM microcontroller architecture. Developed subsystem will collect measuring data in the real time, pre-process it and provide information for the cloud computing. To enhance reliability of the system, it will be powered from industrial battery or accumulator recharged by the solar batteries. From physical/sensor layer point of view, the multimodal data transmission will be implemented covering cable (LAN) and wireless data transmission like GSM and Wi-Fi. Module will automatically change transmission channel accordingly to the availability. In special cases, the satellite data transmission by INMARSAT D/D+/IsatM2M system will be implemented.

For the data transmission system, the specialized, cloud oriented RESTful API web service will be implemented. This web service is the most interoperable and universal, commonly used in IT industry for machine-machine and machine-human interaction. For data protection the specific authorization framework will be implemented according to OAuth 2.0 specification and requirements.

Data processing layer (third layer) will cover verification of data coherence, real time estimation of uncertainty of data as well as data fusion focused on different sources with dependable data coherence. As a result sensors will be the subject of continuous validation on the base of data received from Enterprise Resource Planning (ERP) system concerning e.g. tank refilling quantity or by the comparison among the sensors with similar characteristics. As a result, during the operation, proposed system will develop autonomously the database and rules enabling scoring the quality and accuracy of sensors, indicating in advance possible malfunction or decreases of quality of measurements. Proposed scenario of predictive maintains of measuring systems can be adapted also to other, non-agricultural oriented industrial applications.

The top layer, decision support system (fourth layer) will be based on artificial intelligence prediction algorithms enabling predictive maintenance of agricultural industry infrastructure [7] as well as just-in-time delivery of materials such as liquid fertilizers.

On the base of collected data provided to the cloud, the decision support system will perform the optimisation of delivery system on the base of trends extrapolation in the conjunction with Bayesian learning. As a result the truck driver will get the permission to refill the specified material for the agricultural plot from the supplier and the customer, through the smartphone app. The system will give the farmers the ability to check the quantity and significant changes of liquid fertilizers in the containers, also, based on machine learning algorithms, they will receive the prediction of state of each container, which will give them an opportunity to optimize the supply chain.

Efficiency and functional parameters of developed system will be practically verified in cooperation with Deshen Ha'zafon company, which is the operator of large scale liquid fertilizers tank network as well as different collective communities in Israel. On the Polish side verification will be carried out in cooperation with BioAlt sp. z o.o. as well as AMP Hobda agricultural company.

3   PIAP-UDAM Module Concept

Schematic block diagram of PIAP-UDAM module is given in Fig. 1.

From the point of view of sensor layer, PIAP-UDAM module is equipped in commonly used standard interfaces, such as digital I/Os (0–24 V standards), industrial

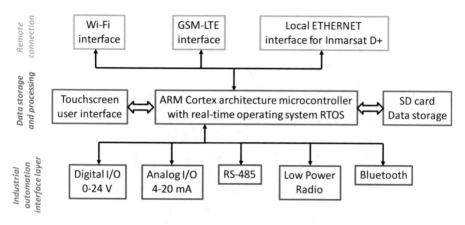

Fig. 1. Schematic block diagram of PIAP-UDAM module.

analog IOs (4–20 mA) as well as in RS-485, Bluetooth and Low Power Radio according to IEEE 802.15.4 standard.

PIAP-UDAM module is equipped in ARM-CORTEX, low power consumption microcontroller STM32L4. However, in spite of energy efficiency, STM32L4 can efficiently realize data processing and aggregation as well as identify sophisticated alarm states based on multiple dependencies among local variables. Moreover, ARM-CORTEX controller is equipped in FreeRTOS+ operating system enabling flexible updates and software development.

PIAP-UDAM provides the multimodal data transmission to the cloud based on MQTT (Message Queuing Telemetry Transport) protocol. Accordingly to the needs also CoAP (Constrained Application Protocol) may be implemented. From data transmission point of view, Internet connection may be established on the base of Wi-Fi interface, GSM-LTE or even INMARSAT D+ connection.

It should be highlighted, that PIAP-UDAM modules may work in the swarm mode. In such a case, only one of modules is communicating with the global Internet network and distributes Wi-Fi Internet access to other modules. Such mode of operation is especially useful for the more expensive internet connections, such as satellite INMARSAT D+. It should be highlighted, that in spite of growing popularity of INMARSAT D+, operation in the swarm mode is still not commonly available in embedded, industrial devices.

Figure 2a presents the view of control touch screen of PIAP-UDAM module. Upper part of the screens presents the data transmission state, right side visualize the local communication protocols, whereas left-central part is presenting the state of local signals for quick process verification. Due to the use of FreeRTOS+ operating system, touch screen can be easily adopted to the needs of specific application.

For the pilot purpose a lot of PIAP-UDAM modules was produced on the base of rapid prototyping and 3D printing technology. The digital model of PIAP-UDAM module (so called digital twin) is presented in Fig. 2b, whereas prototypes of PIAP-UDAM modules (with and without touch screen) are presented in Fig. 2c.

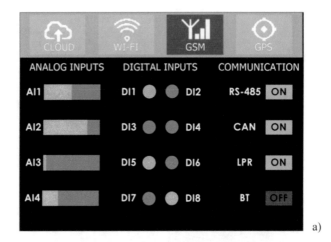

a)

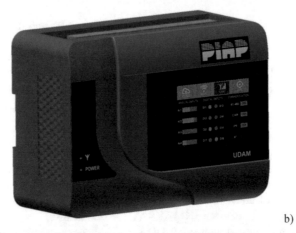

b)

c)

Fig. 2. PIAP-UDAM module: (a) visualization of the touch screen view, (b) digital model of the PIAP-UDAM module for 3D printing, (c) prototypes of the PIAP-UDAM modules with and without the touch screen.

4 Practical Implementation

First practical implementation of PIAP-UDAM module was focused on common problem in agriculture observed both in Israel and in Poland. This problem is monitoring of the state of the large number of fertilizer tanks distributed on large area. For this reason the prototype version of PIAP-UDAM module was integrated with fertilizer tank cup as well as equipped in temperature and ultrasonic liquid level sensor.

Figure 3 presents the result of integration of PIAP-UDAM module with fertilizer tank cup. For this specific purpose, PIAP-UDAM was stored in specialized casing. Due to the environmental requirements, for this application PIAP-UDAM module was not accessorized in touch screen.

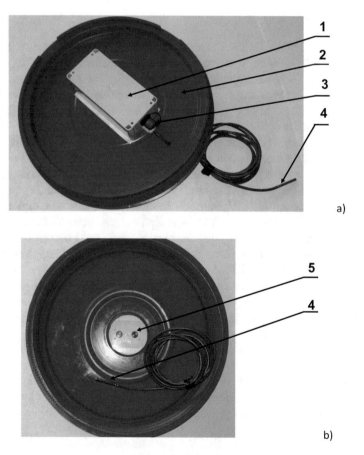

Fig. 3. Integration of PIAP-UDAM module with fertilizer tank cup: (a) top view, (b) bottom view. 1 – PIAP-UDAM module in the special casing, 2 – fertilizer tank cup, 3 – power connector, 4 – fertilizer temperature sensor, 5 – ultrasonic fertilizer level sensor.

Figure 3a presents the top view of PIAP-UDAM module (1) integrated with the fertilizer tank cup (2). In this specific case, power supply is provided for the PIAP-UDAM module (1) via the connector (3). Figure 3b presents the bottom view of the module. Due to specific requirements of fertilizer level measurements, temperature sensor (4) and ultrasonic level sensor (5) is located directly in the centre of the fertilizer tank cup.

The set of the prototypes of the PIAP-UDAM modules integrated with fertilizer tank cups was successfully mounted in the fertilizer system of agricultural cooperative (Kibbutz) in Jordan Valley in Israel. Data collected from the sensors are transmitted to the cloud enabling advanced analyses of fertilizer use, prediction of fertilizing process efficiency as well as enabling more efficient fertilizer distribution in the network.

5 Conclusions

Application of the INDUSTRY 4.0 based solutions to agricultural industry (known also as FARMING 4.0) is directly connected with the implementation of more efficient production without higher exploitation of resources. Due to significant economic advances of this approach as well as positive environmental impact, transformation towards FARMING 4.0 will be the constant and inevitable process in for both Polish and Israeli agriculture.

To enable efficient, practical application of FARMING 4.0 solutions, low cost, flexible and robust data acquisition and control modules are necessary. Such module, oriented on different standards of cloud processing solutions was presented in the paper. PIAP-UDAM enables data transmission to the cloud using MQTT protocol and multimodal data transmission. Module can be easily integrated with agricultural processes control due to wide selection of local data transmission standards.

From functional point of view, developed field version of PIAP-UDAM module is dedicated for cooperation with ultrasonic level sensors and digital temperature sensors. Moreover, all types of PIAP-UDAM modules are is able to work in swarm – in the scenario when selected modules enable Wi-Fi internet access for other modules. Such functionalities are separately available in alternative solutions offered on the global market, however, we have no information about module enabling all these functionalities in the single module.

Practical verification of usability of PIAP-UDAM modules was carried out on the monitoring of fertilizer tank level in agricultural system of agricultural cooperative (Kibbutz) in Jordan Valley in Israel. After adaptation to the environmental conditions, PIAP-UDAM modules enabled cloud-based monitoring and control of fertilizing tanks as a key element of agricultural processes.

Acknowledgements. Module PIAP-UDAM was partially developed within Polish-Israeli bilateral project "CPS-AGRI: Customisable cyber-physical system for distributed monitoring and control in agriculture" and is protected as the utility model no. WP-26300.

The project is co-financed by The National Centre for Research and Development under bilateral cooperation – V Polish-Israel Call.

References

1. Halang, W.A., Unger, H.: Industrie 4.0 und Echtzeit. Springer, Heidelberg (2014)
2. De Silva, P.C.P., De Silva, P.C.A.: Ipanera: an Industry 4.0 based architecture for distributed soil-less food production systems. In: Manufacturing & Industrial Engineering Symposium (MIES), Colombo, pp. 1–5 (2016). https://doi.org/10.1109/mies.2016.7780266
3. Bratek, A., Słowikowski, M., Turkowski, M.: The improvement of pipeline mathematical model for the purposes of leak detection. In: Jabłoński, R., Turkowski, M., Szewczyk, R. (eds.) Recent Advances in Mechatronics. Springer, pp. 561–565 (2007). https://doi.org/10.1007/978-3-540-73956-2_110
4. Hermann, M., Pentek, T., Otto, B.: Design Principles for Industrie 4.0 Scenarios. In: 49th Hawaii International Conference on System Sciences (HICSS) 2016. https://doi.org/10.1109/hicss.2016.488
5. Nukala, R., Panduru, K., Shields, A., Riordan, D., Doody, P., Walsh, J.: Internet of things: a review from 'farm to fork'. In: 27th Irish Signals and Systems Conference (ISSC), Londonderry, pp. 1–6 (2016). https://doi.org/10.1109/issc.2016.7528456
6. Turkowski, M.: Simple installation for measurement of two-phase gas–liquid flow by means of conventional single-phase flowmeters. Flow Meas. Instrum. **15**, 295–299 (2004). https://doi.org/10.1016/j.flowmeasinst.2004.06.001
7. Piotrowski, K., Sojka-Piotrowska, A., Stamenkovic, Z., Kraemer, R.: IHPNode platform as a base for precision farming and remote diagnosis in agriculture. In: 24th Telecommunications Forum (TELFOR), Belgrade, pp. 1–5 (2016). https://doi.org/10.1109/telfor.2016.7818712
8. Marosi, A.C., Farkas, A., Lovas, R.: An adaptive cloud-based IoT back-end architecture and its applications. In: 26th Euromicro International Conference on Parallel, Distributed and Network-based Processing (PDP), Cambridge, pp. 513–520 (2018). https://doi.org/10.1109/pdp2018.2018.00087

New Methods of Power Nodes Automatic Operation in Scope of Voltage Regulation, Reactive Power and Active Power Flow Control

Piotr Kolendo[✉]

Institute of Power Engineering, Research Institute,
Gdańsk Division, Gdansk, Poland
p.kolendo@ien.gda.pl

Abstract. Group control systems (ARNE/ARST) are used for the automatic regulation of voltage, reactive power and active power flow control in polish National Power System (NPS). Those systems are used in polish NPS since '90, however due to constantly expanding electrical grid and new requirements for automation systems, a new solutions and control methods should be elaborated. Conditions for automatic control of power nodes are strongly dependent on requirements of Transmission System Operators (TSO) in each country, therefore solutions presented in paper will be appropriate for polish NPS. Paper describes solution of complex power node, on which automation should be prepared individually. On the basis of its case a new methods of group control were introduced i.e. control algorithms for phase shift transformers, coordination between step-up transformers and generators and control of industrial factories active/reactive power flow in respect to requested power angle. Control methods presented in this paper were implemented on industrial objects. Verification of its correctness is presented in the form of measurements from those industrial objects.

Keywords: Coordinated voltage control · ARNE/ARST systems · Hierarchical voltage control

1 Introduction

Group control systems (ARNE/ARST) are used for the automatic regulation of voltage, reactive power and active power flow control in all nodes of polish National Power System (NPS). Such a systems are used in polish NPS since '90, however due to constantly expanding electrical grid and new requirements for automation systems, complexity of power nodes has significantly increased, makes it necessary to expand range of duties and develop new methods of automatic control.

In polish NPS increasing complexity of nodes is mainly connected with building of new big generation units (i.e. Włocławek 460 MW, Turów 496 MW, Belchatow 858 MW, Opole 2 × 900 MW, Jaworzno 910 MW and Kozienice 1075 MW) and ways of deriving power from these units.

© Springer Nature Switzerland AG 2020
R. Szewczyk et al. (Eds.): AUTOMATION 2019, AISC 920, pp. 286–296, 2020.
https://doi.org/10.1007/978-3-030-13273-6_28

A frequent case is building of new generators near big industrial factories to deliver power from several lines simultaneously. For precise active power exchange phase shift transformers are installed. Such a solutions increase level of complexity of whole node, causing need of new algorithms development. That is the reason that modern automation systems for group control are elaborated for each node individually.

Author of paper took part in preparation of analysis, elaboration of control algorithms as well as implementation of automatic group controllers for newly built big power units in polish NPS and would like to present in article general algorithms used in newly built automation systems in power nodes.

Control logic and algorithms of voltage and reactive power are specific for different countries. They are strongly dependent on requirements of Transmission System Operator (TSO) in each country and they differ significantly. Due to the extent of subject matter paper will present solutions appropriate for polish NPS. General informations about coordinated voltage control in another countries such as Italy, France Belgium etc. can be found in [1].

Second section describes general characteristic of ARNE/ARST group control systems, while Sect. 3 presents description of example node with general algorithms description. Section 4 presents verification of algorithms correctness on the basis of selected measurements from control object. Fifth section concludes paper.

2 Group and Area Voltage Control

Automated voltage control systems ARNE/ARST are obligatory in every High Voltage (HV) transmission node in NPS on the basis of [10]. Aim of those systems is to coordinate work of all transmission/generation nodes in NPS. National dispatch centre (KDM) in coordination with regional dispatch centres (ODM) specify set voltages in individual nodes for safe and optimal work of the NPS. Those set voltages are set directly by dispatchers in ODM and transmitted to ARST systems in HV substations. Set voltages are transmitted from ODM by means of communication module of area voltage control system (MK SORN). Structure of area voltage control in polish NPS is shown on Fig. 1.

ARNE/ARST systems use all available devices in each node such as generators, transformers, capacitor banks etc. to maintain set value of voltage and to even load between each devices. Description of standard ARNE/ARST solutions can be found in [5, 7, 8, 11].

In standard solutions ARNE/ARST system is physically divided into two separate systems: ARNE and ARST. Group controller (ARNE) is responsible of control of all devices located in power plant. ARNE systems coordinate work with ARST systems located on HV substation, which are responsible for transmission to dispatch centres and control of all devices located on HV substation.

ARNE systems in general are responsible for:

- Control of voltage level on substation bus bars,
- Control of generators permissible work range,
- Assure even load between generators,
- Prevent the circulation of reactive power in node,

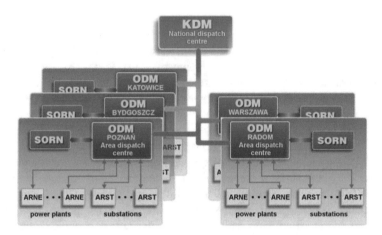

Fig. 1. Structure of area voltage control in polish NPS [3]

ARST systems in general are responsible for running automated control of HV substation in scope of:
- Control of voltage level on upper or lower side of transformers,
- Reactive power flow,
- Even a generation load between each HV systems,

3   Control Methods on the Basis of System Installed in Selected NPS Node

Methods of power node automatic control will be presented on the basis of complex power node, which consists of industrial factory and new big generation unit connected to node in specific way.

In presented example node contains industrial factory, which have to be supplied from several power lines simultaneously. Moreover, industrial factory would like to have a possibility to control amount of active power taken from each line. Another requirement for automation system except from voltage regulation, even load between generators, control of active and reactive power flow – is to keep right value of power factor on power lines, not to exceed value set by TSO (due to it is penalized).

Example below consists of 6 generators of approximate power 60 MVA and one of approximate 700 MVA. The biggest generator G7 should constitute alternative power supply for industrial factory and be able to sell energy to TSO at the same time. To enable such a possibility G7 is connected to R400 kV and R30 kV at the same time, equipped with two phase shift transformers (PST1, PST2). Those PST enables to set desired value of active power flowing from each line (Fig. 2).

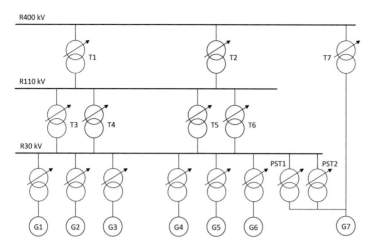

Fig. 2. Topology of complex ARNE/ARST system containing industrial factory under regulation

For this case control algorithms looks as follows:

1. Generators G1–G7 are maintaining set voltage on R400 kV and R30 kV bus bars. ARNE system control excitation current of generators to keep desired level of voltage, by sending pulses "up" and "down" to increase or decrease reactive power produced by generator.

$$U_s - \varepsilon_U > U_m => Q_g \uparrow \tag{1}$$

$$U_s + \varepsilon_U < U_m => Q_g \downarrow \tag{2}$$

where:

U_s – set voltage on bus bar,
U_m– measured voltage on bus bar,
Q_g – reactive power value on each generator,
ε_U – voltage deadband.

2. If generators are equipped with step-up transformers with on-load tap changer (OLTC), ARNE system coordinate also its work. ARNE monitors voltage on generator clamps and if it reach maximum or minimum value, cause change of the tap position to enable further regulation by generator. More detailed description of step-up transformer control algorithm is described in [6]. Control of tap position is shown in a simplified way – it is assumed that tap position "up" will increase voltage on generator clamps, while "down" will decrease it.

$$U_g \geq 1.05\, U_n => T_i \uparrow \tag{3}$$

$$U_g \leq 0.95\, U_n => T_i \downarrow \tag{4}$$

where:

U_g – measured voltage on generator clamps,
U_n – nominal voltage of generator,
T_i – tap position of each transformer,

3. Transformers T1–T2 coordinates work with generators G1–7 and their work mode is dependent from several factors:

- If generators on each systems are off or do not have control possibilities, transformers keep set value of voltage on lower side,

$$U_s - \varepsilon_U > U_m => T_i \uparrow \tag{5}$$

$$U_s + \varepsilon_U < U_m => T_i \downarrow \tag{6}$$

- If there are generators with available control range then transformers keep set value of voltage on upper side,
- If there are generators with available control range on upper and lower side of transformer, then transformers keep set value of reactive power, which flows through transformer (value of set reactive power is calculated automatically by ARNE/ARST system, to balance reactive power range on upper and lower side of transformer – more detailed information about method of calculating set value can be found in [2].

$$Q_s - \varepsilon_Q > Q_m => T_i \uparrow \tag{7}$$

$$Q_s + \varepsilon_Q < Q_m => T_i \downarrow \tag{8}$$

where:

Q_s – set value of reactive power flowing through transformer,
U_n – measured value of reactive power flowing through transformer,
ε_Q – reactive power deadband.

4. Transformers T3–T6 can work in two modes. Work mode is selected depending on fact if on R30 kV there are generators with available control range:

- In case of R30 kV with turned on generators – T3–T6 works in power factor (tg *fi*) mode, controlling right value of Power Factor (PF). Exceeding of PF over 0.4 is penalized by TSO. PF is sum of reactive power, which flows to industrial factory through T3–T6 divided by sum of active power:

$$\text{tg } fi = \frac{\sum_{T=3}^{6} Q_T}{\sum_{T=3}^{6} P_T} \tag{9}$$

For regulation purposes, a transformer with biggest flow of reactive power is selected for tap change:

$$IF \sum_{i=3}^{6} Q_i < 0 \; AND \; |\mathrm{tg}\,fi| > 0.4 => T_i \uparrow \tag{10}$$

$$IF \sum_{i=3}^{6} Q_i > 0 \; AND |\mathrm{tg}\,fi| > 0.4 => T_i \downarrow \tag{11}$$

$$IF |\mathrm{tg}\,fi| \leq 0,4 => \text{no control} \tag{12}$$

- R30 kV without turned on generators- transformers T3–T6 control voltage level on R30 kV similar to (5), (6).

5. Phase shift transformers (PST1, PST2) are responsible for delivering desired active power value to industrial factory. PST have two tap changers: one transversal and second longitudinal. To simplify control algorithms, it can be assumed, that longitudinal is responsible for control of reactive power flow and transversal is responsible for active power flow:

- Transversal tap changer – this tap changer can work in two modes, which are selected by dispatcher.
  First one is keeping set value of active power set by dispatcher. To simplify description it is assumed that tap position up is increasing value of active power, while tap down is decreasing.

$$P_m - \varepsilon_P > P_s => T_i \uparrow \tag{13}$$

$$P_m + \varepsilon_P < P_s => T_i \downarrow \tag{14}$$

where:

P_s – set value of active power flowing through PST,
P_m – measured value of active power flowing through PST,
ε_P – active power dead band.
Second mode is also keeping set value of active power, however this value is calculated by ARNE/ARST and it is a sum of active power flowing through T3–T6. Calculated value will compensate active power from T3–T6, so that all energy is taken from PSTs.

$$\sum_{i=3}^{6} P_i < -\varepsilon_P => T_i \uparrow \tag{15}$$

$$\sum_{i=3}^{6} P_i > \varepsilon_P => T_i \downarrow \tag{16}$$

- Longitudinal – prevents circulation of reactive power in node – in practice it keeps reactive power value close to zero including deadband,

$$Q_m > \varepsilon_Q => T_i \uparrow \qquad (17)$$

$$Q_m < -\varepsilon_Q => T_i \downarrow \qquad (18)$$

Additional description of PST work modes can be found [4, 9].

4 Verification of Algorithms Correctness on the Basis of Selected Measurements from Control Object

This section presents selected measurements of voltage, active and reactive power on the basis, which correctness of prepared method of regulation can be estimated.

Figure 3 shows voltage control of R400 kV, where orange colour shows measured voltage, while blue shows set voltage. For R400 kV deadband $\varepsilon_U = 1.5\,\text{kV}$. Figure 4 presents trend of generator G7 reactive power changes according to voltage changes on R400 kV.

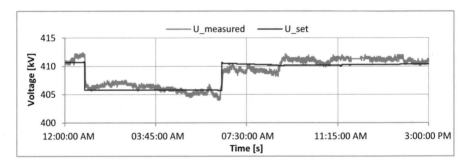

Fig. 3. Measured and set voltage of R400 kV

As it can be seen on Fig. 5, ARNE stopped influencing on generator due to it has reached voltage maximum value (21.8 kV). To enable further regulation, transformer T7 changed tap position.

Figure 6 presents voltage control of R30 kV, where orange colour shows measured voltage, while blue shows set voltage. For R30 kV deadband $\varepsilon_U = 0.2\,\text{kV}$.

Figure 7 presents load distribution between generators working on R30 kV. For considered period of time, only three generators where turned on (G2, G3, G6). Their reactive power control range was sufficient to keep set voltage on bus bars, therefore there was no need to change tap positions on transformers T3–T6. As it can be seen on figures, ARNE/ARST was properly keeping set voltage on bus bars, maintaining an even distribution of reactive power at the same time.

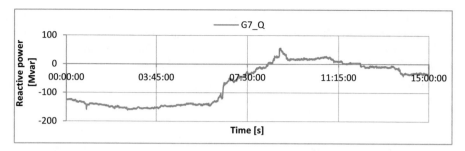

Fig. 4. Distribution of reactive power load on generator G7 connected to R400 kV

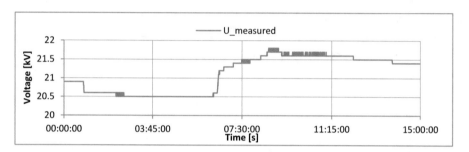

Fig. 5. Voltage level on generator clamps for considered period of time

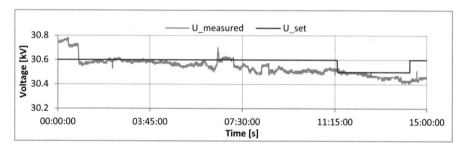

Fig. 6. Measured and set voltage of R30 kV

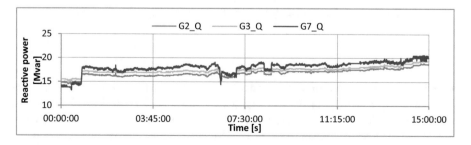

Fig. 7. Distribution of reactive power load between generators connected to R30 kV

Figure 8 shows work of phase shit transformer PST1 concerning control of active power flow. Orange colour shows measured value, while blue set one. As it can be seen on Figs. 8 and 9, active power was maintaining on the right level, by changing tap position of PST1 (Fig. 10).

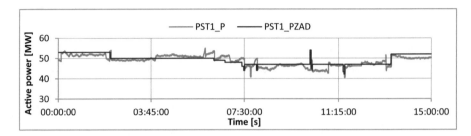

Fig. 8. Distribution of active power on PST1

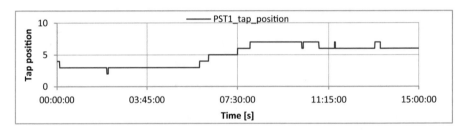

Fig. 9. Transversal tap position of PST1

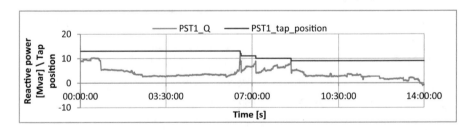

Fig. 10. Reactive power value on PST1 and longitudinal tap position

5 Summary

ARNE/ARST group control systems enables to automatically operate power node in scope of voltage regulation, reactive and active power flow control. Such a systems increases safety of NPS, due to coordination of all nodes work. Voltage control, control of reactive and active power flow is carried out simultaneously in a coordinated manner. Despite the theoretically uncomplicated control algorithms, implementation of automatic regulation for power devices (at size of hundred megawatts) is connected with many difficulties, among others:

- Necessity to ensure safe and stable operation of the power node – algorithms described in the article are presented only in a general way. Very important issue (due to the high power of controlled devices) are algorithms of preventing erroneous control of devices, due to the limited space of paper, such algorithms are not presented. Such algorithms take up to 80% of the overall regulation algorithms (this will be material for another publication),
- Many limitations in the control of high power devices – one example is the limited number of transformers tap position changes. Depending on the type of transformer, number of tap position changes is from a few to several hundred thousand changeovers to the overhaul inspection. Such a feature affects the way the control algorithms are developed, i.e. the accuracy of voltage control is reduced to lower number of tap changes,
- General resistance related to the implementation of automatic control systems for high power devices – due to economic and safety reasons, there is great fear in resigning from manual control – it is associated with a significant expansion of the safety functions in regulation algorithms,
- Necessity to coordinate the work of devices belonging to different owners (i.e. TSO, Industrial factory and Power plant) – not always the interests of individual owners are convergent. For example, the power plant will want to maintain the minimum amount of reactive power produced, to reduce the load of the generators, what may result in too high or too low voltage on the grid. The developed algorithms must be a compromise to ensure safer and optimal work of the node at the expense of the interests of individual owners.

Due to the number of mentioned difficulties in the implementation of this type of automation, it can be concluded that the implementation of at least one new simple algorithm of automatic regulation for high power devices is a great success. As it was mentioned in the paper, due to increasing complexity of nodes, group control systems are developed individually for each node. Not everything can be predicted during the analysis. Often, after starting the node, the idea of its functioning changes. The development of algorithms takes place in an iterative manner, and often the tuning and improvement of algorithms is carried out long after its launch. A novelty in this type of group control systems is automation of phase shifting transformers work (related to this active power flow control), automation of step-up transformers control and Power Factor control. On the basis of presented measurements a proper work of elaborated control algorithms can be claimed.

References

1. CIGRE: Coordinated Voltage Control in Transmission Networks. Task Force C4.602 (2007)
2. Dolny R., et al.: Analiza wpływu zmian programów i algorytmów ARNE ARST na możliwości regulacyjne i bezpieczeństwo pracy KSE, Etap I [Analysis of the impact of changes in ARNE ARST programs and algorithms on the regulatory capacity and operational security of the NPS, Stage I], a study of the Institute of Power Engineering (2008)
3. Institute of Power Engineering, Gdańsk division: Mikroprocesorowy układ automatycznej regulacji napięcia i mocy biernej typu ARNE2 [Microcontroller system of reactive Power and voltage control of ARNE2 type]. Information Card (2010)
4. Kocot, H., et al.: Dobór głównych parametrów przesuwników fazowych dla zachodnich połączeń transgranicznych KSE [Selection of main parameters of phase shifting transformers for NPS western cross-border interconnections]. Przegląd Elektrotechniczny 90(42), 124–127 (2014). https://doi.org/10.12915/pe.2014.04.28. ISSN 0033-2097, R. 90 NR 4/2014
5. Kolendo, P., Szuca, M., Drop, M.: Zastosowanie transformatorów blokowych z regulacją podobciążeniową w układach regulacji grupowej Krajowego Systemu Elektroenergetycznego. Elektroenergetyczna Automatyka Zabezpieczeniowa, [Use of unit transformers with on-load control in group control systems of the National Power System. Electrical power automatic protections in thermal power plants]. Seminarium Kozienice (2014)
6. Kolendo, P., Jendrzejewska, A., Szuca, M., Ogryczak, T.: Current issues of group control in the example of solutions for the Włocławek node. Acta Energetica Number 3/32 (September 2017). https://doi.org/10.12736/issn.2300-3022.2017307
7. Łosiński M.: Regulacja grupowa napięcia w systemie elektroenergetycznym – algorytmy i modelowanie [Voltage group regulation in the power system – algorithms and modelling], Ph.D. thesis, Gdansk University of Technology (2005)
8. Machowski, J., Regulacja i stabilność systemu elektroenergetycznego [Power system adjustment and stability], Warsaw (2007)
9. Opala, K., Rozenkiewicz, P., Kolendo, P., Opracowanie algorytmów lokalnej współpracy automatyki przesuwnika fazowego z układem ARST/ARNE w węźle regulacyjnym, Etap I [Development of algorithms for local interoperation of phase shift transformers automatic controls with ARST/ARNE in a regulation node, Stage I], a study of the Institute of Power Engineering (2015)
10. PSE S.A.: Instruction of Transmission System Operation and Maintenance (2017)
11. Sołtysiak, M., Kolendo, P., Szuca, M., Ogryczak, T.: Adaptive algorithm of generator reactive power control range expansion in national grid system. IEEE Publ. Electron. (2016). https://doi.org/10.1109/icaipr.2016.7585221

Robotics

High Precision Automated Astronomical Mount

Krzysztof Kozlowski[1], Dariusz Pazderski[1], Bartlomiej Krysiak[1(✉)],
Tomasz Jedwabny[1], Joanna Piasek[1], Stanislaw Kozlowski[2], Stefan Brock[3],
Dariusz Janiszewski[3], and Krzysztof Nowopolski[3]

[1] Institute of Automatic Control and Robotics, Poznan University of Technology,
ul. Piotrowo 3a, 61-138 Poznan, Poland
{office_iar,bartlomiej.krysiak}@put.poznan.pl
[2] Nicolaus Copernicus Astronomical Centre, Polish Academy of Sciences,
Rabiańska 8, 87-100 Torun, Poland
stan@ncac.torun.pl
[3] Institute of Control, Robotics and Information Engineering,
Poznan University of Technology, ul. Piotrowo 3a, 61-138 Poznan, Poland
office_cie@put.poznan.pl
http://www.iar.put.poznan.pl, http://www.camk.edu.pl
http://www.cie.put.poznan.pl

Abstract. In this paper, some specific aspects of a new automated astronomical mount of a 0.5 m class telescope is provided. This is the first astronomical mount built on the ground of Polish technical concept. This mount was designed and built by the interdisciplinary team consisting of researchers representing automation and robotics discipline (Poznan University of Technology) and astronomy (Nicolaus Copernicus Astronomical Centre of the Polish Academy of Sciences). The project takes advantage of CAD software for mechanical design, analysis and optimization. Furthermore, the modern control theory is utilized for coping with complicated physical phenomena which are especially troublesome in the low range of angular velocities of the mount. The astronomical mount is able to work with the high accuracy of positioning, greater than one second of arc, and large range of accessible velocities. The mount is dedicated for observations of stars and satellites. Mechanical and electrical parts of the mount and its measuring system are discussed. Furthermore, an algorithm designed for motion control, based on the active disturbance rejection paradigm, is outlined. In order to illustrate the performance of the closed-loop system, experimental results of trajectory tracking in the joint and the task spaces are compared.

Keywords: Astronomical mount · Robotic telescope ·
Control system · Sky observation · Trajectory tracking · ADRC

This work was supported by the National Science Centre (NCN) under the grant No 2014/15/B/ST7/00429, contract No UMO-2014/15/B/ST7/00429.

R. Szewczyk et al. (Eds.): AUTOMATION 2019, AISC 920, pp. 299–315, 2020.
https://doi.org/10.1007/978-3-030-13273-6_29

1   Introduction

This paper considers design and build of an automated astronomical telescopic mount which is capable of supporting and control of an astronomical telescope of a class 0.5 m. It is expected that the mount will ensure accurate pointing of the telescope instruments and tracking the motion of the fixed stars and satellites as the Earth rotates. The design concerns the preparation of the mechanical structure of the mount as well as motion control system along with low and high software level. This article presents a more comprehensive survey about the new astronomical mount in relation to the paper [1].

Nowadays, a lot of effort is put on the design of large telescopes, ranging in aperture from the 6-m Zelentchouk telescope on Mt. Pastukhov [2] to the 24-m Giant Magellan telescopes in Las Campanas [3]. The large telescopes excel in the detailed study of faint objects at the edge of the observable universe and high-resolution examination of specific objects. In the same time, small telescopes (in the 0.5–1.0 m aperture range) continue their valuable role in astronomical research through time series and other observations or tasks which are cost prohibitive for large telescopes. Small telescopes remain to play also a vital role in recruiting and training the next generation of astronomers and instrumentalists, and serve as test beds for developments of novel instruments and experimental methods.

The basic application of telescopes is the observation of objects that move in the sky at a star speed relative to the observer, the one that results from a daily motion of the Earth: $15°/h = 7,27 \cdot 10^{-5}$ rad/s $= 15$ asec/s. Non-star objects, e.g. planets, planetoids or comets within the Solar System, move in relation to the background stars. The closer the object is to the Earth, the higher is its speed in relation to the stars.

A special case is the observation of artificial satellites of the Earth. Since the launch of Sputnik in 1957, it was obvious that it was necessary to monitor the orbits of satellites from the surface of the Earth. Position measurements are carried out using three techniques: laser, radar and optical. The latter and the most common one, consists in measuring the position of a satellite in a plane tangential to the celestial sphere, against the stars in the background. Due to the wide range of technical parameters of telescopes and cameras, this method enables to measure the position of objects in almost all orbits. The exceptions are the lowest orbits in the LEO range (Low Earth Orbits, full range of LEO is 160–2000 km), where the angular velocities of the satellites are significant, this is a challenge in the process of tracking [4]. It is estimated that there are currently approximately 600,000 objects with a size greater than 1 cm in the orbit around the Earth. The American NORAD database provides information about the orbits of about 17,200 objects, the smallest of them have linear size of about 10 cm, and only some of them are active satellites. 95% of the objects in orbit around the earth are space debris. Governmental institutions, military and private companies around the world are responsible for orbit objects tracking and cataloguing. Objects that are not active satellites cannot be controlled in any way from Earth, while large number of active satellites is deprived of this possibility. This means that their orbits require continues monitoring. It is worth noting that for this purpose laser techniques are successfully used i.e. in the station in Borowiec (Poland) [5].

Satellites observation, especially in low orbits, provides a challenge to the control drive system design. At speeds of the order of $2°/s$, it is necessary to achieve the tracking precision better than 1 asec. This is due to the scale of the image projected on the camera sensor - typically one can roughly assume that 1 camera pixel corresponds to about 1 asec (this value, however, depends on the type of camera and focal length of the telescope). The expected precision refers to the measurements in the external space, after taking into account the disturbances (e.g. wind) or mechanical-optical system's bending. In practice, this means that control error of the isolated system should not exceed several hundredth arc seconds. As a result, the drive system's expected dynamic range of motion speed is wide and is about 1:20 000. For such demanding applications, the best solution is to use a direct drive system where a high torque motor is mounted on each axis itself. This approach has been used to good effect on many large alt-az telescopes (examples in [6] and [7]). In contrast to all other drive mechanisms, in which the force is concentrated on a pinion or wheel, direct drives distribute the thrust along the structure, thus minimizing localized deformation and maximizing the structural stiffness. This results in the highest drive stiffness possible. Another key advantage of direct drives usage is their insensitivity to mechanical misalignment, reducing telescope installation time [8].

Several research projects concerning small robotic telescopes are nowadays ongoing. One of them, the TBT project is being developed under ESA's General Studies and Technology Programme (GSTP), and shall implement a test-bed for the validation of an autonomous optical observing system in a realistic scenario [9]. The goal of the project is to provide two fully robotic telescopes: one in Spain and the second one in the Southern Hemisphere, which will serve as prototypes for development of a future network. The telescope is a fast astrograph with a large Field of View of 2.5×2.5 square-degrees and a plate scale of 2.2 asec/pixel. The tube is mounted on a fast direct-drive mount moving with speed up to 20 degrees per second. Detection software and hardware are optimised for the detection of NEOs and objects in high Earth orbits (objects moving from 0.1–40 asec/s). Nominal exposures are in the range from 2 to 30 s, depending on the observational strategy. Telescopes are managed by RTS2 control software, that performs the real-time scheduling of the observation and manages all the devices at the observatory.

In June 2007 the Alt-Az Initiative was established. The Initiative focused, from its inception, on the development of advanced technology, lightweight, low cost, 0.5–1.5 m alt-az research telescopes. Within the framework an 18-Inch Direct Drive Alt-Az Telescope Cal Poly 18 (CP 18) was designed and built [10]. The telescope's drive system has no gears, belts, or friction wheels; instead direct drive motors and high resolution encoders are completely integrated into the bearing assemblies and telescope's structure. The electronic control system has been designed to operate these brushless motors in a high precision mode.

This prototype was later transferred to quantity production in modified version as PlaneWave CDK700 Telescope.

The PlaneWave CDK700 Telescope is a 0.7 m, alt-azimuth mounted telescope system [11]. The CDK700 has dual Nasmyth port outputs at f/6.5, with an image scale of 22 microns per asec. The telescope pointing is controlled by two direct drive motors with high-speed encoders, resulting in a pointing accuracy of 10 asec RMS, a pointing precision of 2 asec, and a tracking accuracy of 1 asec over a three-minute period. Additionally, the focuser is also motor-controlled and can be remotely adjusted, useful for defocusing the telescope when doing photometric work.

The structure of the work is as follows. Section 2 discusses the assumptions for the design of the mechanical part and the design in general. In the next section, the drive system and the solution used for precise position measurement are characterized. Section 4 deals with the selected algorithm of controlling the drive axis with an adaptive structure and presents its properties on the basis of an experimental evaluation. The last section summarizes the work.

2   Mechanical Construction of the Mount

In this section we will at first discuss the kinematic structure of the mount. It will be described with use of the structural model, the CAD model and the picture of prototype mount. Furthermore, some chosen design details of the main part of the mount will be considered with special focus on the stress analysis. At the end of this section the drives and measure system will be explained.

2.1   Kinematic Structure

The schematic structure of the mount is presented in Fig. 1a. The kinematic model of the mount can be simplified to a system with two masses which rotate around two intersecting axes. The first mass (represented by the parameter J_1) is a symmetrical rotational mass where its position is given by φ_1. It rotates around a stationary vertical axis (element no. 5 in Fig. 1b) which passes through the center of J_1. The second mass (represented by the parameter J_2) is also a symmetrical mass where its position is described by φ_2. J_2 rotates around the horizontal axis (element no. 6 in Fig. 1b) and the orientation of J_2 is given by φ_2. This model is a two degree of freedom system with a state described by $\varphi = [\varphi_1\ \varphi_2]^T$. The mount is equipped with two engines which provide the torque for each degree of freedom. Therefore, the actuation of the considered model is defined as $\tau = [\tau_1\ \tau_2]^T$, these are the direct control inputs for φ_1 and φ_2 respectively.

The mechanical design of the mount is presented in Fig. 1b. It consists of five main parts providing various functional features. The first one, the basis unit Z1, is the main supporting column which is rigidly connected with the ground. The unit Z1 is also a mounting socket for the vertical axis unit Z2 which is responsible for rotational movement around the vertical axis. It consists of a

Table 1. Basis mechanical parameters of the mount

Description	Unit	Value
Nominal angular velocity of each axis	[rad/s]	0.52
Time period for reaching the nominal angular velocity	[s]	0.2
Angular acceleration of each axis	[rad/s²]	2.62
Inertia moment of the telescope in the horizontal axis	[kg · m²]	$J_1 = 15.84$
Inertia moment of the telescope in the vertical axis	[kg · m²]	$J_2 = 11.64$
Safety coefficient for calculation of forces and momentums		1.3

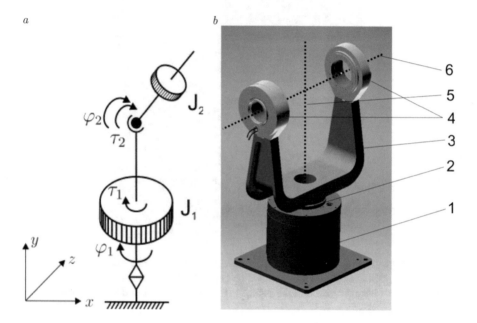

Fig. 1. Model of the astronomical mount a – schematic model, b – CAD model: 1 – basis unit (Z1), 2 – vertical axis unit (Z2), 3 – handler of the horizontal axis unit (Z3), 4 – horizontal axis unit (Z4), 5 – mount vertical axis, 6 – mount horizontal axis

bearing, a driving engine and a measurement system. The next element of the mount is the handler of the horizontal axis unit Z3. It is responsible for fixation of two half–shaft and their connection to Z2. Unit Z3 is designed in the shape of pitchfork which allows for a free rotation of the telescope (presented in Fig. 2) around the horizontal axis. The elements that hold the telescope constitute the horizontal axis unit (Z4) whose two main parts are the aforementioned half–shafts. Each half–shaft is fixed to the arm of the pitchfork. The first half–shaft consists of two bearings and drive engine, whereas the second one consists of two bearings and measurement system.

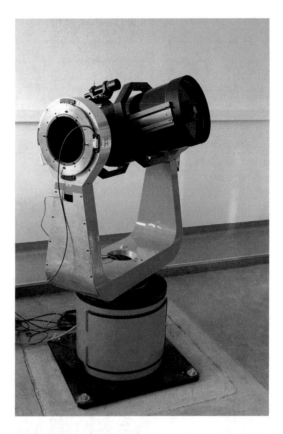

Fig. 2. The prototype astronomical mount with the industrial telescope of a mirror diameter 279,4 mm (11 in.). The system is prepared for realisation of observation experiments in the laboratory room with use of artificial star displayed on the monitors.

2.2   Optimization of the Handler Shape and Mass

The design of the system was performed taking into account a list of assumptions concerning specific properties of movement, durability and functionality presented in Table 1.

In order to optimize the shape and the mass of the handler Z3, a static stress analysis was performed with use of numerical calculations environment. The following set of mechanical loads acting on the structure of Z3 was assumed:

– forces generated from the inertia of the telescope,
– forces and torques generated by the mass of the vertical axis engine,
– forces and torques generated by the inertia momentum of Z3,
– forces generated by a wind acting on the surface of the telescope.

The loads described above are illustrated in Fig. 3. In the simulation, the material stresses and displacement were tested. The analysis was performed iteratively for

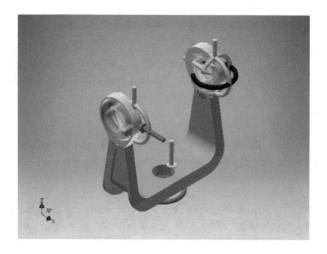

Fig. 3. Forces and momentums acting on Z3

different shapes and masses of Z3. The Von Mises stress of the one of analysed scenarios for Z3 is presented in Fig. 4. In this scenario Z3 element is equipped with the housing elements of Z4, because they were needed for attaching considered forces and torques into the structure of Z3. The simulation tests pointed out the localisation of the maximal loads occurrence and the maximal displacements. The final construction of Z3 ensures that the loads appearing in Z3 are on the safe level with the assumed safety coefficient given in Table 1. Furthermore, the simulations show that the design of Z3 ensures that bending of this element will not disrupt working conditions of Z4 unit's bearings.

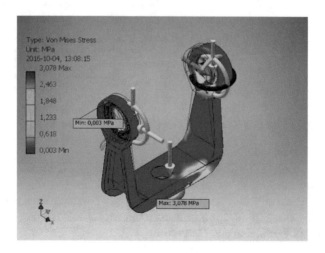

Fig. 4. Structural visualization of Von Mises stress for Z3

2.3   Driving Motors and Power Management Unit

On the basis of literature analysis it was assumed that the telescope mount will be driven directly, without use of gears, by electric permanent magnets synchronous motors (PMSM). The advantage of direct drives concerns its simplicity, the lack of backlash and also the wide range of accessible velocities. The PMSM were provided by Alxion company where its essential parameters are provided in Table 2. The motors are controlled by dedicated electronic unit designed specially for this project. The structure of this unit is presented in Fig. 5. The power supply module provides the DC bus supply voltage for Power Management Unit (PMU) from which the power converters for both axes are supplied. The PMU perform several functionalities. The main one concerns the drive braking phase where the PMU stores and dissipates the electric energy generated by the kinetic energy form the mechanism. Initially small amount of electric energy is stored in the capacitor battery package. In case of high amounts of electric energy the high power resistors are used to dissipate it.

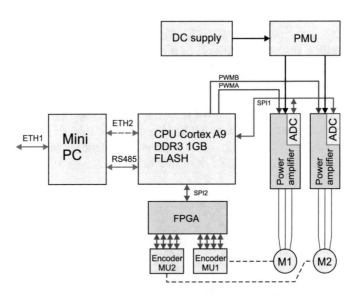

Fig. 5. Structure of the electrical control system

Power electronic converters have been designed in a three-phase Voltage Source Inverter structure, as shown in Fig. 7. The measured signals are supply voltage and supply current as well as output currents from the converter. Because of the low supply voltage usage, the MOSFET transistors have been applied in the power electronic converters.

Table 2. Basic parameters of the electrical drives

Description	Unit	Value
Engine type		300STK1M
Nominal drive torque	[Nm]	50
Maximal drive torque	[Nm]	110
Nominal current	[A]	20
Maximal current	[A]	45.5
Nominal voltage	[V]	33

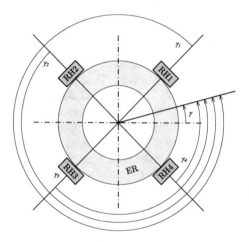

Fig. 6. The placement of reading heads (RH1–RH4) of the measuring system around the measuring ring (ER)

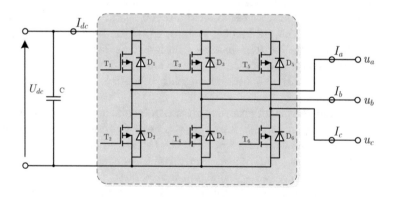

Fig. 7. Structure of a single axis module converter

2.4   Measurement System

The measurement of the electrical drives angular position is performed by measuring system provided by Renishaw company. Two separate measuring units are used to measure coordinates φ_1 and φ_2.

Each system consists of a measuring ring and 4 reading heads, placed symmetrically around the measuring ring (see Fig. 6). This solution helps to eliminate the positioning inaccuracy. The measuring system allows for measuring the angle with accuracy of 1 arcsec. Information from the measuring heads is transmitted to the control system with use of serial communication with a BISS-C protocol. Because the main control system unit does not support the BISS-C communication standard, the additional FPGA module is responsible for communication between the reading heads and the main control system unit. The mean value is calculated from all reading heads and the final absolute position of the measuring ring is marked as γ. This method allows to increase the measuring precision and additionally it performs the elimination of the second and fourth harmonic from the signal of position. The implemented data fusion algorithms take advantage of a simple averaging as well as the Uncented Kalman Filter was used for fusion of signals from individual heads [12]. In the second case the estimation state vector consists of the reconstructed acceleration, speed and position of the motor rotor.

To illustrate the correct operation of the proposed measurement algorithm, the results of a start-up test for a very low speed are presented. The reference speed during the test was set to 1 mrad/s (206 arcsec/s). Figure 8 shows the waveform of the position read directly from one of the heads and the waveform of the position estimated by the Kalman filter. After a time of 0.3 s from the start, the reference speed is reached and the position states to change linearly. Differences between the measured and estimated position on the drawing scale are negligible. Figure 9 shows the speed of motion, obtained by discrete differentiation of the position measured by the reading head. Ripple of the speed signal can be observed as result of measurement noise, that amplitude exceeds 2mrad/s. The signal with such high level of noise cannot be used in the control system, as it would lead to saturation of the current regulator and loss of the stability. It is therefore necessary to filter or estimate the speed signal [13]. Based on the last resonant frequency of the mechanical system, equal to about 100 Hz, a low pass filter with a frequency of 120 Hz has been designed. A 2nd order filter with the same cut-off frequency and critical damping has also been designed. Figure 10 illustrates the comparison of filtered angular velocities $\frac{d\gamma}{dt}$ and the estimated angular velocity $\hat{\omega}$.

It was observed that the first order filter was insufficient as the ripple amplitude of the speed signal exceeds 0.2 mrad/s. The 2nd order filtration results are also not satisfactory as they are similar to the estimated speed but in the same time introduce a delay, that limits the dynamic properties of the closed loop speed control. The estimated speed signal is free of this weakness. It has been shown that simultaneous use of multiple sensors and the Kalman filter based sensors fusion algorithm enables a higher performance in terms of resolution and dynamics.

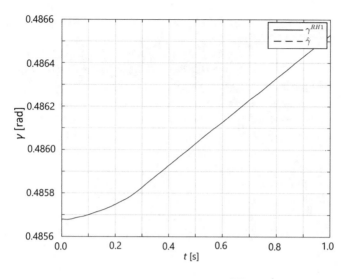

Fig. 8. Angle γ measured by read head no. $1-\gamma^{RH1}$ and estimated angle $\hat{\gamma}$

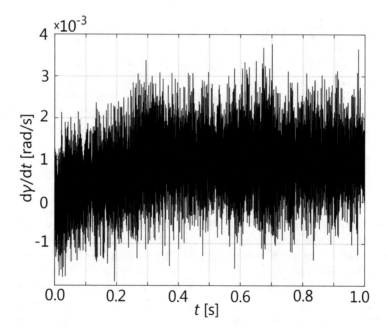

Fig. 9. Angular velocity obtained by as a time derivative of the measured angle

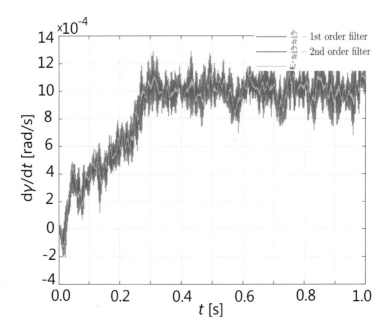

Fig. 10. Comparison of the estimated angular velocity and angular velocities calculated as a time derivative of measured angle with use of low–pass filters

3   Control Algorithm for a Joint Trajectory Tracking

The main control task for the considered astronomical mount is to ensure a precise trajectory tracking in the joint space. For the simplicity of the controller synthesis we assume that the astronomical mount and the telescope has a rigid mechanical structure and its kinematic model is fully known. This assumption allows one to state that a proper trajectory tracking in the joint space is sufficient to ensure a desired quality of a sky objects tracking in the global state space.

An important challenge in control of an astronomical mount is to obtain a control algorithm which would be robust to unmodelled or not fully modelled dynamic phenomena. These phenomena mostly include ripples of the torque introduced by PMSM motors and the occurrence of a friction which is a non-stationary, nonlinear and hard to model effect. Moreover, the also external disturbance such as the wind blows should be taken into account. The requirement of robustness is especially important when the mechanism has to work in the wide range of velocities.

In this paper we take advantage of the active disturbance rejection paradigm to design the motion controller for the mount. The method is based on on–line identification of a matched bounded disturbance and simultaneous application of the computed disturbance estimate in order to partially compensate unknown dynamics by a feedback.

For the simplicity of analysis we assume that the controller is designed for one rotational axis which configuration is marked as $\theta \in \mathbb{S}^1$. The angle θ represents the angle φ_1. The new symbol θ is entered here because of the consistency with the description of the considered control algorithm presented in [1]. Additionally, the inertia momentum of the reduced mechanism is J and its dynamics is described by the following equation:

$$\ddot{\theta} = \delta + J^{-1}\tau, \tag{1}$$

where τ is the torque provided by the engine and δ takes into account the friction phenomenon and a gravity force. For the simplicity we assume that δ is a bounded function with a bounded derivative (at least in a neighbourhood of the operating point).

Assuming that the measurement of the angle θ and the torque τ is accessible, the state of the mechanical system and additive component δ is estimated throughout the high gains linear observer given by:

$$\dot{z} = \begin{bmatrix} z_2 \\ z_3 + J^{-1}\tau \\ 0 \end{bmatrix} + \begin{bmatrix} l_1 \\ l_2 \\ l_3 \end{bmatrix} (\theta - z_1), \tag{2}$$

where $z = [z_1 \ z_2 \ z_3]^T$ represents respectively the estimate of the angle, the angular velocity and the disturbance, whereas l_i, $i = 1, 2, 3$ are gains.

The trajectory tracking controller is designed as a nonlinear feedback using fractional exponents in the feedback (compare with [14, 15]). It makes it possible to increase the sensitivity of the feedback to small errors which is extremely relevant in the considered application. The desired trajectory is defined by $\theta_d \in C^2$. The proposed controller is defined by the following formula:

$$\tau = J\left(-k_p \text{sgn}\left(\theta - \theta_d\right)|\theta - \theta_d|^{\alpha_1} - k_d \text{sgn}\left(z_2 - \dot{\theta}_d\right)\left|z_2 - \dot{\theta}_d\right|^{\alpha_2} - z_3 + \ddot{\theta}_d\right), \tag{3}$$

where k_p, k_d are positive gains, $|\cdot|$ is the absolute value of the real function, sgn is the sign function, $\alpha_1 \in (0, 1)$ and $\alpha_2 = \alpha_1 / (\alpha_1 + 1)$.

It is worth to notice that in a real application torque τ in Eq. 3 cannot be exerted on the axis directly. This is due to the existence of an auxiliary current controller dedicated to the three-phase PMSM motor. In the considered case we take advantage of such a cascade control structure and assume that τ is the desired signal for a simple PI regulator. In addition, a feed-forward term is used in order to improve the performance of the current loop and to attenuate the torque ripples.

4   Experiments

The control algorithm was tested experimentally in the laboratory conditions for the vertical axis only. The following list of parameters was used: $k_p = \omega_n^2$, $k_d = 2\zeta_n\omega_n$ where $\zeta_n = 0{,}8$, $\omega_n = 12$ rad/s, $\alpha_1 = 0{,}8$ and $l_1 = 3\zeta_o\omega_o$, $l_2 = \left(1 + 2\zeta^2\right)\omega_o^2$,

$l_3 = \omega_o^3$ for $\zeta_o = 0.8$ i $\omega_o = 500\,\mathrm{rad/s}$. The value of inertia momentum was defined as $J = 20\,\mathrm{kg \cdot m^2}$. The telescope installed on the astronomical mount was fixed in the position where its vertical axis is parallel to the horizontal axis of the mount. In this configuration the maximal value of the inertia momentum of a whole system around the vertical axis is achieved.

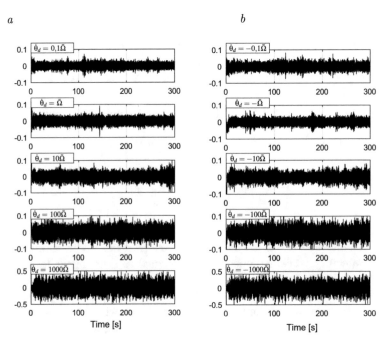

Fig. 11. E1: position error given in arcsec for specific reference velocities: $|\dot{\theta}_d| \in \{10^{-1}, 1, 10, 100, 1000\}\bar{\Omega}$, a – positive reference velocity, b – negative reference velocity

In the first experiment (E1) the angular velocity was constant while in the starting phase the linear velocity ramp was used for the period of 10 s. For each experiment the data was gathered with a frequency of 1 kHz throughout 300 s time period. The reference angular velocities were: $\{\pm 10^{-1}, \pm 1, \pm 10, \pm 100, \pm 1000\}\bar{\Omega}$, where $\bar{\Omega}$ is the daily velocity equal to $7,29212 \cdot 10^{-5}\,\mathrm{rad/s}$. The error plots are presented in Fig. 11 where the values of standard deviation are presented in Fig. 12. It can be noticed, that in the range of low velocities, below $10\bar{\Omega}$, the standard deviation of the tracking error in the steady state does not exceed 0,02 arcsec. The important increase of error is observed for the velocities near $1000\bar{\Omega}$, that is $4°/\mathrm{s}$.

In the second experiment (E2), the velocity was changed according to the sine function defined by $\dot{\theta}_d = \bar{\Omega}\sin\left(\frac{\pi}{15}t\right)$. The data was gathered in the period of 150 s. It can be seen from Fig. 13 that the error of trajectory tracking in the joint space increased in case of velocity sign change and was probably caused by the

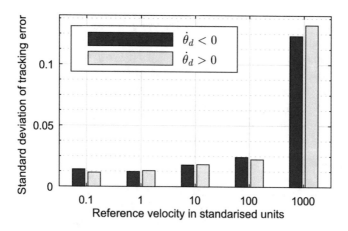

Fig. 12. E1: effective value of position error in the steady state given in arcsec (calculated in time 290 s after the starting phase)

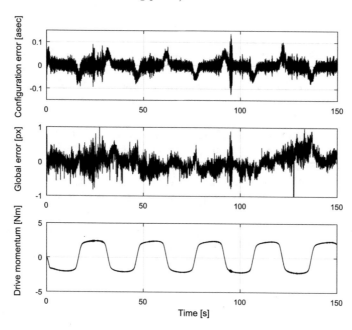

Fig. 13. E2: a – position error in the joint state space, b – position error in the global state space, c – force momentum calculated by the controller

effect of friction. In E2 experiment the image of the artificial star, which was projected on the LCD display, was recorded with use of camera from the telescope. The camera was working with the frequency of 50 fps. On each of 7500 acquired images the position of the artificial star was measured. The used measurement method is the least squares fitting for two-dimensional axially symmetric

Gaussian profile. Further, for calculation the x position the reference function $\dot{\theta}_d$ was fitted with the achieved data. The residual fitted plot is presented in Fig. 13. It should be interpreted as the position error in the global state space. The RMS value of this error is on the level of 0.3 px. Taking into account the focal length of the telescope and the size of the pixel on the camera, the scale of the picture is 0.43 arcsec/px. It means that the RMS value of the error is equal to 0.13 arcsec in time period of 150 s.

5 Summary

The design of an astronomical mount is a non-trivial problem due to high expectation about the positioning precision. This prerequisite is necessary for a proper image acquisition of objects in the sky. Another important requirement for an astronomical mount refers to the wide range of accessible velocities. In case of stars movement tracking very slow velocities are required, whereas for tracking of satellites movement relatively high velocities are needed. In order to meet these requirements new technological solutions in the area of mechanical construction, sensing and control units have to be utilized.

The key issue concerns a proper movement control algorithm selection that is robust to physical phenomena, that are hard to model, such as friction. The control algorithm must also be robust to the influence of the construction stiffness and its vibrations which strongly affect the quality of the image stabilization in the telescope.

On the basis of achieved result it can stated that the presented astronomical mount is able to perform a high quality of trajectory tracking for a stationary object in the range of low velocities. Still, the trajectory tracking in high velocities and stabilization in case of a reverse movement have to be improved. Furthermore, an important challenge poses a damping of mechanical vibrations that results from a mechanism construction. Also the level of noise limitation in the current and the position measurements is an issue that has to be improved.

Despite the possibilities for performing the experimental tests in the laboratory conditions, in the future we are expecting to achieve a fully verifiable system for a real sky observations.

References

1. Kozlowski, K., Pazderski, D., Krysiak, B., Jedwabny, T., Kozlowski, S., Brock, S., Janiszewski, D., Nowopolski, K.: Struktura mechaniczna i uklad sterowania zrobotyzowanym montazem astronomicznym, Krajowa Konferencja Robotyki 2018 (2018)
2. Ioannisiani, B.K., et al.: The Zelenchuk 6M Telescope (BTA) of the USSR Academy of Sciences. In: International Astronomical Union Colloquium, vol. 67. Cambridge University Press, Cambridge (1982)
3. Johns, M., et al.: Design of the giant Magellan telescope. In: Ground-Based and Airborne Telescopes V, vol. 9145. International Society for Optics and Photonics (2014)

4. Hampf, D., Wagner, P., Riede, W.: Optical technologies for the observation of low Earth orbit objects. In: Proceedings of IAC 2014 (2015)
5. Lejba, P., Suchodolski, T., Michalek, P., Bartoszak, J., Schillak, S., Zapasnik, S.: First laser measurements to space debris in Poland. Adv. Space Res. **61**(10), 2609–2616 (2018)
6. Andersen, T., et al.: The Euro50 extremely large telescope, vol. 4840. International Society for Optics and Photonics (2003)
7. Gilmozzi, R., Spyromilio, J.: The European extremely large telescope (E-ELT), The Messenger 127 (2007)
8. Bely, P.: The Design and Construction of Large Optical Telescopes. Springer Science & Business Media, New York (2006)
9. Ocaña, F., et al.: First results of the Test-Bed Telescopes (TBT) project: Cebreros telescope commissioning. In: Ground-Based and Airborne Telescopes VI, vol. 9906. International Society for Optics and Photonics (2016)
10. Genet, R., et al.: An 18-Inch Direct Drive Alt-Az Telescope, Society for Astronomical Sciences Annual Symposium, vol. 27 (2008)
11. Hedrick, R.L., et al.: New paradigms for producing high-performing meter class ground-based telescopes. In: Modern Technologies in Space-and Ground-Based Telescopes and Instrumentation, vol. 7739. International Society for Optics and Photonics (2010)
12. Janiszewski, D., Kielczewski, M.: Kalman filter sensor fusion for multi-head position encoder. In: 19th European Conference on Power Electronics and Applications, Warsaw, pp. P.1–P.7 (2017). https://doi.org/10.23919/EPE17ECCEEurope.2017.8099389
13. Brock, S.: Influence of filters for numerical differentiation on parameter tuning of PI speed controllers. Przeglad Elektrotechniczny **05**(2018), 60–64 (2018). https://doi.org/10.15199/48.2018.05.10
14. Galicki, M.: Finite-time control of robotic manipulators. Automatica **51**, 49–54 (2015)
15. Su, Y., Swevers, J.: Finite-time tracking control for robot manipulators with actuator saturation. Robot. Comput. Integr. Manuf. **30**, 91–98 (2014)

Estimation of Free Space on Car Park Using Computer Vision Algorithms

Mateusz Bukowski[1], Marcin Luckner[1(✉)], and Robert Kunicki[2]

[1] Faculty of Mathematics and Information Science,
Warsaw University of Technology, ul. Koszykowa 75, 00–662 Warsaw, Poland
bukowskim@student.mini.pw.edu.pl, mluckner@mini.pw.edu.pl
[2] Digitalisation Department, City of Warsaw, pl. Bankowy 2, 00-095 Warsaw, Poland
rkunicki@um.warszawa.pl

Abstract. A system for monitoring of vacant parking spots can save drivers a lot of time and costs. Other citizens can benefit from a reduction of pollutions too. In our work, we proposed the computer vision system that estimates free space in a car park. The system uses three separate estimation methods based on various approaches to the estimation issue. The free car park area is recognised on a video frame by as the broadest cohesive area, the largest group of pixels with similar colours, and background for parked cars. The raw results of the estimations are aggregated by a Multi-Layer Perceptron to obtain the final estimate. The test on real data from the City of Warsaw showed that the system reaches 95% accuracy. Moreover, the results were compared with the registers from the parking machines to estimate a gap between covered payment and the accurate number of parked cars.

1 Introduction

Monitoring of free spaces on car parks benefits the parking operator companies and citizens, who can reduce the time spent on searching for a free parking spot. For instance, a smart parking application in the City of Milan communicates if space is free or not. Simulations showed that the technology could help each driver to save an average of 77.2 h every year, 86.5 euros in fuel costs, and the entire city of Milan can reduce CO_2 emissions by 44,470 tons per year [9].

Data to drive such applications can be obtained from dedicated systems based on existing sensors and parking machines. However, if a city lacks such system a computer vision solution using surveillance cameras can be used instead.

We proposed the computer vision system that applies separate algorithms to estimate the free space. The algorithms use various approaches to the estimation issue. The flood fill algorithm calculates a cohesive area representing not occupied places. The k-mean algorithm finds a set of the dominant colours that represents the asphalt area of the car park. Finally, the GrabCut algorithm cuts off the occupied area from the background determined by free space.

© Springer Nature Switzerland AG 2020
R. Szewczyk et al. (Eds.): AUTOMATION 2019, AISC 920, pp. 316–325, 2020.
https://doi.org/10.1007/978-3-030-13273-6_30

The proposed solution utilises the Multi-Layer Perceptron to process the raw results of the algorithms considering global image parameters. Next, the same type of the neural networks is used to aggregate all estimation in a single result.

The system was verified on real data from the City of Warsaw (Poland). The obtained accuracy was 95%. Moreover, the system was compared with payment data from the parking machine system.

The rest of this work is structured as follows. Section 2 briefly describes related works. Section 3 presents used data. Section 4 describes the used clustering model and a data preparation. Section 5 discuss the obtained results. Finally, Sect. 6 presents conclusions and future works.

2 Related Work

Several works discussed the issue of free places detection in car parks. In work [8] a fisheye optics was used to detect free parking lots from a car. The authors stressed that due to occlusions and perspective distortions, the classification performance dropped rapidly at longer distances. However, the authors obtained a relatively good average classification error from 3.2% to 3.6% for near vehicles.

The distortion issue is broader discussed in [12]. The issues of a wide-angle camera are not very important in our case, but the perspective issue is one of the major problems of the recognition of objects in an extensive area.

An image from a surveillance camera is analysed in work [14]. The proposed system tracked in real time the state of the parking complex. The authors investigated 500 frames in five locations. The error varied from 5% to 20%.

In work [1] a bird's-eye view of a car park is analysed to detect free spaces. The recognition was errorless on sunny days when on cloudy days it grew up to 6 per cent. The importance of the weather conditions was also stressed in [6] where the authors reached 98% accuracy but only in the absence of strong shadows.

Finally, work [10] presented a list of the results obtained by state-of-the-art techniques in vacant parking spot detection. The accuracy varies from 88% to over 99% – including the result obtained by the authors that were 98.8%. The results were obtained on various data sets but allow us to estimate that an acceptable accuracy rate should exceed 90%.

In our work, we compared the estimation results with data from the parking machine registers. We did not find any paper that compared such data. However, the extended discussion of parking-related data can be found in work [3].

We also discussed an occupancy of the car park according to a day of the week. This kind of analysis can be used to calculate parking fees automatically. The application of machine learning techniques for this task can be found in [13].

3   Data

3.1   Video

Video data with a car park view was taken from an Internet camera http://www.lookcam.com (formerly http://www.oognet.pl). The observed car park is localised on Constitution Square in centre of Warsaw. The camera view includes the whole car park. Figure 1 presents the view of the whole car park that was used in the tests.

Fig. 1. View of the car park used in tests

Collected data were a quarter long movies collected between 8 am and 6 pm from 03/18/17 to 04/28/17. The data contains some lacks caused mostly by stoppages of the camera. In the discussed methods static pictures were utilised. Therefore, one-thousandths of frames were taken for the learning and testing process. There is no knowledge transmitted among the frames. Each of the estimation methods uses additional filters to prepare the data. The parameters of the filters were fixed using random movies.

Each algorithm took a frame as input. The output is a single number from range $[0, 1]$ that estimates a free space in the car park where zero is an entirely free and one a fully occupied car park.

3.2   Payment Data

Significant shortages of parking spaces characterise unguarded Paid Parking Zone (UPPZ) on public roads in the Capital City of Warsaw. The area covered by the obligation to pay the parking fee increases periodically. Payments are charged working days from 8 am to 6 pm.

The payment for parking the vehicle in the UPPZ should be paid immediately after parking the vehicle in one of the following forms:

1. Purchase of a ticket in a parking machine (cash-coins or bank card).
2. Through the mobile payment carried out by an external operator.
3. Paying a residential subscription fee.

Due to various form of the payment, payment data from the parking machines may not fully correspond to the actual state for various reasons, including:

- the mobile payment is not assigned to the location,
- data of parking with the subscription is not recorded in the parking machine,
- the payment in the parking machine is valid in the whole UPPZ and it is possible to move the vehicle after paying the fee,
- the differences between the actual and declared time of using the zone.

4   Methods

4.1   Flood Fill

The first strategy used the Flood fill algorithm [16]. Firstly, several pixels belonged to the car park were selected. The pixels were not obscured by cars on the learning probe and with a high probability would not be covered on the rest of the frames.

Next, the strategy coloured all pixels connected to the starting pixels and having a similar colour. The procedure was repeated for newly coloured pixels. The pixels were coloured to black.

The algorithm returns value ρ_F from the range $[0, 1]$ defined as the number of the black pixels to the number of pixels in the region of interests.

Before the procedure, noises on an input frame were reduced using several filters. Firstly, a bilateral filter [15]. The filter smooths images while preserving edges, by means of a nonlinear combination of nearby image values. Next, a non-local algorithm for image denoising was used [4]. In the filter, a denoised value at pixel x is a mean of the values of all pixels whose Gaussian neighbourhood looks like the neighbourhood of x. The main difference of the NL-means algorithm concerning local filters or frequency domain filters is the systematic use of all possible self-predictions the image can provide.

Fig. 2. Usage of flood fill algorithm

Figure 2 presents the input frame for the flood fill algorithm and the car park area coloured by the algorithm. We see that not only cars but also people influence the estimation of the free space area.

4.2   K-Means

The second strategy used the K–means algorithm [7]. All pixels were clustered into three groups according to their colours. The biggest cluster was taken as a representation of the car park area. Observations on the learning probe supported this assumption. All pixels from the biggest cluster were coloured to black.

The algorithm returns value ρ_M from the range $[0, 1]$ defined as the number of the pixels from the dominant cluster to the number of pixels in the region of interests.

Initially, the input frame was filtered by the Laplacian filter [16]. The filter is a 2-D isotropic measure of the 2nd spatial derivative of an image. The Laplacian of an image highlights regions of rapid intensity change and is therefore often used for edge detection.

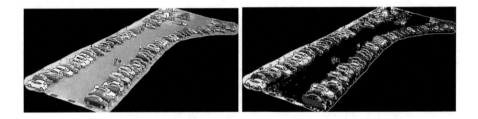

Fig. 3. Usage of K-means algorithm

Figure 3 presents the input frame for the K-means algorithm and the car park area coloured by the algorithm. In comparison to the flood fill algorithm, the result also includes segments of the cars (mostly windows) into the free space area.

4.3   GrabCut

The third strategy used the GrabCut algorithm [11]. The algorithm estimated the alpha-matte simultaneously around an object boundary and the colours of foreground pixels. The algorithm needs a definition of the foreground area.

Firstly, the mask was created to define potential car park area (background) and area potentially occupied by cars (foreground). The aisle between the cars was taken as the background when two rows of cars were taken as the foreground.

However, a statical definition of the areas was not good enough. Therefore the edge detection was used to estimate a location of cars on the mask. When the edges are detected in the foreground area, then the segment outlined by the edges is taken as a reliable foreground and taken as an input for the algorithm.

The algorithm returns value ρ_G from the range $[0, 1]$ defined as the number of the background pixels to the number of pixels in the region of interests.

The input frame is initially filtered using the same denoising filters as the flood fill strategy. Additionally, the image is sharpened. For that, a Gaussian smoothing filter is used. Next, the image is weighted subtracted with the smoothed version from the original image. The edges of the preprocessed image were detected using the Canny edge detectors [5].

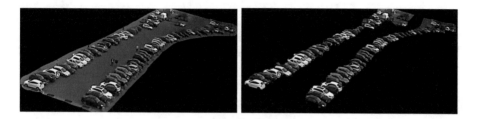

Fig. 4. Usage of GrabCut algorithm

Figure 4 presents the input frame for the GrabCut algorithm and the car park area coloured by the algorithm. In comparison to the previous algorithms, the people did not obscure the free space area.

4.4   Multi Layer Perceptron

Each strategy took as an input a frame with a view of the car park. The output was the surface of detected free space on the car park area divided by the region of the considered area ρ. Additionally, each frame was described by two additional values that described conditions when the photo was taken. For the grey-scale image, the average colour μ and the standard deviation σ were calculated.

The created triple (ρ, μ, σ) is the input for a Multi-Layer Perceptron [2]. The neural network has three layers with 3, 12, and 1 neuron on them. The output is a value from range $[0, 1]$ that multiplied by the maximal number of parked cars n_{max}– estimated as 150 – returns the current number of the parked cars.

The following formula calculates the estimation:

$$n = n_{max} * \mathrm{mlp}\,(\rho, \mu, \sigma), \tag{1}$$

where mlp represents the Multi-Layer Perceptron function.

4.5   Ensemble of Estimators

Additionally to the estimation made by separates strategies a global estimation is calculated. The estimate of the number of parked cars is done by a next Multi-Layer Perceptron that aggregates the estimations obtained by the rests of neural networks.

The following formula calculates the final estimation:

$$n = n_{max} * \mathrm{mlp}\left(\mathrm{mlp}_F\left(\rho_F, \mu, \sigma\right), \mathrm{mlp}_M\left(\rho_M, \mu, \sigma\right), \mathrm{mlp}_G\left(\rho_G, \mu, \sigma\right)\right), \quad (2)$$

where mlp is the perceptron function, $\mathrm{mlp}_F, \mathrm{mlp}_M, \mathrm{mlp}_G$, are the perceptron function trained for the separated estimators, ρ_F, ρ_M, ρ_G, are raw estimations made by the estimators, and μ, σ are descriptors of the image.

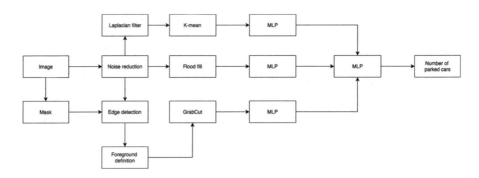

Fig. 5. Summary of the number of cars estimation process

Figure 5 summarises the whole estimation process. Firstly, the input image is used to create filtered versions according to the estimators' needs. Secondly, the raw estimations are calculated separately. Next, the estimations are calculated using Multi-Layer Perceptrons (MLP). Finally, the results are aggregated by the neural network and the final estimation is calculated.

5   Results

5.1   Valuation of the Estimators

The proposed methods were tested on real video data collected between 8 am and 6 pm from 03/18/17 to 04/28/17. The photos were taken in various weather conditions with varying levels of sun exposure and during different precipitation.

We have collected 17325 frames. Among them, 618 frames were chosen randomly for the tests of the estimation accuracy. For all these frames the number of the parked cars were manually counted. After that 430 frames were taken as the learning set for the estimators and 188 frames created the testing set.

Table 1 presents the average error obtained on the learning set. Each of the estimators was validated separately as well as a part of the ensemble. The obtained errors show a variety of the analysed set. Various conditions registered on the frames are displayed in disproportion between the errors obtained on the learning and testing sets.

Table 1. The average errors obtained on the learning and testing sets

Method	Flood fill	K-means	GrabCut	Ensemble
Learning	0.07	0.07	0.06	0.05
Testing	9.89	11.04	9.53	8.16

On the learning set, the difference between the right number of the parked cars and the estimates ones is no higher than one for all estimators. On the testing set, the results vary from 9 to 11 vehicles for various estimators.

The ensemble of the estimators obtained the smallest error of 8 cars. Because the estimated maximal number of the parked vehicles is 150 the obtained error is about 5% of the parking places. This result is in the range determined by the best algorithms compered in [10].

5.2   Comparison with the Parking Machines

To compare our estimation with payment data from the parking machines we collected information on payments from three parking machines situated on Constitution Square (G050120, G050122, and G050123).

The comparison of the results obtained on the working days in the observation period showed that the measures from the parking machines were underestimated. The mean difference between our system and the parking machines data was 30 cars. It is a higher value than the obtained mean error, which was eight cars. However, because of the diversity of the payment methods, the existence of the underestimation was expected.

Let us stress that the observed gap could not be estimated using the parking machines data only. An external company gathers data about mobile payments. Therefore, the access to the data is restricted. However, the alternative estimation presented in our work gives a reliable estimate of the gap.

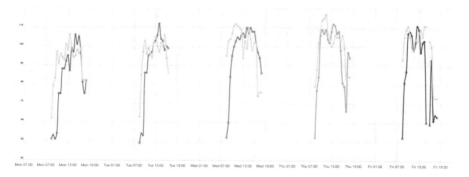

Fig. 6. Comparison of the estimated number of parked cars with data from the parking meters. The value from the parking machines (yellow line) was shifted by 30.

Figure 6 presents the comparison of the estimated number of parked cars with the data from the parking machines. The value obtained from the parking meters registers (yellow line) was shifted by 30 to compare the characteristic of the plots. Both plots look similar without unexpected fluctuations.

5.3   Analysis of Days of Week

Figure 7 compares the estimations from 7 weeks distributed by days of the week. Data on the plots were taken with a half hour resolution. Mostly, the estimations obtained on the same day of the week are similar. We observe exception for the 6th week where the estimated number of cars is lower than at others weeks on Sunday and Monday. It was caused by the Easter holidays.

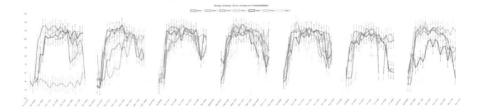

Fig. 7. The estimated number of cars grouped by days of the week

On weekends the number of parked cars stays on a similar level than on the working days. It may undermine the legitimacy of free parking at weekends. It should be noticed that the parking machines data does not allow the local government to estimate a need for parking places at weekends because such data are not collected. This fact stresses the sense of new estimation methods including the computer vision methods proposed in this work.

6   Conclusions

We presented the computer vision system that can estimate free space in a car park. The system uses three separate algorithms to estimate the free space. Next, the system aggregates the raw results considering global image parameters.

We tested the system on real data from the City of Warsaw. The 95% accuracy was obtained. Additionally, we compared the received data with the parking machines registers. That allowed us to estimate the gap in the registered cart and the actual number.

In the future, we want to compare obtained results with dynamic strategies that count arriving and departing cars and create a combined dynamic system with relaxation based on the static methods described in this paper.

Acknowledgements. This research has been supported by the European Union's Horizon 2020 research and innovation programme under grant agreement No. 688380 *VaVeL: Variety, Veracity, VaLue: Handling the Multiplicity of Urban Sensors.*

References

1. Bibi, N., Majid, M.N., Dawood, H., Guo, P.: Automatic parking space detection system. In: 2017 2nd International Conference on Multimedia and Image Processing (ICMIP), pp. 11–15. IEEE (2017)
2. Bishop, C.M.: Pattern Recognition and Machine Learning (Information Science and Statistics). Springer, New York (2006)
3. Bonsall, P.W.: The changing face of parking-related data collection and analysis: the role of new technologies. Transportation 18(1), 83–106 (1991). https://doi.org/10.1007/BF00150560
4. Buades, A., Coll, B., Morel, J.M.: A non-local algorithm for image denoising. In: Proceedings of the 2005 IEEE Computer Society Conference on Computer Vision and Pattern Recognition (CVPR'05), CVPR 2005, vol. 2, pp. 60–65. IEEE Computer Society, Washington, DC (2005). http://dx.doi.org/10.1109/CVPR.2005.38
5. Canny, J.: A computational approach to edge detection. IEEE Trans. Pattern Anal. Mach. Intell. 8(6), 679–698 (1986). http://dx.doi.org/10.1109/TPAMI.1986.4767851
6. Döge, K.: Experiences with video-based incident detection and parking space surveillance systems on motorways in the free state of Saxony. In: TST. Communications in Computer and Information Science, vol. 239, pp. 257–264. Springer (2011)
7. Hartigan, J.A., Wong, M.A.: A k-means clustering algorithm. JSTOR Appl. Stat. 28(1), 100–108 (1979)
8. Houben, S., Komar, M., Hohm, A., Lüke, S., Neuhausen, M., Schlipsing, M.: On-vehicle video-based parking lot recognition with fisheye optics. In: ITSC, pp. 7–12. IEEE (2013)
9. Mangiaracina, R., Tumino, A., Miragliotta, G., Salvadori, G.: Smart parking management in a smart city: costs and benefits. In: SOLI, pp. 27–32. IEEE (2017)
10. Marmol, E., Sevillano, X.: Quickspot: a video analytics solution for on-street vacant parking spot detection. Multimed. Tools Appl. 75(24), 17711–17743 (2016)
11. Rother, C., Kolmogorov, V., Blake, A.: Grabcut -interactive foreground extraction using iterated graph cuts (2004). https://www.microsoft.com/en-us/research/publication/grabcut-interactive-foreground-extraction-using-iterated-graph-cuts/
12. Shih, S., Tsai, W.: A convenient vision-based system for automatic detection of parking spaces in indoor parking lots using wide-angle cameras. IEEE Trans. Veh. Technol. 63(6), 2521–2532 (2014)
13. Simhon, E., Liao, C., Starobinski, D.: Smart parking pricing: a machine learning approach. In: INFOCOM Workshops, pp. 641–646. IEEE (2017)
14. Sukhinskiy, I.V., Nepovinnykh, E.A., Radchenko, G.I.: Developing a parking monitoring system based on the analysis of images from an outdoor surveillance camera. In: MIPRO, pp. 1603–1607. IEEE (2016)
15. Tomasi, C., Manduchi, R.: Bilateral filtering for gray and color images. In: Proceedings of the Sixth International Conference on Computer Vision, ICCV 1998, pp. 839–846. IEEE Computer Society, Washington, DC (1998). http://dl.acm.org/citation.cfm?id=938978.939190
16. Vernon, D.: Machine Vision. Prentice-Hall, Upper Saddle River (1991)

City Bus Monitoring Supported by Computer Vision and Machine Learning Algorithms

Artur Wilkowski[1], Ihor Mykhalevych[2], and Marcin Luckner[2(✉)]

[1] Faculty of Geodesy and Cartography, Warsaw University of Technology,
Pl. Politechniki 1, 00-661 Warsaw, Poland
artur.wilkowski@pw.edu.pl
[2] Faculty of Mathematics and Information Sciences,
Warsaw University of Technology, Koszykowa 75, 00-662 Warsaw, Poland
{i.mykhalevych,mluckner}@mini.pw.edu.pl

Abstract. In this paper there are proposed methods and algorithms supporting city traffic controllers in effective perception and analysis of the visual information from the public transport monitoring system implemented in the City of Warsaw. To achieve this goal, public transport vehicles must be recognised and tracked in camera view. In this work, we describe a structure and give preliminary results for the detection and tracking system proposed. The algorithms discussed in this paper uses background subtraction to extract moving vehicles from the scene and the classification system to reject objects that are not city buses. Furthermore, a custom tracking module is utilized to enable labeling of city buses instances. During the test performed in the City of Warsaw the system was able to successfully detect 89% bus instances giving less than 15% erroneous detections.

Keywords: Computer vision · Detection · Tracking · Traffic monitoring

1 Introduction

Public transport is an area that still needs improvement to satisfy its users [5]. Local government are aware of this fact and apply systems to monitor and supervise the public transport. Typically, a human traffic controller monitors public transport using cameras from city infrastructure. However, quite often an Internet of Things (IoT) systems are introduced to support him or her.

The City of Warsaw monitors position of the public transport vehicles: buses and trams. Their position is reported every 10 s. Recently, these data were made public to improve services offered by the government and addressed to citizens, as well as to enable third–party application development. It was done because recent studies emphasise the role of Open Data also as an enabler of innovation

© Springer Nature Switzerland AG 2020
R. Szewczyk et al. (Eds.): AUTOMATION 2019, AISC 920, pp. 326–336, 2020.
https://doi.org/10.1007/978-3-030-13273-6_31

[2,3]. The data can be used to improve public transport information systems [7] that mostly lacks a real–time information [7].

Our goal in this work is to support the traffic controller in using information from the public transport monitoring system implemented in the City of Warsaw. At the moment, the controller has the view of a street from a camera and the view of a map with superimposed positioning data. Our aim is to integrate both systems by associating information from both data sources - the camera and the map. To achieve this, first public transport vehicles must be recognised and tracked in camera view.

In this work, we present a structure and preliminary results of the tracking system. The algorithm proposed in this paper uses background subtraction to extract moving vehicles. The initial module tracks foreground objects, extracts images of the tracked objects from the original picture, and feds them to the classification system. The classification system rejects objects that are not city buses. Detection and classification results are subsequently filtered by the tracking system.

On one of the main intersection in the City of Warsaw the system proposed was able to successfully detect almost 89% bus instances present in the video while giving less than 15% erroneous detections.

The rest of the paper is structured as follows. Section 2 presents algorithms and methods that were used to track and recognize city buses. Section 3 describes experiments that have been performed to evaluate the system. Finally, Sect. 4 summarizes the results.

2   Methods and Algorithms Used

The algorithms proposed in this paper uses background subtraction to binarize the scene. First, foreground objects are detected and tracked. Next, the images of the tracked objects are extracted from the original picture and fed to the classification system. The classification system rejects objects that are not classified as city buses. Detection and classification results are filtered using the tracker. The following sections give details.

2.1   Application of Background Subtraction

Background Subtraction. In order to perform image segmentation the Adaptive Gaussian Mixture Model (AGMM) (described in [10,11]) is utilized. According to these papers color model for each pixel (both background and foreground) is given by the Gaussian Mixture Model:

$$p(x|BG + FG) = \sum_{m=1}^{m} \hat{\pi}_m \mathcal{N}(x; \hat{\mu}_m, \hat{\sigma}_m^2 I)$$

where $\hat{\mu}_m$ and $\hat{\sigma}_m^2$ are estimations of mean and variance of Gaussian components, and $\hat{\pi}_m$ are non-negative mixing weights.

For each new color sample $x^{(t)}$, we apply the following update rules to our GMM

$$\hat{\pi}_m \leftarrow \hat{\pi}_m + \alpha(o_m^{(t)} - \hat{\pi}_m) - \alpha c_T$$

$$\hat{\mu}_m \leftarrow \hat{\mu}_m + o_m^{(t)}(\alpha/\hat{\pi}_m)\delta_m$$

$$\hat{\sigma}_m \leftarrow \hat{\sigma}_m + o_m^{(t)}(\alpha/\hat{\pi}_m)(\delta_m^T \delta_m - \hat{\sigma}_m^2)$$

where α is the adaptation coefficient (in our application we assume that $\alpha = 1/T$, where T is the predefined time span), $\delta_m = x^{(t)} - \hat{\mu}_m$. $o_m^{(t)}$ is set to 1 for a single selected mixture component compliant with $x^{(t)}$ and to 0 for remaining components. The selected component must be within 3-sigma distance from the new sample $x^{(t)}$ (in the sense of Mahalonobis distance) and have the largest $\hat{\pi}_m$ value. αc_T is responsible for 'smoothing' the distribution of $\hat{\pi}_m$ according to the Dirichlet prior [10] with c_T being a constant. The values $\hat{\pi}_m$ are normalized after update.

If no compliant component could be found for the new sample $x^{(t)}$ a new component is created for this sample with parameters $\hat{\pi}_{M+1} = \alpha$, $\hat{\mu}_{M+1} = x^{(t)}$ and $\sigma_{M+1} = \sigma_0$ (σ_0 is a fixed initial variance). The number of components is limited. If surplus components are present the ones with the smallest $\hat{\pi}_m$ are discarded.

The Gaussian Mixture components constituting the color model can be divided into those representing background and foreground. It is assumed that background components dominate in the model. Therefore, the background model is constituted by the largest Gaussian clusters (with the highest values of $\hat{\pi}_m$). To separate background and foreground components, they are first ordered by descending $\hat{\pi}_m$ values, and the number B of components constituting background model is given by

$$B = \arg\min_b \left(\sum_{m=1}^{b} \hat{\pi}_m > (1 - c_f) \right)$$

where c_f is a fixed constant describing the maximum proportion of the data that can belong to foreground objects. The background model is thus a fraction of the full color model and can be given as

$$p(x|BG) \sim \sum_{m=1}^{B} \hat{\pi}_m \mathcal{N}(x; \hat{\mu}_m, \hat{\sigma}_m^2 I)$$

This model then make decisions if the pixels belongs to the background or foreground by thresholding $p(x^{(t)}|BG)$.

$$p(x^{(t)}|BG) > c_{thr}$$

and c_{thr} is a fixed threshold value.

Applying Background Subtraction to City Bus Detection. Similarly to [9] we apply the Hue-Saturation-Value (HSV) color space as found to be giving the best relation between true and false foreground detections.

Binary Image Processing. The output of the background subtraction module is a binary image with foreground pixels marked as 1 and background pixels marked as 0. The resulting binary image is first preprocessed to ameliorate pure color detection results. To do so n-times morphological erosion followed by m-times morphological dilation is applied. The first operation removes false detection patches and the second connects close detections that were accidentally separated. In our approach, we set $m \geq n$.

After obtaining a list of disconnected image patches, we apply hole-filling algorithm [6] to obtain the final list of candidates to track.

Prefiltering. To early distinguish between promising track candidates and accidental movement detections, a prefiltering step is applied to the list of candidates. The filtering step uses two features: the minimum bounding box r, the center of the bounding box c and the size of the image patch s (effectively - the number of pixels). Early rejection of the candidate is applied when **at least one** of the following conditions hold

$$\begin{cases} s > minArea \\ r.height < maxHeight \\ r.width < maxWidth \\ IStreetMask(c.x, c.y) = 0 \end{cases}$$

where $IStreetMask$ is the mask containing area when vehicles are expected to appear (street, bus stops). The resulting list of observations is subsequently supplied to object tracker and recognizer modules.

The output list of observation for frame t can be described as a list of pairs $o_i^t = (o_i^t.x, o_i^t.y)$, where (x, y) stand for the bounding box center coordinates for each patch detected.

2.2   Bag of Words Features

In order to efficiently distinguished between city buses and other vehicles, we utilize the bag-of-words feature set [1]. In our application a number of SIFT keypoints and features [4] is first extracted from each training patch. Then k-means clustering algorithm is applied to the set composed of SIFT descriptors from all training patches. In results, we obtain a set of clusters grouping similar descriptors available in the training set. Each cluster is thus represented by a *codeword* which is its geometrical center.

Then, from each training, validation or test patch once again some SIFT keypoints and descriptors are extracted and each descriptor is matched against the existing set of codewords. For each descriptor in the patch closest codeword is selected. The resulting set (or bag) of codewords assigned to a single image patch is transformed into a codeword histogram. A codeword histogram forms a feature used for subsequent image patch classification.

2.3   Recognition of City Buses

A classifier used in this work distinguishes between city buses and other street vehicles. The implementation is based on Support Vector Machines (SVM) [8].

The basic setup is as follows: we have two classes labelled -1 and $+1$ and the space of features is d-dimensional (R^d), and each object occurrence is described by a feature vector $x_i \in R^d$ labelled as y_i, for $i = 1, 2, \ldots, N$ and N being the cardinality of the learning set. For a non linear separation of the two classes, the space R^d is mapped into a space of higher dimension using kernel function $K(x, x')$. In this paper a classic SVM decision function is utilized

$$f(x) = \text{sgn} \left(\sum_{i=1}^{N} y_i * \alpha_i * K(x, x_i) + b \right) \tag{1}$$

where coefficients α_i and b are computed by maximization of the following convex quadratic programming (QP):

$$\sum_{i=1}^{N} \alpha_i - \frac{1}{2} \sum_{i=1}^{N} \sum_{j=1}^{N} \alpha_i * \alpha_j * y_i * y_j * K(x_j, x_i) \tag{2}$$

subject to the following constrains:

$$\bigwedge_{i \in \{1, 2, \ldots, N\}} 0 \leqslant \alpha_i \leqslant C \ \wedge \ \sum_{j=1}^{N} \alpha_i * y_i = 0 \tag{3}$$

This formulation aims at finding a hyperplane separating positive and negative samples that maximizes minimum distance of the training samples from the plane (the margin).

The regularization coefficient C in Eq. 3 controls trade–off between margin width and misclassification errors to avoid overfitting. The training set is thus divided into proper training and validation subsets. The parameter C is adjusted to provide best recognition results on the validation subset.

As the input data are generally not linearly-separable, the Gaussian kernel function $K(x, x')$ is utilized

$$\exp \left(-\frac{1}{d} ||x - x'||^2 \right), \tag{4}$$

where d is the number of features.

2.4 Tracking of City Buses

Object Tracking. In the application of city bus tracking we follow to some extent methods proposed in [9]. The system assumes a possibility of tracking multiple objects (both buses and other vehicles) at the same time. The observations obtained from the image are associated with the *tracked objects* by a minimum distance criterion. Tracked objects that were not observed for too many frames are dropped. The list of tracked objects is updated using newly created associations. The detailed description of the process is given below. For the purpose of tracking two sets need to be maintained:

- The output set of observation for frame t described as a set of pairs $o_i^t = (o_i^t.x, o_i^t.y)$. Let us denote this set as O^t.
- The set of *tracked object entries* A^t with elements $a_j^t = (a_j^t.x, a_j^t.y, a_j^t.k)$, where $a_j^t.x$ and $a_j^t.y$ is the last known position of the tracked object, and $a_j^t.k$ is the last frame when it was observed.

For each frame processed, conceptually, a Cartesian product of both sets $A^t \times O^t$ is first computed. Resulting pairs (a_j^t, o_i^t) are then filtered leaving only those for which the following conditions hold

$$\begin{cases} a_j^t.k < \varepsilon_{age} \\ dist(a_j^t, o_i^t) < \varepsilon_{dist} \end{cases}$$

where *dist* is a Euclidean distance function defined as

$$dist(a, o) = \sqrt{(a.x - o.x)^2 + (a.y - o.y)^2)}$$

The second filtering steps removes repeating a_j^t and o_i^t. The resulting set of pairs (a_j^t, o_i^t) is sorted according to the value of the *dist* function. Then, while the list is traversed starting from the smallest values of the *dist* function, there are left only entries with so far unused a_j^t, o_i^t values. In effect we enforce one-to-one *tracking associations* (a_j^t, o_i^t) between objects currently tracked and the new observations.

The new set A^{t+1} of tracked object entries is initially set to A^t then updated in the following way:

- If for the observation o_i^t there exist a valid tracking association in the form (a_j^t, o_i^t) then the tracked object entry a_j^t is only updated according to o_i^t
- If for the observation o_i^t there do not exist any valid tracking association (a_j^t, o_i^t) a new tracked object entry (say a_l^{t+1}) is added to A^{t+1} and then its values are set according to o_i^t).

A newly created $a_{\bullet}^{t+1}$ inherits the observation position (x and y attributes from associated observation). Attribute k is set to the current frame t.

Sample results of the tracking procedure together with background subtractor output are given in Fig. 1.

Fig. 1. Sample output of background subtractor and the tracker

Filtering of Recognition Results. The SVM image patch recognition engine tracks all moving objects. Each image patch is recognized as a city-bus or other vehicle, and individual recognition results are then projected into the classification of the whole tracked object.

In order to do so, in addition to tracked object entries A^t, a set of recognition memory C^t is maintained. For each entry of A^t there is exactly one entry in C^t. Each entry in C^t is an 2-tuple $c_i^t = (c_i^t.n, c_i^t.n_b)$. $c_i^t.n$ - denotes the total number of frames in which the given tracked object was observed, while $c_i^t.n_b$ denotes the excess of frames ("classification rate") in which the tracked object i was classified as a city bus or an other vehicle. When tracking begins, both $c_i^t.n$ and $c_i^t.n_b$ are set to 0. $c_i^t.n$ is incremented with each frame in which object is successfully tracked, $c_i^t.n_b$ is incremented when the object is recognized as a city bus in frame t otherwise it is decremented. In the moment when $c_i^t.n_b$ passes through 0, the recognition decision is inverted.

For each frame t, and for each C^t entry there is made a decision whether a tracked object associated with some entry c_j could be classified as a city bus. The decision is positive when it is recognized so and two conditions regarding the length of successful tracking and the classification rate are satisfied:

$$\begin{cases} c_j^t.n > \varepsilon_{fcount} \\ c_j^t.n_b > 0 \end{cases}$$

3   Experiments

3.1   Recognition of City Buses

Classification of City Buses. In the first experiment, there was evaluated the performance of SVM classifier distinguishing city buses from other vehicles. The classifier was trained on a five minutes sequence. From the sequence, some tracked objects were extracted by the algorithms discussed, and from these, a total of 16216 vehicle samples were obtained with 6651 manually classified as city buses and 9565 classified as other objects. These samples were directly used for training the SVM classifier. Randomly selected samples were used as the training (30%) and validation (70%) sets. Using these two sets SVM parameters C and γ were tuned and SVM classifier was eventually trained.

Fig. 2. Samples of city buses

Fig. 3. Samples of other vehicles

Figure 2 demonstrates samples of city buses, while Fig. 3 demonstrates samples of other vehicles.

The classifier gave no errors on the training set. Therefore, in the next step the classifier was verified on the test set. The test set was created basing on another 5-min video, from which 10623 image samples were extracted and manually labelled as city buses (1129 samples) and other vehicles (9494 samples).

The ROC curve giving the results of the classification on the test set is given in Fig. 4. The results are worse than those obtained on the test set. However, they can be still evaluated as very promising.

It must be noted that these are raw results before application of filtering of the recognition results based on the assumption of motion consistency.

Per-frame Detection of City Buses. In the second experiment, there was evaluated about 1 h of recording. Each frame of the recording was manually labelled as containing a city bus or not. At the same time the system using methods given in Sects. 2.1, 2.2 and 2.3 was asked to label each frame as containing city bus or not. The frame was marked as containing a city bus if at least one city bus was recognized in the frame. The total of 111465 frames was analyzed with 39683 samples containing buses and 71782 samples that do not contain buses. Resulting confusion matrices are given in Tables 1 and 2.

The detector turns out to be quite conservative in its decisions. It has a minimal rate of false acceptance (about 2.1%) at the cost of the limited sensitivity (about 60%). However, these figures can be adjusted by appropriately calibrating classifier to specific conditions.

Table 1. Confusion matrix for bus detection expressed in total numbers

		True label	
		P	N
Recognized as	P	24147	1537
	N	15536	70245

Table 2. Confusion matrix for bus detection expressed in percentages

		True label	
		P	N
Recognized as	P	60.9	2.1
	N	39.1	97.9

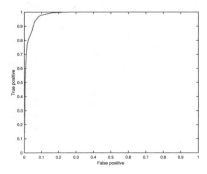

Fig. 4. ROC curve for classification into 'buses' and 'other vehicles' categories

3.2 Tracking City Buses' Instances

The objective of the system is to detect and track instances of city buses in image sequences, so they could be matched against data retrieved from the public transport positioning system. To evaluate our solution, with respect to this specific goal, the following experiment was undertaken.

One hour video was human-processed to extract information regarding the presence of buses. For each bus present in the video, there was noted the first frame in which the bus appeared and the frame in which the bus disappeared, thus providing ground truth information. Similar annotations were performed automatically by the application using methods described in Sect. 2. For each tracked object that was recognized as city bus using methods from Sect. 2.4 there was noted the frame of bus appearance and its disappearance.

In cases where an object appeared, then the track was lost (e.g. due to occlusion) and the object was rediscovered in a new position, the whole object trajectory was treated as a single entity. We treat this only as a slight and acceptable abuse to our test scenario since, in the scenario of matching detections against positions from external source, the re-appearing bus would still obtain the same label.

When matching ground truth with tracking/detection results, we denoted positive match when time spans of single bus presence in the ground truth and the tracking/detection results overlapped more than by threshold t. In the experiments $t_{overlap} = 10\%$ and $t_{overlap} = 80\%$ were used. For each bus present in the

Fig. 5. Application interface showing the detection of a city bus and the detection of some other vehicle

Table 3. Confusion matrix for bus instance matching for $t_{overlap} = 10\%$. Precision: 85.9%, Recall: 88.7%

		True label	
		P	N
Recognized as	P	55	8
	N	7	-

Table 4. Confusion matrix for bus instance matching for $t_{overlap} = 80\%$. Precision: 65.6%, Recall: 67.7%

		True label	
		P	N
Recognized as	P	42	22
	N	20	-

ground truth but absent in the tracking/detection results a single instance of False Negative was counted and for reverse case a single instance of False Positive was counted. Since it is difficult to establish the total number of negative examples in this scenario, the True Negatives were not counted at all. The confusion matrices both threshold are given in Tables 3 and 4. The results obtained reveal that the system can effectively detect city buses instances. Especially for more lenient overlap assessment thresholds, the system can successfully detect almost 89% bus instances present in the video while giving less than 15% erroneous detections. Therefore, we claim the tracking and detection in conjunction with the external bus positioning system could provide reliable graphical annotation regarding buses identification in the video. An example screenshot from the application is given in Fig. 5.

4 Conclusions

In this paper we proposed the system for monitoring public transport vehicles. The algorithm proposed in this paper used background subtraction to extract moving vehicles and the classification system to reject objects that are not city buses. On the test done in the City of Warsaw the system successfully detected 89% bus instances giving less than 15% erroneous detections.

In the future, we want to firmly integrate the public transport vehicle recognition with the information system created for the City of Warsaw in the VaVeL project. The system gives information not only on a position of the vehicle but also calculates its delay and detects critical events such as groups of stopped vehicles. The system combined with a camera view will allow the traffic controller to understand a critical situation and quickly handle incidents.

Acknowledgements. This research has been supported by the European Union's Horizon 2020 research and innovation programme under grant agreement No. 688380 *VaVeL: Variety, Veracity, VaLue: Handling the Multiplicity of Urban Sensors.*

References

1. Csurka, G., Dance, C.R., Fan, L., Willamowski, J., Bray, C.: Visual categorization with bags of keypoints. In: Workshop on Statistical Learning in Computer Vision, ECCV, pp. 1–22 (2004)
2. Grabowski, S., Grzenda, M., Legierski, J.: The adoption of open data and open API telecommunication functions by software developers. In: Proceedings of the Business Information Systems: 18th International Conference, pp. 337–347. Springer, Poznań, 24–26 June 2015
3. Lakomaa, E., Kallberg, J.: Open data as a foundation for innovation: the enabling effect of free public sector information for entrepreneurs. IEEE Access **1**, 558–563 (2013)
4. Lowe, D.G.: Distinctive image features from scale-invariant keypoints. Int. J. Comput. Vis. **60**(2), 91–110 (2004). https://doi.org/10.1023/B:VISI.0000029664.99615.94
5. Nesheli, M.M., Ceder, A.A., Estines, S.: Public transport user's perception and decision assessment using tactic-based guidelines. Transp. Policy **49**, 125–136 (2016). http://www.sciencedirect.com/science/article/pii/S0967070X16301998
6. Suzuki, S., Abe, K.: Topological structural analysis of digitized binary images by border following. Comput. Vis. Graph. Image Process. **30**(1), 32–46 (1985). http://www.sciencedirect.com/science/article/pii/0734189X85900167
7. Tyrinopoulos, Y.: A complete conceptual model for the integrated management of the transportation work. J. Public Transp. **7**(4), 101–121 (2004)
8. Vapnik, V.: Statistical Learning Theory. Wiley, New York (1998)
9. Wilkowski, A., Luckner, M.: Low-cost canoe counting system for application in a natural environment. In: Szewczyk, R., Zieliński, C., Kaliczyńska, M. (eds.) Challenges in Automation, Robotics and Measurement Techniques, pp. 705–715. Springer (2016)
10. Zivkovic, Z.: Improved adaptive Gaussian mixture model for background subtraction. In: Proceedings of the 17th International Conference on Pattern Recognition, ICPR 2004, vol. 2, pp. 28–31, August 2004
11. Zivkovic, Z., van der Heijden, F.: Efficient adaptive density estimation per image pixel for the task of background subtraction. Pattern Recogn. Lett. **27**(7), 773–780 (2006). https://doi.org/10.1016/j.patrec.2005.11.005

A Description of the Motion of a Mobile Robot with Mecanum Wheels – Dynamics

Zenon Hendzel[(✉)]

Faculty of Mechanical Engineering and Aeronautics, Department of Applied
Mechanics and Robotics, Rzeszow University of Technology, Rzeszów, Poland
zenhen@prz.edu.pl

Abstract. This paper formulates dynamic equations of motion of a 4-wheel
mobile robot equipped with mecanum-type wheels. This new approach towards
the formulation of equations has been applied by the usage of Maggi's mathematical formalism. When describing the dynamics of the mobile robot,
dynamic equations of motion have been designed by the application of
Lagrange's equations with multipliers. Lagrange's multipliers occurring in
dynamic equations of motion cause difficulties in the application of such a form
to the real-time steering synthesis of the analyzed object. Maggi's mathematical
formalism has been applied to eliminate the need for multipliers. Numerical
simulations of the inverse dynamics task have been conducted for the obtained
dynamic motion parameters stemming from the inverse kinematics task.

Keywords: Mobile wheeled robot · Mecanum wheels · Dynamics

1 Introduction

The motivation for analytical considerations of the behaviour of a mobile robot with
Swedish wheels, referred to in the literature as mecanum wheels [1, 6, 7, 10] comes from
the fact that there is a relatively small amount of literature in this area, especially with
regards to the impact of resistance to motion and variable operating conditions on the
quality of motion and its control in real time. An analysis of topics related to the
dynamics of mobile wheeled robots is conducted to aid in selecting the correct solution of
the problem of steering for this type of system. When describing the dynamics of a
system, one often applies simple models, in which there is no taking into consideration
the weights of many individual movable elements. The authors describing the dynamics
of these systems often use mostly classical equations stemming from general mechanics.
One may also encounter works describing the motion of these systems with the assistance of Lagrange's equations of the second kind. For nonholonomic systems, these are
Lagrange's equations with multipliers. This is a form of dynamic equations of motion
inconvenient for modeling the steering of these objects in real-time. Therefore, various
methods of Lagrange's multiplier decoupling are applied [12]. The following paper
applies a new approach towards the formulation of dynamic equations of motion of a
4-wheeled mobile robot equipped with mecanum (WMR_4M) type wheels by the
application of Maggi's mathematical formalism [1, 8]. In Sect. 2, dynamic equations for

© Springer Nature Switzerland AG 2020
R. Szewczyk et al. (Eds.): AUTOMATION 2019, AISC 920, pp. 337–345, 2020.
https://doi.org/10.1007/978-3-030-13273-6_32

WMR_4M motion are formulated. Computer simulations of the inverse dynamics task have also been conducted in Sect. 3. The article ends with a summary and a bibliography.

2 Dynamic Robot Equations of Motion

In dynamics, we are interested mainly in the values of the driving moments for the respective wheels, therefore we analyze the inverse dynamics task. If we are describing the dynamics of mobile wheeled robots, we are mainly aiming at providing dynamic equations of motion. For this purpose, we most often apply Lagrange's equations with multipliers for nonholonomic objects [6, 10–12]. However, due to the entangled form of the equations describing the motion of the system, their application in the synthesis of the steering of these objects is complex. It is then necessary when introducing Lagrange's multiplier decoupling transformation to the moments, to effect equations describing the motion of the system to the so-called reduced form of motion description with nonholonomic links. This reduced form makes the determination of the values of the moments which are of interest to us easier, and the knowledge of these values subsequently allows us to determine the value of Lagrange's multipliers, and therefore the dry friction forces occurring within the plane of contact of the wheels and the road. The knowledge of these values gives a basis to solve the issue of the steering of the mobile robot tracking motion. An analysis of the issues regarding the dynamics of mobile wheeled robots is conducted mainly to obtain the correct solution in terms of the steering of the motion of such types of systems [3, 4, 5, 9].

The motion of the model, which is shown in Fig. 1 is described via the application of Lagrange's equations, which for the nonholonomic system we write in the following vector form [11, 12]:

$$\frac{\mathrm{d}}{\mathrm{dt}}\left(\frac{\partial \mathrm{E}}{\partial \dot{\mathrm{q}}}\right)^{\mathrm{T}} - \left(\frac{\partial \mathrm{E}}{\partial \mathrm{q}}\right)^{\mathrm{T}} = \mathrm{Q} + \mathrm{J}^{\mathrm{T}}(\mathrm{q})\lambda \tag{1}$$

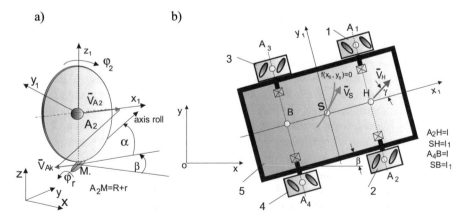

Fig. 1. The WMR_4M model [6]

where q is the vector of the generalized coordinates, $E = E(q, \dot{q})$ is the kinetic energy of the system, Q is the vector of the generalized forces, $J(q)$ is the Jacobian, and λ is the Lagrange multiplier vector. In the analyzed WMR_4M model, these are the forces of friction within the plane of contact of the driving wheel rollers and the road. Taking into consideration that the analyzed system has 3 degrees of freedom, the description of the kinematics of the system [4], (21) has been transformed into three equations in the following form:

$$
\begin{aligned}
\dot{x}_S[\cos(\beta + \alpha)] + \dot{y}_S[\sin(\beta + \alpha)] &= \left(\omega_3 + \tfrac{1}{2}(\omega_4 - \omega_1)\right)k \\
\dot{x}_S[\cos(\beta - \alpha)] + \dot{y}_S[\sin(\beta - \alpha)] &= \left(\omega_4 - \tfrac{1}{2}(\omega_4 - \omega_1)\right)k \\
\dot{\beta} &= \tfrac{1}{2c}(\omega_4 - \omega_1)k
\end{aligned}
\tag{2}
$$

where $k = (R + r)\cos\alpha$, $c = (l\cos\alpha + l_1\sin\alpha)$, $q = [x_S, y_S, \beta, \varphi_1, \varphi_3, \varphi_4]^T$. The above kinematics equations can also be written in the following form:

$$
J(q)\dot{q} = 0 \tag{3}
$$

in such a case, the Jacobian will be determined as:

$$
J(q) = \begin{bmatrix}
\cos(\beta + \alpha) & \sin(\beta + \alpha) & 0 & k/2 & -k & -k/2 \\
\cos(\beta - \alpha) & \sin(\beta - \alpha) & 0 & -k/2 & 0 & -k/2 \\
0 & 0 & 1 & k/2c & 0 & -k/2c
\end{bmatrix}
\tag{4}
$$

When determining via vector the generalized Q forces, the following have been taken into consideration: the forces of gravity of the individual parts, resistance when rolling the wheels, and driving moments. The forces of friction within the plane of contact of the individual wheel rollers with the road are introduced into the system as Lagrange's multipliers. Other resistances occurring within the internal pairs have been omitted. The generalized force vector can be determined from the following dependency [1, 8]:

$$
Q_j\delta q_j = \sum_{i=1}^{6} F_i^T \begin{bmatrix} \delta r_i \\ \dots \\ \delta \varphi_i \end{bmatrix} \qquad j = 1, .., 6 \tag{5}
$$

where j – the number of generalized coordinates. F_i is the generalized vector of the set of forces influencing the i- part, reduced to a point being the beginning of an i-set of reference, δr_i is the vector of shift of the virtual point being the i- set of reference, $\delta \varphi_i$ is the vector of virtual rotation of the i-part. The relations between the virtual shift vector δr_i and the virtual rotation $\delta \varphi_i$, and the appropriate generalized virtual shifts, stem from the equations in (2). When completing Eq. (5) the generalized forces Q_j have been determined, and the part with the Lagrange multipliers has been added λ. The results were the right-hand sides of Eq. (1) marked as Z_j

$$
\begin{bmatrix} Z_1 \\ Z_2 \\ Z_3 \\ Z_4 \\ Z_5 \\ Z_6 \end{bmatrix} = \begin{bmatrix} \{(M_1 - N_1 f_1)\cos(\beta - \alpha) + (M_3 - N_3 f_3)\cos(\beta + \alpha) + (M_4 - N_4 f_4)\cos(\beta - \alpha)\}1/k \\ + \cos(\beta + \alpha)\lambda_1 + \cos(\beta - \alpha)\lambda_2 \\ \{(M_1 - N_1 f_1)\sin(\beta - \alpha) + (M_3 - N_3 f_3)\sin(\beta + \alpha) + (M_4 - N_4 f_4)\sin(\beta - \alpha)\}1/k \\ + \sin(\beta + \alpha)\lambda_1 + \sin(\beta - \alpha)\lambda_2 \\ \{-(M_1 + N_1 f_1) - (M_3 - N_3 f_3) - (M_4 - N_4 f_4)\}c/k + \lambda_3 \\ M_1 - N_1 f_1 + k/2\lambda_1 - k/2\lambda_2 + k/2c\lambda_3 \\ M_3 - N_3 f_3 - k\lambda_1 \\ M_4 - N_4 f_4 - k/2\lambda_1 - k/2\lambda_2 - k/2c\lambda_3 \end{bmatrix} \tag{6}
$$

where M_k – the wheel driving moments, N_k – the forces of the wheel pressing forces, and f_k – the friction parameters of the appropriate wheel rolling, $k = 1...4$.

However, the system's kinetic energy, omitting the roller mass, has been determined on the basis of the following relation:

$$
E = \left(\frac{1}{2}m_r\right)\left(\dot{x}_S^2 + \dot{y}_S^2\right) + \frac{1}{2}I_S\dot{\beta}^2 + \frac{1}{2}m_k\sum_{j=1}^{4}V_{Aj}^2 + 2I_z\dot{\beta}^2 + \frac{1}{2}I_k\left(\omega_1^2 + \omega_2^2 + \omega_3^2 + \omega_4^2\right) \tag{7}
$$

where m_r – frame mass, $m_1 = m_2 = m_3 = m_4 = m_k$ – driving wheel mass, V_{Aj} – velocity of the wheel center of weight, I_S – mass frame moment of inertia calculated with regard to the axis from the frame going through to the center of mass and perpendicular to the frame plane, $I_{zA1} = I_{zA2} = I_{zA3} = I_{zA4} = I_z$ – mass moment of inertia of the wheel calculated in relation to axis z_{Aj} going through the wheel's center of mass and perpendicular to the plane of motion, and $I_{A1} = I_{A2} = I_{A3} = I_{A4} = I_k$ – mass moment of inertia of j-wheel calculated in relation to the axis going through the wheel's center of mass and perpendicular to the wheel plane.

Upon the completion of the required mathematical operations stemming from Lagrange's equations (1), the motion of the analyzed robot model is described by the following set of differential equations:

$$
\begin{aligned}
(m_r + 4m_k)\ddot{x}_S &= Z_1 \\
(m_r + 4m_k)\ddot{y}_S &= Z_2 \\
\left(I_S + 4I_z + m_k\left(l^2 + l_1^2\right)\right)\ddot{\beta} &= Z_3 \\
I_k\ddot{\varphi}_1 &= Z_4 \\
I_k\ddot{\varphi}_3 &= Z_5 \\
I_k\ddot{\varphi}_4 &= Z_6
\end{aligned} \tag{8}
$$

The Lagrange multipliers which can be seen on the right sides of the set of Eqs. (8) make it impossible to apply these dynamic equations of motion for example for the steering motion of the analyzed object in real-time. Therefore, in the later part of the analysis, a different mathematical formalism was used by the application of Maggi's equations [1, 5, 8], which allow the elimination of Lagrange's multipliers. These equations which describe the motion of the systems in generalized coordinates are determined in the following form:

$$
\sum_{j=1}^{n} c_{ij}\left[\frac{d}{dt}\left(\frac{\partial E}{\partial \dot{q}_j}\right) - \frac{\partial E}{\partial q_j}\right] = \Theta_i \qquad i = 1,....s \tag{9}
$$

s is the number of independent system parameters within the generalized coordinates $q_j (j = 1, \ldots, n)$, which represents the degrees of freedom. In such a case we write the generalized velocities as follows:

$$\dot{q}_j = \sum_{i=1}^{s} c_{ij} \dot{e}_i + G_j \tag{10}$$

The values $\dot{e}_i$ are called the characteristics or system kinetic parameters within the generalized coordinates, and c_{ij} as well as G_j are the dependent values in the general case regarding time and the q_j generalized coordinates. The right sides of the set (9) are the coefficients with variations δe_i in the expression for the outer system force prepared work. These coefficients can be determined from the following:

$$\sum_{i=1}^{s} \Theta_i \delta e_i = \sum_{i=1}^{s} \delta e_i \sum_{j=1}^{n} c_{ij} Q_j \tag{11}$$

The provided Maggi's equations will be applied for the preparation of the dynamic equations of motion of WMR_4M. If we assume that the motion of the mobile robot, the model of which has been shown in Fig. 1b, takes place within one plane, then a straightforward setting of this model requires providing point S, and therefore providing the x_S and y_S coordinates of this point, as well as the orientation of the frame, and therefore the temporary rotation angle β, as well as the angles describing the driving wheels own rotations, which we will respectively mark as φ_i, $i = 1, 3, 4$. Then, the generalized coordinates and generalized velocity vector will be as follows:

$$q = [x_S, y_S, \beta, \varphi_1, \varphi_3, \varphi_4]^T \tag{12}$$

$$\dot{q} = [\dot{x}_S, \dot{y}_S, \dot{\beta}, \dot{\varphi}_1, \dot{\varphi}_3, \dot{\varphi}_4]^T \tag{13}$$

Taking into consideration kinematics Eqs. (2) and writing out the relation (10) for $G_j = 0$, the following generalized values have been obtained:

$$
\begin{aligned}
\dot{q}_1 &= \dot{x}_S = c_{11}\dot{e}_1 + c_{21}\dot{e}_2 + c_{31}\dot{e}_3 = k/2[\sin(\beta - \alpha) + \sin(\beta + \alpha)]\dot{e}_1 \\
&\quad - k\sin(\beta - \alpha)\dot{e}_2 - k/2[\sin(\beta - \alpha) - \sin(\beta + \alpha)]\dot{e}_3 \\
\dot{q}_2 &= \dot{y}_S = c_{12}\dot{e}_1 + c_{22}\dot{e}_2 + c_{32}\dot{e}_3 = -k/2[\cos(\beta - \alpha) + \cos(\beta + \alpha)]\dot{e}_1 \\
&\quad + k\cos(\beta - \alpha)\dot{e}_2 - k/2[\cos(\beta + \alpha) - \cos(\beta - \alpha)]\dot{e}_3 \\
\dot{q}_3 &= \dot{\beta} = c_{13}\dot{e}_1 + c_{23}\dot{e}_2 + c_{33}\dot{e}_3 = -k/2c\dot{e}_1 + 0\dot{e}_2 + k/2c\dot{e}_3 \\
\dot{q}_4 &= \dot{\varphi}_1 = c_{14}\dot{e}_1 + c_{24}\dot{e}_2 + c_{34}\dot{e}_3 = 1\dot{e}_1 + 0\dot{e}_2 + 0\dot{e}_3 \\
\dot{q}_5 &= \dot{\varphi}_3 = c_{15}\dot{e}_1 + c_{25}\dot{e}_2 + c_{35}\dot{e}_3 = 0\dot{e}_1 + 1\dot{e}_2 + 0\dot{e}_3 \\
\dot{q}_6 &= \dot{\varphi}_4 = c_{16}\dot{e}_1 + c_{26}\dot{e}_2 + c_{36}\dot{e}_3 = 0\dot{e}_1 + 0\dot{e}_2 + 1\dot{e}_3
\end{aligned}
\tag{14}
$$

The generalized forces determined from relation (11), taking into consideration the coefficients c_{ij} determined in system (14), lead to the final result of:

$$\Theta_1 = c_{11}Q_1 + c_{12}Q_2 + c_{13}Q_3 + c_{14}Q_4 + c_{15}Q_5 + c_{16}Q_6 = 2M_1 - 2N_1f_1$$
$$\Theta_2 = c_{21}Q_1 + c_{22}Q_2 + c_{23}Q_3 + c_{24}Q_4 + c_{25}Q_5 + c_{26}Q_6 = 2M_3 - 2N_3f_3 \quad (15)$$
$$\Theta_3 = c_{31}Q_1 + c_{32}Q_2 + c_{33}Q_3 + c_{34}Q_4 + c_{35}Q_5 + c_{36}Q_6 = 2M_4 - 2N_4f_4$$

Knowing the c_{ij} factors, two left sides of the relation (9) have been determined, which allow for the formulation of Maggi's equations, and which, within the case in consideration, have been written in a vector-matrix form:

$$M(\upsilon)\ddot{\upsilon} + C(\upsilon, \dot{\upsilon})\dot{\upsilon} + F(\dot{\upsilon}) = u \quad (16)$$

where the matrices and vectors take on the following form:

$$M(\upsilon) = \begin{bmatrix} a_1 \sin\beta + a_2 \cos(\beta - \alpha) & a_2 \sin(\beta - \alpha) - a_1 \cos\beta & -a_3 \\ a_2 \cos(\beta + \alpha) - a_8 \sin(\beta - \alpha) & a_8 \cos(\beta - \alpha) + a_2 \sin(\beta + \alpha) & -a_4 \\ a_1 \cos\beta + a_2 \cos(\beta - \alpha) & a_2 \sin(\beta - \alpha) + a_1 \sin\beta & a_3 \end{bmatrix}$$

$$C(\upsilon, \dot{\upsilon}) = \begin{bmatrix} -a_2 \sin(\beta - \alpha)\dot{\beta} & a_2 \cos(\beta - \alpha)\dot{\beta} & 0 \\ -a_2 \sin(\beta + \alpha)\dot{\beta} & a_2 \cos(\beta + \alpha)\dot{\beta} & 0 \\ -a_2 \sin(\beta - \alpha)\dot{\beta} & a_2 \cos(\beta - \alpha)\dot{\beta} & 0 \end{bmatrix}$$

$F(\dot{\upsilon}) = [a_5\mathrm{sgn}(\omega_1) \quad a_6\mathrm{sgn}(\omega_3) \quad a_7\mathrm{sgn}(\omega_4)]^T$, $u = [M_1 \quad M_3 \quad M_4]^T$, $\upsilon = [x_S \quad y_S \quad \beta]^T$, ω_1, ω_3, ω_4 have been determined from the relation [6], (21).

The meaning of the a_i parameters is as follows:

$$\begin{bmatrix} a_1 = (m_r + 4m_k)k \cos\alpha/2 & a_2 = I_k/2k & a_4 = I_kc/2k \\ a_3 = [I_S + 4I_z + m_k(l^2 + l_1^2)]k/4c + I_kc/2k & a_5 = N_1f_1 & a_6 = N_3f_3 \\ a_7 = N_4f_4 & a_8 = (m_r + 4m_k)k/2 \end{bmatrix}$$

Maggi's equations (16) provide an advantageous form of dynamic equations of movement for the completion of the simple and inverse dynamics task. In dynamics, we are mainly interested in the values of the driving moments of the appropriate wheels, therefore we are analyzing an inverse dynamics task. The knowledge of these values gives the basis for the solution of the issue regarding the tracking motion of the analyzed WMR_4M. The following remarks stem from the conducted considerations.

1. The analyzed system has 3 degrees of freedom, and the 4 driving wheels are driven by drive systems. In the literature, these types of systems are referred to as over-actuated [2]. They are characterized by a larger number of executive systems than of degrees of freedom. Solutions with additional executive systems find application, apart from mobile robotics, in various spheres of technology, e.g. in aviation, and the shipbuilding industry. To determine additional steering signals, one most often apply the complex method of optimal separation of steering (control allocation) [2].

2. In this paper, for the solution of the over-actuated type of system, it has been assumed that the powers of the driving modules are the same. Without taking into consideration the energy losses in the transfer of this power, one has obtained additional equations, on the basis of which the missing wheel 2 driving moment M_2 has been determined.

3. Taking into consideration the WMR_4M kinematics equations [4], (21) and assuming that the robot's frame 5 is in translational motion, therefore $\dot{\beta}(t) = 0$, then the angular velocities of, respectively, wheels {2 and 3}, and {3 and 1}, are identical. The equality of the power of the driving modules of these wheels shall be written as $\omega_2 M_2 = \omega_3 M_3$, and $\omega_1 M_1 = \omega_4 M_4$. When adding by sides the given relations, we will obtain the missing wheel 2 driving moment: $M_2 = M_3 + M_4 - M_1$.

4. However, if the robot's frame 5 is in plane motion, $\dot{\beta}(t) \neq 0$, which means that the angular velocities of, respectively, wheels 2 and 4, and 1 and 3 are the same. In such a case, we have relations $\omega_2 M_2 = \omega_4 M_4$ and $\omega_1 M_1 = \omega_3 M_3$. Proceeding similarly, we will determine the driving moment as $M_2 = M_3 + M_4 - M_1$.

5. As per the conducted research, it seems that it is not important what type of motion the robot's frame is, assuming consistency in the power of the driving sets. We determine the missing driving moment from the relation $M_2 = M_3 + M_4 - M_1$, therefore we obtain a simple solution to the over-actuated type of problem.

3 Example

Analyzing the inverse dynamics task WMR_4M, on the basis of the obtained Eq. (16), a computer simulation of the inverse dynamics task has been conducted. As before, various stages have been taken into consideration, such as motion, start, driving with an unchanging speed for the characteristic point S, stopping, driving in a straight line, and driving in a loop. For the simulation of the inverse dynamics task, the value of the a_i parameters occurring in Eq. (16) has been given in Table 1. The remaining data was given in [4].

Table 1. The values of the mobile robot's a_i parameters.

a_1	a_2	a_3	a_4	a_5	a_6	a_7	a_8
0.044	11.5708	2.4558	2.4545	3.001	3.001	3.001	0.0622

In this example there is a solution of the inverse dynamics task for the case of motion of point S as per the requested trajectory in the form of a loop, shown in [4 – Fig. 4 and 4d], when, during the time of the motion, there is no change in the angular velocity of the self-rotation of the robot, Fig. 2b and with change in the angular velocity of the robot Fig. 2a.

The obtained moment patterns stem from the structure of the trajectory of motion of point S. Within the initial stage of the movement, there is an increase in the values of

a)   b)

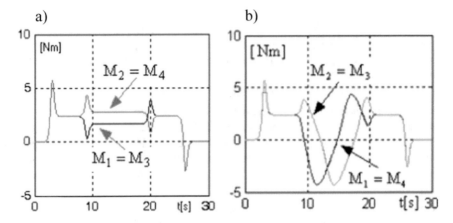

Fig. 2. The results of the inverse dynamics task simulation.

the moments, which is caused by the start-up stage, and then the values of the driving moments decrease since there is steady motion with the velocity of point S, $V_S = V_S^* = $ const. For time t $\geq$ 7.5 s the robot's point S begins to move through a circular trajectory for $\dot\beta(t) \neq 0$, Fig. 2a and $\dot\beta(t) = 0$, Fig. 2b. It is then that the value of the wheel 2, Fig. 2b, driving moment increases, and the simultaneously the value of the wheel 1 driving moment decreases. This stems from the robot's structure, shown in Fig. 1b, and from the requested trajectory of motion. A change of these values is shifted in time. A change in the value of the moments has confirmation in the change of the wheel angular velocities [4 – Fig. 4b], and in the progress of the wheel angular accelerations [4 – Fig. 4c]. When point S proceeds with an angle $\pi/2$ rad, then the moments have equal values, after which there is a change of the values and in the turns of their vectors.

4 Summary

This paper provides dynamic equations of a 4-wheel mobile robot equipped with mecanum type wheels. A new approach has been applied for the formulation of dynamic equations of motion of the robot, by applying Maggi's mathematical formalism. By applying such an approach, a mathematical model of a mobile robot has been formulated, which can be used in the synthesis of the real-time steering of this type of objects [5, 9]. Computer simulations have been conducted for the assumed solutions. The solutions obtained in kinematics [4] have been applied in the inverse dynamics task when determining the wheel driving moments implementing the requested tracks of motion of a chosen point of the robot. Assuming equality of the power of the drive modules, this was a simple solution for a mechanical system characterized by a larger number of executive systems than the number of the degrees of freedom. The simulation research has been conducted within a specified class of mathematical models, and the selection of the best mathematical model within this class requires the parametrical identification with the application of a real object.

References

1. Amengonu, Y., Kakad, Y.: Dynamics and Control of Constrained Multibody Systems modeled with Maggi's equation: application to Differential Mobile Robots Part I. IOP Conf. Ser. Mater. Sci. Eng. **65**(1), 012017 (2014)
2. Bodson, M.: Evaluation of optimization methods for control allocation. J. Guid. Control Dyn. **25**(4), 703–711 (2002)
3. Hendzel, Z.: Robust neural networks control of Omni-Mecanum wheeled robot with Hamilton-Jacobi inequality. J. Theoret. Appl. Mech. **56**(4), 1193–1204 (2018)
4. Hendzel, Z.: A description of the motion of a mobile robot with Mecanum wheels-kinematics. In: Advances in Intelligent Systems and Computing. Springer (2019). (accepted to the printing)
5. Kurdila, A., Papastavridis, J.G., Kamat, M.P.: Role of Maggi's equations in computational methods for constrained multibody systems. J. Guid. Control Dyn. **13**(1), 113–120 (1990)
6. Lin, L.-C., Shih, H.-Y.: Modeling and adaptive control of an Omni-Mecanum-Wheeled robot. Intell. Control Autom. **4**, 166–179 (2013)
7. de Wit, C.C., Siciliano, B., Bastin, G.: Theory of Robot Control. Springer, London (1996)
8. Papastavridis, J.G.: Maggi's equations of motion and the determination of constraint reactions. J. Guid. Control Dyn. **13**(2), 213–220 (1990)
9. Szuster, M., Hendzel, Z.: Intelligent Optimal Adaptive Control for Mechatronic Systems. Springer (2018)
10. Taheri, H., Qiao, B., Ghaeminezhad, N.: Kinematic model of a four Mecanum wheeled mobile robot. Int. J. Comput. Appl. **113**(3) (2015)
11. Zimmermann, K., Zeidis, I., Abdelrahman, M.: Dynamics of mechanical systems with Mecanum wheels. In: Applied Non-Linear Dynamical Systems, pp. 269–279. Springer International Publishing (2014)
12. Żylski, W.: Kinematics and Dynamics of Wheeled Mobile Robots (in Polish). Rzeszow University of Technology Publishing Mouse, Rzeszow (1996)

A Description of the Motion of a Mobile Robot with Mecanum Wheels – Kinematics

Zenon Hendzel[(✉)]

Faculty of Mechanical Engineering and Aeronautics,
Department of Applied Mechanics and Robotics,
Rzeszow University of Technology, Rzeszów, Poland
zenhen@prz.edu.pl

Abstract. This paper formulates kinematic equations of motion of a 4-wheel mobile robot equipped with mecanum-type wheels. It has been assumed that the motion of the robot occurs within the x-y plane. A classical approach applied in mechanics has been used for the formulation of the robot's equations of motion. Computer simulations have been conducted with the obtained kinematics equations of motion, assuming motion through straight line tracks and tracks in the shape of a loop, and also assuming a constant robot frame orientation angle.

Keywords: Mobile wheeled robot · Mecanum wheels · Kinematics

1 Introduction

In the literature, one may encounter various approaches to the analytical considerations regarding the kinematics for a mobile robot equipped with omni wheels [2], or alternatively known as mecanum-type wheels [1–6]. When describing the motion of mobile wheeled robots, we are interested in topics related to the kinematics and dynamics of these systems [8, 9]. When describing the kinematics of mobile wheel robots, kinematics equations are generally provided from which one may determine the linear parameters of motion, such as: distance, velocity or acceleration of any chosen point; or the angular parameters of motion, such as: angle of rotation, angular velocity and angular acceleration of solids. In the case of mobile wheeled robots, we are analyzing an inverse kinematics task in which we assume the movement of a robot's characteristic point and then determine the motion parameters. The following paper applies a new approach towards the formulation of kinematic equations of motion of a 4-wheeled mobile robot equipped with mecanum type wheels. Section 2 formulates the robot's kinematic equations of motion. In Sect. 3 computer simulations have been conducted which assume driving through straight line tracks and a track in the shape of a loop, assuming a constant robot frame orientation angle. The article ends with a summary and a bibliography.

© Springer Nature Switzerland AG 2020
R. Szewczyk et al. (Eds.): AUTOMATION 2019, AISC 920, pp. 346–355, 2020.
https://doi.org/10.1007/978-3-030-13273-6_33

2 Robot Kinematics

In the following section, we analyze the inverse kinematics task of a 4-wheeled mobile robot equipped with mecanum wheels (WMR_4M). When describing WMR_4M kinematics, a model such as that presented in Fig. 1 has been applied. In this representation x, y, z are motionless system axes. The basic components of this model are frame 5 and the driving units. Wheels 1–4 are the elements of the driving unit, along with the points located at their centers of symmetry A_1, A_2, A_3, A_4, set on semi-axes which are set into motion by the driving module related to a given wheel. These wheels rotate along their own axes which do not change their position relative to the frame. Rollers are located on the wheel perimeter, set at an angle of $\alpha = \pi/4$ [rad] to the driving wheel axis. Figures 1 and 2 demonstrate the appropriate geometric measurements and the characteristic points of the system. Point S is the center of mass of the frame, and point H is at the midpoint lying on the $A_1 A_2$ axis. As far as B is concerned, it is a point belonging to the frame equivalent to point H but on the other wheel axis. The β angle is an angle of the temporary frame rotation frame temporary rotation. In Fig. 2, as an example, one has marked the angular velocity vectors for, respectively, the wheel A_2, $\overline{\omega}_2$ and the roller $\overline{\omega}_{r2}$. It has been assumed that the driving wheel radiuses are equal, therefore $R_1 = R_2 = R_3 = R_4 = R$. For the description of the kinematics of any given point of the system, it is advantageous to determine the equation of motion. A right-handed rectangular frame of reference is assigned to each of the movable parts of the system, with its origin being the center of mass of a given part. In addition, we assume that the motion of the robot occurs within the xy plane. It is advantageous to provide an equation of motion for the description of the kinematics of any point of the set. We can determine such an equation with knowledge of the geometry of the set, and by applying homogeneous coordinates and transformation matrices [7], but one may also apply classical methods form general mechanics [9]. In this spirit, we will apply the classical approach for the description of the robot's kinematics. First it can be determined from the system geometry that the coordinates of points H and A_2 in the xy system are related by the following dependency:

$$
\begin{aligned}
x_{A_2} &= x_H + l\sin\beta \\
y_{A_2} &= y_H - l\cos\beta
\end{aligned}
\tag{1}
$$

The frame of the mobile wheeled robot is in planar motion, therefore the speed of point A_2 can be written as:

$$
\overline{V}_{A_2} = \overline{V}_H + \overline{V}_{A_2 H}
\tag{2}
$$

The circle with its center point A_2 is in compound motion, and therefore we can determine the velocity of this point A_2 with the equation

$$
\overline{V}_{A_{2b}} = \overline{V}_{A_{2u}} + \overline{V}_{A_{2w}}
\tag{3}
$$

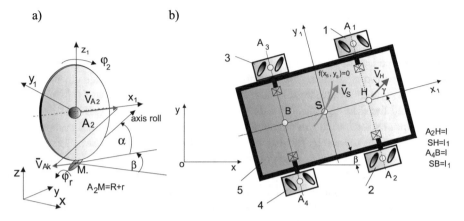

Fig. 1. The MRW_4M model.

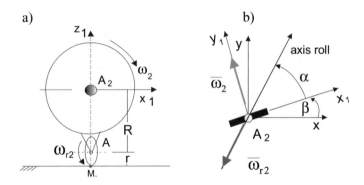

Fig. 2. The geometry and kinematic parameters of the A_2 circle.

where $\overline{V}_{A_{2w}} = 0$ is the velocity in relative motion, $\overline{V}_{A_{2u}} = \overline{V}_{A_2}$ is the velocity in raising motion, and therefore the absolute velocity of point A_2, which we will further mark as $\overline{V}_{A_2}$. Let us note the velocity of point A_2 in an analytical form is

$$\overline{V}_{A_2} = \dot{x}_{A_2}\overline{i} + \dot{y}_{A_2}\overline{j} \tag{4}$$

and we will determine the velocity of point M shown in Fig. 2a from the relation

$$\overline{V}_M = \overline{V}_A + \overline{V}_{MA} \tag{5}$$

where the velocity of point A is

$$\overline{V}_A = \overline{V}_{A_2} + \overline{V}_{AA_2} \tag{6}$$

We can then determine the $\overline{V}_{AA_2}$ relative velocity vector as

$$\overline{V}_{AA_2} = \overline{\omega} \times \overline{R} = \begin{bmatrix} \overline{i} & \overline{j} & \overline{k} \\ \omega_x & \omega_y & \omega_z \\ 0 & 0 & -R \end{bmatrix} \tag{7}$$

The following relation stems from the outline of the angular velocity vectors of circle A_2, $\overline{\omega}_2$ and the frame angular velocity $\overline{\omega}_r$:

$$\overline{\omega} = \overline{\omega}_2 + \overline{\omega}_r \tag{8}$$

where $\overline{\omega}_2 = \dot{\varphi}_2 \overline{j}_1$ $\overline{\omega}_r = \dot{\beta} \overline{k}_1$. The markings appearing in Eq. (8) are, respectively $\dot{\varphi}_2$ - the value of the A_2 circle angular velocity, $\dot{\beta}$ - the value of the frame angular velocity, $\overline{i}_1, \overline{j}_1, \overline{k}_1$ - the versors of the system related to the robot's frame. We will obtain the following from Eq. (7):

$$\begin{aligned} V_{AA_{2x}} &= -R\omega_y \\ V_{AA_{2y}} &= R\omega_x \\ V_{AA_{2z}} &= 0 \end{aligned} \tag{9}$$

Projections of the vector of the angular velocity of the circle with its center point A_2 are:

$$\begin{aligned} \omega_x &= -\omega_2 \sin \beta \\ \omega_y &= \omega_2 \cos \beta \end{aligned} \tag{10}$$

Introducing the relation (10) into Eq. (9), we get:

$$\begin{aligned} V_{AA_{2x}} &= -R\omega_2 \cos \beta \\ V_{AA_{2y}} &= -R\omega_2 \sin \beta \\ V_{AA_{2z}} &= 0 \end{aligned} \tag{11}$$

After introducing the relation (6) to (5), we can note the speed of point M in the following form:

$$\overline{V}_M = \overline{V}_{A_2} + \overline{V}_{AA_2} + \overline{V}_{MA} \tag{12}$$

The relative velocity stemming from the rotation of point M with regard to point A will be determined by us in the form of the following vector product:

$$\overline{V}_{MA} = \overline{\theta} \times \overline{r} \tag{13}$$

where θ is the absolute angular velocity of the roller, which is equal to

$$\bar{\theta} = \bar{\omega}_2 + \bar{\omega}_r + \bar{\omega}_{r2} \tag{14}$$

and $\bar{\omega}_{r2} = \dot{\psi}\bar{e}_1$, $\dot{\psi}$ - the value of the rotation angular velocity of the roller itself, $\bar{e}_1$ - the roller axis versor. From Eq. (13), also taking into consideration Eq. (14), we get the following results:

$$\begin{aligned}
V_{MAx} &= -r(\omega_2 \cos\beta - \omega_{r2} \sin(\beta + \alpha)) \\
V_{MAy} &= -r(\omega_2 \sin\beta + \omega_{r2} \cos(\beta + \alpha)) \\
V_{MAz} &= 0
\end{aligned} \tag{15}$$

Assuming that the roller is rolling without slipping, the following equation is fulfilled:

$$\overline{V}_M = 0 \tag{16}$$

since point M is the point of contact of the roller with the road. In such a case, we will write Eq. (12) as follows:

$$\bar{0} = \overline{V}_{A_2} + \overline{V}_{AA_2} + \overline{V}_{MA} \tag{17}$$

Projecting Eq. (17) into the xyz system axes and taking into consideration the relation (9), (10) and (15), we will get

$$\begin{bmatrix} \dot{x}_{A_2} \\ \dot{y}_{A_2} \end{bmatrix} = \begin{bmatrix} (R+r)\cos\beta & -r\sin(\beta+\alpha) \\ (R+r)\sin\beta & r\cos(\beta+\alpha) \end{bmatrix} \begin{bmatrix} \omega_2 \\ \omega_{r2} \end{bmatrix} \tag{18}$$

Applying the geometrical relations of point A_2 and H (1), and next of point S, the Eq. (18), we will get

$$\dot{x}_S[\cos(\beta+\alpha)] + \dot{y}_S[\sin(\beta+\alpha)] + \dot{\beta}(l\cos\alpha + l_1\sin\alpha) = \omega_2(R+r)\cos\alpha \tag{19}$$

$$-\dot{x}_S \sin\beta + \dot{y}_S \cos\beta + \dot{\beta}l_1 = \omega_{r2}r\cos\alpha \tag{20}$$

where the coordinates of point S are $x_S = x_{A_2} - l\sin\beta - l_1\cos\beta$, $y_S = y_{A_2} + l\cos\beta - l_1\sin\beta$. Equations (19) and (20) allow us to determine, respectively, the angular rotation velocity of the A_2 wheel and that of the roller of this wheel in the requested function of the velocity $\overline{V}_S$ of point S, and the rotation angle of the robot's frame $\beta(t)$. Assuming indexes in the markings of the angular velocities of wheels in accordance with their numbering provided in Fig. 1b and proceeding similarly in the case of the other wheels, the WMR_4M kinematics equations have been determined in the following form:

$$\begin{aligned}
\dot{x}_S[\cos(\beta-\alpha)] + \dot{y}_S[\sin(\beta-\alpha)] - \dot{\beta}(l\cos\alpha + l_1\sin\alpha) &= \omega_1(R+r)\cos\alpha \\
\dot{x}_S[\cos(\beta+\alpha)] + \dot{y}_S[\sin(\beta+\alpha)] + \dot{\beta}(l\cos\alpha + l_1\sin\alpha) &= \omega_2(R+r)\cos\alpha \\
\dot{x}_S[\cos(\beta+\alpha)] + \dot{y}_S[\sin(\beta+\alpha)] - \dot{\beta}(l\cos\alpha + l_1\sin\alpha) &= \omega_3(R+r)\cos\alpha \\
\dot{x}_S[\cos(\beta-\alpha)] + \dot{y}_S[\sin(\beta-\alpha)] + \dot{\beta}(l\cos\alpha + l_1\sin\alpha) &= \omega_4(R+r)\cos\alpha
\end{aligned} \tag{21}$$

Based on the obtained WMR_4M kinematics Eqs. (21) and on the equations of the inverse task of the kinematics, computer simulations have been conducted. These simulations consider motion through tracks in the shape of a loop, and also consider a constant robot frame orientation angle. When completing the inverse kinematics task, the angular parameters of the driving wheels have been determined. The obtained solutions will constitute the requested kinematics parameters for the implementation of the inverse dynamics task. Three characteristic stages of motion have been assumed, an example of which is shown in Fig. 3a.

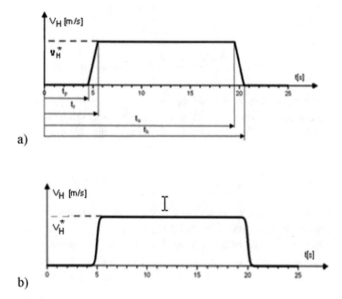

Fig. 3. The robot's point S velocity profile.

The characteristic stages of motion are:

- driving through a straight-line track, start-up:

$$V_H = \frac{V_H^*}{t_r}(t - t_p) \qquad t_p \leq t \leq t_r, \quad \dot{\beta} = 0,$$

where t_p – motion initial time, t_r – time of the completion of the start-up,

- motion with a determined speed, when $V_H = V_H^* = const$

$$t_r \leq t \leq t_u \qquad \dot{\beta} = 0$$

where t_u – steady motion time,

- motion through a circular track with radius RR, a change of the robot's configuration for:

$$V_H = V_H^* = \text{const}, \quad t_1 \leq t \leq t_2, \quad \dot{\beta} = \frac{V_H^*}{RR}$$

where t_1 – time of entering the arch, t_2 – time of driving through the circular track,

- exiting the arch taking into consideration the transitional period, then driving through a straight line track with a constant velocity ($V_H = V_H^* = \text{const}$) in the time $t_2 \leq t \leq t_u$, where t_u is the steady motion time,
- braking:

$$V_H = V_H^* - \frac{V_H^*}{t_h}(t - t_u), \quad t_u \leq t \leq t_k, \quad \dot{\beta} = 0$$

where t_k – the final time, t_h – braking time.

Within the entire scope of the analyzed motion, the assumption was the approximation of the velocity of point S, Fig. 3b, in accordance with the relation:

$$V_H = V_H^* \left(\frac{1}{1 + e^{-c(t-b)}} - \frac{1}{1 + e^{-c(t-b1)}} \right)$$

with the constants c, b, and b_1. A change in the robot's orientation angle $\dot{\beta}(t)$ in the period of motion through a circular track taking into consideration the transition period of entering onto the wheeled track, followed by motion through the track and an exit from the wheeled track has been marked as $\dot{\beta}_a(t)$ and can be described by the relation

$$\dot{\beta}_a(t) = \dot{\beta}(t) \left(\frac{1}{1 + e^{-ca(t-ba)}} - \frac{1}{1 + e^{-ca(t-b1a)}} \right), \quad i = 1, 2$$

with constants ca, ba, b1a. The introduction of such an approximation makes it possible to implement the system's movement with a smooth change of angular parameters such as velocity and acceleration. In the conducted simulations, it was assumed that the braking and start-up times are the same and there are $t_r = t_h = 1$ s. A maximum linear velocity of the mobile robot's point H was assumed at the level of $V_H^* = 0.4$ [m/s], $\gamma = 0$ and the approximation of point H velocity was determined by the following values c = 5, b = 3, and b1 = 26.

3   Example

In the example, a simulation of the suggested solution has been conducted for the motion of point H for a requested trajectory in the form of a straight-line track and loop. Motion through a circular track is considered with a radius of $RR = 0.7$ [m], for a time of $t_1 \leq t \leq t_2$, where $t_1 = 8$ [s], $t_2 = 21$ [s], exiting an arch with taking into consideration the transition period with constant values ca = 5, ba = 9, b1a = 20, driving through a straight-line track with a constant velocity and braking. For these periods of motion, Fig. 4 shows the desired motion of the robot's wheels and the motion track of point S assuming the initial motion conditions: $x_s(0) = 1$ [m], $y_s(0) = 5$ [m], $\beta(t) = 0$. Within the timeframe of motion, there is no change in the angle of the robot frame self-rotation. The coefficients occurring in the equations which describe the system geometry correlate to the chosen WMR_4M and are as follows: $\alpha = \pi/4$ [rad], R = 0.15 [m], r = 0.01 [m], $l_1 = l = 0.15$ [m].

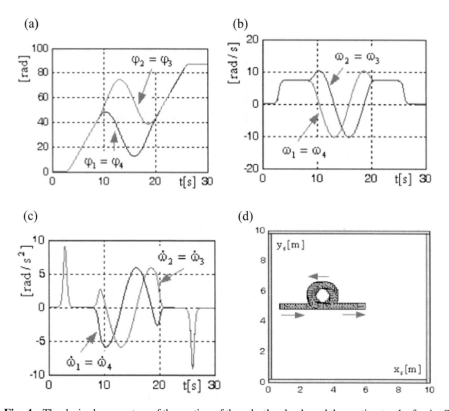

Fig. 4. The desired parameters of the motion of the robot's wheels and the motion track of point S

For time $t \geq 7.5$ [s], the robot's point S starts moving through a circular trajectory with a radius of RR = 0.7 [m]. Then, the self-rotation angle of wheel 2 increases as the self-rotation angle of wheel 1 decreases. This stems from the robot's structure, which is shown in Fig. 4a. A change of these values is shifted in time. A change in motion is confirmed in the change of the wheel angular velocities, shown in Fig. 4b, and in the progress of the wheel angular accelerations, shown in Fig. 4c. When point S proceeds with an angle of $\pi/2$ [rad], then the angular velocities and angular accelerations have identical values, after which there is a change in the values and in the turns of the vectors of these kinematic parameters.

The different parameters of the movement of the robot were put in Fig. 5, putting on, that the angle of the orientation of the robot is changing. The angle of the turn and the angular velocity of the frame of the robot were put in Fig. 5a but the change of the projections of the linear velocity of the point S was showed in Fig. 5b.

a)   b)

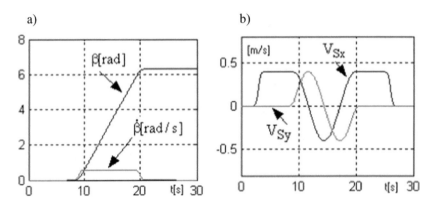

Fig. 5. The desired parameters of the motion of point S.

4   Summary

This paper provides kinematics equations of a 4-wheel mobile robot equipped with mecanum type wheels. When formulating the kinematics equations for the robot, the classical approach from general mechanics has been applied. Computer simulations have been conducted for the assumed solutions. Within the sphere of research of kinematics equations, there has been a determination of the angular parameters of the movement parameters of a set of solids for a number of scenarios. These include driving through straight line tracks and through a track in the shape of a loop, and assuming a constant robot frame orientation angle.

References

1. Becker, F., Bondarev, O., Zeidis, I., Zimmermann, K., Abdelrahman, M., Adamov, B.: An approach to the kinematics and dynamics of a four-wheeled mecanum vehicles. Sci. J. IfToMM Prob. Mech. Special Issue **2**(55), 27–37 (2014)
2. de Wit, C.C., Siciliano, B., Bastin, G.: Theory of Robot Control. Springer, London (1996)
3. Han, K.-L., Choi, O.-K., Kim, J., Kim, H., Lee, J.S.: Design and control of mobile robot with Mecanum-Wheel. In: ICROS-SICE International Joint Conference, Fakuoka International Congress Center, Japan, pp. 2932–2937 (1996)

4. Hendzel, Z., Rykała, Ł.: Modelling of dynamics of a wheeled mobile robot with Mecanum wheels with the use of Lagrange equations of the second kind. Int. J. Appl. Mech. Eng. **22**(1), 81–99 (2017)
5. Hendzel, Z.: Robust neural networks control of Omni-Mecanum wheeled robot with Hamilton-Jacobi inequality. J. Theor. Appl. Mech. **56**(4), 1193–1204 (2018)
6. Lih-Chang, L., Hao-Yin, S.: Modeling and adaptive control of an Omni-Mecanum-Wheeled robot. Intell. Control Autom. **4**(02), 166 (2013)
7. Taheri, H., Qiao, B., Ghaeminezhad, N.: Kinematic model of a four Mecanum wheeled mobile robot. Int. J. Comput. Appl. **113**(3) (2015)
8. Szuster, M., Hendzel, Z.: Intelligent Optimal Adaptive Control for Mechatronic Systems. Springer (2018)
9. Żylski, W.: Kinematics and Dynamics of Wheeled Mobile Robots. Rzeszow University of Technology Publishing Mouse, Rzeszow (1996). (in Polish)

Using the Raspberry PI2 Module
and the Brain-Computer Technology
for Controlling a Mobile Vehicle

Szczepan Paszkiel[✉]

Faculty of Electrical Engineering, Automatic Control and Informatics,
Department of Biomedical Engineering, Opole University of Technology,
Proszkowska 76, 45-271 Opole, Poland
s.paszkiel@po.opole.pl

Abstract. This paper describes the execution process of a four-wheeled robot controlled by a user via an Emotiv EPOC+ NeuroHeadset device. The following, inter alia, was described for this purpose - the issue of selecting a controller with additional modules necessary to create a robot; execution of a four-wheeler prototype; connecting the devices: Raspberry PI2 and Emotiv EPOC+ NeuroHeadset in a network, which allows the transfer of data grouped in packs. An original control algorithm, presented in this paper was developed and calibration with an Emotiv EPOC+ NeuroHeadset device was conducted for the purposes of the research.

Keywords: EEG · Brain-computer interfaces · Mobile vehicle

1 Introduction

Many activities can be simplified in the 21st century by using smart technological solutions. The electroencephalograph is certainly one of them. It studies the bioelectrical brain activity on the basis of a non-invasive technique. The study involves appropriate placement of electrodes on the surfaces of the scalp, which record the electrical potential on the skin surface, resulting from the activity of the cerebral cortex neurons and after adequate strengthening, form a record - the electroencephalogram [1]. There are many fields in which these devices can find their application. The electroencephalograph was created mainly for medical purposes, disease diagnosis, but that is not its only application. When studying the brain activity and using relevant signal filtration methods, it is possible to also control many devices, such as a TV, computer and even a mobile robot [2]. It is that last solution, on which this paper focuses. The concept of a mobile robot is understood as a device, which can perform a certain movement, both, unmanned and manned. In order to simulate the operation of the device for research purposes, a four-wheeled robot driven by two DC motors and turning with two front wheels via servo-mechanisms was constructed. The device is small in size and weighs about one kilogram. The very idea of the device is, however, similar to the applications used in, for example, wheelchairs.

© Springer Nature Switzerland AG 2020
R. Szewczyk et al. (Eds.): AUTOMATION 2019, AISC 920, pp. 356–366, 2020.
https://doi.org/10.1007/978-3-030-13273-6_34

2   The Brain-Computer Interface Technology

The brain-computer interface allows direct communication between the brain and an appropriate external device. The device used in the conducted research was created in the laboratories of Emotiv System Inc. The software delivered by the manufacturer reads and interprets conscious and unconscious emotions, head movement and even facial expressions. The Emotic EPOC+ NeuroHeadset consists of 14 electrodes + 2 reference electrodes, which read brain wave signals [3]. It operates on the principle of an electroencephalograph. Electrodes placed on the head measure electrical potential changes, which appear as a result of neuron activity in the cerebral cortex [4]. A signal is formed as a result of brain activity associated with a specific action [5]. Based on the measurement of this signal and assigning it to an appropriate activity, it is then possible to control a given device. As the conducted analyses show, it is currently impossible to read human thoughts [6]. The works on this technology are developing on a daily basis, however, the degree of their advancement, does not make the broadly understood though identification possible. Classifying all of the mental possibilities itself, seems very impeded. This is why, controlling involves a human teaching the control software, which EEG signal is responsible for a given activity. After classifying individual activities, they can be assigned relevant actions. An Emotiv EPOC+ NeuroHeadset device shown in Fig. 1 was used for the purposes of the conducted research. After installing the devices software, a user gains a very broad spectrum of possibilities in the scope of the EEG signal analysis. The manufacturer provides appropriate software, which enables real-time access to the signals from each of the electrodes installed in the device. The device communicates with a PC via a wireless Bluetooth 4.0 adapter.

An alternative solution for the device from Fig. 1 may be a custom designed and executed device. Figure 2 shows biomedical electrodes, which can be used for EEG, EKG and EMG tests. Their disadvantage, however, is single-use and no frame. The electrodes can be connected to the computer using a 3.5 mm jack connection and additional cables.

3   A Four-Wheeled Mobile Robot

When beginning the design works, it was necessary to select a controller, which would perform relevant computations, had the option to connect necessary module and would operate according to a developed algorithm. The Raspberry Pi2 was selected for the purposes of the work [7]. The controller operates on any Linux system. It will act as a computer, computation unit of the entire device, to which all movement commands will be sent. They will be processed with appropriate algorithms and a signal controlling the drive motors and servo-mechanisms will be sent on their basis. Raspbian, normally installed on Raspberry PI2 was selected as the operating system. The main algorithm was written in Python.

Fig. 1. Emotiv EPOC+ NeuroHeadset device.

Fig. 2. An example of biomedical electrodes.

A quad-core Raspberry PI2 processor with 1GB RAM memory guarantees sufficient computing power [8]. The Raspbian system has an option of additionally separating the processes onto individual processor cores. Hence, the operating system may reliably work on one core, while the other three will be available for other processes. It also has a dedicated camera output and expandable memory. Besides a very big computing power, the option to connect additional modules is very important [9]. The Raspberry Pi2 has two independent I2C buses, SPI, UART and 26 programmable GPIO I/Os. The controller has 4 USB 2.0 ports. Hence, there is an option to connect one of them to a Wi-Fi module, a computer mouse and keyboard and even an additional disk to one of them. Controlling the rotational speed of the DC motor and the steering angle of the servos is executed via changing the signal fill level. In the initial research phase, software PWM Raspberry PI2 were used, however, due to large inaccuracy, an additional module was used, the Adafruit Mini Kit, a 16-channel PWM I2C controller - Servo Hat (Fig. 3).

Fig. 3. A 16-channel PWM controller Adafruit Mini Kit.

A TP-Link TL-WN722N network card together with a TL-ANT2405CL antenna was used to work on the device. Its features include the option to connect a separate antenna. Broadband speed in the N standard up to 150 Mbps ensures smooth data transfer, video streaming, using internet telephony and online games. 64/128 bit encryption WEP, WPA/WPA2-WPA-PSK/WPA2-PSK(TKIP/AES), compatibility with the IEEE 802.1X standard. The standard antenna was replaced with the TL-ANT2405 model for research purposes, in order to increase the range of the module and decrease lags. The gain of a 5 dBi omnidirectional antenna allows the coverage of a larger area with a wireless signal. Operating frequency is 2.4 GHz and is compatible with 11 b/g/n devices. The vehicle is powered by two Lego 8882 motors. Each motor is connected independently and is responsible for the rotating speed of one rear wheel. A big advantage of the motor is a planetary gear installed inside and a high torque in relation to the weight of the motor. In order to additionally increase the rotational speed of the motors a 1–3 transmission gear was installed, which ensures a maximum speed of 600 rpm. Two TowerPro SG-92 Carbon servo-mechanism were used in the device. They were set up in such a way, so that each servo would change the steering angle of one of the front wheels. In the first stage, the wheel was installed directly on the axis of a servo-mechanism. The selection of a battery was strictly associated with the previously selected sub-assemblies. The number of battery chambers and power supply voltage are very important elements. L298N controlled module (Fig. 4) enables the change of the rotational direction and speed of two motors. By adjusting the values on four Inputs, the motor rotational direction can be changed. Two additional inputs (A-B Enable) are used to change the rotational speed of a motor. A big advantage is the 5 VDC output, which can be also used to power the robots electronics.

In order to ensure stable operation of the Raspberry PI2, it needs to be powered by 5 VDC. The battery voltage is 11.1 V. A step-down inverter was used for that purpose. In the first phase, the Raspberry PI2 was powered from an additional output of the H-bridge module, however, due to large energy consumption, an additional inverter was used, which is plugged in directly between the electronics and the battery. The frame of the four-wheeler was build using

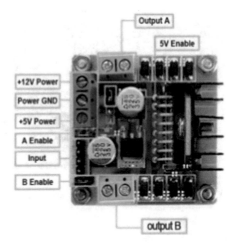

Fig. 4. L298N controller module.

Lego Technic Modules. Thanks to the use of modules, default design of the dimensions and the frame shape is possible. Additionally, through a wide range of blocks, it is possible to modify the frame of the device during operation. Another advantage is the gears, which can be used to form a transmission. The device also uses motors, which perfectly fit the frame, forming together a stable structure. The used 62.4 × 20 cm slick wheels ensure excellent grip. A complete device weighs about 1 kg. The construction of the frame started with building a transmission from the motors onto the wheels. Next, modelling servos were remade in such a way, so as to fit the frame. In the first stage, the wheels were steered directly by the servo. Unfortunately, a disadvantage of that solution is the fact that a travelling wheel tilts and transfers the vibrations onto the servo-mechanism, which causes vibrations and speeds up the wear of the servo. The problem was solved by installing a 1–3 transmission gear. The Raspberry Pi2 controller together with additional modules was mounted in the centre of the vehicle. The inverter is in a clearance between the motors. The batteries are placed at the rear. It is the heaviest sub-assembly of the four-wheeler. The aim of installing it on the drive axles is to ensure additional downforce. The H-bridge was installed at the top (Fig. 5).

Figure 6 shows the electrical diagram of the device. Since the Servo Hat extension plugged into a Raspberry PI2 occupies all additional inputs, the module (H-bridge) was plugged into the exit from the second plate. The system is powered by a 11.1 VDC 2200 mAh lithium-polymer battery. A Wi-Fi module is used in order to programme the system and for further transmission, to set appropriate movement parameters. Electrical motors were used as a drive in the car, while the servomechanisms for steering. Since the motors operate at maximum 9 VDC voltage and the system is powered with 11.1 VDC, the PWM signal maximum fill was limited in the control software, which translates into

Fig. 5. Final structure of a four-wheeled mobile vehicle.

output voltage. This way, the H-bridge will never receive a higher fill, which would mean a voltage higher than 9 VDC. The PWM signal maximum fill value was measured before the motor was connected to the system.

4   Controlling a Robot via an Emotiv EPOC+ NeuroHeadset

Controlling the devices requires the creation of a network in which all the necessary components will be able to operate. The Bluetooth and Wi-Fi technologies were selected for the purposes of this work. All of the parameters downloaded from the Emotiv EPOC+ NeuroHeadset device as sent to a PC via Bluetooth 4.0. The computer operates in a network to which the four-wheeled mobile robot was also connected. By using the SSH protocol, a network user has access to a virtual console of the device and is able to send appropriate commands this way. The control algorithm was divided into two parts. The first one was implemented in the robot and it consists of the devices mobile features, which means they specify what should the device be doing after receiving a specific command. The second part of the algorithm was implemented on a PC. In this case, all commands are generated on the basis of pulses gathered from the Emotiv EPOC+ NeuroHeadset device, and are sent to the mobile robot. The entire system is shown in Fig. 7.

The works with the Emotiv EPOC+ NeuroHeadset device were commenced with soaking the electrodes - the felt items installed on them - with saline solution. The measurement accuracy depends on both, the correct placement of an electrode, as well as its contact with the skin. After soaking the electrodes, it was necessary to put the device on the head as shown in Fig. 8.

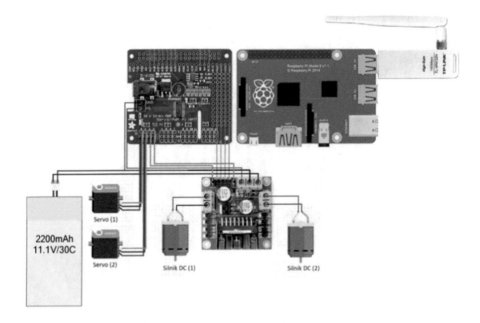

Fig. 6. The electrical diagram of the circuit used in the studies.

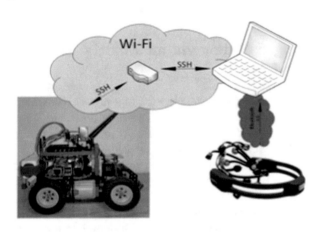

Fig. 7. A flow diagram of the network structure.

After starting the EPOC Control Panel software and communicating with the device, the software interface signals whether the electrodes are correctly correlated with the skin. The system itself, needed about 1–2 min to detect and identify appropriately the electrodes. A correctly configured device is shown in Fig. 9.

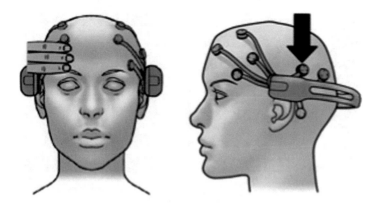

Fig. 8. A properly fitted Emotic EPOC+ NeuroHeadset device.

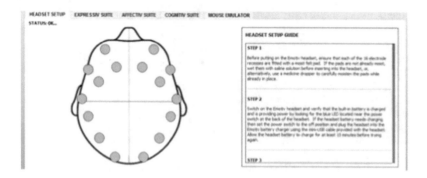

Fig. 9. EPOC Control Panel with correct fitting of the device.

After a connection was established and the electrodes were properly placed, signals were identified on the basis of potentials generated by facial expressions [10]. The software delivered with the device enables a preview of the virtual face (avatar) of the controlling person. As shown in Fig. 10, gesture or motion sensitivity could be set individually the identification algorithm is so accurate that it identifies even a blink of an eye, which is shown in Fig. 10.

Figure 11 shows, inter alia, the condition of a standard activity, in which a tested person does not perform any specific actions.

Moreover, Fig. 11 shows the changes of an EEG signal during eyelid movement. We can see than electrodes AF3, AF4, F7 and F8 recorded a significant hanged of the signal amplitude. In Fig. 11, it is also possible to count that a tested person performed 13 eyelid blink cycles. Figure 12 shows an EEG signal of a person clenching their teeth. It can be clearly seen that all electrodes recorded a significant change of the EEG signal. Such a behaviour results mainly from the fact that when a person clenches their jaw, all of the muscles of the head are tensioned. When muscles are tensioned, they begin to tremble and also

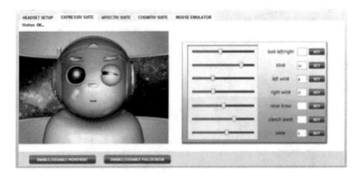

Fig. 10. Virtual avatar.

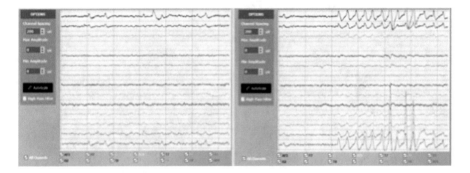

Fig. 11. An EEG signal for the condition of standard-normal activity (left side) and eye blinking (right side) of a tested person on 14 transmission channels.

generate a pulse, which was recorded by the device. Figure 12 also presents an EEG signals adequate for eyebrow movement. It can be also noticed that only some electrodes recorded certain signal amplitude changes.

5   Signal Identification and Classification

The device was controlled through detecting appropriate, certain behaviours observed in the EEG signal - artifacts, induced by facial expressions. Classified artifact and their assigned movements are presented in Fig. 13.

Selecting an appropriate sensitivity of the measured signal in relation to the one generated for the four-wheeled vehicle is important for such control. To frequent blinking or an intensive smile may cause lack of comfort noticeable during the controlling process. The data receiving software was developed in such a way, so that it is not necessary to maintain a characteristic artifact, in order to maintain the required movement. Unfortunately, the Putty software granting access to a virtual console requires confirming each command with Enter. Therefore, a programme was created, which sends only one character at a time, without the need of confirming with Enter.

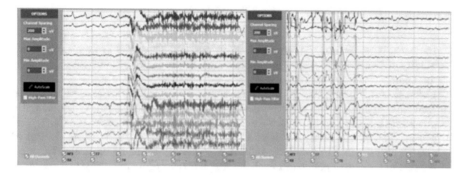

Fig. 12. The EEG signal for teeth clenching (left side) and eyebrow movement (right side) of a person subjected to testing on 14 transmission channels.

Classified artifact	Four-wheeled robot movement	Assigned character
Eye blinking	Forward travel	„w"
Right eye blinking	Turning right	„d"
Left eye blinking	Turning left	„a"
Teeth clenching	Stop	„"
Smiling	Backwards travel	„s"
	Exit program	„r"

Fig. 13. Classification of received signals relative to the generated command.

6   Conclusion

A four-wheeled vehicle controlled by a user with the use of an Emotic EPOC+ NeuroHeadset device operating in the BCI technology was created for the purposes of the research. The Raspberry PI2 was the main computation heart of the mobile robot. Modules responsible for generating the PWM signal and a Wi-Fi module were added to it. The designed device operates in a Wi-Fi network. By using the Putty software, the use acquired access to a virtual console of the controller installed on-board the four-wheeler. The software was developed in such a way, as to be able to receive characters at different time intervals, without the need to confirm with Enter. The main aim of the studies was to develop a miniature version prototype of a mobile robot controlled with the use of a brain-computer interface. The manner presented in this paper enables the control, inter alia, of a wheelchair. Currently constructed are also devices supporting identification of correct perception of reality - appropriate concentration, for example, when driving a car [11]. The used Wi-Fi network can be easily replaced with a cable solution. Signals received with the use of an Emotiv EPOC+ NeuroHeadset device for the purposes of controlling processes were classified in the course of the conducted tests.

References

1. Badcock, N.A., et al.: Validation of the Emotiv EPOC EEG system for research quality auditory event-related potentials in children. Peer J. **3**, 907 (2015). http://dx.doi.org/10.7717/peerj.907
2. Delorme, A., Makeig, S.: EEGLAB: an open source toolbox for analysis of single-trial EEG dynamics including independent component analysis. J. Neurosci. Methods **134**(1), 9–21 (2004). https://doi.org/10.1016/j.jneumeth.2003.10.009
3. Mathewson, K.E., Lleras, A., Beck, D.M., Fabiani, M., Ro, T., Gratton, G.: Pulsed out of awareness: EEG alpha oscillations represent a pulsed-inhibition of ongoing cortical processing. Front. Psychol. (2011). https://doi.org/10.3389/fpsyg.2011.00099
4. Ghaemi, A., Rashedi, E., Pourrahimi, A.M., Kamandar, M., Rahdari, F.: Automatic channel selection in EEG signals for classification of left or right hand movement in BCI using improved binary gravitation search algorithm. Biomed. Sign. Process. Control **33**, 109–118 (2017). https://doi.org/10.1016/j.bspc.2016.11.018
5. Lin, Y., Breugelmans, J., Iversen, M., Schmidt, D.: An adaptive interface design (AID) for enhanced computer accessibility and rehabilitation. Int. J. Hum. Comput. Stud. **98**, 14–23 (2017). https://doi.org/10.1016/j.ijhcs.2016.09.012
6. Gareis, I.E., Vignolo, L.D., Spies, R.D., Rufiner, H.L.: Coherent averaging estimation autoencoders applied to evoked potentials processing. Neurocomputing **240**(31), 47–58 (2017). https://doi.org/10.1016/j.neucom.2017.02.050
7. Kuziek, J.W.P., Shienh, A., Mathewson, K.E.: Transitioning EEG experiments away from the laboratory using a Raspberry Pi 2. J. Neurosci. Methods **277**(1), 75–82 (2017). https://doi.org/10.1016/j.jneumeth.2016.11.013
8. Bolaños, F., LeDue, J.M., Murphy, T.H.: Cost effective raspberry pi-based radio frequency identification tagging of mice suitable for automated in vivo imaging. J. Neurosci. Methods **276**(30), 79–83 (2017). https://doi.org/10.1016/j.jneumeth.2016.11.011
9. Arcidiacono, C., Porto, S.M.C., Mancino, M., Cascone, G.: Development of a threshold-based classifier for real-time recognition of cow feeding and standing behavioural activities from accelerometer data. Comput. Electron. Agric. **134**, 124–134 (2017). https://doi.org/10.1016/j.compag.2017.01.021
10. Paszkiel, S.: Characteristics of question of blind source separation using Moore-Penrose pseudo inversion for reconstruction of EEG signal. In: Szewczyk, R., Zieliski, C., Kaliczyska, M. (eds.) Recent Research in Automation, Robotics and Measuring Techniques. Series: Challenges in Automation, Robotics and Measurement Techniques, Advances in Intelligent Systems and Computing. Springer, Cham (2017)
11. Paszkiel, S., Hunek, W., Shylenko, A.: Project and simulation of a portable proprietary device for measuring bioelectrical signals from the brain for verification states of consciousness with visualization on LEDs. In: Szewczyk, R., Zieliski, C., Kaliczyska, M. (eds.) Recent Research in Automation, Robotics and Measuring Techniques. Series: Challenges in Automation, Robotics and Measurement Techniques, Advances in Intelligent Systems and Computing, vol. 440, pp. 25–36. Springer, Cham (2016). https://doi.org/10.1007/978-3-319-29357-8

A Self-driving Car in the Classroom: Design of an Embedded, Behavior-Based Control System for a Car-Like Robot

Mateusz Mydlarz and Piotr Skrzypczyński[✉]

Institute of Control, Robotics, and Information Engineering,
Poznań University of Technology, ul. Piotrowo 3A, 60-965 Poznań, Poland
{mateusz.mydlarz,piotr.skrzypczynski}@put.poznan.pl

Abstract. In this paper we study the design of a small mobile robot that resembles a self-driving car. The robot has been designed on the basis of a cheap, radio controlled toy car, but equipped with an embedded controller, a vision system and range sensors. To keep the robot affordable, low-cost, off-the-shelf components are used in the design. The developed robot is used to demonstrate the behavioral paradigm of the control system design, which is commonly used in mobile robots. The small autonomous car becomes an attractive educational tool, that may give the students an insight in both the hardware and software aspects of mobile robotics.

Keywords: Mobile robot · Vision · Behavior-based control · Education

1 Introduction

The importance of robotics and – in a broader sense – AI-related education is increasing constantly due to the high demand for engineers in the developing "Industry 4.0". A good example is the automotive industry, that quickly absorbs information technologies on its road to the extended safety and autonomy of the manufactured vehicles. Thus, self-driving cars are an interesting inspiration for more advanced educational robots. Real self-driving cars require a pretty complicated perception system to avoid arbitrary obstacles and to find some predefined elements of the environment, e.g. the traffic signs [14]. According to the recent report by the US-based University Professional and Continuing Education Association [15], the development of autonomous cars gives the higher education institutions an opportunity to enhance their curricula in the areas related to mobile robots autonomy and related industries. As so far, few Universities offer courses focusing on programming and based on the self-driving car concept [6]. Although open-source hardware projects of autonomous racing car models are available [13], the costs of the proposed hardware seem to be too high for regular courses offered to large groups of students.

© Springer Nature Switzerland AG 2020
R. Szewczyk et al. (Eds.): AUTOMATION 2019, AISC 920, pp. 367–378, 2020.
https://doi.org/10.1007/978-3-030-13273-6_35

Therefore, we present in this paper an approach to develop a car-like mobile robot that is intended to be used for educational purposes, focusing on the aspects of hardware-software integration and autonomy. The robot is based on a commercial, radio-controlled (RC) car model, low-cost sensors, and a typical microcontroller, which makes it affordable. The car-like kinematics offers a possibility to observe in practice how much different is to control a car from the control algorithms of two-wheeled, differential drive robots that are by far most popular in education [8]. The contribution of this paper is threefold:

- the design of a mobile robot from low-cost, off-the-shelf components is outlined, and the hardware-software integration steps are detailed;
- the integration of the CMUCam5 Pixy module as the vision system of the robot is demonstrated;
- the software implementation of two different behavioral control architectures on the developed robot is described, and the resulting behaviors are compared in the context of autonomous obstacle avoidance.

2   Self-driving Car Robot

2.1   Car Model as the Robot Base

The basis of the developed mobile robot is a 1/10 scale car model toy (Fig. 1a). The model has a DC electric motor that drives the rear wheels, whereas the passive front wheels are steered by a servo. Originally, the toy was radio controlled using a simple remote control device. For the self-driving version an entirely new controller was designed and fabricated. All electronics of the new controller are divided into three parts:

- The main board controls the basic components, like the distance sensors, the CMUCam5 Pixy module, the incremental optical encoder mounted on the DC motor shaft, the servos that steer the robot, the Bluetooth HC-05 module responsible for communication, the power and charging module, the DC motor driver module, and the external lightings. Moreover, it controls a LCD TFT 3.2″ display mounted on the car's roof, which displays the most important internal parameters of the robot. The main board employs a fast and efficient microcontroller from the STMicroelectronics Cortex-M4 family – STM32F407VG, which sits on the STM32F4-Discovery evaluation board. The whole main board is designed in the form of a sandwich to make it compact.
- The DC motor driver module controls the DC motor that propels the entire robot. It is based on the popular L298 chip.
- The power and charging module controls the automatic charging process of the battery. This module measures the voltage and current of the battery, and sends this information to the main board via the I2C bus. It provides 5 V and 3.3 V voltages necessary for operation of the logic circuits. The power source consists of nine 18650 Li-Ion cells.

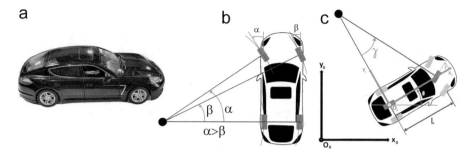

Fig. 1. RC model of a Porsche car (a), Ackerman steering (b), and a simplified kinematics model of a four-wheeled vehicle (c)

2.2 Vehicle Kinematics and Control

In order to create an autonomous mobile robot, it was necessary to determine the kinematics of the vehicle. In a wheeled autonomous robot the inverse kinematics is less useful due to the non-holonomic constraints that prevent such a robot from performing some motions, e.g. the robot cannot move sideways. Hence, one can't specify an arbitrary pose and just compute the control inputs (i.e. the linear velocity and steering angle in the car-like robot) to get there.

Therefore, we model only the direct kinematics, which is required to compute the pose of the robot from known control inputs, assuming that the vehicle moves without slippages. The approach described in [10] is adopted, that defines the generalized coordinates of a car-like robot as: $\mathbf{q} = [x, y, \theta, \gamma]^T$, where x and y are the Cartesian coordinates of the rear axle mid-point, θ gives the robot orientation (heading), and γ defines the steering angle (Fig. 1c). Using these coordinates a set of non-holonomic constraints for a car-like robot is defined:

$$\dot{x}_f \sin(\theta + \gamma) - \dot{y}_f \cos(\theta + \gamma) = 0, \quad \dot{x} \sin(\theta) - \dot{y} \cos(\theta) = 0, \qquad (1)$$

where x_f and y_f are the Cartesian coordinates of the front axle mid-point.[1] Applying the rigid-body constraint $x_f = x + l \cos(\theta)$ we obtain the formula:

$$\dot{x} \sin(\theta+\gamma) - \dot{y} \cos(\theta+\gamma) - \dot{\theta} L \cos(\gamma) = 0, \dot{x} \sin(\theta+\gamma) - \dot{y} \cos(\theta+\gamma) - \dot{\theta} L \cos(\gamma) \quad (2)$$

where L is the distance between both axles of the car. From the constraints given by (1) and (2) the kinematic model of a car that has rear-wheels driving can be derived:

$$\dot{\mathbf{q}} = \begin{bmatrix} \dot{x} \\ \dot{y} \\ \dot{\theta} \\ \dot{\gamma} \end{bmatrix} = \begin{bmatrix} \cos(\theta) \\ \sin(\theta) \\ \frac{\tan(\gamma)}{L} \\ 0 \end{bmatrix} v_m + \begin{bmatrix} 0 \\ 0 \\ 0 \\ 1 \end{bmatrix} v_s, \qquad (3)$$

where v_m is the motor (driving) velocity, and v_s is the steering velocity.

---

[1] They are coordinates of the physical front wheel in the tricycle car configuration.

In an actual four-wheeled car the orientation angles of the front wheels are governed by the Ackerman steering mechanism, which ensures that inner front wheel orientation angle is slightly larger than the orientation angle of the outer wheel, in order to eliminate geometrically induced slippage (Fig. 1b). Unfortunately, the car model toy used in this research shows a lot of slack in the mechanisms (including the Ackerman steering) and poor quality of the steering. Therefore, it was decided to consider only the velocity of the rear axle motor, which is measured by an optical encoder added to the model [7], and the front wheels steering angle that is read from the servo. This simplification leads to a mathematical model of a two-wheeled vehicle, which has the wheels in the mid-points of both exes [4]:

$$\dot{\theta} = \frac{v}{L}\tan(\gamma), \quad \dot{x} = v\cos(\theta), \quad \dot{y} = v\sin(\theta), \tag{4}$$

where v is the robot velocity measured using the encoder, and γ is given by the servo.

To control the robot we specify the desired v and γ values that together form the control vector of the car-like robot. This vector is computed instantaneously by the behavioral control architecture. Details of the computation depend on the implemented behaviors and the arbitration mechanism. This mechanism ensures that a single $[v, \gamma]^T$ vector is passed to the actuators. The velocity v is threated as the reference value PID controller for the driving motor, while the angle γ is directly set by the servo on the steering front wheels.

3   Sensory System

The self-driving car robot needs a sensory system that enables to detect obstacles and visually salient objects. For obstacle avoidance five optical distance sensors of the SHARP GP2Y0A21YK0F type are employed. They detect obstacles in the front of the robot within a much larger angular sector than the single sonar used previously [7], whereas their measurement range is between 10 and 80 cm. The task of the robot is to reach a visually tagged destination, therefore the easiest way to locate the target is to employ a vision system that detects the predefined, characteristic object in the image.

To accomplish this task, the CMUCam5 Pixy image sensor is used. Its advantage is that the image processing is carried out on the sensor. The main control unit receives only the information about the location of the target in the image and the size of the target object. Such a solution does not burden the embedded main control module with image processing, and makes it possible to dedicate computing power of the controller to the autonomous driving tasks.

In order to mount the optical distance sensors on the robot housing, a laminate handle has been designed and fabricated. The five distance sensors have been arranged on a semi-circle arc and offset from each other by 45°. The image sensor is attached to a pan/tilt mechanism. This mechanism allows the camera to rotate horizontally from −90° to 90°, and to tilt in the range from 0° to 35° to the back. The use of the pan/tilt mechanism makes it possible to actively search for objects through a much larger area of the environment without having to change the orientation of the car-like robot. The whole structure is mounted at the front hood of the vehicle (Fig. 2a), as the size of the sensors prevents them from being mounted inside the streamlined car chassis.

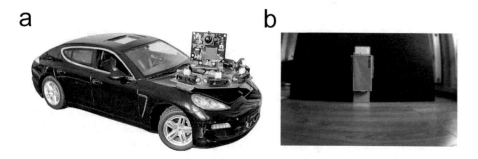

Fig. 2. Car-like robot in the final configuration (a) and a view from the CMUCam5 Pixy camera with a salient object bounding box (b)

The CMUCam is a self-containing sensor [9], as its microcontroller processes the images, extracts the information required by the user, and sends this information to the higher-level system (in our case another microcontroller) using a serial interface. The self-driving car model currently uses only the color blob tracking function to find the visually salient target object, and to obtain the image coordinates of its centroid. The blob search algorithm finds groups of pixels that have the RGB values within the bounds specified by the higher-level system. The image coordinates of the blob centroid, the bounding box, and the number of pixels belonging to the blob are sent to the higher-level system. The area of an object that is significantly different from the background (i.e. salient enough) can be selected by the user to define the bounding box for the target. Data describing the selected object are stored in the memory of the CMUCam5 Pixy, and then used by the robot. Figure 2b shows an example view from the camera with a salient object and the bounding box marked as a rectangle.

The developed pan/tilt mechanism helps the CMUCam sensor to detect and track the target object from a moving vehicle. The main controller receives from the CMUCam the coordinates of the center of the object in the image, and controls the pan/tilt two servos in such a way that the target remains roughly in the central part of the image (Fig. 3a). This helps the CMUCam to track the selected color blob, as when the car model takes a turn, a tracked object located close to the image border can easily move out of the camera's field of view. The same could happen on an uneven ground due to the tilting motion of the car. The mechanism rotates right and left up to the extreme positions ($\pm 90°$) searching for a target object. Whenever the target is spotted, the controller enters the tracking mode (Fig. 3a), trying to keep the centroid of the target object within

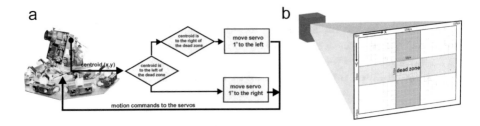

Fig. 3. Block diagram of the target tracking method (a) and the CMUCam image acquisition area controlled by the pan/tilt mechanism (b)

the "dead zone" in the center of the image (Fig. 3b). The implementation of the tracking algorithm runs in a loop, updating the camera position every 20 ms. If the centroid of the object is on the left side of the "dead zone", the servo rotates by 1° until the centroid reaches the "dead zone". The same happens when the object is on the right side of the image, but the direction of rotation is reversed. Same scheme is applied to track the centroid in the vertical axis (not shown in Fig. 3a).

4   Software of the Robot

4.1   Behavior-Based Control

The main idea behind the low-cost self-driving car was to use it to demonstrate and teach control paradigms specific to autonomous robots [11]. To make the robot capable of real-time autonomous operation the behavior-based control paradigm is chosen. Behavior-based control is popular in mobile robotics, and used even in quite large outdoor mobile robots [5]. In the self-driving car robot two different behavioral control algorithms are implemented directly in the main microcontroller. They are stored in the memory and activated at the user's request.

The idea of behavioral control for mobile robots dates back to the seminal work of Brooks [3], who proposed to discard symbolic representations in robot control in favour of physical embodiment in the real-world. The robot architecture proposed by Brooks is layered, as the individual behaviors that constitute the whole controller are stacked in a hierarchy, depending on their priority in the access to the actuators. As several behaviors can be active at the same time depending on the current sensory input, an arbitration mechanism is required to ensure deterministic control of the actuators. Brooks proposed hard-wired priorities between the stacked layers, where a layer located higher in the hierarchy subsumes the lower layers, suppressing their output to actuators. Thus, this concept is known as the *subsumption architecture*.

In the case of the simple, car-like robot, the subsumption architecture consists of three layers: wandering, pursuing the goal, and avoiding obstacles (Fig. 4a). The layers are wired together so that the top layer has the highest priority. The task of the *wandering* layer is to generate forward driving behavior at a specific speed. This behavior is clock-initiated and provides just a constant input to the driving motor's PID controller that drives the car at the specified speed observing the output of an optical encoder mounted on the motor's shaft. The *pursuing the goal* layer generates a more specific driving behavior towards the target at a constant speed. It is activated once a visually salient target previously defined by the user is detected by the CMUCam5 Pixy sensor. The steering angle γ of the front axle is determined based on the rotation of the pan mechanism and the location of the center of the detected target object with respect to the image coordinates. The *avoiding obstacles* layer generates the behavior of avoiding obstacle that are detected by the optical sensors in front of the robot. Depending on which distance sensor has detected the object, the encountered obstacle is avoided from the left or right side, making the detour as short as possible. However, the instantaneous change of the steering angle is limited to the range of $\pm 22°$, due to mechanical limitations of the car model. The maximum speed of the robot during the execution of this behavior can be set by the user.

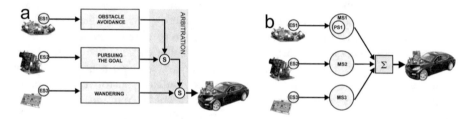

Fig. 4. Subsumption-based control system (a) and Motor-Schema-based control system (b) with the sensory input and the output arbitration mechanisms

Another similar idea has been proposed by Arkin, who took however a point of view more inspired by biology and cognitive science [1]. This resulted in the *motor-schema-based architecture*, which defines motor schema, that are a close analogy of individual behaviors. The Arkin's architecture defines also modules that can aggregate or abstract the data from physical sensors – *perceptual schemas*, that have no direct analogy in the subsumption architecture. However, the major difference is that the behavior arbitration in the Arkin's architecture is cooperative, in contrast to the hard-wired competitive scheme in the Brooks concept. Namely, all motor schema modules contribute their outputs in the form of vectors that define the desired direction and speed of the robot to a vector summation module. This mechanism produces the final output in the form of an aggregated direction and speed vector, which is then interpreted by the actuators to drive the robot.

The motor-schema architecture of the self-driving car robot contains three motor schemas that generate individual control vectors (Fig. 4b). The meaning of the behavior implemented by these schemas is the same as for the three behavioral modules in the subsumption architecture. Therefore, the resulting behavior

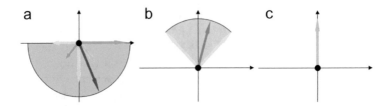

Fig. 5. Individual control vectors generated by the three motor schemas: MS1 (a) with summation of control vectors, MS2 (b), and MS3 (c)

of the robot should be similar as well. The motor schema MS1 is responsible for obstacle avoidance. It uses the distance measurements from the SHARP sensors, which are processed by the perceptual schema PS1 that allows to treat the group of rangefinders as a single external sensors. Each sensor generates a vector directed in the opposite direction than the direction of the individual measurement. This allows to generate a repulsive behavior from the encountered obstacle. Before executing this behavior, the vectors are scaled in such a way, that their magnitude corresponds to the maximum permitted speed of the robot whenever the acquired distance is in the smallest range of measurements (<10 cm). The magnitude decreases linearly with the increasing distance from the obstacle. All generated vectors are summed up to form the single control vector of the motor schema (Fig. 5a). The MS2 schema computes a control vector directly towards the detected salient object, using the values of the pan mechanism rotation angle and the location of the object centroid in the image yielded directly by the CMUCam sensor through the controller. The obtained angular values are summed together with the current value of the steering angle γ considering the $22°$ limit of the front wheels steering mechanism (Fig. 5b). The magnitude is constant and can be set by the user. The motor schema MS3 generates a vector directed along the robot's longitudinal axis, with the magnitude determined by the maximum speed set by the user (Fig. 5c). This motor schema is active only if there is no target object within the CMUCam's field of view. The final vectors from the individual motor schemas are added together to form the single control vector for the mobile robot. The direction of this vector determines the direction of motion, but the instantaneous change of the steering direction is limited to the range of $\pm 22°$, due to mechanical limitations of the steering system.

4.2   User Interface

As the self-driving car is fully autonomous and self-containing with respect to both sensing and control, the user interface is implemented as a remote supervision system on a separate Windows-based personal computer (PC). This computer exchanges data with the robot using a standard Bluetooth interface.

The user interface program has been designed focusing on the insight the operator (student) should have in the internal states of the robot and its control algorithms. Hence, the user interface contains three separate modes. The main

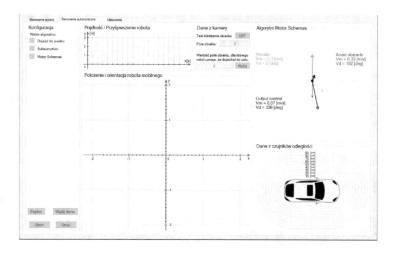

Fig. 6. User interface – example visualization in the self-driving mode

mode (Fig. 6, note that the interface is in Polish) allows the user to select and configure one of the two possible control algorithms (subsumption or motor schemas) and graphically shows the robot's internal parameters, the speed and coordinates of the robot, displaying the current control vectors, and the incoming sensory data in real time. The two other modes make it possible to manually control the robotic car from the keyboard, showing selected internal parameters in the form of a car dashboard, and to configure and test communication with the robot.

5   Experimental Results

The self-driving car robot was evaluated in a series of simple experiments aimed at comparing the performance of the subsumption and motor schema architectures. Moreover we wanted to evaluate the suitability of the new suite of exteroceptive sensors for obstacle avoidance and target seeking. Four tasks of increased complication have been defined. The simplest one was to each a target represented by the red box avoiding a single obstacle (Fig. 7a). A slightly harder task was to reach the same target going in between two boxes (Fig. 7b). Then, the robot had to find the target avoiding three obstacles located between the starting position and the goal (Fig. 7c). The last task was to reach the target avoiding a U-shaped object open towards the approaching vehicle (Fig. 7d).

All the experiments were registered by a ceiling-mounted, wide-lenses camera. Then, the path of the robot was extracted using image processing and drawn on the last image from the sequence as a series of red boxes bounding the instantaneous poses of the car (Fig. 8).

From these results one can conclude that the perception system of the robot works properly and finds the target from a considerable distance. The simple

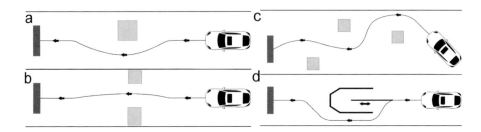

Fig. 7. Four tasks of the self-driving car in the experimental evaluation

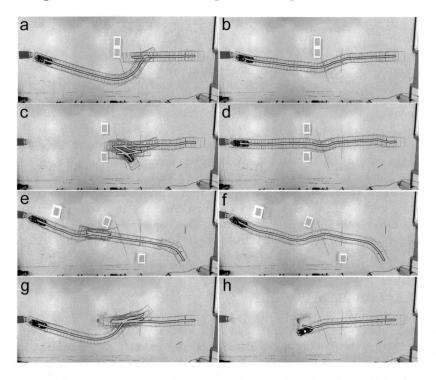

Fig. 8. Self-driving car robot trajectories registered in the four experiments for the subsumption (left column) and motor-schema (right column) architectures

infra-red distance sensors in most cases provide enough information to avoid the box-shaped objects. In general, both the subsumption architecture and the motor schemas architecture managed to guide the self-driving car through the tests. In general, the output motion of the motor schemas was much smoother, while the car controlled by competing behaviors of the subsumption system dodged a lot when avoiding obstacles. This is related to the non-holonomic constraints and the minimal turning radius of the car-like vehicle. If a behavior was activated too late, it could not circumvent the object and had to drive backward to avoid a collision (Fig. 8c). However, both architectures have their specific pitfalls. The vector summation mechanism in the motor-schema-based architecture could not manage to avoid the U-shaped object, and eventually caused a collision (Fig. 8h). This is a known problem in all mobile robot control approaches that employ vector summation, as the vectors generated from several behaviors (three in our case) can sum to zero immobilizing the robot [1]. However, due to the uncertainty in sensor measurements, the sum can also oscillate around a zero magnitude, while the direction is towards an obstacle. In such a case the robot can crash into the object.

6 Conclusions

The focus of our project was on making the control software principles transparent and easy to understand for students that have only background in control engineering and a limited knowledge of robotics issues.

The new sensory subsystem developed for the car-like robot, consisting of the CMUCam5 Pixy and the SHARP infrared range sensors turned out to be an effective source of information for obstacle avoidance and target seeking. As in is a separate module it can be replicated and used also on other educational mobile robots, such as our older, differential drive SanBot [8].

The main limitations of the current design can be circumvented by upgrading the controller to one of the new embedded controller platforms, which make possible to run even convolutional neural networks, as demonstrated by the very recent DeepPicar project [2] using a RaspberryPi 3 quad-core board.

References

1. Arkin, R.C.: Behavior-Based Robotics. The MIT Press, Cambridge (1998)
2. Bechtel, M.G., McEllhiney, E., Kim, M., Yun, H.: DeepPicar: a low-cost deep neural network-based autonomous car. arXiv:1712.08644v4 (2018)
3. Brooks, R.A.: Intelligence without representation. Artif. Intell. **47**, 139–159 (1991)
4. Corke, P.: Robotics, Vision and Control. Springer, Heidelberg (2011)
5. Doroftei, D., Colon, E., De Cubber, G.: A behaviour-based control and software architecture for the visually guided ROBUDEM outdoor mobile robot. J. Autom. Mobile Robot. Intell. Syst. **2**(4), 19–24 (2008)
6. Karaman, S., Anders, A., Boulet, M., Connor, J., Gregson, K., Guerra, W., Guldner, O., Mohamoud, M., Plancher, B., Shin, R., Vivilecchia, J.: Project-based, collaborative, algorithmic robotics for high school students: programming self-driving race cars at MIT. In: IEEE Integrated STEM Education Conference (ISEC 2017), Princeton (2017)
7. Mydlarz, M.: Sonarowy system nawigacji dla robota mobilnego ze sterowaniem Ackermana. BSc. Thesis, University of Zielona Góra (2017). (in Polish)
8. Rostkowska, M., Topolski, M., Skrzypczyński, P.: A modular mobile robot for multi-robot applications. Pomiary Automatyka Robotyka **2**, 283–287 (2013)
9. Rowe, A., Nourbakhsh, I.R., Rosenberg, C.: A second generation low cost embedded color vision system. In: IEEE Conference on Computer Vision and Pattern Recognition (CVPR 2005) - Workshops, San Diego, p. 136 (2005)

10. Siegwart, R., Nourbakhsh, I.R.: Introduction to Autonomous Mobile Robots. The MIT Press, Cambridge (2004)
11. Zieliński, C., Kornuta, T., Winiarski, T.: A systematic method of designing control systems for service and field robots. In: IEEE International Conference on Methods and Models in Automation and Robotics, pp. 1–14 (2014)
12. CMUCam5. http://www.cmucam.org/projects/cmucam5/wiki
13. Donkey Car. An opensource DIY self driving platform for small scale cars. http://www.donkeycar.com
14. Here's how Tesla's self-driving cars see the world (2016). http://www.businessinsider.com/how-tesla-driverless-cars-see-world-2016-11?IR=T
15. University Professional and Continuing Education Association, The Effect of Autonomous Vehicles on Education (2018). https://upcea.edu/autonomous-vehicles/

Capabilities of the Additive Manufacturing in Rapid Prototyping of the Grippers' Precision Jaws

Piotr Falkowski[(✉)] ⓘ, Bogumiła Wittels ⓘ, Zbigniew Pilat ⓘ,
and Michał Smater ⓘ

Industrial Research Institute of Automation and Measurements PIAP,
Warsaw, Poland
pfalkowski@piap.pl

Abstract. Nowadays, additive manufacturing is becoming increasingly popular, cheap and easily accessible. It got particularly useful in industry. While the engineering designs need to be verified as quickly as it is possible, it is hardly unmanageable due to their high costs and a long time of production. 3-D printing enables a totally different approach towards these problems. This technique lets the mechanic apply the changes in a design on spot and test it under the real conditions. Even though it gives the crucial advantage of the price and the save of time, it brings some major problems as well. The 3-D printers offer the accuracy and precision limited to much lower standards than traditional machining. Moreover, the filaments used for the additive manufacturing have usually much worse material parameters than the most common metals. This study assesses the capabilities of the additive manufacturing in the production of the small elements for the tests. It is based on the real-life case of rapid prototyping of the grippers' precision jaws used for assembling the cylindrical objects in the casings. The paper contains an overview of grippers, 3-D printers and filaments, finite element strength tests and the summary of the method's potential. Its result is a method of creating models and testing fragile elements so to guarantee their coherence and to verify their applicability for the particular intent.

Keywords: Additive manufacturing · Industrial robotics · Precision gripping

1 Introduction

The following paper describes the tests of the additive manufacturing's capabilities for the precise parts. Its main purpose is to assess whether the innovative rapid prototyping is applicable in the real design processes of the small and accurate objects under assumed constraints. These are among the others size limitations (according to the operational space) and strength/stiffness criteria (based on the chosen material parameters and the designed part's geometry). Conducted research is based mostly on the FEM analysis but includes also a printout of the prototype and its evaluation.

The paper is divided into six sections. They contain respectively: short introduction, description of the considered case, choice of the gripper's model, analysis of the additive manufacturing method's application for the given case, presentation of the results and a short discussion over them.

2 Case Description

The case described in the paper is based on the experience of the designing and constructing an automatic assembling machine. The aim of its main process is accurate, precise and repeatable gripping of a cylindrical metal object by a robot and then placing it in a casing. The object is 39.9 mm long and has a diameter of 12 mm. As the operational space of the gripper's jaws is significantly limited by the flat surfaces of the casing (see Fig. 1), a stroke of the gripper and its jaws' thickness need to be minimised.

Assembling process is currently carried manually; however, the objects are typically gripped by its cylindrical surface and placed in a casing nonparallel to its base. Due to the lack of flexible elements (low durability) and vision control system (high costs) in a considered solution, replication of this method is impossible.

Moreover, the cylindrical object consists of two subparts. One of them is constrained to another with a cylindrical joint; therefore, the application of the additional force is necessary to keep it still. For this reason, there is no alternative but to grip the object by its flat surfaces.

As a machine with a robot is meant to run more assembling processes simultaneously, the longest of them states for the whole cycle's time. The task of placing the described object in the casing is significantly faster than the others, so its time is not relevant; however, it is estimated at 16.4 s.

Pose repeatability of the robot used for this project (KUKA KR10 [6]) is determined as ±0.01 mm and it generally states for the machine's precision. The other factor which might affect it is positioning of the casing. To mitigate the risk of inaccuracies, every casing is locked in the special socket while placing the object in it. This guarantees accurate positions of the surfaces which limit the operational area.

For the presented case, a parallel gripper was chosen as a robot's tool (see Sect. 2). Its jaws are the most problematic elements to design. They need to be relatively thin (a sum of one's thickness and a stroke of the gripper per side has to be smaller than 2 mm, see Fig. 1), durable enough to carry the weight of the element as well as the force caused by the closing gripper and rough so to eliminate slip of an object. Due to the high number of different versions and expected modifications of the jaws being tested, rapid prototyping method including 3D printing is involved. This enabled faster *manufacturing – testing – modifying* cycle and lowered production costs [5]. To create the accurate precision jaws meeting the design intents mentioned above, different materials and technologies were considered. Their comparison, potential applications and main advantages and disadvantages are presented in the following sections.

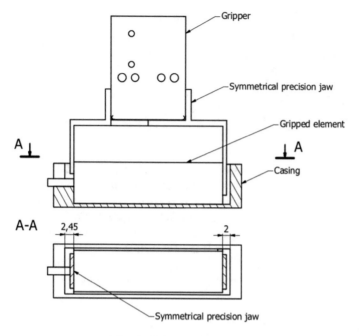

Fig. 1. Scheme of gripping.

3 Choice of the Gripper

As the grippers for the robotic applications are designed to carry different objects, they vary in terms of their principle of work, type of their drives, construction, size and numerous related parameters [1]. To ensure the fully controlled grip while not damaging the object, the following parameters were considered:

- a principle of work – choice of either mechanical, vacuum or electro-magnetic gripper;
- type of drive – choice of either a pneumatic [3] or an electric drive for the mechanical gripper;
- construction of the gripper – choice of either radial or parallel gripper with one or two movable jaws (as only grippers with two fingers are considered due to their application); consideration of additional construction modifications;
- size of the gripper – a choice based on the dimensions of a gripper, force occurring on its every jaw and its stroke.

The mechanical parallel grippers are the most suitable to place a cylindrical object in the casing of the given geometry. Only the pneumatic grippers are considered as electric ones are much more expensive and usage of the multiple gripping distances are not necessary for this case.

There are not many standard industrial grippers with a satisfying stroke (see Table 1). Moreover, there is only 2 mm of free space on one of the object's sides and

Table 1. Considered grippers' parameters [9–11].

Producer	Model	Stroke per jaw	Gripping force	Opening force
Schunk	MPG-plus 12	1.2 mm	10 N	8 N
Zimmer	MGP801N	1 mm	8 N	6 N
Festo	DHPS-6-A	2 mm	30 N	25 N
Schunk	MPG-plus 25	3 mm	38 N	32 N
Zimmer	GP-403N-C	3 mm	100 N	85 N

the manufactured jaw of the gripper must work without any contact with the casing. To reconcile this design intent with a need of achieving a satisfying gripping force and choosing the one with the size appropriate to catch the object, grippers with the stroke bigger than 2 mm per jaw are considered as well (see Table 1). Additionally, their operation area is limited by the jaws designed in the way that their gripping surfaces contact the object while the gripper is not fully closed.

Because of the small sizes of the grippers with the smallest strokes, only two mentioned above with the stroke of 3 mm per jaw are involved in the study.

4   Analysis of the Method

4.1   Printing Technique and Devices

All the jaws' printouts are manufactured with a usage of two 3D printers. Their parameters are compared in Table 2.

The consecutive jaws' series are printed to adjust their shapes to the gripped object, minimise their thickness regarding the strength of the material (including some strengthening modifications) and test the full operational capabilities of the robot's end effector.

To reduce the printouts' irregularities their manufacturing process is slowed down so to cools down every layer before applying the next one.

Table 2. Printers comparison.

Printer	Prusa i3 MK3 [12] with Melzi 2.0 control	Modified hypercube
Layer thickness	0.2 mm	0.16 mm
Printing precision	±0.011 mm	
Nozzle diameters	0.3 mm	0.5 mm
Printing speed	35 mm/s	2000 mm/min
Compatible filaments	All with the extruder's temperature below 270 °C	All with the extruder's temperature below 250 °C
Standard print size	200 × 200 × 180 mm	200 × 200 × 350 mm

4.2   Materials

Rapid prototyping is involved in the case to design the final jaws within a shorter time and to decrease production costs. Therefore, mostly the cheapest and most common filaments are considered (see Table 3).

Table 3. Considered materials' mechanical parameters [4, 7, 8].

Material	ABS	PLA	PC
Young's Modulus [GPa]	1.1–2.9	3.5	2.0–2.44
Poisson's Ratio	0.36–0.38		0.29–0.39
Shear Modulus [GPa]	0.58–0.74	2.4	0.54–0.67
Density [Mg/m^3]	1.3–1.8	1.25	1.14–1.21
Yield Strength [MPa]	18.5–51		59–70
Tensile Strength [MPa]	27.6–55.2	36–55	60–72.4

The sample prototypes (Fig. 2) manufactured with the different materials differ mostly in terms of their physical parameters. On the other hand, their structure and the typical inaccuracies (mostly in the region of the holes for the assembling screws and the irregular gripping surface's shape) of the printouts are generally very similar.

Fig. 2. ABS (orange) and PLA (black) sample printouts.

4.3   Correlation of the Internal Stress and the Printouts' Parameters

To assess the minimum thickness of the jaws not colliding with the functional trials, the simulation tests are held in 3D modelling FEM environment of *Autodesk Inventor 2015*.

Variations of models (different in terms of thickness and strength modifications) are loaded with the reaction force equal to the maximum gripping force in their points of contact with an object and fully constrained with the surface connected to the gripper's fingers. The simulations are run for the two series of jaws – designed for Schunk MPG-plus 25 and Zimmer GP-403N-C grippers.

The factors taken into consideration are the maximum internal stresses (yield criterion) and the deviation of the points of contact with an object (stiffness criterion) (Fig. 3). The simulations are prepared for the different designs and two materials – ABS and PLA (outcomes presented in Tables 4 and 5).

Table 4. Simulation results for ABS.

Thickness	Modification	Schunk MPG-plus 25		Zimmer GP-403N-C	
		Max. stress	Deviation	Max. stress	Deviation
1.0 mm	–	370 MPa	20.2 mm	876.3 MPa	83.4 mm
0.8 mm	–	536.2 MPa	36.4 mm	1336 MPa	209.8 mm
1.0 mm	Fillet 5 mm	221 MPa	8.4 mm	590.9 MPa	51.6 mm
0.8 mm	Fillet 5 mm	340.3 MPa	11.2 mm	902.1 MPa	85.3 mm
1.0 mm	Rib 5 × 45°	209.2 MPa	4.4 mm	551.1 MPa	41.4 mm
0.8 mm	Rib 5 × 45°	329.5 MPa	6.8 mm	801.3 MPa	74.6 mm

Table 5. Simulation results for PLA.

Thickness	Modification	Schunk MPG-plus 25		Zimmer GP-403N-C	
		Max. stress	Deviation	Max. stress	Deviation
1.0 mm	–	410.2 MPa	13.6 mm	933.4 MPa	56.3 mm
0.8 mm	–	626.1 MPa	24.6 mm	1414 MPa	85.1 mm
1.0 mm	Fillet 5 mm	230.9 MPa	6.3 mm	627 MPa	47.8 mm
0.8 mm	Fillet 5 mm	353.5 MPa	7.5 mm	954.8 MPa	57.8 mm
1.0 mm	Rib 5 × 45°	226.2 MPa	2.9 mm	580.7 MPa	27.8 mm
0.8 mm	Rib 5 × 45°	351.8 MPa	3.4 mm	874.6 MPa	37.8 mm

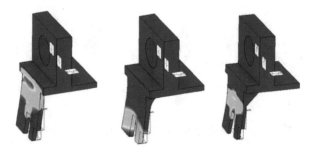

Fig. 3. Stress distribution in the shorter jaws with different modifications, designed for the MPG-plus 12 gripper.

Table 6. Simulation results for the MPG-plus 12 gripper's jaws manufactured with ABS.

Thickness	Modification	Max. stress	Deviation
1.0 mm	–	49.2 MPa	0.6 mm
0.8 mm	–	49.4 MPa	0.7 mm
1.0 mm	Fillet 3 mm	36.9 MPa	0.3 mm
0.8 mm	Fillet 3 mm	57.4 MPa	0.5 mm
1.0 mm	Rib 3 × 45°	33.7 MPa	0.15 mm
0.8 mm	Rib 3 × 45°	50 MPa	0.2 mm

As the results are not meeting the stiffness criterion for the grippers with the bigger stroke, the simulations are also run for Schunk MPG-plus 12 gripper and its shorter jaws (see Table 6).

4.4   Risk Assessment

As may be seen in Tables 4 and 5 the bigger grippers' jaws, printed with both materials are not strong enough to be applied in this case study. Acquired results deduce the certain break of the thin parts of the printed elements. To make the models more resistant to the loads, strengthening modifications should be applied. Also, minimisation of the jaws' length is desirable as well. However, even then only the modified designs for the smallest gripper may be tested with the full closing force.

Besides the problems regarding the strength of materials, the precision of the printouts may be an issue as well. Fortunately, they are soft enough to apply the additional manual amendments, so to make the gripper's assembly operate within the assumed scope. Moreover, small defects may occur in the small and irregular geometries where trails of filaments are carried unparallel to the edge (see Fig. 4). They may be reduced analogically to the problems with precision.

Fig. 4. Defect of the prototyped jaw from ABS.

5   Results

Different jaws are printed out to be tested in two aspects. These were held in two separate trials and involved different grippers and jaws' designs.

Primarily, they are verified in terms of their geometry (whether they are not colliding with the casing and if they may realise the gripping in the operational space). As none of the materials nor designs gave an opportunity to test the gripping with either Schunk MPG-plus 25 or Zimmer GP-403N-C (the grippers meant to be finally used), this trial was run without pneumatic supply (with the manual movements of grippers' fingers). The models used were printed from ABS and had a thickness of 1 mm (see Fig. 5). As a result, only the general jaws' application was assessed. Gripping effectiveness and strength of the elements were not tested at this stage.

Fig. 5. Examples of the tested grippers with the jaws' printouts.

Then, the jaws, with the same geometry of the part being in contact with the gripped object, may be printed for the gripper with a smaller stroke. They are manufactured from ABS of 0.8 mm thick as well and they have strengthening modifications applied (the rib). As the smaller gripper's closing force is much lower, the jaws are stiff enough to be tested with the pneumatic control of the system. This leads to the verification of the gripping effectiveness (assuming the final force is not less than the tested gripper's and it would not damage the gripped object). However, even though the gripper's stroke is much smaller it needs to be limited to 0.6 mm (therefore, the sum of this stroke, jaw's thickness and the maximum deviation equals 1.6 mm). Furthermore, the grippers may be also tested with the reduced pneumatic supply. This enables the functional tests of the full gripping set with a lower possibility of its damage (due to the reduced closing force). The result of this approach is a design which meets the requirements, guarantees the appropriate grip of an object and may be manufactured from metal.

6 Summary

Rapid prototyping method presented in the paper may be applied for the precision jaws. However, the materials typically used for 3D printing have too low strength parameters to be used for the thin elements while the big closing force is applying to them. Such prototypes may be assessed in terms of their geometry and then be redesigned for the smaller grippers so to have the same gripping facet's shapes. This combination of approaches towards the complex problem gives an opportunity to test the new models of jaws quicker. It enables verification of the designs even though theoretically the material's parameters are not sufficient to guarantee the safe work of the gripping elements. Its use reduces the number of designs manufactured with metal, which results in the time and money savings [2].

It is worth noticing, that the presented method considers an application of only common types of 3-D printers and the most basic materials. It does not require any additional complicated nor expensive technologies. Therefore, it may be used even for the low budget projects conducted by the small teams.

An outcome of a method involved in this case is a choice of the most appropriate gripper, development of the jaws' projects and a possibility to test the other components placement or even to program the industrial robot for the given task.

References

1. Klimasara, W., Pilat, Z.T.: Basics of Automatic Control and Robotics. Wydawnictwa Szkolne i Pedagogiczne, Warszawa (2006). (in Polish)
2. Chua, C.K., Leong, K.F., Lim, C.S.: Rapid Prototyping: Principles and Applications (with Companion CD-ROM). World Scientific Publishing Company, Singapore (2010)
3. Pneumatic driven grippers. Pomiary Automatyka Robotyka, vol. 2, pp. 22–31 (2013, in Polish)
4. Cantrell, J., et al.: Experimental characterization of the mechanical properties of 3D-printed ABS and polycarbonate parts. Rapid Prototyping J. **23**(4), 811–824 (2017)
5. Muita, K., Westerlund, M., Rajala., R.: The evolution of rapid production: how to adopt novel manufacturing technology. IFAC-PapersOnLine **48**(3), 32–37 (2015)
6. KUKA Robots KR AGILUS Specification. https://www.kuka.com/-/media/kuka.../spez_kr_agilus_en.pdf. Accessed 1 Oct 2018
7. University of Cambridge Engineering Department. Materials Databook (2003). http://www-mdp.eng.cam.ac.uk/web/library/enginfo/cueddatabooks/materials.pdf. Accessed 16 Sept 2018
8. Comparison of typical 3D printing materials. http://2015.igem.org/wiki/images/2/24/CamJIC-Specs-Strength.pdf. Accessed 16 Sept 2018
9. SCHUNK's catalogue webpage. https://schunk.com/de_en/gripping-systems/product/66092-0340007-mpg-plus-12/. Accessed 15 Sept 2018
10. Zimmer's catalogue webpage. http://www.zimmer-group.de/pl/product/%24mn-som-%24plc-v-%24pg-gre-%24sr-mgp800_3765_1/mgp801n#tecData. Accessed 15 Sept 2018
11. Festo's catalogue webpage. https://www.festo.com/cat/pl_pl/products_DHPS. Accessed 15 Sept 2018
12. Prusa i3 M3K printer's manual. https://www.prusa3d.com/downloads/manual/prusa3d_manual_175_en.pdf. Accessed 16 Sept 2018

Real-Time 3D Mapping with Visual-Inertial Odometry Pose Coupled with Localization in an Occupancy Map

Jacek Szklarski[1(✉)], Cezary Ziemiecki[2], Jacek Szałtys[2], and Marian Ostrowski[2]

[1] Department of Intelligent Systems, Institute of Fundamental Technological Research, Polish Academy of Sciences, Pawińskiego 5b, 02-106 Warsaw, Poland
jszklar@ippt.pan.pl
[2] Invenco sp. z o.o., Al. Żwirki i Wigury 93, 02-089 Warsaw, Poland

Abstract. Recent research has shown that visual and inertial measurements can serve as a powerful, robust and accurate odometry source when processed by state-of-the-art algorithms. One of the main benefits of such approach is short latency, even for on-board computers working on Miniature Autonomous Vehicles (MAV). However, depending on environmental conditions or sensor motion patterns, this type of odometry may be prone to drift or even divergence. In the presented work, it is shown that employing occupancy maps can limit such undesirable behaviour while still providing pose estimate at high frequencies. This is of particular importance for highly dynamical MAV control with limited on-board numerical capabilities.

Keywords: 3D maps · Visual-inertial odometry · Aerial systems · MAV control · Occupancy maps

1 Introduction

Planning and performing any autonomous actions by robotic platforms require knowledge about its surrounding environment. Of particular importance is information regarding distance to obstacles in order to avoid collisions. This remains true even if such a platform performs in a "semi-automatic" way, that is in the case when human supervises and remotely operates the device. Even if the operator has a general control on the larger spatial scale, still it might be beneficial that, e.g., the device will keep a safe distance from any obstacles or provide some specific obstacle-related information to the human.

In this paper we discuss a system to map surroundings of Micro Aerial Vehicles (MAV) which is expected to perform complex tasks in vicinity of obstacles like walls, pillars, beams and other structural elements related to civil engineering. The main requirements regarding the final system can be stated as follows:

- keep track of distances to obstacles in range of $d_{max} \approx 20\,\mathrm{m}$ around the MAV (the environments is assumed to be static) with accuracy of order of 1–$10\,\mathrm{cm}$,

© Springer Nature Switzerland AG 2020
R. Szewczyk et al. (Eds.): AUTOMATION 2019, AISC 920, pp. 388–397, 2020.
https://doi.org/10.1007/978-3-030-13273-6_37

- real-time operation with all calculations performed on-line on the on-board computer,
- used sensors should be light and should require low power,
- provide – independently of GPS – 6D position in the frame of the human operator,
- allow to gather data necessary to perform off-line optimization (like loop closing) so that the resulting maps can be used for efficient localization in subsequent missions,
- any software component used in the system should be licensed in the way enabling its application in commercial products.

The on-board computer is assumed to be used by other sub-systems related to MAV operation, in particular thrust control and control of a robotic arm mounted on the platform. Therefore, the mapping sub-system has a limited CPU/GPU power and the implemented algorithms should be tuned in the way that a proper balance between computational power and accuracy can be achieved. In the course of design process, it was decided that the main sensors will be: a visual camera (wide-field preferred; possibly more than one), a depth camera and an Inertial Measurement Unit (IMU; possible more than one). Particularly, using camera and the depth sensors makes it possible to utilize information of a very distinct nature, making the system less prone to instabilities or drift.

One of the obvious first steps in order to tackle this mapping problem would be to choose one of the many SLAM (Simultaneous Localization And Mapping; see e.g. [1]) algorithms which are described in robotic literature. A typical 6D-SLAM optimization is not suitable in the discussed case, mainly due to the high motion dynamics and requirement of low latency with calculations to be performed on the computationally constrained platform. Another alternative would be to use Kinect-Fusion-like approach [2] which utilizes only depth measurements in order to construct local 3D models and estimate 6D pose. However, besides intensive GPU computational requirements, this dense reconstruction is not expected to work well, for example, in front of large, uniform planar surfaces. Similar argument hold against using closed-source software, ZEDfu [3].

After a thorough survey, due to the requirements stated above, the focus was on solution based on monocular Visual-Inertial Odometry (VIO) and occupancy 3D maps constructed using depth images. Regarding VIO, modern algorithms are very robust and have sufficient accuracy. However, under certain circumstances the drift can become a problem. In the solution proposed in the paper, it is shown that pose estimate can be improved and the drift can be limited. This is realized by simultaneously constructing a 3D map of environment, localizing the robot in it, and subsequently using the estimated pose as a feedback to the VIO filter.

1.1 Visual-Inertial Odometry

Combining complementary information from visual sensors and IMUs has been shown to be of a particular use in 6DoF state estimation of highly dynamical robotic platforms, including aerial vehicles. Such systems have already proofed to be robust, accurate and reliable. Research in this field reached a mature stage, and now there exist open-source libraries which implement various VIO approaches, like Rovio [4,5] or Okvis [6]. For comparison of various monocular VIO algorithms, bearing in mind applications for MAV, see the recent review [7].

Visual odometry (VO) algorithms are usually based on standard feature tracking methods or direct measurements if image intensities. For a properly calibrated visual system, it is possible to estimate pose which is fused with IMU measurements using, e.g., EKF. Incorporating IMU in the VO results in higher pose estimation frequency, more accurate velocities and rotation estimates. Additionally, IMU provides the direction of gravity vector which is very useful for human readable visualisation. It should be noted that a proper fusion of an output of a VIO with yet another IMU can lead to further improvement of the filter, including estimation of position with frequency equal to the one of the second IMU [8,9]. This is of special importance for control algorithms.

Although VIO algorithms have been already proved to be suitable and robust for many applications, still there are significant problems with design-space exploration and parameter optimization [7]. With each VIO, there is a vast number of parameters which need to be tuned, and they can have a dramatic influence on the performance. Also, pure VIO will almost inevitably suffer from a drift. These undesirable effects can be reduced by fusing VIO with some kind of mapping and localization processes.

1.2   Occupancy Maps

One of the ways to represent 3D environment is to use a volumetric representation of space divided into voxels where each voxel describes some spatial property – *occupied* and *free* being the most straightforward. Since the space in which a robot moves can be relatively large, it is important that the environment representation will fit in memory and can be accessed efficiently. Data structure known as oct-tree suits best for this task – the space is hierarchically divided into boxes in a tree-like structure, with leaves representing a certain property. Such structure has many desirable properties: specific nodes can be located quickly in $O(\log n)$, traversing the entire tree is $O(n)$. The most widely used and extensively tested library for 3D occupancy mapping is OctoMap [10]. Apart of utilizing oct-trees, the library uses sensor fusion which makes use of certain sensor characteristics. Therefore the space representation in nodes is probabilistic. After querying a node state one gets three states based on which trajectory planning can be realized: empty, occupied and unknown.

One should note that there exist other methods of environment representation which are suitable for local trajectory planning, obstacle avoidance and visualisation. Most notably, there is a class of approaches based on Signed Distance Fields (SDF). For example, Euclidean STF are voxel representation where each voxel contains Euclidean distance to a nearest obstacle. Recent research shows that this approach is suitable for online planning, incremental mapping, and convenient environment visualization [11]. Nevertheless, due to maturity, reliability, and efficiency of OctoMap it was decided that this solution fits better goals of the discussed system.

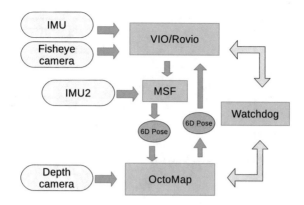

Fig. 1. Schematic representation of the discussed mapping system.

2   System Design

The schematic representation of the system, consisting of VIO and OctoMap processing, is depicted in Fig. 1. Currently, the main sensor is Intel Realsense ZR300 equipped with fisheye camera, depth sensor and IMU. Using the fisheye-type optics is preferred for most visual odometry systems [12]. Optionally, information from additional IMU from Pixhawk is utilized. The entire system is calibrated using *Kalibr* [13]. The details related directly to MAV hardware are not relevant in this paper and will be discussed elsewhere.

As stated above, there is a large number of parameters controlling the Rovio algorithm. Properties like noise-characteristics for all of the input variables, various thresholds, number of images patches can profoundly alter accuracy and stability of the VIO. Moreover, chosen values for a particular environment and lighting might not work well in different conditions.

An example of VIO failure is shown in Fig. 2. The data were gathered by a hand-held motion in front of a white wall towards and away from it along a line perpendicular to the wall. It should be noted that for a properly calibrated and tuned VIO system such large errors should rarely happen in practice. Here, however, the unfavorable conditions were chosen deliberately in order to exaggerate wrong estimate.

Along the output of Rovio, the figure contains a plot of pose estimation using the classical Iterative Closest Point (ICP) procedure. The depth sensor provides point clouds at the rate of 30 Hz and these clouds can, in principle, serve as an input of ICP odometry. However, since the point clouds contain 170k points, ICP can not be performed with full accuracy in real time. Low resolution ICP works well in these type of motion (towards/away planar surfaces), however it will fail in complex, unstructured environments or with fast rotations. ICP-like approach may be improved in many ways: utilizing point-to-plane correspondences, matching planar patches [14], making use of semantic information [15],

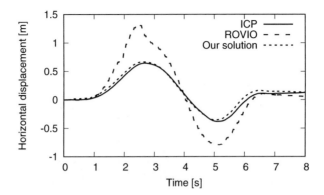

Fig. 2. An example of failure of Rovio during a horizontal translation along line perpendicular to a wall lacking distinct features. Results of a typical ICP and the presented solution are shown for comparison (c.f. Fig. 3).

and many more [16]. However, currently the only real-time efficient approaches are similar in nature as the dense Kinect fusion mentioned above.

As can be seen in Fig. 1 our approach is somehow different and can be summarized in the following steps.

1. Use VIO Rovio to get the 6D pose estimate x in the world frame.
2. Use x as the initial guess for ICP-like pose refining in the OctoMap M.
3. If the refined pose and measurements are consistent with the existing map, the pose is fed back to the EKF filter of Rovio and the OctoMap is updated accordingly.

Additionally, all voxels which are further from platform center than the given distance threshold, $d_{\max}$, are removed from the tree. The *watchdog* subsystem monitors VIO estimates and pose as a result of localisation in the map. If discrepancies becomes above a certain threshold, or VIO velocities or covariances become too large, the occupancy map is wiped out and the entire system is reset (with initial pose taken from GPS, if available).

All results related to operations on point clouds were performed with the use of the PointCloud library [17]. Steps 2 and 3 require additional, more detailed explanations.

2.1 Pose Refinement and Map Update

First, the measured point cloud I is down-sampled to a voxel grid with voxel size corresponding to the one used in the oct-tree map representation. All points falling into a voxel are replaced by a single point – their centroid, and the set of these points will be denoted as P. It should be noted that voxels in OctoMap have a constant size defined at the moment of oct-tree initialization. Obviously, increasing resolution (i.e. using smaller voxels) leads to increase in

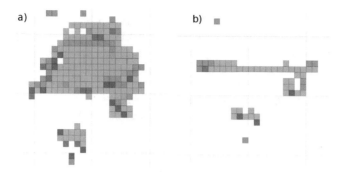

Fig. 3. Map as viewed from the top (orthogonal projection) for the case of motion along a line perpendicular to a wall, as shown in Fig. 2. (a) Rovio, (b) Rovio corrected with use of the OctoMap.

memory requirements and processing power needed for the tree update or traversal. Additionally, in OctoMap voxel grid is always coaligned with the coordinate system. This influences, e.g., accuracy of reconstructed planes. For example, it would be preferred to start map building being exactly perpendicular to a wall. Then the voxels faces would be perfectly aligned with the wall, which would improve visualization and the ICP-like alignment. However, for obvious reasons this usually will not be the case.

Next, voxel centers of occupied nodes in the map are converted into a point cloud Q, and Q is matched to P, so that the pose related transformation $\mathbf{t}_{\mathrm{icp}}$ minimizes distances between corresponding points in Q and P. Having analogous transformation from Rovio, $\mathbf{t}_{\mathrm{vio}}$, both transformations are compared by taking into account several factors. This is realised by considering how $\mathbf{t}_{\mathrm{icp}}I$ and $\mathbf{t}_{\mathrm{vio}}I$ would alter the existing map, in particular the consistency with the previous measurements. Additionally, if rotational or translational parts of $\mathbf{t}_{\mathrm{icp}}$ are above a certain threshold (when compared to $\mathbf{t}_{\mathrm{vio}}$), $\mathbf{t}_{\mathrm{icp}}$ is ignored since it is a common manifestation of a clearly bad ICP match. After deciding which transformation is preferred, it is used to update the map with I. Additionally, if $\mathbf{t}_{\mathrm{icp}}$ is better, it is used as correction in the EKF filter of Rovio. This is done with a raycast method, i.e., all voxels on the line between the sensor and an occupied voxel are treated as empty space measurements.

2.2 Parameter Tuning

As mentioned earlier, VIO algorithms require a large number of parameters which have a great influence on the performance. Additionally, when coupling with the map there is even more. In particularly, one has to assume some values of covariances for translational and rotational components of $\mathbf{t}_{\mathrm{icp}}$. We used a Stewart platform to fine-tune parameters in various types of motions and motion patterns in front of various textures, shapes and lighting conditions. Utilizing the platform makes it possible to easily and quickly test the system in very different conditions with accurate ground-truth pose as a reference.

VIO algorithms are prone to diverge or drift especially if the system is not excited enough [4]. Therefore, these type of motions – e.g. simple translations – were of special importance during the tunning procedure. Except where sensors properties were known (e.g. IMU), all the parameters were chosen manually, using common sense in order to obtain best outcome for the expected conditions. Regarding the map update, there are also parameters associated with the sensor model: hit/miss probabilities, probability clamping, and occupancy threshold (Fig. 4).

Fig. 4. Stewart platform with sensors mounted on the top of it. The platform has been used in the parameter-tuning procedure.

3   Experimental Results

After calibrating and tunning the system, several experiments were performed, mostly in indoor environment. As can be seen in Figs. 2 and 3, the erroneous position of VIO is greatly corrected and the resulting map becomes usable.

Regarding the drift, Fig. 5 depicts a situation where sensor was placed in a vicinity of a wall and a floor with minimal motion. Due to the slow downward drift of Rovio, the representation of floor becomes incorrect, here depicted after 4 s since initialization.

More complex experiments related to indoor flying were performed as well. The resulting maps are significantly improved, walls and angles between walls are represented more accurately which leads to the conclusion that the pose estimate is clearly better. One should keep on mind, however, that the system is not aimed at a typical mapping problem. Instead, it should provide a reliable information regarding pose and near obstacles during MAV operation in vicinity

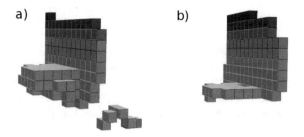

Fig. 5. Result of a slow motion in close vicinity of a wall and floor. Making use of the map significantly reduces drift leading to a consistent map. (a) Rovio, (b) The discussed solution.

of some infrastructure and this should be done with sufficient frequency. In all the experiments, an i7@2.5 GHz CPU was used to run the presented pipeline at 30 Hz without any issues for voxel size of 0.1 m. The limiting, practical resolution will depend on the final CPU and number of other algorithms to be run simultaneously on it (Fig. 6).

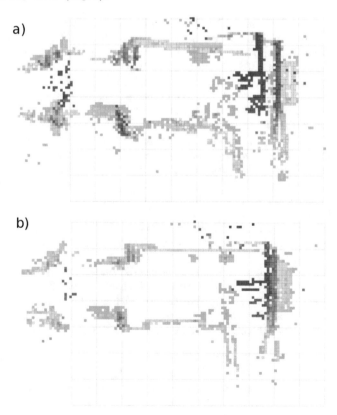

Fig. 6. A view from above on a map made after three consecutive passes in an indoor corridor (on the right side, there is visible a recess in the wall). (a) Rovio, (b) the discussed solution.

4 Conclusions

In the work it was shown how the state-of-the-art VIO system – Rovio – can be integrated with the probabilistic 3D occupancy mapping framework, OctoMap. The main benefits can be summarized as follows: (i) pose correction – especially in cases where environmental conditions are not in favor of visual tracking, however, at the same time allow pose estimate with depth images; (ii) limiting pose drift.

Future work will be focused on integration the discussed pose estimate system with control algorithms running on-board on a MAV.

Acknowledgments. This work was completed as part of the project titled: *"DIAGSTAR – dynamically stabilized universal manipulator for unmanned aerial vehicles"*. It was funded by research grant no. POIR.01.02.00-00-0084/16 supported by the National Centre for Research and Development (NCBiR), Poland.

References

1. Yang, A., Luo, Y., Chen, L., Xu, Y.: Survey of 3D map in SLAM: localization and navigation. In: Fei, M., Ma, S., Li, X., Sun, X., Jia, L., Su, Z. (eds.) Advanced Computational Methods in Life System Modeling and Simulation, pp. 410–420. Springer, Singapore (2017). ISBN 978-981-10-6370-1
2. Newcombe, R.A., Izadi, S., Hilliges, O., Kim, D., Davison, A.J., Kohli, P., Shotton, J., Hodges, S., Fitzgibbon, A.: Kinectfusion: real-time dense surface mapping and tracking. In: IEEE ISMAR. IEEE, October 2011
3. https://www.stereolabs.com/developers/
4. Bloesch, M., Omari, S., Hutter, M., Siegwart, R.: Robust visual inertial odometry using a direct EKF-based approach. In: 2015 IEEE/RSJ International Conference on Intelligent Robots and Systems (IROS), pp. 298–304. IEEE (2015)
5. Bloesch, M., Burri, M., Omari, S., Hutter, M., Siegwart, R.: Iterated extended kalman filter based visual-inertial odometry using direct photometric feedback. Int. J. Robot. Res. **36**(10), 1053–1072 (2017)
6. Leutenegger, S., Lynen, S., Bosse, M., Siegwart, R., Furgale, P.: Keyframe-based visual-inertial odometry using nonlinear optimization. Int. J. Robot. Res. **34**(3), 314–334 (2015)
7. Delmerico, J., Scaramuzza, D.: A benchmark comparison of monocular visual-inertial odometry algorithms for flying robots. Memory **10**, 20 (2018)
8. Sa, I., Kamel, M., Burri, M., Bloesch, M., Khanna, R., Popovic, M., Nieto, J., Siegwart, R.: Build your own visual-inertial drone: a cost-effective and open-source autonomous drone. IEEE Robot. Autom. Mag. **25**(1), 89–103 (2018). ISSN 1070-9932
9. Lynen, S., Achtelik, M.W., Weiss, S., Chli, M., Siegwart, R.: A robust and modular multi-sensor fusion approach applied to MAV navigation. In: 2013 IEEE/RSJ International Conference on Intelligent Robots and Systems (IROS), pp. 3923–3929. IEEE (2013)
10. Hornung, A., Wurm, K.M., Bennewitz, M., Stachniss, C., Burgard, W.: OctoMap: an efficient probabilistic 3D mapping framework based on octrees. Auton. Robot. **34**(3), 189–206 (2013)

11. Oleynikova, H., Taylor, Z., Fehr, M., Siegwart, R., Nieto, J.: Voxblox: incremental 3D Euclidean signed distance fields for on-board MAV planning. In: 2017 IEEE/RSJ International Conference on Intelligent Robots and Systems (IROS), pp. 1366–1373. IEEE (2017)
12. Zhang, Z., Rebecq, H., Forster, C., Scaramuzza, D.: Benefit of large field-of-view cameras for visual odometry. In: 2016 IEEE International Conference on Robotics and Automation (ICRA), pp. 801–808. IEEE (2016)
13. Rehder, J., Nikolic, J., Schneider, T., Hinzmann, T., Siegwart, R.: Extending kalibr: calibrating the extrinsics of multiple IMUs and of individual axes. In: 2016 IEEE International Conference on Robotics and Automation (ICRA), pp. 4304–4311. IEEE (2016)
14. Pathak, K., Vaskevicius, N., Poppinga, J., Pfingsthorn, M., Schwertfeger, S., Birk, A.: Fast 3D mapping by matching planes extracted from range sensor point-clouds. In: IEEE/RSJ International Conference on Intelligent Robots and Systems, IROS 2009, pp. 1150–1155. IEEE (2009)
15. Borkowski, A., Siemiatkowska, B., Szklarski, J.: Towards Semantic Navigation in Mobile Robotics, pp. 719–748. Springer, Heidelberg (2010). ISBN 978-3-642-17322-6
16. Pomerleau, F., Colas, F., Siegwart, R., Magnenat, S.: Comparing ICP variants on real-world data sets. Auton. Robot. **34**(3), 133–148 (2013)
17. Rusu, R.B., Cousins, S.: 3D is here: point cloud library (PCL). In: IEEE International Conference on Robotics and Automation (ICRA), Shanghai, China, 9–13 May 2011 (2011)

Optimal Tuning of Altitude Controller Parameters of Unmanned Aerial Vehicle Using Iterative Learning Approach

Wojciech Giernacki[✉]

Institute of Control, Robotics and Information Engineering,
Faculty of Electrical Engineering, Poznan University of Technology,
Piotrowo Street 3a, 60-965 Poznan, Poland
wojciech.giernacki@put.poznan.pl
http://uav.put.poznan.pl

Abstract. Dynamics and flight stabilization of a multirotor unmanned aerial vehicle (UAV) can be shaped by appropriate mechanisms of tuning parameters of its position and orientation controllers. In the article, the attention is focused on a fixed-parameters altitude controller. Its gains can be tuned optimally and automatically according to the expected criterion, and the search process takes place during the UAV short-time flight. For this purpose, it is proposed to use the auto-tuning method based on the bootstrapping technique and zero-order optimization using Fibonacci-search algorithm. The theoretical basis of the proposed method and discussion of the results from conducted simulation experiments for the exemplary quadrotor model, are presented in the paper.

Keywords: Unmanned aerial vehicle · Zero-order optimization ·
Controller tuning · Fibonacci-search algorithm ·
Bootstrapping technique

1 Introduction and Problem Statement

In recent years, there has been a strong trend in the development of machine learning methods [1,2]. With the dynamic increase in the number of unmanned aerial vehicles [3], appears the possibility of using computational intelligence algorithms to improve the efficiency of such machines [4–6]. This trend is particularly evident in the aspects of improving the quality of autonomous flight control of the multirotor UAV by optimization the work of the control systems of its position and orientation. There are numerous approaches to the search for the optimal controller gains, such as Reinforcement Learning [7], Bayesian Learning [8], Artificial Neural Networks [9] and Deep Learning [10], as well as metaheuristic optimization algorithms [11] and iterative learning control (ILC) [12]. Particularly noteworthy are those that can be successfully used during UAV

© Springer Nature Switzerland AG 2020
R. Szewczyk et al. (Eds.): AUTOMATION 2019, AISC 920, pp. 398–407, 2020.
https://doi.org/10.1007/978-3-030-13273-6_38

flight due to potential applications – such as transporting of payload or dynamic change of flight characteristics (for example, to save energy or to produce a large force of actuators attached to the UAV in manipulation tasks).

Using machine learning methods in the mentioned applications it is desired to quickly and optimally adjust the controller's gains to the expectations expressed by the cost function minimized over time. The proposed solution for Naze/Multiwii flight controllers (CleanFlight G-Tune) an Bayesian learning algorithm SAFEOPT [13] are (according to the best knowledge of the author) currently the only available in the context of the iterative in-flight auto-tuning of orientation controllers. Moreover, only the SAFEOPT algorithm guarantees optimal tuning and insight into the optimizer code. Literature studies, however, indicate the lack of approaches to the problem of the optimal, iterative auto-tuning of the altitude controller during the UAV flight. This issue is addressed in the current article. For a defined research problem, the main contributions of this paper are the following:

- presentation of a novel, iterative learning approach to real-time zero-order optimization in the context of auto-tuning of altitude controller of the multirotor UAV based on the mechanisms of bootstrapping technique & Fibonacci-search algorithm (which are fast and have low computational complexity),
- giving a study of performance of the proposed method for altitude controller by simulation experiments for the well-known dynamical model of the quadrotor from [14].

Some of proposed mechanisms of optimization strategy used in this paper has primarily been successfully used for model-based off-line optimal auto-tuning of the fixed-wing UAV orientation controllers [15], where its high quality performance was confirmed.

The paper is organized as follows – in Sect. 2 the altitude controller description is presented. In Sect. 3, the basics of proposed tuning method are included. In this section the optimization mechanisms based on Fibonacci-search algorithm and bootstrapping technique are explained. The results of simulation experiments are shown in Sect. 4. In Sect. 5 conclusions are presented.

2   Altitude Controller

For a more transparent form of presentation of research results (especially the details of the proposed optimization method), the author decided to use in this work the specific structure of altitude controller commonly used in UAVs, i.e. proportional-derivative (PD) type. A modified version of this controller was used in control system from Fig. 1. It can be written by the equation:

$$T = K_p e\left(t\right) + T_D \frac{d}{dt} e\left(t\right) + \omega_0, \tag{1}$$

where: T – generated thrust (in order to change the UAV altitude), $e(t)$ – control error. Parameters: K_P (proportional gain) and T_D (time required by the proportional term to repeat the output provided by the derivative term), correspond to the searched values of the controller gains in the optimal auto-tuning procedure. In the further part of the article, they are marked as: k_P and k_D.

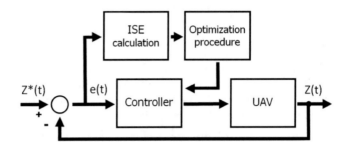

Fig. 1. A block diagram of the considered control system.

Control error is defined as:

$$e(t) = Z^*(t) - Z(t),\tag{2}$$

where: $Z^*(t)$ and $Z(t)$ are desired and actual UAV's altitudes, respectively. In Eq. (1) ω_0 is a rotor speed that needs to be generated to provide a thrust value equal (corresponding) to the total weight of the UAV:

$$\omega_0 = \sqrt{\frac{mg}{4b}},\tag{3}$$

where: m – quadrotor mass, g – gravitational acceleration, b – lift constant for particular UAV ($b > 0$).

3   Method of Optimal Auto-Tuning of the Altitude Controller Parameters

3.1   Idea of Proposed Method

Let us consider an example in which, drawing from the general idea of Iterative Learning Control, one analyzes the problem of tracking repeated reference primitives in the UAV control system from Fig. 1 – in the considered case – the repetitical change of flight altitude. Due to the fact that the control system works in a repetitive manner, it is possible to record periodically the predefined index, e.g. based on control error $e(t)$, which can be minimized iteratively with the zero-order optimization algorithm for each controller parameter. Minimization is carried out repetitively for all parameters of the controller to achieve expected tolerance ϵ or till the execution of the required, total number of iterations (reference primitives) N_{total} – which can be understand as

$$N_{total} = N \cdot N_b \cdot N_{Par},\tag{4}$$

Algorithm 1. Fibonacci-search algorithm [18].

Step 1. Choose lower and upper bounds (values: a and b) and calculate $L = b - a$. Set the number of calculations N for the cost function value. Set $k = 1$.

Step 2. Compute $L_k^* = (1 - \rho_{N-k+1})...(1 - \rho_1) = \frac{F_{N-k+1}}{F_{N+1}}$. Set $x_1 = a + L_k^*$ and $x_2 = b - L_k^*$.

Step 3. From $f(x_1)$ and $f(x_2)$ get the value that was not previously obtained. Eliminate the region according to the fundamental method for region elimination. Set new a and b.

Step 4. Is k equal to N? If no, set $k = k + 1$ and go to the **Step 2**; otherwise, **Terminate**.

where: N – number of calculations for the cost function value in Fibonacci-search algorithm, N_b – number of bootstraps, N_{Par} – number of controller gains, which must be tuned.

Process of particular gains optimization of the altitude controller using the Fibonacci-search algorithm and bootstrapping mechanism is explained below.

3.2 Fibonacci-Search Algorithm

As it has been shown in [16], there are many methods of finding the minimum. Among the optimization methods of one-dimensional cost function, especially the region elimination methods (REMs) are very fast and effective, e.g. Fibonacci-search method devised by Kiefer [17]. In the task of in-flight tuning of altitude controller with the use of Fibonacci numbers (see Algorithm no. 1), the minimized cost function can be the Integral of the Squared Error (ISE):

$$ISE = \int_0^\infty [e(t)]^2 \, dt. \tag{5}$$

For each of the altitude controller gains (k_P and k_D) Fibonacci-search algorithm is used separately according to bootstrapping technique and to the fundamental rule for region elimination methods.

Fundamental Rule for Region Elimination Methods [18]. Let us consider two points x_1 and x_2, which lie in interval (a, b), and satisfy $x_1 < x_2$. For a unimodal nature of the optimized function, it is true that:

(a) If $f(x1) > f(x2)$, then the minimum does not lie in $(a, x1)$,
(b) If $f(x1) < f(x2)$, then the minimum does not lie in $(x2, b)$,
(c) If $f(x1) = f(x2)$, then the minimum does not lie in $(a, x1)$ and $(x2, b)$.

Proportions of Intervals Reduction Using Fibonacci Numbers. The optimization algorithm, by narrowing the range where the searched optimum for a given controller gain is located, is based on the ISE values recorded for two neighboring primitives from the UAV flight. In the rule of regions elimination, according to Fibonacci numbers, the optimum search interval is narrowed in such a way that from iteration to iteration it should be decreased as much as possible.

Introducing ρ_k, and expressing the proportion by which the interval in the k-th iteration decreases, in the Fibonacci-search algorithm the following relationship applies:

$$\frac{1 - \rho_{k+1}}{1} = \frac{\rho_k}{1 - \rho_k}. \tag{6}$$

This leads to a conclusion that values $\rho_k \in (0, 0.5]$, where $k = 1, ..., N$, which minimize the expression $(1 - \rho_1)(1 - \rho_2)...(1 - \rho_N)$ and satisfy (6) are the numbers:

$$\rho_1 = 1 - \frac{F_N}{F_{N+1}},$$

$$\rho_2 = 1 - \frac{F_{N-1}}{F_N},$$

$$...,$$

$$\rho_k = 1 - \frac{F_{N-k+1}}{F_{N-k+2}},$$

$$...,$$

$$\rho_N = 1 - \frac{F_1}{F_2},$$

where F_k are the subsequent Fibonacci numbers:

$$F_1 = 1,$$

$$F_2 = 1,$$

$$F_k = F_{k-1} + F_{k-2}$$

for $k \geq 3$ and $k \in \mathbb{N}$.

3.3   Bootstrapping Technique

For an example of a altitude controller, a mechanism of alternate tuning of its gains (called bootstrapping) is shown in the Fig. 2.

The introduction of such a simple solution (keeping a constant value of one of the parameters, while the other is searched and vice versa for a given number of bootstraps N_b) is to ensure a fast exploration of the two-dimensional space (k_P, k_D) in the vicinity of primary gains in order to tune them optimally.

The duration time of a single bootstrap t_b determines the relationship:

$$t_b = t_{rp} \cdot N \cdot N_{Par}, \tag{7}$$

where the set value t_{rp} is the duration time of a single reference primitive:

$$t_{rp} = T_s \cdot N_{max}, \tag{8}$$

T_s – set value of the sampling interval – needed to calculate the ISE value based on the set N_c (number of samples when performance index is collected) and N_{max} (total number of samples during a single reference primitive for the T_s).

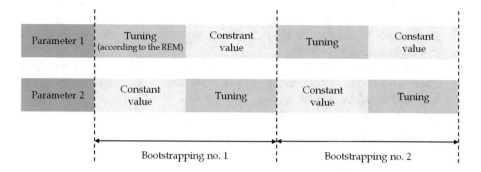

Fig. 2. Idea of the bootstrapping mechanism.

Remark 1. The values of N_c and N_{max} result from the adopted concept, in which the value of the ISE index is calculated during only the fragment of reference primitive, and the time resulting from $N_{max} - N_c$ is a time interval to decay transients after changing controller parameters in every single trajectory primitive.

The low computational complexity and reduction to the smallest number of iterations of the optimization algorithm (N_{total}) are crucial in tuning the altitude controller during the UAV flight, where the number of trajectory primitives to be carried out is limited by the type of power source (flight duration of multirotor UAV is usually only from a few to dozen minutes) [3]. The bootstrapping technique also allows the zero-order optimization algorithm to give a global character of searches, which is important if there are many local optimum of the minimized function.

4 Simulation Experiments

The results of five representative tests from numerous experiments conducted in MATLAB/Simulink 2015a environment for the quadrotor dynamics model from [14] are presented below. They illustrate the work and effectiveness of the proposed method of autotuning of the altitude controller in the structure from Sect. 2. The configuration parameters of each test, the obtained optimal controller gains and the corresponding ISE value are shown in Table 1.

In each test the task of the UAV model was to start the procedure of automatic tuning of the altitude controller parameters after obtaining the 3 m starting at the exemplary initial gains ($k_P = 150$ and $k_D = 2$). The space for searching for the controller's gains was created from $k_P \in <50, 150>$ and $k_D \in <0.1, 10>$. In each experiment, following parameters had to be set each time:

(a) flight time (t_h), in which the tuning and verification procedures were performed,

(b) number of bootstraps (N_b) during which the UAV flying in the X–Y plane periodically increased the altitude (0.5 m per single trajectory primitive as in Fig. 3) drawing a spiral,

(c) number of calculations for the cost function value in Fibonacci-search algorithm (N) equal to the number of primitive retries for each bootstrap (for a single, tuned parameter),

(d) duration time of a single reference primitive (t_{rp}),

(e) set value of the sampling interval (T_s),

(f) number of samples when performance index ISE is collected (N_c).

Table 1. Results of simulation experiments

Parameter	Test no. 1	Test no. 2	Test no. 3	Test no. 4	Test no. 5
t_h [sec/min]	430/7.17	190/3.17	230/3.83	270/4.5	115/1.92
ϵ	0.1	0.15	0.1	0.05	0.1
N_{total}	40	16	20	24	20
N	10	8	10	12	10
N_b	2	1	1	1	1
N_{Par}	2	2	2	2	2
t_{rp} [sec]	10	10	10	10	5
t_b [sec]	200	160	200	240	100
T_s [sec]	0.4	0.4	0.4	0.4	0.1
N_{max}	25	25	25	25	50
N_c	20	20	20	20	40
k_P	145.2381	95.1923	93.2234	91.0364	93.2234
k_D	0.8615	1.3375	0.8615	0.5714	0.8615
ISE	0.9268	1.5862	1.0341	0.8303	6.8082

Remark 2. From inequality:

$$F_{N+1} \geq \frac{1}{\epsilon} \tag{9}$$

the estimated accuracy ϵ in optimization procedure for the set value of N (see Table 1), can be calculated.

The next chapter contains comments to the results from Table 1. In Figs. 3–5 and in the video (https://youtu.be/_a8ZiCVl8Yo) the time courses for simulation experiments are also illustrated.

5   Conclusions

The main goal of the conducted research works was to check in simulation experiments does the proposed method has the potential to automatically optimize the altitude controller gains during the UAV flight, which is very time-limited. For this purpose, the results obtained from Test no. 1 and 3 (see Table 1) were compared. They were recorded for two and one bootstraps, respectively. For the analyzed case, it turns out that the increase in the ISE value by 11.58% allows one to shorten the tuning procedure by half (up to 200 s) and to 20 iterations (Fig. 4). The recorded time courses are very smooth in this case and have no

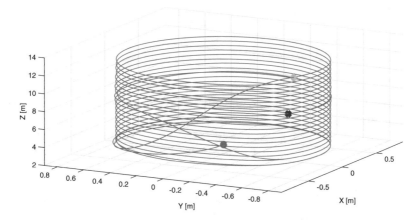

Fig. 3. Flight trajectory (Test no. 3): Start (red circle), end of the tuning process (yellow diamond), end of the flight (purple X).

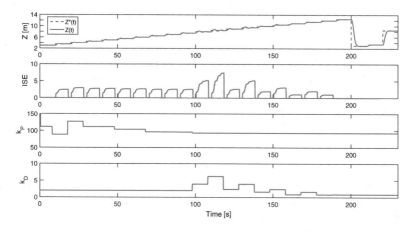

Fig. 4. Simulation experiment results (Test no. 3) – from the top: $Z^*(t)$ and $Z(t)$, ISE, k_P and k_D time courses.

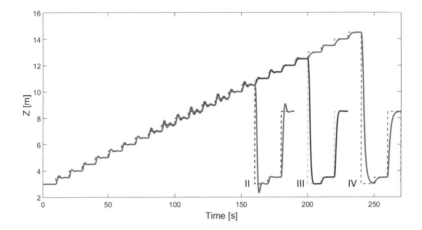

Fig. 5. $Z^*(t)$ and $Z(t)$ time courses comparison for Tests no. 2–4.

overshoot (Fig. 5). This is due to a smaller value of k_P, which for Tests no. 2–4 is very similar.

Comparative analysis of test results (tests no. 2–4) from Table 1 and Fig. 5, allows to assess how the number of iterations is correlated with the accuracy ϵ. Increase of the ISE value for Test no. 2 in comparison to Test no. 3 (as a result of overshoots emerging) (Fig. 5) confirms that the number N can not be too small.

In the last test, the effectiveness of the proposed method was tested with respect to the sampling frequency $f = 1/T_s$ and duration time of a single reference primitive t_{rp}. Comparing the results from Table 1 for Tests no. 3 and no. 5, it can be seen that the same optimal gains were obtained at a 4 times increased sampling frequency for a 2 times shorter duration time of a single reference primitive. As a result, after just 100 s the auto-tuning procedure has been completed.

In further research, the results from a real-world experiments will be investigated. First of all, it is important to verify how the in-flight tuning method can be used in a stochastic environment, how to minimize the impact of disturbances and use the full potential of available, simple on-board drone sensors to determine the ISE values.

References

1. Bonaccorso, G.: Mastering Machine Learning Algorithms. Expert Techniques to Implement Popular Machine Learning Algorithms and Fine-Tune Your Models. Packt Publishing, Birmingham (2018)
2. Mohammed, M., Khan, M.B., Bashier, M.B.E.: Machine Learning: Algorithms and Applications. CRC Press, Boca Raton (2016)
3. Valavanis, K., Vachtsevanos, G.V. (eds.): Handbook of Unmanned Aerial Vehicles. Springer, Heidelberg (2015)

4. Koch, W., Mancuso, R., West, R., Bestavros, A.: Reinforcement Learning for UAV Attitude Control. arXiv (2018). https://arxiv.org/abs/1804.04154
5. Ramirez-Atencia, C., Rodriguez-Fernandez, V., Gonzalez-Pardo, A., Camacho, D.: New artificial intelligence approaches for future UAV ground control stations. In: 2017 IEEE Congress on Evolutionary Computation (CEC), pp. 2775–2782. IEEE Press (2017). https://doi.org/10.1109/CEC.2017.7969645
6. Imanberdiyev, N., Fu, C., Kayacan, E., Chen, I-M.: Autonomous navigation of UAV by using real-time model-based reinforcement learning. In: 2016 14th International Conference on Control, Automation, Robotics and Vision (ICARCV), pp. 1–6. IEEE Press (2017). https://doi.org/10.1109/ICARCV.2016.7838739
7. Rodriguez-Ramos, A., Sampedro, C., Bavle, H., de la Puente, P., Campoy, P.: A deep reinforcement learning strategy for UAV autonomous landing on a moving platform. J. Intell. Robot. Syst. 1–16 (2018). https://doi.org/10.1007/s10846-018-0891-8
8. Berkenkamp, F., Schoellig, A.P., Krause, A.: Safe controller optimization for quadrotors with Gaussian processes. In: 2016 IEEE International Conference on Robotics and Automation (ICRA), pp. 491–496. IEEE Press, Sweden (2016). https://doi.org/10.1109/ICRA.2016.7487170
9. Muliadi, J., Kusumoputro, B.: Neural network control system of UAV altitude dynamics and its comparison with the PID control system. J. Adv. Transp. 1–18 (2018). https://doi.org/10.1155/2018/3823201
10. Carrio, A., Sampedro, C., Rodriguez-Ramos, A., Campoy, P.: A review of deep learning methods and applications for unmanned aerial vehicles. J. Sens. 1–13 (2017). https://doi.org/10.1155/2017/3296874
11. Giernacki, W., Fraire, T.E., Kozierski, P.: Cuttlefish optimization algorithm in autotuning of altitude controller of unmanned aerial vehicle (UAV). In: Third Iberian Robotics Conference. ROBOT 2017. AISC, vol. 693, pp. 841–852. Springer, Cham (2018). https://doi.org/10.1007/978-3-319-70833-1_68
12. Alikhani, H.: PID type iterative learning control with optimal variable coefficients. In: 2010 5th IEEE International Conference Intelligent Systems, pp. 479–484. IEEE Press, London (2010). https://doi.org/10.1109/IS.2010.5548329
13. Berkenkamp, F., Krause, A., Schoellig, A.: Bayesian optimization with safety constraints: safe and automatic parameter tuning in robotics. arXiv (2018). https://arxiv.org/pdf/1602.04450.pdf
14. Corke, P.: Robotics, Vision and Control: Fundamental Algorithms in MATLAB® Second, Completely Revised, Extended and Updated Edition (Springer Tracts in Advanced Robotics). Springer, Heidelberg (2017)
15. Giernacki, W., Horla, D., Espinoza Fraire, T.: Strategy for Optimal Autotuning of the Fixed-Wing UAV Controllers by the Use of Zero-order Optimization Algorithm (in review)
16. Horla, D.: Computational Methods in Optimization, 2nd edn, p. 358. Publishing House of Poznan University of Technology, Poznan (2016). (in Polish)
17. Kiefer, J.: Sequential minimax search for a maximum. Proc. Am. Math. Soc. 4(3), 502–506 (1953). https://doi.org/10.2307/2032161
18. Lewandowski, M.: Optimization methods – theory and selected algorithms. https://web.sgh.waw.pl/~mlewan1/Site/MO_files/mo_skrypt_21_12.pdf

Measuring Performance in Robotic Teleoperation Tasks with Virtual Reality Headgear

Mateusz Maciaś[(✉)], Adam Dąbrowski, Jan Fraś, Michał Karczewski,
Sławomir Puchalski, Sebastian Tabaka, and Piotr Jaroszek

Przemysłowy Instytut Automatyki i Pomiarów PIAP,
Al. Jerozolimskie 202, 02-486 Warsaw, Poland
{mmacias,adabrowski,jfras,spuchalski,pjaroszek}@piap.pl
http://www.piap.pl

Abstract. With the current rise of a new wave of Virtual Reality technologies, wider range of their applications are to be expected. Head Mounted Displays (HMDs) are starting to find use in some of human-machine interfaces for robots. We seek to measure and evaluate the impact that these devices can have on performance in standard robotic teleoperation tasks such as driving, observation and manipulation. In a study conducted with two real robots and standardized testing environment, including a total of 17 operators and 5 different tasks, an interface consisting of HMD and stereo vision module is compared to a more traditional one.

Keywords: Virtual reality · Headgear · Robotics · Telepresence ·
Teleoperation

1 Introduction

The concept of using Head Mounted Display (HMD) to achieve virtual experience had been known for a long time [1,6], with applications so far limited by technical shortcomings and the quality of experience. Currently, virtual reality headsets market is experiencing an unprecedented rise, which is expected to continue in the near future [3,16], marking 2016 as the year when the Virtual Reality (VR) really took hold [9]. With multiple models developed for consumer applications, the technology is driven to other fields as well.

HMDs can be used to achieve a high level of immersion in telepresence. In such applications, stereo video feed rather than a virtual reality environment is displayed on the headgear's screen. This telepresence technology applied to the field of human-machine interfaces provides a great potential to reshape the future of robot operation. With prices of headgear units dropping and display parameters improving, HMDs use for mobile robot teleoperation is becoming more feasible [4]. Virtual reality headsets can facilitate natural, immersive interfaces for tasks such as flexible grasping control [5]. Another study indicates the

© Springer Nature Switzerland AG 2020
R. Szewczyk et al. (Eds.): AUTOMATION 2019, AISC 920, pp. 408–417, 2020.
https://doi.org/10.1007/978-3-030-13273-6_39

need to prove that immersion pays off in telerobotics, while also describing an experiment where significant improvements in visual search task were observed after introducing stereo vision [13].

The aim of our work described in this paper was to test and evaluate the use of virtual reality headgears for teleoperation and telemanipulation tasks in robotics. Research described in this paper was conducted using Oculus Rift Development Kit 2 [7,8].

2   System Description

Our system consists of mobile robot and operator station. For scheduling reason two similar robots were used for tests: for manipulation tests we used PIAP Gryf [11], Fig. 1a, and for driving and observation PIAP Scout [12], Fig. 1c. The manipulator of Gryf robot, operated in telemanipulation tests, has five degrees of freedom plus the gripper. Both robots are equipped with pan/tilt/zoom camera controlled by operator and three fixed cameras at the gripper, robot's front and back. The arrangement of the cameras is same for both robots as presented in Fig. 1b. Both robots were tested in classic and stereovision configurations.

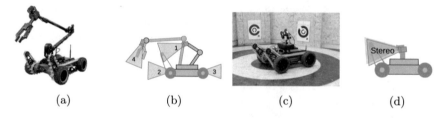

(a)   (b)   (c)   (d)

Fig. 1. PIAP Gryf in standard configuration, (a) actual robot, (b) standard camera arrangement: 1-PTZ camera, 2-front camera, 3-rear camera, 4-gripper camera. PIAP Scout in stereovision configuration as used in driving tests, (c) actual robot, (d) stereovision camera arrangement.

In classic configuration control system, also called Operator Control Unit (OCU), consists of a gaming pad [15], capable of controlling all degrees of movement, and a screen capable of displaying up to four video feeds simultaneously. The operator can arrange videos freely on the screen to adapt the interface for current task. One of cameras Pan/Tilt/Zoom camera (PTZ) that can be controlled by user. This configuration is shown in Fig. 2a.

In the stereovision configuration the robot was equipped with set of stereo cameras, mounted on pan/tilt/roll bracket assembly, see Fig. 1d. On the operator side, the interface included a HMD, capable of displaying video feeds from these stereo cameras. Head position tracking was used to control pan, tilt and roll of cameras. Other robot cameras were not used. Stereovision configuration for driving tests is shown in Fig. 1c and the one used for manipulation tests is shown in Fig. 6b. Operator is using same type of gaming pad as in the classic configuration. It is worth noting that while using the HMD, the operator could not see the pad. The configuration is shown in Fig. 2b.

(a)   (b)

Fig. 2. Operator performing test (a) with the classic configuration, (b) in the stereo-vision configuration.

The vision systems differs not only in camera arrangements and number of available observation points but also in the image parameters itself. The streams have different resolution, frame rate, field of view and transmission delay. That is important to note, that the field of view at the stereo-vision mode was fairly limited when compared to the default configuration due to cameras limitations. The images provided by both systems are presented in Fig. 3.

(a)   (b)

Fig. 3. Vision systems used for the tests: (a) screenshot from the operators tablet with front, PTZ and rear camera displayed, (b) Image displayed in Oculus headgear and controlled by operator's head movement.

3   Tests Procedures

The use of Human Machine Interfaces (HMI) by many different operators, multiple times, in several time-sensitive procedures leads to detecting and fixing issues with the current software and hardware, found both as errors and through user feedback. The quality of HMI is tested and improved. Ideas for improvements can be gathered and used to direct further development effort. Performance of operators with test cases can be compared, which informs us on difference between novice and expert operators, variance between attempts, and the learning curve.

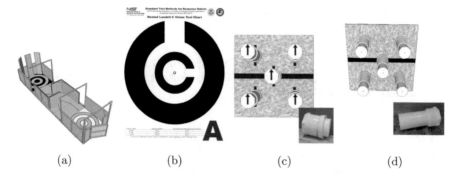

(a)   (b)   (c)   (d)

Fig. 4. Test apparatus used in tests: (a) visualization of test stand used in driving/following line procedure, (b) a marker used in observation tests, (c) cups used in turn objects tests, (d) cups set used in telemanipulation test. Source: NIST Apparatus Assembly Guide for Operator Training and Proficiency Evaluation.

For test apparatus, test stands from DHS-NIST-ASTM International Standard Test Methods [2] were used as a base. Visualisation of test stand in basic configuration is shown on Fig. 4a.

Since testing was focused on HMI and not on the robot itself, moving focus away from issues like platform connectivity, wireless connection quality and differences in video transfer protocols, we have used network cable to connect robot with control station, making it a bit different from the normal robot configuration. The gain was that disturbances did not affected the test outcome and needed not to be controlled for through other means such us increasing the number of test runs.

For observation tests, Landolt Tumbling "C" visual acuity charts were used as specified in the standard. All accessories can be seen in Fig. 4b.

First two proposed end executed test procedures (driving/following line and observation), were also prepared and executed first. We decided that manipulation tasks will be prepared after feedback from the first two procedures. After reviewing results and factoring in the operators comments, additional driving scenario has been added with a more complex terrain. This was due to the fact that in the first driving procedure time differences between test runs were minimal, which implied that they were too easy to complete. On top of performance measured as the time to complete each task, feedback from users that were participating in testing was noted and reported.

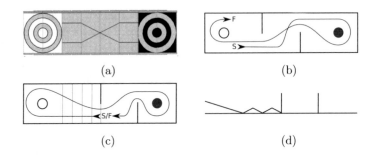

Fig. 5. Test stand layout, (a) top view, (b) path for procedure 1, (c) path for procedure 3, (d) ramps configuration for procedure 3 - side view. Start points are marked with letter S, finish positions with letter F.

- Procedure 1: Driving/following line - robots drives on test stand visualised on Figs. 4a and 5a, according to the path is shown in Fig. 5b.
- Procedure 2: Observation - in this procedure each operator has to read aloud letters on markers, shown on the left of the Fig. 4, in proper order. Operator is free to move the robot inside test stand as he sees fit. Some such movement is in fact required in order to see all the markers. The robot starts from the centre of white circles. Markers are shown on Fig. 4b.
- Procedure 3: Driving/slopes - objective of this test is to drive the robot along the path showed on Fig. 5c in the shortest time possible. This time, instead of going around circles, the robot had to go around poles that were placed in the centres of circles. Slopes were added to one side of test track, as shown on Fig. 6a.
- Procedure 4: Manipulation/turn objects - the task was to turn the lids by 90 degrees each. The lids had arrows printed on them, as shown on the Figs. 4c and 6c. Each operator executed the test only once, but involved a process of rotating five ids. This way the learning curve was already factored in the task.
- Procedure 5: Manipulation/place objects - the task was to cover all the tubes, shown on Fig. 4d, with lids eployed on the floor as shown on Fig. 6b. One lid was placed behind the test board, and the other two on each side of it.

4   Test Results

Operators are identified by numbers (from 0 to 16), and those numbers are coherent across all procedures, for example operator 4 was performing procedures 1, 2, 4 and 5 but not 3. This is due to operator availability.

Results for first driving procedure (line following) are shown in Fig. 7a. This is the only procedure for which teleoperation with stereovision gave consistently worse results than for classic configuration. The differences are slight though. Results for the second procedure, shown in Fig. 7b don't allow to declare one of

(a)   (b)   (c)

Fig. 6. Pictures taken during tests: (a) driving on slopes during test 2, (b) general view of robot placing object in manipulation tests, (c) view from robot gripper camera while rotating objects.

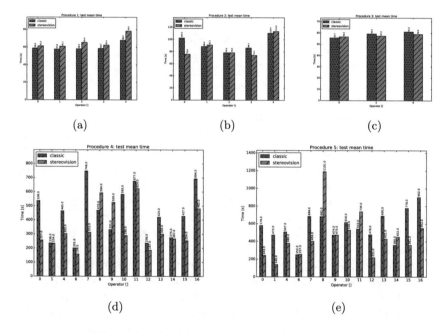

Fig. 7. Results for each procedures (from 1 to 5).

configurations superior, with some operators having better results with stereovision and others with classic configuration. Differences are also insignificant. The third procedure, involving driving with slopes, is also not showing huge results variation, and is not conclusive, as shown in Fig. 7c.

Manipulation results, shown in Figs. 7d and e, show a clear advantage of stereovision configuration. Almost all test runs show better results.

For every procedure, mean time required to complete was calculated and is shown in Fig. 8a and b. It was calculated together for all operators that were performing this procedure, separated into robot configurations.

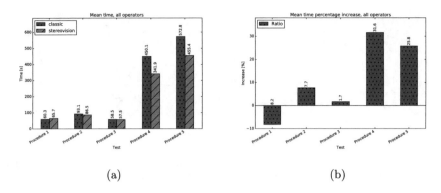

(a)   (b)

Fig. 8. All procedures (a) mean times, (b) mean percentage increase.

Percentage increase was calculated using formula:

$$\frac{t_{classic} - t_{stereovision}}{t_{stereovision}} \cdot 100\% \tag{1}$$

5   Analysis and Discussion

Several informative perspectives can be used to analyse the collected data. One apparent question is why results from driving and observation tests are similar for both configurations, while there is a substantial difference in results of manipulation tasks. A plausible explanation for this phenomena is that depth perception gained from the use of headgear and stereo camera plays an important role in the static and precise tests of manipulation, but is negligible for other tasks.

Human distance perception research [10] shows that in simulated object perception, when self-motion or relatively constant observed object motion were present, study participants were able to quite accurately estimate the distance. The human brain has a way of estimating the distance based on optic flow divergence and motion parallax. Another argument for why depth perception is not significant in driving tasks comes from experience and studies of monocular car drivers. The capability of such people to drive cars is well-known, and drivers need not to inform regulating bodies of their monocular vision. The study [18] examines one-eyed drivers in the context of car racing and concludes that after a period of adaptation the performance is surprisingly good, also noting that there are obvious deficiencies in the extent of the central and peripheral field due to a missing eye.

From our experience with two different field-of-view configurations and operators feedback, sufficient field of view is undoubtedly a more important factor than depth perception in driving and observation tests. A well-studied concept of useful field of view [14] for car drivers can be employed for determining the impact and optimal sensor and screen configurations.

In manipulation tasks, neither self-motion or object motion is present and depth perception is crucial for correct distance evaluation. We can confidently attribute the substantial performance increase of VR headset versus the standard configuration to stereo vision. The results agree fully with notes from observation of operators in action: with VR headset, operators quickly place the gripper over the object, showing that they can correctly perceive the distance. Operators with a single camera vision often struggled and tried to grasp the object while the gripper was positioned in front or behind it.

The other benefit that VR headset interface can provide is an intuitive camera control, as it is following the head motion. Feedback from several operators pointed out that it is superior to traditional control. The effect of this way of control on the measured performance on average was not noticed, possibly diminished or offset by other differences in compared configurations, such as the camera placement, field of view, stabilization or VR headset focus issues.

The operators who conducted the tests differ by experience level in robot teleoperation (operators 0, 8 and 15 were most experienced, and 2, 3 and 4 had some experience).

While in general the experience level was reflected in scores, there were outliers such as operators 1 and 6 excellent performance in manipulation procedures (procedures 4 and 5). With the scope and focus of our data, we are unable to reach statistically significant and definite conclusions on how the teleoperation experience affects performance in these tasks.

The use of HMD is know to have a potential for side effects. During the experiments, there were individual differences in the range and degree of discomfort reported by the operators using the VR headset. Out of 17 operators that took part in the research, one reported a significant degree of discomfort. It is important to note that constraints of the test sessions limited the continuous use of the HMD to the range of 10 min per operator, and that the study of side effects of VR headset was not in the focus of our research. These discomforts, however, certainly should be taken into consideration for HMD applications. The effects of long-term VR use are not yet well-studied. Among published results, there is a self-experimentation research [17] indicating that side effects such as disorientation and nausea can be substantial after extended HMD use.

6   Conclusions

Immersive configuration with headgear was shown to perform significantly (20–25%) better than default configuration in manipulation tasks, slightly better in observation tasks, and slightly worse in driving tasks. Both observation and driving results seemed to vary between operators depending on personal preference, while results from manipulation were uniformly better across different operators.

We have shown that the improvement in chosen teleoperation tasks is large enough to make virtual reality headgear a major consideration for human-robot interface designers, especially for competitive or operation time sensitive applications. It is advisable to test how the proposed design improves over alternatives, which can be done in a way similar (perhaps, scaled down) to the one described in this document.

Our work offers insight into factors influencing both the task performance and operator satisfaction. When designing a teleoperation system employing virtual reality headgear, it is important to provide a well-placed point of view, view angle that is wide enough, stereovision system that can be calibrated, and network capable of low latency. If there is a selection of operators who can work with the system, it is advisable to check their skills and satisfaction with the headgear interface, since our results indicate that there are significant individual differences.

In future tests, hardware should be improved to reduce reported issues that could have had negative impact on results. Better quality cameras, with specialised compression (or no compression at all) should be tested. Also mechanical structure of pan/tilt/roll assembly should be improved to have better angular resolution.

Acknowledgements. The authors gratefully acknowledge that part of the work presented was carried out within the European Research Programme ARTEMIS (Advanced Research and Technology for Embedded Intelligence and Systems), project R5-COP (Reconfigurable ROS-based Resilient Reasoning Robotic Co-operating Systems), co financed by Polish National Centre for Research and Development.

We are thankful to Artur Kaczmarczyk who helped with tests execution, ensuring all data is recorded properly and to our colleagues who participated in tests as operators, always trying to achieve best possible results.

References

1. Brooks Jr., F.P.: What's real about virtual reality? IEEE Comput. Graph. Appl. **19**(6), 16–27 (1999)
2. Jacoff, A., Messina, E., Huang, H.M., Virts, A., Norcross, A.D.R., Sheh, R.: Guide for evaluating, purchasing, and training with response robots using DHS-NIST-ASTM international standard test methods. Technical report, Intelligent Systems Division, Engineering Laboratory, National Institute of Standards and Technology (2009)
3. Gaudiosi, J.: Over 200 million VR headsets to be sold by 2020 (2016). http://fortune.com/2016/01/21/200-million-vr-headsets-2020
4. Kot, T., Novák, P.: Utilization of the oculus rift HMD in mobile robot teleoperation. In: Applied Mechanics and Materials, vol. 555, pp. 199–208. Trans Tech Publication (2014)
5. Krupke, D., Einig, L., Langbehn, E., Zhang, J., Steinicke, F.: Immersive remote grasping: realtime gripper control by a heterogenous robot control system. In: Proceedings of the Symposium on Virtual Reality and Software Technology (VRST) (Poster Presentation), November 2016
6. Morton, H.L.: Stereoscopic-television apparatus for individual use. US Patent 2,955,156 (1960)
7. Oculus VR, LLC. Oculus rift DK2 documentation (2016). http://developer.oculus.com/documentation
8. Oculus VR, LLC. Oculus rift website (2016). http://www.oculus.com

9. Lamkin, P.: HTC vive VR headset sales revealed (2016). http://www.forbes.com/sites/paullamkin/2016/10/21/htc-vive-vr-headset-sales-revealed
10. Peh, C.-H., Panerai, F., Droulez, J., Cornilleau-Pérès, V., Cheong, L.-F.: Absolute distance perception during in-depth head movement: calibrating optic flow with extra-retinal information. Vis. Res. **42**(16), 1991–2003 (2002)
11. PIAP GRYF product website. PIAP Gryf® product webpage (2017). http://www.antiterrorism.eu/portfolio-posts/piap-gryf/
12. PIAP Scout product website. PIAP Scout® product webpage (2017). http://www.antiterrorism.eu/portfolio-posts/piap-scout/
13. Rodrigues, F.A.C.: Immersive telerobotic modular framework using stereoscopic HNDs (2015)
14. Seya, Y., Nakayasu, H., Yagi, T.: Useful field of view in simulated driving: reaction times and eye movements of drivers. i-Perception **4**, 285–298 (2013)
15. Sony. Sony Dualshock ii controller manual (2017). https://www.playstation.com/manual/pdf/scph-10010u.pdf
16. Charara, S.: Sony could sell 1.4 million playstation VR headsets in less than 3 months (2016). https://www.wareable.com/sony/sony-playstation-vr-headset-sales-estimates-2016
17. Steinicke, F., Bruder, G.: A self-experimentation report about long-term use of fully-immersive technology. In: Proceedings of the ACM Symposium on Spatial User Interaction (SUI), pp. 66–69. ACM Press (2014)
18. Westlake, W.: Is a one eyed racing driver safe to compete? formula one (eye) or two? Br. J. Ophthalmol. **85**, 619–624 (2001)

Construction and Preliminary Testing of the Force Feedback Device for Use in Industrial Robot Control Based on the BCI Hybrid Interface

Arkadiusz Kubacki[✉], Tymoteusz Lindner,
and Arkadiusz Jakubowski

Institute of Mechanical Technology, Poznan University of Technology,
ul. Piotrowo 3, 60-965 Poznań, Poland
{arkadiusz.kubacki,tymoteusz.lindner,
arkadiusz.jakubowski}@put.poznan.pl

Abstract. The article describes design process of building force feedback device for use in hybrid brain-computer interface based on Electrooculography (EOG) and center eye tracking. In first paragraph authors presented information about built hybrid Brain-Computer Interface (BCI). The interface was built with used of the bioactive sensors mounted on the head. Research on both the model and the industrial robot has been described In second paragraph presented construction and test of force feedback device. The authors checked the proportionality of the input force to the output one. They also conducted research on the positioning of the robot's tip with both on and off force feedback.

Keywords: Force feedback · Electrooculography · EOG ·
Brain-computer interface · BCI

1 Introduction

Nowadays, brain-computer interfaces are becoming more and more widespread. Brain-computer interfaces (BCI) can recognize some activities directly from the brain [1]. EEG is developing rapidly thanks to decreasing price of the headsets [2]. EEG is a non-invasive method used to record the activity of the brain from the skull through the electrodes. Number of possible commands and the effectiveness of their detection are limited by using only one method of BCI. Therefore, more and more often there are used hybrid BCI [3]. Their classification is shown in Fig. 1. Thanks to using more interfaces is to increase the number of the classification commands. Actually more and more often BCI is also linked with other interfaces such as speech recognition or electromyography (EMG) [3]. There are many ways to connect interfaces. It is possible to connect steady-state visual evoked potential (SSVEP) into the conventional P300 event-related potential [4], electromyographic (EMG) with electroencephalographic (EEG) activity [5], SSVEP with ERD/ERS method [6, 7].

© Springer Nature Switzerland AG 2020
R. Szewczyk et al. (Eds.): AUTOMATION 2019, AISC 920, pp. 418–427, 2020.
https://doi.org/10.1007/978-3-030-13273-6_40

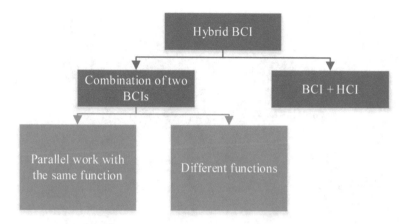

Fig. 1. Classification of hybrid BCIs [3].

2   Hybrid Brain-Computer Interface

It is worth noting that the following article is a continuation of research on the hybrid BCI and part of the research was not presented again. In the article [8], the authors presented research on BCI based on the EOG. In the next article [9] the authors introduced the vision system and carried out additional research. The results were compared with the results from the article [10]. The authors concluded that the use of the vision system significantly improved the results. Additionally, such a system is immune to many interferences thanks to the synergistic connection. Thanks to the camera, the authors have eliminated artifacts in the EEG signal originating from e.g. blinking eyes. On the other hand, the vision system cannot work when the eyes are closed. The EOG makes it possible. The built-in hybrid BCI interface was adapted to the virtual robot model and tested in the article [11]. After positive tests, the interface was tested on a real industrial robot, and the results were presented by the authors in the article [10]. The authors of the article have built a hybrid BCI system based on Electrooculography (EOG) and a vision system. With it, they controlled the movement of the virtual robot's tip. Movement of the eyeballs to the right resulted in moving the robot's tip in the positive direction. Movement of the eyeballs to the left resulted in movement of the robot tip in the negative direction. Double-closed eyes changed the selected axis. In the Unity 3D environment, a mathematical model of the robot inverse kinematics and the actual appearance of the Mitsubishi RV-12sl robot has been implemented (Fig. 2).

The subjects were supposed to transport the virtual block from one table to another in such a way that they would avoid the obstacle. For the obstacle, the authors chose the virtual wall. Touching the wall with a block or any of the robot's elements resulted in the interruption of the attempt. The study involved three people aged from 20 to 35 years. Each of the subjects carried out 10 tests. Satisfactory results were obtained. About 80% of the trials were successful. Figure 3 shows the trajectory of movement of the robot tip for selected trials (Fig. 4).

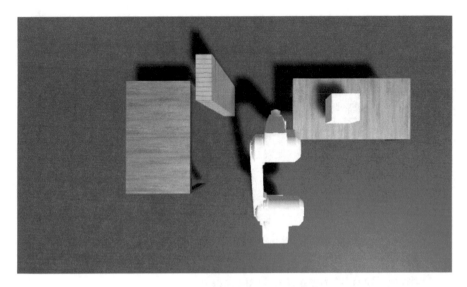

Fig. 2. Robot model Mitsubishi RV-12sl.

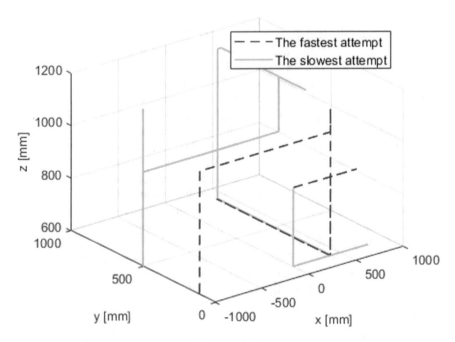

Fig. 3. Movement of the end point of virtual robotic arm.

After obtaining the desired effects, the authors adapted the interface to the real Mitsubishi RV-12sl industrial robot. The subjects were tasked with overturning the upper carton in such a way as to bypass the cardboard posts. In addition, the lower carton could not tip over. The fall of the post or lower carton resulted in the discontinuation of the trial. The way of moving the robot has not changed.

The study involved three people aged from 20 to 35 years. Each of the subjects carried out 10 tests. Satisfactory results were obtained. About 90% of the trials were successful. Figure 5 shows the trajectory of movement of the robot tip for selected trails.

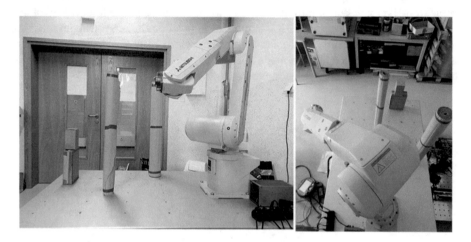

Fig. 4. Test with a real industrial robot.

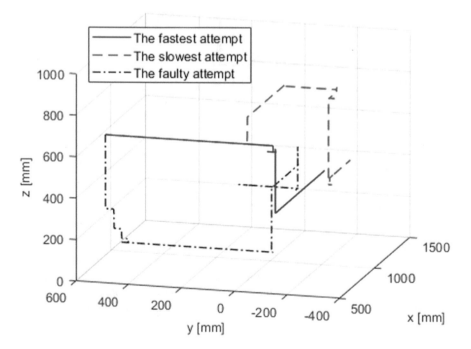

Fig. 5. Movement of the end point of virtual robotic arm.

3   Construction and Testing of the Force Feedback Device

Due to the fact that the tip of the robot is controlled by means of eyeballs, it is not always in the field of view of the subject. Errors associated with this are particularly evident when manipulating the tip of the robot in close proximity to the obstacle. The authors decided to introduce a force feedback. The purpose of the device is to create a sensation of force proportional to the signal from the tip of the robot. This signal can come from the force sensor if for the test it is important to recognize force at the robot's tip or from the distance sensor if it is important to position the robot without touching the surrounding elements. The microcontroller collects data from the load cell or distance sensor mounted on the robot's tip. For this purpose, a prototype of a device for force feedback on the arm was designed and built (Figs. 6 and 7).

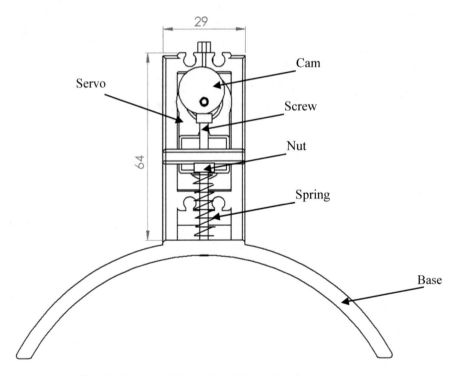

Fig. 6. Diagram of the device with the described components.

The method of mounting the device on arm is shown in Fig. 8. Proportionally to the input signal at the tip of the robot, the rotation angle at the servo is set. The servo engages the cam, which moves the element that exerts pressure on the skin. The return is realized by means of a spring. A study was conducted on the proportionality between the angle of rotation of the servo and the force exerted. The results are shown in Fig. 9.

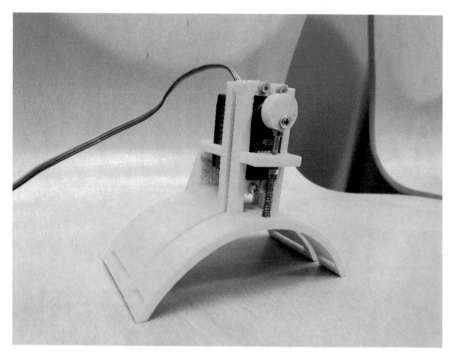

Fig. 7. Movement of the end point of virtual robotic arm.

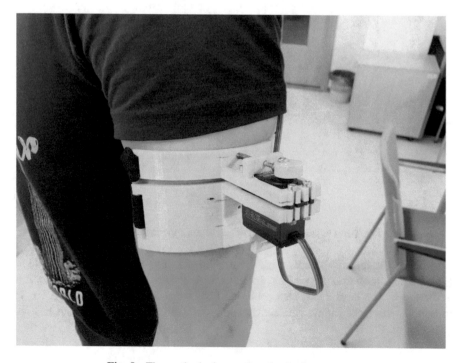

Fig. 8. The method of mounting the device on arm.

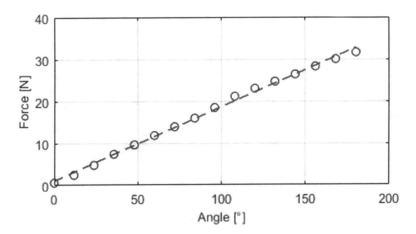

Fig. 9. Force graph depending on the angle.

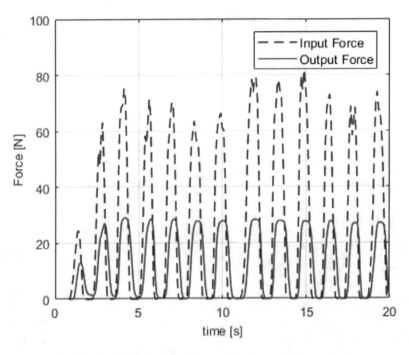

Fig. 10. Graph of the dependence of the input and output force.

Very good linearity results have been achieved. The R^2 coefficient was 0.9967 (Fig. 9). The proportionality of the force exerted by the tip of the robot to the force generated by the device was also checked. The results are presented in Fig. 10. The time delay that was achieved in the construction of this device is 80 ms. It has been presented in Fig. 11.

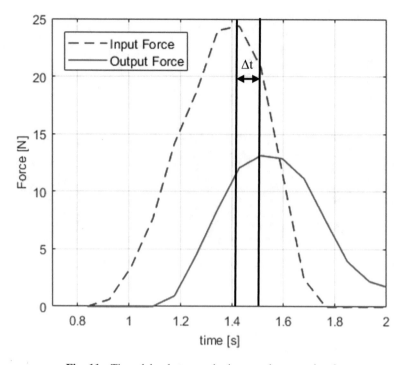

Fig. 11. Time delay between the input and output signal.

4   Research of the Force Feedback Device in the Hybrid Brain-Computer Interface System

The device described above has been implemented in the already existing BCI system described in the article [9]. In the Unity 3D environment, a robot with inverse kinematics was modeled. Additionally, a force sensor was simulated at its end, which sent information to the feedback system. The view of the camera positioned in such a way that the examined person could not determine the distance between the robot's tip and the object (Fig. 10). The subject's task was to move the end of the robot as close to the virtual block as possible without touching it. Three people aged 25 to 35 participated in the study. Each of the subjects tested from 5 to 10 attempts. The output force was proportional to the distance between the robot tip and the block (Figs. 12 and 13).

In the last phase of the movement, the block completely covers the robot's tip. Two tests were performed. In the first scenario the feedback loop was turned off, and the second one was provided with feedback. In both cases, the tests were carried out with an accuracy of 1 mm. The results obtained are presented in Table 1.

Almost four times improved results. In addition, the results are presented in Fig. 11.

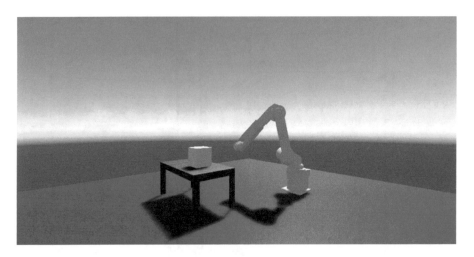

Fig. 12. Simulation view to test force feedback.

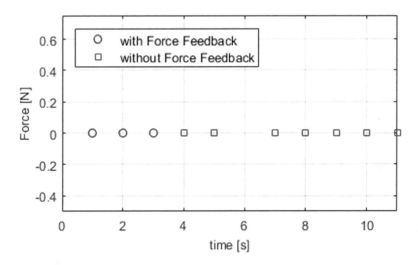

Fig. 13. Distance between the robot tip and the target depending on the force feedback on or off.

Table 1. Results of the robot's tip distance from the object.

	Without force feedback	With force feedback
Minimum distance	4 mm	1 mm
Maximum distance	11 mm	3 mm
Arithmetic average	8 mm	2.27 mm
Standard deviation	2.26 mm	0.65 mm

5 Conclusion

The article describes the design process and research of the created force feedback device for use in hybrid brain-based interface based on Electrooculography (EOG) and center eye tracking. Authors has proved that it is able to obtain satisfactory results from the Emotiv EPOC + headset and standard web camera. The investigations have shown that the use of feedback in such an interface improves the positioning quality of the robot arm and can enable more precise manipulation of objects in the future. In subsequent studies, the authors want to use the described device for controlling a real industrial robot. The next stage will be the extension of the hybrid brain-computer system with an additional system based on SSVEP.

Acknowledgement. The work described in this paper was funded from 02/23/DS-PB/1434.

References

1. Kubacki, A., Jakubowski, A., Sawicki, L.: Detection of artefacts from the motion of the eyelids created during EEG research using artificial neural network. Adv. Intell. Syst. Comput. **440**, 267–275 (2016)
2. O'Regan, S., Faul, S., Marnane, W.: Automatic detection of EEG artefacts arising from head movements. In: 2010 Annual International Conference of the IEEE Engineering in Medicine and Biology Society (EMBC), pp. 6353–56 (2010)
3. Cudo, A., Zabielska, E., Bałaj, B.: Wprowadzenie w zagadnienie interfejsów mózg-komputer. Wydawnictwo KUL (2011)
4. Yin, E., Zhou, Z., Jiang, J., Chen, F., Liu, Y., Hu, D.: A novel hybrid BCI speller based on the incorporation of SSVEP into the P300 paradigm. J. Neural Eng. **10**(2), 026012 (2013)
5. Leeb, R., Sagha, H., Chavarriaga, R., del R. Millán, J.: Multimodal fusion of muscle and brain signals for a hybrid-BCI. In: 2010 Annual International Conference of the IEEE Engineering in Medicine and Biology, pp. 4343–4346 (2010)
6. Pfurtscheller, G., Solis-Escalante, T., Ortner, R., Linortner, P., Muller-Putz, G.R.: Self-Paced operation of an SSVEP-based orthosis with and without an imagery-based brain switch; a feasibility study towards a hybrid BCI. IEEE Trans. Neural Syst. Rehabil. Eng. **18**(4), 409–414 (2010)
7. Allison, B.Z., Brunner, C., Kaiser, V., Müller-Putz, G.R., Neuper, C., Pfurtscheller, G.: Toward a hybrid brain–computer interface based on imagined movement and visual attention. J. Neural Eng. **7**(2), 026007 (2010)
8. Kubacki, A., Owczarek, P., Lindner, T.: Use of Electrooculography (EOG) and facial expressions as part of the Brain-Computer Interface (BCI) for controlling an electric DC motor. In: Automation 2018, pp. 82–92 (2018)
9. Kubacki, A.: Hybrid Brain-Computer Interface (BCI) based on Electrooculography (EOG) and center eye tracking. In: Automation 2018, pp. 288–297 (2018)
10. Kubacki, A., Jakubowski, A.: Controlling the industrial robot model with the hybrid BCI based on EOG and eye tracking. In: ITM Web Conference (2018)
11. Kubacki, A., Jakubowski, A.: Controlling the industrial robot model with the hybrid BCI based on EOG and eye tracking. In: AIP Conference Proceedings, vol. 2029, no. 1, p. 020032, October 2018

Lane Finding for Autonomous Driving

Łukasz Sztyber[✉]

Systems Research Institute, Polish Academy of Sciences, Warsaw, Poland
lukasz.sztyber@ibspan.waw.pl

Abstract. The problem of lane finding is one of the main components of scene understanding for autonomous driving. This paper presents the application of computer vision to lane finding on motorways. This technique is used to transform an image captured by a camera into binary form and apply a moving histogram window to derive the most likely position of the lane markers. It is proved to be fast enough to operate in real-time conditions.

Keywords: Autonomous driving · Lane finding · Computer vision ·
Sobel filter · Sliding histogram window technique

1 Introduction

Autonomous driving deals with the development of vehicles capable of recognizing the surrounding environment and navigating without human input. The significant increase in the number of cars in use results in a rising number of car accidents involving injuries and fatalities. Reports indicate [1] that a major cause of car accidents is human error. Therefore, the development of Advanced Driving Assistance Systems (ADAS) may significantly reduce the number of car accidents and improve driving safety.

The main tasks in autonomous driving can be placed in one of three categories: Sensing, Perception or Decision-Making [2] (Fig. 1). Computer vision is one of the main tools used to detect the details of the AV's surrounding environment. Camera-based systems are used to distinguish lane markings on the road. Lane detection is crucial to local trajectory planning and lane-changing decisions. There is a substantial body of lane detection research [3–8], and a wide variety of algorithms have been proposed, including various representations (including fixed-width line pairs, spline ribbon, and deformable-template model), detection and tracking techniques (from Hough transform to probabilistic fitting and Kalman filtering), and modalities (stereo or monocular).

In [3] Wang et al. propose similar solution to one presented in this article but instead of histograms they use longest straight lines within the horizontal segments with the height of segments decreasing bottom up. In [4] Ieng et al. present algorithm based on the image line level with analysis of gradient change. Another interesting approach is presented in [5] by Taylor et al. They demonstrate an algorithm based on the images from stereo camera. Each image is analyzed separately and we look for local maxima (line by line) and after that we compute the correlation between two images. A slightly different approach is described in [6] by Broggi. It also analyze image line by line but it looks for change in the brightness of the points on black and white input image and it validates candidate points by computing horizontal distance between the points.

© Springer Nature Switzerland AG 2020
R. Szewczyk et al. (Eds.): AUTOMATION 2019, AISC 920, pp. 428–444, 2020.
https://doi.org/10.1007/978-3-030-13273-6_41

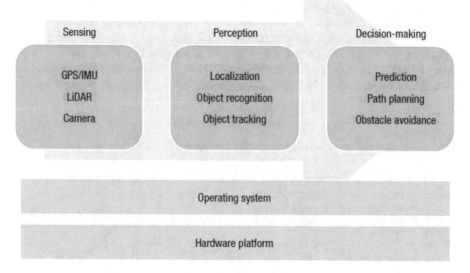

Fig. 1. Main flow of tasks in autonomous driving [2].

Detecting lane marking from information collected through visual sensors is a critical step and primary requirement to support driver assistance and automated driving safety features like Lane Departure Warning (LDW), Lane Keep Assist (LKA), Forward Collision Alert (FCA) among others [7–10]. As these features are directly associated with pedestrian and driver safety, and are not merely a nice-to-have feature, lane detection algorithm with accuracy close to 100% is absolutely essential.

The main goal of this paper is to draw up a methodology that could be applied to series of raw images captured by an RGB video camera to detect lane markings in real time with high accuracy regardless of weather and lighting conditions. There is a number of potential challenges that need to be addressed. Images coming from most types of modern cameras are distorted. Lane markings can be encountered in various shades of white or yellow, and can be solid, dashed or dotted lines. Their visibility can be significantly affected by the weather conditions such as: clear sky, clouds, rain, snow and lighting conditions i.e. day, night, shadows, artificial light. On top of that, it's not uncommon to encounter irregular patterns on the asphalt that can be the result of some road works.

This paper presents an effective technique for lane detection based on conventional computer vision methods and algorithms. Section 2 formulates the lane finding problem and describes the algorithm chosen to tackle it. Details of this algorithm, implementation and numerical examples are provided in Sect. 3. Section 4 provides a summary and final outlook.

2 Lane Finding Algorithm

The algorithm identifying the location of the lane marks based on the images acquired by car's camera consists of several steps including the smoothing operation, transformation to a bird's eye view perspective, the detection of lane edges using Canny algorithm, the use of histogram window method to identify lane pixels and then a line fitting algorithm to calculate the location of mark lanes. The main steps of the proposed algorithm are summarized in Fig. 2 and discussed in the subsequent paragraphs.

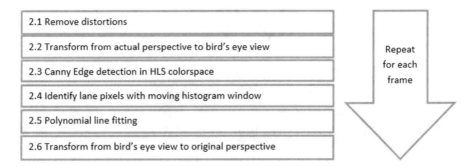

Fig. 2. Overview of the applied methodology.

2.1   Removing Distortions

Modern cameras use curved lenses to form an image. The light rays often bend too much or too little at the edges of these lenses. This creates an effect that distorts the edges of images, so that lines or objects appear more or less curved than they actually are (see Fig. 3). This is called radial distortion, and it's the most common type of distortion. Another type of distortion is tangential distortion, which occurs when a camera's lens is not aligned perfectly parallel to the imaging plane, where the camera film or sensor is. This makes an image look tilted so that some objects appear farther away or closer than they actually are. Fortunately, the relationship between the distorted image and the curvature of photographed objects is constant and can be corrected by applying calibration and remapping.

Let $P = P(x, y)$ denote a point of the distorted image having coordinates (x, y) where $x \in [X_{min}, X_{max}]$ and $y \in [Y_{min}, Y_{max}]$ and X_{min}, X_{max}, Y_{min}, Y_{max} are given real constants. By r we denote the known distance between a point (x_c, y_c) in an undistorted, corrected image and the center (x_c, y_c) of the image distortion. The latter point is often the center of the image. Let us denote the radial distortion coefficients by k_1, k_2, k_3. These coefficients are real given constants. Similarly, the constants p_1, p_2 will denote tangential distortion coefficients. There are three real coefficients required to correct the radial distortion: $k_1, k_2,$ and k_3. To correct the appearance of radially distorted points in an image, one can use a correction formula. The distortion coefficient k_3 is required to accurately reflect major radial distortion such as that which occurs in wide-angle lenses. However, for minor radial distortion, which most regular

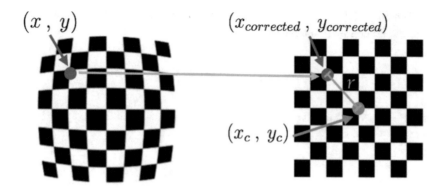

Fig. 3. Point mapping from a distorted to an undistorted (corrected) image.

camera lenses have, k_3 has a value close to or equal to zero and is negligible. This coefficient can be ignored; this is why it appears at the end of the distortion values vector: $[k_1, k_2, p_1, p_2, k_3]$. We will use it in all calibration calculations so that our calculations apply to a wider variety of lenses and can correct for both minor and major radial distortion. Radial distortion correction is governed by the relations:

$$x_{radial} = x\left(1 + k_1 r^2 + k_2 r^4 + k_3 r^6\right) \tag{1}$$

$$y_{radial} = y\left(1 + k_1 r^2 + k_2 r^4 + k_3 r^6\right) \tag{2}$$

The tangential distortion correction formula is given by:

$$x_{tangential} = x + \left[2p_1 xy + p_2\left(r^2 + 2x^2\right)\right] \tag{3}$$

$$y_{tangential} = y + \left[p_1\left(r^2 + 2y^2\right) + 2p_2 xy\right] \tag{4}$$

Summing (1)–(2) and (3)–(4) yields:

$$x_{corrected} = x\left(1 + k_1 r^2 + k_2 r^4 + k_3 r^6\right) + \left[2p_1 xy + p_2\left(r^2 + 2x^2\right)\right] \tag{5}$$

$$y_{corrected} = y\left(1 + k_1 r^2 + k_2 r^4 + k_3 r^6\right) + \left[p_1\left(r^2 + 2y^2\right) + 2p_2 xy\right] \tag{6}$$

To find parameters k_1, k_2, k_3, p_1, p_2 we need the camera to observe a planar pattern shown at a minimum of two different orientations. This calculation is based on the algorithm proposed by Zhang in [11].

2.2 Removing Distortions

Before computing the histograms we make warp transformation in order to get from the actual perspective image (farther points get skewed towards the center of an image) to a bird's-eye view (parallel lane lines). In order to do that we use The Direct Linear

Transformation (DLT) algorithm is described in [12] and [13]. DLT algorithm is used to find the homography matrix H:

$$c \begin{pmatrix} u \\ v \\ 1 \end{pmatrix} = H \begin{pmatrix} x \\ y \\ 1 \end{pmatrix} \tag{7}$$

where x' = (u, v) is a target point corresponding to original point x = (x, y), c is a non-zero constant and H is a 3 by 3 real matrix given as:

$$H = \begin{pmatrix} h_1 & h_2 & h_3 \\ h_4 & h_5 & h_6 \\ h_7 & h_8 & h_9 \end{pmatrix} \tag{8}$$

From (7) it follows:

$$-h_1 x - h_2 y - h_3 + (h_7 x + h_8 y + h_9)u = 0 \tag{9}$$

$$-h_4 x - h_5 y - h_6 + (h_7 x + h_8 y + h_9)u = 0 \tag{10}$$

Equations (9) and (10) can be written in a matrix format (11):

$$A_i h = 0 \tag{11}$$

where the matrix A_i, i = 1,...,4, and the vector h, are given by:

$$A_i h = \begin{pmatrix} -x_i & -y_i & -1 & 0 & 0 & 0 & u_i x_i & u_i y_i & u_i \\ 0 & 0 & 0 & -x_i & -y_i & -1 & v_i x_i & v_i y_i & v_i \end{pmatrix} \tag{12}$$

$$h = \begin{pmatrix} h_1 & h_2 & h_3 & h_4 & h_5 & h_6 & h_7 & h_8 & h_9 \end{pmatrix}^T. \tag{13}$$

For every two-point correspondence (x, x') there are two equations. Therefore, four corresponding points are sufficient to calculate H for the 8° of freedom.

2.3   Canny Edge Detection in HLS Color Space

In a perfect situation, in which the color of the lane markings, were ideally consistent we could rely on RGB color space and easily obtain marking lines by filtering out all other colors. However, in practice they can be observed in all possible shades of white or yellow. Therefore, to find them we need to rely on more advanced image features than simple RGB intensity. To address this problem we will use HLS color space which represents colors in 3 dimensions such as Hue (H), Saturation (S) and Lightness (L). We convert the RGB image into HLS. First we define the following constants [14]:

$$V_{max} \leftarrow \max\{R, G, B\} \text{ and } V_{min} \leftarrow \min\{R, G, B\} \tag{14}$$

These are the maximum and minimum values across all three RGB values for a given color. Consider first H channel conversion equations: there are three different equations; which one is used depends on the value of V_{max} and whether it is R, G, or B.

$$H \leftarrow 30(G - B)/(V_{max} - V_{min}), \text{ if } V_{max} = R \tag{15}$$

$$H \leftarrow 60 + 30(B - R)/(V_{max} - V_{min}), \text{ if } V_{max} = G \tag{16}$$

$$H \leftarrow 120 + 30(R - G)/(V_{max} - V_{min}), \text{ if } V_{max} = B \tag{17}$$

The L and S channel conversion equations take the forms:

$$L \leftarrow (V_{max} + V_{min})/2, \tag{18}$$

and

$$S \leftarrow (V_{max} - V_{min})/(V_{max} + V_{min}), \text{ if } L < 0.5, \tag{19}$$

$$S \leftarrow (V_{max} - V_{min})/(2 - V_{max} + V_{min}), \text{ if } L \geq 0.5. \tag{20}$$

Canny Edge Detection Algorithm

Canny edge detection is a technique based on functional minimization to extract useful structural information from different visible objects. Among the edge detection methods developed so far, Canny edge detection algorithm is one of the most strictly defined methods. and provides good and reliable detection. Let us briefly recall from [15] this algorithm. It consists of five main steps:

Step 1: Smoothing: blurring of the image to remove noise.

In order to smooth the image and reduce the effects of obvious noise on the edge detector, a Gaussian filter is applied. The equation for a Gaussian filter kernel of size $(2k + 1) \times (2k + 1)$ is given by:

$$H_{ij} = \frac{1}{2\pi\sigma^2} \exp\left(-\frac{(i - (k+1))^2 + (j - (k+1))^2}{2\sigma^2}\right) \tag{21}$$

for 1 <= i, j <= (2k + 1)

Step 2: Finding gradients: the edges should be marked where the gradients of the image have large magnitudes.

For image function $I(x, y)$ its gradient G with respect to x and y is denoted by $G = [G_x(x, y), G_y(x, y)]$. The angle θ between vectors G_x and G_y is calculated using function arctangent with two arguments denoted as atan2. This gradient is calculated using the Sobel kernel.

$$G = \sqrt{G_x^2 + G_y^2} \qquad \theta = atan2(G_y, G_x) \tag{22}$$

The edge direction angle is rounded to one of four angles representing vertical, horizontal and the two diagonals $\{0°, 45°, 90°, 135°\}$.

Step 3: Non-maximum suppression: Only local maxima should be marked as edges.

Non-maximum suppression is an edge thinning technique used to suppress all the gradient values (by setting them to 0) except the local maxima, which indicate locations with the sharpest change of intensity value. The algorithm for each pixel in the gradient image is:

- Compare the edge strength of the current pixel with the edge strength of the pixel in the positive and negative gradient directions.
- If the edge strength of the current pixel is larger than the other pixels in the mask with the same direction (i.e., the pixel that is pointing in the y-direction, it will be compared to the pixel above and below it in the vertical axis), the value will be preserved. Otherwise, it will be suppressed.

Step 4: Double thresholding: Potential edges are determined by thresholding.

After non-maximum suppression has been applied, the remaining edge pixels provide a more accurate representation of the real edges in an image. However, some edge pixels remain as the result of noise and color variation. These edge pixels can be filtered out with a weak gradient value while preserving edge pixels with a high gradient value. This is accomplished by selecting high and low threshold values. If an edge pixel's gradient is higher than the high threshold, it is marked as a strong edge pixel. If it is smaller than the high threshold and larger than the low threshold, it is marked as a weak edge pixel. If an edge pixel's value is smaller than the low threshold value, it will be suppressed. The two threshold values are empirically determined and their definition will depend on the content of a given input image.

Step 5: Edge tracking by hysteresis.

The strong edge pixels should certainly be involved in the final edge image, as they are extracted from the true edges in the image. However, so-called weak edge pixels can be extracted from either the true edge or the noise/color variations. To achieve an accurate result, the weak edges caused by noise/color variations should be removed. Usually a weak edge pixel caused from true edges will be connected to a strong edge pixel while noise responses are unconnected. To track the edge connection, blob analysis is applied by looking at a weak edge pixel and its 8-connected neighborhood pixels. As long as there is one strong edge pixel that is involved in the blob, the weak edge point can be identified as one that should be preserved.

2.4   Identify Lane Pixels with Moving Histogram Window

At this stage we can apply a histogram technique to identify road surface markers. Using a binary image after applying gradient transformation we can identify pixels that comprise lane lines. For that purpose we start analysis of the image from the bottom and focus on a horizontal window of a given height (i.e. 100 pixels) and covering the full width. Within that window we compute histograms for each horizontal position (each x value) and assume that two points for which the value of histograms are the highest belong to the lane lines (left and right). Then we slide the window towards the top of the input image and repeat the calculation of the histograms and store the coordinates of each pair of points where the histograms have the biggest values.

2.5   Polynomial Line Fitting

Based on 2 vectors of points, one for each lane line, representing pairs of points (x) at which histograms have the highest values, we compute the best polynomial fit of the second degree $y = Ax^2 + Bx + C$ for 2 curves going through these points. The least square method is used to approximate the lanes. Having the best polynomial fits representing lane lines we clearly know where the lane markings lie, which area belongs to our lane and which is outside of it. We can apply this knowledge to decision-making i.e. with regard to local path planning to optimize the car's path versus center of the lane.

2.6   Transform from Bird's Eye to Original Perspective for Visualization

Though we have full knowledge about detected lanes, some cases might require visualizations on the original input image. For that purpose we need to perform inverse transformation applied to a mask representing our lane and get from birds-eye view to original perspective and merge it with the original input image. To do this, we apply inverse warp transformation using the same mapping matrix described in Subsect. 2.2.

3   Numerical Implementation and Results

The algorithm described in Fig. 2 has been used to identify the location of lane marks. First the distortion parameters in Eqs. (1)–(6) have been calculated. Using the 10 pictures of chessboards we obtained the following distortion parameters: k1 = –0.24688572,   k2 = –0.02372824,   k3 = –0.00109832,   p1 = 0.00035104, p2 = –0.00260433. These parameters can be calculated only once and they will be used for removing distortions from each frame. Next we experiment with the Canny Edge detection algorithm in various channels of various color spaces i.e. RGB, HLS. We look for binary filtering method that would allow us to distinguish lane markings and remove all the noise caused by other patterns on the road. Figures 4, 5, 6 and 7 present the effect of thresholding applied to various channels in RGB and HLS color space.

As the above images show, we get the best result with the least noise on the road surface and clearly exposed lane marking in channel S of HLS color space (see Fig. 7). This proves that regardless of misleading or irrelevant patterns on the road, the best result can be obtained by applying thresholding to Saturation (S) channel in HLS color space. Therefore in our case examples we will search for line markers that rely strictly on Saturation (S) channel. This would allow us to make the solution regardless of the color (white or yellow) of the lane markers.

The warp transformation H from the actual perspective to bird's eye view has been calculated using the frame presented in Fig. 8a. We arbitrarily select 4 points in the perspective view: left bottom point of sample lane marking, left top, right bottom point for the other marking and right top. Assuming that left and right marking are parallel

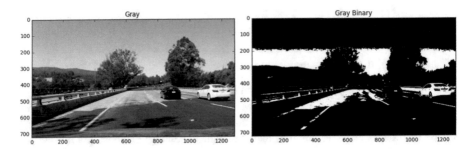

Fig. 4. Image in grayscale with corresponding binary representation after thresholding.

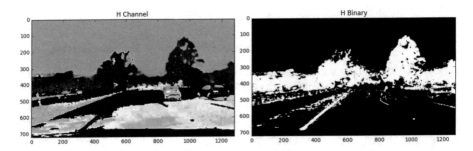

Fig. 5. Hue channel of the image with corresponding binary representation after thresholding.

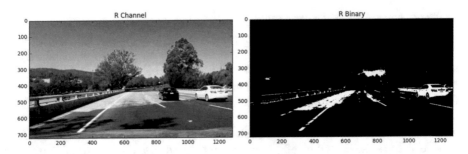

Fig. 6. Red channel of the image with corresponding binary representation after thresholding.

and that the terrain is perfectly flat, we calculate accordingly four target points for those lines and use them to compute a conversion matrix for the DLT algorithm. Table 1 provides the exact points we used for the computation. The effect of the DLT algorithm is presented below on Fig. 8b.

The homography matrix calculated for the points in Table 1 is:

$$
H = \begin{pmatrix} h_1 & h_2 & h_3 \\ h_4 & h_5 & h_6 \\ h_7 & h_8 & h_9 \end{pmatrix} = \begin{pmatrix} -0,7129 & -1,5155 & 1091,1671 \\ 0 & -2,1389 & 1026,6246 \\ 0 & -0,0024 & 1,0 \end{pmatrix}
$$

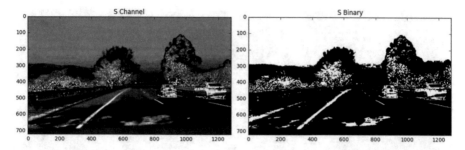

Fig. 7. Saturation channel of the image with corresponding binary representation after thresholding.

Fig. 8. Perspective transform from (a) actual perspective (left) to (b) bird's eye view (right).

Table 1. Original and warp coordinates.

	Original coordinates (x, y)	Warp transformation coordinates (xt, yt)
Left bottom	(200, 720)	(200, 720)
Left top	(550, 480)	(200, 0)
Right bottom	(1104, 720)	(1104, 720)
Right top	(730, 480)	(1104, 0)

Figure 8 displays a warping transformation showing the original source picture (left) and one after the warping transformation (right).

The lane pixels in step 2.4 of the algorithm are identified using the moving histogram window technique. Figure 9 displays the original bird's eye view of the lanes while Fig. 10 presents the corresponding histogram window.

Once the lane pixels have been identified, in step 2.5 the least squares method is used to calculate the coefficients of the second order polynomial passing through the identified pixels. Figure 11 presents sample polynomial fitting for each lane marking as a curve going through points from respective histogram windows.

The computations were performed using a PC with an Intel i7700 k CPU, 64 GB RAM and two NVIDIA TITAN X graphic cards. The algorithm runs at a rate of 35 frames/second, which is sufficient for high frequency real-time processing. In the next

Fig. 9. Bird's eye binary view of lane lines.

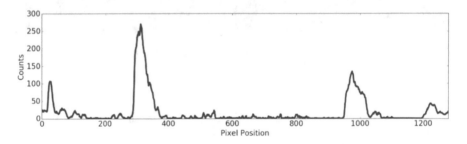

Fig. 10. Histogram for the bird's eye view lane lines presented above.

subsection, two examples of the application of this algorithm to identify lane markers are presented and discussed.

3.1 Numerical Examples

Example 1 deals with the lane finding on the Polish motorway A2 between Warsaw and Łódź. Lane markings are very good. Road is straight and terrain is flat. This example is good for initial calibration. We proceed with the input frame according to the pipeline described in the previous subsection.

First we remove distortions for the input frame using calibration matrices (see Fig. 12). We then transform the original perspective into the bird's eye view. As we can observe for the straight, flat road, on the perspective view the lines converge to the center of the frame. On the bird's eye view perspective the lane lines are parallel (see Fig. 13). In Step 3 of the algorithm on the bird's eye view perspective we apply Canny edge detection in HLS color space to initially distinguish candidate pixels that might

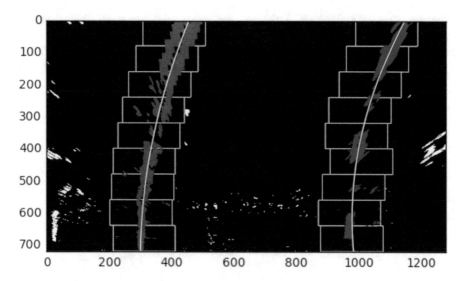

Fig. 11. Visualization of search boxes and polynomial line fitting $y = Ax^2 + Bx + C$ applied to points identified with sliding window histogram technique.

Fig. 12. Example 1. Calibration.

Fig. 13. Example 1. Bird's eye view perspective.

belong to lane lines. As Fig. 14 shows, after this operation the lane lines are left on the binary image and all other information from the input frame is filtered out. After moving the histogram window and the polynomial line fitting, we can see, in Fig. 15, that the algorithm deals very well with a binary image that is the output of Canny edge detection. Indeed, it marked all the line segments very accurately. Lane lines are approximated by the quadratic equations $y = Ax^2 + Bx + C$. For the left curve, the parameters of the quadratic equation are determined as $A_L = -0.00002$, $B_L = -0.0069$, $C_L = 219.4746$ and, respectively, for the right curve as $A_R = 0.00004$, $B_R = -0.0423$, $C_R = 1115.3165$.

Fig. 14. Example 1. Canny edge detection.

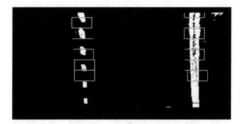

Fig. 15. Example 1. Polynomial line fitting.

In the final step of the algorithm, we merge the original frame with the space identified as the current lane. As we can see in Fig. 16 the marked space falls precisely within the lane space.

Fig. 16. Example 1. Lane visualization after applying full pipeline algorithm.

3.1.1 Error Calculation

In order to measure the accuracy of our algorithm, we introduce the following set of measures calculated using the bird's eye view image. For that purpose we assume we have 4 control points where we check the distance of the lines identified by algorithm and identified by a human analyzing the image. Those four points would be the ones where the errors might be the biggest, which means top-left (xTL) and bottom-left (xBL) for left line and similarly top-right (xTR) and bottom-right (xBR) for the right line. All x variables with A in lower index represent results achieved by algorithm and all variables with H in lower index represent the benchmarking base achieved by a human.

$$E_{TL} = \sqrt{\left(x_{TLH} - x_{TLA}\right)^2} \quad \text{and} \quad E_{TR} = \sqrt{\left(x_{TRH} - x_{TRA}\right)^2} \qquad (23)$$

$$E_{BL} = \sqrt{\left(x_{BLH} - x_{BLA}\right)^2} \quad \text{and} \quad E_{BR} = \sqrt{\left(x_{BRH} - x_{BRA}\right)^2} \qquad (24)$$

To prove the algorithm to be valid and accurate enough for real world application we assume that none of ETL, ETR, EBL, EBR should be greater than 5 cm. It has been experimentally determined that 1 pixel in front of the car is equivalent to 0.53 cm. Therefore for further benchmarking between the human and the algorithm we can easily transform from pixel to centimeters and vice versa. For example, the error values are as follows: ETL = 1.59 cm, ETR = 1.59 cm, EBL = 1.06 cm, EBR = 1.06 cm, i.e. they are satisfactory.

Example 2 deals with a much more challenging scene at Aleja Krakowska in Warsaw. We can see that lane markings are worn off. Sunny weather makes the contrast of the lines lower. There is also some noise in the form of color patterns on the surface of the road. The trees along the road on the right and tram rails on the left make the scene even more complicated.

We remove distortions using the same calibration matrices as for Example 1. The result is displayed in Fig. 17. In step 2 of the algorithm we transform the original perspective into a bird's eye view. As the road is curved to the right we can see in Fig. 18 that the lines get skewed to the right as we move more from the bottom to the top of the bird's eye output.

Fig. 17. Example 2. Calibration.

Fig. 18. Example 2. Eye bird's view perspective.

The application of the Canny edge detection in HLS color space provides the solution displayed in Fig. 19. In this Example we see that some of the patterns on the road don't get filtered out equally well, as in Example 1, which might be challenging for the next steps in the algorithm.

Fig. 19. Example 2. Canny edge detection result.

After the identification of the lane pixels using the moving histogram window and the polynomial line fitting of the identified pixels, we obtain lines as displayed in Fig. 20. The lane curves are approximated by quadratic equations $y = Ax^2 + Bx + C$. For the left curve the parameters are determined as $A_L = 0.00008$, $B_L = -0.1414$, $C_L = 225.6067$ and for the right line as $A_R = 0.00002$, $B_R = -0.132$, $C_R = 121.7785$.

Fig. 20. Example 2. Polynomial line fitting.

As we can see in this figure the noise from the strange patterns on the road is ignored as the true line markings have a much higher number of pixels for each step of moving window. After transforming from the bird's eye to the original perspective for visualization, in step 6 of the algorithm we obtain the scene displayed in Fig. 21. Merging the original input frame captured by the dash camera and lane space identified by our algorithm proves it to work well not only in optimal conditions but also in challenging scenes with interfering objects, misleading patterns on the road, in the presence of shadows and an inconsistent road structure.

Fig. 21. Example 2. Visualisation after applying full pipeline algorithm.

As in Sect. 3.1.1, we now calculate the accuracy error for this example using measures (22)–(23). In Example 2, the error values are as follows: E_{TL} = 1.59 cm, E_{TR} = 2.12 cm, E_{BL} = 1.06 cm, E_{BR} = 1.06 cm, i.e., they are less than 5 cm and are also satisfactory.

4 Conclusions

The algorithm described above demonstrates an efficient methodology for lane-finding in various road conditions. It works well even in the occurrence of confusing patterns on the asphalt which are not a part of the road's markings. The algorithm has also proved to be effective enough to run in real time on modern workstations by achieving a performance of over 30 frames per second on a modern workstation.

References

1. Kim, D.U., Park, S.H., Ban, J.H., Lee, T.M., Do, Y.: Vision-based autonomous detection of lane and pedestrians. In: 2016 IEEE International Conference on Signal and Image Processing (ICSIP), pp. 680–683 (2016). https://doi.org/10.1109/siprocess.2016.7888349
2. Liu, S., Tang, J., Zhang, Z., Gaudiot, J.: Computer architectures for autonomous driving. IEEE Comput. **50**(8), 18–25 (2017)
3. Wang, Y., Teoh, E.K., Shen, D.: Lane detection and tracking using B-snake. Image Vis. Comput. **22**(4), 269–280 (2004)

4. Ieng, S.-S., Tarel, J.-P., Labayrade, R.: On the design of a single lane-markings detectors regardless the on-board camera's position. In: Proceedings of the IEEE Intelligent Vehicles Symposium, pp. 564–569 (2003)
5. Taylor, C.J., Malik, J., Weber, J.: A real-time approach to stereopsis and lane-finding. In: Proceedings of the IEEE Intelligent Vehicles, pp. 207–212 (1996)
6. Broggi, M.: Real-time lane and obstacle detection on the GOLD system. In: Proceedings of the IEEE Intelligent Vehicles, pp. 213–218 (1996)
7. Hsiao, P., Yeh, C., Huang, S., Fu, L.-C.: A portable vision-based real-time lane departure warning system: day and night. IEEE Trans. Veh. Technol. **58**(4), 2089–2094 (2009)
8. Viswanath, P., Swami, P.: A robust and real-time image based lane departure warning system. In: 2016 IEEE International Conference Consumer Electronics (ICCE), Las Vegas, pp. 819–827 (2016)
9. Hillel, A.B., Lerner, R., Levi, D., Raz, G.: Recent progress in road and lane detection: a survey. Mach. Vis. Appl. **25**, 727–745 (2014). https://doi.org/10.1007/s00138-011-0404-2
10. Mammar, S.S., Netto, M.: Time to line crossing of lane departure avoidance: a theoretical study and an experimental setting. IEEE Trans. Intell. Transp. Syst. **7**(2), 226–241 (2006)
11. Zhang, Z.: A flexible new technique for camera calibration. IEEE Trans. Pattern Anal. Mach. Intell. **22**(11), 1330–1334 (2000)
12. Hartley, R., Zisserman, A.: Multiple View Geometry in Computer Vision. Cambridge University Press, Cambridge (2004)
13. Dubrofsky, E.: Homography Estimation. The University of British Columbia, Vancouver (2009)
14. HSL and HSV, 20 January 2018. https://en.wikipedia.org/wiki/HSL_and_HSV
15. Canny, J.: A computational approach to edge detection. IEEE Trans. Pattern Anal. Mach. Intell. **PAMI-8**(6), 679–698 (1986)

Using Multiple RFID Readers in Mobile Robots for Surface Exploration

Marcin Hubacz$^{(\boxtimes)}$, Bartosz Pawłowicz$^{(\boxtimes)}$ (iD),
and Bartosz Trybus$^{(\boxtimes)}$ (iD)

Rzeszow University of Technology, al. Powstancow Warszawy 12,
35-959 Rzeszow, Poland
marcin.hubacz@outlook.com,
{barpaw,btrybus}@prz.edu.pl

Abstract. The article presents selected aspects related to the implementation of RFID readers in a mobile robot that tests a given surface. The advantages of such a solution are given, which significantly improves the accuracy of the location. Issues related to energy flow in RFID systems are discussed, including inductive and propagation coupling. Details on software handling of multiple RFID readers are given. Test results of two modes of operation are presented, i.e. using polling and interrupts. The algorithm for reader prioritization is also shown.

Keywords: RFID · Mobile robot · Robot group · Surface exploring

1 Introduction

Autonomous navigation robots are currently the subject of research and many scientific studies and the RFID technique [1–3] is used often in such experimental solutions as a part of the robot navigation system. The introduction of automatic identification systems for contactless RFID brings a number of advantages. They are used to identify objects with passive or semi-passive transponders (or identifiers) equipped with a writable memory with an information capacity definitely exceeding currently used barcodes [1]. RFID system may be applied in many areas, especially in the field of ISM (industrial, scientific, medical). Exploration of buildings or surfaces using such robots is also a promising research area [4, 5].

The great advantage of RFID systems is that it is possible to read information from many objects at the same time. A robot equipped with an RFID reader can use information and coordinates stored in RFID transponders embedded in walls, floors, doors and furniture [6]. Unfortunately, reading the information does not give knowledge about the location of an object on the surface or in the space. For this reason, quickly reaching the right object can be a problem. Hence, in the scope of RFID systems, the problem of location of an object on which the identifier is mounted, is an important research area [7–12].

It was also noticed that it is possible to reverse this issue and attempt to determine the position of a mobile robot equipped with an RFID reader based on readings and analysis of signals received during reading information from transponders placed in

R. Szewczyk et al. (Eds.): AUTOMATION 2019, AISC 920, pp. 445–456, 2020.
https://doi.org/10.1007/978-3-030-13273-6_42

known locations [13–15]. Many of the existing studies are based on the application of the RSSI method (received signal strength indicator) [15, 16]. This is sometimes extended with methods related to odometry and application of EKF filtering to increase the accuracy of position estimation [17]. There are also original solutions based, e.g., on phase measurement of the signal reflected from the transponder and carrying the information read from the transponder [18].

However, the above-mentioned studies mostly feature a very extensive mathematical apparatus and almost completely ignore the issue of localization support with data stored in the transponder memory. This gap was very quickly noticed [6], what resulted in the creation of various solutions using RFID tags to determine the location and orientation of a robot equipped with an RFID reader, starting from simple solutions with information about how the robot should move written the transponders [19], to advanced systems using the power measurement of the signal reflected from RFID transponders arranged in a grid [15, 20, 21]. It is significant, however, that due to interference [22], the most commonly used solution is the use of a single reader in a mobile robot.

As shown in [23], a swarm of mobile robots with a main control unit can be used to discover and create a surface map. The first generation of robots, having a single reader, placed centrally, was characterized by a very limited measurement accuracy what imposed great limitations. To improve the estimated location of a robot, we have introduced the concept of multiple readers. Each increment of the number of readers improved the accuracy of the position approximation. The tests have been performed using up to four readers. In the considered grid types, i.e. triangular and square, it proved that the most universal solution is the four-reader version, which allowed correct and effective work with both types of grid. Increasing the number of readers makes it possible to cover them with the contour of the robot, which can be seen in larger dispersed transponder grids. Increasing the number of readers makes it possible to place them on the contour of the robot, which can be useful in dispersed grids of transponder. The reading of the position from RFID transponders takes place before the robot is in direct contact with, e.g. an obstacle or the end of the explored area. A single reader located centrally in the robot does not provide such a possibility. Another important aspect to pay attention to is the size of the robot in terms of the number of readers. Laboratory tests were carried out for a small robot with dimensions not exceeding 20×20 cm. However, in the case of large industrial robots, the size relative to the density and range of readers is more important. For this reason, a much better solution is to place readers as far as possible on the contour of the robot housing. However, the higher number of readers introduces additional challenges related to the flow of energy and software handling of data reading from multiple devices connected to a single bus.

2 Energy Flow During RFID Exploration

Consider a robot with four readers and components shown in Fig. 1. Since it can be treated as a radio communication device [24], one of the most important aspects that must be taken into account during the is to ensure uninterrupted operation of RFID readers. This is important due to specifics of RFID systems, which must take into account the fact that undisturbed operation of readers is necessary to provide power to transponders and to correctly read information allowing robots to navigate.

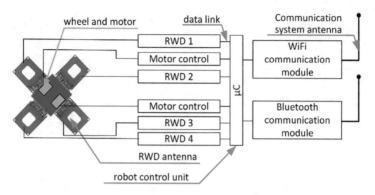

Fig. 1. Block diagram of an autonomous robot.

Due to the principle of operation, one can distinguish two types of solutions, that define the transmission of energy in RFID systems. The first are systems operating on the basis of inductive coupling, and the second one is propagation coupling. The inductive coupling is used in the LF and HF bands. The energy transmitted between the reader and the transponder is transmitted via a magnetic field (Fig. 2a), and its quantity depends on the explored surface and the mutual position of the receiving and transmitting antennas. For correct operation of the system, it is necessary to activate the antenna of the RFID transponder with its resonant frequency, because it causes the flow of the maximum current in the antenna circuit.

Passive transponders are most commonly used in systems using inductive coupling. In a common scenario, each transponders sends its identification number as long as it is located in the interrogation zone i.e. the area of correct operation of the reader, with an appropriate frequency and sufficient field strength. In the case of inductive coupling, the maximum range is obtained when the magnetic field lines generated by the reader antenna are perpendicular to the plane of the transponder antenna coil windings. If the field lines are parallel to the transponder coil, the coupling does not occur and the supply of transponders in the interrogation zone is not provided. The maximum range of readers is most limited by administrative restrictions, concerning the maximum allowed intensity of the magnetic field produced by the reader's antenna. These restrictions vary from country to country and are described in ETSI standards.

The propagation coupling (Fig. 2b) is used for communication in the UHF band. Here, unlike in inductively coupled systems, the area of the far-field is used, where it is assumed that the intensity of the electromagnetic field is independent of the presence of transponders in the field of operation of the reader's antenna. The exchange of information between the passive transponders and the reader relies on the modulation of the carrier's reflection coefficient (backscatter).

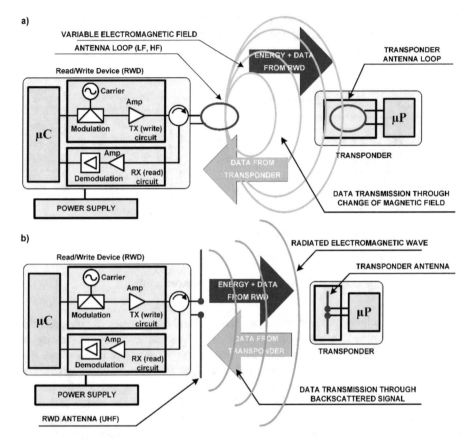

Fig. 2. General scheme of the RFID identification system: (a) an inductive coupled system; (b) a system using propagation coupling.

This effect is obtained by changing the load on the antenna of the transponders. This results in a partial reflection of the energy of the wave produced by the reader antenna in the direction opposite to the direction of the wave radiating on the transponders. The reader receiving the changes in field strength can signal this demodulation and reproduce the data. In order to properly transfer data between system elements, it is necessary to use appropriate modulation so that the energy sent from the reader to the transponders is sufficient to supply them.

The efficiency of RFID system can be comprehensively understood through the interrogation zone. It describes both energy and communication issues of RFID system elements [24, 25], and the estimation of this area allows for a wider implementation of multiple identification of objects. Due to the different ways of operation, the interrogation zone of system is determined differently for induction and propagation systems. In both cases, however, the energy is transmitted via the electromagnetic field, so each of them must comply with acceptable radiation standards based on the CEPT/ERC

70-03 recommendations. Based on these guidelines, one can specify the minimum field strength needed to power the transponders [26].

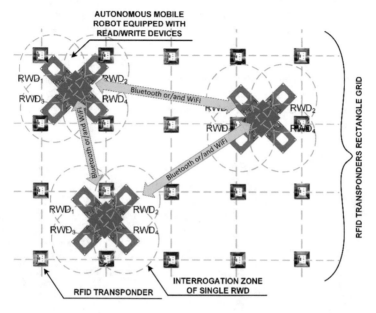

Fig. 3. Group of robots on a RFID transponders grid.

Various communication systems including popular networks such as Bluetooth or Wi-Fi, and other specialized methods of communications are used to exchange data between robots in the group (Fig. 3) [23]. Such a group of robots often collaborates with a master control unit that manages the entire group. Individual robots use radio communications to connect with the master control unit.

As any electrical/electronic equipment, the robot is a potential source of electromagnetic disturbances and must be constructed in accordance with requirements of the EMC Directive 2014/30/EC in such a way that the disturbances it generates are tolerated in the electromagnetic surrounding environment and that it has immunity to exposure occurring in this environment. It is obvious that every robot in the swarm must build and tested according to directive to assure continuous reliable operation of swarm. Unfortunately there are no subject standards for this class of equipment, the assessment for the moment must carried out in accordance with requirements of the general standard IEC 61000-6-3 [26]. This standard assures that measurements are taken in the operating state of the object for which it generates the highest level of disturbances.

3   Software Handling of Multiple RFID Readers

Equipping a robot with four readers required an appropriate communication scheme. The readers have been arranged on the vertices of the square, allowing testing two types of mesh, i.e. triangular and square. Order of the readings is also an important aspect due to the robot movement. The time needed for reading generates an offset of the approximated position, which should be taken into account in the calculations [27–29].

A device equipped with certain peripherals requires a suitable algorithm for communication with the components. Different modes of operation and notifications are used to minimize the time of the communication and to maintain the appropriate time frame. Single-core microcontrollers have various hardware interfaces for communicating with the peripherals. However, without additional hardware solutions, such as DMA or simultaneous access to the interfaces, they do not allow using more than one medium at the same time. In most cases, components are supported sequentially, due to the single-core nature of simple microcontrollers and the number of physical hardware buses, such as SPI.

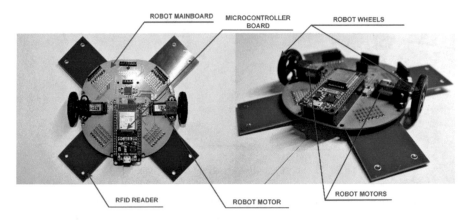

Fig. 4. Autonomous robot with four RFID readers designed and built for the purpose of work.

The presented mobile robot (Fig. 4) uses one hardware bus to communicate with RFID readers based on NXP MFRC522 chip [30]. The rectangular reader antenna dimension are 39 mm × 42 mm. The tests have been carried out at a 2 MHz SPI clock. The SPI interface is running at 2 MHz clock and assumes the use of a master and slave device. Here, slaves are the RFID the readers connected to the common bus. Each reader can be enabled using a separate line (Chip Select). The slave devices cannot start transmission by themselves, and thus return data, without a proper query from the master device. Two modes of operation can be configured:

- polling
- interrupt-based.

Polling is the basic form of communication with a periphery, using the principle of continuous monitoring. It is associated with high resource-consuming processor time and bus occupancy. However, it does not require additional control lines between the microcontroller and the device. It is particularly useful in simple devices that return the desired information in one communication cycle. In the case of RFID reader, polling is based on continuous queries in which the microcontroller attempts to read the transponder data. Cyclic polling of subsequent readers may adversely affect the response time of a robot encountering an obstacle.

In order to determine performance of the read modes in the robot, tests have been performed with an attempt to read a transponder. Two cases have been considered, i.e. when the transponder is in the range of antenna and when it is in the range.

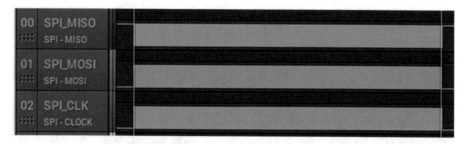

Fig. 5. SPI lines using the polling mode: start reading, if an transponder is detected, its data are read.

Figure 5 shows usage of three SPI bus lines during reading of a transponder using polling operation mode. The lines are constantly busy and changing their state, so they look like bars in the chart. The tests revealed the results shown in Table 1. The difference between the above cases, i.e. the transponder within or not within the range results from the reading which is automatically performed when the transponder is in the range.

Table 1. Polling mode test results

Activity	Duration
1. Attempt to read the transponder	25 ms
2. Transponder not within the range	
3. No reading	
1. Attempt to read the transponder	32 ms
2. Transponder within the range	
3. Reading transponder data	

Interrupt mode is characterized by automatically generated impulse when an event occurs. In the case of RFID readers, after the appropriate interrupt configuration, only the activation of the receiving block is required. If the transponder is within the range of the reader, a falling edge is generated in the interrupt line (IRQ) to inform the microcontroller about the event. The readers can generate the interrupts signalling different events. In addition, this mode allows to use DMA while receiving data.

The interrupt mode in the RC522 reader requires additional triggering of the read block, i.e. a command to trigger the reader should be sent with the selected frequency. Without this command, the reader will not react to the transponders appearing in the range. The read-block trigger command takes up to 6 bytes of data.

The state of the SPI lines and the IRQ line between read-block triggering and signalling the detection of a transponder with the interrupt is shown in Fig. 6. The first part of the chart (between the vertical lines) takes about 58 us. If a transponder has been detected and read, a falling edge is generated (left part) after about 360 us (part of the chart has been dropped for readability). Then a sequence of reading the transponder data is carried out (Fig. 7), which takes about 51 ms. The results for the interrupt mode in the scenario when the transponder is within the range of the reader have been gathered in Table 2.

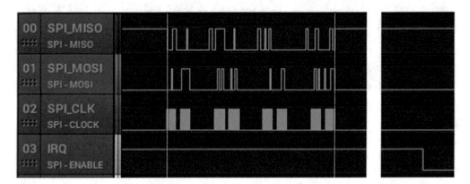

Fig. 6. SPI and IRQ lines using the interrupt mode: read-block triggering and interrupt signaling of detection of the transponder within the range.

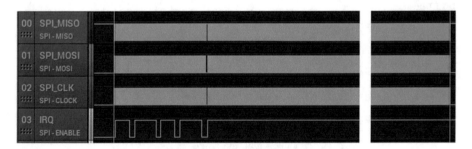

Fig. 7. SPI and IRQ lines using the interrupt mode: reading data from the transponder and reconfiguring the interrupt mode.

Table 2. Interrupt mode test results.

Activity	Duration
1. Read block triggering	58 us
2. Transponder within the range	360 us
3. IRQ signal received	
4. Reading transponder data	51 ms (max)

As seen, the use of the interrupt mode results in a noticeable increase in software performance, which, along with the algorithm used, allows the robot to move smoothly when reading the transponder data.

The density of transponders on the grid is another important factor that must be taken into account. In the tests, the distance between them was about 42 mm, while a transponder had a diameter of 25 mm. The range of reader antenna is about 33 mm from the center (the diameter of the range is about 66 mm). It means that the grid is quite densely covered with transponders, and the reading ranges overlap. The tests showed, that the maximum speed for the polling mode is 0.5 m/s, while for the interrupt mode it as about 0.3 m/s. This means, that the interrupt mode gives poorer results in terms of robot speed for dense grids of transponders. This must be taken into account when the reading mode is selected by the designer.

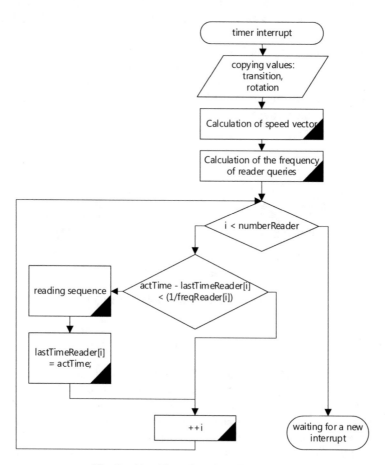

Fig. 8. Algorithm of reader prioritization.

To improve the robot's response to possible obstacles that may occur in its surroundings, an algorithm that prioritizes reading from the RFID readers has been introduced. It is shown in Fig. 8. The algorithm has to balance the maximum reading frequency between the respective readers depending on the speed vector. This is based on two components of the speed vector, i.e. translation and rotation. The timer interrupt is generated at a specified maximum frequency. The instantaneous speed, being the basic parameter of the algorithm, is determined using the translation and rotation of the robot. Then a starting frequency is determined for each reader. If the specified period has elapsed, the reading sequence is started. This operation is repeated for all the readers in the robot.

To illustrate how the algorithm works, we will consider a moving robot with four readers, R1 and R2 at the front, and two others (R3, R4) at the rear (Fig. 9). The algorithm has to properly divide the maximum reading frequency between the readers (R1, R2, R3, R4). The maximum frequency in Fig. 9 is 12 Hz. The robot moves with the momentary velocity vector v and the instantaneous velocity of the wheels v_L, v_R. If the rotation is zero, along with the translation, the front readers (R1, R2) will be interrogated more frequently than the rear readers (R3, R4), e.g. front at 4 Hz, rear at 2 Hz. At the maximum speed of 0.5 m/s only R1 and R2 will be used. If no translation occurs, all the readers will be handled with a fixed cycle of 3 Hz. Figure 9a shows the calculated frequencies for a slight turn. The reader R2 is interrogated more frequently than R1, and R4 more frequently than R3. If the turn is sharper (Fig. 9b), R1 is handled even less frequently in favour of R2 and R4.

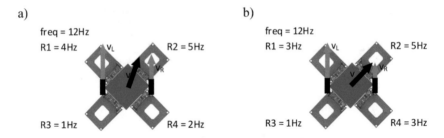

Fig. 9. Balancing the interrogation frequency of the RFID readers: (a) slight turn (b) sharp turn.

4 Summary

The use of multiple readers in a robot exploring the surface is justified by many factors. It helps to improve the accuracy of mapping and better react to obstacles. However, it is associated with increased requirements regarding consideration of energy radiation in the electromagnetic terms. It also affects the load on the hardware bus between the microcontroller and the readers. This load is significantly lower in case of using the interrupt mode. The tests have shown that polling takes up to 100 times more time when there is no RFID transponder in the interrogation zone than in the case of interrupt mode. When an transponder is already present in the zone, the interrupt handling requires about 30% more time, but it is caused by necessity of reenabling interrupts in the NXP MFRC522 reader used in the tests. However, the interrupt mode gives worse results in the case of dense grids, not allowing the robot to reach high movement speeds. Nevertheless, the interrupt mode seems mode seems to have much more potential, especially in the case of less populated grid. In addition, the interrupt mode provides additional flexibility related to the ability to prioritize data readings.

Acknowledgment. Results of Grant No. PBS1/A3/3/2012 from Polish National Centre for Research and Development as well as Statutory Activity of Rzeszow University of Technology were applied in this work. The work was developed by using equipment purchased in Operational Program Development of Eastern Poland 2007–2013, Priority Axis I Modern Economics, Activity I.3 Supporting Innovation under Grant No. POPW.01.03.00-18-012/09-00 as well as Program of Development of Podkarpacie Province of European Regional Development Fund under Grant No. UDA-RPPK.01.03.00-18-003/10-00.

References

1. Finkenzeller, K.: RFID Handbook, 3rd edn. Wiley, New York (2010)
2. ISO/IEC 15693. Identification cards - Contactless integrated circuit cards - Vicinity cards
3. ISO/IEC 14443-3. Identification cards - Contactless integrated circuit cards - Proximity cards (2016)
4. Sanpechuda, T., Kovavisaruch, L.: A review of RFID localization: applications and techniques. In: 5th International Conference on Electrical Engineering/Electronics, Computer, Telecommunications and Information Technology, May 2008, pp. 769–772 (2008)
5. Jian, M.-S., Wu, J.-S.: RFID applications and challenges. In: Radio Frequency Identification from System to Applications. InTech, Vienna (2013)
6. Willis, S., Helal, S.: RFID information grid for blind navigation and wayfinding. In: Ninth IEEE International Symposium on Wearable Computers (ISWC 2005), October 2005, pp. 34–37 (2005)
7. Lionel, M.N., Liu, Y., Lau, Y.C., Patil, A.P.: LANDMARC: indoor location sensing using active RFID. Wirel. Netw. **10**(6), 701–710 (2004)
8. Subramanian, S.P., Sommer, J., Schmitt, S., Rosenstiel, W.: RIL - reliable RFID based indoor localization for pedestrians. In: 16th International Conference on Software, Telecommunications and Computer Networks, September 2008, pp. 218–222 (2008)
9. Chia-Yu, Y., Wei-Chun, H.: An indoor positioning system based on the dual-channel passive RFID technology. IEEE Sens. J. **18**(11), 4654 (2018)
10. Pomárico-Franquiz, J.J., Shmaliy, Y.S.: Accurate self-localization in RFID tag information grids using FIR filtering. IEEE Trans. Ind. Inform. **10**(2), 1317 (2014)
11. Soltani, M.M., Motamedi, A., Hammad, A.: Enhancing cluster-based RFID tag localization using artificial neural networks and virtual reference tags. In: 2013 International Conference on Indoor Positioning and Indoor Navigation, 28–31 October 2013
12. Han-Yen, Y., Jiann-Jone, C., Tien-Ruey, H.: Design and implementation of a real-time object location system based on passive RFID tags. IEEE Sens. J. **15**(9), 5015 (2015)
13. Zhou, J., Shi, J.: RFID localization algorithms and applications - a review. J. Intell. Manuf. **20**(6), 695–707 (2009)
14. Saab, S.S., Nakad, Z.S.: A standalone RFID indoor positioning system using passive tags. IEEE Trans. Ind. Electron. **58**(5), 1961–1970 (2011)
15. Han, S., Lim, H., Lee, J.: An efficient localization scheme for a differential-driving mobile robot based on RFID system. IEEE Trans. Ind. Electron. **54**(6), 3362–3369 (2007)

16. Deyle, T., Nguyen, H., Reynolds, M.S., Kemp, C.C.: RFID-guided robots for pervasive automation. IEEE Pervasive Comput. **9**(2), 37–45 (2010)
17. Boccadoro, M., Martinelli, F., Pagnotelli, S.: Constrained and quantized Kalman filtering for an RFID robot localization problem. Auton. Robots **29**(3–4), 235–251 (2010)
18. DiGiampaolo, E., Martinelli, F.: Mobile robot localization using the phase of passive UHF-RFID signals. IEEE Trans. Ind. Electron. **61**(1), 365–376 (2014)
19. Gueaieb, W., Miah, M.S.: An intelligent mobile robot navigation technique using RFID technology. IEEE Trans. Instrum. Meas. **57**(9), 1908–1917 (2008)
20. Choi, B.S., Lee, J.W., Lee, J.J., Park, K.T.: A hierarchical algorithm for indoor mobile robot localization using RFID sensor fusion. IEEE Trans. Ind. Electron. **58**(6), 2226–2235 (2011)
21. DiGiampaolo, E., Martinelli, F.: A passive UHF-RFID system for the localization of an indoor autonomous vehicle. IEEE Trans. Ind. Electron. **59**(10), 3961–3970 (2012)
22. Papapostolou, A., Chaouchi, H.: RFID-assisted indoor localization and the impact of interference on its performance. J. Netw. Comput. Appl. **34**(3), 902–913 (2011)
23. Hubacz, M., Pawlowicz, B., Trybus, B.: Exploring a surface using RFID grid and group of mobile robots. In: Szewczyk, R., Zieliński, C., Kaliczyńska, M. (eds.) Automation 2018 Advances in Automation, Robotics and Measurement Techniques. Advances in Intelligent Systems and Computing, vol. 743, pp. 490–499. Springer (2018)
24. Jankowski-Mihułowicz, P., Węglarski, M.: Determination of 3-dimentional interrogation zone in anticollision RFID systems with inductive coupling by using Monte Carlo method. Acta Phys. Pol. A **121**(4), 936–940 (2012)
25. Jankowski-Mihułowicz, P., Kalita, W., Pawłowicz, B.: Problem of dynamic change of tags location in anticollision RFID systems. Microelectron. Reliab. **48**(6), 911–918 (2008). https://doi.org/10.1016/j.microrel.2008.03.006
26. IEC 61000-6-3 Electromagnetic Compatibility - Part 6-3: Generic Standards - Emission Standard For Residential, Commercial And Light-industrial Environments
27. Wang, J., Wang, D., Zhao, Y., Korhonen, T.: Fast anti-collision algorithms in RFID systems. In: International Conference on Mobile Ubiquitous Computing, Systems, Services and Technologies - UBICOMM 2007, 4–9 November 2007, pp. 75–80 (2007)
28. Jia, L., Bin, X., Xuan, L., Lijun, C.: Fast RFID polling protocols. In: 45th International Conference on Parallel Processing (ICPP). https://doi.org/10.1109/icpp.2016.42. Accessed 16–19 Aug 2016. ISBN 978-1-5090-2823-8
29. Liu, J., Po, Y.: A Localization algorithm for mobile robots in RFID system. In: 2007 International Conference on Wireless Communications, Networking and Mobile Computing (2007). https://doi.org/10.1109/wicom.2007.527. Accessed 21–25 Sept 2007. ISBN 1-4244-1311-7
30. MFRC522 Standard performance MIFARE and NTAG frontend, NXP Semiconductors. https://www.nxp.com/docs/en/data-sheet/MFRC522.pdf

Proposition of the Methodology of the Robotised Part Replication Implemented in Industry 4.0 Paradigm

Michał Nazarczuk[1] , Maciej Cader[2]([⊠]) , Michał Kowalik[3] ,
and Mikołaj Jankowski[1]

[1] Imperial College London, Exhibition Road, London SW7 2AZ, UK
{michal.nazarczuk17,
mikolaj.jankowski17}@imperial.ac.uk
[2] Industrial Research Institute for Automation and Measurements PIAP,
Al. Jerozolimskie 202, 02-468 Warsaw, Poland
mcader@piap.pl
[3] The Institute of Aeronautics and Applied Mechanics,
Warsaw University of Technology, Nowowiejska 24, 00-665 Warsaw, Poland
mkowalik@meil.pw.edu.pl

Abstract. In this article, we present a proposition of the methodology of the robotised part replication, designed to be implemented according to Industry 4.0 paradigm. The replication process relies on a digital model which establishes the base for robotised 3D scanning, the preparation of the CAD model and robotised milling. The main goal of the article was to estimate errors that potentially may occur in predefined key stages of the proposed approach. We present the results of analysing the errors arising from numerical controlled machining of the representative part – from collecting the digital point cloud model to a computer simulation of the machining process based on the CAD model. The article also contains propositions for future work in robotised part replication.

Keywords: 3D scanning · Robotised milling · Replication methodology ·
Rapid prototyping · Process automation and robotisation · Industry 4.0

1 Introduction

Industry 4.0 is a concept which highly affects the utilisation of robots in the industry [1, 2]. It is predicted that the production management processes in their current form will undergo significant changes to the acquisition and processing of large volumes of data generated by control systems [3, 4]. Nowadays, these data are mainly used for monitoring the state of the technological processes. However, in the future they will enable prediction of the processes' behaviour and quality parameters of the products and will allow global production control. In addition, Machine to Machine (M2 M) [5] communication technologies, application of the Industrial Internet of Things (IIoT) [6, 7] and advanced methods of information processing will be a key factor in the digital transformation. The production will become more flexible and manufacturers will be able to process more complex orders faster and cheaper than before, e.g. using properly

© Springer Nature Switzerland AG 2020
R. Szewczyk et al. (Eds.): AUTOMATION 2019, AISC 920, pp. 457–472, 2020.
https://doi.org/10.1007/978-3-030-13273-6_43

programmed industrial robots capable of quick adaptation to a given task. One of the key aspects in the field of automation and robotics that is currently undergoing a revolution is the process of manufacturing the parts based on fully digital Computer Aided Design (CAD) models. Automation and robotisation of production based on digital CAD models allows production of components that meet the highest quality requirements while guaranteeing very high efficiency. Therefore, the acquired CAD models, which form the basis of the manufacturing tools functioning and the replication methodology, should meet high standards of dimensional accuracy [8]. The implementation of the part replication process according to Industry 4.0 means intelligent automation of the process of acquiring reference geometry, and its postprocessing to a form suitable for subsequent use as a basis for manufacturing processes, e.g. machining [9]. Therefore, it is crucial to develop a method of replication of the part and its verification that is carried out with respect to Industry 4.0 principles, and to examine the utility of the methodology for the industry requirements [10]. Reverse engineering is a process that responds to modern replication needs, either obtaining reference geometry, or learning about the functioning of existing objects [11].

In this paper, we propose a reverse engineering approach for a model built of freeform surfaces [12]. Analyses and comparisons of the accuracy of the real object representation at different stages of the replication process were performed. In order to acquire and verify geometrical data, a structured-light 3D scanning technique was used [13]. The scanning process allowed acquisition of a very dense cloud of points, which underwent a polygonisation process, i.e. filtering and triangulation, to build a triangular model of the part. The model in the form of a triangle mesh was reconstructed into a CAD model. The reconstruction process was carried out using a method based on NURBS surfaces [14], i.e. the most general parametric geometries in CAD programs. Further, the parameters were calculated and the tools selected to model the milling process of the object. At this stage, machining paths were planned using a Computer Aided Manufacturing (CAM) software and exported as Computerized Numerical Control (CNC) code.

Ultimately, the milling process is to be carried out using an industrial robot. Therefore, it was necessary to simulate the movements of the milling tool. The work was done in the Laboratory of Rapid Prototyping and Numerical Calculations of the PIAP Institute by using research infrastructure for robotised milling of large-scale objects based on 3D documentation obtained from precise scanning of micro and small details, purchased under the Ministry of Science and Higher Education grant, decision No. 6468/IA/SN/2015. The proposed methodology (Fig. 1) assumes that robotic devices are the process effectors, both at the stage of acquisition and verification of measurement data (Fig. 2), and production (Fig. 3). The aim of the research in terms of the proposed methodology is to automate the replication process of machine parts using industrial robots.

The analysis of the related work in the area of 3D scanning suggests an insufficient number of publications regarding the use of 3D scanners as tools for generating CAD models used for the robotic milling process. The state of the art is mainly focused on reverse engineering, i.e. acquiring parametric geometry of the model. However, those approaches do not include the further analysis of acquired 3D CAD models as the basis of manufacturing processes. In this respect, this paper presents a unique approach focused not only on the acquisition of parametric geometry, but also the verification of the effects of each mathematical operation and their impact on the final process - robotic milling based on obtained 3D model.

In the robotised part replication process, which is implemented in the Industry 4.0 convention, the basis is a digital model of reference geometry, which due to the characteristics of the reverse engineering process and programming of the manufacturing is modified by mathematical algorithms [15]. As long as the accuracy of the processing of reference models at individual stages of the process is not known, the replication will be a vague task. This publication is a beginning of work related to the automation of the manufacturing process with the use of industrial robots as effectors.

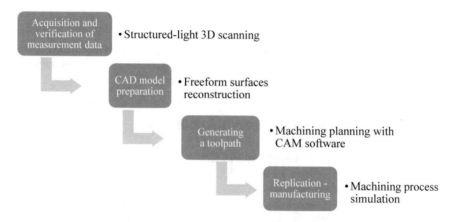

Fig. 1. Diagram of the proposed methodology of robotised part replication carried out according to Industry 4.0 principles.

Fig. 2. Industrial 3D scanners produced by German Producer - GOM mounted on industrial robots (system effectors) are responsible for obtaining measurement data and reference geometries.

The structure of this paper corresponds to the workflow of obtaining the CAD representation of the physical part with 3D scanning, followed by robotised milling. Firstly, the paper focuses on presenting the equipment and the reference model that were used. Thereafter, the accuracy of subsequent mathematical operations related to

Fig. 3. Industrial robot with a milling head in the laboratory of the Industrial Research Institute for Automation and Measurements PIAP in Warsaw as the system effector responsible for the manufacturing.

the transition from the point cloud (representation of the scanned model) to the parametric geometry of the CAD format is examined. Finally, an analysis of the robotised milling simulation is presented. The milling process is performed on the rescaled CAD model that has been obtained as a result of 3D scanning.

2   Reference Model and Research Apparatus

The reference model is a small body made of steel with main dimensions: 70 L × 50 W × 25 H mm, built of freeform surfaces (Fig. 4). The model has been scanned twice. The first scan was captured with the ATOS Core 80 scanner (Fig. 6a) and was adopted as a reference measurement, performed with the most accurate of currently available methods on the market. All accuracy comparisons in this paper were made with respect to this measurement. The second scan was acquired with the ATOS III Triple Scan device (Fig. 6b) and was used to examine the accuracy of structured-light scanning (Fig. 5) in the reverse engineering process and to generate the digital model for the subsequent manufacturing process. Both scanners use the ATOS Professional software for the acquisition and preprocessing of data.

Fig. 4. Reference model – 3D solid built of freeform surfaces.

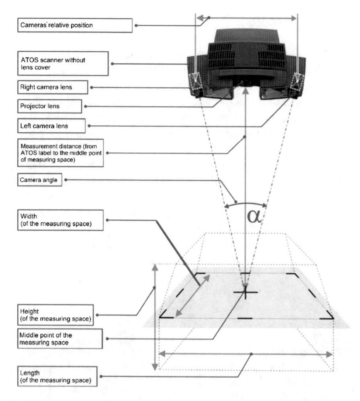

Cameras' relative position

ATOS scanner without
lens cover

Right camera lens

Projector lens

Left camera lens

Measurement distance (from
ATOS label to the middle point
of measuring space)

Camera angle

Width
(of the measuring space)

Height
(of the measuring space)

Middle point of the
measuring space

Length
(of the measuring space)

α

Fig. 5. Diagram of the 3D scanning process with the use of the GOM system [16].

a)   b)

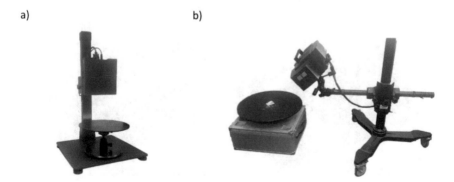

Fig. 6. 3D scanner for reference geometry acquisition - ATOS Core 80 (a); 3D scanner for
measurement data verification - ATOS II Triple Scan (b).

3 Methods and Results

Exposure time is an essential parameter that can be modified during scanning and which has a direct impact on the time and the accuracy of the scanning process. For the software employed, it is possible to perform scans with one, two or three exposure times. A single exposure time is designed for simply shaped objects that reflect light well. Setting the three exposure times results in the execution and assembly of three exposures into one scan, and is intended for objects that reflect light poorly and have gaps and pockets. The figure below (Fig. 7) shows the result of performing 4 test scans with single exposure times: 5 ms, 10 ms, 15 ms and 20 ms. It can be observed (Fig. 7) that the exposure time of 20 ms results in significantly worse scan quality than in the remaining samples. This is due to the coating of the object with titanium oxide 33 which is very reflective. Nevertheless, exposure times of 5 ms, 10 ms and 15 ms all provide satisfactory results. The repetitive procedure of performing scans with two exposure times (5 ms and 12 ms) together with using the rotary table was introduced to facilitate further automation of the process and to ensure the correctness of acquired scans.

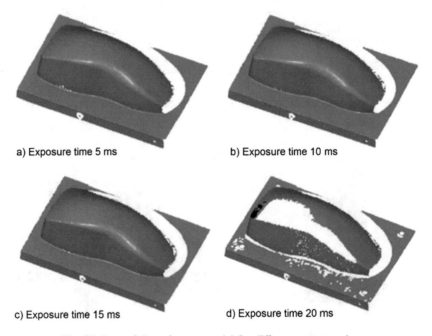

a) Exposure time 5 ms   b) Exposure time 10 ms

c) Exposure time 15 ms   d) Exposure time 20 ms

Fig. 7. Scan of the reference model for different exposure times.

The reference scan was captured in two sessions, one for each of the upper and lower part of the element. Both sessions were performed in the form of single "photo" series of the tested object. In order to match subsequent scans (made within one session) in space with each other, it is necessary to recognise at least 3 common, previously captured reference markers. The combination of two scanning sessions also requires selecting a minimum of 3 recognised markers common to both. Selection of a larger number of markers results in averaging the relative position of the point clouds with each other by minimising the mean square error, which usually leads to an increase in the accuracy of mutual positioning. As a result of the above operations, a dense cloud consisting of several million points was created and initially presented as a surface consisting of triangles stretched between the three points nearest to each other.

The result is shown in the figure below (Fig. 8). Nevertheless, these data are not suitable for further processing and require another operation – polygonisation, which means further processing with a number of tools for filtration of the point cloud and generation of a surface.

Fig. 8. Point cloud visualisation of the reference measurement.

The data acquisition process generates a large volume of data points, which is a big problem in further processing due to the limited computing capabilities of computers. To be able to perform further reconstruction procedures, the point cloud should be converted into a set of surfaces, without reducing the accuracy of the mapping. The entire polygonisation process is performed automatically by the software provided by the scanner manufacturer. In the first part, the polygonisation includes operations for point cloud filtration, which involves removing noise and small point clusters. Noise is described as single points placed at a threshold distance from the rest of the cloud [11]. The software algorithm selects the value of the acceptable thickness of the point cloud layer so that almost all points belonging to the scan of the reference model can be placed in this range. Further, the software looks for points with no neighbours in the distance equal to the selected thickness. These points are then removed, considered as the noise. Small clusters are sets of close points separated from the rest of the cloud. They arise as a result of the accidental capture of unnecessary data in the work area not belonging to the scanned object. Small clusters are also removed from the point cloud using analogous algorithms, as in the case of the measurement noise. Next, to reduce the size of the resulting data, the unnecessary points are removed from the cloud, which is called the cloud density reduction. The cloud density is reduced in an adaptive way, i.e. more points are removed within flat areas than in high curvature areas (as opposed to the regular method, where the density is evenly reduced over the entire surface). There are several algorithms used in the process of cloud density reduction with the adaptive method. The software producer – the German company GOM – does not disclose the details of the algorithm, but below we present the basic methods for cloud density reduction from which individual algorithms originate, such as [15]: difference in surface vectors, analysis of standard deviations of normal vectors and surface patch analysis. During the polygonisation of the point cloud, the triangulation algorithm is used to construct triangles on the cloud and to connect their common sides. The algorithm builds a grid of triangular surfaces (called triangular or faceted areas) according to three main principles:

1. Each triangle must be described by three vertices and a normal vector,
2. Each triangle must have two common points with all adjacent triangles,
3. The inside of the excircle of each triangle must not contain the vertex of any other triangle.

This type of model is a new class of description in relation to the point cloud because the object is entirely represented by surfaces. Such a model can be expressed in the form of an STL file [17], which contains the vertices and normal vectors of subsequent triangles.

To create a correct model, it is necessary to repair the mesh. The ATOS Professional software, for example, automatically removes markers that would be visible on the grid as small protrusions in the model. The removal process is based on mesh smoothing tools. Such procedures approximate the correct position of the offset surface based on the coordinates of the surrounding triangles. However, the area approximated under the marker may not adequately reflect the actual surface, especially for small objects. The scanner software also allows patching of the mesh holes, but in the case of larger holes, another scan is recommended. Small holes are filled based on continuity of the mesh elements around the hole. The last step in building a faceted model is to analyse its correctness. In the case study, the Geomagic Wrap software was used for this purpose. The correctness of the generated reference model grid was verified by observing the potential occurrence of:

- edges not connected to any surface,
- intersecting surface fragments,
- too-sharp edges,
- spikes,
- triangles with too small an area,
- duplicated triangles,
- small holes.

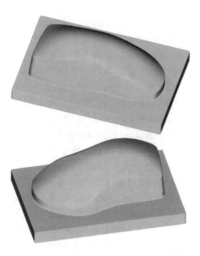

Fig. 9. Triangular model obtained by 3D scanning.

The procedures described above allowed generation of a point cloud which was used to create a triangular model (Fig. 9) to measure the accuracy of the scanning operation and, in the next step to create a CAD model.

The measurement performed on the ATOS II Triple SCAN scanner was compared to the reference measurement made on the ATOS Core 80 scanner to determine the errors introduced as a result of the operations mentioned above. CloudCompare software was used to compare point clouds and triangle grids, and to compute the errors. The results are presented below (Fig. 10). For numerical comparison, several values were gathered to present the accuracy of the measurement using a 3D scanner: RMSE: 5.91 μm, average error: 0.58 μm with a standard deviation of 7.55 μm. In the figures below (Figs. 10 and 12), a map of the geometrical deviations of the tested model with respect to its reference is shown. The presented legend and the corresponding

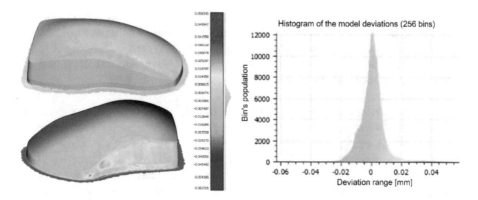

Fig. 10. Comparison of the measurement performed on the ATOS II Triple SCAN with the reference measurement [mm] along with the deviation histogram [mm].

histogram show the distribution of deviation values expressed in millimetres. The positive deviations are marked with red and the negative ones with blue.

The above data indicate the very high accuracy of model mapping in the 3D scanning process. In addition, more than 97% of the results are within the range of the scanner's accuracy (provided by the manufacturer), which indicates that the measurement has been carried out correctly for the given object.

The next step of the replication methodology was the reconstruction of the CAD model. There are several basic methods for reconstructing the CAD model of an object, the selection of the appropriate method depends mainly on the shape of the object, as well as on its intended use and the way it was manufactured.

In the presented case, the method of creating a model composed of Non-Uniform Rational B-Spline (NURBS) surfaces was selected. Therefore, it was necessary to perform several operations to reconstruct the surface using Geomagic Design X software. In order to create a NURBS surface, it is necessary to fit a so-called grid of NURBS curves on the model, between the edges extracted previously [18]. The software allows creation of a grid with either a manually selected or an automatically estimated number of elements, which forms the basis for transforming the surface into a parametric CAD model (Fig. 11).

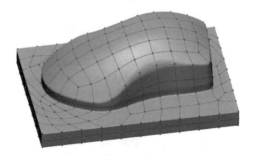

Fig. 11. NURBS grid fitted to the triangular model.

The parametric CAD model was compared with the reference model. The CloudCompare software was used, performing identical procedures as in the case of measuring the accuracy of scanning. The results are presented below (Fig. 10). For numerical comparison, several values were printed showing the accuracy of CAD model reconstruction: RMSE: 7.2 µm, average error: 0.31 µm with a standard deviation of 9.54 µm.

An accurate parameterized CAD model can be used in a further reverse engineering process. It is worthwhile to mention that in the process of building the Finite Elements Methods (FEM) model, archiving geometric data and inspection tasks, the step of creating a parametric model can be omitted. However, the CAD model is necessary for the process of designing toolpaths in CAM systems. In the further part of the work, an analysis of the accuracy of the milling process simulation based on the obtained parametric model is presented. It was assumed that the milling will take place for a model six times larger than the reference. Scaling was possible due to the model

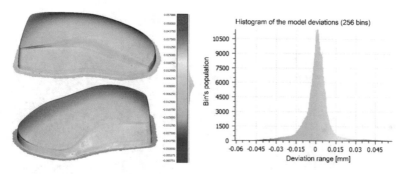

Fig. 12. Comparison of the parametric CAD model with the reference measurement [mm] along with the deviation histogram [mm].

parametrisation. Machining of NURBS surfaces enforces the use of 5-axis milling which requires toolpath planning based only on the CAD parametrised geometry of the model. Therefore, inaccuracies resulting from object scaling have a direct impact on the accuracy of the entire milling process. The milling was planned with the use of Mastercam 2017 software. At the toolpath generation stage, all the necessary parameters of the process were determined and are summarised in Tables 1 and 2 (Fig. 13).

Table 1. Parameters of the milling process (flat end mill).

Parameter	Value
Mill diameter	30 mm
Number of teeth	6
Overall length	166 mm
Flute length	90 mm
Cutting speed	450 m/min
Spindle speed	4775 rpm
Feedrate	8000 mm/min (rough)
	5600 mm/min (finishing)

Table 2. Parameters of the milling process (ball end mill).

Parameter	Value
Mill diameter	12 mm
Number of teeth	4
Overall length	110 mm
Flute length	53 mm
Cutting speed	450 m/min
Effective tool diameter	7.9 mm
Spindle speed	18000 rpm
Feedrate	10000 mm/min

a) Toolpath – face milling

b) Toolpath – rough milling

c) Toolpath – profile milling

d) Toolpath – finishing milling (flat end mill)

e) Toolpath – finishing milling of the top surface

f) Toolpath – finishing milling of the side surface

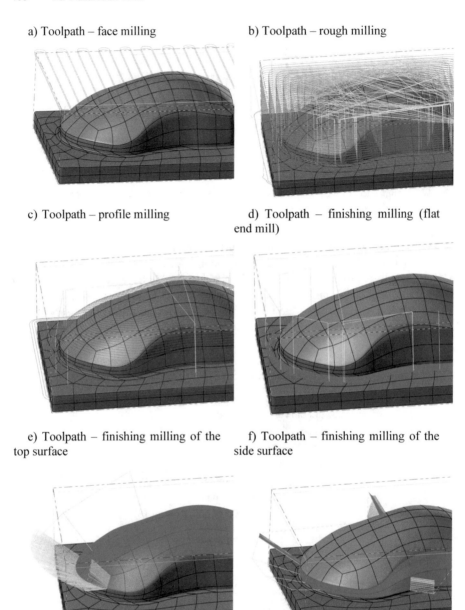

Fig. 13. Consecutive stages of milling generated on the basis of the CAD model.

As before, an error analysis was carried out. The model obtained from CAM simulation (scaled) was compared to the scaled reference by using the CloudCompare software as previously. The results are presented below (Fig. 14). For numerical comparison, several values were printed showing the accuracy of toolpath planning:

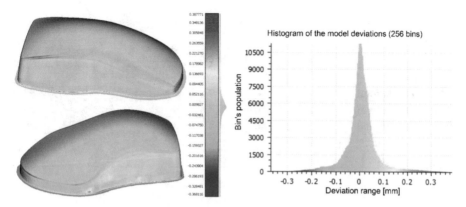

Fig. 14. Comparison of the model obtained from CAM simulation (scaled) with scaled reference measurement [mm] along with the deviation histogram [mm].

Root-mean-square error (RMSE): 48 µm, average error: 5.58 µm with standard deviation: 65.57 µm.

Based on the figure and error values, it can be observed that RMSE and standard deviation remained relatively small in comparison to the previous stages of the process (taking the scaling into consideration). Therefore, it can be inferred that the process of generating toolpaths did not introduce a large error to the reconstruction process. From the histogram shape and percentile range, it can be deduced that the machining planned in the CAM software increased the error in the areas where the biggest deviations on previous stages were observed.

4   Conclusions

At every key stage of the replication process an analysis of the accuracy of the digital model, which is the basis of the whole process, was performed. The table below (Table 3) provides a comparison of all collected statistical data from the conducted tests.

All data contained in Table 3 is compared w.r. to the reference model. The biggest disadvantage of the process is the accumulation of errors from previous stages. Nevertheless, one can observe that the reconstruction preserves satisfactory accuracy, which allows for proper replication of machine parts in the Industry 4.0 convention, i.e. based on a digital model and using industrial robots as effectors in the applied methodology. The preliminary tests achieved a satisfactory accuracy of replication for potentially large production batches, in a relatively short preparation time. The process of replication seems to be particularly sensitive to the high-curvature surfaces reconstruction, which requires special attention in the context of a comprehensive approach to the process – both obtaining CAD models and preparing toolpaths.

Table 3. Comparison of the replication accuracy at key stages of the process.

Statistics	Stage		
	3D scanning	CAD model reconstruction	Toolpaths planning
RMSE [μm]	5.91	7.20	48
97th percentile [μm]	<−17.8, 18.14>	<−26.92, 26.84>	<−185.05, 189.94>
95th percentile [μm]	<−14.72, 15.07>	<−21.82, 21.28>	<−152.57, 151.56>
90th percentile [μm]	<−10.94, 11.28>	<−14.41, 13.86>	<−102.37, 101.36>
Standard deviation [μm]	7.55	9.54	65.57
Average error [μm]	0.58	0.31	5.58

Based on the conducted process, further research was undertaken to generate and simulate the paths of the milling tool mounted on a KUKA KR 500 industrial robot. A virtual robotic milling station is shown in the Fig. 15. The milling process was programmed using the KUKA.CNC module provided by the manufacturer. The module allows interpretation and execution of a program written in the form of G-code. Before loading the model built of free-form surfaces, an analogous process was carried out for a simpler geometric model (Fig. 15). Detailed results of milling will be presented in a separate publication, which is the next stage of the research in the field of part replication in the Industry 4.0 convention.

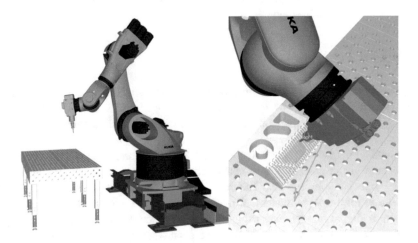

Fig. 15. Simulation model of a milling station equipped with an industrial robot (on the left) and preliminary tests of robotic milling of a simple geometric model (on the right).

The research shows that the processing of point cloud into the parametric model does not introduce any significant inaccuracies. This means that the scaled CAD model obtained as a result of 3D scanning can be the basis for the Direct Digital

Manufacturing (DDM) for a robotised machining. The research shows that the process of robotised milling (including large-scale) can be classified as a part of Industry 4.0 paradigm being a technology that allows the individualised production. In the proposed approach, this means that the CAD representation of the model obtained from 3D scanning process can be considered as a digital label containing the necessary geometrical data for the reproduction. In addition, due to the repeatability of the effectors – industrial robots, the process may be scaled to perform large volume production. Therefore, the research results suggest that the robotised process of obtaining 3D geometry and following milling is feasible within acceptable numerical errors. Nevertheless, further work should be carried on full automation of the process. Further research should focus on two aspects – performing more trial on different shapes of reference geometries and using the most recent achievements of computer science and numerical methods to automate decision process in terms of converting point cloud into a parametric CAD model, including automation of tool path generation.

References

1. Rojko, A.: Industry 4.0 concept: background and overview. Int. J. Interact. Mobile Technol. **11**(5), 77–90 (2017)
2. Lasi, H., Fettke, P., Kemper, H.-G., Feld, T., Hoffmann, M.: Industry 4.0. Bus. Inf. Syst. Eng. **6**(4), 239–242 (2014)
3. Khan, A.R., Schioler, H., Kulahci, M., Knudsen, T.: Big data analytics for industrial process control. In: 22nd IEEE International Conference on Emerging Technologies and Factory Automation (ETFA), Limassol (2017)
4. Kumar, R., Singh, S.P., Lamba, K.: Sustainable robust layout using big data approach: a key towards industry 4.0. J. Cleaner Prod. **204**, 643–659 (2018)
5. Verma, P.K., Verma, R., Prakash, A., Agrawal, A., Naik, K., Tripathi, R., Alsabaan, M., Khalifa, T., Abdelkader, T., Abogharaf, A.: Machine-to-Machine (M2M) communications: a survey. J. Netw. Comput. Appl. **66**, 83–105 (2016)
6. Wang, L., Törngren, M., Onori, M.: Current status and advancement of cyber-physical systems in manufacturing. J. Manuf. Syst. **37**(2), 517–527 (2015)
7. Jeschke, S., Brecher, C., Meisen, T., Özdemir, D., Eschert, T.: Industrial Internet of Things and cyber manufacturing systems. In: Industrial Internet of Things, pp. 3–19. Springer, Cham (2017)
8. Huangfu, D., Pei, X.: Research on the key technologies of CAD/CAM integrated system. In: International Conference on Mechanic Automation and Control Engineering, Wuhan (2010)
9. Huiying, Q.: Research on NC simulation technology integrating CAD/CAM/CAPP. In: 7th International Conference on Intelligent Computation Technology and Automation, Changsha (2014)
10. Vaidya, S., Ambad, P., Bhosle, S.: Industry 4.0 – a glimpse. Procedia Manuf. **20**, 233–238 (2018)
11. Wulf, O., Wagner, B.: Fast 3D scanning methods for laser measurement systems. In: International Conference on Control Systems and Computer Science (CSCS14), pp. 2–5 (2003)
12. Sarkar, B., Meng, C.-H.: Smooth surface approximation and reverse engineering. Comput. Aided Des. **23**(9), 623–628 (1991)

13. Stjepan Jecić, N.D.: The assessment of structured light and laser scanning methods in 3D shape measurements. In: 4th International Congress of Croatian Society of Mechanics, Bizovac (2003)
14. Liang, K.M., Rajeswari, M., Khoo, B.E.: Free form shape representation using NURBS. In: The 10th International Conference in Central Europe on Computer Graphic, Plzen-Bory (2002)
15. Trstenjak, M., Cosic, P.: Process planning in Industry 4.0 environment. Procedia Manuf. **11**, 1744–1750 (2017)
16. GOM GmbH. https://www.gom.com. Accessed Oct 2018
17. Szilvśi-Nagy, M., Mátyási, G.: Analysis of STL files. Math. Comput. Model. **38**, 945–960 (2003)
18. Dan, J., Lancheng, W.: An algorithm of NURBS surface fitting for reverse engineering. Int. J. Adv. Manuf. Technol. **31**(1), 92–97 (2006)

Optimal Control of a Wheeled Robot

Zenon Hendzel and Paweł Penar[(✉)]

The Faculty of Mechanical Engineering and Aeronautics,
Rzeszow University of Technology,
Powstancow Warszawy 12, 35-959 Rzeszów, Poland
ppenar@prz.edu.pl

Abstract. The article presents the experimental verification of the zero-sum differential game in the mobile robot tracking control task in changing environmental conditions. The solution allows to generate an adaptive optimal control and solve the H_∞ control problem. The adopted solution is based on the Hamilton-Jacobi-Bellman principle of optimality and is a generalization of the minmax optimization problem. The verification of adaptive optimal control is a new approach to this problem in the context of currently optimal solutions in mobile robotics in real time. The article presents the experimental results, which were used to verify the solutions adopted and confirmed the high accuracy of mobile robot tracking control.

Keywords: Mobile robot · Differential game · H_∞ control problem

1 Introduction

Modern control theory exposes the problem of optimal control of nonlinear objects [1–3,5]. The interest in such objects is related to industrial application since mobile robots are nonlinear objects.

In the classic approach, control is optimal when it minimizes the function value [8,9]. In the case of a zero-sum differential game, the value functions are influenced by two players who are understood as control and disturbance. One minimizes and the other maximizes the value functions. This approach combines game theory and optimal control theory and is a generalization of the minmax optimization problem [10,11].

From the perspective of optimal control theory, the zero-sum differential game solution is based on Bellman's principle of optimality. At the same time, from the perspective of game theory, the solution of differential game is the Nash saddle point. Finding it is equivalent to solving the problem of H_∞ control [12]. This fact combines the theory of differential games with the theory of dissipative system [10,11]. In the nonlinear case, an analytical zero-sum differential solution is not available. For this reason, the solution is approximated using the approximation dynamic programming methods. Such approach has been described in various aublications; for a continuous [1–4] and discrete case [4,13,14].

© Springer Nature Switzerland AG 2020
R. Szewczyk et al. (Eds.): AUTOMATION 2019, AISC 920, pp. 473–481, 2020.
https://doi.org/10.1007/978-3-030-13273-6_44

The article presents the experimental research on the optimal control algorithm of a wheeled mobile robot (WMR) with the use of zero-sum differential game. The achieved numerical solution was verified on the real object using the rapid prototype method.

2 H_∞ Optimal Control Problem

The equation of dynamic nonlinear control object is given $[1, 10, 11]$ as

$$\dot{\mathbf{x}} = f(\mathbf{x}) + g(\mathbf{x})\mathbf{u} + k(\mathbf{x})\mathbf{d} \tag{1}$$

where $\mathbf{x} \in \mathbf{R}_x^N$ is the state space vector of the system, f, k, d are nonlinear functions;

With the control object (1) the function was related, given by $V \geq 0$

$$V(\mathbf{x}) = \int_0^\infty r(\mathbf{x}, \mathbf{u}, \mathbf{d}) = \int_0^\infty \left[\mathbf{x}^T \mathbf{Q} \mathbf{x} + \mathbf{u}^T \mathbf{R} \mathbf{u} - \gamma \mathbf{d}^T \mathbf{d} \right] dt \tag{2}$$

which form is derived from the theory of the dissipative systems $[10, 11]$, $\mathbf{Q}, \mathbf{R}$ are design matrices. The $\mathbf{u}$ and $\mathbf{d}$ signals are, respectively, the control signal being a minimizing player and the maximizing player signal that acts as a disturbance.

For the adopted indicator for optimization, the Hamiltonian has the form

$$H(\mathbf{x}, \nabla V, \mathbf{u}, \mathbf{d}) = r(\mathbf{x}, \mathbf{u}, \mathbf{d}) + (\nabla V)^T (f(\mathbf{x}) + g(\mathbf{x})\mathbf{u} + k(\mathbf{x})\mathbf{d}) \tag{3}$$

where $\nabla V = \frac{\partial V}{\partial x}$ is a gradient. For the control object $0(1)$, a gain L_2 can by specified that is less than or equal to γ if the following inequality is met

$$\frac{\int_0^\infty \left[\mathbf{x}^T \mathbf{Q} \mathbf{x} + \mathbf{u}^T \mathbf{R} \mathbf{u} \right] dt}{\int_0^\infty \left[\mathbf{d}^T \mathbf{d} \right]} \leq \gamma^2 \tag{4}$$

H_∞ control problem comes down to determining the smallest value $\gamma^* > 0$ such that for any γ the inequality

$$\gamma > \gamma^* \tag{5}$$

is satisfied.

With $[10, 11]$ it is known that the determination of gain L_2 for the control object described by the dependency (1) is equivalent to finding the optimal function V. Therefore, the L_2 stability problem can, among other things, be reduced to the zero-sum differential games theory.

2.1   Continuous Zero-Sum Differential Game

As given in the paper $[12]$, the zero-sum differential game solution which the control object is given by (1) and the value function is of the form (2) are such signals $\mathbf{u}^*$ and $\mathbf{d}^*$, which solve the Nash saddle equation, i.e.

$$V^*(x(0)) = min_{\mathbf{u}} max_{\mathbf{d}} r(\mathbf{x}, \mathbf{u}, \mathbf{d}) = min_{\mathbf{d}} max_{\mathbf{u}} r(\mathbf{x}, \mathbf{u}, \mathbf{d}) \tag{6}$$

where V^* is the optimal value of the function value. Which solve a pair of $(\mathbf{u}^*, \mathbf{d}^*)$ signals, solving the Nash saddle, is equivalent the solution of the Hamilton-Jacobi-Issac equation

$$H(\mathbf{x}, \nabla V, \mathbf{u}, \mathbf{d}) = 0 \tag{7}$$

Therefore the signals are the following form

$$\mathbf{u}^*(t) = -\frac{1}{2}\mathbf{R}^{-1}g(\mathbf{x})^T \frac{dV^*(\mathbf{x})}{d\mathbf{x}} \tag{8}$$

and

$$\mathbf{d}^*(t) = \frac{1}{2\gamma^2}k(\mathbf{x})^T \frac{dV^*(\mathbf{x})}{d\mathbf{x}} \tag{9}$$

In articles [1,2,14] it was pointed out that the solution of zero-sum game based on (8) and (9) is very difficult. The solution is possible only for the linear case. In the nonlinear case, the frequently used solution is the approximation of the value function by neural networks (NN). In this case, the NN adaptation is based on the minimization of the $EHJI$ error, which results from the (7) equation and has the form

$$EHJI = r(\mathbf{x}, \mathbf{u}, \mathbf{d}) + (\nabla V)^T(f(\mathbf{x}) + g(\mathbf{x})\mathbf{u} + k(\mathbf{x})\mathbf{d}) \tag{10}$$

In the article [1] the authors propose the solution of nonlinear differential game using the actor-critic structure. In this approach, the critic approximates the value function $V(\mathbf{x})$ and the actor generates signals $(\hat{\mathbf{u}}, \hat{\mathbf{d}})$ approximating $(\mathbf{u}^*, \mathbf{d}^*)$.

2.2 Approximation of the Value Function. Critic Structure

The value function $V(\mathbf{x})$ can be written as [1]

$$V(\mathbf{x}) = \mathbf{W}^T \varphi(\mathbf{x}) + \epsilon(\mathbf{x}) \tag{11}$$

where $\epsilon(\mathbf{x})$ is an approximation error, the expression $\mathbf{W}^T \varphi(\mathbf{x})$ describes NN of M neurons in which $\mathbf{W} \in \mathbf{R}^M$ is the output layer weight vector and $\varphi(\mathbf{x}) \in \mathbf{R}^M$ is the vector of the activation function. When $M \to \infty$, approximation error $\epsilon \to 0$. The structure of the critique approximating the value function has the form

$$\hat{V}(\mathbf{x}) = \hat{\mathbf{W}}^T \varphi(\mathbf{x}) \tag{12}$$

where $\hat{\mathbf{W}}$ are weights that approximate the unknown values of the weight $\mathbf{W}$.

From (12) and (3) we have approximations

$$\hat{\mathbf{u}}(t) = -\frac{1}{2}\mathbf{R}^{-1}g(\mathbf{x})^T \left[\frac{d\varphi(\mathbf{x})}{d\mathbf{x}}\right]^T \hat{\mathbf{W}} \tag{13}$$

and

$$\hat{\mathbf{d}}(t) = \frac{1}{2\gamma^2}k(\mathbf{x})^T \left[\frac{d\varphi_1(\mathbf{x})}{d\mathbf{x}}\right]^T \hat{\mathbf{W}} \tag{14}$$

From (12)–(14) and (10) we obtain

$$E_{HJI} = r(\mathbf{x}, \hat{\mathbf{u}}, \hat{\mathbf{d}}) + (\nabla V \varphi(\mathbf{x})^T \hat{\mathbf{W}})^T (f(\mathbf{x}) + g(\mathbf{x})\mathbf{u} + k(\mathbf{x})\mathbf{d}) \qquad (15)$$

The equation of (15) determines the error of NN traning, which is used in the adaptation of SN weights. The update of NN is based on

$$\dot{\mathbf{W}} = \eta \frac{\sigma}{(1 + \sigma^T \sigma)^2} E_{HJI} \qquad (16)$$

where η is the learning rate and $\sigma = \nabla V \left[f(\mathbf{x}) + g(\mathbf{x})\hat{\mathbf{u}} + k(\mathbf{x})\hat{\mathbf{d}} \right]$

The approximation solution of the zero-sum differential-game is based on the weighted adaptation low in the form (16), which is a modified version of the Levenberg-Marquart algorithm where the normalizing factor is $(1 + \sigma^T \sigma)^2$ instead $(1 + \sigma_1 \sigma)$. Proof of convergence of the proposed adaptation law is given in [1].

3   Control of a Wheeled Mobile Robot

Zero-sum differential game was used to generate suboptimal control in the tracking task, in which the control object is a WMR. Its scheme is shown in Fig. 1.

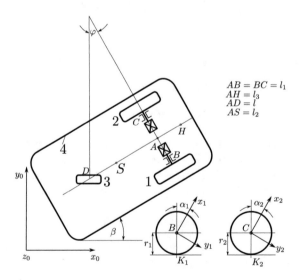

Fig. 1. WMR schematics

WMR consists of a frame 4, two driver wheels (1 and 2), whose drives are independent, and a free rolling castor driving wheel 3.

3.1   Tracking Control Problem

The WMR dynamics equation can be written in a vector-matrix form as [6,7]

$$\mathbf{M}\ddot{\alpha} + \mathbf{C}(\dot{\alpha})\dot{\alpha} + \mathbf{F}(\dot{\alpha}) + \mathbf{F}_\Delta(\dot{\alpha}) = \mathbf{U} \tag{17}$$

where $\boldsymbol{\alpha} = [\alpha_1, \alpha_2]^T$ is the vector of generalized coordinates whose elements α_1, α_2 are the angles of rotation of driving wheels, $\mathbf{U} = [M_1, M_2]^T$ is a control vector whose elements are the driving torques. Matrix $\mathbf{M}, \mathbf{C}$ and vectors $\mathbf{F}, \mathbf{F}_\Delta$ are

$$\mathbf{M} = \begin{bmatrix} a_1 + a_2 + a_3 & a_1 - a_2 \\ a_1 - a_2 & a_1 + a_2 + a_3 \end{bmatrix}, \mathbf{C}(\dot{\alpha}) = \begin{bmatrix} 0 & 2a_4(\dot{\alpha}_2 - \dot{\alpha}_1) \\ -2a_4(\dot{\alpha}_2 - \dot{\alpha}_1) & 0 \end{bmatrix}, \tag{18}$$

$$\mathbf{F}(\dot{\alpha}) = \begin{bmatrix} a_5 sign(\dot{\alpha}_1) \\ a_6 sign(\dot{\alpha}_2) \end{bmatrix}, \mathbf{F}_\Delta(\dot{\alpha}) = \begin{bmatrix} \Delta a_5 sign(\dot{\alpha}_1) \\ \Delta a_6 sign(\dot{\alpha}_2) \end{bmatrix} \tag{19}$$

whereas the *sign* function (in the experiment approximated by a sigmoidal bipolar function) allows taking into account the direction of motion. The coefficients $a_1 - a_6$ are grouped parameters related to geometry, mass or moment of inertia WMR. Vector $\mathbf{F}_\Delta$ is a disturbance vector due to friction changes in the wheel-road kinematics pair.

In article [6] a synthesis of WMR tracking control task was presented. In this task the character point H WMR is tracking the desired path. Defining the tracking error of angle coordinate

$$\mathbf{e} = \boldsymbol{\alpha}_d - \boldsymbol{\alpha} \tag{20}$$

and a generalized error

$$\mathbf{x} = \begin{bmatrix} x_1 \\ x_2 \end{bmatrix} = \dot{\mathbf{e}} + \mathbf{\Lambda}\mathbf{e} \tag{21}$$

where $\mathbf{\Lambda} > 0$ is a design matrix, $\dot{\alpha}$ is a desired angles of rotation. The tracking control task can be turned into a problem of stabilizing the error $\mathbf{x}$ around zero. The WMR dynamics in the generalized error $= \mathbf{x}$ are written in the form

$$\dot{\mathbf{x}} = f(\mathbf{x}) + g(\mathbf{x})\mathbf{U} + k(\mathbf{x})\mathbf{F}_\Delta \tag{22}$$

where $f(\boldsymbol{x}) = \mathbf{M}^{-1}[-\mathbf{C}(\boldsymbol{\alpha})\boldsymbol{x} + \mathbf{M}\dot{\boldsymbol{v}} + \mathbf{C}(\boldsymbol{\alpha})\boldsymbol{v} + \boldsymbol{F}]$, $g(\boldsymbol{x}) = -\mathbf{M}^{-1}, k(\boldsymbol{x}) = \mathbf{M}^{-1}$ a $\boldsymbol{v} = \boldsymbol{\alpha}_d + \mathbf{\Lambda}\mathbf{e}$.

4   Test of Adopted Solutions

Test of optimal control solution generated by zero-sum differential game included verifications of numerical simulation of the real object and checking the condition of H_∞ control problem. The results obtained from the experimental study were used for verification analytical and simulation solutions. The scheme of the control system in the case of simulations and tests on the real object is shown in Fig. 2. For evaluation of the quality of the numerical simulation control and verification on the real object, the mean square error (RMSE) of the form was defined

$$\varepsilon_j^{sim} = \sqrt{\frac{1}{N}\sum_{i=1}^{N}[e_j^{sim}(i)]^2}, \dot{\varepsilon}_j^{sim} = \sqrt{\frac{1}{N}\sum_{i=1}^{N}[\dot{e}_j^{sim}(i)]^2}, j = 1, 2 \qquad (23)$$

where $e^{sim}, \dot{e}^{sim}$ are, respectively, to angle tracking error and velocity tracking error, $N = 2501$ is the number of discrete data. Similarly quality indicators can be defined for the error following the verification, i.e. e^{ver}.

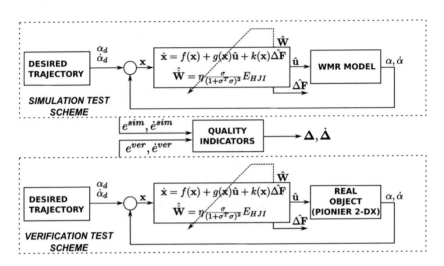

Fig. 2. Scheme of the control system in the case of simulation and experimental tests on the real object with the evaluation of the quality of the numerical solution

The second goal of the presented experiment is to verify the obtained numerical results. Therefore, it was compared with the results of experimental research, using the quality indicators, i.e

$$\Delta_j = \sqrt{\frac{1}{N}\sum_{i=1}^{N}(e_j^{ver}(i) - e_j^{sim}(i))^2}, \dot{\Delta}_j = \sqrt{\frac{1}{N}\sum_{i=1}^{N}(\dot{e}_j^{ver}(i) - \dot{e}_j^{sim}(i))^2}, j = 1, 2$$
$$(24)$$

4.1   Numerical Experiment

In the numerical test, optimal control was generated for the tracking control task. An WMR motion was assumed on a straight line with the velocity of point A, $v_a = 0.35\,[\text{m/s}]$.

The control used an approximation of the differential game which is related to the equation of (22).

As the structure of the function approximating the value of V was chosen NN, in which the vector of the activation function is a quadratic form, i.e. $\varphi(\mathbf{x}) = [x_1^2, x_1 x_2, x_2^2]$.

Assuming data $r = 0.0825[m]$, $\boldsymbol{a} = [0.2187, 0.0158, 0.1104, 0.0039, 3.0836,$
$3.0836]^T$, $\Lambda = 3\boldsymbol{I}_{2\times2}$, $\boldsymbol{W}(0) = 3\boldsymbol{I}_{3\times1}$, $\boldsymbol{Q} = 1\boldsymbol{I}_{2\times2}$, $\boldsymbol{R} = 16\boldsymbol{I}_{2\times2}$, $\gamma = 4$, $\eta = 9$
numerical simulation of the minmax control was performed. Its results are shown
in Fig. 3.

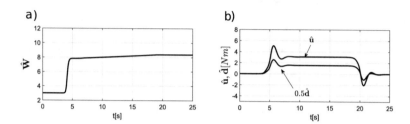

Fig. 3. Numerical test result: adaptation on weight $\hat{\boldsymbol{W}}$ (a); $\hat{\boldsymbol{u}}$ and $\hat{\boldsymbol{d}}$ signals (b)

The approximated control signal $\hat{\boldsymbol{u}}$, which is a minimizing player, is shown
in Fig. 3b. Its figure is compared to the signal $\hat{\boldsymbol{d}}$, which acts as a maximizing
player, and approximates the change in the friction of motion of the kinematic
pair wheel-ground. In the Fig. 3 process of adaptation of the weight was shown.
The values of which stabilize and reach constant values ($\hat{\boldsymbol{W}} = const$).

4.2   Experimental Test

In order to verify the simulation results, conducted experimental test (in real
object) solutions resulting from game theory (13)–(16) (selecting these constants,
which in the case of the numerical procedure). Its results are shown in Fig. 4.

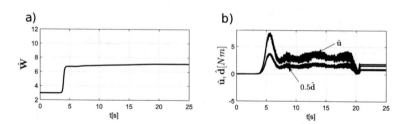

Fig. 4. Verification test result: adaptation on weight $\hat{\boldsymbol{W}}$ (a); $\hat{\boldsymbol{u}}$ and $\hat{\boldsymbol{d}}$ signals (b) for
minmax process

The obtained results of experimental tests are close to the results of numerical
simulations. This fact indicates a good selection of coefficients in the appropriate
class of the mathematical model. The nonzero control signals that occurs for
$t \in [20, 25]$ is the result of a non-zero tracking error at that time.

Table 1 presents the values of RMSE error of tracking WMR for numerical simulation and experimental testing. In addition, the values of quality indicators were determined using the equation (24). Their small values confirm the high accuracy of the robot's movement.

Table 1. Indicators of the quality of experimental test

	ε_1	ε_2	$\dot{\varepsilon}_1$	$\dot{\varepsilon}_2$	Δ_1	Δ_2	$\dot{\Delta}_1$	$\dot{\Delta}_2$
Numeric. (sim)	0.5885	0.5885	0.2815	0.2815	0.3321	0.3023	0.3082	0.3226
Experim. (ver)	0.8151	0.7806	0.4260	0.4354				

4.3 H_∞ Control Problem

The next step in the verification of zero sum differential game was to check the condition H_∞ control problem. For this purpose, a disturbance involving the change in friction coefficient of kinematic pairs of wheel-ground, which was carried out by changing the ground.

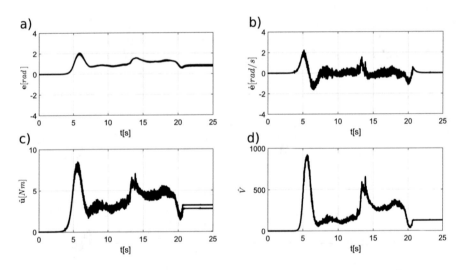

Fig. 5. Experimental test result in changing environmental conditions: tracking errors: e (a) and $\dot{e}$ (b) if the trajectory of the WMR is line; control signal $\hat{u}$ (c) and value function $\hat{V}$ (d)

Figure 5 presents the results of the verification of the approximate solution zero-sum differential game. Figure 5a and b shows tracking errors. About 13 [s] there was a change in the friction of the kinematics pair of the wheel-ground. The answer of the control system to the occurring disturbance is the increase in the signal $\hat{u}$ (Fig. 5c), which increases the value function (Fig. 5d). Based on the solutions (Fig. 5), it can be concluded that the solution is stable. This is due to the limited signals.

5 Summary

The article presents the experimental verification of the use zero-sum differential game in the WMR control task in changing environmental condition. To approximate the solution of the HJI equation, an actor-critic type structure was used in which the value functions were approximated by SN with square basic functions. The obtained verification results confirm the high accuracy tracking for desired path in changing. Verification carried out (in real time) confirms the applicability of adaptive optimal control in mobile robotics.

References

1. Vamvoudakis, K.G., Lewis, F.L.: Online solution of nonlinear two-player zero-sum games using synchronous policy iteration. Int. J. Robust Nonlinear Control **22**, 1460–1483 (2012)
2. Yasini, S., Sistani, M.B.N., Karimpour, A.: Approximate dynamic programming for two-player zero-sum game related to H_∞ control of unknown nonlinear continuous-time systems. Int. J. Control Autom. Syst. **13**(1), 99–109 (2015)
3. Zhu, Y., Zhao, D., Li, X.: Iterative adaptive dynamic programming for solving unknown nonlinear zero-sum game based on online data. IEEE Trans. Neural Netw. Learn. Syst. **28**(3), 714–725 (2017)
4. Jiang, H., et. al.: Iterative adaptive dynamic programming methods with neural network implementation for multi-player zero-sum games. Neurocomputing **30**, 54–60 (2018)
5. Szuster, M., Hendzel, Z.: Intelligent Optimal Adaptive Control for Mechatronic Systems. Springer International Publishing, Cham (2018)
6. Giergiel, M., Hendzel, Z., Zylski, W.: Modeling and control of wheeled mobile robots (in polish). PWN (2002)
7. Zylski, W.: Kinematics and dynamics of wheels mobile robots. Oficyna Wydawnicza Politechniki Rzeszowskiej, Rzeszów (1996)
8. Prokhorov, D.V., Wunsch, D.C.: Adaptive critic designs. IEEE Trans. Neural Netw. **8**(5), 997–1007 (1997)
9. Wang, F., Zhang, H., Liu, D.: Adaptive dynamic programming: an introduction. IEEE Comput. Intell. Mag. **4**(2), 39–47 (2009)
10. Abu-Khalaf, M., Huang, J., Lewis, F.L.: Nonlinear H_2/H_∞ Constrained Feedback Control. A Practical Design Approach Using Neural Networks. Springer (2006)
11. van der Schaft, A.J.: L_2-gain analysis of nonlinear systems and nonlinear state feedback H_∞ Control. IEEE Trans. Autom. Control **37**(6), 770–784 (1992)
12. Starr, A.W., Ho, Y.C.: Nonzero-sum differential game. J. Optim. Theory Appl. **3**(3), 184–206 (1969)
13. Zhang, H., Qin, C., Jiang, B., Luo, Y.: Online adaptive policy learning algorithm forH_∞ state feedback control of unknown affine nonlinear discrete-time systems. IEEE Trans. Cybern. **44**(12), 2706–2718 (2014)
14. Liu, D., Li, H., Wang, D.: Neural-network-based zero-sum game for discrete-time nonlinear systems via iterative adaptive dynamic programming algorithm. Neurocomputing **110**, 92–100 (2013)

The Use of Force Feedback to Control the Robot During Drilling

Marcin Chciuk[1(✉)] and Andrzej Milecki[2]

[1] Faculty of Mechanical Engineering, University of Zielona Góra,
Zielona Góra, Poland
m.chciuk@iibnp.uz.zgora.pl
[2] Institute of Mechanical Technology, Poznan University of Technology,
Poznan, Poland
andrzej.milecki@put.poznan.pl

Abstract. The article proposes the use of a joystick with the so-called force feedback, called in the English-language terminology as "haptic", to control a small robot. First, the haptic-type joysticks are briefly described and then their applications in robotics and in teleoperation application are presented. A laboratory stand designed to control a small robot performing drilling operations was described. Signal changes during the drilling process performed by the robot in the conditions of teleoperation controlled by a joystick with magnetorheological brakes were recorded. These signals were compared when the force feedback system was used and when it was not used.

Keywords: Force feedback · Haptic · Control · Manipulator · Joystick

1 Introduction

In some machines controlled by human operators, e.g. excavators, cranes, lifts, etc. different types of joystics, usually uniaxial, are used. Each of them usually controls one drive and therefore it is necessary to use several joysticks to control the whole arm. The operator, by changing the position of the joystick, sends a signal to the drive controller which changes the position of the operating element. The device movement is monitored with the help of eyesight. In Ergonomic requirements also mean that the operator's cabin is optimally silenced and air-conditioned which isolates the operator from the environment. The use of such solutions means that when controlling the device only the eyesight is used to a limited extent depending on the parameters of the camera and monitor and the other senses are not used at all. Problems arise when visual observation is difficult for example, due to the dustiness or significant distance of the operator from the machine's actuators. Due to the poor visibility of the work area the operator can hit the working part of the machine with objects in its surroundings, causing their destruction or damage to the machine itself.

A solution to this situation may be the introduction of additional force feedback and the use of a joystick with adjustable movement resistance. This force depends on the resistance force acting on the controlled drive. This system, the joystick and all the technology is called "haptic". Using it the user can feel the moment of collision with an

© Springer Nature Switzerland AG 2020
R. Szewczyk et al. (Eds.): AUTOMATION 2019, AISC 920, pp. 482–491, 2020.
https://doi.org/10.1007/978-3-030-13273-6_45

obstacle with the sense of touch and prevent possible damage in case he cannot observe the controlled object. This method is also commonly used in the case of teleoperation in which a system of cameras transmitting the image to the operator's screen is used to observe the controlled object. The system presented in Fig. 1 was the starting point for the construction of the test stand described in this article.

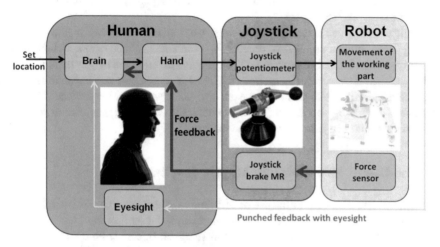

Fig. 1. Block diagram of human cooperation with the machine using force feedback.

2 Haptic Technology in Teleoperations

The use of computers and digital serial transmission of electrical signals made by internet communication enabled the construction of systems for controlling devices from a considerable distance [29]. Solutions of this type are also used when there is a need to "move" people away from hazardous or arduous zones. Such control is called teleoperation and aims at improving the safety, comfort and convenience of employees. The separation of the controlled machine from the operator, however, creates some problems and even dangers caused by the lack of the possibility of close and direct observation of the work process. In most cases of teleoperation the operator can only view the image from the camera placed near the working element of the machine on the monitor [26].

Due to the fact that in teleoperations man does not have direct contact with the machine and does not feel directly with the sense of touch what is happening to it during work, it would be very helpful to introduce elements providing additional information to the operator. Manually controlled machines often generate enormous power and the operator does not receive important information about resistance to the movement or encountering an obstacle. It may damage the device or surrounding objects. Managing machines by humans can be improved by using intensively developed joysticks and force feedback systems in recent years. Their task, in addition to controlling, is also to provide the operator with various tactile stimuli from controlled devices [6, 11, 13, 14, 18, 24, 25, 29].

3   Haptic Interfaces in Robotics

The issue of human interaction with the robot using the sense of touch is described in many publications. It is considered in two basic aspects. The first of these covers the issues of conveying a touch experience from the robot to a human being [1, 5, 7, 10, 15, 16, 20–22, 27, 28, 30–32] and the other one concerns equipping the robot with artificial sense of touch [12, 16, 19, 23].

Until recently haptic type joysticks were used mainly to deliver tactile sensations coming from human-controlled objects in computer games or in virtual reality. Recently, research has been carried out on the use of this type of joysticks to control real objects. In most cases, electric motors are used to generate resistance to motion in these devices which allows these joysticks to act also as active devices. Currently, research is also being conducted on the use of tactile devices in medicine in specialties such as ophthalmology, radiotherapy or surgery [13].

4   The Use of MR Fluids in Haptic Joysticks

Descriptions and examples of the use of MR fluid in joysticks can be found, inter alia, in the publication [17] which describes the construction of a semi-active joystick with two degrees of freedom in which two magnetorheological brakes were applied. The construction of another uniaxial joystick is described in publications [3] and [4]. On its shaft, a torque sensor with the handle is used. To measure the joystick angular position an encoder placed in the housing of the DC motor, was mounted. The presented in papers [3, 4] research results on the usage of this joystick, are focused on the detection of obstacles in virtual reality. The next publication [2] presents an active touch device with two brakes with MR fluid and with two DC motors. Additional interesting descriptions of solutions using MR fluids in joysticks can be found in publications [8, 9].

5   Description of the Test Stand

The conducted by authors research are focused on a control of a robot performing drilling operations. The view of the test bench is shown in Fig. 2. It consisted of a small robot (1) with a three-axis force sensor (7) mounted on the working tip and a mini milling machine (8). A three-axis joystick served to set the motion. In this joystick three brakes with magnetorheological fluid (2) are used. Its kinematic structure was similar to the structure of a controlled robot. For creating a controlled resistance in the joystick, magnetorheological brakes were assembled in each joystick wrist.

Electronic systems complemented the control system (3) is shown in Fig. 2. The controller of the system was implemented on a PC (4) on whose monitors (5) and (6) it was possible to additionally observe the movement of the robot based on images received from cameras (10) and (11). The material in which it was drilled or milled (9) was mounted to the ground in the robot's working space.

Figure 2 also shows the monitors used at the station. The monitor on the left (5) displayed the image from the camera (10) mounted on the last wrist of the robot and

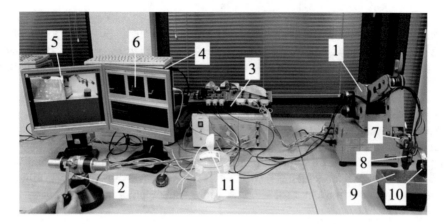

Fig. 2. View of the test station.

the image from the second camera observing the robot from a distance (11). On the monitor (6) it was possible to observe graphs showing changes in the angular position of the joints q_1, q_2, q_3 of the joystick and the robot. The window at the bottom of the screen showed changes in X, Y, Z forces measured by a three-axis force sensor. They corresponded to the forces of motion resistance occurring during processing.

The view of the joystick is shown in Fig. 3a and its construction in Fig. 3b. The joystick had the same range of angular movement in individual wrists as the robot. The difference between them, lies only in the proportional reduction of the length of individual links in the joystick to the robot, while maintaining the same scale in the whole system.

Three rotary MR (1), (2) and (3) brakes of similar design were used in the joystick (Fig. 3). The entire joystick construction is mounted on the brake main axis (1) located in the middle of the base. The brake (2) is connected directly to the first arm. A toothed wheel (5) has been connected on the third brake shaft (3) by a toothed belt (6) to the same wheel (7) attached to the forearm. Measurements of the swing angles of all axis of the joystick were made by potentiometers (4).

a)   b)

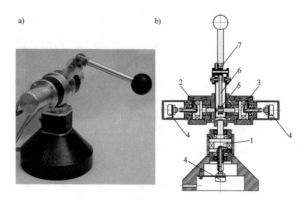

Fig. 3. Three-axis joystick MR: (a) external view; (b) section.

Electronic systems complemented the control system (3) is shown in Fig. 2. The controller of the system was implemented on a PC (4) on whose monitors (5) and (6) it was possible to additionally observe the movement of the robot based on images received from cameras (10) and (11). The material in which it was drilled or milled (9) was mounted to the ground in the robot's working space.

The block diagram of the control system built in the MATLAB/Simulink environment is shown in Fig. 4. The main elements of this system are: (1) measurement block for joystick position, (2) robot position signal measurement block, (3) robot control block, (4) the control block of the joystick brakes. The joystick arm joints angles were measured using potentiometers. The signals originating from them were fed to the analog inputs of the RT-DAC card. The measured values of the load forces from the three-axis forces sensor were fed to the same type of input.

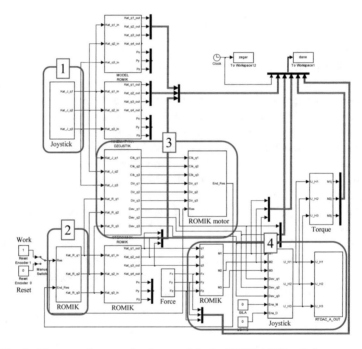

Fig. 4. The control system implemented in the MATLAB/Simulink program.

These signals were converted to digital values by the A/D converter on this card. Positions of the manipulator's angle were measured by incremental encoders connected to the RT-DAC card counters inputs. All input signals were subject to scaling by multiplying them by appropriate coefficients. In this way correspondingly adjusted angular position values for the joystick and the robot joints in radians, and forces in the N, were obtained.

6   Study of the Influence of Force Feedback Application on Control Parameters During Drilling

The experimental study with the use of a mini milling machine consisted of piercing a 1 mm thick plastic plate. The operation was performed under teleoperation conditions, i.e. the robot was invisible to the operator. He could only observe the robot on the screens of the monitors on which images from two cameras were displayed: the first camera showed a view of the robot's workspace and the second camera showed the image of a drill mounted in a miniature milling machine. The operator's task was to first bring the milling machine vertically down using the arm (articulation q_2) and the forearm (articulation q_3) of the robot to the workpiece and then perform the drilling operation up to the drilling. The tests were carried out for two control variants, i.e. with the force feedback in the joystick is switched off (I) and switched on (II). The attempts were performed one by one by 10 participants in the experiment.

Figure 5 shows the recorded example waveforms when the force feedback was turned off. Because the quality of the images from the camera presented on the screen, was not good the operator was unable to precisely move the milling machine and usually when it comes in contact with the workpiece and the beginning of drilling, the force Fz in the Z axis with which the robot operated on the mill was from about 20 N to 30 N (Fig. 5). Such a significant force occurring during the first initiation of drilling caused the drill (motor) to stop. This was a situation repeated in 80% of the tests carried out. The operator, seeing this i.e. stopping of drilling in the camera image, withdrew the milling machine, and after a while made the second attempt to start drilling again, moving the joystick's arm a little bit more slowly this time. This allowed the operator to drill material with an average force in the Z axis equal to approx. 10 N which no longer caused the drill to stop. On the course of changes in Fy and Fx, forces it can be seen that very large values of these forces occurred while pulling the stopped drill from the material and immediately after piercing the object "through". During the drilling the force values were small but different from zero. This was mainly due to the fact that the operator was not able to move the joystick arm perfectly along the Z axis (vertically down).

Figure 5 shows the waveforms that also show a large overshoot $E_{q\,2}$ that arose after the drilling operation when the operator moved the joystick arm too quickly when the drill was removed from the hole.

If using II. control variant (Fig. 6) the force feedback in the joystick was switched on. The feedback signal to the joystick was generated basing on the ex control error and on the M_{qn} torques in all robot joints qn. As in the case of the I-st control variant the operator had at his disposal an image from the same cameras.

The results of the research show that in this case, the operator felt the resistance of the joystick from the very beginning of contact between the miller and the object, prompting him to move the joystick arm and thus the robot arm more slowly.

This was the result of the use of force feedback from the ex-regulation error. It allowed to accelerate the reaction of the operator who, with a delay of about 0.3 s, felt resistance to the movement on the joystick coming from the torques occurring in the robot joints. This control allowed for almost "smooth" drilling of the workpiece

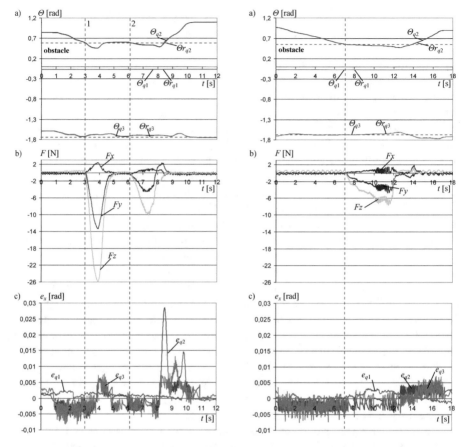

Fig. 5. Signals recorded during experimental investigations of drilling for I. Control variant in three axes.

Fig. 6. Signals recorded during experimental investigations of drilling for II. Control variant in three axes.

without blocking the drill. The main pressure force F_z of the robot on the workpiece vertically, i.e. in the Z axis, for this algorithm did not exceed 7 N. This algorithm allowed to limit the overshoot value which can be seen on the waveforms for the ex control error shown in Fig. 6.

The oscillations visible in the graphs of all Fx, Fy and Fz forces come from the vibrations created in the miniature milling machine. Drilling experiments were carried out for two control variants. For comparison, to see whether the force feedback in the joystick helps in controlling the robot, tests were also carried out using the I. control variant in which no force feedback was applied. The research was carried out with the participation of 10 operators. All of them, based on subjective assessments, concluded that the force feedback allowed them to control better when controlling the robot during drilling.

7 Summary

As part of this work research was undertaken on the three-axis arms robot control systems using a three-axis touch joystick with brakes with a magnetorheological fluid. The research was carried out during drilling, while the operator's control was performed in the teleoperation.

The work focused on the comparison of control without force feedback and control using force feedback in the joystick. The investigations have shown that, thanks to the applied force feedback, the operator had the ability to better control the drilling process by using the sense of touch. Such a control system, by increasing the resistance of movement of the joystick arm, allowed to prevent the drill from blocking in the material being processed. Also the overloading the robot joint drives are increased. In the tests the feedback signals are taken as force signals occurring at the robot working end as well as a signals originating from a regulation position error between the joystick and the robot.

Thanks to the use of a joystick with controllable force of its arm movement, the operator controlled the robot with a greater convergence of the angular position of the individual joystick and robot joints and thus the position error decreased. The research was carried out with the participation of 10 operators. All of them, based on subjective assessments, concluded that the force feedback allowed them to control better when controlling the robot during drilling.

References

1. Ahmadkhanlou, F.: Design, modeling and control of magnetorheological fluid-based force feedback dampers for telerobotic systems. Master thesis, Ohio State University, Columbia (2008)
2. An, J., Kwon, D.-S.: Control of multiple DOF hybrid haptic interface with active/passive actuators. In: International Conference on Intelligent Robots and Systems (2005)
3. An, J., Kwon, D.-S.: Haptic experimentation on a hybrid active/passive force feedback device. In: Proceedings of IEEE International Conference on Robotics and Automation (2002)
4. An J., Kwon, D.-S.: In Haptics, the influence of the controllable physical damping on stability and performance. In: Proceedings of International Conference Intelligent Robots and Systems, Sendai, Japan (2004)
5. Bardorfer, A., Munih, M.: Connecting haptic interface with a robot. In: Melecon 2000 10th Mediteranean Electrotechnical Conference, Cyprus (2000)
6. Basañez, L., Rosell, J., Palomo, L., Nuño, E., Portilla, H.: A framework for robotized teleoperated tasks. In: ROBOT 2011 Robótica Experimental, Sevilla, España (2011)
7. Bruder, J., Klimentjew, D., Zhang, J.: Haptic interactive telerobotics control architecture with iterative solving of inverse kinematics. In: Proceedings of the 2011 IEEE International Conference on Robotics and Biomimetrics, Phuket, Thailand (2011)
8. Chciuk, M., Milecki, A., Bachman, P.: Comparison of a traditional control and a force feedback control of the robot arm during teleoperation. In: Innovations in Automation, Robotics and Measurement Techniques - AUTOMATION 2017. Warsaw, Polska. Advances in Intelligent Systems and Computing, vol. 550, pp. 277–289. Springer (2017)

9. Chciuk, M., Myszkowski, A.: Wykorzystanie trzy-osiowego dżojstika dotykowego z cieczą magnetoreologiczną i siłowym sprzężeniem zwrotnym do sterowania ramienia robota. Archiwum Technologii Maszyn i Automatyzacji, Poznań (2008)

10. Chotiprayanakul, P., Wang, D., Kwok, N., Liu, D.: A haptic base human robot interaction approach for robotic grit blasting. In: Proceedings of the International Symposium on Automation and Robotics in Construction, Vilnius (2008)

11. El Saddik, A., Orozco, M., Eid, M., Cha, J.: Haptics Technologies - Bringing Touch to Multimedia. Springer, Heidelberg (2011)

12. Fumagalli, M.: Increasing Perceptual Skills of Robots Through Proximal Force/Torque Sensors. Ph.D. thesis, Department of Informatics, Systems and Telecommunication, University of Genoa, Springer (2014)

13. Goethals, P.: Tactile feedback for robot assisted minimally invasive surgery: an overview. Division PMA Department of Mechanical Engineering Katholieke Universiteit Leuven (2008)

14. Kern, T.A.: Engineering - Haptic devices. Springer, Heidelberg (2009)

15. Lee, J.-B., Lim, J.-H., Park, C.-W.: Development of ethernet based tele-operation systems using haptic devices. Inf. Sci. **172**, 263–280 (2005)

16. León, B., Morales, A., Sancho-Bru, J.: From Robot to Human Grasping Simulation. Springer, Heidelberg (2014)

17. Liu, B.: Development of 2-DOF haptic device working with magnetorheological fluids. MEng thesis, Faculty of Engineering, University of Wollongong (2006). http://ro.uow.edu.au/theses/136/

18. Lin, M.C., Otaduy, M.A.: Haptic Rendering: Foundations, Algorithms, and Applications. A K Peters, Ltd., Wellesley (2008)

19. Kim, M., Yoon, S.-S., Kang, S., Kim, S.-J., Kim, Y.-H., Yim, H.-S., Lee, C.-D., Yeo, I.-T.: Safe arm design for service robot. In: Proceedings of The Second IARP -IEEE/RAS Joint Workshop on Technical Challenge for Dependable Robots in Human Environments, Toulouse, France, pp. 88–95 (2002)

20. Marchese, A., Hoyt, H.: Force sensing and haptic feedback for robotic telesurgery, Raport z badań. Worcester Polytechnic Institute (2010)

21. O'Malley, M.K., Gupta, A.: HCI: Beyond the GUI. In: Haptic Interfaces. Morgan-Kaufman (2007)

22. Mishima, M., Kawasaki, H., Mouri, T., Endo, T.: Haptic teleoperation of humanoid robot hand using three-dimensional force feedback. In: Proceedings on the 9th International Symposium on Robot Control (SYROCO'09) The International Federation of Automatic Control Nagaragawa Convention Center, Gifu, Japan (2009)

23. Miyashita, T., Tajika, T., Ishiguro, H., Kogure, K., Hagita, N.: Haptic communication between humans and robots. In: Robotics Research, vol. 28, pp. 525–536. Springer, Heidelberg (2007)

24. Molen van der, K.: The vicinity sensor: exploring the use of hapticons in everyday appliances. Master's thesis, Technische Universiteit Eindhoven Department Mechanical Engineering, Dynamics and Control Technology Group, Eindhoven (2005)

25. Otaduy, M., Lin, M.: High Fidelity Haptic Rendering. Morgan & Claypool, Los Altos (2006)

26. Park, H.-J., Lee, S., Kang, S.-K., Kang, M.-S., Song, M.-S., Han, C.: Experimental study on hydraulic signal compensation for the application of a haptic interface to a tele-operated excavator. In: Proceedings of the 28th ISARC, Seoul, Korea (2011)

27. Radi, M., Reinhart, G.: Industrial haptic robot guidance system for assembly processes. In: IEEE International Workshop on Haptic Audio visual Environments and Games, Lecco, pp. 69–74 (2009)

28. Rozo, L., Jiménez, P., Torras, C.: Learning force-based robot skills from haptic demonstration. In: Artificial Intelligence Research and Development 2010, volume 220 of Frontiers in Artificial Intelligence and Applications, pp. 331–340 (2010)
29. Stone, R.J.: Haptic feedback: a potted history from telepresence to virtual reality. Robotica **10**, 461–467 (1992)
30. Takamuku, S., G´omez, G., Hosoda, K., Pfeifer, R.: Haptic discrimination of material properties by a robotic hand. In: IEEE 6th International Conference on Development and Learning (ICDL 2007), London (2007)
31. Horie, T., Abe, N., Tanaka, K., Taki, H.: Controlling two remote robot arms with direct instruction using hapticmaster and vision system. In: w: VSMM 2000: 6th International Conference on Virtual Systems and MultiMedia, pod. red. Hal Thwaites, Ogaki, Gifu, Japan (2000)
32. Zhengyi, Y., Chen, Y.: Haptic rendering of milling. Mechanical Engineering Department, The University of Hong Kong, Hong Kong, China (2003)

Hierarchical Petri Net Representation of Robot Systems

Maksym Figat$^{(\boxtimes)}$ and Cezary Zieliński

Institute of Control and Computation Engineering,
Warsaw University of Technology, Nowowiejska 15/19, 00-665 Warsaw, Poland
{M.Figat,C.Zielinski}@ia.pw.edu.pl

Abstract. The paper presents a holistic robot system specification methodology taking into account both the system structure and its activities. It is based on the concept of an embodied agent. Each agent is decomposed into cooperating subsystems. Previously subsystem activities were defined by a hierarchical finite state machine (HFSM). In that approach communication between subsystems was not specified explicitly. This paper utilises a Hierarchical Petri Net (HPN) with conditions as an alternative modelling tool. HPN can be obtained by transformation of the HFSM into HPN. The resulting HPN consists of consecutive layers: subsystem layer, behaviour layer and communication layer. The proposed methodology not only organizes in a systematic and holistic manner the development of the robot system, but also introduces a comprehensive description of concurrently acting subsystems. The HPN description can be utilised to automatically generate the robot controller code.

Keywords: Robot System Specification Methodology ·
Hierarchical Finite State Machine · Hierarchical Petri Net

1 Introduction

Each robot within a robotic system consists of hardware (i.e. receptors – collecting data from the environment, and effectors – affecting the environment) and an onboard computer (running the robot control software, i.e. robot controller). The robot controller must be developed in a such way that it guarantees that the robot will perform the entrusted class of tasks. To ensure that, one of the software engineering development approaches should be followed, e.g.: waterfall, incremental, iterative, agile. All of them include major steps: formulation of requirements, specification, implementation, validation and deployment [1]. Requirements define what services should be provided by the system and identify the constraints imposed on the system activities. Based on the provided requirements, the model of the system emerges, i.e. its specification. This model is transformed into implementation and is further validated on test examples. This methodology is suitable for developing robotic systems, however domain specific knowledge should be taken into account.

© Springer Nature Switzerland AG 2020
R. Szewczyk et al. (Eds.): AUTOMATION 2019, AISC 920, pp. 492–501, 2020.
https://doi.org/10.1007/978-3-030-13273-6_46

Many general robot control system architectures have been proposed, e.g.: Sense-Plan-Act [2], multi-tier [3] (two-tier (2T) or three-tier (3T)), subsumption [4], robot schema [5–7]. Nevertheless, in many existing robotic systems it is difficult to determine their architecture type [8]. Although robot control systems are inherently complex, rarely their authors formally specify their structure and activities. Lack of formal system specification usually results in obscure architecture, that is difficult to modify, extend or integrate with other systems. The process of developing robotic system requires not only the description of its architecture, but also a general methodology leading to the specification of the robotic system. Moreover additional tools facilitating the transformation of the specification into implementation are required.

This paper introduces the Robot System Specification Methodology (RSSM), which is based on the concepts embedded in the robotics domain, i.e. embodied agents composed of effectors, receptors and control system, exhibiting behaviours defined in terms of transition functions [9–15]. The specification of a robot system takes into account both its structure and activities [9–11]. Since the embodied agent consists of a set of subsystems, its activities result from the activities of its subsystems and their interdependence. Previously the activities of each subsystem were expressed by a two-layered hierarchical finite state machine (HFSM) [13–15]. The upper HFSM layer switched subsystem behaviours specified as FSMs of the lower layer. Those FSMs iteratively executed the associated transition functions, which used the input information to the subsystem as arguments and produced their output information. However, the inter-subsystem communication was not defined explicitly.

Alternatively robot system activities can be expressed using Petri nets with conditions [16–18]. A Petri net can be used throughout the whole development cycle of the robot system, including: detailed description of subsystem activities and inter-subsystem communication models, verification of system specification, code generation, verification of compliance with requirements and testing the developed robot controller [19,20]. This paper proposes Hierarchical Petri Nets (HPN) with conditions as an alternative to HFSM.

Section 2 presents an introduction to RSSM based on HPN. RSSM requires the determination of the robot system structure (Sect. 2.1) and activities (Sect. 2.2). The definition of HPN is introduced in Sect. 3. HPN is obtained partially by transforming the HFSM specifying the subsystem activities. HPN consists of three layers: subsystem layer – expressing how the sybsystems switch behaviours (Sect. 4.1), behaviour layer – expressing each behaviour of the subsystem (Sect. 4.2) and communication layer – defining the communication models utilised by a behaviour to interact with other subsystems (Sect. 4.3). The proposed methodology is summarised in Sect. 5.

2   Robot System Specification Methodology

Following the Robot System Specification Methodology (RSSM) based on FSMs [12], the robot system developer obtains a formal specification of the robotic system encompassing both its structure and activities. The structure specifies how the system is decomposed into smaller components, i.e. agents, their subsystems and their interconnections. Activities are specified in terms of behaviours, which iteratively execute transition functions.

2.1   Specification of Robot System Structure

A robot system consists of a set of cooperating embodied agents [10,21]. An embodied agent executes its task by affecting the environment using its effectors and acquiring data from its surroundings using receptors [9–11]. An embodied agent a_j, where j is its name, is decomposed into a set of communicating subsystems (Fig. 1): control subsystem c_j, real effectors $E_{j,h}$, virtual effectors $e_{j,n}$, real receptors $R_{j,l}$, virtual receptors $r_{j,k}$, where h, n, l and k indicate specific real/virtual effectors/receptor. Real entities directly interact with the environment, i.e. they contain hardware devices, while the virtual entities act as drivers transforming data into a form accepted/produced by the control subsystem. The control subsystem c_j of the embodied agent a_j acquires data about the current state of the environment using its virtual receptors $r_{j,k}$, which provide the aggregated data obtained from the real receptors $R_{j,l}$. This data and the internal imperative to realise the agent's task is utilised to produce control commands that are delivered to the virtual effectors $e_{j,n}$, which transform them into control signals for the real effectors $E_{j,h}$. The loop closes through the environment. Agent subsystems communicate with each other through buffers, as presented in Fig. 1. The buffers are named in a systematic way, e.g. [11]. The letter in the center determines the type of the subsystem s, where $s \in \{c, e, r, E, R\}$. The compound right subscript specifies the names of: the agent, its subsystem and optionally a buffer element. The left subscript determines type of the buffer: input buffer uses x, output buffer uses y, while internal does not use this subscript. The right superscript determines a discrete time, e.g. i, and the left superscript specifies the subsystem type from which the data arrives or to which the data is dispatched. For example, ${}^r_x c^i_{j,k}$ is the input buffer of the control subsystem c_j receiving the data from the virtual receptor $r_{j,k}$ at the discrete time i, while ${}^T_x c^i_{j,j'}$ is the input buffer of c_j receiving the data from the agent $a_{j'}$.

2.2   Specification of Activities of an Embodied Agent

Previously the activities of subsystem $s_{j,v}$ were determined by a two-layered hierarchical FSM (HFSM) [13–15]. The first layer described how the subsystem $s_{j,v}$ switches between behaviours ${}^s\mathcal{B}_{j,v,\omega}$, where ω is the name of behaviour. Those behaviours are associated with the states (nodes of the graph) of the upper layer FSM ${}^s\mathcal{F}_{j,v}$. The nodes of ${}^s\mathcal{F}_{j,v}$ are connected to each other by directed arcs labeled by initial conditions ${}^s f^\sigma_{j,v,\alpha}$, where α indicates the name of condition. Hereinafter no distinction is made between automaton ${}^s\mathcal{F}_{j,v}$ and its graph, due to contextual obviousness. An exemplary FSM ${}^s\mathcal{F}_{j,v}$ is presented in Fig. 2a. The second layer is composed of FSMs ${}^s\mathcal{F}^\mathcal{B}_{j,v,\omega}$, which define the behaviours ${}^s\mathcal{B}_{j,v,\omega}$. The general structure of behaviour was presented in Fig. 2b.

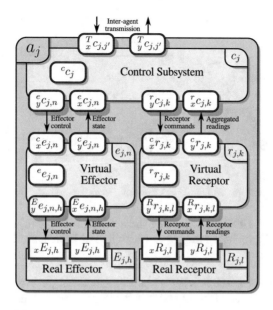

Fig. 1. The general structure of embodied agent.

Each behaviour ${}^{s}\mathcal{B}_{j,v,\omega}$ causes the subsystem $s_{j,v}$ to execute iteratively the following sequence of steps. It calculates the associated transition function ${}^{s}f_{j,v,\omega}$

$$\left({}^{s}s_{j,v}^{i+1}, {}_{y}s_{j,v}^{i+1}\right) := {}^{s}f_{j,v,\omega}\left({}^{s}s_{j,v}^{i}, {}_{x}s_{j,v}^{i}\right), \tag{1}$$

sends the result to the connected subsystems, increments the discrete time index i, and receives the data from the connected subsystems. Once this sequence is completed two conditions are evaluated: terminal condition ${}^{s}f_{j,v,\xi}^{\tau}$ (ξ indicates the name of terminal condition), error condition ${}^{s}f_{j,v,\beta}^{\varepsilon}$ (β indicates the name of error condition). If neither ${}^{s}f_{j,v,\xi}^{\tau}$ nor ${}^{s}f_{j,v,\beta}^{\varepsilon}$ is fulfilled, then $s_{j,v}$ executes the next iteration of behaviour ${}^{s}\mathcal{B}_{j,v,\omega}$. Otherwise, the behaviour is terminated and depending on the values of initial conditions ${}^{s}f_{j,v,\zeta}^{\sigma}$, where ζ is the initial condition designator, the subsystem $s_{j,v}$ switches to another state of ${}^{s}\mathcal{F}_{j,v}$.

The FSM based specification method did not introduce the inter-subsystem communication model. Thus, in fact, both subsystems act independently of each other. Communication had to be dealt with at the implementation stage [13–15]. An alternative way of describing the activities of the agent, enabling both the description of its activities and the communication model specifying explicitly interaction between communicating subsystems uses a hierarchical Petri net.

3 Hierarchical Petri Net

A Petri net is a bipartite graph containing transitions t and places p alternatively connected by directed arcs [16]. In a HPN some places can be substituted by

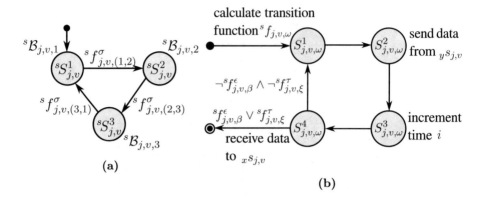

Fig. 2. Two-layered HFSM determining the activities of $s_{j,v}$, (a) An exemplary FSM $^s\mathcal{F}_{j,v}$, (b) FSM $^s\mathcal{F}^{\mathcal{B}}_{j,v,\omega}$ being the general template of behaviour $^s\mathcal{B}_{j,v,\omega}$

pages $\mathcal{P}$. A page is a HPN with a distinguished single input place p_{in} and a single output place p_{out}. Herein HPNs with conditions are used, thus with each transition a condition $\mathcal{C}$ is associated. When markers are assigned to places the net $\mathcal{H}$ becomes a marked HPN with conditions. With each place a single operation $\mathcal{O}$ is associated. Places from different nets can be fused with each other. Fused places p^{fusion} are in principle the same place appearing in two or more nets [18]. These places are indistinguishable from each other and thus contain the same markers. Fusing places of different nets combines those nets into a single net of a more complex structure. This article uses a graphical representation of HPNs, in which places are represented by single circles, pages by double circles, transitions by rectangles, directed arcs by arrows and markers by black filled circles. Conditions $\mathcal{C}$ associated with transitions are placed within square brackets. If a condition is always fulfilled (i.e. is True) then the label may be omitted. Association of two places: $p^{\text{fusion}}_{\mathcal{H}_1,\alpha}$ and $p^{\text{fusion}}_{\mathcal{H}_2,\beta}$, belonging respectively to nets: $\mathcal{H}_1$ and $\mathcal{H}_2$ is represented by a single place $p^{\text{fusion}}_{(\mathcal{H}_1,\mathcal{H}_2),(\alpha,\beta)}$. The definition of a safe HPN requires that at the most one marker can reside in a place.

4   Subsystem Activities Represented by a HPN

Activities of subsystem $s_{j,v}$ are expressed by a HPN $^s\mathcal{H}_{j,v}$ having three layers: (1) subsystem layer – Petri net $^s\mathcal{H}_{j,v}$, which is obtained by a transforming $^s\mathcal{F}_{j,v}$ (Sect. 4.1), contains pages $^s\mathcal{P}^{\mathcal{B}}_{j,v,\omega}$ defining behaviours $^s\mathcal{B}_{j,v,\omega}$ of subsystem $s_{j,v}$, (2) behaviour layer – composed of Petri nets $^s\mathcal{H}^{\mathcal{B}}_{j,v,\omega} \equiv {}^s\mathcal{P}^{\mathcal{B}}_{j,v,\omega}$ defining behaviours $^s\mathcal{B}_{j,v,\omega}$ (Sect. 4.2), (3) communication layer – composed of two Petri nets defining communication models utilised by behaviour $^s\mathcal{B}_{j,v,\omega}$, i.e. $^s\mathcal{H}^{\mathcal{B}}_{j,v,\omega,\text{snd}} \equiv {}^s\mathcal{P}^{\mathcal{B}}_{j,v,\omega,\text{snd}}$ describes how the data is sent to the other subsystems and $^s\mathcal{H}^{\mathcal{B}}_{j,v,\omega,\text{rcv}} \equiv {}^s\mathcal{P}^{\mathcal{B}}_{j,v,\omega,\text{rcv}}$ describes how the data is received from the other subsystems (Sect. 4.3). An exemplary three-layered HPN $^s\mathcal{H}_{j,v}$ specifying the activities of subsystem $s_{j,v}$ is presented in Fig. 3.

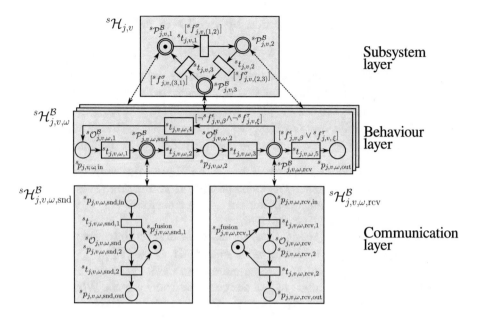

Fig. 3. Three-layered HPN $^s\mathcal{H}_{j,v}$ specifying the activities of the subsystem $s_{j,v}$.

4.1 Subsystem Layer – Transformation of $^s\mathcal{F}_{j,v}$ into $^s\mathcal{H}_{j,v}$

Let $^s\mathcal{F}_{j,v}$ be the FSM expressing the activities of subsystem $s_{j,v}$. Transformation of $^s\mathcal{F}_{j,v}$ into $^s\mathcal{H}_{j,v}$ requires the following steps:

1. Compose $^s\mathcal{H}_{j,v}$ of $|^s\hat{S}_{j,v}|$ pages $^s\mathcal{P}^\mathcal{B}_{j,v,\omega}$, $\omega = 1,\ldots,|^s\hat{S}_{j,v}|$, where $^s\hat{S}_{j,v}$ is the set of states of $^s\mathcal{F}_{j,v}$ and $|^s\hat{S}_{j,v}|$ is its cardinality. Page $^s\mathcal{P}^\mathcal{B}_{j,v,\omega}$ represents the activities of behaviour $^s\mathcal{B}_{j,v,\omega}$ associated with the state $^sS^\omega_{j,v}$ of $^s\mathcal{F}_{j,v}$,
2. For each pair of states $(^sS^\gamma_{j,v}, {}^sS^\omega_{j,v})$ connected by a directed arc starting in $^sS^\gamma_{j,v}$ and ending in $^sS^\omega_{j,v}$, add to $^s\mathcal{H}_{j,v}$ a transition $^st_{j,v,\zeta}$ with condition $^s\mathcal{C}_{j,v,\zeta} = {}^sf^\sigma_{j,v,(\gamma,\omega)}$, where ζ is a consecutive number of transition within the $^s\mathcal{H}_{j,v}$ and $^sf^\sigma_{j,v,(\gamma,\omega)}$ is the initial condition associated with the directed arc connecting $^sS^\gamma_{j,v}$ and $^sS^\omega_{j,v}$,
3. For each transition $^st_{j,v,\zeta}$ connect $^s\mathcal{P}^\mathcal{B}_{j,v,\gamma}$ and $^st_{j,v,\zeta}$ as well as $^st_{j,v,\zeta}$ and $^s\mathcal{P}^\mathcal{B}_{j,v,\omega}$ by directed arcs,
4. Add a single marker to the page corresponding to the initial state of $^s\mathcal{F}_{j,v}$.

The transformation of the exemplary $^s\mathcal{F}_{j,v}$ (presented in Fig. 2a) into $^s\mathcal{H}_{j,v}$ is presented in the subsystem layer in Fig. 3.

4.2   Behaviour Layer – Specified by Petri Net $^s\mathcal{H}^{\mathcal{B}}_{j,v,\omega}$

The behaviour $^s\mathcal{B}_{j,v,\omega}$ specified by the lower layer FSM $^s\mathcal{F}^{\mathcal{B}}_{j,v,\omega}$ is transformed into Petri net $^s\mathcal{H}^{\mathcal{B}}_{j,v,\omega}$ using the same algorithm as presented in Sect. 4.1. The activities of $^s\mathcal{H}^{\mathcal{B}}_{j,v,\omega}$ start when a marker appears in its input place $^sp_{j,v,\omega,\text{in}}$ and cease when it reaches the output place $^sp_{j,v,\omega,\text{out}}$. $^s\mathcal{H}^{\mathcal{B}}_{j,v,\omega}$ is the general pattern of behaviour. The refinement of page $^s\mathcal{P}^{\mathcal{B}}_{j,v,\omega}$ is presented in the subsystem layer in Fig. 3. The page $^s\mathcal{P}^{\mathcal{B}}_{j,v,\omega}$ executes the behaviour $^s\mathcal{B}_{j,v,\omega}$ in the following way:

1. A marker appearing in the place $^sp_{j,v,\omega,\text{in}}$ initiates the operation $^s\mathcal{O}^{\mathcal{B}}_{j,v,\omega,1}$ associated with this place. The value of transition function (1) is calculated and inserted into the output buffer $_ys_{j,v}$. The transition $^st_{j,v,\omega,1}$ will not fire until the calculations are completed,
2. When the transition $^st_{j,v,\omega,1}$ fires, a marker is inserted into the input place of the page $^s\mathcal{P}^{\mathcal{B}}_{j,v,\omega,\text{snd}}$ initiating its execution. The page $^s\mathcal{P}^{\mathcal{B}}_{j,v,\omega,\text{snd}}$ represents the net $^s\mathcal{H}^{\mathcal{B}}_{j,v,\omega,\text{snd}}$ defining the communication model used to send the data from the $s_{j,v}$ output buffer $_ys_{j,v}$ to the connected subsystems. When the page $^s\mathcal{P}^{\mathcal{B}}_{j,v,\omega,\text{snd}}$ terminates its execution, the transition $^st_{j,v,\omega,2}$ fires,
3. The operation $^s\mathcal{O}^{\mathcal{B}}_{j,v,\omega,2}$, associated with the place $^sp_{j,v,\omega,2}$, increments the subsystem $s_{j,v}$ discrete time i,
4. When the transition $^st_{j,v,\omega,3}$ fires, it activates the page $^s\mathcal{P}^{\mathcal{B}}_{j,v,\omega,\text{rcv}}$, i.e. net $^s\mathcal{H}^{\mathcal{B}}_{j,v,\omega,\text{rcv}}$ defining the communication model utilised to receive data from the connected subsystems,
5. When the page $^s\mathcal{P}^{\mathcal{B}}_{j,v,\omega,\text{rcv}}$ finishes its activity, the terminal condition $^sf^\tau_{j,v,\xi}$ and the error condition $^sf^\epsilon_{j,v,\beta}$ are checked. If none of them is fulfilled then the next iteration of behaviour $^s\mathcal{B}_{j,v,\omega}$ is started. Otherwise, the execution of page $^s\mathcal{P}^{\mathcal{B}}_{j,v,\omega}$ is terminated, i.e. a marker appears in the output place $^sp_{j,v,\omega,\text{out}}$ and thus control returns to the net $^s\mathcal{H}_{j,v}$.

4.3   Communication Layer – Communication Models Used by $^s\mathcal{B}_{j,v,\omega}$

The communication layer specifies communication models utilised within $^s\mathcal{B}_{j,v,\omega}$. In each iteration of $^s\mathcal{B}_{j,v,\omega}$ subsystem $s_{j,v}$ uses two communication models. One specifying how the calculated data is sent to the connected subsystems and another specifying how data is received from the associated subsystems. Each of two communicating subsystems, where one is the producer and the other is the consumer, can act either in a blocking mode or a non-blocking mode. The producer in the blocking mode suspends its execution until it receives the confirmation of the receipt of data from the consumer, while in the non-blocking mode it dispatches the data and instantaneously starts executing its further activities. The consumer in the blocking mode waits until it receives the data, while in the non-blocking mode if the data is not available it resumes its operation without waiting for data arrival. Selection of the mode for each pair of subsystems determines the communication model. Two communication models are presented in this article: (1) both subsystems act in the non-blocking mode (communication layer in Fig. 3) and (2) both subsystems act in the blocking mode (Fig. 4a).

The behaviour layer contains two pages ${}^s\mathcal{P}^{\mathcal{B}}_{j,v,\omega,\text{snd}}$ and ${}^s\mathcal{P}^{\mathcal{B}}_{j,v,\omega,\text{rcv}}$ executing the communication models used by behaviour ${}^s\mathcal{B}_{j,v,\omega}$ of subsystem $s_{j,v}$. Those models are represented by Petri nets defining the communication modes of communicating parties. Each of those nets has to be divided into two nets with fused places. One net defines the behaviour of the data producer, while the other the behaviour of the data consumer – it should be noted that those two nets are created for two different subsystems. For instance, let subsystem $s_{j,v}$, executing the behaviour ${}^s\mathcal{B}_{j,v,\omega}$, send data to subsystem $s_{j,h}$, which executes the behaviour ${}^s\mathcal{B}_{j,h,\omega'}$. Let those two subsystems both use the blocking mode (Fig. 4a) to send and receive data, respectively. The Petri net representing this model is divided into two nets: ${}^s\mathcal{H}^{\mathcal{B}}_{j,v,\omega,\text{snd}}$ (Fig. 4b) to send the calculated data by $s_{j,v}$ and ${}^s\mathcal{H}^{\mathcal{B}}_{j,h,\omega',\text{rcv}}$ (Fig. 4c) to receive this data by $s_{j,h}$, thus synchronous communication results. When the data is ready to be sent by $s_{j,v}$ to $s_{j,h}$, a marker appears in the place ${}^s p_{j,v,\omega,\text{snd,in}}$ and thus transition ${}^s t_{j,v,\omega,\text{snd,1}}$ fires, producing two markers, one in the place ${}^s p_{j,v,\omega,\text{snd,2}}$ and the other in the fused place ${}^s p^{\text{fusion}}_{j,v,\omega,\text{snd,1}}$. The latter marker appears also in the fused place ${}^s p^{\text{fusion}}_{j,h,\omega',\text{rcv,1}}$, thus $s_{j,h}$ is informed that data has arrived, so it sends the confirmation of receiving the data to $s_{j,v}$ (by inserting a marker into the place ${}^s p^{\text{fusion}}_{j,h,\omega',\text{rcv,3}}$) and resumes its further activities. When the marker appears in the fused place ${}^s p^{\text{fusion}}_{j,v,\omega,\text{snd,3}}$, the transition ${}^s t_{j,v,\omega,\text{snd,2}}$ fires, $s_{j,v}$ terminates the execution of ${}^s\mathcal{P}^{\mathcal{B}}_{j,v,\omega,\text{snd}}$ and resumes further activities of ${}^s\mathcal{P}^{\mathcal{B}}_{j,v,\omega}$. The Petri net ${}^s\mathcal{H}^{\mathcal{B}}_{j,v,\omega,\text{rcv}}$ is defined analogically.

If subsystem $s_{j,v}$ sends or receives data from two or more subsystems, for each such subsystem a Petri net is defined in the above-mentioned way. Those nets can be connected either sequentially or in parallel.

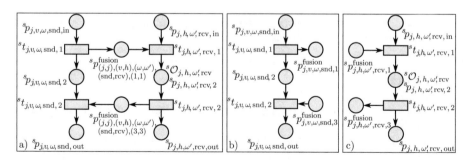

Fig. 4. (a) Communication model where both $s_{j,v}$ and $s_{j,h}$ use the blocking mode. (b) Petri net ${}^s\mathcal{H}^{\mathcal{B}}_{j,v,\omega,\text{snd}}$ defining the blocking communication mode of $s_{j,v}$ executing behaviour ${}^s\mathcal{B}_{j,v,\omega}$. (c) Petri net ${}^s\mathcal{H}^{\mathcal{B}}_{j,h,\omega',\text{rcv}}$ defining the blocking communication mode of $s_{j,h}$ executing behaviour ${}^s\mathcal{B}_{j,h,\omega'}$.

5   Recapitulation

This paper presents the Robot System Specification Methodology (RSSM) which can be used for developing robot system controllers. The methodology indicates the methods for specifying both the robot control structure and its activities. The structure of the robot system is specified using well-known robotics concepts, i.e. concepts associated with embodied agents, while activities are defined by HPNs describing the activities of each subsystem within the embodied agent. The HPN describes in a holistic manner the activities of each subsystem, enabling explicit definition of the communication model utilised by two communicating subsystems. The two independent HPNs determining the activities of two communicating subsystems are connected by a fused place forming a single HPN. The proposed modeling approach based on HPNs can be extended to specify the activities a multi-agent robot system.

Petri nets can be utilised in the whole robot control software development cycle: specification, implementation and validation. A graphical tool has been developed for specifying the multi-agent robot systems using HPNs. This tool can also be utilised to automatically generate the source code of the robot controller based on formal specification formulated by the developer. Utilising the RSSM and the developed tool specification of a rudimentary robot executing the follow-the-line task has been created. The HPN of the developed robot controller has been verified – no deadlocks existed. To this end, each layer of HPN was analysed using the place invariants method. Because both lower layers of HPN, i.e. behaviour and communication layers, are created out of reusable patterns, the verification of succeeding robot systems is limited only to the verification of the subsystem layer. The verified model has been then automatically transformed into the robot controller software, which subsequently has been verified in the simulation. Further research will be directed at utilising RSSM in the development of much more complex multi-agent robot systems.

Acknowledgments. This work was supported by the National Science Centre, Poland (grant number 2017/25/N/ST7/00900).

References

1. Sommerville, I.: Software Engineering, 9th edn. Addison-Wesley Publishing Company, New York (2010)
2. Nilsson, N.: Shakey the robot, technical note 323. Technical report, AI Center, SRI International, Menlo Park (1984)
3. Beetz, M., Mösenlechner, L., Tenorth, M.: CRAM – a cognitive robot abstract machine for everyday manipulation in human environments. In: IEEE/RSJ International Conference on Intelligent Robots and Systems, IROS, October 18–22, 2010, Taipei, Taiwan, pp. 1012–1017. IEEE (2010)
4. Brooks, R.A.: A robust layered control system for a mobile robot. IEEE J. Robot. Autom. **2**(1), 14–23 (1986)

5. Arbib, M.: Perceptual structures and distributed motor control. In: Handbook of Physiology – The Nervous System II. Motor Control, pp. 1449–1480. Wiley Online Library (1981)
6. Lyons, D.M., Arbib, M.A.: A formal model of computation for sensory-based robotics. IEEE Trans. Robot. Autom. **5**(3), 280–293 (1989)
7. Arkin, R.C.: Motor schema-based mobile robot navigation. Int. J. Robot. Res. **8**, 92–112 (1989)
8. Kortenkamp, D., Simmons, R.: Robotic systems architectures and programming. In: Khatib, O., Siciliano, B. (eds.) Springer Handbook of Robotics, pp. 187–206. Springer, Heidelberg (2008)
9. Zieliński, C., Winiarski, T.: General specification of multi-robot control system structures. Bull. Pol. Acad. Sci. Tech. Sci. **58**(1), 15–28 (2010)
10. Brooks, R.A.: Intelligence without reason. Artif. Intell. Crit. Concepts **3**, 107–163 (1991)
11. Kornuta, T., Zieliński, C.: Robot control system design exemplified by multi-camera visual servoing. J. Intell. Robot. Syst. **77**(3–4), 499–524 (2013)
12. Zieliński, C., Figat, M.: Robot system design procedure based on a formal specification. In: Recent Advances in Automation, Robotics and Measuring Techniques. Advances in Intelligent Systems and Computing (AISC), vol. 440, pp. 511–522. Springer (2016)
13. Figat, M., Zieliński, C., Hexel, R.: FSM based specification of robot control system activities. In: 2017 11th International Workshop on Robot Motion and Control (RoMoCo), pp. 193–198 (2017)
14. Zieliński, C., Figat, M., Hexel, R.: Robotic systems implementation based on FSMs. In: Szewczyk, R., Zieliński, C., Kaliczyńska, M. (eds.) Automation 2018, pp. 441–452. Springer, Cham (2018)
15. Zieliński, C., Figat, M., Hexel, R.: Communication within multi-FSM based robotic systems. J. Intell. Robot. Syst., 1–19 (2018). https://doi.org/10.1007/s10846-018-0869-6
16. Peterson, J.L.: Petri Net Theory and the Modeling of Systems. Prentice Hall, Upper Saddle River (1981)
17. Girault, C., Rudiger, V.: Petri Nets for Systems Engineering. Springer, Heidelberg (2003)
18. Huber, P., Jensen, K., Shapiro, R.M.: Hierarchies in Coloured Petri Nets. In: Rozenberg, G. (ed.) Advances in Petri Nets. Lecture Notes in Computer Science, vol. 483, pp. 313–341. Springer, Heidelberg (1991)
19. Montano, L., García, F.J., Villarroel, J.L.: Using the time petri net formalism for specification, validation, and code generation in robot-control applications. Int. J. Robot. Res. **19**(1), 59–76 (2000)
20. Billington, D., Estivill-Castro, V., Hexel, R., Rock, A.: Requirements Engineering via Non-monotonic Logics and State Diagrams, pp. 121–135. Springer, Heidelberg (2011)
21. Jennings, N.R., Sycara, K., Wooldridge, M.: A roadmap of agent research and development. Auton. Agents Multi Agent Syst. **1**(1), 7–38 (1998)

Laser-Based Localization and Terrain Mapping for Driver Assistance in a City Bus

Michał R. Nowicki$^{(\boxtimes)}$, Tomasz Nowak, and Piotr Skrzypczyński

Institute of Control, Robotics, and Information Engineering,
Poznań University of Technology, ul. Piotrowo 3A, 60-965 Poznań, Poland
{michal.nowicki,tomasz.nowak,piotr.skrzypczynski}@put.poznan.pl

Abstract. High costs of labor and personnel training in public transport lead to increased interest in the advanced driver assistance systems for city buses. As buses have to execute precise maneuvers when parking in a limited and cluttered environment, they need accurate localization and reliable terrain perception. We present preliminary results of a project aimed at equipping an electric city bus with localization and terrain mapping capabilities. The approach is based on 3-D laser scanners mounted on the bus. Our system provides the bus pose estimate and elevation map to the motion planning algorithm that in turn provides the human driver with steering suggestions through a human-machine interface.

Keywords: ADAS · Bus · Laser scanner · Localization · Elevation map

1 Introduction

Although the interest in hybrid and electric cars brings a promise of a cleaner and more quiet vehicles, the problem of congestion in urban traffic may be only solved by an increased use of public transportation. The electric city buses, which are recently introduced by major bus manufacturers, such as Solaris Bus & Coach [10], are zero-emission, quiet vehicles, produced in the single- and multi-body variants, that are capable of accommodating more than 100 passengers for a ride. They provide reduction in air pollution and improve traffic flow. However, the developing fleet of large buses requires an increased number of bus drivers.

Hence, various assistive technologies are integrated into these vehicles to reduce the effort and stress in human drivers. These solutions are known as Advanced Driver Assistance Systems (ADAS). The variant of ADAS we consider in this work focuses on providing the driver with clear cues how to operate the steering wheel in order to perform the desired maneuvers. The maneuvers are performed with respect to an external coordinate system that may be attached to a chosen object, e.g. a charging station or a parking slot. Obstacles including movable objects and other vehicles have to be considered while planning and

© Springer Nature Switzerland AG 2020
R. Szewczyk et al. (Eds.): AUTOMATION 2019, AISC 920, pp. 502–512, 2020.
https://doi.org/10.1007/978-3-030-13273-6_47

executing the maneuvers, to avoid any collisions. The maneuvers are planned and executed using an approach originally conceived for vehicles with n-trailers [7], because buses having two or even three segments are considered in the project. The trajectory planning and control algorithms are out of the scope of this paper, however, they define the requirements for the perception and localization system of the ADAS-augmented bus.

The chosen sensors for localization and mapping are the Sick MRS6124 laser scanners. A 3-D laser scanner (called also LIDAR) is a common sensor in self-driving cars. In particular, the high-performance Velodyne LIDARs mounted on a car's roof are often employed [6]. An interesting survey of large-scale 3-D laser localization methods is presented in [3]. Recent LIDAR-based localization methods make it possible to handle even non-stationary scenes [11]. Although specific representations of the occupied space have been developed for vision-based self-driving cars [8], we consider semi-static environments where the map accuracy is more important than rapid updates. Hence, for terrain mapping, we re-use our proven elevation map concept [1], developed originally for legged rescue robots [2].

This paper presents our approach to the problems of accurate, large-scale localization and terrain mapping using a 3-D laser scanner, and describes the software, which leverages the Robot Operating System (ROS) as the backbone of the ADAS. We present the results of preliminary experiments performed using an outdoor mobile robot equipped with two laser scanners. Although the experiments did not involve a real bus, or even a car, their scale with respect to the covered area and the length of the paths are fully matching the scale of the planned maneuvers.

2 Laser-Based Perception System

In the experiments, we use scans coming from two 3-D laser scanners: Sick MRS 6124 and Velodyne VLP-16. These sensors differ when it comes to the parameters, as compared in Table 1. The most obvious difference between these two sensors is the horizontal field of view. Velodyne VLP-16 provides measurements within the full 360°, while Sick MRS 6124 measures distances in the 120° field of view. This makes VLP-16 suitable for personal cars, as it can observe obstacles in every direction. On the other hand, mounting Velodyne VLP-16 without obstructing its field of view on an electric bus is troublesome, as other elements are already mounted on top of this vehicle. Despite the difference in the horizontal measurement angle, both sensors provide a roughly similar number of points, as the measurements from Sick MRS 6124 are captured more densely. For each scan, Velodyne VLP-16 provides 16 independent scanning lines while Sick MRS 2164 captures 24 scanning lines in 4 groups. Each group contains 6 scanning lines, and scanning in a new group starts once all measurements from the previous group are completed. Both sensors differ when it comes to the number of evaluated responses from the obstacles (echoes) that can be useful to select proper measurements, i.e. in snow or heavy rain.

Table 1. Comparison of Sick MRS 6124 and Velodyne VLP-16 parameters

Parameter	Sick MRS 6124	Velodyne VLP-16
Horizontal angle	120°	360°
Vertical angle	15°	30°
Horizontal resolution	0.13°	0.2°
Vertical resolution	0.625°	2°
Scanning lines	24	16
Scanning groups	4	1
Points per second	221 520	288 000
Number of echos	4	2
Maximal range	200 m	100 m
Weight	2.2 kg	0.83 kg

In our experiments, both sensors were mounted on a mobile platform as presented in Fig. 1A, and the exemplary scans from both sensors taken at the same time are presented in Fig. 1B and C. Velodyne VLP-16 was mounted on the top of the platform to not obstruct the 360° field of view. The Sick MRS 6124 was mounted upside down in the front of the robot to see the ground in front of the robot on several scanning lines. The sensor was tilted up by 6°, so it could see obstacles further away from the robot, as it was not possible to mount it at the assumed 3 m height. The Ethernet interfaces of both sensors were connected to a router on the robot. A laptop connected to this router was used to log the incoming laser scans to ROS *.bag files. To communicate with sensors we utilized open source ROS drivers: *velodyne*[1] and *Sick_scan*.[2]

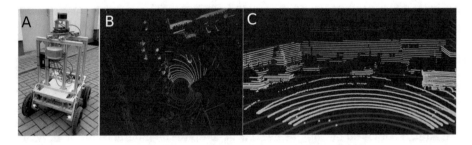

Fig. 1. Sick MRS 6124 and Velodyne VLP-16 were mounted on a mobile robot (A) to capture scans from both sensors in similar conditions. Exemplary captured scans from Velodyne VLP-16 (red) and Sick MRS 6124 (green) are presented from the top view (B) and from the viewpoint of the robot (C)

---

[1] https://github.com/ros-drivers/velodyne.
[2] https://github.com/SickAG/Sick_scan.

3 Localization Method

In the experiments we wanted to verify accuracy of the recent open source localization systems available in ROS. The first system is LOAM [12],[3] that is a state-of-the-art LIDAR-based solution, which is ranked first on the car localization KITTI dataset [5]. Processing of the laser scans in LOAM is divided into several steps: feature extraction, odometry, mapping, and integration. The system operates on the detected planar and edge features that make it possible to downsample the extensive datastream from a laser scanner. The 3-D features are then matched between the current scan and the previous one. In this step, the 6 d.o.f. pose $\mathbf{x} = [x\ y\ z\ \theta\ \phi\ \psi]^T$ is estimated using constraints defined by the associated features. The system considers that the scanning is performed continuously while the scanner is moving. Therefore, after the odometry step, the measured points are projected to the coordinate system related to the start point of the scanning process, and a geometrically corrected point cloud is obtained. The corrected cloud is the input to the mapping procedure that optimizes the alignment between this cloud and the global map, yielding a more accurate pose estimate than the one obtained in the odometry step. The mapping optimization is performed less frequently (2 Hz) than odometry (10 Hz) as this step is more computationally demanding. The final localization step fuses the pose from map optimization with the odometry, to provide the best pose estimate available at the moment.

The LOAM system works without any assumptions as to the type of motion performed by the LIDAR. But in the case of a wheeled vehicle, we can assume that it moves on the ground. This assumption led to the most recent extension of the LOAM system called LeGO-LOAM [9] that is also publicly available.[4] The LeGO-LOAM introduces a procedure to determine the ground plane and distinguishes between features found on the ground and on obstacles. In this system, the planar features are used to determine the elevation z from the ground, and the pitch and roll angles (ϕ, ψ). These partial pose estimates help to obtain correspondences between edge features that are then used to determine the position on the 2-D plane and the yaw angle: $\mathbf{x}^{2D} = [x\ y\ \theta]^T$. Owing to the modifications, LeGO-LOAM utilizes prior belief about the vehicle motion, which makes the localization results more stable in challenging scenarios.

Both systems, LOAM and LeGO-LOAM, are prepared to be used with the Velodyne VLP-16. In our comparison, we captured the scans also from the Sick MRS 6124 and therefore we modified the systems to operate in the case of scans coming also from that sensor. The introduced modifications included the parameters concerned with the different horizontal and vertical observation angles and the necessary changes to the assumptions about the way the points are captured in each scan to correctly represent the measurements at the beginning of the scan based on the current estimate of the motion.

---

[3] https://github.com/laboshinl/loam_velodyne.
[4] https://github.com/RobustFieldAutonomyLab/LeGO-LOAM.

4   Terrain Mapping

The driver assistance system requires to register the acquired range data into a map that serves mainly the trajectory planner, but also enables the system to detect objects appearing in the vicinity of the vehicle, including people. A dual-grid elevation map is used (Fig. 2) as in [1], which employs a local grid that moves with the vehicle and is directly updated from the range data, and a global grid describing the whole scene, which is updated indirectly from the small grid content.

The implemented terrain mapping system is based on *grid map* and *elevation mapping* ROS packages.[5] The *grid map* is a library implementing grid maps that can have multiple data layers. The *elevation mapping* package is responsible for building a robot-centric elevation map of the environment considering the uncertainty of range measurements and robot/sensor pose. Measurements are registered in the local *raw_map*. Each cell in this map is described by elevation, elevation variance, horizontal variances, last update time, and parameters required for visibility cleanup based on ray tracing. In a parallel thread, at a lower frequency, a fusion process is performed in order to combine elevation values considering the sensor pose uncertainty. The result of this step is the global *fused_map* that contains information about upper and lower bounds of elevation, a current estimate of the elevation and its uncertainty. Another parallel thread is responsible for removing invalid elevation values according to the line-of-sight visibility constraints, i.e. considering occlusions.

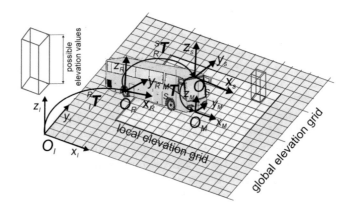

Fig. 2. Dual elevation grid concept with visualization of the coordinate systems

We define the coordinate system of the local map O_M relatively to the vehicle and the laser sensor coordinate system O_S. The map origin coincides with the origin of the sensor coordinates in the xy plane, and the x and y axes of the map

---

[5] https://github.com/ANYbotics/elevation_mapping.

frame match the respective axes of the sensor frame. The z-axis of the map frame is aligned with the gravity vector, while the sensor may rotate slightly in the pan and tilt angles, due to the suspension of the vehicle. The origin of the map z axis is located on the ground plane, while the sensor is assumed to be elevated about 3 m above the ground, as it is mounted to the roof of the bus. The sensor coordinate frame is rigidly related to the vehicle coordinate frame O_R, which for the convenience of the motion planner is usually set to the midpoint of the driving axle of the bus. The relation (rototranslation) ${}_R^S\mathbf{T}$ between O_R and O_S is determined by calibration. The coordinates of the global map are expressed in the inertial frame O_I, which is fixed in the environment, and is set at the start of the ADAS-aided maneuvers. All motion planning is accomplished with respect to this frame.

Each range measurement from the 3-D laser scanner is transformed into a point $\mathbf{p}_s = [x_s, \; y_s, \; z_s]^T$ in the sensor coordinate system. Next, these points are transformed to the elevation map coordinates O_M: $\mathbf{p}_m = {}_S^M\mathbf{T}\mathbf{p}_s$ with the homogeneous matrix ${}_S^M\mathbf{T}$ describing the transformation between those two coordinate systems, which is defined by the pan ϕ_r and tilt ψ_r angles measured by inertial sensors of the vehicle or obtained from laser-based localization, and the elevation offset calibrated beforehand. The spatial uncertainty of an acquired 3-D point $\mathbf{p}_m$, given by the covariance matrix $\mathbf{C}_{p_m}$, depends on two factors: the uncertainty of the range measurement $\mathbf{C}_p$, and the uncertainty $\mathbf{C}_s$ of the ${}_S^M\mathbf{T}$ transformation, which accounts only for the uncertainties in the measured pan and tilt angles: $\mathbf{C}_{p_m} = \mathbf{J}_s\mathbf{C}_s\mathbf{J}_s^T + \mathbf{J}_p\mathbf{C}_p\mathbf{J}_p^T$, where $\mathbf{J}_s$ and $\mathbf{J}_p$ are Jacobians of the point transformation with respect to the sensor pose $\mathbf{x}_s$ and the point coordinates $\mathbf{p}_s$, respectively.

The 3-D points acquired in consecutive scans are integrated into the local elevation grid applying a static Kalman filter per each grid cell:

$$\mathbf{K} = \mathbf{C}_{h(k)}^{[i,j]}\left(\mathbf{C}_{h(k)}^{[i,j]} + \mathbf{C}_{p_m(k+1)}\right)^{-1}, \tag{1}$$

$$\mathbf{p}_{h(k+1)}^{[i,j]} = \mathbf{p}_{h(k)}^{[i,j]} - \mathbf{K}\left(\mathbf{p}_{h(k)}^{[i,j]} - \mathbf{P}_{p_m(k+1)}\right), \; \mathbf{C}_{h(k+1)}^{[i,j]} = (\mathbf{I} - \mathbf{K})\mathbf{C}_{h(k)}^{[i,j]},$$

where $\mathbf{p}_{h(k)}^{[i,j]} = [x_h \; y_h \; z_h]^T$ represents the elevation measurement in the given cell, its uncertainty is represented by $\mathbf{C}_{h(k)}^{[i,j]}$, while $\mathbf{K}$ is the Kalman gain, and the k index denotes time instances.

The position of the local map depends on the 2-D position of the vehicle $\mathbf{x}_r^{2D} = [x_r \; y_r \; \theta_r]^T$, because the transformations between the vehicle, the sensor, and the local map are rigid. Hence, the elevation estimates in the global map are affected by the vehicle position drift accumulated along the trajectory and given by the covariance matrix $\mathbf{C}_r^{2D}$. This effect is taken into account by the global map elevation update procedure that follows the approach introduced in [4]. The global elevation estimates are computed from the formulas:

$$\hat{h}^{[i,j]} = \frac{\sum_{a,b\in\mathcal{C}} w_{a,b}h_{a,b}}{\sum_{a,b\in\mathcal{C}} w_{a,b}}, \quad \hat{\sigma}_h^{2[i,j]} = \frac{\sum_{a,b\in\mathcal{C}} w_{a,b}\left(\sigma_{h(a,b)}^2 + h_{a,b}^2\right)}{\sum_{a,b\in\mathcal{C}} w_{a,b}} - \hat{h}^{2[gi,j]}, \tag{2}$$

where $h_{a,b}$ is the elevation stored in a global map cell located in the neighborhood $\mathcal{C}$ of the considered cell $[i,j]$, while $\sigma^2_{h(a,b)}$ is the elevation variance. The weights $w_{a,b}$ define the influence the neighboring cells have on the $[i,j]$ cell:

$$w_{a,b} = \left(\Phi_x \left(d_x + \frac{b}{2} \right) - \Phi_x \left(d_x - \frac{b}{2} \right) \right) \left(\Phi_y \left(d_y + \frac{b}{2} \right) - \Phi_y \left(d_y - \frac{b}{2} \right) \right), \quad (3)$$

where b is the cell side length, d_x and d_y are distances between the centers of the $[i,j]$ and $[a,b]$ cells, while Φ_x and Φ_y are cumulative normal distribution functions with the covariances $\sigma^2_{x_r}$ and $\sigma^2_{y_r}$, respectively. These functions define the probability that elevation measurements from a cell located in the neighborhood overlap with the cell of interest for the given position uncertainty given by $\sigma^2_{x_r}$ and $\sigma^2_{y_r}$ taken from $\mathbf{C}^{2D}_r$.

5   Mapping Performance in Simulation

The *elevation mapping* package was designed for legged robots, so performance tests were needed to verify that the modified software can be employed in the bus application. ADAS application requires real-time performance with larger map size and faster motion of the robot. Therefore, we focused on experiments concerning cell size in the map (resolution) and the overall size of the map.

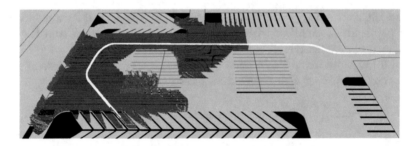

Fig. 3. Simulated parking lot with the generated elevation map and bus trajectory

Performance of the package was evaluated in a simulated environment. A realistic parking scenario (Fig. 3) was used in the simulator. The environment is represented as a ROS *OccupanyGrid* using the *map_server* package. With the information about the pose of the vehicle, we generate simulated scans of the environment (*Point Cloud*) using a custom *scan_simulator* package, configured to simulate the Sick MRS 6124 sensor. The simulation involved an approximately 270 m long trajectory on the provided map of the bus parking lot. We measured the time of updating the local map *raw_map*, the time of fusing local map into the global map *fused_map*, and the visibility cleanup time. These times were measured with changing parameters of the maximal scanner range, cell size, and map sizes. All simulations were performed on a computer with i7-5820K 3.30 GHz CPU.

Results for different cell sizes are presented in Fig. 4A. Here the scan range equal to 40 m and the global map size equal to 200 m were fixed parameters. As expected, the time needed for processing incoming measurements (update time) increases when the cell size of the map decreases. Similarly, the fusion and visibility cleanup times increase with greater resolution as more cells need to be considered. The curve flattens for cell size greater than 15 cm as the cleanup, fusion and update times are decreasing slowly. For very small cell sizes the update time increases exponentially preventing the package from producing a dense map. Figure 4B plots the performance as a function of the map size for fixed resolution. The relation between the global map size and computing times is almost linear rather than quadratic. For larger maps, the update time increases significantly while no significant change can be observed for the fusion time. From these results, we conclude that the best parameters for our application are 15 cm cell size and 200 m global map size. For these parameters, the reported update time is below 0.2 s, so we can include laser scans with 5 Hz and the fusion time is below 2 s.

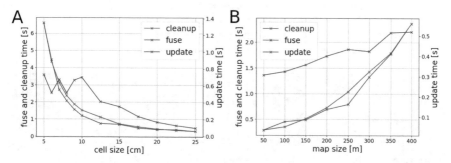

Fig. 4. Local map update (update time), fusing local map into global map (fuse time) and visibility cleanup (cleanup time) depending on cell size (A) and map size (B)

6 Preliminary Experimental Results

The laser-based localization system was verified using a mobile robot equipped with Velodyne VLP-16 and Sick MRS 6124 (cf. Fig. 1A). The robot was moving in a typical parking scenario at PUT campus, which can be observed in more detail using the Google Earth view presented in Fig. 5D. The obtained trajectories for the LOAM and the LeGO-LOAM systems and both evaluated sensors are presented in Fig. 5.

The least accurate estimate of the trajectory was obtained with the LOAM system when the Velodyne VLP-16 was used (Fig. 5A) as the trajectory drifts in the vertical direction. This is caused by the lack of constraints in some of the 6 d.o.f. Surprisingly, the trajectory obtained with the LOAM system and Sick MRS 6124 (Fig. 5B) lacks the vertical drift, which can be attributed to the greater density of the measurements from the sensors. The trajectories obtained using the LeGO-LOAM system are similar for both sensors and outperform the LOAM system due to employing additional constraints provided by the detected ground plane.

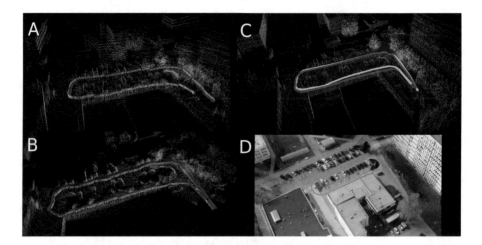

Fig. 5. Comparison between the trajectories obtained with LOAM using Velodyne VLP-16 (A) and Sick MRS 6124 (B), LeGO-LOAM (C) using Velodyne VLP-16 (red) and Sick MRS 6124 (green), and satellite view from similar observation angle from the Google Earth with rough estimation of the robot movement

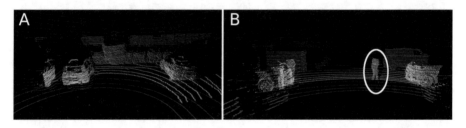

Fig. 6. Exemplary views from Velodyne VLP-16 (red) and Sick MRS 6124 (green) during two typical scenarios: parking maneuver (A) and when a person (marked by the white ellipse) appears in front of the robot (B)

When performing the experiments we were also interested in scenarios that are typical for the application of the city buses. In Fig. 6A, an exemplary view from both scanners is shown when the robot performed a parking maneuver between cars. Here the advantage of MRS 6124 is the better density of the acquired point cloud, that allows the robot to map obstacles with greater accuracy. On the other hand, its limited horizontal field of view poses a challenge once the robot is close to a wall, because the lateral constraints disappear. Another aspect is a detection of intrusions in the scene, i.e. a person walking in front of the vehicle. Figure 6B shows that the MRS 6124 registers the person with a much larger number of scan lines, while VLP-16 hardly detects the silhouette.

The LeGO-LOAM provides accurate localization that can be combined with the point clouds to map the surrounding environment. Figure 7 depicts elevation maps created with the previously chosen parameters without visibility cleanup.

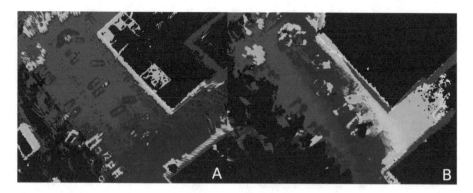

Fig. 7. Comparison between the elevation maps obtained from Velodyne VLP-16 (A) and Sick MRS 6124 (B) data registered with trajectories from LeGO-LOAM system

The map generated from Sick scanner data is large enough to map the terrain for motion planning. There is no need to acquire data at elevations above the scanner. However, the elevation map generated with point clouds from the Velodyne scanner represents objects that are above the scanner level. While the Velodyne map may look more natural, processing these points consumes resources, but does not contribute to better terrain representation in this application.

7   Final Remarks

We presented the concept, implementation, and preliminary results for a laser-based localization and mapping system, which is a vital part of the ADAS in a modern city bus. The contribution of this work lies in the modification and integration of several ROS packages into a working system and customization of the software for the recently introduced Sick MRS 6124 laser scanners. The results of the experiments confirmed that the MRS 6124 is suitable for localization in spite of the smaller horizontal field of view, while its more dense scans give it an advantage when mapping the terrain. However, adding to the system a pair of simple 2-D scanners to perceive close obstacles on the sides of the bus are considered for further research.

Acknowledgement. This work was funded by the National Centre for Research and Development grant POIR.04.01.02-00-0081/17.

References

1. Belter, D., Labecki, P., Skrzypczyński, P.: Estimating terrain elevation maps from sparse and uncertain multi-sensor data. In: IEEE International Conference on Robotics and Biomimetics, Guangzhou, pp. 715–722 (2012)
2. Belter, D., Nowicki, M., Skrzypczyński, P., Walas, K., Wietrzykowski, J.: Lightweight RGB-D SLAM system for search and rescue robots. In: Szewczyk, R., et al. (eds.) Progress in Automation, Robotics and Measuring Techniques. AISC, pp. 11–21. Springer, Heidelberg (2015)

3. Bedkowski, J., Röhling, T., Hoeller, F., Shulz, D., Schneider, F.E.: Benchmark of 6D SLAM (6D simultaneous localization and mapping) algorithms with robotic mobile mapping systems. Found. Comput. Decis. Sci. **42**(3), 275–295 (2017)

4. Fankhauser, P., Bloesch, M., Gehring, C., Hutter, M., Siegwart, R.: Robot-centric elevation mapping with uncertainty estimates. In: Mobile Service Robotics, pp. 433–440. World Scientific, Singapore (2014)

5. Geiger, A., Lenz, P., Urtasun, R.: Are we ready for autonomous driving? The KITTI vision benchmark suite. In: Conference on Computer Vision and Pattern Recognition (CVPR), Rhode Island, pp. 3354–3361 (2012)

6. Levinson, J., Thrun, S.: Robust vehicle localization in urban environments using probabilistic maps. In: IEEE International Conference on Robotics and Automation, Anchorage, pp. 4372–4378 (2010)

7. Michałek, M.M., Kiełczewski, M.: The concept of passive control assistance for docking maneuvers with n-trailer vehicles. IEEE/ASME Trans. Mechatron. **20**(5), 2075–2084 (2015)

8. Pfeiffer, D., Franke, U.: Efficient representation of traffic scenes by means of dynamic stixels. In: IEEE Intelligent Vehicles Symposium, pp. 217–224. San Diego (2010)

9. Shan, T., Englot, B., LeGO-LOAM: lightweight and ground-optimized lidar odometry and mapping on variable terrain. In: IEEE/RSJ International Conference on Intelligent Robots and Systems, Madrid, pp. 4758–4765 (2018)

10. Solaris Bus & Coach. Alternative powertrains, product catalogue (2018)

11. Withers, D., Newman, P.: Modelling scene change for large-scale long term laser localisation. In: IEEE International Conference on Robotics and Automation, Singapore, pp. 6233–6239 (2017)

12. Zhang, J., Singh, S.: Low-drift and real-time lidar odometry and mapping. Auton. Robots **41**(2), 401–416 (2017)

Motion Planning and Control of Social Mobile Robot – Part 1. Robot Hardware Architecture and Description of Navigation System

Marcin Słomiany[1]([✉]), Przemysław Dąbek[2], and Maciej Trojnacki[2]

[1] Institute of Vehicles, Warsaw University of Technology, Warsaw, Poland
m.slomiany93@gmail.com

[2] Industrial Research Institute for Automation and Measurements (PIAP),
Warsaw, Poland
{pdabek,mtrojnacki}@piap.pl

Abstract. This two-part paper is concerned with the problem of motion planning and control of PIAP IRYS social robot mobile platform, which is the adapted NOMAD 200 robot. In the first part of the work the developed hardware and software architectures of the research object, as well as its kinematic structure are described. The robot consists of two independent components: (1) a head with a trunk and (2) a mobile platform. As the mobile platform the modernized version of NOMAD 200 robot with synchronously driven and steered wheels was used. The mobile platform was equipped among others with modern low-level miControl motor controllers and a high-level controller working under GNU/Linux operating system with Robot Operating System framework. Kinematics of the three-wheeled robot is discussed and software architecture including ROS-based navigation system is described.

Keywords: Hardware architecture · Software architecture · Wheeled mobile robot · Social robot · Motion planning · Robot Operating System

1 Introduction

Mobile robots usually must be able to operate in the environment which involves moving to the goal point while avoiding obstacles. This is true in case of vast majority of service robots, including social robots that have to operate hand in hand with humans in natural environment of the humans. Navigation systems used for this purpose can be divided into behavioral-based and map-based [2,8]. Both methods are in constant developing but in recent years the map-based system seems to be more popular for autonomous mobile robots [5]. Great example of mobile robot that use this type of navigation system is robot SPENCER [11]. This social mobile robot operates in very difficult environment of airport where

R. Szewczyk et al. (Eds.): AUTOMATION 2019, AISC 920, pp. 513–523, 2020.
https://doi.org/10.1007/978-3-030-13273-6_48

serving as a guide helping travelers find their way to the gate. The map-based navigation system that ensure autonomous operation of robot in that environment consists of the following subsystems: perception, mapping, localization, motion planning and drive control. Motion planning could be divided between the global and local path planning. During global path planning a whole path from starting point to goal point are computed, based on entire map of environment [2,8]. Local path planning, also often called collision avoidance [2,8,10] focuses on generating trajectories that bring the robot closer to the goal while avoiding obstacles unknown at the time of global planning activity and often moving when the planned motion is executed. These trajectories are converted into desired angular velocities of individual wheels and then executed by control system. The robot discussed in this article is designed to operate in natural human environment, therefore there is need to develop a navigation system capable of operating in such environment.

2   PIAP IRYS Social Robot

The robot consists of two main parts: upper part of the robot comprising upper tank and head, and bottom part comprising mobile platform. The robot's upper trunk shown in Fig. 1a was described in detail in another article [3] and will not be described in the present paper due to text length limitations. In contrast, mobile platform shown in Fig. 1b will be described in details later in this paper. PIAP IRYS robot together with mobile platform will be used as a robot teacher for children between 7 and 10 years old. It is planned to combine the functions described in [3] with the possibility of moving around children, transporting small items – such as notebooks, toys, etc., or playing hide and seek.

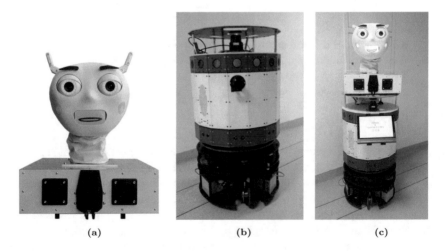

(a)   (b)   (c)

Fig. 1. PIAP IRYS social robot: a – PIAP IRYS upper trunk, b – PIAP IRYS mobile platform, c – PIAP IRYS mobile platform with upper trunk

As a mobile platform, the modernized NOMAD 200 robot was used. The modernization consisted in replacing inefficient and obsolete elements of power supply, control and scanning systems, which forced the design of a new hardware architecture of the robot. The main design assumption during the modernization of the robot was to obtain as much hardware modularity as possible which allows for easy integration or exchange of individual elements, so that the platform can be quickly adapted to perform other tasks.

The structure can be divided into two parts: a mobile base and a rotating tower connected to mobile base by a rotary connector. The robot mobile platform contains three synchronously driven wheels which can move forward or backward and also make a turn at any angle in each direction synchronously – also when standing still. In addition, the tower can rotate independently of the mobile base. The hardware architecture of the mobile platform is illustrated in Fig. 2.

The modernization of the robot included design of a new robot hardware architecture and replacement of ultrasonic sensors and the addition of a 2D laser scanner. The robot control system includes a high-level controller (Single Board Computer: UP Board with x5-Z8350 CPU, 4 GB RAM) responsible for planning the robot's movements and low-level controllers to control the robot's drive units. The mobile base is equipped with three DC motors with encoders, and each motor is responsible for a different type of robot movement: synchronous rotating of all wheels, synchronous steering of all wheels and rotating of the tower. Each motor is powered and controlled by a separate miControl controller. The mobile base can be fitted with mechanical bumpers – incorporating 20 contact sensors connected with controller responsible for driving and can be used for emergency stop of the robot's movement. At the top of the mobile base there is a rotary connector, which allows the rotation of the tower through any angle, independent of the mobile base. It also provides a safe connection between the tower and the mobile base to signal and power cables that would otherwise be damaged by the tower's rotation. The tower structure can be divided into four connected segments. The first segment is located at the very bottom and connects to the base of the mobile rotary connector. At the perimeter of this segment there are ultrasonic distance sensors that are operated by the Arduino microcontroller board. The second segment has four DC-DC converters which power most of robot's components, a single board computer and in the geometric center of the bottom plate of this segment ADIS IMU is placed and around it, four iNEMO IMU's (connected with SBC by USB hub). The third segment of the tower has a power connector, an emergency stop button and a main switch. All these elements were mounted in the side panels of the third segment. The fourth segment includes a Wi-Fi router and a 2D laser scanner. On the top of the fourth segment the bust of PIAP IRYS robot can be placed and attached to the mobile platform, which together with the mobile platform is a mobile social robot.

2.1   Hardware Architecture

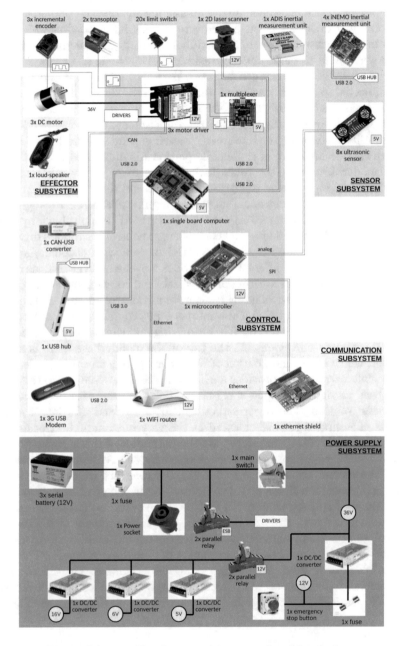

Fig. 2. Scheme of hardware architecture of mobile platform

2.2 Software Architecture

The software architecture consists of distributed programs (called nodes) which communicate with each other by means of the popular robotic middleware called Robot Operating System (ROS). ROS computation graph of designed system is shown in Fig. 3. The system can be divided into subsystems: navigation, control, data aggregation and communication. The navigation system is guiding the robot from the starting position to the goal position based on the environment map. The map of the environment is transformed into so-called cost maps, providing data about obstacles. The global motion planner determines the shortest path based on the current global cost map. In turn, the local motion planner generates trajectories leading to the goal point, taking into account the (global) path and distances to the obstacles. The trajectory is processed by the drive control subsystem, which controls the individual drives, resulting in the movement of the robot. During the system operation, the odometry subsystem takes data from the encoders of individual motors and calculates the current position of the robot and its components relative to the starting point. As part of Simultaneous Localization And Mapping (SLAM) based on odometry data and data from sensors scanning the environment an up-to-date map of the surrounding is created on an ongoing basis. This means that each time the robot is started map of environment is created and robot's location on that map is estimated.

The control system consists of the drive control subsystem, which controls the operation of the motors depending on the trajectory set by the navigation system, and provides odometry data from the encoders. The data aggregation system acquires and delivers data about the environment from ultrasonic sensors and a laser scanner to the navigation system. The communication system is responsible for communication of operator with the robot via a Wi-Fi network, which allows robot control and data exchange. For motion planning we used the `Navigation` metapackage, which generates global path and trajectories published to the ROS network. Trajectories are published through `cmd_vel` topic and when obtained by the `nomad_driver` node are they are processed using inverse kinematics on the rotational speeds of individual robot's movements and then executed by individual drive controllers. This node also provides odometry data from encoders mounted on motor shafts. Data from the laser scanner are processed and published by the `hokuyo_node`, and from ultrasonic sensors by the `sonars_node`. The SLAM method comes from the `hector_mapping` package from the metapackage `hector_mapping`, which requires data from the laser scanner and the odometry source for operation. Remote communications with the robot was based on Wi-Fi network formed between operator's laptop and WiFi router on the robot. On the software side, we used the `node_manager_fkie` package to connect the ROS on the mobile platform with ROS on the operator's platform, which makes this operation easy and efficient from the user point of view. The available graphical interface allows to set a target pose as well as visualize and record the state of the system from the operator's station.

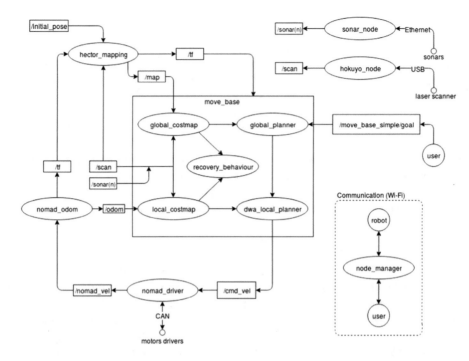

Fig. 3. ROS computation graph of designed navigation system

3   Kinematics of Mobile Platform

The kinematic structure of the mobile robot platform is illustrated in Fig. 4.

It is possible to distinguish the following main components of the mobile platform: 0 – body, 1–3 – synchronously driven and steered wheels (for which 2 DC motors are used), 4 – rotating tower (driven by DC motor), on which the upper part of the robot is placed. For the mobile platform are introduced two

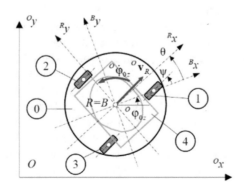

Fig. 4. Kinematic structure of mobile platform

coordinate systems: associated with its upper part (tower), to which the upper body with robot's head is attached to – denoted with symbol R, and reference system connected with its lower part, i.e. body – denoted with symbol B.

The current state of the mobile robot platform includes its pose $^O\mathbf{q}$, the angle of rotation of the tower relative to the lower part of the mobile platform θ, the angular speed of rotation of wheels ω and the steering angle of these wheels Ψ. In the actual construction, the steering axis is shifted relative to the axis of wheel own rotation by a distance of about 0.02 m. The work omits the influence of this shift on kinematics and robot motion control.

It is assumed that robot's motion is realized in O_{xy} plane of the fixed coordinate system $\{O\}$. Desired pose of the robot (connected with tower 4) is described by the vector of generalized coordinates:

$$^O\mathbf{q} = [^O x_R, {}^O y_R, {}^O \varphi_z]^T, \tag{1}$$

where: $^O x_R$, $^O y_R$ are coordinates of point R belonging to the robot, and $^O \varphi_z$ denotes angle of spin of the robot about z axis with respect to fixed coordinate system $\{O\}$, that is course of the robot.

Between the angle of the robot's course (including its tower 4) and the lower part of the mobile platform, i.e. body 0, there is a dependency:

$$^O \varphi_z = {}^O \varphi_{0z} + \theta, \tag{2}$$

where θ is the angle of rotation of the tower relative to the mobile base, as indicated earlier.

Therefore, for velocities of robot's point R in coordinate systems $\{B\}$ and $\{R\}$ the following relationships can be applied:

$$^R\mathbf{v}_R = {}^R_B\mathbf{R}\ ^B\mathbf{v}_R \quad \text{and} \quad {}^B\mathbf{v}_R = {}^B_R\mathbf{R}\ ^R\mathbf{v}_R, \tag{3}$$

where $^R_B\mathbf{R}$ and $^B_R\mathbf{R}$ are rotation matrices from $\{B\}$ to $\{R\}$ coordinate system and vice-versa.

These matrices have the following forms:

$$^R_B\mathbf{R} = \begin{bmatrix} c_\theta & s_\theta & 0 \\ -s_\theta & c_\theta & 0 \\ 0 & 0 & 1 \end{bmatrix} \quad \text{and} \quad {}^B_R\mathbf{R} = \begin{bmatrix} c_\theta & -s_\theta & 0 \\ s_\theta & c_\theta & 0 \\ 0 & 0 & 1 \end{bmatrix}, \tag{4}$$

where $c_\theta = \cos(\theta)$, $s_\theta = \sin(\theta)$.

It should be noted that as a result of identical and simultaneous turning of all wheels, the orientation of the body 0 is not substantially changed, i.e. $^O \varphi_{0z} = const$, therefore the angular velocity of the tower own rotation relative to the fixed reference system $\{O\}$ is equal to the angular velocity with which this tower rotates relative to body 0, i.e.

$$^O \dot{\varphi}_z = \dot{\theta} \tag{5}$$

In turn, vectors of generalized velocities respectively in $\{O\}$ and $\{R\}$ coordinate systems can be written as:

$$^O\mathbf{v} = {}^O\dot{\mathbf{q}} = [^O \dot{x}_R, {}^O \dot{y}_R, {}^O \dot{\varphi}_z]^T, \quad {}^R\mathbf{v} = {}^R\dot{\mathbf{q}} = [^R \dot{x}_{Rx}, {}^R \dot{y}_{Ry}, {}^R \dot{\varphi}_z]^T \tag{6}$$

where: $^{O}v_{R_x} = {}^{O}\dot{x}_R$, $^{O}v_{R_y} = {}^{O}\dot{y}_R$, $^{O}\omega_z = {}^{O}\dot{\varphi}_z$, $^{R}\omega_z = {}^{R}\dot{\varphi}_z$.

Those two vectors satisfy the relationship:

$$^{O}\dot{\mathbf{q}} = {}^{O}\mathbf{J}^{R}{}^{R}\dot{\mathbf{q}} \tag{7}$$

Matrix $^{O}\mathbf{J}^{R}$ has the following form:

$$^{O}\mathbf{J}^{R} = \begin{bmatrix} \cos(^{O}\varphi_z) & -\sin(^{O}\varphi_z) & 0 \\ \sin(^{O}\varphi_z) & \cos(^{O}\varphi_z) & 0 \\ 0 & 0 & 1 \end{bmatrix} \tag{8}$$

If it is assumed that the robot's wheels roll without slipping, the velocity module with which the characteristic robot's point R travels is: $||^{O}v_R|| = ||^{R}v_R|| = \omega r$, where r is the radius of the driving wheels. In order to solve forward kinematics problem for the robot, one has to determine relationship $^{R}\mathbf{v} = \mathbf{F}(\mathbf{x})$, which allow calculation of generalized velocities $^{R}\mathbf{v} = [^{R}v_{R_x}, {}^{R}v_{R_y}, {}^{R}\omega_z]^{T}$ for known kinematic parameters of the robot in configuration space $\mathbf{x} = [\omega, \psi, \dot{\theta}]$. For this purpose, the following equations can be used:

$$^{B}v_{R_x} = \omega r \cos(\psi), \quad ^{B}v_{R_y} = \omega r \sin(\psi), \quad ^{O}\dot{\varphi}_z = \dot{\theta}, \tag{9}$$

where the dependencies on the velocities of the robot's R point in the $\{R\}$ and $\{B\}$ coordinate systems result from the transformation described by Eq. (3).

These equations can be used to estimate the robot's generalized velocities based on measured kinematic parameters of the robot in configuration space. In turn, to control the robot's movement, it is necessary to be able to solve the inverse kinematics problem. For this purpose, a dependence $\mathbf{x}_d = F^{-1}(^{R}\mathbf{v}_d)$, should be set to allow finding the desired kinematic parameters of the robot in configuration space $\mathbf{x}_d = [\omega_d, \psi_d, \dot{\theta}_d]$. For desired generalized velocities $^{R}\mathbf{v}_d = [^{R}v_{Rxd}, {}^{R}v_{Ryd}, {}^{R}\omega_{zd}]^{T}$ In the analyzed case of the robot, the equations that enable solving the inverse kinematics problem are as follows:

$$\omega_d = \frac{\sqrt{(^{R}v_{Rxd})^2 + (^{R}v_{Ryd})^2}}{r}, \quad \dot{\theta}_d = {}^{O}\dot{\varphi}_{zd},$$

$$\psi_d = \begin{cases} \psi_d^{t-\Delta t} & \text{for } ^{R}v_{Rxd} = 0 \,\&\, ^{R}v_{Ryd} = 0 \\ \theta_d + \arctan2(^{R}v_{Ryd}, {}^{R}v_{Rxd}) & \text{otherwise} \end{cases} \tag{10}$$

Where arctan2 is a two-argument arcustangent function that allows determining the steering angle of driven wheels ψ_d in the full range of this angle. In the case when the $^{R}v_{Rxd} = 0 \,\&\, ^{R}v_{Ryd} = 0$ steering angle of the wheels does not change, it takes the value from the previous time moment, i.e. $\psi_d^{t-\Delta t}$.

4   Motion Planning and Control

The motion planning methods assume that the phase limitations at each point of the configuration space of a system that is a mobile robot are subject to non-holonomic constraints. This allows the determination of the permitted system motion velocity without reducing the size of the available configuration space [9]. A motion planning and control are subsystems of the navigation system, which can be divided into behavioral-based and map-based. First one uses a set of predefined behaviors in response to stimulus according to unchanging principles. This assumption is based on the conviction that there is a specific procedure that provides a solution to the task of realization of the desired robot movement. This type of system avoids explicit motion planning and robot position determination. Second one are based on the use of the environment model, usually in the form of an environment map. In this approach, actions are taken regarding the robot's location in environment, as well as planning the path to the goal point based on the available environment map [8]. In this paper we focus on the map-based system which [1] performs tasks of scanning of surrounding, mapping, localization and finally navigation (motion planning and control).

Let us focus on navigation, which first step is to determine the path connecting the starting point with the goal point based on the representation of the environment in the form of e.g. a map. This step is called global path planning and for this purpose graph searching algorithms like A* or Dijkstra [6] can be used. Next through the time-based parameterization of the geometrical path, a trajectory is determined, which allows to defining the velocity of movement along the path. This stage is called local path planning and to generate trajectories it uses methods like Dynamic Window Approach [4] or Timed Elastic Band [7]. The last stage is execution of desired trajectory. This is realized by the drive control system which by application of inverse kinematics on the desired trajectories then compute and control desired velocities for each wheel. In addition on an on-going basis data from the laser scanner and odometry are processed and

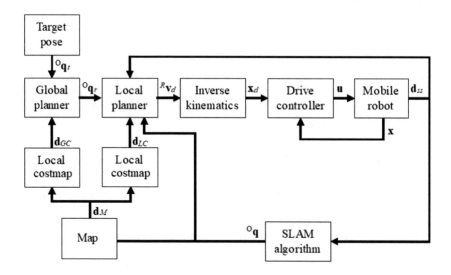

Fig. 5. Scheme of control system used during experimental research

used by SLAM algorithm to create an up-to-date map of the environment and to estimate the actual location of the robot relative to it.

The navigation system developed by us works in the manner described earlier and also as follows. The goal point is the pose of robot – that is the position and orientation of the robot base in the environment. We also assumed that each time the map of the environment is created from scratch and localization is carried out on it. It is worth pointing out that there are only static obstacles on the global map but on the local map there are also dynamic obstacles. Scheme of that system is shown in Fig. 5.

5   Conclusion and Future Work

In the present work a hardware and software architectures of social mobile robot with a mobile base with synchronous drive were presented. Also general description of autonomous navigation system, especially in the field of motion planning and control was presented. The second part of the work is concerned with experimental research of the developed navigation system in natural environment of human and focus on impact of various local path planning algorithm parameters on avoidance of static and dynamic obstacles. Effectiveness of particular configurations will be evaluated using the introduced quality indexes.

Directions of future works will include:

- development of solutions of motion planning where artificial intelligence techniques like artificial neural networks or systems with fuzzy logic will be used,
- simulation and experimental research of various local path planning methods for other types of kinematic structures, to find optimal solution for the mobile platform of PIAP IRYS social robot.

Acknowledgements. The work was carried out as a part of the FB2517 project funded by Own Research Fund of Industrial Research Institute for Automation and Measurements (PIAP). We also gratefully acknowledge the technical assistance as well as important comments and suggestions provided by Mr. Piotr Jaroszek of PIAP Defence and Security Systems Division and Mr. Michał Pełka of the PIAP Interdisciplinary Research Team.

References

1. Choset, H., Lynch, K.M., Hutchinson, S., Kantor, G.A., Burgard, W., Kavraki, L.E., Thrun, S.: Principles of Robot Motion: Theory, Algorithms, and Implementations. A Bradford Book, Cambridge (2005)
2. Corke, P.: Robotics, Vision and Control: Fundamental Algorithms in MATLAB, 1st edn. Springer, Berlin (2011)
3. Dąbek, P., Trojnacki, M., Jaroszek, P., Zawieska, K.: Concept, physical design and simulator of IRYS social robot head (in press)
4. Fox, D., Burgard, W., Thrun, S.: The dynamic window approach to collision avoidance. IEEE Robot. Autom. Mag. 4(1), 23–33 (1997)

5. Janiak, M., Zieliński, C.: Control system architecture for the investigation of motion control algorithms on an example of the mobile platform Rex. Bull. Polish Acad. Sci. Tech. Sci. **63**(3), 667–678 (2015)
6. Li, D., Niu, K.: Dijkstra's algorithm in AGV. In: 9th IEEE Conference on Industrial Electronics and Applications, pp. 1867–1871, June 2014
7. Rösmann, C., Feiten, W., Wösch, T., Hoffmann, F., Bertram, T.: Efficient trajectory optimization using a sparse model. In: European Conference on Mobile Robots, pp. 138–143, September 2013
8. Siegwart, R., Nourbakhsh, I.R., Scaramuzza, D.: Introduction to Autonomous Mobile Robots, 2nd edn. The MIT Press, Cambridge (2011)
9. Tchoń, K.: Manipulatory i roboty mobilne: modele, planowanie ruchu, sterowanie. Akademicka Oficyna Wydawnicza, Warszawa (2000)
10. Thrun, S., Burgard, W., Fox, D.: Probabilistic Robotics. Intelligent Robotics and Autonomous Agents. The MIT Press, Cambridge (2005)
11. Triebel, R., Arras, K., Alami, R., Beyer, L., Breuers, S., Chatila, R., Chetouani, M., Cremers, D., Evers, V., Fiore, M., Hung, H., Ramírez, O.A.I., Joosse, M., Khambhaita, H., Kucner, T., Leibe, B., Lilienthal, A.J., Linder, T., Lohse, M., Magnusson, M., Okal, B., Palmieri, L., Rafi, U., van Rooij, M., Zhang, L.: SPENCER: a socially aware service robot for passenger guidance and help in busy airports. In: Wettergreen, D.S., Barfoot, T.D. (eds.) Field and Service Robotics: Results of the 10th International Conference. Springer Tracts in Advanced Robotics, pp. 607–622. Springer, Cham (2016)

Motion Planning and Control of Social Mobile Robot – Part 2. Experimental Research

Marcin Słomiany[1]([✉]), Przemysław Dąbek[2], and Maciej Trojnacki[2]

[1] Institute of Vehicles, Warsaw University of Technology, Warsaw, Poland
m.slomiany93@gmail.com
[2] Industrial Research Institute for Automation and Measurements (PIAP),
Warsaw, Poland
{pdabek,mtrojnacki}@piap.pl

Abstract. The paper is the second part of the work concerned with the motion planning and control of PIAP IRYS social robot mobile platform, which is the adapted NOMAD 200 robot with synchronously driven and steered wheels. In the first part of the work developed hardware and software architectures of the research object, as well as its kinematic structure were described. In the part two of the paper, state of the art in mobile robot navigation systems is described and the problem of motion planning and control of a mobile platform is considered. The experiments were conducted with a real robot in a real environment. For experimental research the most representative and relevant local motion planning algorithm was chosen. During each experimental traverse of the robot a map of indoor environment was created using Hector SLAM algorithm. For robot localization the same SLAM algorithm was used. The experiments were carried out for the selected local motion planning algorithm. During some experiments additional static and dynamic obstacles were introduced, that were not present before. Results of selected experiments for various environments and additional obstacles were presented and discussed.

Keywords: Wheeled mobile robot · Social robot · Motion planning · Robot operating system · Experimental research

1 State of the Art

In the literature there is often a division of navigation systems to based on the map and behavioral ones. The system developed by us belongs to the first category, therefore we will focus on the state of knowledge for this type of systems. In this case, mobile robot motion planning can be divided into three stages [13,15]:

© Springer Nature Switzerland AG 2020
R. Szewczyk et al. (Eds.): AUTOMATION 2019, AISC 920, pp. 524–537, 2020.
https://doi.org/10.1007/978-3-030-13273-6_49

- Generating a path from starting point to goal point – global planning.
- Generating a trajectory – local planning.
- Motion control – execution of trajectory including tracking control.

Global motion planning is the first stage of motion planning during which a path is generated based on the provided map from the starting point to a given goal point, avoiding obstacles. The main criterion of optimization that is taken into account by this type of algorithms it is above all the length of the path, but there are also methods that choose the optimal route in terms of time needed to traverse it, energy consumption [9] or some other parameter. In the literature you can meet many algorithms that solve these tasks and the most common ones are:

- A*, Dijsktra [10],
- rapidly-exploring random tree [4],
- probabilistic roadmap [4],
- ant colony [3],

Local motion planning, often referred to as the obstacle avoidance algorithm [13], consists in generating a trajectory on which the robot is supposed to move. In contrast to the global planner, these types of methods operate on the map of the nearest environment, which is less computationally demanding and can generate new trajectories and thus react to changes in the environment much faster than the global planner that generates entire path based on the entire map. In the literature there can be local planners that can operate without the participation of a global planner leading the robot to the goal, however, these algorithms often can easily get stuck in the local minimum and make it impossible to proceed in the case of more complicated obstacles or longer routes. Therefore, we focused on algorithms that interact with the global planner, and in the literature can be found the following examples:

- Dynamic Window Approach [6],
- Trajectory Rollout [7],
- Elastic Band [13],
- Timed Elastic Band [12],
- Tentacles Clothoid [1],
- VFH+ [13].

The mobile robot motion control system executes the trajectory desired by the local planner, which in the simplest case is associated with the task of setting appropriate rotational velocities of the mobile base kinematics pairs, for which the forward kinematics task is used. However, this approach works well only in the case of constant resistances of movement and the occurrence of small slides on the robot's wheels. For this reason, an approach was developed where the desired trajectory is compensated by an additional element compensating for the non-linearity of the object and a stabilizer (a controller) that eliminates the error resulting from the noise. The following approaches to controlling robot movement are distinguished [5,8,16]:

– Robust control,
– Adaptive control,
– Artificial-Intelligence-methods based control.

In this paper we assume a scenario in which we set the desired pose and the navigation system tries to realize it. The current pose and speed of the robot are known thanks to odometry measurements, which in combination with readings from the laser scanner and the use of the SLAM method allows locating the robot surroundings. The global motion planner, based on the current map, determines the path to the destination point, and the local planner determines the trajectory following the path to the destination. This trajectory is then executed by the drive controller, which causes the robot to move. Our research is focused on finding the optimal parameters of the planner for a robot moving in the natural environment of human, that is containing static and dynamic obstacles.

2   Analyzed Algorithm

In our research we focus on ROS map-based local motion planning algorithms Dynamic Window Approach and Trajectory Rollout. During the analysis, we evaluated selected methods for obstacle avoidance and the generation of a trajectory for a robot whose kinematic structure can be treated as holonomic – in the case of a synchronous drive it is possible to move forward and then to the side. Change of the movement's direction does not happen in place, but the turning radius is small enough (in the case of the robot NOMAD 200 it is 0.02 m) that it can be neglected, therefore we can call this a holonomic kinematic structure. Dynamic Window Approach method is based on choosing the set of velocities (translational and rotational velocities of a mobile platform) such that they can be attained from the point of view of robot dynamics and the resulting trajectory does not lead to collision with any obstacles. The algorithm for this method is as follows [6,7]:

– Searching a configuration space for potential free spaces available at the velocities achievable by the robot.
  - Generating a trajectory in the form of arcs – calculated on the basis of set of pairs of linear and angular velocities $(^{R}v_{R_x}, {}^{R}\omega_z)$. As a result of this operation, a set of velocities is obtained which are searched.
  - Determining permissible velocities $\mathbf{V}_a$ in given simulation time: limit calculated in the previous step velocities to those that stop the robot before it collide with the obstacle.

$$\mathbf{V}_a = \left\{ {}^R v_{R_a}, {}^R \omega_a \right\}$$

$$({}^R v_{R_x}, {}^R \omega_z) \in \mathbf{V}_a \iff \begin{cases} {}^R v_{R_x} \leq \sqrt{2 \cdot \text{dist}({}^R v_{R_x}, {}^R \omega_z) \cdot {}^R \dot{v}_{xb}} & (1) \\ \\ {}^R \omega_z \leq \sqrt{2 \cdot \text{dist}({}^R v_{R_x}, {}^R \omega_z) \cdot {}^R \dot{\omega}_{zb}} \end{cases}$$

where: ${}^R \dot{v}_{xb}, {}^R \dot{\omega}_{zb}$ – braking accelerations, dist – distance to the closest obstacle on the trajectory.

- Dynamic window $\mathbf{V}_d$ is a set of permissible velocities that can be achieved in a simulation time depending on the limits of accelerations achievable by the robot.

$$\mathbf{V}_d = \left\{ {}^R v_{R_d}, {}^R \omega_d \right\}$$

$$({}^R v_{R_x}, {}^R \omega_z) \in \mathbf{V}_d \iff \begin{cases} {}^R v_{R_x} \in [{}^R v_a - {}^R \dot{v}_{R_x} \cdot t, {}^R v_a + {}^R \dot{v}_{R_x} \cdot t] & (2) \\ \\ {}^R \omega_z \in [{}^R \omega_a - {}^R \dot{\omega}_z \cdot t, {}^R \omega_a + {}^R \dot{\omega}_z \cdot t] \end{cases}$$

– optimization, that is, searching for the objective function with the maximum value, taking into account the current position and orientation of the robot.

$$G({}^R v_{R_x}, {}^R \omega_z) = \sigma(\alpha \cdot \text{heading}({}^R v_{R_x}, {}^R \omega_z) + \beta \cdot \text{dist}({}^R v_{R_x}, {}^R \omega_z) \\ + \gamma \cdot \text{vel}({}^R v_{R_x}, {}^R \omega_z)) \quad (3)$$

where: $G({}^R v_{R_x}, {}^R \omega_z)$ – optimization function, α, β, γ – weight values

$$\text{heading} = 180° - \theta_{gc} \quad (4)$$

where: θ_{gc} – angle between goal and robot's course.

ROS implementation – `dwa_local_planner` and `base_local_planner` (described in documentation [11]) work on the principle described above, with the difference that the optimization function takes different parameters to evaluate a trajectory:

$$cost = \text{path}_{bias} \cdot \text{path}_{dist} + \text{goal}_{bias} \cdot \text{goal}_{dist} + \text{obstcl}_{scale} \cdot \text{obstcl}_{cost} \quad (5)$$

where: path_{bias}, goal_{bias} and obstcl_{scale} – are weight values of path_{dist} – distance to global path from the endpoint of trajectory, goal_{dist} – distance to local goal from endpoint of trajectory, obstcl_{cost} – maximum obstacle cost (0–254) along the trajectory.

An important difference between the two planners is the so-called Trajectory Rollout, which is different from the Dynamic Window Approach in way in which the control space is sampled. In Trajectory Rollout set of permissible velocities in each simulation is calculated including acceleration limits for the entire simulation time, while in Dynamic Window Approach it is limited to one simulation step. This difference results in the fact that Dynamic Window Approach is less computationally demanding, and Trajectory Rollout is more suitable for robots with low value of acceleration [7]. Both available planners could be used on robot with non-holonomic and holonomics kinematic structure and because the accelerations achieved by the NOMAD 200 mobile platform are not low, we decided to conduct experimental research only for the Dynamic Window Approach method as less computationally demanding.

3   Experimental Research

Research Conditions and Environment. The experimental research were carried out with the real NOMAD 200 mobile platform (as described in [14]) in real environment of the PIAP office areas. Localization of the robot was possible with the SLAM algorithm. The major aim of the research was to study the influence of selected local motion planner parameters on the realization of movement in known static environment with additional unknown static and dynamic obstacles. Secondary goal was to check readiness of the developed system for autonomous operation in real environment such as offices, educational facilities or shopping centers. Therefore, the experiments were carried out in a close-to-real environment in which unnecessary obstacles were not removed and the surroundings were not adjusted to reduce the occurrence of measurement errors. The research object – that is PIAP IRYS mobile robot – was described in the part 1 of the work [14].

Based on the conducted experiments it is also possible to recommend some changes of the planner parameters, so that the navigation can run on a single board computer (SBC). All traverses were made in the buildings of the Industrial Research Institute for Automation and Measurements (PIAP), which resulted in a map visualizing the research scene and proving a good calibration of the system components responsible for collecting data on the surroundings, odometry and localization. Created map is shown in Fig. 1 where **A** and **B** are PIAP office spaces, **P** is a narrow corridor connecting room **B** with a staircase **KL**. The letter **K** denotes the corridor through which one can get from room **A** to the staircase. During each experiment goal point was set by operator from operator's station and the state of navigation system was saved in .bag file on robot's SBC.

The research was divided into several stages. The first stage involved examining the influence of selected local motion planner parameters on the realization of movement in a known static environment. During next stages, the effect of adding unknown static and dynamic obstacles on the operation of the system was checked. The last stage of the research was to check how the developed system will work while overcoming various types of obstacles during realization of a complex route.

Quality Rates. In order for the robot to successfully complete the traverse, it is crucial to avoid obstacles and also to reach the target pose – that is desired position and orientation of a robot – in the shortest possible time, by the shortest possible route or with the possibly constant speed – which translates into greater energy efficiency. We used the following quality rates for a comparison of traverses with various local path planner parameters [2]:

- The time T that takes for the robot to reach the target pose.
- Robot's average velocity ${}^R v_M$ during traverse (within a time from 0 to T).
- The sum of squares of the robot's distance to the goal E

$$E = \sum_{i=0}^{N} e(i \cdot \Delta t)^2 \tag{6}$$

where: i – iteration number, e – the robot's distance to the goal point for time instant $t = i \cdot \Delta t$, N – the number of iterations till robot reaches target pose.
- Standard deviation of the robot's speed S

$$S = \sqrt{\frac{\sum_{i=0}^{N}({}^R v_R(i \cdot \Delta t) - {}^R v_M)^2}{N-1}} \tag{7}$$

where: ${}^R v_R$ – robot's actual linear velocity

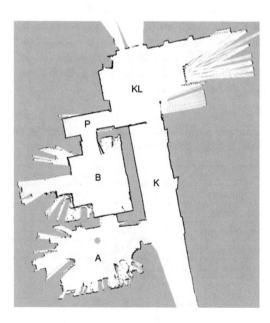

Fig. 1. The map of the surroundings was created during the experimental research

– Approximate length of the path from the starting pose to target pose s

$$s \approx \sum_{i=0}^{N} {}^{R}v_R(i \cdot \Delta t)\Delta t. \tag{8}$$

3.1   Experiment Case 1 – The Influence of Local Planner Parameters on Avoiding Obstacles

During this stage, we checked the influence of the DWA planner parameters on the efficiency of avoiding known and unknown static obstacles. The optimization function based on which the trajectories realized by the mobile base are selected, takes into account parameters described earlier such as $path_{bias}$ (*path_distance_bias*), $goal_{bias}$ (*goal_distance_bias*) and $obstcl_{scale}$ (*occdist_scale*), additionally the simulation time (*sim_time*) significantly influences the generated trajectory, therefore the impact of these parameters on the system operation was investigated. Another important parameters determining overall performance of motion planning are: acceleration limit on a given axis of motion (*acc_lim*), maximum/minimum velocity on a given axis of motion (*max/min_vel*), maximum/minimum absolute translational velocity of robot (*max/min_trans_vel*), absolute value of the maximum/minimum rotational velocity for the robot (*max/min_rot_vel*), the tolerance for the controller in yaw/rotation when achieving its goal (*yaw_goal_tolerance*), number of velocity samples used during exploration of given axis of motion velocity space (*vx/vy/vtheta_samples*). Parameters that values were changed are shown in the Table 2. In turn, in Table 1 some additional reference values of parameters recommended for use of the navigation package with a single board computer are shown – those parameters turned out to be appropriate for the SBC used in our experimental setup ([14]). Default values of parameters (not presented in this paper) can be found in the documentation of `dwa_local_planner` in [11].

Table 1. Values of changed parameters of costmap and `hector_mapping` package recommended for use with a single board computer

costmap		hector_mapping	
update_frequency	2.0	*map_resolution*	0.020
publish_frequency	2.0	*map_size*	2500
transform_tolerance	0.5		
local_costmap (height/width)	6/6		
global_costmap (height/width)	20/20		

Table 2. Values of changed `dwa_local_planner` parameters

acc_lim_x	1.5	sim_time	1.7
acc_lim_th	1.8	$yaw\_goal\_tolerance$	0.2
max_vel_x	0.3	vx_sample	8
min_vel_x	−0.1	vy_sample	4
max_trans_vel	0.3	$vtheta\_samples$	30
min_trans_vel	0.0	$occdist\_scale$	0.02
max_rot_vel	0.5	$goal\_distance\_bias$	20
min_rot_vel	0.0	$path\_distance\_bias$	32
$sim\_granularity$	0.02	$controller\_frequency$	5.0

The realization of the traverse is significantly affected by the global path generated by the planner, which is not identical for subsequent attempts, therefore we decided to compare the impact of parameters on the course of subsequent traverses based on the described quality rates. In this research scenario, we decided to make traverses from room **B** to **A** through the door in the examined environment – that is, the robot moved from room **A** to **B**, set in a given pose, then driven back to room **A**. For this reason, the starting point slightly differs for subsequent attempts, but the goal point is always the same.

During the traverses with the known static obstacle, the effect of the path and goal distance bias parameters was in line with the expectations – that is, when path distance bias was higher than goal distance bias, the actual path was close to the path, while in the opposite case it was very different from the path. We can say that this effect depends on the ratio of these parameters – the higher the value of path distance bias compared to goal distance bias, the more actual path resembles the generated path. The $occdist\_scale$ scale parameter which defines how much the controller has to avoid obstacles has a minimal impact on the actual path – only a 100-fold increase in its value causes a noticeable change in the robot's passage. Increasing the parameter specifying the simulation time sim_time causes that the actual path traveled is smoother, and its reduction results in apparently greater sticking to the generated path.

During the traverses with an unknown static obstacle, we checked the operation of the system in the same way as in the previous stage, but in room **A** we placed an additional obstacle in the form of a box of width comparable to width of the robot located on the direct actual path from the door to the destination point. This obstacle was added after the robot entered room **B** and it was invisible to robot until it passed through the door. For comparative purposes, we made analogous traverses, however, with a previously known obstacle. During the traverses we noticed that the higher the value of the goal distance bias parameter in relation to path distance bias, the faster the robot is able to overcome the previously unknown obstacle. From a value of 40/32 respectively, the obstacle is overcome without stopping, and below this ratio the robot stops in front of the obstacle, but still is able to avoid it. Comparsions of actual path and path from global planner for selected traverses are shown in Fig. 3 and time histories

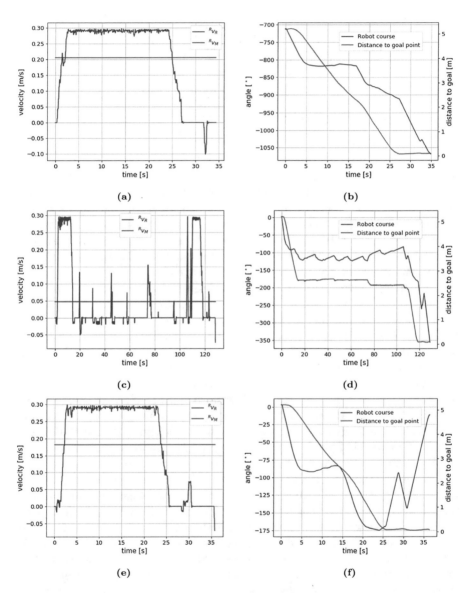

Fig. 2. Results of experiment 1 with *goal_distance_bias* = 50 (a, b), 10 (c, d), default (e, f); known (a, b, c, d) and unknown (e, f) obstacle. Note: a, c, e – time history of the linear velocity of robot during a traverse; b, d, f – time history of the direction of movement during a traverse

Table 3. Results of experiment

path$_{bias}$ [-]	goal$_{bias}$ [-]	T [s]	S [-]	s [m]	E [-]	$^{R}v_M$ [$\frac{m}{s}$]
Passings with unknown obstacle						
32	10	126.75	0.101	6.19	18298	0.05
32	20	90.45	0.114	6.07	12179	0.06
20	20	43.6	0.131	6.23	6641	0.14
12	20	38.22	0.134	7.12	5812	0.16
32	40	35.03	0.132	8.04	5264	0.18
32	50	29.92	0.125	7.78	5235	0.20
Passings with known obstacle						
32	10	35.99	0.129	6.66	5919	0.16
32	20	33.44	0.133	6.51	5363	0.18
32	50	32.34	0.135	6.43	5210	0.19

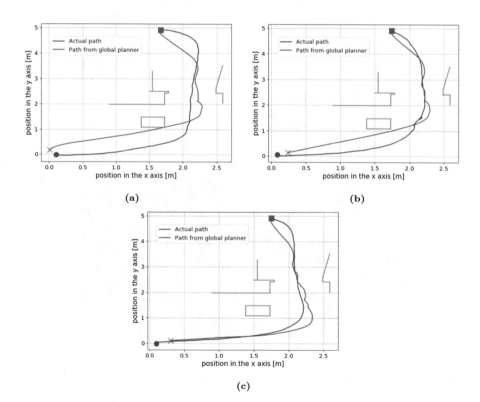

Fig. 3. Results of experiment 1 with *goal_distance_bias* = 50 (a), 10 (b), default (c); known (c) and unknown (a, b) obstacle. Note: a, b, c – comparison of actual path and desired path from global planner.

of robot's velocities, course and distance to the goal point for these traverses are shown in Fig. 2. Increasing value of goal distance bias during traverses with unknown obstacle results in that the quality rates ^Rv_M, T and E achieve better values, and quality rates s and S achieve worse values. Similar correlation occurs during traverses with known obstacle but with exception that quality rate s is slightly better what is consistent with the assumptions. The Table 3 presents a comparison of parameter values and reached quality rates.

3.2 Experiment Case 2 – Avoidance of Dynamic Obstacles

Studies of system operation in the presence of dynamic obstacles were performed in corridor K and consisted in checking how the system behaves when avoiding a dynamic obstacle moving towards the robot and moving across the path of the robot's movement. For this case, the effectiveness of the traverse is assessed, not the quality rates. During the traverses where the dynamic obstacle moves in close proximity to the robot, it turned out that bypassing the dynamic obstacle is very much dependent on its speed – which should not exceed about 1/3 of the maximum speed of the robot, which means that in real world the robot will not be able to avoid such an obstacle. When the robot is in close proximity to an obstacle that moves in a collision fashion, the robot will attempt to pass it, but with given speed limitations, this maneuver will be successful for obstacles moving slowly or at the right distance from the robot's path. During the traverses in the narrow corridor our robot was able to bypass the dynamic obstacle moving in its direction if it did not overlap the path, and the dynamic obstacle moving across the path which is visualized in Fig. 4. Velocities and course during this traverse are presented in Fig. 5.

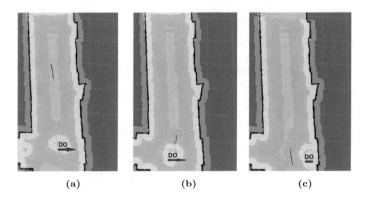

(a) (b) (c)

Fig. 4. Visualization in RViZ of traverse with dynamic obstacle moving across the path, where DO is dynamic obstacle

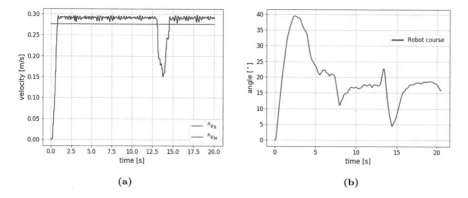

(a)   (b)

Fig. 5. Velocities (a) and course (b) during traverse with dynamic obstacle moving across the path

3.3   Experiment Case 3 – Passings in a Complex and Unknown Environment

The last stage of the research was to check whether the system could bring the robot to its destination in an unknown and complicated environment. These traverses were carried from room **A** to room **B** with the door closed between these rooms, which forced the passage through **K**, **KL** and **P**, and as a result of their execution a map was created (Fig. 1). During these traverses, it turned out that setting the destination point in an unknown place for such a complicated route in most cases results in failure. On the other hand, setting goals gradually – from room to room, where the environment is sometimes already scanned by the laser – in most cases results in an efficient passage. It should be noted that the return path through the environment already researched always resulted in success. An interesting case occurred during some traverses while passing through corridor P, where due to the highly reflective surface of the wall, artifacts appeared (Fig. 6) in the corridor in the form of obstacles of the size of one cell that prevented passage.

Fig. 6. RViZ visualisation of artifact blocking possibility of moving through corridor P

4   Conclusion and Future Work

In the present work the results of experimental research on the developed navigation system have been shown, during which impact of local motion planner parameters on generated trajectories was tested. Control quality indexes were introduced and calculated based on results of each experimental trial.

The most important conclusions of the work are summarized below.

- Localization, mapping and control systems for synchronously driven and steered mobile robot was implemented and successfully tested.
- The created autonomous navigation system is able to act in simple unknown environment (e.g. negotiation of a single previously unknown static obstacle at once).
- The created autonomous navigation system is capable of navigating in a complex environment provided that this environment in known (that is, mapped before autonomous navigation is conducted) and contains only one static unknown obstacle for the navigation system to deal with at once.
- System is able to avoid dynamic obstacles on condition they move with low velocities, that is, about one third of the velocity of the robot.
- It is possible to use the ROS navigation stack with a single board computer, however the default parameters must be relaxed to obtain acceptable speed of operation.

Directions of future works can cover among others:

- Implementation of a method that divides a path in an unknown complex environment to waypoints located in a known robot environment that allows exploration of that environment.
- Implementation of local motion planning methods that take into account to a greater extent the movement possibilities of the robot with the holonomic kinematic structure.
- Inclusion of other type of sensors than 2D laser scanner, e.g. a video camera, to reduce impact of laser measurement errors on system performance.

Acknowledgements. The work was carried out as part of the FB2517 project funded by Own Research Fund of Industrial Research Institute for Automation and Measurements (PIAP).

We also gratefully acknowledge the technical assistance as well as important comments and suggestions provided by Mr. Piotr Jaroszek of PIAP Defence and Security Systems Division and Mr. Michał Pełka of the PIAP Interdisciplinary Research Team.

References

1. Alia, C., Gilles, T., Reine, T., Ali, C.: Local trajectory planning and tracking of autonomous vehicles, using clothoid tentacles method. In: 2015 IEEE Intelligent Vehicles Symposium (IV), pp. 674–679, June 2015
2. Bigaj, P., Bartoszek, J., Trojnacki, M.: The analysis of influence of sensors' failure on the performance of mobile robot autonomy. J. Autom. Mob. Robot. Intell. Syst. **8**(4), 31–39 (2014)
3. Brand, M., Masuda, M., Wehner, N., Yu, X.H.: Ant Colony Optimization algorithm for robot path planning. In: 2010 International Conference on Computer Design and Applications, vol. 3, pp. V3-436–V3-440, June 2010
4. Corke, P.: Robotics, Vision and Control: Fundamental Algorithms in MATLAB, 1st edn. Springer, Heidelberg (2011)
5. Fahimi, F.: Autonomous Robots: Modeling, Path Planning, and Control. Springer, New York (2009)
6. Fox, D., Burgard, W., Thrun, S.: The dynamic window approach to collision avoidance. IEEE Robot. Autom. Mag. **4**(1), 23–33 (1997)
7. Gerkey, B.P., Konolige, K.: Planning and control in unstructured terrain. In: Workshop on Path Planning on Costmaps, Proceedings of the IEEE International Conference on Robotics and Automation (ICRA) (2008)
8. Hendzel, Z., Żylski, W., Giergiel, M.: Modelowanie i sterowanie mobilnych robotów kołowych. Wydawnictwo Naukowe PWN, Warszawa (2002)
9. Jaroszek, P., Trojnacki, M.: Model-based energy efficient global path planning for a four-wheeled mobile robot. Control. Cybern. **43**(2), 337–363 (2014)
10. Li, D., Niu, K.: Dijkstra's algorithm in AGV. In: 2014 9th IEEE Conference on Industrial Electronics and Applications, pp. 1867–1871, June 2014
11. ROS Community: Documentation – ROS Wiki. http://wiki.ros.org/
12. Rösmann, C., Feiten, W., Wösch, T., Hoffmann, F., Bertram, T.: Efficient trajectory optimization using a sparse model. In: 2013 European Conference on Mobile Robots, pp. 138–143, September 2013
13. Siegwart, R., Nourbakhsh, I.R., Scaramuzza, D.: Introduction to Autonomous Mobile Robots, 2nd edn. The MIT Press, Cambridge (2011)
14. Słomiany, M., Dąbek, P., Trojnacki, M.: Motion Planning and Control of Social Mobile Robot - Part 1. In: Robot Physical Design and Software Architecture Description; Theoretical Considerations. Advances in Intelligent Systems and Computing, Springer (2018)
15. Tchoń, K.: Manipulatory i roboty mobilne: modele, planowanie ruchu, sterowanie. Akademicka Oficyna Wydawnicza, Warszawa (2000)
16. Trojnacki, M.: Modelowanie dynamiki mobilnych robotów kołowych (Dynamics modeling of wheeled mobile robots), January 2013

Unified CAMELOT Interoperability Adapters for Existing Unmanned Command and Control Systems

Jan Piwiński[1(✉)], Rafał Kłoda[1], Mateusz Macias[1], and Francisco Pérez[2]

[1] Industrial Research Institute for Automation and Measurements, Al. Jerozolimskie 202, 02-486 Warsaw, Poland
{jpiwinski,rkloda,mmacias}@piap.pl
[2] Universitat Politècnica de València, Camino de Vera, s/n, 46022 Valencia, Spain
frapecar@upvnet.upv.es

Abstract. In this paper we present the CAMELOT project Command Control (C2) framework, which aim is to provide seamless interoperability with each of the UxV platform domains, by a single standard for message formats and data protocols, without necessitating a re-structuring of existing interface protocols and underlying architectures. This will lead to evolution to one unified standard that can be applied for development of all unmanned systems types in the future. The CAMELOT framework will include different service modules for controlling the C2 Infrastructure and Operation, for data management and running analytics, for supporting communications and network connectivity between blocks, and for providing interfaces to external systems. Special care will be taken to maximise interfaces with external information frameworks and to ensure future implementation and adoption by end users.

Keywords: Interoperability · Adapter · Command & control

1 Introduction

One of the most cutting-edge and challenging aspects of autonomous systems is enabling systems that operate in different domains work together as a heterogeneous whole, controlled by single control station. Examples of such vehicles considered here are Unmanned Aerial Vehicles (UAVs), Unmanned Surface Vehicles (USVs), and Unmanned Underwater Vehicles (UUVs), but one could imagine other domains.

As autonomous systems become more important to military operators, and especially as they are used for more diverse and complex missions, the issue of command and control (C2) of cross-domain unmanned vehicles (UxVs) will become more important. Today, while the performance of UxV in all domains has improved dramatically, the C2 issues of controlling UxVs in multiple domains simultaneously remains an area requiring additional research, modeling and simulation, and operational testing [1].

© Springer Nature Switzerland AG 2020
R. Szewczyk et al. (Eds.): AUTOMATION 2019, AISC 920, pp. 538–548, 2020.
https://doi.org/10.1007/978-3-030-13273-6_50

The combat development community is calling for interoperability as a critical element to the future unmanned systems fleet. The ability for manned and unmanned systems to share information will increase combat capability, enhance situational awareness, and improve flexibility of resources. Interoperability will improve the ability for unmanned systems to operate in synergy in the execution of assigned tasks [2]. Properly stabilized, implemented, and maintained, interoperability can serve as a force multiplier, improve warfighter capabilities, decrease integration timelines, simplify logistics, and reduce total ownership costs [3].

To address above challenges the CAMELOT project will provide a cross-organizational, cross-domain, cross-level interoperability between the involved C2 Systems. Although individual standards and specifications are usually adopted in C2 Systems separately, there is no common, unified interoperability specification to be adopted in an emergency situation, which creates a crucial interoperability challenge for all the involved organizations.

The CAMELOT framework will ensure interoperability with each of the UxV platform domains, without necessitating a re-structuring of existing interface protocols and underlying architectures. Providing seamless interoperability between UAVs with its UGVs, by a single standard for message formats and data protocols is needed where two such standards, STANAG 4586 and JAUS exist today. Widely accepted or approved standards are often too broadly defined with varying options and inadvertently allow compliance but not necessarily interoperability [3].

CAMELOT partners will contribute in evolution to a unified standard, which lead to one interoperability standard that can be applied for development of all unmanned systems types in the future. The work on such integration of these two standards was pursued by NATO Industrial Advisory Group (NIAG) in Study Groups 157 and 202 (SG157 and SG202), starting to converge on identification of a set of Internet Protocol-based development schemas and software development. The most flexible solution for CAMELOT is service-oriented, layered approach which is consistent with the common approach started by SG202.

The CAMELOT framework will need to discover, control, and receive data from a wide variety of UxVs and their sensors. As such, the data model must provide a common description for entities within the system, their relationships, and the messages between them. A number of architectures/data models have been developed for the Unmanned Vehicle domain and the MDCS (Multi-Domain Unmanned Platform Control System Services) architecture has been selected as the baseline for CAMELOT.

The information contained in the Conceptual Data Model of MDCS framework is designed to support complex, multi-system deployments in the context of a wide variety of established architectures and interface standards. In another way, system which is built on such concept is able to interoperate by leveraging capabilities from components that already conform to existing interoperability standards, such as STANAG 4586, JAUS and UCS. This is the basic assumption of CAMELOT project therefore MDCS architecture standard will play major role and certainly should be based on the NATO MDCS architecture [4, 5].

Currently, there is no widespread standard for multi-service, multi-domain C2 systems. The CAMELOT architecture and modules will help build critical support in both the industrial and the practitioner communities leading to the adoption of these technologies. The CAMELOT project aims to develop and demonstrate different advanced command and control service modules for multiple platform domains customisable to the user needs.

This paper focuses on the part of the CAMELOT architecture, which is adaptor layer, which will allow interoperability of systems designed to operate using external protocols. This layer converts bi-directionally between the external protocol of the other system and the Logical Data Model (LDM) of an instantiation of the MDCS architecture. The advantage of adapting to a single LDM is that a single MDCS implementation can easily support multiple platform protocols.

The MDCS architecture relies on protocol adaptors, which stand between the platform and payload devices and services in the vehicle domains. The adaptors provide the service interfaces to the MDCS architecture services at the MDCS side and the vehicle and payload protocol at the other side, external to the MDCS architecture.

2 Camelot High Level Architecture

The CAMELOT system architecture will be based on a distributed platform as shown in Fig. 1. This platform will offer the scalability, availability and security demanded as main design requirement. The core of the platform, will be designed in a flexible and uncoupled way to easily implement new functionalities and to adapt to user requirements changes. It consists of specific modules that will be developed along the different stages of the project.

- CAMELOT Adaptor Layer (CAL): Adapts the different data models and information available in connected assets to the CAMELOT Data Model.
- Automatic Asset Tasking & Control (AATC): Exposes different methods to control the different assets connected to the platform.
- Data Manager and analytics (DMA): Brings to the platform a set of advanced analysis and data management functions to perform information operations.
- Middleware: Allows the interaction and the information exchange among the different submodules and services deployed in CAMELOT platform using the Publish-subscribe paradigm. This component is able to decouple all the interactions providing services and information according to CAMELOT data model.
- CAMELOT Core Layer: Contains all internal functionalities required by the platform to perform CAMELOT main capabilities. This module includes Mission-related Services, Visualization & Display Services and Sensing & Detection Services.

Additionally, the platform will expose different interoperability services accessible by a REST API that will allow the different C2 Systems to interact with the platform to get and feed information or to consume the provided services.

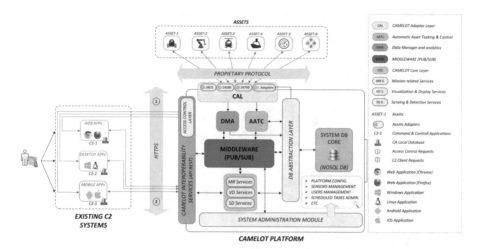

Fig. 1. The CAMELOT architecture

2.1 Generic Logic Layers in the Architecture

This chapter aims to define generic interfaces in CAMELOT architecture to external components and systems like UxV, payloads sensors, assets or external C2 systems. Since the CAMELOT architecture data model is quite complex, therefore it is important to show the generic block diagram (Fig. 2), which illustrates the main logic layers, which enrich the main goals and ambition of CAMELOT project. The main layers are as followed:

- CAMELOT adaptor layer – which includes interface services based on approved standards e.g. JAUS, STANAG 4586, 4748 etc. to heterogeneous assets to providing seamless interoperability between them.
- CAMELOT application layer, which represents its service modules e.g. Mission related services and corresponds to MDCS framework capabilities and adopt its conceptual data model.
- CAMELOT implementation specifics, which demonstrates new functionalities to Multi Service Domain for Command and Control frameworks e.g. complex Visualization services including augmented/mixed reality.
- CAMELOT platform layer, which represents middleware and security and privacy issues of the project.

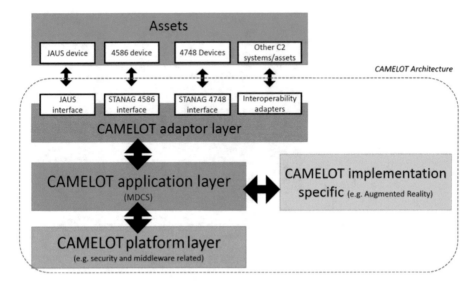

Fig. 2. The main logic layers in the CAMELOT architecture

3   CAMELOT Specific Protocol Adapters

The proposed architecture of CAMELOT is an extension of existing candidate architecture and data model from NATO NIAG SG.202 Conceptual Data Model for Multi-
Domain Unmanned Platform Control System Services, which includes suggested
application domains in the application layer to support multi-domain control of UxVs.

The proposed architecture includes an adaptor layer to allow interoperability of
systems designed to operate using external protocols. This layer converts bi-
directionally between the external protocol of the other system and the Logical Data
Model (LDM) of an instantiation of the MDCS architecture.

Centralized C2 nodes ashore and afloat, as well as individual cross-domain
(air/sea/ground) unmanned systems, must be able to exchange information seamlessly
for networked operations that support distributed control and flexible hierarchies
conforming to evolving tactical scenarios. The overwhelming number of unmanned air
systems relative to sea or ground vehicles supported early establishment of the NATO
Standard Agreement (STANAG) 4586 messaging protocols as a candidate for networking unmanned systems [6].

Fortunately a limited number of these standards and protocols have been identified,
so an implementation should be able to develop this limited set of protocol adaptors
and attach them to the required communication device(s) used by the specific instance
of the platform. The details of manipulating the communication device and the
implementation of the platform specific protocol (e.g. 4586, JAUS or 4748) are the
responsibility of the protocol adaptor.

Each protocol adaptor must be configured and must enable discovery and initialization of the protocol specific devices and must convey this information to instances of
the CAMELOT services so that they can be properly configured to use the protocol

specific assets. This is enough to implement protocol adapters for STANAG 4586, JAUS and STANAG 4748, providing discovery and initialisation functions. Below sections briefly present each standard, which are the most widely adopted and the most interesting for the scope of CAMELOT project.

3.1 Joint Architecture for Unmanned Systems (JAUS)

Motivation to JAUS was to establish a common set of message formats and data protocols for UGVs made by various manufacturers. The JAUS messaging architecture enables communication with and control of unmanned systems across the entire unmanned system domain. It provides a common language enabling internal and external communication between unmanned systems. It incorporates a component-based, message-passing architecture specifying data formats that promote the stability of capabilities by projecting anticipated requirements as well as those currently needed. JAUS addresses unmanned system capabilities including payload control, autonomous systems, and weapons systems. In order to ensure that the component architecture is applicable to the entire domain of current and future unmanned systems, JAUS is built on five principles: vehicle platform independence; mission isolation; computer hardware independence; technology independence; and operator use independence. Architecture of JAUS is open and scalable to the unmanned systems community's needs [7].

3.2 STANAG 4586

STANAG 4586 is an unmanned aerial equivalent of JAUS. Traditionally UAVs comprise system elements and functions that are generally unique to that nation as well as being generally vehicle-specific. The result is a variety of non-interoperable "stovepipe" vehicles and systems. To address this issue, the NATO STANAG 4586 was developed to provide a level of unmanned aerial vehicle (UAV) interoperability across different entities to allow for the ability to quickly task available assets, to mutually control these vehicles and their payloads, and to rapidly disseminate tactical information to the collective force as required. STANAG 4586 specifies interfaces to be implemented to achieve the level of interoperability (LOI) that is operationally required per each respective UAV concept of operations (CONOPS). This can be achieved by implementing standard interfaces in the UCS to communicate with different UAVs and their payloads, and with different C4I systems. In large part, STANAG 4586 is an Interface Control Definition (ICD), which defines the architectures, interfaces, communication protocols, data elements, message formats, and related STANAGs that must be complied with in order to operate and manage multiple legacy and future UAVs [7].

3.3 STANAG 4748 (JANUS)

JANUS is the protocol used by UAVs for initial contact establishment between UAV and operator. It is an open-source robust signalling method for underwater communications, freely distributed under the GNU General Public License.

Underwater (UW) communication capabilities are currently manufacturer-specific, generally using proprietary digital coding technologies. There exists no general interoperable capability for digital UW communication between assets using modems from different manufacturers Interoperability is an essential feature as maritime operators seek to integrate an increasingly heterogeneous mix of maritime assets. Currently there are no existing means to discover other communicating assets to permit the formation of ad-hoc networks [8, 9]. The establishment of an UW digital communications standard therefore has wide application in both military and civilian contexts.

The proposed physical layer standard, named JANUS, has been designed to minimise the changes required to bring existing UW communications equipment into compliance, leveraging the inherent flexibility of modern digital communications systems and existing acoustic frequencies and bandwidths. JANUS is currently moving in the process of a NATO Standardization Agreement (STANAG) promulgation and a NATO Industry Advisory Group is addressing issues, future protocol extensions and compliance to the standard [10].

4   CAMELOT Interoperability Adapters with Existing C2 Systems and Assets

This section describes the concept for implementation of the interface modules and protocols (other than widely known and approved-mentioned above in chapter "CAMELOT specific protocol adapters" called Interoperability adapter (IA) that represent CAMELOT interfaces to involved end users assets and systems in border protection management. Since the end user's tools/assets are a-priory not interoperable and each of them "speaks its own language", the IAs foster the interoperability by connecting these systems to the CAMELOT framework and provide necessary transformation to standard presentations.

The Interoperability adapter are application modules that have been implemented to allow CAMELOT interfacing with the external C2 systems and other assets of local users that are involved in the demonstration. Each IA has two main interfaces:

- On the one side, IA interfaces directly with the system of the local stakeholder and therefore takes into account the standard used by this system, implementing features that allow interaction with it;
- On the other side, IA interfaces with CAMELOT framework and exchanges information with the CAMELOT network of components.

Between these two interfaces, IA assures that the data is transformed to the right format and sent to the right system or stakeholder, in accordance with the workflow description from the relevant technical specification profiles. Each IA, thus, allows unidirectional communications between CAMELOT and the local user system with no or minimal changes to the local system and to the local operational procedures.

The IAs are interchangeable modules complying with a common API. Their role is to ensure the communication with a legacy/proprietary systems at the application protocol level, based on the hypothesis that underlying low-level communication protocol is TCP/IP or UDP/IP.

4.1   Planned Implementation of SOA Concept

To address interoperability needs, CAMELOT will expose the functionality of already existing C2 systems, assets and applications by adopting a Service-Oriented Architecture (SOA) concept.

Resources of different systems are required to be packaged as web services that may contain data, business operations or models. The interoperability between a variety of resources requires appropriate data representations and a subsequent data exchange between services wrapping resources. For designing and implementing communication between mutually interacting software applications in a Service-Oriented Architecture (SOA) is used a shared messaging layer (which is one of the possibility) named Enterprise Service Bus (ESB). Once data is received by the ESB all the other applications/systems can consume it within their permission.

An ESB implements a SOA through middleware that offers virtualization and management of service interactions between communication participants.

In order to minimize effort and maximize added value for the CAMELOT system, the project will deploy a middleware solution, namely a message-oriented middleware solution that supports the transmission of messages between distributed systems, allows service modules (applications) to be shared over disparate platforms, and reduces the complexity and effort required for the development of applications for multiple operating systems and network protocols. Given the fact that there are multiple solutions available in the industry, in order to choose the most suitable for CAMELOT, a benchmark had to be done to understand the pros and cons of the usage of each solution.

CAMELOT system has a list of requirements towards future middleware, namely the fact that the system will be networked, using Ethernet, but with direct visibility of all hardware items. This means that all external components will be connected using adapters, with the suitable data link/connection/encryption is required.

When writing this paper several middleware solutions were analysed and no consent was reached. In this sense, a middleware benchmark has performed, being the solutions under analysis DDS, ECOA, JAUS, Kafka, RabbitMQ with JSON, and ZeroMQ with Protocol Buffers.

In communication processing, chosen middleware should allow many message delivery models, such as request/response, publish/subscribe, and a large number of transport protocols as well as synchronous and asynchronous mechanisms. For example ESB, via the bus mode, integrates all the services needed for CAMELOT, enabling developers no longer to consider each other's message format or protocol between services, and improve system adaptability and scalability.

The concept of an Enterprise Service Bus was first described as "a new architecture that exploits Web services, messaging middleware, intelligent routing, and transformation" [11]. Since then, ESB technologies have become the subject of countless papers, works, and have generally known a widespread level of adoption across enterprise, SOA and Web-oriented contexts.

ESB is the increase prevalent structure in service-oriented architecture. In a typical SOA environment, there is a service provider and a service consumer. In order to work, we also need a mechanism allowing them to communicate with each other and also describe and discover one another. The fundamental structure of SOA is shown in Fig. 3 below:

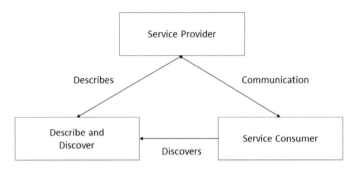

Fig. 3. The fundamental structure of SOA

W3C has defined an open standard for web services in order to make SOA work. A web service uses the XML-based Simple Object Access Protocol (SOAP) for communication between service provider and service consumer. The service provider describes the services as interfaces defined by e.g. Web Service Definition Language (WSDL). Universal Description, Discovery and Integration (UDDI), a platform-independent, XML-based registry, is used for looking up the services.

An Enterprise Service Bus can formally be described as providing a set of infrastructure capabilities, implemented by middleware technology, that enable the integration of services in an SOA. A variety of ESB capabilities has been identified and summarized in many public works, but not all of them are required in every possible ESB-based solution. A set of minimum capabilities that fulfil the most basic needs for an Enterprise Service Bus that are consistent with the principles of SOA has also been defined in the literature [12]. Identifying these minimum capabilities enables the adopter to identify which existing technologies can be used to implement an ESB. By considering how the requirements of a specific situation dictate the need for additional capabilities, it is possible to choose the most appropriate implementation technology for that situation.

Summarizing, making use of an ESB, an application/sensor/integrator will communicate via the bus, which acts as a message broker between systems/actors. Such an approach has the primary advantage of reducing the number of point-to-point connections required to allow systems/actors to communicate. This, in turn, makes impact analysis for major software changes simpler and more straightforward.

5 Conclusions and General Recommendations

One of the main goal of the CAMELOT project is to provide seamless interoperability between heterogeneous assets, which are conformant to widely accepted or approved standards in unmanned vehicles domain, by a single standard for message formats and data protocols.

The CAMELOT project will provide a cross-level interoperability between the involved C2 Systems. This will be done by implementing the CAMELOT framework that will ensure interoperability with each of the UxV platform domains, without necessitating a re-structuring of existing interface protocols and underlying architectures. This requires a heavy utilization of protocol adaptors and data integrators.

Data is essential to the day-to-day operations of every security organization responsible for border protection. Unfortunately, the ad-hoc development of many legacy systems that contain redundant and inconsistent data. As stated in the previous sections, SOA provides a framework for integrating these information and leveraging existing systems, applications or sensors to create more flexible, agile security framework.

Within the CAMELOT project applicative aspects will be designed using a novel approach, which addresses all the layers of the communication stack in security field to empower Command and Control (C2) systems with a harmonized view of information originating from different source providers (like physical sensors, legacy security systems and applications), which is one of the key aspects of the CAMELOT project.

Acknowledgements. This project has received funding from the European Union H2020 Programme for research, technological development and demonstration under the Grant Agreement No. 740736.

References

1. Pastore, T., Galdorisi, G., Jones, A.: Command and control (C2) to enable multi-domain teaming of unmanned vehicles (UxVs). In: OCEANS 2017 - Anchorage, pp. 1–7, September 2017
2. Joint Publication 1-02. Department of Defense Dictionary of Military and Associated Terms, 12 April 2001 (as amended through 17 March 2009). https://marineparents.com/downloads/dod-terms.pdf
3. Unmanned Systems Integrated Roadmap, FY2013-2038. UnderSecretary of Defense Acquisition, Technology & Logistics, 3010 Defense Pentagon, Washington, DC, 20301-3010. https://apps.dtic.mil/dtic/tr/fulltext/u2/a592015.pdf
4. Unmanned Systems Roadmap FY2007-2032. https://www.globalsecurity.org/intell/library/reports/2007/dod-unmanned-systems-roadmap_2007-2032.pdf
5. NATO Industrial Advisory Group: On Development of Conceptual Data Model for Multi-Domain Unmanned Platform Control System Services. (Report NATO Unclassified)
6. Incze, M.L., et al.: Communication and collaboration among heterogeneous unmanned systems using SAE JAUS standard formats and protocols. IFAC-PapersOnLine **48–5**, 007–010 (2015)

7. Blais, C.L.: Unmanned systems interoperability standards. Calhoun: The NPS Institutional Archive, Monterey, CA (2016). https://core.ac.uk/download/pdf/81222182.pdf
8. Heidemann, J.: Research challenges and applications for underwater sensor networking. In: IEEE Wireless Communications and Networking Conference, pp. 228–235 (2006)
9. Chitre, M., et al.: Underwater acoustic communications and networking: recent advances and future challenges. (2008). https://stuff.mit.edu/people/millitsa/resources/pdfs/mandar.pdf
10. Potter, J., et al.: The JANUS underwater communications standard. In: Underwater Communications and Networking (UComms) (2014)
11. Schulte, R.: Predicts 2003: enterprise service buses emerge (2012)
12. Keen, M., et al.: Patterns: implementing an SOA using an enterprise service bus. IBM Redbooks (2004).https://www.redbooks.ibm.com/redbooks/pdfs/sg246346.pdf

Measuring Techniques and Systems

Magnetic Induction Measurements as an Example of Evaluating Uncertainty of a Vector Output Quantity

Adam Idźkowski[(⊠)] [ID]

Faculty of Electrical Engineering, Białystok University of Technology,
Wiejska 45D st., 15351 Białystok, Poland
a.idzkowski@pb.edu.pl

Abstract. This paper presents the way of estimating results in the case of indirect multivariate measurements performed using electronic measuring instrument. The simultaneous measurements of several quantities, which are usually correlated, are performed by using electronic devices in automatic control applications. The aim of paper is to discuss the propagation of measurement uncertainty in measuring channels. The presented model equation contains a vector measurand and the covariance matrix. The covariance of two random variables is a measure of their mutual dependence. A novel application is presented. The uncertainties of output quantities as well as matrix of correlation coefficients are computed. A comparison of two time series is another problem which is discussed in this paper. A simple technique of finding correlation between two signals produced by the automatic system with two magnetic field sensors, which is used for the purpose of vehicle's length estimation, is presented.

Keywords: Standard uncertainty · Uncertainty propagation ·
Magnetic induction measurement · Normalized cross-correlation ·
Pearson's correlation coefficient

1 Introduction

Propagation of measurement uncertainty using covariance matrix is one of the issues presented in Guide to the expression of uncertainty in measurement [1]. The input measurand is formulated by a random vector $X = [X_1, X_2, ..., X_m]$ and the output measurand by a random vector $Y = [Y_1, Y_2, ..., Y_n]$. These vectors of dimensions m and n represent the multi-dimensional distributions of X and Y, respectively [2]. If f is a linear operator, then $n \leq m$. The relation between them is presented by the generalized formula [3]

$$Y = f'(h) = f(X), \tag{1}$$

where f describes the system equation as a function of the input quantities, while f' is a function of the uncorrelated subsystems.

© Springer Nature Switzerland AG 2020
R. Szewczyk et al. (Eds.): AUTOMATION 2019, AISC 920, pp. 551–559, 2020.
https://doi.org/10.1007/978-3-030-13273-6_51

The relations between covariance matrices of the output and input measurand are expressed in [2, 3]

$$c_Y = Sc_X S^T, \tag{2}$$

where $S \equiv \partial Y/\partial X$ is the matrix of sensitivity coefficients. The model (2) is true if it is either linear or can be linearized in the working point using a Taylor series. Usually, it is assumed that the uncertainties for the input parameters are small. The other requirement is that the probability density functions (PDFs) should be symmetric and common [3].

2   Vehicle Length Estimation with Two Magnetic Sensors

In Fig. 1 we can see a part of intelligent transportation system which consists of two three-axis digital magnetic field sensors LIS3MDL. Their full-scale range is ±0.4 mT. The sensors are placed at small distance on a roadway. The data from sensors are sampled and processed to evaluate speed and length of a traveling vehicle [4]. Soft iron effect, caused by metallic moving parts, e.g. car engine, influences the Earth's magnetic field [5, 6] (Fig. 1).

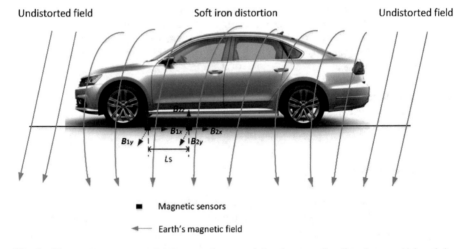

Fig. 1. Two sensors mounted on a roadway and local magnetic disturbances induced by traveling vehicle.

The input measurand in the system in Fig. 2 consists of six components of magnetic induction, measured in two points on a roadway, and the output measurand has three quantities

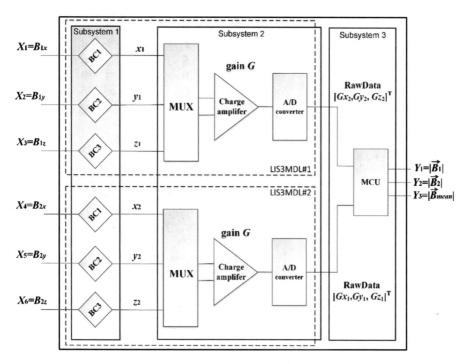

Fig. 2. The measuring system based on two LIS3MDL magnetic sensors and microcontroller.

$$Y = \left[\left| \overrightarrow{B_1} \right| \quad \left| \overrightarrow{B_2} \right| \quad \left| \overrightarrow{B}_{mean} \right| \right]^T,$$ (3)

where

$$\left| \overrightarrow{B_1} \right| = \sqrt{B_{1x}^2 + B_{1y}^2 + B_{1z}^2}, \ \left| \overrightarrow{B_2} \right| = \sqrt{B_{2x}^2 + B_{2y}^2 + B_{2z}^2},$$ (4)

$$\left| \overrightarrow{B}_{mean} \right| = \sqrt{\left(\frac{B_{1x} + B_{2x}}{2}\right)^2 + \left(\frac{B_{1y} + B_{2y}}{2}\right)^2 + \left(\frac{B_{1z} + B_{2z}}{2}\right)^2}.$$ (5)

As it is depicted in Fig. 2 an AMR sensor consists of two subsystems. The first one includes three classic magneto-resistive bridge-circuits BC1-BC3. The second one is responsible for multiplexing, amplifying and converting into RAW data. The third one – a microcontroller (MCU) is responsible for arithmetic operations.

Assuming that the uncertainties of every component of magnetic field in both sensors are equal $u_{X1} = u_{X2} = u_{X3} = \sigma_1$, $u_{X4} = u_{X5} = u_{X6} = \sigma_2$, uncorrelated, have uniform distributions, the input and output variance-covariance matrices are presented as follows

$$c_X\left(B_{1x}, B_{1y}, B_{1z}, B_{2x}, B_{2y}, B_{2z}\right) = \begin{bmatrix} \sigma_1^2 & 0 & 0 & 0 & 0 & 0 \\ 0 & \sigma_1^2 & 0 & 0 & 0 & 0 \\ 0 & 0 & \sigma_1^2 & 0 & 0 & 0 \\ 0 & 0 & 0 & \sigma_2^2 & 0 & 0 \\ 0 & 0 & 0 & 0 & \sigma_2^2 & 0 \\ 0 & 0 & 0 & 0 & 0 & \sigma_2^2 \end{bmatrix} \tag{6}$$

$$
\begin{aligned}
c_Y\left(\left|\overrightarrow{B_1}\right|, \left|\overrightarrow{B_2}\right|, \left|\overrightarrow{B}_{mean}\right|\right) &= \begin{bmatrix} c_{11} & c_{12} & c_{13} \\ c_{21} & c_{22} & c_{23} \\ c_{31} & c_{32} & c_{33} \end{bmatrix} \\[2mm]
&= \begin{bmatrix} \sigma_1^2 & 0 & \dfrac{2b_1 \cdot \sigma_1^2}{\left|\overrightarrow{B}_{mean}\right| \cdot \left|\overrightarrow{B_1}\right|} \\[4mm] 0 & \sigma_2^2 & \dfrac{2b_2 \cdot \sigma_2^2}{\left|\overrightarrow{B}_{mean}\right| \cdot \left|\overrightarrow{B_2}\right|} \\[4mm] \dfrac{2b_1 \cdot \sigma_1^2}{\left|\overrightarrow{B}_{mean}\right| \left|\overrightarrow{B_1}\right|} & \dfrac{2b_2 \cdot \sigma_2^2}{\left|\overrightarrow{B}_{mean}\right| \left|\overrightarrow{B_2}\right|} & 4(\sigma_1^2 + \sigma_2^2) \end{bmatrix}
\end{aligned} \tag{7}
$$

where:

$$b_1 = B_{1x}(B_{1x} + B_{2x}) + B_{1y}(B_{1y} + B_{2y}) + B_{1z}(B_{1z} + B_{2z}) \tag{8}$$

$$b_2 = B_{2x}(B_{1x} + B_{2x}) + B_{2y}(B_{1y} + B_{2y}) + B_{2z}(B_{1z} + B_{2z}) \tag{9}$$

Equation (7) is the simplified expression for the propagation of uncertainties from one set of variables X onto another Y. Additionally, it is assumed that gain and offset errors (uncertainties) of subsystem 2 are negligible, with the above mentioned assumptions c_{12} and c_{21} components are equal 0. The coverage region is a hyper-ellipsoid, with coverage factor k_p for stated probability, usually 95% [2].

Similar equations to (7–9) were presented in [7], where the modulus of difference $\left|\overrightarrow{B_1} - \overrightarrow{B_2}\right|$ was measured in two points in the space by using a positioning probe. The above presented model can be expanded. It can contain more output parameters, e.g. x and y components and their modules ratio [8].

2.1 Numerical Example of Evaluating Type A Uncertainties and Correlation Coefficients

The xyz components of magnetic induction measured in two points, their magnitudes and mean magnitude are presented in Table 1. They represent means of 10000 samples. Sampling rate was equal to 1 ms. The data was not filtered.

Table 1. The xyz components of magnetic field, their magnitudes and mean magnitude in mT.

B_{1x}	B_{1y}	B_{1z}	B_{2x}	B_{2y}	B_{2z}	$\left|\overrightarrow{B_1}\right|$	$\left|\overrightarrow{B_2}\right|$	$\left|\vec{B}_{mean}\right|$
0.010082	0.005997	0.005797	−0.005796	0.0000002	−0.014398	0.0149	0.0169	0.0159

As you can see the c_{12} and c_{21} components in variance-covariance matrix (10) are not equal 0. However, they are very small when comparing to the other values. This is slight effect of multiplexing and amplifying by the subsystem 2

$$c_Y\left(\left|\overrightarrow{B_1}\right|, \left|\overrightarrow{B_2}\right|, \left|\overrightarrow{B}_{mean}\right|\right) = 10^{-8} \cdot \begin{bmatrix} 193 & -11 & 91 \\ -11 & 272 & 131 \\ 91 & 131 & 1862 \end{bmatrix}. \tag{10}$$

The correlation matrix shows weak relationship between output quantities

$$k\left(\left|\overrightarrow{B_1}\right|, \left|\overrightarrow{B_2}\right|, \left|\overrightarrow{B}_{mean}\right|\right) = \begin{bmatrix} 1 & -0.048 & 0.152 \\ -0.048 & 1 & 0.183 \\ 0.152 & 0.183 & 1 \end{bmatrix}. \tag{11}$$

Standard deviations represent type A uncertainties of the measured magnitudes of magnetic induction and their mean value. It is also observed that the uncertainty of mean magnitude is about 3 times larger: $\sigma_{B1} = 0.00044$ mT, $\sigma_{B2} = 0.00052$ mT, $\sigma_{Bmean} = 0.00133$ mT. The first two values are a bit higher than the RMS noise (0.00032 mT) [9] which is typical resolution of LIS3MDL sensor.

2.2 Computing a Time Series Correlation and Estimating a Vehicle's Length

The numerical example in Sect. 2.1 is evaluated in the case when magnetic field is constant. Computing a mean magnitude of magnetic induction entails higher uncertainty. In practice we have to deal with the changes of magnetic field and to analyze time series.

The system in Fig. 2 measures two magnetic induction magnitudes and their mean. The sample delay between two signals $\left|\overrightarrow{B_1}(n)\right|$ and $\left|\overrightarrow{B_2}(n)\right|$ represents a vehicle's speed [8], n is sample number, the signals are discrete and sampling frequency is 2 kHz.

In ideal situation on a road, while a vehicle follows the line of two sensors at a constant speed, the shapes of both signals are the same and their areas under the curves are equal. In this case the estimation of vehicle's length is executed basing on the $\left|\overrightarrow{B_1}(n)\right|$ curve and the selected threshold [4]. The threshold can be set lower or higher (e.g. basing on m-time value of standard deviation calculated before a vehicle passage) and this fact have influence on the estimated length value.

However, in many observations the highest peaks of both signals are not the same and the curves are not so highly correlated. This is caused by a non-standard passage of a vehicle. Then, the length estimation can be performed basing on $\left|\overrightarrow{B}_{mean}(n)\right|$ curve. It is the distance in samples between n_b and n_e (Fig. 3). In most cases the accuracy of such estimation is less in comparison to the simpler estimation, which is based on the $\left|\overrightarrow{B_1}(n)\right|$ curve only.

Below presented algorithm (Fig. 4) was tested using a dataset of 60 vehicles. There were buses (Setra, 12.3 m) and trucks (tractor units with 13.6 m semitrailers) traveling

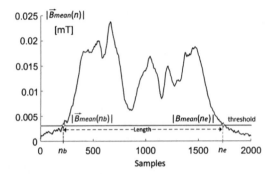

Fig. 3. The change in mean magnetic field magnitude observed while a truck traveled at 80 km/h.

in a traffic at different speed. It was helpful in exploring relationship between two time series for all registered vehicle passages. It significantly allowed to detect the non-standard passage cases when both curves were weakly correlated or were not very highly correlated (the positive correlation coefficient was less than 0.95). Additionally, mean values from 2 matrices B_1 and B_2 and their ratio c were computed. The results are presented in Table 2.

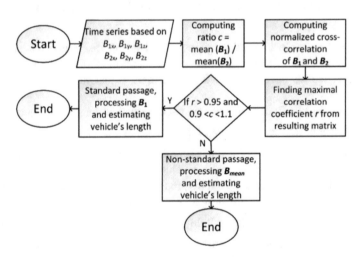

Fig. 4. The algorithm for making decision in estimating a vehicle's length using only $\left| \overrightarrow{B_1}(n) \right|$ magnitude or using mean of two magnitudes $\left| \vec{B}_{mean}(n) \right|$.

In Fig. 5 the exemplary curves from two sensors and their mean are presented. The normalized cross-correlation (NCC) is computed [11–13] as well as ratio c. The maximal value of correlation coefficient in the resulting matrix 0.9863 signifies that two signals are highly correlated. However, c is 1.51. Therefore, in this case the decision is that the $\left| \vec{B}_{mean}(n) \right|$ curve should be thresholded in order to estimate a vehicle's length.

Table 2. The real and estimated values of length based on the dataset of 60 passages.

Type of vehicle	Real length (m)	Correlation coefficient r	Ratio c	Estimated mean length based on B_1 (m)	Uncertainty for $k_p = 2$, $p = 0.95$ (m)	Estimated mean length based on B_{mean} (m)	Uncertainty for $k_p = 2$, $p = 0.95$ (m)
Bus	12.30	$r > 0.95$	$0.9 < c < 1.1$	13.06	0.50	13.31	0.90
Truck (TIR)	16.50	$r > 0.95$	$c > 1.1$	17.12	1.62	16.46	1.60

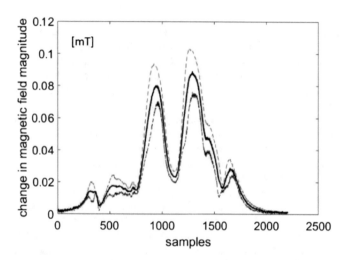

Fig. 5. A pair of magnetic field signals (dashed lines) and their mean (constant black line). The maximal correlation coefficient r is 0.9863, the ratio of means is 1.51, which means the non-standard passage of a vehicle.

3 Conclusions

The uncertainty evaluation can be performed using the methods of uncertainty propagation or propagation of distributions. The first method is presented in this paper, the alternative way is the use of the Monte Carlo method [14]. It can be applied for multichannel measuring systems [15].

The measuring system has been presented. It consists of the bridge-circuits, the multiplexing, amplifying and sampling blocks and microcontroller. The presented model of uncertainty propagation, using Eqs. (3)–(9), is simplified. However, it shows the way of evaluating uncertainty from one set of variables onto another. Basing on the real output data, the type A uncertainties reach the values up to 4-time sensor resolution (when low magnetic inductions are measured, sensor range is ± 0.4 mT).

Recently, vehicle detection and identification techniques have been widely applied in parking and traffic scenes [5, 16]. In this work the normalized cross-correlation between two time series (dynamic signals) has been studied.

The original algorithm, which bases on computing the NCC coefficient between two signals from AMR sensors, is proposed. This helps at start to detect the standard and non-standard vehicle passages as well as to estimate a vehicle's length by means of two magnetic field sensors. The uncertainty of estimated length of trucks (TIRs) was about ± 1.6 m for $k_p = 2$ and $p = 0.95$.

The paper was prepared within a framework of S/WE/2/2018 project at Bialystok University of Technology sponsored by Ministry of Science and Higher Education.

References

1. JCGM: Evaluation of measurement data – Supplement 2 to the Guide to the expression of uncertainty in measurement – Extension to any number of output quantities, http://www.bipm.org/utils/common/documents/jcgm/JCGM_102_2011_E.pdf. Accessed 31 Aug 2018
2. Fotowicz, P.: Coverage Region for the Bidimensional Vector Measurand. In: Szewczyk, R., Zieliński, C., Kaliczyńska, M. (eds.) ICA 2017. AISC, vol. 550, pp. 401–407. Springer, Cham (2017). https://doi.org/10.1007/978-3-319-54042-9_37
3. Hampel, B., Liu, B., Nording, F., Ostermann, J., Struszewski, P., Langfahl-Klabes, J., Bieler, M., Bosse, H., Guttler, B., Lemmens, P., Schilling, M., Tutsch, R., et al.: Approach to determine measurement uncertainty in complex nanosystems with multiparametric dependencies and multivariate output quantities. Meas. Sci. Technol. **29**, 1–12 (2018). https://doi.org/10.1088/1361-6501/aa9d70
4. Markevicius, V., Navikas, D., Idzkowski, A., Valinevicius, A., Zilys, M., Andriukaitis, D.: Vehicle speed and length estimation using data from two Anisotropic Magneto-Resistive (AMR) sensors. Sensors **17**(8), 1778 (2017). https://doi.org/10.3390/s17081778
5. Nguyen, P., Nguyen, H., Nguyen, D., Dinh, T.N., La, H.M., Vu, T.: ParkSense: automatic parking positioning by leveraging in-vehicle magnetic field variation. IEEE Access **5**, 25021–25033 (2017). https://doi.org/10.1109/ACCESS.2017.2767029
6. Vala, D.: Advanced AMR sensor using spread spectrum technology. IFAC-PapersOnLine **49** (25), 511–516 (2016). https://doi.org/10.1016/j.ifacol.2016.12.068
7. Warsza, Z., Puchalski, J.: Estymacja macierzowa niepewności wieloparametrowych pomiarów pośrednich z przykładami - Matrix Estimation of Uncertainty of Indirect Multiparameter Measurements with Examples. Pomiary Automatyka i Robotyka **22**(2), 31–40 (2018). https://doi.org/10.14313/PAR_228/31
8. Daubaras, A., Markevicius, V., Navikas, D., Zilys, M.: Analysis of magnetic field disturbance curve for vehicle presence detection. Elektronika ir Elektrotechnika **20**(5), 80–83 (2014). https://doi.org/10.5755/j01.eee.20.5.7104
9. ST Microelectronics Inc.: LIS3MDL three-axis digital output magnetometer. AN4602 Application note (2014)
10. Markevicius, V., Navikas, D., Idzkowski, A., Andriukaitis, D., Valinevicius, A., Zilys, M.: Practical methods for vehicle speed estimation using a microprocessor-embedded system with AMR Sensors. Sensors **18**(7), 1–12 (2018). https://doi.org/10.3390/s18072225
11. Lewis, J.P.: Fast normalized cross-correlation. Vis. Interface, 120–123 (1995)

12. Jalili, M., Barzegaran, E., Knyazeva, M.G.: Synchronization of EEG: bivariate and multivariate measures. IEEE Trans. Neural Syst. Rehabil. Eng. **22**(2), 212–221 (2014). https://doi.org/10.1109/TNSRE.2013.2289899

13. Gajda, J., Mielczarek, M.: Automatic vehicle classification in systems with single inductive loop detector. Metrol. Meas. Syst. **21**(4), 619–630 (2014). https://doi.org/10.2478/mms-2014-0048

14. Kroese, D.P., Taimre, T., Botev, Z.I.: Handbook of Monte Carlo Methods. Wiley, New York (2011)

15. Wojtkowski, W., Zankiewicz, A.: Parallel RF power amplifiers temperature monitoring system based on FPGA device. IFAC PapersOnLine **51**(6), 414–419 (2018). https://doi.org/10.1016/j.ifacol.2018.07.118

16. Dong, H.H., Wang, X.Z., Zhang, C., He, R.S., Jia, L.M., Qin, Y.: Improved robust vehicle detection and identification based on single magnetic sensor. IEEE Access **6**, 5247–5255 (2018). https://doi.org/10.1109/ACCESS.2018.2791446

Evaluation Methods for the Ergatic System Reliability Operator

Igor Korobiichuk[1]([✉]), Andriy Tokar[2], Yuriy Danik[2],
and Vadim Katuha[2]

[1] Industrial Research Institute for Automation and Measurements PIAP,
Warsaw, Poland
ikorobiichuk@piap.pl
[2] Zhytomyr Military Institute n.a. S.P. Korolyov, Zhytomyr, Ukraine
tapir@i.ua, zhvinau@ukr.net, vadkatyukha@ex.ua

Abstract. In modern society, multipurpose hardware, which together with a human operator represents ergatic system, is commonly used to perform various tasks. Effectiveness of such systems depends on reliability of both components. Modern science helps to find ways to improve reliability of hardware at design and production stages, which leads to their effective functioning. However, despite high hardware performance, quality selection, staff training and coaching, the accidents, disasters, disruptions and drop in ergatic task performance continue to take place. 40% to 80% accidents and emergencies in various fields of activity happen due to human error as a result of lack of preparation, adverse psychological factors and fatigue. Therefore, search for ways of assessing reliability of ergatic system operators is an urgent scientific challenge aimed at determining a critical moment of its deterioration. The article suggests methods for assessing reliability of ergatic system operator based on the theory of fuzzy logic, which allows for assessment in terms of individual character and variability of psychological, physiological and professional capabilities and characteristics of operator and his/her sensitivity to the effects of external and internal factors.

Keywords: The reliability operator · The efficiency of work ·
The membership function · Ergatic systems

1 Introduction

The modern society is characterized by wide application of multi-functional technical means to perform a lot of various tasks. These means present the human-machine system and are considered as the ergatic systems. The efficiency of such system functioning depends on the reliability of both components. The modern science allows successful finding the ways of improving the reliability of technical means on the stages of their design and production and it causes their effective functioning. At the same time accidents, disasters, failures and the decrease in the quality of task performance within the ergatic systems take place despite the high characteristics of technical means, quality selection, instructing and training of personnel. From 40 up to 80% of

© Springer Nature Switzerland AG 2020
R. Szewczyk et al. (Eds.): AUTOMATION 2019, AISC 920, pp. 560–570, 2020.
https://doi.org/10.1007/978-3-030-13273-6_52

all the accidents and emergencies in different spheres of activity [1–6] happen due to the errors of humans as a result of their bad training, psychological factors, fatigue, etc. Therefore, the search for the ways to evaluate the reliability of the operators of ergatic systems with the purpose to determine the moment of its critical decline is the scientific task of current interest.

The results of researches of the psychological and physiological mechanisms of the reliability regulation prove that the given quality of human-operator is caused by his/her level of work efficiency [1, 2, 6]. The main direction of this level determination is connected with the evaluation of the functional state by means of calculating the composite index by input vector of private performance with the application of convolutions of different types [7–14]. The convolution application requires knowledge of clear quantitative boundaries of indexes and their optimal values. The determination of the mentioned above characteristics is complex and can cause the significant errors while determining the level of the operator work efficiency [7–10] in the conditions of the individual character and variability of psychological, physiological and professional abilities and features of a human-operator, his/her sensitivity to the impact of factors of the external and internal environment.

The formulation of the purposes of the article. Taking into consideration all the mentioned above, the purpose of the given article is to develop the evaluation methods of an ergatic system reliability operator. These methods will allow determining the level of work efficiency in conditions of fuzzy qualitative changes of the input index vector.

2   Materials and Methods of Research

Taking into account the fact, that reliability is closely connected with the state and dynamics of operator work efficiency [1, 2, 6], the evaluation methods of the egotine system reliability operator will be based on the determination of the level of his/her work efficiency. It will be performed by measuring the physiological indexes which enable us to estimate the functional state of operator. The necessity to continuously monitor the state of the operator currently on shift brings some limitations to the quantity and quality of indexes which are being measured. Having analyzed the literature, taking into account the limitations, the following physiological indexes of a human state have been selected [4, 10]: heart rate $-X_1$; heart rate stability $-X_2$; skin resistance $-X_3$; body temperature $-X_4$. The external environment temperature $-X_5$, time of day $-X_6$ and the duration of shift $-X_7$ are analyzed additionally.

It is recommended to apply the theory of fuzzy logic and namely the method described in [15–17] because of the individual character of changes of the presented indexes and the absence of clear quantitative boundaries of their changes. The idea of the above mentioned method is to design and configure fuzzy base of knowledge which presents itself as a set of linguistic statements IF <input> THEN <output>. The essence of this method is to find the non-linear dependencies with the required precision [15, 16] by configuring the fuzzy base of knowledge.

The operator's work efficiency evaluation is based on a mathematical model which defines specific relations between the input variables (measured values of physiological indexes and the external environment) and the output variable (work efficiency level). To design the model we have to: form the matrix of knowledge, obtain fuzzy logical equations, determine the membership functions.

In order to form the matrix of knowledge, let us enumerate the possible levels of the ergatic system reliability operator state states N_i, $i \in \{1, \ldots, 5\}$ [1, 2, 6]: N_1 – work (W); N_2 – compensations (C); N_3 – sub-compensations (SC); N_4 – de-compensations (DC); N_5 – failure (F).

Besides, let us determine the approximate boundaries where the measured indexes change: X_1 – (45–180) beats/min, X_2 – (0–5) beats/min, X_3 – (10–30) kOhm, X_4 – (35, 5–40) °C, X_5 – (10–35) °C, X_6 – (0–24) h, X_7 – (0–6) h. The task of work efficiency evaluation is focused on the finding one of the possible solutions N_i in compliance with every single combination of indexes. The structure of the model for estimation of work efficiency is presented in Fig. 1 in the form of hierarchical inference tree, where T, P – are intermediate variables, $N_i \in (N_1, N_2, N_3, N_4, N_5)$ – is the state of work efficiency of operator.

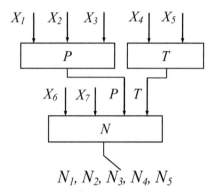

Fig. 1. The inference tree.

Here

$$T = f(X_4, X_5), \tag{1}$$

$$P = f(X_1, X_2, X_3), \tag{2}$$

$$N = f(X_6, X_7, T, P). \tag{3}$$

The indexes $X_1 - X_7$ will be considered as linguistic variables and the scale of qualitative terms presented in Table 1. will be used for evaluation.

We will use the following terms to define (characterize) the linguistic variables T and P:

$T = \{norm (N), out of norm (ON)\}$; $P = \{repose (R), optimal work (OW), work with maximum mobilizationp (WMM), stress (S)\}$. Every introduced term is regarded as a fuzzy value. Let us show the ratio (1–2) in the form of Tables 2, 3 and 4 using the introduced terms and the knowledge of the experts.

Table 1. The scale of qualitative terms.

Linguistic variable	Range	(linguistic) estimation term
X_1 – heart-beat rate	(45–180) beats/min	High (H) Increased (I) Normal (N) Reduced (R) Low (L)
X_2 – heart-beat stability	(0–5) beats/min	Stable (S) Variable (V)
X_3 – skin resistance	(10–30) kOm	Low (L) Reduced (R) Normal (N)
X_4 – body temperature	(35, 5–40) °C	Dangerous (D) Increased (I) Normal (N) Reduced (R)
X_5 – the temperature of external environment	(18–35) °C	Hot (H) Comfortable (C) Cold (C)
X_6 – the time of day	(0–24) h	Optimal (O) Reduction of work efficiency (RWE)
X_7 – the duration of shift	(0–6) h	Beginning (B) Middle (M) End (E)

Table 2. The scale of qualitative terms of variable T.

The terms of variable T	The terms of variable X_4	The terms of variable X_5
ON term	The terms of X_4 for which the variable T belongs to ON term	The terms of X_5 for which the variable T belongs to ON term
N term	Terms X_4 for which the variable T belongs to N term	Terms X_5 for which the variable T belongs to N term

The presented tables form the knowledge base for the identification of the operator's work efficiency state. By using the tables, operations (AND)$\wedge$ and (OR)$\vee$, let us write the systems of logical equations, which connect the solutions about the intermediate variables T, P and the variable N with the functions of membership (FM) of the corresponding indexes in the form:

$$\mu^{T_Y}(Y) = \mu^{T_{X_{i1}}}(X_{i1}) \wedge \mu^{T_{X_{j1}}}(X_{j1}) \ldots \wedge \mu^{T_{X_{m1}}}(X_{m1}) \vee$$
$$\vee \mu^{T_{X_{i2}}}(X_{i2}) \wedge \mu^{T_{X_{j2}}}(X_{j2}) \ldots \wedge \mu^{T_{X_{m2}}}(X_{m2}) \vee \qquad , \qquad (4)$$
$$\vee \mu^{T_{X_{in}}}(X_{in}) \wedge \mu^{T_{X_{jn}}}(X_{jn}) \ldots \wedge \mu^{T_{X_{mn}}}(X_{mn})$$

where $\mu^{T_Y}(Y)$ – MF of the variable Y to term T_Y, $Y = (T, P, N)$; $\mu^{T_{X_{mn}}}(X_{mn})$ – MF of index X_{mn} to term $T_{X_{mn}}$; n – the number of indexes for the variable Y; m - the number of combinations of n indexes where the variable Y belongs to term T_Y.

Table 3. The scale of quantitative terms of the variable P.

Variable P terms	Variable X_1 terms	Variable X_2 terms	Variable X_3 terms
Term R	Terms X_1 for which the variable P belongs to term R	Terms X_2 for which the variable P belongs to term R	Terms X_3 for which the variable P belongs to term R
Term OW	Terms X_1 for which the variable P belongs to term OW	Terms X_2 for which the variable P belongs to term OW	Terms X_3 for which the variable P belongs to term OW
Term WMM	Terms X_1 for which the variable P belongs to term WMM	Term X_2 for which the variableпри P belongs to term WMM	Terms X_3 for which the variable P belongs to term WMM
Term S	Terms X_1 for which the variable P belongs to term S	Terms X_2 for which the variable P belongs to term S	Terms X_3 for which the variable P belongs to term S

Table 4. The scale of qualitative terms of the variable N.

The terms of variable N	The terms of variable P	The terms of variable T	The terms of variable X_6	The terms of variable X_7
Term W	Terms P for which the variable N belongs to term W	Terms T for which the variable N belongs to term W	Terms X_6 for which the variable N belongs to term W	Terms X_7 for which the variable N belongs to term W
Term C	Terms P for which the variable N belongs to term C	Terms T for which the variable N belongs to term C	Terms X_6 for which the variable N belongs to term C	Terms X_7 for which the variable N belongs to term C
Term S	Terms P for which the variable N belongs to term S	Terms T for which the variable N belongs to term S	Terms X_6 for which the variable N belongs to term S	Terms X_7 for which the variable N belongs to term S
Term DC	Terms P for which the variable N belongs to term DC	Terms T for which the variable N belongs to term DC	Terms X_6 for which the variable N belongs to term DC	Terms X_7 for which the variable N belongs to term DC
Term F	Terms P for which the variable N belongs to term F	Terms T for which the variable N belongs to term F	Terms X_6 for which the variable N belongs to term F	Terms X_7 for which the variable N belongs to term F

In order to determine MF of output parameters $X_1 - X_7$ by fuzzy terms, let us apply the method of rank estimations [14]. Let us consider the method on the example of parameter X_1, which is determined on interval 45–180. By using the fuzzy terms from Table 1, let us form the matrix which shows the pair wise comparisons of various parameter values from the point of view of their proximity to the term "LOW":

$$\mu^H(X_1) =$$

	45	55	65	75	95	120	180
45	1	7/9	5/9	3/9	2/9	1/9	1/9
55	9/7	1	5/7	3/7	2/7	1/7	1/7
65	9/5	7/5	1	3/5	2/5	1/5	1/5
75	9/3	7/3	5/3	1	2/3	1/3	1/3
95	9/2	7/2	5/2	3/2	1	1/2	1/2
120	9	7	5	3	2	1	1
180	9	7	5	3	2	1	1

In order to form the given matrix, the range of parameter X_1 change was split into six quantum and it provided the conversion of continuous magnitude $X_1 = [\underline{x}, \bar{x}]$, into discrete seven element one:

$$X_1 = \{X1_1, X1_2, X1_3, X1_4, X1_5, X1_6, X1_7\},$$

where $X1_1 = \underline{x}$ – the lower boundary of range, $X1_7 = \bar{x}$ – is the upper boundary of range.

Line 7 of the above matrix was found experimentally, the elements in lines 1–6 were calculated with the use of the following equalities [15, 16]:

$$\mu_{ii} = 1, \quad \mu_{ji} = \frac{1}{\mu_{ji}}, \quad \mu_{ij} = \frac{\mu_{kj}}{\mu_{ki}}$$

where k – the number of known line; i, j – the number of line and column where the element which is being searched.

The 9 point scale of Saaty [15, 16] was used to set the ranks. This scale is presented in Table 5.

Table 5. The rank scale.

The intensity of the relative importance I	Determinations
1	Equal proximity
3	Moderate (low) excess of one over the other
5	Essential (significant) excess of one over the other
7	Obvious excess of one over the other
9	Absolute excess of one over the other
2, 4, 6, 8	Intermediate solutions between two neighboring estimations

3 The Results of Research

Taking into account the matrix properties, all its diagonal elements are equal to one. Matrix $\mu^H(X_1)$ gives us the opportunity to obtain the degrees of the membership of the elements to the term "LOW" by summing line elements:

$$\mu^H(45) = \frac{1}{1+\frac{7}{9}+\frac{5}{9}+\frac{3}{9}+\frac{2}{9}+\frac{1}{9}+\frac{1}{9}} = 0,32, \quad \mu^H(55) = \frac{1}{\frac{9}{7}+1+\frac{5}{7}+\frac{3}{7}+\frac{2}{7}+\frac{1}{7}+\frac{1}{7}} = 0,25,$$

$$\mu^H(65) = \frac{1}{\frac{9}{5}+\frac{7}{5}+1+\frac{3}{5}+\frac{2}{5}+\frac{1}{5}+\frac{1}{5}} = 0,18, \quad \mu^H(75) = \frac{1}{\frac{9}{3}+\frac{7}{3}+\frac{5}{3}+1+\frac{2}{3}+\frac{1}{3}+\frac{1}{3}} = 0,11,$$

$$\mu^H(95) = \frac{1}{\frac{9}{2}+\frac{7}{2}+\frac{5}{2}+\frac{3}{2}+1+\frac{1}{2}+\frac{1}{2}} = 0,07, \quad \mu^H(120) = \frac{1}{9+7+5+3+2+1+1} = 0,04,$$

$$\mu^H(180) = \frac{1}{9+7+5+3+2+1+1} = 0,04.$$

Similarly the matrixes were formed which showed the pair wise comparisons of various values of parameter X_1 from the point of view of their proximity to the terms "LOWER" (L), "REDUCED" (R), "NORMAL" (N), "INCREASED" (I) and "HIGH" (H).

The obtained values of MF are normalized by means of division into the highest degree of membership. As a result, we obtain fuzzy multiplicities of parameter X_1, which are presented in the form:

$$\text{parameter } X_1 \text{ "L"} = \left\{ \frac{1}{45}, \frac{0,77}{55}, \frac{0,55}{65}, \frac{0,33}{75}, \frac{0,22}{95}, \frac{0,11}{120}, \frac{0,11}{180} \right\},$$

$$\text{parameter } X_1 \text{ "R"} = \left\{ \frac{0,77}{45}, \frac{1}{55}, \frac{0,77}{65}, \frac{0,55}{75}, \frac{0,33}{95}, \frac{0,22}{120}, \frac{0,11}{180} \right\},$$

$$\text{parameter } X_1 \text{ "N"} = \left\{ \frac{0,11}{45}, \frac{0,66}{55}, \frac{1}{65}, \frac{1}{75}, \frac{0,55}{95}, \frac{0,33}{120}, \frac{0,11}{180} \right\},$$

$$\text{parameter } X_1 \text{ "I"} = \left\{ \frac{0,11}{45}, \frac{0,39}{55}, \frac{0,55}{65}, \frac{0,72}{75}, \frac{1}{95}, \frac{0,89}{120}, \frac{0,77}{180} \right\},$$

$$\text{parameter } X_1 \text{ "H"} = \left\{ \frac{0,11}{45}, \frac{0,33}{55}, \frac{0,44}{65}, \frac{0,55}{75}, \frac{0,77}{95}, \frac{0,89}{120}, \frac{1}{180} \right\}.$$

The obtained fuzzy multiplicities in the graphical form are presented in Fig. 2.

As a result of conducting the similar calculations, we obtained MF for parameters: X_2 – (Fig. 3), X_3 – (Fig. 4), X_4 – (Fig. 5), X_5 – (Fig. 6), X_6 – (Fig. 7), X_7 – (Fig. 8).

The knowledge of MF of the estimated parameters gives the possibility to find the degree of membership of the operator's state to every level N_i, $i \in \{1,\ldots,n\}$. In order to evaluate the reliability of operator, let us calculate the reliability coefficient by using the convolutions by non-linear scheme of compromise of expression (5), where a_i the significant coefficients which characterize the degree of the influence of the corresponding level on the decrease of operator reliability and are determined by the expression (6), where f_i – the evaluation of importance of i- level of work efficiency and it is estimated by the expert with the band [6]

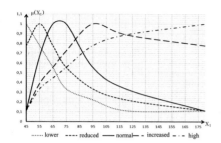

Fig. 2. MF of parameter X_1

Fig. 3. MF of parameter X_2

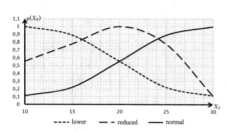

Fig. 4. MF of parameter X_3

Fig. 5. MF of parameter X_4

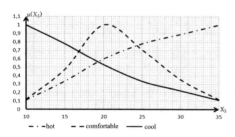

Fig. 6. MF of parameter X_5

Fig. 7. MF of parameter X_6

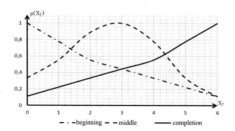

Fig. 8. MF of parameter X_7

$$K = 1 \frac{1}{\sum\limits_{i=1}^{n} a_i [1 \; N_i]^{-1}}, \tag{5}$$

$$a_i = \frac{f_i}{\sum\limits_{j=1}^{n} f_i}. \tag{6}$$

Therefore, the methods of operator reliability of ergatic system consists of the following steps:

1. Calculate the values of MF to fuzzy terms, which are used in the knowledge base of Tables 2, 3 and 4 for the fixed vector of measured indexes $X = \{X_1, X_2, X_3, X_4, X_5, X_6, X_7\}$.
2. Apply the logical Eq. (4) it is recommended to calculate the multi-dimensional MF of vector X_4, X_5 for all the values of intermediate variable T. Here the logical operations $(AND) \wedge$ and $(OR) \vee$ are replaced by the operations min and max correspondingly:

$$\mu \; (x) \wedge \mu(y) = \min\{\mu(x), \mu(y)\}, \; \mu(x) \vee \mu(y) = \max\{\mu(x), \mu(y)\}. \tag{7}$$

3. Calculate the multi-dimensional MF of vector X_1, X_2, X_3 for all the values of intermediate variable P by applying the logical Eq. (4) and the rule (7).

By applying the logical Eq. (4) and the rule (7) it is also recommended to calculate the multi-dimensional MF of vector T, P, X_6, X_7 for all N_i values of input variable.

4. Find the reliability coefficient for the vector of input parameters $X = \{X_1, X_2, X_3, X_4, X_5, X_6, X_7\}$ by applying the expression (5).

4   Conclusions

The developed methods for the evaluation of operator reliability of ergatic system will allow to determine the level of operator reliability taking into account the individual character of a human operator and variability of psychological, physiological and professional qualities of a human operator, his/her sensitivity to the impact of factors of external and internal environment at fuzzy quantitative changes of input index vector.

It can significantly contribute to the further research on the evaluation of effectiveness of ergatic systems, where the focus is on the operator's reliability level.

References

1. Bodrov, V.A., Orlov, V.Y.: Psychology and Reliability: A Human Being Within the System of Technique Control. Publishing house "The Institute of Psychology of ASR", Moskva, 288 p. (1998). (in Russian)
2. Islam, R., Yu, H., Abbassi, R., Garaniya, V., Khan, F.: Development of a monograph for human error likelihood assessment in marine operations. Saf. Sci. **91**, 33–39 (2017)
3. Korobiichuk, I., Nowicki, M., Szewczyk, R.: Design of the novel double-ring dynamical gravimeter. J. Autom. Mob. Rob. Intell. Syst. **9**(3), 47–52 (2015). https://doi.org/10.14313/JAMRIS_3-2015/23
4. Voronin, A.N., Ziatdinov, Y.K., Kozlov, A.I., Cabanyuk, V.S.: Vector Optimization of Dynamic Systems, 284 p. Techique, Kyiv (1999). (in Russian)
5. Korobiichuk, I., Ladanyuk, A., Shumyhai, D., Boyko, R., Reshetiuk, V., Kamiński, M.: How to Increase Efficiency of Automatic Control of Complex Plants by Development and Implementation of Coordination Control System. In: Recent Advances in Systems, Control and Information Technology. Advances in Intelligent Systems and Computing, vol. 543, pp. 189–195 (2017). https://doi.org/10.1007/978-3-319-48923-0_23
6. Dushkov, B.A., Lomov, B.F., Rubahin, V.F.: The Fundamentals of Engineering Psycology: Text-Book for Technical High Schools, 2nd edn, p. 448. High School, Moskva (1986). (in Russian)
7. Brahman, T.P.: Multi Criteria and Selection in Technique, 288 p. Radio and communication, Moskva (1984). (in Russian)
8. Jahan, A., Edwards, K.L., Bahraminasab, M.: Multi-criteria Decision Analysis for Supporting the Selection of Engineering Materials in Product Design, 2nd edn, p. 238. Butterworth-Heinemann, Oxford (2016)
9. Korobiichuk, I., Fedushko, S., Juś, A., Syerov, Y.: Methods of determining information support of web community user personal data verification system. In: Advances in Intelligent Systems and Computing, vol. 550, pp. 144–150 (2017). https://doi.org/10.1007/978-3-319-54042-9_13
10. Smith, K.M., Larrieu, S.: Mission Adaptive Display Technologies and Operational Decision Making in Aviation, pp. 1–355. IGI Global, Pennsylvania (2015)
11. Gorgo, Y.P., Popduha, Y.A., Tokar, A.M.: The coefficient of evaluation for the quality of operator work in complexes "human-machine". Abstract scientific and practical conference "The problems of creation, development and application of information systems of special purpose". Zhitomir military Institute named after S.P. Korolyov at National Aviation University (2009). (in Ukranian)
12. Korobiichuk, I., Ladanyuk, A., Shumyhai, D., Boyko, R., Reshetiuk, V., Kamiński, M.: How to increase efficiency of automatic control of complex plants by development and implementation of coordination control system. In: Advances in Intelligent Systems and Computing, vol. 543, pp. 189–195 (2017). https://doi.org/10.1007/978-3-319-48923-0_23
13. Solodkov, A.S., Buharin, V.A., Melnikov, D.S.: Working abilities of sportsmen: its criteria and the ways of correction. The scientific and theoretic journal "The Scientific Notes", **3**(25), 74–79 (2007). https://cyberleninka.ru/article/n/rabotosposobnost-sportsmenov-ee-kriterii-i-sposoby-korrektsii
14. Pomeshchikova, I.P., et al.: Influence of exercises and games with ball on coordination abilities of students with disorders of muscular skeletal apparatus. J. Phys. Educ. Sport **16**(1), 146–155 (2016)

15. Gerasimov, B.M., Divipnyuk, M.M., Subach, I.Y.: The systems of support and decision-making: design, application, efficiency assessment. Sevastopol, 320 p. (2004). (in Russian)
16. Pettang, C., Manjia, M.B., Abanda, F.H.: Decision Support for Construction Cost Control in Developing Countries, p. 385. Pennsylvania, IGI Global (2016)
17. Gerasimov, B.M., Kamyshyn, V.V.: Organizational ergonomics: methods and the algorithms of investigation and design, 212 p. Infosystem, Kyiv (2009). (in Ukranian)

Information Support the Operative Control Procedures of Energy Efficiency of Operation Modes of Municipal Water Supply System Facilities

Igor Korobiichuk[1](✉), Liudmyla Davydenko[2],
Volodymyr Davydenko[3], and Nina Davydenko[3]

[1] Industrial Research Institute for Automation and Measurements PIAP,
Warsaw, Poland
ikorobiichuk@piap.pl
[2] Lutsk National Technical University, Lutsk, Ukraine
l.davydenko033@gmail.com
[3] National University of Water and Environmental Engineering, Rivne, Ukraine
vd19688691@gmail.com, ninadavydenko1992@gmail.com

Abstract. The principles of information support organization of control procedures, which take into consideration the features of the research object functioning for obtaining the correct control results, have been considered in the article. The energy efficiency monitoring system is considered as a tool for forming a database. Complex analysis of daily water consumption graphs has been provided for taking into consideration of water consumption unevenness and influence of external factors during the water supply planning. Its results are the basis of planning and control procedures. An architecture of the information support system of the control, which reflects the connection between its procedures, has been proposed. The formalization of energy efficiency control has been accomplished based on object-oriented technology. Classes that take into consideration the characteristic of the facility, water consumption, environment, and energy consumption, methods of planning and control of energy consumption have been formed. Examples of calculated procedures realization have been demonstrated. Using the proposed principles gives an opportunity to take into consideration the actual operation conditions of the research object and provide information for correct control.

Keywords: Municipal water supply system · Energy efficiency ·
Operative control

1 Introduction

The issue of the energy performance improvement is one of the priority tasks in the conditions of reduction energy carriers and growth their market value. IPPC Directive 2008/1/EU [1] requires efficient use of energy during the running of any equipment. Energy efficiency is one of the criteria used to determine the best available techniques. Creation of the energy management system (EnMS) on the enterprise is one of the directions of energy performance improvement. Its functioning is based on the

© Springer Nature Switzerland AG 2020
R. Szewczyk et al. (Eds.): AUTOMATION 2019, AISC 920, pp. 571–582, 2020.
https://doi.org/10.1007/978-3-030-13273-6_53

principle of continuous improvement [2, 3] and the main mission is the purposeful improvement of the energy consumption efficiency. At present, research of efficiency energy consumption issues in various branches are carried out by specialists and scientists. Many publications on the introduction the EnMS [4–6], forming of energy performance indicators of complex facilities [7], description of energy consumption mode [8, 9], monitoring of indicators and energy performance [10, 11], increasing the efficiency of management the operation modes of various complexes [12, 13], energy efficiency control [14–16], etc. are the result of these researches.

Recently, a high degree of detailing, energy efficiency control of the end user and reaction on deterioration of energy consumption efficiency is carried out in energy management methodology [3]. It requires the improvement of control procedures and their integration into EnMS [12, 14]. Energy efficiency control procedure should give the opportunity to identify the moments of increasing or decreasing the energy efficiency, and the reasons these changes were occurred [7, 10]. Control provides information to decide on the need to energy performance improvement. At the same time, the control procedure also requires input information about the mode and the actual operation conditions of the facility, the external factors, etc. Part of the information is obtained based on measurements. However, significant part of the data is obtained based on calculations or mathematical modeling [17, 18]. Correct obtaining of such information is the basis of correct results. The research aim is improving the effectiveness of control the energy efficiency operation modes the water supply facilities by forming the principles of information support of control procedures to detect and prevent negative trends and improve energy consumption efficiency in municipal water supply system (MWSS).

2   Research Methodology

The energy efficiency control system provides [2]: control of technological processes; identifying the energy performance indicators and methods of their measurement and control; documenting and analysis abnormal situations to identify and remove their causes. Control of energy efficiency is a part of the energy efficiency monitoring system, which is the EnMS subsystem of the enterprise (Fig. 1).

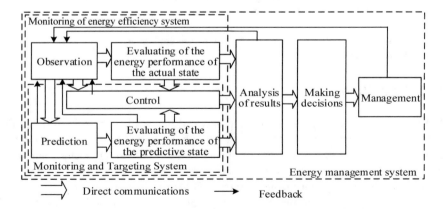

Fig. 1. Components of energy management system of the enterprise.

The current measures of controlled parameters are determined to control the state of the research object and its operation modes. Their deviation from the defined standard value are evaluated. In general, controlled parameters have to satisfy the limit:

$$\underline{x}_i \leq x_i \leq \bar{x}_i, \qquad i = \overline{1, N} \tag{1}$$

Estimated values of controlled of the functioning parameters are described with the equation:

$$x_i^*(t) = F\big(x_i(t_0), u_{[t_0, t]}, K_i\big) \tag{2}$$

where $x_i(t_0)$ – vector of calculated technical parameters of the research object; $u_{[t_0, t]}$ – conditions of functioning in the time interval $[t_0, t]$; K_i – vector of the parameters characterizing the operation mode.

Separate units, technological equipment, industrial processes, etc. are the objects of control. Description the objects, choice necessary calculated algorithms, correct computation of target variables are the basis of energy efficiency control procedure and getting correct results to decide on energy performance improvement.

The introduction of energy efficiency monitoring system provides a possibility: (1) the creation of large databases that contain information on operation modes of the MWSS facilities; (2) the use of intellectual data analysis methods to explore information and reveal hidden regularities that determine the forming of the technological modes. Dependencies with small number of input and output variables can be received by processing large amounts of data.

Water consumption is one of the factors that affect on efficiency of energy consumption mode. It is uneven and is influenced by many factors. Incompleteness and inaccuracy of initial information leads to errors of planning the water supply mode. It causes irrational operation mode of pumping stations (PS) and electricity overspending [11]. Water supply has to be maximum appropriate to water consumption. Therefore, it is necessary to analyze water consumption modes, identify trends of its changes depending on the season and climatic conditions and form the water supply graphs based on it [9]. Water consumption graph (WCG) is the main mode indicator of the water supply process. Based on the database it is possible to research of graph parameters for searching similarities in water consumption and construction of similar WCG set. WCG set has to be differentiated to the seasons and reflect the specificity of water consumption during working, weekend, pre-holiday and holiday days. Daily WCG are described by absolute characteristics of the water supply mode and the classical indicators of unevenness in order to reveal their similarity by a season [9]. Daily WCG can be represented as circular time-pie-chart – the chart radar type (CRT). This representation allows using the analysis tool of figures of various form – morphometric approach [8]. The influence of social factors causes the difference in the form of daily WCG for different days. WCG are described by morphometric parameters to detect the similarity by the day type [8, 9]. Consideration the seasonal and social factors influence is a prerequisite for the planning of efficient water supply and power consumption modes for each characteristic days.

Modern EnMS contain a subsystem operative management of energy consumption efficiency – the so-called Monitoring and Targeting System [16]. The identification of energy consumption dependency from significant indicators – construction of energy baseline (EB) is in its base.

Thus, the operative control system of energy efficiency of water supply facilities has to provide control of [13]: water supply mode; energy performance indicators; EB observance. The definition of EB in the MWSS consists of two stages: (1) planning of the water supply mode with taking into consideration social and climatic factors; (2) construction of power consumption models with taking into consideration factors that affect on the power consumption efficiency.

Organization of the energy efficiency control system provides standards determination for controlled parameters and adjustment the alarm tools – information to energy manager about exceeding the standard. Current trends in the energy efficiency management in accordance with ISO 50001 [3] provide the introduction of automated energy accounting systems, technological processes management (which is a part of the monitoring system) and energy management information systems that are integrated with systems ACS TP and SCADA-systems. Monitoring system data allow determining the standard of controlled parameter, attaching to him alarm and controlling deviations from the standard.

Hierarchy of production system causes hierarchy of the research problem, which is directed at improving this system [7]. Stratified representation can be used for construction of information support system of energy efficiency control of MWSS [7]. It simplifies describing its facilities and procedures-algorithms, but keeps their subordination to a single goal. That is, all of its subsystems regardless of their functions are consolidated by common information space. Forming the unified knowledge base for decision concerning on energy performance improvement of the MWSS have to be a result of their operation. The architecture of information support of procedures of operational control of energy efficiency modes of MWSS facilities is shown in Fig. 2.

The set of the existing structural and functional relationships in MWSS indicates the sequence of sample information, order of the necessary calculations and control procedures [14].

It is necessary to setup procedures for the exchange of input and output data to ensure the functioning of the proposed architecture. Availability of automated control systems simplifies the collection and processing of information for each MWSS facility. Their connection to the Internet provides communication between control points and the central enterprise server. This will allow realizing the consolidation of data in a single information system.

3   Results

Procedure formalization of operational control of energy efficiency of MWSS facilities modes has been performed using object-oriented technology. Information field objects are modeled by the classes with combined properties and existence rules - sets that have the same characteristics and qualities. The class contains properties of the facility (defines the data structure of the facility, the rules, which facilities act), as well as

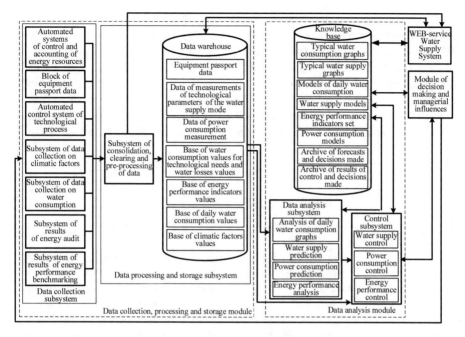

Fig. 2. Architecture of information support of energy efficiency operational control procedures.

methods (functions) that have access to the facility data, process them and perform certain operations and tasks. The architecture of the procedure of operational control of energy efficiency is shown in Fig. 3.

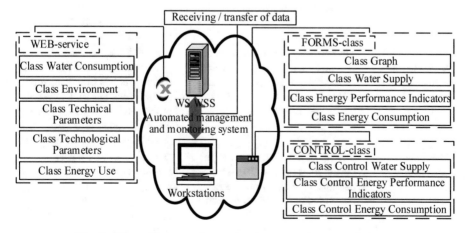

Fig. 3. The architecture of operational control of energy efficiency.

Three categories of classes have been formed: WEB-service – a set of classes that are combined by procedure of obtaining initial information about the research object; FORMS-class – a set of classes that are combined by computational procedures and models; CONTROL-class – a set of classes that are combined by procedures of the accomplishment of the energy efficiency control. Class properties are quantitative characteristics of the research object. Class methods are the algorithms calculations, procedures, communications, actions, etc. Description of the properties and methods of classes with taking into consideration their tasks, described formally, are shown in the Table 1.

Table 1. Properties and methods of classes.

Class	Properties of classes	Methods of classes
Category WEB- service		
Class Water consumption	The set of data that characterize water consumption mode: season; date; type of day; value of water consumption	Obtaining data on hourly water consumption. Forming daily WCG. Determining the volume of water consumption. Data transferring for analysis WCG
Class Environment	The set of data that characterize environment: season; date; temperature; precipitation	Obtaining data on value of daily temperature, amount and duration of precipitation. Fixing the date, type of day, season. Data transferring for calculation of climatic parameters
Class Technical parameters	The set of technical parameters: passport data of units; characteristic of pipeline network	Obtaining data on passport data of units; characteristic of pipeline network. Data transferring for calculation of energy performance indicators
Class Technological parameters	The set of technological parameters of water supply mode: the volume of water that is pumped PS; head in dictating points of the network; excess heads in the network	Obtaining data on the volume of water that is pumped PS, head in dictating points in the network, excess heads in dictating points of the network. Data transferring for organization of control procedures of water supply mode
Class Energy consumption	The set of data that characterize power consumption. The set of data that characterize of water consumption for industrial needs	Obtaining data on power consumption. Obtaining data on picking out of water for industrial needs. Data transferring for organization of control procedures of energy consumption mode

(*continued*)

Table 1. (*continued*)

Class	Properties of classes	Methods of classes
Category FORMS-class		
Class Graph	WCG. Main indicators. Auxiliary indicators. Morphometric parameters	Obtaining data on daily WCG. Calculation of main and auxiliary indicators WCG. Transformation of daily WCG in CRT. Calculation of morphometric parameters. Obtaining data on climatic factors. Calculation of climatic indicators. The combined analysis of indicators WCG. Forming of the set of similar daily WCG. Data transferring for forming of the set of typical graphs of water supply
Class Water supply	WCG. Climatic indicators. Water supply	Getting the set of daily WCG. Forming of the set of typical graphs of water supply. Adjustment of typical graphs with taking into consideration climatic conditions. Modelling of water supply mode. Data transferring for modelling of EB. Data transferring for forming of standards and alarms
Class Energy performance indicators	Technical and technological energy performance indicators. Specific power consumption	Obtaining data on water supply mode. Calculation of technical and technological energy performance indicators with taking into consideration typical graph of water supply. Data transferring about energy performance indicators for modelling of EB. Calculation of specific power consumption
Class Energy consumption	Power consumption. Picking out of water for technological needs	Obtaining data on water supply mode. Obtaining data on energy performance indicators. Modelling of power consumption with taking into consideration of water supply graph. Definition of EB

(*continued*)

Table 1. (*continued*)

Class	Properties of classes	Methods of classes
Category CONTROL-class		
Class Control of water supply	The standard of water supply. Permissible values for water supply. The standard of excess heads. Permissible values for excess heads. The standard of pressure in dictating points of the network	Obtaining predictive values of water supply. Definition of standards. Definition of permissible parameters values. Setting alarms. Organization of control procedures. Organization of procedures of identification of emergency and abnormal situations and informing person on duty. Data transferring into the system of management of technological process. Forming reports. Data transferring into the knowledge base
Class Control of energy performance indicators	The standard of i's energy performance indicator. Permissible values for i's energy performance indicator	Obtaining theoretical values of energy performance indicators. Definition of standards. Definition of permissible parameters values. Setting alarms. Organization of control procedures. Organization of procedures of identification of inefficient mode and informing energy manager. Data transferring into the EnMS. Forming reports. Data transferring into the knowledge base
Class Control of energy consumption	The EB. Permissible values for power consumption. The standard of picking out of water for technological needs. Permissible values for picking out of water for technological needs	Obtaining predictive values of power consumption. Definition of permissible values of power consumption changing. Setting alarms. Organization of control procedures. Organization of procedures of identification of inefficient mode and informing energy manager. Data transferring into the EnMS. Forming reports. Data transferring into the knowledge base

The realization of each class involves the sequence of several calculated procedures. The result is the information that is collected in the database. Part of the information is the initial for the following procedures. The other part is used to make a decision on the energy efficiency of the operation mode of the MWSS facility.

Consider the accomplishment sequence of some procedures of FORMS-class category that have been developed for the water supply PS. Consider the Class Graph. The first step is calculation the main and additional parameters of WCG, transformation of the daily WCG into the CRT and the calculation of the morphometric parameters. An example of the results of this procedure is shown in Table 2.

The second step is joint analysis of WCG indicators and forming a set of similar daily WCG. The procedure involves the identifying of hidden regularity in the water consumption character. It has been performed in two stages: (1) the influence of seasonality has been revealed; (2) the influence of social factors has been revealed. Consistent using of cluster analysis and discriminant analysis methods has been solved many issues. The number of clusters has been determined and groups of similar WCG have been formed by means of cluster analysis. Thus, an educational sample has been formed. Then a discriminant analysis has been applied and classification functions have been constructed. They have been used to determine the appliance of WCG to one of the classes [19].

The results of WCG analysis have been showed the following. For the first stage, there are four classes of WCG: 1st class - "summer"; 2nd class - "winter"; 3rd class - "spring-summer-autumn"; 4t class - "irregular days" (WCG of different months and seasons). For the second stage, there are three classes: 1st class - "working days"; 2nd class - "weekend"; 3rd class - "irregular days" (WCG of different type days) [19].

Table 2. Parameters of the daily water supply graph for 1 January.

Classic parameters		Morphometric parameters	
Graph		CRT	
Q_{min}	1436	Perimeter	786,19
Q_{max}	2786	Area	40 673
Q_{Σ}	50076	Circularity	0,66
Q_{AM}	2086,5	Compactness	0,83
σ	433,1	Elongation	0,77
K_{form}	1,02	Main axle of elongation	260,41
K_{max}	1,33	Additional axle of elongation	199,8
K_{fill}	0,75	Angle of elongation	127
K_{unev}	0,69	Convexity	0,97

The obtained results have been used to implement Class Water supply procedures. Formation of the similar WCG groups has been allowed describing the water supply mode (Fig. 4), in particular, the definition of averaged characteristics of water consumption from the water supply network for each season (Table 3), as well as the profiles of its daily graphs for the typical days, which are the basis of the water supply process planning, and its parameters.

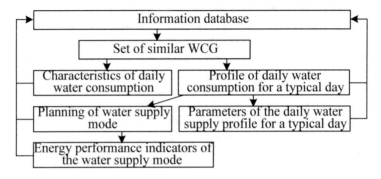

Fig. 4. Procedure "Characteristics of the water supply mode".

Table 3. Value of water consumption.

Cluster	Season	Daily water supply Q, м³	
		Average value	Confidence interval ($p = 0.99$)
1	«summer»	42992.86	($41928.85 < \overline{Q} < 44313.65$)
2	«winter»	47850.1	($47241.9 < \overline{Q} < 48458.3$)
3	«spring-summer-autumn»	49860.44	($48466.25 < \overline{Q} < 51254.63$)

The average value of the water supply is accepted as planned, and confidence interval, which is constructed with a confidence probability p = 0.99, is accepted as the permissible limits of changing the seasonal water consumption [Q_{min}, Q_{max}]. These results are used in the water supply control procedure.

Implementation of the described procedures provides taking into consideration the cyclic changes of the technological process that are caused by the external factors influence, as well as determine the time ranges for defining (observing) relevant variables and construction the EB. The obtained results are the basis for construction the energy consumption model (Class Energy consumption), which is adapted to the water supply mode for a typical day type of the appropriate season. It provides the ability to determine the EB of water supply PS taking into consideration its actual operation conditions during a specific time interval.

4 Conclusions

Energy performance control requires constant analysis of deviations of actual parameter values of control facility from the expected theoretical values and notifying about exceeding of established standards. Defining standards of energy performance indicators and EB for the selected research object have to be carried out by being based on statistics, which is accumulated in the data-base monitoring system, take into consideration the real conditions of its functioning, as well as the influence on the character of the daily water consumption of social and weather factors. It is appropriate to use Web-oriented systems to transfer of information flows between objects of subject area and the central server company. This will allow creating a single information space and ensuring the ability to process information about the modes parameters and energy performance indicators of water supply facilities. The result of realization of energy efficiency control has to be creating a single information knowledge base. It will provide integration of all the complex of operational control procedures into EnMS of the enterprise and facilitate effective management making decisions.

References

1. Directive 2008/1/EC of the European Parliament and of The Council of 15 January 2008 concerning integrated pollution prevention and control. Official Journal of the European Union L 24
2. Reference Document on Best Available Techniques for Energy Efficiency. Institute for Prospective Technological Studies, European Commission IPPC Bureau, Seville (2009)
3. ISO 50001:2011(E) Energy management systems - requirements with guidance for use. ANSI (2011)
4. Karcher, P., Jochem, R.: Success factors and organizational approaches for the implementation of energy management systems according to ISO 50001. TQM J. **27**(4), 361–381 (2015)
5. Marvin, T.: Howell Effective Implementation of an ISO 50001 Energy Management System (EnMS). ASQ Quality Press, Milwaukee (2014)
6. Majerník, M., Bosák, M., Štofová, L., Szaryszová, P.: Innovative model of integrated energy management in companies. Qual. Innov. Prosperity **19**(1), 22–32 (2015)
7. Davydenko, L.: Indicators system creation for the energy efficiency benchmarking of municipal power system facilities. Problemele energeticii regionale **1**(27), 58–70 (2015)
8. Komenda, T., Komenda, N.: Morphometrical analysis of daily load graphs. Electr. Power Energy Syst. **42**, 721–727 (2012). https://doi.org/10.1016/j.ijepes.2012.03.028
9. Rosen, V., Davydenko, N.: Formation of the set of characteristics of the actual regime of water consumption in municipal water supply systems. Energy: Econ. Technol. Ecol. **3**, 85–92 (2015). [in Ukrainian]
10. Flizikowski, J., Bielinski, K.: Technology and Energy Sources Monitoring: Control, Efficiency, and Optimization. IGI GLOBAL, Hershey (2013)
11. Davydenko, L.: Principles of constructing an integrated energy efficiency monitoring system for a water supply and sewage company. Power Eng.: Econ. Technol. Ecol. **3**, 107–115 (2015). [in Ukrainian]

12. Korobiichuk, I., Ladanyuk, A., Shumyhai, D., Boyko, R., Reshetiuk, V., Kamiński, M.: How to increase efficiency of automatic control of complex plants by development and implementation of coordination control system. In: Recent Advances in Systems, Control and Information Technology. Advances in Intelligent Systems and Computing, vol. 543, pp. 189–195 (2017). https://doi.org/10.1007/978-3-319-48923-0_23

13. Korobiichuk, I., Lysenko, V., Reshetiuk, V., Lendiel, T., Kamiński, M.: Energy-efficient electrotechnical complex of greenhouses with regard to quality of vegetable production. In: Advances in Intelligent Systems and Computing, vol. 543, pp. 243–251 (2017). https://doi.org/10.1007/978-3-319-48923-0_30

14. Davydenko, L., Rozen, V., Davydenko, V., Davydenko, N.: Formalization of energy efficiency control procedures of publicwater-supply facilities. In: Advances in Intelligent Systems and Computing, vol. 543, pp. 196–202 (2017)

15. Kulcsar, T., Balatonb, M., Nagyb, L., Abonyi, J.: Feature selection based root cause analysis for energy monitoring and targeting. Chem. Eng. Trans. 39, 709–714 (2014)

16. John, P.: Quick Start Guide to Energy Monitoring & Targeting (M&T). Effective Energy Management Guide (2005). http://www.oursouthwest.com/SusBus/susbus9/m&tguide.pdf

17. Korobiichuk, I., Osadchuk, R., Fedorchuk, D., Nowak, P.: Approach to determination of parameters of probability density function of object attributes recognition in space photographs is considered within statistical method. In: Advances in Intelligent Systems and Computing, vol. 550, pp. 425–432 (2017). https://doi.org/10.1007/978-3-319-54042-9_40

18. Korobiichuk, I., Kuzmych, L., Kvasnikov, V., Nowak, P.: The use of remote ground sensing data for assessment of environmental and crop condition of the reclaimed land. In: Advances in Intelligent Systems and Computing, vol. 550, pp. 418–424 (2017). https://doi.org/10.1007/978-3-319-54042-9_39

19. Davydenko, L., Davydenko, N.: Construction of discrimination rules of daily water consumption graphs from the water supply network with consideration of seasonal and social factors. Trans. Kremenchuk Mykhailo Ostrohradskyi Nat. Univ. 3/2018(110), 58–64 (2018). [in Ukrainian]

The Use of Brain-Computer Interface to Control Unmanned Aerial Vehicle

Szczepan Paszkiel$^{(\boxtimes)}$ and Mariusz Sikora

Faculty of Electrical Engineering, Automatic Control and Informatics,
Department of Biomedical Engineering, Opole University of Technology,
Proszkowska 76, 45-271 Opole, Poland
s.paszkiel@po.opole.pl

Abstract. The article discusses the capabilities of Emotiv EPOC+ NeuroHeadset in the context of quadcopter control. It also presents project and the practical implementation of the inertial measurement unit based on the extended Kalman filter and a low-pass filter. Data produced by the inertial measurement unit are used by the quadcopter stability control system that consist of 3 PID controllers. Effectiveness of the stability control system was examined. Based on that examination, the areas for further development was pointed. Additional control systems can be implemented in order to make the control of quadcopter via BCI less demanding for the pilot.

Keywords: EEG · Brain-computer interfaces ·
Unmanned aerial vehicle

1 Introduction

In 1973 Jacques J. Vidal described the concept of Brain-Computer Interface (BCI). Even then the author was considering a possibility of controlling with its help a flying object – namely, a spaceship. A great deal of progress has been made in the field of BCI since 1973. New types of electrodes has been developed, which can allow to use BCI more comfortable. With the development of new signal processing algorithms it is possible to reduce the training time and to improve accuracy and reliability [1]. Devices like OpenBCI Ultracortex, Emotiv EPOC+ Neuroheadset or NeuroSky MindWave are commercially available and can be used with a variety of different projects. Other branch of technology which has recently been developing very dynamically are unmanned aerial vehicles (UAVs). Simple stability systems are capable of maintaining desired attitude of a rotorcraft. More complicated are able to maintain also its position in reference to the ground. Stability controllers can be based on a device dynamics model and utilize different control algorithms e.g.: Proportional–Integral–Derivative (PID) controller, Linear–Quadratic Regulator (LQR) or fuzzy control system. Modern UAVs can be equipped with expensive and complicated systems that allows them to perform some tasks autonomously. For devices that are not equipped with

© Springer Nature Switzerland AG 2020
R. Szewczyk et al. (Eds.): AUTOMATION 2019, AISC 920, pp. 583–598, 2020.
https://doi.org/10.1007/978-3-030-13273-6_54

sophisticated sensors and data processing algorithms, there is still a necessity for receiving a control signal from the pilot. It is possible to connect those two technologies together, in which BCI provides a control signal for UAV. LaFleur et al. first described the usage of BCI for quadcopter control [2]. In that research a 64 channels EEG system with BCI2000 software was used. EEG signal was filtered in the context of motor imaginations: clenching of the left hand, clenching of the right hand, clenching of both hands. Obtained signals were linked with the quadcopter steering allowing to control the quadcopter in terms of turning left or right and changing its altitude. Forward motion of the quadcopter was constant and wasn't controlled with BCI. Within a few years the technology evolved to such a level, that at the University of Florida worlds 1st Brain-Drone race took place [3]. 16 pilots took part in the competition which involved flying a quadcopter with the help of BCI through a track prepared on an indoor basketball court.

2   Basics of the BCI Technology

BCI allows for direct communication between a brain and an external device. The technology is based on Hans Bergers invention – EEG, that was first described by him in 1929 [4]. EEG is a record of electrical activity of the cerebral cortex which can be measured on the surface of the scalp. It is believed that the most of the electrical signals that are visible in EEG are caused by postsynaptic potentials, also long depolarization can be registered in EEG. In a healthy brain an action potential travels along the axon to the axon terminals, where a neurotransmitter is released. Neurotransmitters are affecting postsynaptic membrane by changing its conductivity and polarization. If the signal stimulates the neuron, there is a local depolarization of the cells membrane and Excitatory Postsynaptic Potential (EPSP) appears. There also exist an Inhibitory Postsynaptic Potential (IPSP) – it is causing a local hyperpolarization. Combination of EPSP and IPSP induces electrical current in a neuron and around it. The electrical field have enough power to be registered on the surface of the scalp. As it turned out, a typical time of duration for the postsynaptic potential is about 100 ms, which strongly correlates with the time of duration for an average alpha wave. Alpha-rhythm consist of sinusoidal or rhythmic alpha waves and it is basic rhythmic frequency in a healthy brain of an adult human [5]. The concept of BCI was first described by Vidal in 1973 [6]. In his article the author is pointing that there is a possibility of extracting useful information from EEG. Evoked responses can be embedded in the electrical signals acquired from the surface of the scalp. They can appear as en effect of an external stimulation that can be delivered via different paths: sight, hearing, and somatosensory system. EEG signal can carry information about eye movement, and muscle potentials (e.g. facial expressions), as well as it can deliver information about the motor decisions. According to Wolpaw et al. basic elements of BCI system can be as follows: signal acquisition subsystem, signal processing subsystem, the output device [7]. In order to process EEG signals, first they have to be registered with the help of electrodes. Since the signals are not very powerful – they are measured in μV – they need to be amplified. Finally, they can be digitalized with the usage of analogue-to-digital converters (ADC) and from that moment the signals are represented as quantized discrete time series. Signal processing phase aims to extract the user commands from EEG, translate them and deliver to the output device [26]. The feature extraction can be done by applying different procedures to the digitalized signals. Broad spectrum of available methods exists working both in time domain, as well as in frequency domain, e.g.: spatial filtering, voltage amplitude measurements, spectral analyses. Recorded EEG signals aside from useful information can also contain artefacts. Numerous different reasons can cause artefacts, such as: poor connection between the electrode and a scalp, movement of facial muscles, tongue movement, blinking, eyeballs movement, pulse and many others. The computer screen can act as an output device. However, a large number of other possibilities exist. Mobile robots can be one example [8,9], following with neurogaming [10] or neuroprosthesis [11]. EMOTIV Epoc+ Neuroheadset can be given as an example of a contemporary BCI. The hardware part of this device can be described as a wireless EEG system with 14 channels, it was presented in the Fig. 1. Electrodes are made as a saline soaked felt pads. Signals collected by the electrodes are sampled internally with the rate of 2048 per second. These data are later filtered and downsampled to 128 or 256 samples per second for each channel. Resolution of the data can be set to either 14 (0.51 µV per least significant bit) or 16 (0.13 µV per least significant bit) bits. Data are transferred from the headset to the host device with the usage of Bluetooth 4.0 technology. Aside from the part that is responsible for EEG, the device also includes motion sensors: accelerometer, gyroscope, magnetometer. EMOTIV Epoc+ Neuroheadset can work with most of the popular platforms: Windows 7, 8, 10, Linux, Mac OS X, iOS 5, Android. The data analysis can be performed in a 3rd party software: EEGLAB (toolbox for Matlab), LabView (there is Emotiv Toolkit available), OpenViBe (open source software), BCI2000 (general-purpose software system for BCI). Emotiv also offers its own software, as well as the Software Development Kit (SDK) for application developers.

Emotiv Xavier Controlpanel application is allowing to detect facial expressions in the context of: blinking, smiling, laughing, teeth clenching, looking right or left, raising brows and others. An example is presented in the Fig. 2 - a detected facial expression was the Right Smirk. It is possible to calibrate the sensitivity for the facial expressions listed above.

The software also offers "The Performance Metrics Detection Suite" for detection of the user subjective emotions. The results are presented on a chart as in the Fig. 3. There are 5 performance metrics available: engagement/boredom, frustration, meditation, instantaneous excitement, long-term excitement. Precision of the detection will increase with the amount of data gathered from the user. Stored data are processed and used for rescaling of the results.

"The Mental Commands detection suite" is another feature that is available for the user. It is based on a real time EEG processing and is designed to recognize brainwave patterns that are related up to 13 different physical actions: 6 linear movements, 6 rotational movements, and disappearing. In order to use this

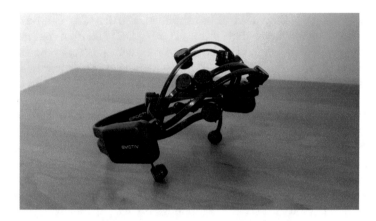

Fig. 1. Emotiv EPOC+ NeuroHeadset.

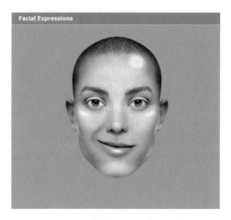

Fig. 2. Emotiv Xavier Controlpanel – facial expression detection.

feature, user has to go through the training phase. During this phase brainwave pattern for neutral state is recorded, as well as patterns related to other actions. It is possible to choose the number of actions that will be detected, but the difficulty increases with the number of actions that the user wants to perform. The Push action was presented in the Fig. 4 - a virtual cube was pushed away.

Functionality that allows to control a mouse pointer with the movement of the head is also available [12].

3   Applications of UAVs

Currently, UAVs are receiving a lot of attention. A number of different applications have been developed. Some of the non-military applications have been presented in this chapter. UAVs are appropriate tool for the air pollution monitoring [13]. They can provide measurements with higher spatial and temporal

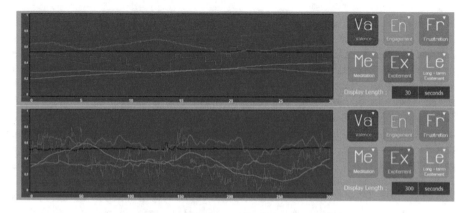

Fig. 3. The Performance Metrics Detection Suite.

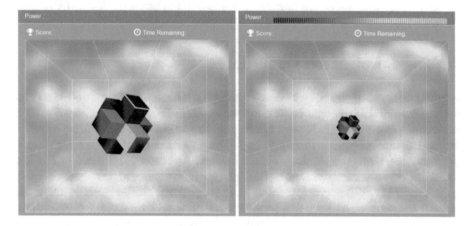

Fig. 4. Example of detected mental command: neutral state (left), push action (right).

resolution in comparison to manned aircraft or satellite, they are also less expensive. UAVs are well suited for local and regional applications. Moreover, they can take measurements in areas where it would be too risky for manned aircraft. Both fixed-wing aircrafts and rotary wing aircrafts (quadcopters, hexacopters, octocopters) are capable of carrying measurement equipment. Sensors for dust monitoring, CO_2, SO_2, radiation and others can be mounted on board. Other type of monitoring can be carried out when there is a camera mounted on UAV. Rafał Czapaj-Atłas and Bogumił Dudek described a usage of UAVs in the field of power engineering [14]. Both regular cameras and thermal imaging cameras can be used for inspection of overhead power lines. Based on this data, it is possible to indicate the places with a different failures, e.g.: places with higher resistance, insulation failures, loose or corroded connections. UAVs can be especially convenient in the areas that are difficult to reach – forests or swamps. Not only overhead power lines can be inspected with UAVs. It is also possible for wind turbines or photovoltaic power stations. Some applications exist also for agriculture [15]. UVAs paired with Global Positioning System (GPS) and cameras can be used for detection of irrigation issues, pest and fungal infestations and crop management. Combining data gathered from camera and infrared camera, it is possible to create images that will show plants chlorophyll levels. Amazon Prime Air is a service for rapid package delivery via UAVs. Another example of delivery UAV can be given as DHL Parcelcopter [16,17]. The limitations of delivery UAVs are their range and the weight that they can carry. Nevertheless, their advantage is short delivery time. The delivery of goods is quicker in comparison to the delivery by car, especially when the terrain is difficult (a mountains during winter). Worth noting is also Swiss search-and-rescue device that won United Arab Emirates Drones for Good Award in 2015. The "GimBall" is a collision-proof search-and-rescue UAV, that can be used in complicated and hazardous sites [18]. Because of its construction, it won't be damaged by the collisions with obstacles, it is also capable of rolling across ceilings and floors. Its protective cage allows to use it in close proximity to people.

4   The Constructed Quadcopter

In order to shorten the time of development for mechanical construction a commercially available quadcopter frame was selected. It is made mostly of carbon fibre tubes and plates. It is very important for the construction of the frame to be rigid as the vibrations are undesired. The used frame has a place for mounting 4 motors, electronic equipment and battery. As presented in the Fig. 5, there are 4 brushless motors mounted that can produce 830 g of thrust each when paired with 11.1 V power supply and 11 in propeller with the pitch of 4.7 in. As well as the frame, the propellers also should be rigid to reduce the vibrations. What's more, they should be well balanced. To meet this requirements carbon fibre propellers were used. Electrical energy is fed to the motors through the devices called Electronic Speed Controllers (ESC) – each motor requires one ESC. On the input side of the device, there is a power supply from the battery and pulse-width modulation (PWM) signal. Output of each ESC consist of 3 lines, that ought to be connected to the brushless motor. Rotational velocity of the motor is proportional to the duty cycle of the PWM signal. Used battery is produced in Lithium Polymer (LiPo) technology. Its voltage is rated at 11.1 V and capacity equals 5000 mAh. For quadcopter weighting 1.6 kg in theory such battery should provide a flight time of about 10 to 15 min.

The main part of electronic system is a microcontroller (μC) with the Cortex M4F core. The used μC is equipped with 1 MB of available flash memory, 192 KB of random access memory (RAM) and it is clocked at 168 MHz. It has floating point unit (FPU), that can accelerate the calculations (addition, divide, multiply, square-root, subtract). Among other features it also provides: ADC, timers, Serial Peripheral Interface (SPI), Universal Synchronous Asynchronous Receiver Transmitter (USART). The μC is connected to sensors that create an inertial measurement unit (IMU): gyroscope, accelerometer and magnetometer.

Fig. 5. The constructed quadcopter.

Data received from those sensors are processed with the algorithm that was described in the Sect. 6 of this article in order to calculate the quadcopter attitude - roll, pitch and yaw angles. 3 axis gyroscope (ST L3G4200D) provides angular velocity data. Its measurement range is set to 500° per second (dps), it incorporates low pass filter and high pass filter. The gyroscope is communicating with the μC via SPI and the measurement results are coded on 16 bits for each axis. Accelerometer is a sensor that provides information about the linear acceleration. In the measurement results also gravitational acceleration is included. Based on the information about the gravitational acceleration it is possible to determine the orientation. The device (ST LSM303D) is communicating with the μC via SPI. The measurement results consist of 3 values – one for each axis. Together with accelerometer, in the same chip also magnetometer is included. It is used to measure the strength and direction of Earth magnetic field. It is worth to note that the magnetometer should be placed as far as possible from the elements that can cause magnetic interference. Two separate communication channels were designed. Radio Control (RC) transmitter is used as a device that allows to change the drone attitude. With its help it is possible to set roll, pitch and yaw angles as well as to control the throttle (total amount of power that is delivered to the motors). Second communication channel is based on Bluetooth technology and it is used mainly for telemetry. It provides data rate of 115200 bits per second (bps). Transmission with such speed allows to deliver the telemetry data with the frequency of 100 Hz without the usage of Direct Memory Access (DMA). Attitude data, setpoints for the attitude, battery voltage as well as the control signals for the motors will be transmitted within this channel to the Ground station.

5 Ground Station Software

To make a full usage of the communication channel that is based on Bluetooth technology an application for the ground station was created.

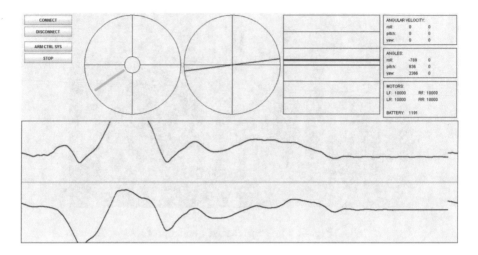

Fig. 6. Ground station software graphical user interface (GUI).

The software that is presented in the Fig. 6 can be run on a PC and its main task is to visualize the attitude data in near real-time (without significant delay). The application consist of two threads with one infinite loop in each thread. First is responsible for the communication with the quadcopter, and the 2nd for displaying the data. The home screen consist of: 4 buttons, heading indicator, roll angle indicator, pitch angle indicator, text fields and graph area. Buttons allows the user to establish or close down the link with the quadcopter, as well as activate and deactivate the stabilization system. Heading indicator provides an information about the quadcopter orientation with respect to the magnetic north. Roll angle and pitch angle indicators shows information about the angles that are measured with respect to the gravity of Earth. Values of those angles are also presented in the text form on the right side of the screen, with one more important value placed there – the battery voltage. On the graph area the roll and pitch angles are presented as well as setpoint values for those angles. Aside from the visualization, all of the received data are also stored in a text file and are available for further analysis after the flight.

6   Digital Data Processing and Stability Control Algorithm Description

Digital Low Pass Filter. A substantial amount of noise is embedded in the measurement results from the accelerometer. It can be caused by the vibrations generated by the motors and propellers. Also, the sensors itself produces noisy results even when it is laying steady on the desk. In order to reduce this noise a digital low pass filter (LPF) was implemented. It is based on the equation:

$$\overline{M}_k = \alpha \overline{M}_{k-1} + (1 - \alpha) M_k \tag{1}$$

where: $\overline{M}_k$ – value after filtration at time step k, M_k – value before filtration at time step k, α – design factor, a constant in the range of $0 < \alpha < 1$, related to the filter cut-off frequency.

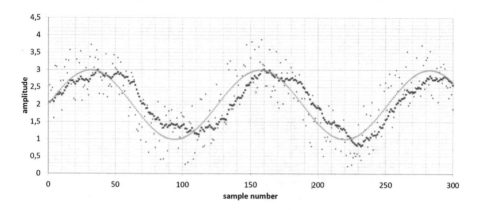

Fig. 7. LPF results for $\alpha = 0.1$.

Green points on the chart in the Fig. 7 are representing sinusoidal signal without the noise. Signal represented by blue points was created on the base of the sinusoidal signal with random noise added to it. The signal marked by red points is produced based on the noisy signal (blue) with the usage of Eq. 1. Based on the above chart it can be stated that the noise in the signal was considerably reduced. On the other hand, some delay was introduced to the filtered data.

Extended Kalman Filter. The data processing algorithm known as the Kalman Filter was first described by Kálmán in 1960 [19]. It is used for estimating the state of the process based on a linear model of the process, the measurement data, the process noise covariance and the measurement noise covariance [20]. For processes that are described by nonlinear equations Extended Kalman Filter (EKF) was developed. There are number of documented applications of EKF, e.g.: Apollo spacecraft, radar tracking and Global Positioning System (GPS) [21]. General form of equations for EKF is given by the literature [22]. The equations that were used for these project assumes that there is no information available about the process inputs, as well as, that the process and measurement noise are additive. EKF equations can be divided into 2 phases: the prediction phase and the correction phase. Equations for the prediction phase are given below:

$$x_k^- = f(x_{k-1}, \dot{x}_k) \tag{2}$$

$$P_k^- = A_k P_{k-1} A_k^T + Q \tag{3}$$

where: x_k^- - prediction of the process state, x_{k-1} - process state estimation from the last time step, $f(x_{k-1}, \dot{x}_k)$ - nonlinear function $A_k = \frac{df_{[i]}}{dx_{[j]}}(x_{k-1})$- Jacobian of nonlinear function with respect to the states, P_k^- - prediction of error covariance matrix, P_{k-1} - error covariance matrix for process state estimation from the last time step, Q - process noise covariance matrix.

Based on the process model and on the previous state estimation, the state prediction is calculated, as well as prediction for the error covariance matrix. With this values calculated it is possible to proceed to the correction phase with the following equations:

$$K_k = P_k^- H^T (H P_k^- H^T + R)^{-1} \tag{4}$$

$$x_k = x_k^- + K_k(z_k - h(x_k^-)) \tag{5}$$

$$P_k = P_k^- - K_k H P_k^- \tag{6}$$

where: K_k - the Kalman gain, H - the linearized measurement model, R - the measurement noise covariance matrix, z_k - the vector of measurements, $h(x_k^-)$ - the measurement model (transition between state variables and measurements).

With the equations presented above the state estimate can be calculated. For this project state variables vector x consist of 3 elements:

where: ϕ - roll angle, θ - pitch angle, ψ - yaw angle.

The model of the process can be explained as a relationship between the current angular position, position in previous time step and the current angular velocity. It is given by the Eq. 8:

As it was stated in Sect. 4 of this article, the gyroscope is used for angular velocity measurements. The sensor output data are given in respect to the local

$$x = \begin{bmatrix} \phi \\ \theta \\ \psi \end{bmatrix} \tag{7}$$

$$f(x_{k-1}, \dot{x}_k) = x_{k-1} + \dot{x}_k * dt \tag{8}$$

reference frame of the gyroscope, on the contrary, state variables are measured in respect to the static reference frame related with the Gravity and the magnetic north. For that reason $\dot{x}$ is given as:

$$\dot{x} = \begin{bmatrix} \dot{\phi} \\ \dot{\theta} \\ \dot{\psi} \end{bmatrix} = \begin{bmatrix} 1 & sin\phi tan\theta & cos\phi tan\theta \\ 0 & cos\phi & -sin\phi \\ 0 & sin\phi sec\theta & cos\phi sec\theta \end{bmatrix} \begin{bmatrix} p \\ q \\ r \end{bmatrix} \tag{9}$$

where: p, q, r are angular velocities registered by the gyroscope and dt is the sampling time. Because the measurement vector is stated as the orientation angles that are calculated based on the data from accelerometer and magnetometer, the measurement model H is given as an identity matrix. New estimates are produced by the EKF with the frequency of 400 Hz.

Proportional-Integral-Derivative Controller. After the attitude estimates are calculated, a control algorithm is executed. Based on the current attitude and on the received setpoints, with the frequency of 400 Hz control signals for the motors are calculated. One PID controller for each of axis was implemented in the software running on the μC. Control signal calculated with the PID algorithm is given by the equation:

$$u(n) = K_p e(n) + K_i \sum_{k=0}^{n} e(k) + K_d[e(n) - e(n-1)] \tag{10}$$

where: u(n) - control signal, K_p - proportional gain, e(n) - control error, K_i - integral gain, K_d - derivative gain. PID control is the most popular control algorithm in industry and is described with details in the literature [23,24].

Control system presented on the Fig. 8 was design for this project. Its primary task is to maintain the desired angular position. It is presented in a simplified form, a block responsible for splitting 3 control signals to 4 motors was omitted.

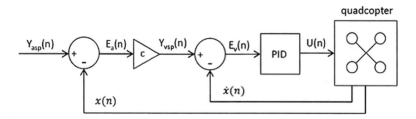

Fig. 8. Simplified block diagram of the control system.

As it is Multi Input Multi Output (MIMO) system the signals are given as vectors. $Y_{asp}(n)$ is a setpoint vector for the angles - this values are transmitted to the quadcopter with the help of RC transmitter. The difference between the setpoint value and the measured value is known as the control error – $E_a(n)$. Marked as c is a coefficient, allowing to calculate the angular velocity $Y_{vsp}(n)$ based on the angle error. $E_v(n)$ is the control error vector for the angular velocity and U(n) is the calculated control signal vector.

7   The Concept of Controlling a Quadcopter with BCI

Traditionally, a quadcopter is controlled with the help of a RC transmitter. The concept of controlling a quadcopter with BCI will be presented below. Such approach will have influence on the used hardware and software for the quadcopter and for the Ground station. RC transmitter used to control a quadcopter should have minimum 4 channels available. With 4 channels of a RC transmitter, it is possible to independently modify 4 values. One of the channels is used for control of the throttle, and other 3 channels are used to change the setpoints for roll, pitch and yaw angles. When quadcopter changes its roll angle to different than zero it moves sideways. When pitch angle changes it moves forward or backward, and the change of yaw angle results in rotation around the vertical axis. With the throttle the user is able to control quadcopter descent or ascent. When controlled by RC transmitter, the quadcopter roll and pitch angles as well as the throttle are proportional to the movement of the control sticks. With movement of the control stick for the yaw angle, we can control velocity of the rotation. The signal sent from the RC transmitter is accepted by the receiver. RC receiver is connected to the μC via 4 lines. With those lines 4 independent PWM signals are transmitted. Based on the measurement results of the duty cycle of those signals, it is possible to calculate setpoint values for the attitude and throttle.

The software for the ground station was created in Java programming language. A package named Emotiv SDK v2.0.0.20−LITE provides the library that allows to use the Emotiv EPOC+ Neuroheadset in programs created in Java. After the library is added to the Java project, it is possible to use methods for reading data regarding the facial expression, emotions, mental commands, and head movement [25]. Methods for reading mental commands data are named: ESCognitivGetCurrentAction() and ESCognitivGetCurrentActionPower().

Those methods returns the name of the action related to the users current brainwave pattern as well as its strength. Also, methods that allows to conduct the training phase (save brainwave patterns) are available. Methods: EECognitivSetTrainingAction() and EECognitivSetTrainingControl() can be used in the program. Due to the fact that minimum of 4 different information is required to control the quadcopter, it seems appropriate to choose mental commands as the source of the information. The telemetry data channel can be used as a medium for transmitting setpoints to the quadcopter. It is possible to realize the control in 2 variants: as proportional or as static. What regards the setpoints for angles, they can be proportional to the strength of the detected brainwave pattern. Other approach would be to set a threshold for the strength of a detected brainwave pattern and make a static assignment of a specific angle whenever a strength threshold is exceeded. The proportional control would provide more flexibility. On the other hand, it would be more demanding for the user. The static approach would not allow for such a flexibility in terms of quadcopter control, but it could be easier for the user, especially during first flights.

Hardware modification should be made in the sensor area of the quadcopter. In the current state, it would be extremely hard to control angles of the quadcopter and constantly tune the throttle in order to maintain a desired altitude. Barometric sensor should be added to the sensor system in order to provide the air pressure data, which in turn can be converted into altitude. In order to increase the measurement accuracy in the proximity to the ground, a ultrasonic rangefinder or a laser rangefinder pointed in the direction of the ground should be added. Sensor fusion of the listed sensors along with the data from accelerometer should be done in order to introduce 4th state variable – the altitude. To estimate this new state variable Kalman filter can be used. The altitude controller will accept this data as an input. This controller would be responsible for maintaining the altitude at a setpoint level. At that point, Emotiv EPOC+ should provide information only about the changes of the altitude and not the changes of the throttle.

Controller responsible for maintaining of the quadcopter altitude should be extended with safety features, that would not allow to crash quadcopter into the ground. Altitude and its rate of change should be constantly measured and monitored. In a close proximity to the ground the descend speed should be automatically decreased in order to allow safe landing and to avoid accidental crash into the ground. Another safety feature that should be present in the ground station system is a failsafe button. Pressing this button would switch the control from BCI to RC transmitter. Bluetooth modules used in quadcopter and in ground station are class 1 modules, which means that their maximum range is 100 m. The value gives enough safety margin for transmitting telemetry data, but for transmitting control data there should be used a communication medium with greater range. Used Bluetooth technology can be changed to communication based on nRF24L01+ modules, that should increase the range as well as broaden a bandwidth of the link for telemetry data.

8 Results of the Quadcopter Test Flight in the Context of BCI

The flight took place in a room, with the altitude of about 1 m above the floor. Quadcopter was controlled by the RC transmitter in order to maintain a steady position. Data for roll angle, roll setpoint, pitch angle and pitch setpoint were collected with the frequency of 100 Hz. Fragments of the collected data was presented below (Fig. 9).

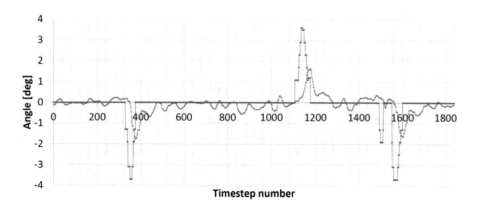

Fig. 9. Comparison of roll angle with the setpoint value for roll angle.

Based on the chart presented above, it can be concluded that roll angle (blue chart) was stable with minor variation when setpoint (red chart) was equal to 0 (max. variation = 0.5°). Such variation of the roll angle do not influence the quadcopter position greatly as it generates the lateral force that is proportional to the sine of the angle. When the RC transmitter is used as the source of control signal a small variation of the position can be easily compensated. On the other hand, when the quadcopter is to be controlled with BCI event a little changes of the quadcopter position can be difficult to compensate. It is related to the limited bandwidth of BCI – a limited amount of the commands can be issued via BCI in the unit of time. When the setpoint was changed briefly, the rate of change for roll angle as well as for pitch angle was not fast enough to follow the setpoint value.

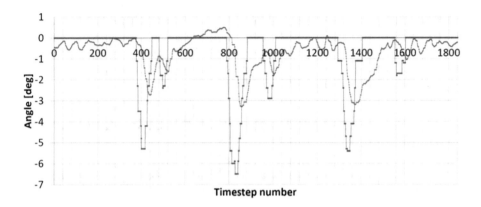

Fig. 10. Comparison of pitch angle with the setpoint value for pitch angle.

In the Fig. 10 there is a chart for pitch angle (blue) and pitch setpoint (red) presented. It is visible that the adjustments of setpoint for that angle was larger and more frequent than for the roll angleIt is possible that the mechanical imperfections are responsible for the higher control effort. Motors might be mounted not perpendicular to the frame and the printed circuit board containing the sensors might be mounted not in parallel with the frame. Angles should follow the setpoint values with less delay. A presence of the delay makes it difficult to control the quadcopter with the RC transmitter and even more with BCI. In addition to control delay, also another delay is introduced. The other delay is related to the processing time of an EEG signals. In general, during the quadcopter flight a situation can change very dynamically and delays are not desired. It would be hard to concentrate on maintaining the proper state of mind and in the same time to think about how early we have to react in order to complete a maneuver. Overall, it is clear that the changes should be made to the quadcopter, so it requires less control action. Also, during the outdoor flights, it is possible to use GPS as an additional source of data for the position control. For indoor flights the quadcopter might be equipped with a vision sensor that is pointing to the ground. Based on the image analysis, it would be possible to calculate quadcopter direction of movement.

9 Conclusion

Considering the presented information it can be concluded, that it would be extremely difficult to use BCI with a quadcopter equipped only with the attitude stability system. A certain improvement in that situation is possible after introducing altitude controller within the quadcopter control system. It would eliminate the necessity for constant throttle tuning in order to maintain desired altitude. Further development of the control system is possible in order to control the direction of movement. Based on the information from vision sensor or GPS, a controller would be able to maintain desired quadcopter position in reference to the ground. An application that would simulate the behaviour of the quadcopter should be created. After BCI is connected to that application, it should be possible to train manoeuvres like: starting, landing, flying in different directions and turning. The simulation can be built with the usage of quadcopter dynamics model. The application should allow to choose the method of control – proportional or static and threshold value. It should be possible to choose the smaller number of controlled variables for the beginners, and increase it for more advanced users.

References

1. Wei-Yen, H.: Brain-computer interface connected to telemedicine and telecommunication in virtual reality applications. Telematics Inform. **34**(4), 224–238 (2017). https://doi.org/10.1016/j.tele.2016.01.003
2. LaFleur, K., Cassady, K., Doud, A., Shades, K., Rogin, E., He, B.: Quadcopter control in three-dimensional space using a noninvasive motor imagery-based brain-computer interface. J. Neural Eng. **10**(4), 046003 (2013)
3. Dearen, J.: University of Florida. https://www.eng.ufl.edu/newengineer/news/mind-controlled-drones-race-to-the-future. Accessed 21 Jan 2017
4. Berger, H.: Uber das Elektrenkephalogramm des Menschen. Archiv für Psychiatrie und Nervenkrankheiten 87, 527–570 (1929)
5. Ghaemia, A., Rashedia, E., Mohammad Pourrahimib, A., Kamandara, M., Rahdaric, F.: Automatic channel selection in EEG signals for classification of left or right hand movement in Brain Computer Interfaces using improved binary gravitation search algorithm. Biomed. Sig. Process. Control **33**, 109–118 (2017). https://doi.org/10.1016/j.bspc.2016.11.018
6. Vidal, J.J.: Toward direct brain-computer communication. Annu. Rev. Biophys. Bioeng. **2**, 157–80 (1973)
7. Wolpaw, J.R., Birbaumer, N., McFarland, D.J., Pfurtscheller, G., Vaughan, T.M.: Brain-computer interfaces for communication and control. Clin. Neurophysiol. **113**(6), 767–91 (2002)
8. Moreno, R.J., Aleman, J.R.: Control of a mobile robot through brain computer interface. INGE CUC **11**(2), 74–83 (2015)
9. Paszkiel, S., Blachowicz, A.: Zastosowanie BCI do sterowania robotem mobilnym. Pomiary Automatyka Robotyka PAR 02/2012, s. 270–274, Warszawa 2012
10. Forbes, F.A.: http://www.forbes.com/sites/singularity/2013/06/03/the-future-of-gaming-it-may-be-all-in-your-head. Accessed 21 Jan 2017

11. Handa, G.: Neural prosthesis - past, present and future. Indian J. Phys. Med. Rehabil. **17**(1), 460 (2006)
12. Emotiv. https://emotiv.zendesk.com/hc/en-us/categories/200089979-EPOC-EPOC-. Accessed 21 Jan 2017
13. Villa, T.F., Gonzalez, F., Miljievic, B., Ristovski, Z.D., Morawska, L.: An overview of small unmanned aerial vehicles for air quality measurements: present applications and future prospectives. Sensors (Basel) **16**(7), 1072 (2016)
14. Netoa, J.R.T., et al.: Performance evaluation of unmanned aerial vehicles in automatic power meter readings. Ad Hoc Netw. **60**(15), 11–25 (2017). https://doi.org/10.1016/j.adhoc.2017.03.003
15. Anderson, C.: Technology review. https://www.technologyreview.com/s/526491/agricultural-drones. Accessed 01 Jan 2017
16. Amazon Prime Air. https://www.amazon.com/Amazon-Prime-Air. 01 Jan 2017
17. Parcelopter. http://www.dpdhl.com/en/mediarelations/specials/parcelcopter. 21 Jan 2017
18. Flyability. http://www.flyability.com. Accessed 21 Jan 2017
19. Kalman, R.E.: A new approach to linear filtering and prediction problems. J. Basic Eng. **82**(1), 35–45 (1960)
20. Klavins, E., Zagursky, V.: Unmanned aerial vehicle movement trajectory detection in open environment. Proc. Comput. Sci. **104**, 400–407 (2017). https://doi.org/10.1016/j.procs.2017.01.152
21. Grewal, M.S., Andrews, A.P.: Applications of Kalman Filtering in aerospace 1960 to the present. IEEE Control Syst. **30**(3), 69–78 (2010)
22. Bell, J., Stol, K.A.: Tuning a GPS/IMU Kalman Filter for a robot driver. In: Australasian Conference on Robotics and Automation (2006)
23. Astrom, K.J., Hagglund, T.: PID Controllers: Theory, Design, and Tuning. Instrument Society of America, Pittsburgh (1995)
24. Visioli, A.: Practical PID Control. Springer, London (2006)
25. Emotiv Software Development Kit User Manual for Release 2.0.0.20
26. Cegielska, A., Olszewski, M.: Nieinwazyjny interfejs mózg–komputer do zastosowań technicznych, Pomiary Automatyka Robotyka – PAR 03/2015, s. 5–14, Warszawa 2015

Increasing Imaging Speed and Accuracy in Contact Mode AFM

Andrius Dzedzickis$^{(\boxtimes)}$, Vytautas Bučinskas, Tadas Lenkutis,
Inga Morkvėnaitė-Vilkončienė, and Viktor Kovalevskyi

Vilnius Gediminas Technical University, J. Basanaviciaus Street 28,
Vilnius, Lithuania
{andrius.dzedzickis,vytautas.bucinskas,
tadas.lenkutis,inga.morkvenaite-vilkonciene,
viktor.kovalevskyi}@vgtu.lt

Abstract. Atomic force microscope (AFM) is a promising tool in micro and nano size objects researches. Contact mode AFM has advantages comparing to non-contact modes: the scanning speed is higher, and atomic resolution can be achieved. The main limiting factor in contact mode AFM is scanning speed. At high scanning speed the 'loss of contact' phenomenon occurs, and probe in this case cannot follow the surface. In order to ensure a constant interaction force (stable contact) between the probe and scanned surface, the additional force created by air flow was applied. Proposed method is based on the idea to apply additional controllable nonlinear force on the upper surface of the AFM cantilever, which will help to keep the probe in contact with sample surface. It was found that dynamic characteristics of various AFM sensor cantilevers can be controlled using proposed method. It has been determined that the use of aerodynamic force has the greatest influence on the scanning results deviation from the real sample in the horizontal direction then scanner z axis goes down. With a compressed air pressure of 7 kPa and a scanning speed 847.6 µm/s, this deviation decreases by 20% comparing to the case when compressed air flow is not used.

Keywords: AFM · Dynamics · Aerodynamic force · Scanning speed

1 Introduction

Atomic force microscopy (AFM) is a powerful tool widely used in researches of different kinds of surfaces by determining their topography, friction, adhesion, and viscoelasticity on nanometric scale [1–3]. AFM is extensive technique based on the measurement of the interaction force between the probe and sample surface. AFM can work in several operating modes: contact, tapping, and non-contact [4–6]. Scanning mode should be chosen considering the properties of the sample's surface. Contact mode AFM is mainly used to image hard surfaces when the presence of lateral forces is not expected to modify the morphological features [7], for example characterization of semiconductors surface topography. Tapping and non-contact modes usually are used for imaging of soft biological samples [8]. Contact mode AFM has some advantages among other modes as:

© Springer Nature Switzerland AG 2020
R. Szewczyk et al. (Eds.): AUTOMATION 2019, AISC 920, pp. 599–607, 2020.
https://doi.org/10.1007/978-3-030-13273-6_55

high scan speeds; high (atomic) resolution images; high resolution when measuring rough samples with extreme changes in vertical topography [9, 10]. The main limitation of contact mode AFM, which increases labor costs is low scanning speed, which depends on multiple dynamic factors, such as the bandwidth of the probe detection system [11], characteristics of the sample positioning system [12], the probe type [13], the probe and surface material [14, 15], etc. Various attempts are made trying to resolve this problem, but all methods are based on improvement of AFM control system. In our group the new method to ensure tip-sample contact was proposed [15–18]. It is based on the idea to implement controllable nonlinear force which will act on the upper surface of the cantilever. This additional force allows to control stiffness of the cantilever and prevent contact loss between probe and sample. In practice additional force is created using stream of compressed air. Such engineering solution can be applied to various AFMs and does not require major changes in AFM hardware. Detail explanation of our idea, research of aerodynamic parameters, results of mathematical modelling and results of initial experimental research presented in our previous reports [4, 15–18].

The aim of this research was to investigate the effect of aerodynamic force, created by air stream, when surface is scanned at various speeds in contact mode AFM, and compare the results with ideal surface profile.

2   Experimental Research of Modified AFM Sensor

2.1   Experimental Setup

Experimental research was performed in the Laboratory of Center for Physical Sciences and Technology. Scanning were performed using home-made AFM setup and modified AFM sensor with cantilever, type NCHV from Bruker (MA, USA) (Table 1) and home-made compressed air supply system.

Table 1. Parameters of cantilever and holder.

Parameter	Value
Cantilever	
Length, L	117 μm
Mass, m	7.96×10^{-11} kg
Resonant frequency, f	320 kHz
Spring constant, k	40 N/m
Thickness, t	3.5 μm
Width, w	33 μm
Young's modulus, E	310 GPa
Manufacturer	Bruker
Type	NCHV
Holder	
Size of the initial gap, Δ_0	0.4 mm
Diameter of air duct	0.2 mm

AFM setup was built using standard components from reliable producers. Main components of AFM setup are: 3d printed housing, AFM head based on a DVD optical pickup, manual micrometer translation stage, closed-loop 3D Nano positioning stage (Nano-M350, Mad City Labs, USA). For rough positioning manual translation stage was used; for precise positioning nanopositioning stage Nano-M350 was used. Nanopositioning stage have its own control system and allows precise positioning under closed loop control with resolution 0.4 nm in Z axis and 0.1 nm in XY axes. AFM head with attached cantilever and a holder with installed micro air duct is mounted on the top of AFM housing. This experimental setup was also used in other researches and is presented in more detail in [15, 19].

For the supply of clean compressed air was designed, a special mechatronic system (Fig. 1) which allows precisely control air output pressure.

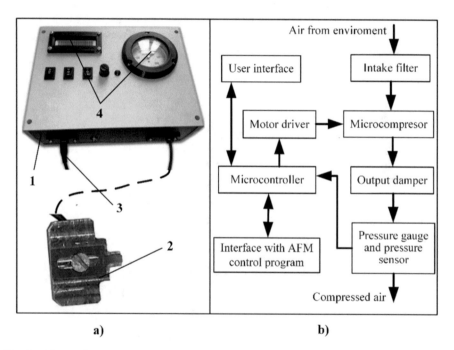

a)   b)

Fig. 1. Air supply system: (a) view physical system, 1 – compressed air supply system, 2 – holder with installed air duct, 3 – USB port to connect to AFM controller, 4 – LCD screen and a pressure gauge; (b) structural scheme of the compressed air supply system.

The pressure of compressed air, applied to the cantilever, is controlled using PID controller which changes the efficiency of the micro compressor according to the desired pressure and output pressure (Fig. 1b). Output pressure is measured using the pressure sensor MPXV5050GP (Freescale Semiconductor, Austin, TX, USA). For manual control, air supply system has mechanical pressure gauge and LCD screen for easier system adjustment. The automatic control is performed by computer through USB interface (Fig. 1). The more detailed system explanation is presented in our previous report [15].

2.2   Research Methodology

Experimental research was performed by scanning the rectangular cross-section calibration grating APCS-0099 with structure height 240 nm (2 μm pitch) from Bruker (MA, USA). Horizontal scanning data was obtained using NCHV cantilever (Table 1) by scanning calibration grating in circular trajectory of 30 μm in diameter. It was decided to use circular trajectory because such solution allows to scan long trajectory of 94.2 μm long with smooth linear speed. Horizontal scanning was performed at 94.2 μm/s, 282.6 μm/s, 471 μm/s, 659.4 μm/s, 847.8 μm/s scanning speeds in contact mode AFM. It was decided to use calibration grating for this experiment due to the regular structure of the surface roughness which makes easier to process the obtained results, in particular to evaluate the results with numerical parameters. Scanning results were evaluated using three parameters as shown in Fig. 2.

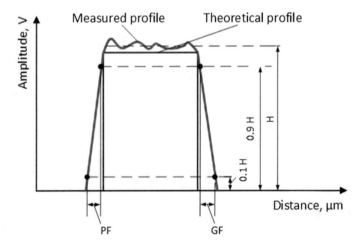

Fig. 2.   Scheme of experimental results evaluation.

Results of the experiments are evaluated by comparing the measured height of the structure H, the steepness of forward signal front PF (horizontal deviation from theoretical profile when probe goes up) and the steepness of backward signal front GF (horizontal deviation from theoretical profile when probe goes down).

Comparison of amplitude allows to detect changes in the vertical axis due to changes in scanning speed and aerodynamic forces. The steepness of the forward and backward fronts of the signal is evaluated on the basis of assumption that the theoretically plateaus and valleys of the calibration grate have ideally perpendicular surface shape (Fig. 2). Both fronts of the signal are evaluated in order to define the effect of the aerodynamic force when the cantilever rises on the plateau and moves from it. In order to ensure the reliability of the measurement data according to the described method, 5 freely chosen plateaus from the curves obtained by scanning the calibration grating were evaluated. Parameters PF and GF evaluated using OriginPro 8.0's "rise/fall" tool for signal analysis. The statistical analysis of the received data was performed using the Microsoft Office 2016 software package.

2.3 Results and Discussion

In this chapter are presented results in graphic form summarized according the above described methodology. Comparison of measured amplitudes (H) is presented in Fig. 3.

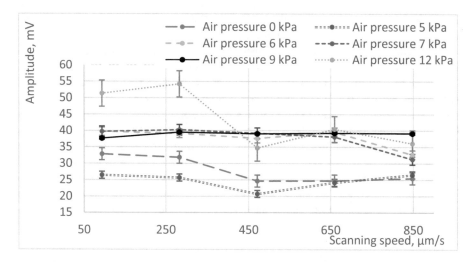

Fig. 3. Comparison of measured amplitudes (H).

In Fig. 3, which shows the dependencies between amplitude and scanning speed, it is seen that the amplitude increases then air flow is used in all cases, except when the air pressure was 5 kPa and the scanning speed did not exceed 650 µm/s. The difference in amplitude values can be explained by the fact that the compressed air flow decreases the cantilever's bounce at the beginning of the plateau of the sample's structure and by fact that the cantilever takes lower position faster and more accurately estimates the surface at the bottom of the valley. The hypothesis that the probe is not always able to reach the bottom of the valley is confirmed by the dependence between the amplitude and the scan speed when aerodynamic force is not used, in this case then scanning speed increases up to 450 µm/s, the amplitude value decreases from 33 mV to 25 mV and this tendency continues when scanning speed is increasing. This indicates that at this speed a threshold is reached when the speed of the feedback or the piezo scanner z-axis is not sufficient to respond adequately to surface roughness. The decrease in amplitude obtained using a 5 kPa pressure flow of compressed air can be explained this way: under such scanning conditions the air flow effects are not effective enough because of the lack of air flow and low speed the surface forms "air bags" which interfere with the probe when it tries to reach bottoms of the surface valleys. This effect is reduced when scanning speed exceeding limit of the 450 μm/s. The stable amplitude values are obtained when the compressed air pressure was 6 kPa and 7 kPa. At such pressures, the increase in amplitude occurs at a scanning speed exceeding the limit of 650 μm/s. The amplitude increase from 33 mV to 40 mV. With a pressure of 9 kPa, the incidence of amplitude presented in previously discussed cases is not noticeable. In contrast, the increase in scanning speed from 100 μm/s to 300 μm/s shows an amplitude increase from 38 mV to 40 mV. This phenomena is explained this way: under such pressure at low scan speeds, corresponds to the case with air pressure 5 kPa, air stream on the sample surface creates an aerodynamic force that prevents the probe from reaching the sample surface. A large scattering of measured points when pressure of compressed air is 12 kPa indicates that in this case the air flow effect is very unstable, thus, it could be concluded, that this pressure is too high. Summarizing the results presented in Fig. 3 it can be stated that the positive effect of the aerodynamic force on the measured amplitude size (H) in the range of scanning speeds from 100 μm/s to 650 μm/s is observed when the compressed air pressure is 6–7 kPa. A positive effect of 9 kPa pressure of the compressed air is detected in the scanning speed range from 300 μm/s to 840 μm/s. Comparison of measured height of unequal structures allows us to evaluate only the effect of aerodynamic force for scanning sufficiently smooth structures when the probability of loss of contact between the probe and the sample surface is low.

The influence of aerodynamic forces on the scanning structures with sharp changes in surface topology was evaluated by parameters PF and GF (Fig. 2). The values of parameter (PF), which describes the steepness of the signal's forward front are shown in Fig. 4.

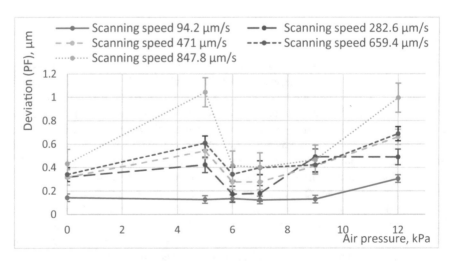

Fig. 4. Dependencies of parameter PF on scanning speed and pressure of compressed air.

The curves presented in Fig. 4 describes the effect of aerodynamic force on scanning structures with unequal surface height when the probe has to "climb" on a high plateau. The minimum deviation is obtained by scanning at speed 94 µm/s. In this case, the effect of an aerodynamic force is only visible when air pressure exceeds value of 9 kPa, moreover, this effect is negative because deviation increases. This can be explained by the fact that the high pressure creates force on the upper surface of the cantilever, which prevents it from "jumping" on the plateau. As the scanning speed increases, the pressure range of the compressed air in which aerodynamic force creates positive effect becomes noticeable. When scanning speed increases up to 282 µm/s and the air pressure is 6–7 kPa, the deviation from known geometric shape of the calibration grate decreases from 0.31 µm to 0.17 µm compared to the case when compressed air flow is not used. With a scanning speed of 471 µm/s, the deviation decreases from 0.31 µm, when compressed air flow is not used, to 0.17 µm. In addition, increasing the scan speed from 659.7 µm/s to 847.8 µm/s, the deviation decreases from 0.35 µm to 0.34 µm and from 0.43 µm to 0.41 µm respectively. Summarizing the results presented in Fig. 4, it can be stated that when compressed air pressure is in range from 6 kPa to 7 kPa a positive effect of aerodynamic force is seen on all scanning speeds tested in this experiment. The use of 5 kPa, 9 kPa and 12 kPa pressure in the researched cases increases the deviations (PF), this can be explained by low air stream velocity at 5 kPa pressure and excessive compression air flow at 9 kPa and 12 kPa air pressures.

The results presented in Figs. 3 and 4 allows to state that the use of aerodynamic force produces a positive effect and creates preconditions for increasing AFM scanning speed. However, in order to fully evaluate this effect, GF parameter, which defines the steepness of signal rear front should be defined. The dependency of the scanning speed and the pressure of compressed air on parameter GF shown in Fig. 5.

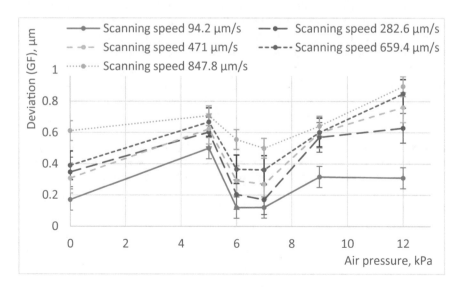

Fig. 5. Dependencies of parameter GF on scanning speed and pressure of compressed air.

From Fig. 5 it is seen that positive effect created by aerodynamic force are observed when the compressed air pressure is from 6 kPa to 7 kPa. In this case, the deviation from theoretical profile is reduced from 0.18 µm to 0.11 µm at scanning speed 94 µm/s. Increasing scanning speed up to 280 µm/s, a deviation decreases from 0.34 µm to 0.2 µm compared to the case when compressed air flow is not used. By further increasing the scanning speed, the trend of deviation decrease remains. With a scanning speed 470 µm/s and 670 µm/s, the deviation is reduced from 0.32 µm to 0.27 µm and from 0.39 µm to 0.36 µm, respectively. By increasing the speed to 847 µm/s, the deviation value decreases from 0.61 µm to 0.49 µm. In cases when air pressure is 5 kPa, 9 kPa and 12 kPa deviations increases due to the above discussed reasons (Fig. 4).

Summarizing the results of experimental research, it can be stated that the effect of aerodynamic force is experimentally evaluated. Positive effect of the aerodynamic force according analyzed parameters is defined scanning a calibration grate at all tested speeds, the when pressure of compressed air was from 6 to 7 kPa.

3 Conclusions

It was determined that significant positive effect of the aerodynamic force is achieved when pressure of the compressed air is 6–7 kPa. It has been determined that the use of aerodynamic force has the greatest influence on the parameter GF, which defines the deviation of measured results from the real sample in the horizontal direction then scanner z axis goes down. With a pressure of compressed air 7 kPa and a scanning speed 847.6 µm/s, this deviation decreases by 20% compared to the case when compressed air flow is not used.

The effect of different aerodynamic forces were evaluated, and results show that that the deviations PF (deviation of the measured result from the theoretical profile in the horizontal axis then probe goes upward) also were lower compared to the results obtained without additional aerodynamic force. The positive effect of aerodynamic force remains even if scanning speed is increased up to 847 µm/s.

Results of experimental research confirms that the proposed AFM scanning speed enhancement method is effective. Selection of suitable air pressure value according to the type of structure being scanned, creates conditions which allows to increase AFM scanning speed and improve quality of the results.

Acknowledgement. This research was funded by the European Social Fund according to the activity "Development of Competences of Scientists, other Researchers and Students through Practical Research Activities" of Measure No. 09.3.3-LMT-K-712. Project No 09.3.3-LMT-K-712-02-0137.

References

1. Morkvenaite-Vilkonciene, I., Ramanavicius, A., Ramanaviciene, A.: Atomic force microscopy as a tool for the investigation of living cells. Medicina **49**, 155–164 (2013)
2. Fujii, S.: Atomic force microscope. In: Compendium of Surface and Interface Analysis. Springer, Singapore (2018)
3. Haase, K., Pelling, A.E.: Investigating cell mechanics with atomic force microscopy. J. R. Soc. Interface **12**(104), 20140970 (2015)
4. Bučinskas, V., Dzedzickis, A., Šešok, N., Šutinys, E., Iljin, I.: Research of modified mechanical sensor of atomic force microscope. In: Dynamical Systems: Theoretical and Experimental Analysis. Springer, Cham (2016)
5. Yang, H. (ed.): Atomic Force Microscopy (AFM): Principles, Modes of Operation and Limitations. Nova Science Publishers, Hauppauge (2014). Incorporated
6. Voigtländer, B.: Scanning Probe Microscopy. Springer, Berlin (2016)
7. Gonnelli, R.S.: Atomic force microscopy in the surface characterization of semiconductors and superconductors. Phil. Mag. B **80**(4), 599–609 (2000)
8. Schillers, H., Rianna, C., Schäpe, J., Luque, T., Doschke, H., Wälte, M., Dumitru, A.: Standardized nanomechanical atomic force microscopy procedure (SNAP) for measuring soft and biological samples. Sci. Rep. **7**(1), 5117 (2017)
9. Quénet, D., Dimitriadis, E.K., Dalal, Y.: Atomic force microscopy of chromatin. In: Atomic Force Microscopy Investigations into Biology-from Cell to Protein. InTech (2012)
10. Materials and Structure Property Correlation Assignment No. 2. http://www.mecheng.iisc. ernet.in/~bobji/mspc/assign_2011/Atomic%20force%20microscopy.pdf. Accessed 15 Oct 2018
11. Adams, J.D., Nievergelt, A., Erickson, B.W., Yang, C., Dukic, M., Fantner, G.E.: High-speed imaging upgrade for a standard sample scanning atomic force microscope using small cantilevers. Rev. Sci. Instrum. **85**(9), 093702 (2014)
12. Tien, S., Zou, Q., Devasia, S.: Iterative control of dynamics-coupling-caused errors in piezoscanners during high-speed AFM operation. IEEE Trans. Control Syst. Technol. **13**(6), 921–931 (2005)
13. Richter, C., Burri, M., Sulzbach, T., Penzkofer, C., Irmer, B.: Ultrashort cantilever probes for high-speed atomic force microscopy. SPIE Newsroom (2011)
14. Butt, H.J., Cappella, B., Kappl, M.: Force measurements with the atomic force microscope: technique, interpretation and applications. Surf. Sci. Rep. **59**(1–6), 1–152 (2005)
15. Dzedzickis, A., Bucinskas, V., Viržonis, D., Sesok, N., Ulcinas, A., Iljin, I., Morkvenaite-Vilkonciene, I.: Modification of the AFM sensor by a precisely regulated air stream to increase imaging speed and accuracy in the contact mode. Sensors **18**(8), 2694 (2018)
16. Dzedzickis, A., Bučinskas, V., Eok, N., Iljin, I.: Modelling of mechanical structure of atomic force microscope. Solid State Phenom. **251**, 77–82 (2016)
17. Bučinskas, V., Dzedzickis, A., Šutinys, E., Lenkutis, T.: Implementation of different gas influence for operation of modified atomic force microscope sensor. Solid State Phenom. **260**, 99–104 (2017)
18. Bučinskas, V., Dzedzickis, A., Šutinys, E., Šešok, N., Iljin, I.: Experimental research of improved sensor of atomic force microscope. In: International Conference on Systems, Control and Information Technologies, pp. 601–609. Springer, Cham (2016)
19. Ulčinas, A., Vaitekonis, Š.: Rotational scanning atomic force microscopy. Nanotechnology **28**(10), 10LT02 (2017)

Rotation Speed Detection of a CNC Spindle Based on Ultrasonic Signal

Grzegorz Piecuch[✉]

Faculty of Electrical and Computer Engineering,
Rzeszow University of Technology, Rzeszów, Poland
gpiecuch@kia.prz.edu.pl

Abstract. This paper describes the methodology and results of research on distinguishing the rotation speed of a CNC machine spindle with the help of ultrasonic signal as well as a classifier created with the help of neural networks. Tests were carried out on laboratory object in real-time. Achieved research results are very good, and developed possible solutions for use in industry and education. The article describes the problems and the methodology of achieved research, indicates the used hardware and software solutions, as well as an analysis of the results.

Keywords: Rotation speed · CNC spindle · Ultrasonic signal · Neural network

1 Introduction

In the factories, detail production is frequently covered by CNC machines. They are often equipped with the spindles to which tools are attached. Despite of the fact, that the tool has direct contact with the object, it is the spindle that made the tool move. To make a particular element accordingly with the presumption and project, it is necessary to retain certain machine parameters [1] for example rotation speed of a spindle or feed speed.

Sensors are used to control the machine parameters. It is usually checked whether the machine performs the process with the given parameters and whether the tools and machine components are not degraded. In terms of TCM (Tool Condition Monitoring) and PCM (Process Condition Monitoring) research, vibration measurement is most commonly used, as evidenced by numerous publications [2–9]. Unfortunately, in the case of ultrasound signal, its usefulness is negligible when detecting wear of the tool, and it is only valid when detecting catastrophic blunting of the blade [2]. According to the popular strategy of predictive maintenance, this information seems to be too late. The question then arises as to whether this signal can be effectively used for other purposes. Despite the existence of many other methods (laser speed sensors, inductive sensors, etc.), the author decided to check whether it is possible to use an ultrasonic sensor for speed detection. It turns out that yes and it is quite a successful method. However, it was not tested on the actual machine during processing but on a laboratory testbed.

© Springer Nature Switzerland AG 2020
R. Szewczyk et al. (Eds.): AUTOMATION 2019, AISC 920, pp. 608–615, 2020.
https://doi.org/10.1007/978-3-030-13273-6_56

2 Preparation of Test Station

In the case of the imbalance detection problem, authors of many publications point out a vibration, current, force and acoustic emission signals usefulness. Single-handedly performed examinations, described in [10], confirm the usefulness a vibration signal but do not confirm this feature for an ultrasonic signal. The other signals were not checked. In the case of the rotation speed detection issue, own trials proved, that measured signal from ultrasonic sensor is absolutely sufficient for the speed evaluation.

For the investigations, the same station as used in imbalance detection examinations was used (Fig. 1). The examined element was low power CNC spindle Teknomotor C41/47-C-3822-400 featuring maximum rotation speed of 24000 rpm, controlled with Siemens Sinamics G110 inverter. The ultrasonic signal was measured with the UE Systems UE Ultra-Trak 750 sensor (Fig. 1) connected to the EL3742 analog input module with oversampling, which through the expansion module EK1100 and EtherCAT protocol, sent measuring data to Beckhoff C6920 IPC. All properties (e.g. operating range) of sensor, inverter, spindle and Beckhoff devices are available in datasheets.

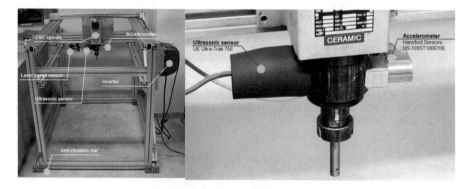

Fig. 1. The laboratory testbed (on the left) and sensors attachment to the spindle (on the right) [10].

3 Measuring Data Registration and Speed Detection

For gathering learning data, a Simulink scheme (Fig. 2), running in the external mode and executed on Beckhoff C6920 IPC was used. The program cycle was set to 2 ms and measuring data was being saved to file every 320 cycles. During the measurement oversampling was set to 50. This translates into a sampling frequency of 25 kHz. Measurements were performed for rotation speed in the range of 40–200 Hz with 20 Hz interval for spindle with no tool attached. From the gathered data, 5 indicators were appointed: RMS, standard deviation, kurtosis, bias and mean average. The analysis of the graphs for different indicators pointed out that only RMS coming from the ultrasonic sensor can provide satisfying results. Good example of useless factor is chart of kurtosis viewed in the Fig. 4.

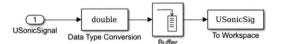

Fig. 2. Simulink scheme providing ultrasonic sensor data registering.

The other indicators were characterizing by a similar time waveforms for different speeds, therefore the distinguishing based on them would be impossible. As a consequence, using only RMS attribute in further steps was decided. An example chart of RMS values for different speeds is depicted in Fig. 3.

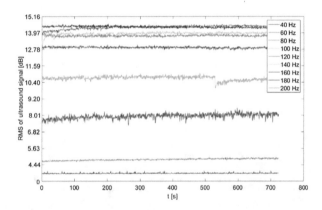

Fig. 3. RMS value of ultrasound signal in [dB] for different rotational frequency of spindle – the distinction is possible.

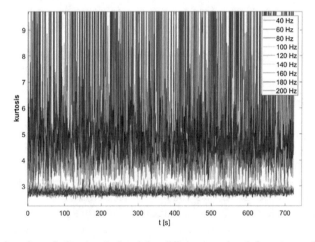

Fig. 4. Kurtosis value of ultrasound signal for different rotational frequency of spindle – the distinction is impossible. For better illustration the graph is enlarged.

Analysis of Fig. 3 indicates that speeds from range of 140–200 Hz can be poorly distinguishable among each other, therefore in trials 5th and 6th only speeds from range 40–120 Hz was taken into consideration. In further step, an impact of position and attachment approach of the sensor (position, tightening) for received results was checked. The outcome of this experiment is presented in Fig. 5. A set of measurements was made in which, after every trial, the position or the tightening of the sensor was changed. To check, whether the changes were not random, between the 4th and 6th trial no changes were made.

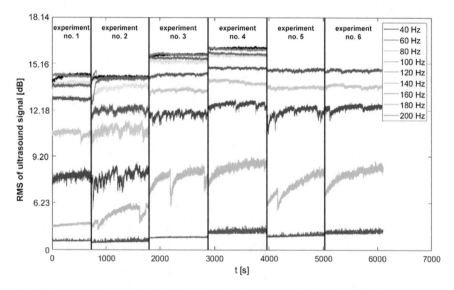

Fig. 5. Comparison of the behavior of ultrasonic sensor signal effected by attachment changes.

For performing classification, 5 classes of speed were distinguished: class 1 – 40 Hz, class 2 – 60 Hz, class 3 – 80 Hz, class 4 – 100 Hz, class 5 – 120 Hz. Although the value of the RMS indicator for the speed of 60 Hz increases in time, it was decided to include the speed in the classification. Given the fact that the rest of the speed classes in the graph are at a considerable distance from the discussed 60 Hz. In addition, observing the behavior of the indicator in days where the mechanical configuration of the sensor was not changed, showed that the indicator despite the change over time, is still in the same value range.

As the learning data, the ones gathered in experiments 4–6 (details in Table 1) were decided to use. With the available MATLAB/Simulink tool Neural Network Pattern Recognition Tool (nprtool), the neural network was learned and a block containing its model, which then was used in project of online speed recognition scheme was generated (details in Table 2). The architecture of the network is depicted in Fig. 6. It is a 2-layer, feed-forward, backpropagation network with sigmoidal hidden and output neurons, where w is a weight and b is a bias of the neural network [11]. The choose was dictated by its features of nonlinear approximator.

Table 1. Information about data used in the learning of neural network process.

Parameter	Value
Number of classes	5
Amount of files containing registered data per class	5043
Amount of all files	25215
Experiment time	3226.88 s ($\sim$54 min)
Buffer capacity	4000 samples
Learning data	17651 files (70%) – 282 416 000 samples
Validation data	3782 files (15%) – 60 512 000 samples
Test data	3782 files (15%) – 60 512 000 samples

Table 2. Information about a structure of the network.

Parameter	Value
Number of layers	2
Number of neurons in hidden layers	10
Number of inputs	1 (RMS)
Data normalization	Embedded in network structure; function mapminmax

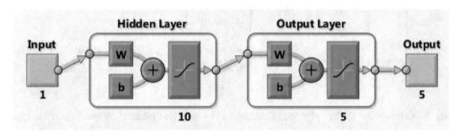

Fig. 6. The architecture of 2-layer neural network used in spindle rotation speed classification [11].

The used tool has an embedded algorithm providing a data division for learning data, validation data and test data. This was the reason for using the hold-out technique, that provides a random division of the data set into 3 independent sets: learning data, test data and validation data, usually in proportion of 70/15/15%. For learning data, either accuracy, sensitivity or specificity had a value of 100% what proves received confusion matrix depicted in Fig. 7.

On-line classifier tests were carried out in real time on the same laboratory testbed, constantly changing the spindle speed. Of course, during the test, the speeds were known to the classifier (5 classes defined earlier). For the real-time tests, a Simulink scheme was created (Fig. 8). To the network inputs a continuously calculated RMS value were being put, and the sample match-to-class result was being gotten on the output. A value close to 1, in any row of the window 'Class' informs of the recognition of the sample as class 1 to 5 (from the top).

Confusion Matrix

	1	2	3	4	5	
1	5043 20.0%	0 0.0%	0 0.0%	0 0.0%	0 0.0%	100% 0.0%
2	0 0.0%	5043 20.0%	0 0.0%	0 0.0%	0 0.0%	100% 0.0%
3	0 0.0%	0 0.0%	5043 20.0%	0 0.0%	0 0.0%	100% 0.0%
4	0 0.0%	0 0.0%	0 0.0%	5043 20.0%	0 0.0%	100% 0.0%
5	0 0.0%	0 0.0%	0 0.0%	0 0.0%	5043 20.0%	100% 0.0%
	100% 0.0%	100% 0.0%	100% 0.0%	100% 0.0%	100% 0.0%	100% 0.0%

Output Class / Target Class

Fig. 7. Confusion matrix after learning and testing network off-line.

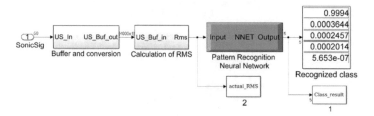

Fig. 8. Real-time Simulink classifier.

The real-time tests results confirmed virtually 100% accuracy achieved in learning process. A result confusion matrix is attached in Fig. 9.

Confusion Matrix

	1	2	3	4	5	
1	2481 20.0%	0 0.0%	0 0.0%	0 0.0%	0 0.0%	100% 0.0%
2	3 0.0%	2484 20.0%	0 0.0%	0 0.0%	0 0.0%	99.9% 0.1%
3	0 0.0%	0 0.0%	2484 20.0%	0 0.0%	0 0.0%	100% 0.0%
4	0 0.0%	0 0.0%	0 0.0%	2484 20.0%	0 0.0%	100% 0.0%
5	0 0.0%	0 0.0%	0 0.0%	0 0.0%	2484 20.0%	100% 0.0%
	99.9% 0.1%	100% 0.0%	100% 0.0%	100% 0.0%	100% 0.0%	100% 0.0%

Output Class / Target Class

Fig. 9. Real-time tests confusion matrix for the rotation speed detection neural network.

The real-time tests were performed for 2482 samples of each class – overall, 12420 records. Achieved indicators of classifier efficiency are gathered in Table 3. Acc (ac-

Table 3. Classifier efficiency comparison based on appointed indicators.

	Offline test (gathered data)			Real-time test		
	Acc [%]	Sen [%]	Spe [%]	Acc [%]	Sen [%]	Spe [%]
Class 1 (40 Hz)	100	100	100	99.98	99.88	100
Class 2 (60 Hz)	100	100	100	99.98	100	99.97
Class 3 (80 Hz)	100	100	100	100	100	100
Class 4 (100 Hz)	100	100	100	100	100	100
Class 5 (120 Hz)	100	100	100	100	100	100

curacy), Sen (sensitivity), Spe (specificity) indicators are determined based on the binary classification from the following formulas [12]:

$$ACC = \frac{TP + TN}{TP + FN + TN + FP} \qquad (1)$$

$$SEN = \frac{TP}{TP + FN} \qquad (2)$$

$$SPE = \frac{TN}{TN + FP} \qquad (3)$$

where: TP – True Positive, TN – True Negative, FP – False Positive, FN – False Negative.

As the time domain classification achieved very good results, no frequency and time-frequency signals were tested, but only in the time domain.

4 Conclusions

Performed investigations proved, that signal from ultrasonic sensor of which the RMS indicator is calculated, being the only input in the neural network is sufficient for almost faultless recognition of class, which is assign to specific spindle rotation speed. It was confirmed with real-time test performed on the laboratory station. The developed method is easy to implement and possible to use in the industry, and its main advantages are the ability to act in real-time and very high rate class of rotation speed detection of a CNC spindle. In addition, using the laboratory testbed it is possible to use with prepared software solutions for educational purposes, aimed at teaching students of Computational Intelligence methods and use them in industrial solutions.

References

1. Mączka, T., Żabiński, T.: Platform for intelligent manufacturing systems with elements of knowledge discovery. In: Manufacturing System, pp. 183–204. InTech, Croatia (2012)
2. Jemielniak, K.: Automatyczna diagnostyka stanu narzędzia i procesu skrawania. Oficyna wydawnicza Politechniki Warszawskiej, Warszawa (2012)
3. Sokołowski, A.: Automatyzacja wytwarzania. Zastosowania sztucznej inteligencji w diagnostyce obrabiarek i procesu skrawania. Wydawnictwo Politechniki Śląskiej, Gliwice (2013)
4. Teti, R., Jemielniak, K., O'Donnell, G., Dornfeld, D.: Advanced monitoring of machining operations. CIRP Ann. - Manuf. Technol. **59**, 717–739 (2010)
5. Linxia, L., Radu, P.: Machinery time to failure prediction - case study and lesson learned for a spindle bearing application. In: IEEE Prognostics and Health Management (2013)
6. Dayong, J., Taiyong, W., Yongxiang, J., Lu, L., Miao, H.: Reliability assessment of machine tool spindle bearing based on vibration feature. In: International Conference on Digital Manufacturing and Automation, China (2010)
7. Abu-Mahfouz, I.: Drilling wear detection and classification using vibration signals and artificial neural networks. Int. J. Mach. Tools Manuf **43**, 707–720 (2003)
8. Jantunen, E., Jokinen, H.: Automated on-line diagnosis of cutting tool condition. Int. J. Flex. Autom. Integr. Manuf. **4**, 273–287 (1996)
9. Jantunen, E.: A summary of methods applied to tool condition monitoring in drilling. Int. J. Mach. Tools Manuf **42**, 997–1010 (2002)
10. Piecuch, G., Żabiński, T.: Implementation of computational intelligence methods for CNC machine spindle imbalance prediction. In: Szewczyk, R., Zieliński, C., Kaliczyńska, M. (eds.) Automation 2018, AUTOMATION (2018)
11. Matlab & Simulink: Neural Network Pattern Recognition Tool – Manual
12. Mączka, T.: Zastosowanie metod inteligencji obliczeniowej i wspomagania decyzji w systemach produkcyjnych. Rozprawa doktorska, Rzeszów (2016)

Neon Oscillator Megohmmeter

Al Julanda Hashim Salim Al Nabhani[1], Klaudia Biś[2],
Michał Nowicki[3(✉)], and Marcin Safinowski[4]

[1] Institute of Micromechanics and Photonics,
Warsaw University of Technology, Warsaw, Poland
[2] Institute of Automation and Robotics,
Warsaw University of Technology, Warsaw, Poland
[3] Institute of Metrology and Biomedical Engineering,
Warsaw University of Technology, Warsaw, Poland
nowicki@mchtr.pw.edu.pl
[4] Industrial Research Institute of Automation and Measurements PIAP,
Warsaw, Poland

Abstract. This article is about the application of the neon lamp in oscillator circuit to the measurement of high electrical resistance. It has been found that the relationship between the period of the oscillator and the resistance in the circuit is linear to a high degree of accuracy. This linear relation are applied to the measurement of high resistance, with use of simple phototransistor voltage divider and popular Arduino platform. Thus extremely basic and inexpensive, but relatively obscure electronic circuit can be utilized for high resolution measurements of high resistance, while microcontroller provides necessary calculations, error corrections and PC connection for optional data storage and visualization.

Keywords: Neon lamp oscillator · High resistance · Megohmmeter · Arduino

1 Introduction

Through the use of different equipment many methods have been used in the measurements of high resistance. Measurement of high resistance could be problematic. In order to avoid possible errors the measurement of resistance greater than 10 MΩ requires the use of some specific techniques, as well as some instruments such as an electrometer, SMU instrument or picoammeter [1]. A constant voltage method or constant current method is usually used to measure the high resistance by the use of electrometer [2].

In this article high resistance was measured by detecting the flashes of a neon lamp in oscillator circuit, which gave a regular period of time between short flashes. The device built used very few basic components. The result however presented very good accuracy-to-cost ratio.

The neon lamp can be used in a wide range of temperatures, brightness and voltages, which allows its usage in harsh environmental conditions.

When we connect a neon lamp with a capacitor in parallel, charging this circuit through a resistor, our result reveal the simplest oscillating electronic circuit that makes

© Springer Nature Switzerland AG 2020
R. Szewczyk et al. (Eds.): AUTOMATION 2019, AISC 920, pp. 616–623, 2020.
https://doi.org/10.1007/978-3-030-13273-6_57

measurement of high resistance possible. The higher the resistance the slower the capacitor will charge, and this will result in an lowering the frequency of the neon lamp flashes. By utilizing modern phototransistors and microcontrollers, the measurement of this frequency, and following calculation of the measured resistance, is exceptionally easy.

2 Theoretical Background

The neon lamp flashes at regular intervals of time in a circuit that has a high resistance connected in a series and a capacity shunted across its electrodes, which will interrupt the current through the lamp. The time between flashes is equal to the time required for the condenser to charge added to the time necessary for it to discharge through the lamp [3].

The drop of cathode potential appears when the steady discharge is passing through the lamp, which is greater than the potential that falls in the rest of the tube. Therefore, the ions which travel through the dark space will cause ionisation by collision. There must be a high voltage source to start the discharge and keep the lamp glowing, and to keep the average field between the lamp electrodes sufficient to cause ionization by collision. The voltage across the electrode may be reduced to the lower critical voltage so that the potential drop across the tube from the confines of the dark space is just sufficient to supply energy to carry ions from the cathodes glow to the anode.

If the cathode potential is constant for different voltages across the lamp electrodes, then there will be a constant value of the field in the dark space. Consider a small element of area at the confines of the dark space which is normal to the current direction; a constant number of ions will be produced per second at the point where the area has been situated since the field in the dark space is constant, and this number of ions passing over to the anode per second through the area will be proportional and is given by n $(V - V_A)$ δA, where n is number of ions, V_A is the cathode fall of potential, V is the voltage across the electrodes and δ_A is the area. Therefore, the small element of current contributed through the area is given by $\delta_i = ne (V - V_A)$ δA, where 'e' is the charge carried by the individual ions, and the total current I is given by

$$I = e(V - V_A)\int_0^A ndA \qquad (1)$$

where the integration is performed over the whole of the glow area.

Condenser flashes charge from lower to higher critical voltage, then discharges through the lamp down to the lower critical voltage again, where the condenser flashes are shunted across the lamp electrodes. The total time between charge and discharge is given by $T = t_1 + t_2$, where t_1 is the time required for the condenser to charge and t_2 is the time to discharge.

The discharge through the lamp begins when the condenser plates have attained a potential difference of high critical voltage. Simultaneously, the condenser is connected to a battery and consequently is charging.

The potential difference across the lamp electrodes at any instant is defined by voltage V, therefore, V is the potential difference between the condenser plates. Then the quantity of electricity passing through the lamp in time is given by

$$dq = k(V - V_A)dt \qquad (2)$$

a quantity $\frac{E-V}{R}dt$ has flowed into the condenser from the battery at the same period.

Therefore, the net change of condenser charge is given by

$$-dQ = \left[k(V - V_A) - \frac{E - V}{R} \right] dt \qquad (3)$$

And solving this equation by

$$t_2 = \frac{CR}{KR + 1} log_e \frac{V_C - D}{V_B - D} \qquad (4)$$

where D is equal to

$$\frac{E + KRV_A}{KR + I} \qquad (5)$$

Therefore, the total period is given by [4]:

$$T = CR \left[log_e \frac{E - V_B}{E - V_C} + \frac{I}{KR + I} log_e \frac{V_C - D}{V_B - D} \right] \qquad (6)$$

The whole area of the lamp electrode is employed during discharge and a large quantity of electricity passes through the lamp during the short luminous period t_2. The K which is proportional to the electrode area employed, is such that KR is large compared with unity and the above expression may be reduced to the approximate form

$$T = CR log_e \frac{E - V_B}{E - V_C} + \frac{C}{K} log_e \frac{V_C - V_A - \frac{E}{KR}}{V_B - V_A - \frac{E}{KR}} \qquad (7)$$

An interesting idea is provided by this equation, as it suggests methods of comparison of capacities and high resistance. These have been experimentally proven to be both quick and accurate methods [5].

3 Project Design

The basic Neon Lamp Oscillator was assembled with measured resistor connected as series resistance charging the ultra-low leakage capacitor with neon lamp shunted across it. A high 100 V DC voltage source was used, which is slightly higher than the breakdown voltage comprised of thirteen 9 V batteries. A 10 k resistor is used in series with voltage source to protect the user against shock (Figs. 1 and 2).

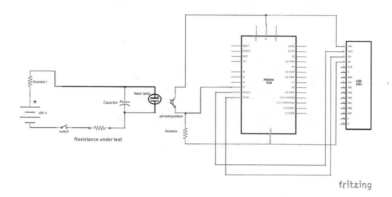

Fig. 1. Neon Oscillator Megohmmeter diagram.

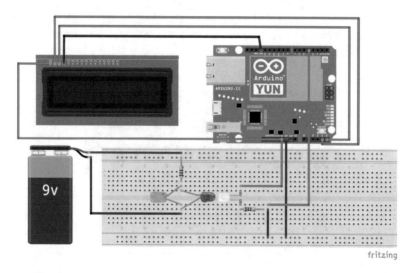

Fig. 2. Fritzing diagram of the Neon Oscillator Megohmmeter prototype.

The capacitor will charge until it reaches the neon lamp breakdown voltage, then quickly discharge. The voltage across the capacitor will cross back below the neon lamp breakdown voltage, therefore, the capacitor will take around Δt to charge and discharge, so the neon lamp will flash at around Δt interval.

The photoresistor's job is to receive the flashes from the neon lamp and to send it as signal to the Arduino [6]. A code is implemented in the Arduino to measure the time between pulses using stop watch function. An LCD is used to give the output of this measurement. The code was written in standard C programming language. The finished device is shown in Fig. 3.

Fig. 3. Finished Neon Oscillator Megohmmeter.

4   Investigation

The finished device was investigated by taking multiple readings of various resistors, and comparison with reference, high accuracy multimeter Tonghui TH1961 (Fig. 4). The 1 MΩ resistor used was polish made wire-wound laboratory standard, 10 MΩ was generic Chinese metal film THT resistor, while the 100 MΩ was carbon film USSR made surplus (Fig. 5).

Fig. 4. The measurement test stand.

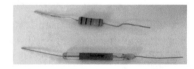

Fig. 5. Typical resistors measured in the project.

The graph in Fig. 6 presents the linearity of the obtained function from the relation between the result measurements read using digital multimeter and neon megohm-meter. The coefficient of determination R^2 equals 0.999, which proves linearity of the Neon Megohmmeter characteristic (Fig. 7 and Table 1).

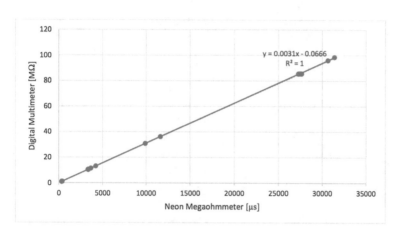

Fig. 6. Graph of the relation between resistance and neon megohmmeter oscillation period.

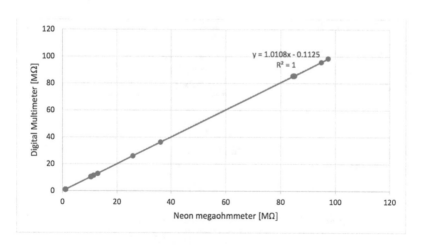

Fig. 7. Graph of the relation between measurement reading using digital multimeter and neon megohmmeter.

Table 1. Comparison of measurement between digital multimeter and neon megohmmeter.

Resistor [MΩ]	Digital Multimeter [MΩ]	Neon Megohmmeter [MΩ]	Relative errors %
1 MΩ	0.999 MΩ	1.060 MΩ	6.1%
1.1 MΩ	1.009 MΩ	1.076 MΩ	6.6%
1.2 MΩ	1.019 MΩ	1.090 MΩ	6.9%
1.3 MΩ	1.029 MΩ	1.097 MΩ	6.6%
1.4 MΩ	1.039 MΩ	1.106 MΩ	6.4%
1.5 MΩ	1.049 MΩ	1.119 MΩ	6.6%
1.9 MΩ	1.089 MΩ	1.151 MΩ	5.6%
10 MΩ	10.36 MΩ	10.38 MΩ	0.19%
10.1 MΩ	10.454 MΩ	10.44 MΩ	0.13%
10.2 MΩ	10.464 MΩ	10.451 MΩ	0.12%
10.4 MΩ	10.482 MΩ	10.47 MΩ	0.11%
10.9 MΩ	10.53 MΩ	10.51 MΩ	0.18%
11 MΩ	11.44 MΩ	11.40 MΩ	0.34%
15 MΩ	12.86 MΩ	12.90 MΩ	0.31%
30 MΩ	25.97 MΩ	25.91 MΩ	0.23%
40 MΩ	36.16 MΩ	36.10 MΩ	0.16%
100 MΩ	85.41 MΩ	84.75 MΩ	0.77%
100.1 MΩ	85.46 MΩ	84.47 MΩ	1.15%
100.2 MΩ	85.48 MΩ	84.49 MΩ	1.15%
100.4 MΩ	85.51 MΩ	84.62 MΩ	1.04%
100.6 MΩ	85.53 MΩ	84.87 MΩ	0.77%
100.9 MΩ	85.55 MΩ	85.04 MΩ	0.59%
110 MΩ	95.70 MΩ	94.81 MΩ	0.92%
115 MΩ	98.5 MΩ	97.30 MΩ	1.21%

The main drawback of the presented circuit is the time between readings, highlighted in Table 2:

Table 2. Time taken between each reading of the resistors.

Resistor [MΩ]	Time [s]
1 MΩ	2 s
10 MΩ	6.7 s
20 MΩ	13.1 s
25 MΩ	15.2 s
30 MΩ	16.8 s
35 MΩ	21.6 s
100 MΩ	54.9 s
110 MΩ	60.01 s

In order to shorten the time between each reading, a capacitor with smaller capacitance would have to be used – the device would then have higher measurement range, but lower resolution.

5 Conclusion

The main goal of this project was to develop megohmmeter which would be truly inexpensive, have low power consumption and data logging capability, and accuracy better than 1% in the 10 to 100 MΩ range, which is outside of the typical handheld DMM range. All of these goals have been met by applying exotic measurement technique utilizing neon lamp oscillator circuit. Presented solution can be used for continuous measurement of high resistance, or capacitance after small hardware and software modification. Thus the presented circuit can be also utilized as cheap high-resolution transducer for sensors utilizing high resistance or capacitance. It should be stressed then, that there is full galvanic isolation between the measured resistance, and the microcontroller circuits, which allows i.e. operating voltages in the kV range.

References

1. Low Level Measurement Handbook, 7th edn. Keithley, Tektronix Company
2. Edwards, D.F.A.: Electronic Measurement Techniques. Butterworth-Heinemann, Oxford (2014)
3. Miller, W.G.: Using and Understanding Miniature Neon Lamps. Howard W. Sams & Co. Inc., Indianapolis (1969)
4. Taylor, J., Clarkson, W.: The application of the neon lamp to the comparison of capacities and high resistances. J. Sci. Instrum. **1**(6), 173 (1924)
5. Taylor, J., Clarkson, W., Stephenson, W.: A further study of the comparison of capacities and high resistances by the neon lamp. J. Sci. Instrum. **2**(5), 154 (1925)
6. Boxall, J.: Arduino. Helion, Gliwice, Poland (2014)

Estimation of Uncertainties of Multivariable Indirect Measurements of Two DC Electrical Circuits

Zygmunt L. Warsza[1(✉)] and Jacek Puchalski[2]

[1] Industrial Research Institute of Automation and Measurements (PIAP),
Warsaw, Poland
zlw@op.pl
[2] Central Office of Measures (GUM), Warsaw, Poland
jacek.puchalski@gum.gov.pl

Abstract. Signal processing in a multi-variable indirect measurement system and its uncertainties is considered. It was proposed to extend the vector method of estimating uncertainties, given in Supplement 2 to GUM, to the statistical description of the instrumental systems accuracy in indirect multivariable measurements. Formula for the covariance matrix of relative uncertainties is given. The covariance matrixes for measurement parameters of few DC electrical circuits are found, i.e. for measurement of the star circuit resistances from its terminals, measurement of three resistances with using them for three variants of Wheatstone bridge. Formulas for absolute and relative uncertainties and their correlation coefficients are given. The general conclusion is that for the description accuracy of multivariable measurement instrumental systems relative uncertainties are preferable then absolute ones and uncertainties of parameters of the main measurement functions have be considered.

Keywords: Estimation · Absolute and relative uncertainty ·
Indirect multivariable measurements · Covariance matrix

1 Introduction

Some of physical quantities or parameters characterizing the object under the test have no any technical possibilities for direct measurements. Then indirect methods must be applied, i.e. other quantities should be measured directly and from their results the tested quantities (observables) should be determined. It applies to measure the set of jointed parameters, named the multivariate or vector measurand. In general case of multivariable measurements, the relation between values X of quantities measured in input and data of Y quantities obtained on output after processing is a functional $F(X, Y) = 0$. Usually the function can be formulated as the following equation

$$Y = F(X) \tag{1}$$

The dependence (1) is multivariable and can be linear or non-linear. The general flow chart of multi-variable measurement system is shown in Fig. 1.

© Springer Nature Switzerland AG 2020
R. Szewczyk et al. (Eds.): AUTOMATION 2019, AISC 920, pp. 624–635, 2020.
https://doi.org/10.1007/978-3-030-13273-6_58

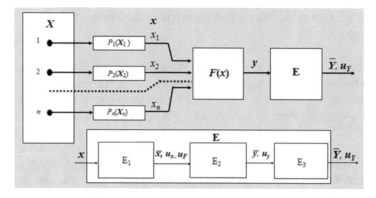

Fig. 1. Signal processing in multi-variable indirect measurement system.

In Fig. 1 and in the text next symbols are used: $x = [x_1, x_2, \ldots x_n], y = [y_1, y_2, \ldots y_m]$ – vectors of input x_i and output y_i signals; $\bar{x}, \bar{y}, u_x, u_{\delta x}, u_y, u_{\delta y}$ – vectors of estimators of values of input x and output y signals and their absolute and relative standard uncertainties; $U_X, U_Y, U_{\delta x}, U_{\delta y}$ – covariance matrixes; $F(x), u_F, u_{\delta F}, U_F, U_{\delta F}$ – function of processing x to y, their absolute and relative uncertainties and its covariance matrixes, E – processing unit of y to obtain $\bar{Y}, u_y, u_{\delta y}$ and $U_Y, U_{\delta Y}$.

The relations of covariance matrixes of absolute [1] and relative uncertainties [6–8] are

$$U_Y = SU_X S^T \tag{2}$$

and

$$U_{\delta Y} = S_\delta U_{\delta X} S_\delta^T \tag{3}$$

where: S, S_δ – matrixes obtained by linearized function $F(x)$ by first partial derivatives of input quantities, defined as

$$S = \begin{bmatrix} \frac{\partial y_1}{\partial x_1} & \cdots & \frac{\partial y_1}{\partial x_n} \\ \cdots & \cdots & \cdots \\ \frac{\partial y_m}{\partial x_1} & \cdots & \frac{\partial y_m}{\partial x_n} \end{bmatrix}, \quad S_\delta = \begin{bmatrix} \frac{x_1}{y_1}\frac{\partial y_1}{\partial x_1} & \cdots & \frac{x_n}{y_1}\frac{\partial y_1}{\partial x_n} \\ \cdots & \cdots & \cdots \\ \frac{x_1}{y_m}\frac{\partial y_m}{\partial x_1} & \cdots & \frac{x_n}{y_m}\frac{\partial y_m}{\partial x_n} \end{bmatrix} \tag{3a, b}$$

The vector $\bar{Y}$ of the estimations of results of indirectly obtained quantities $\bar{y}_i$ (named observables [2, −4]) and covariance matrices $U_Y, U_{\delta Y}$ of their absolute and relative uncertainties $u_Y, u_{\delta y}$ are obtained by direct measurements of n-dimensional signal x, and depends on whether the measurement functional $F(x)$ is linear or non-linear. All, or some results of components of the measurand $\bar{Y}$ can be used further separately or jointly. In the latter case it is necessary to take in considerations the correlation coefficients between the uncertainties of its components y_i, which are in nondiagonal elements of its covariance matrices U_Y or $U_{\delta Y}$.

2   State of Description of the Accuracy of Measurement Systems

In many cases the realization of indirect measurements of m – components of the multivariate measurand y can be made by automatic instrumental measurement systems. When this is not possible, then a collection of individual quantities x is separately measured, and finally from above data both, the vector of estimators $\bar{Y}$ and covariance matrixes $U_{\delta X}$, $U_{\delta Y}$ are calculated. The recommendation of this method of estimation uncertainties for practical applications is described in Supplement 2 to the GUM [1]. This Supplement is designed only for description of indirect multivariate measurements of individual values of tested quantities and use the Eq. (2) for absolute uncertainties. Supplement 2 does not cover situations in instrumental systems, when realization of functional $F(x)$ is not accurate. Such inaccuracy can be due to approximation of transfer functions and limited their frequency ranges, using in signal processing an A/D converters, analogue multipliers, and other functional elements, necessary measurements. Therefore $F(x)$ is also saddled with own uncertainties $u_{\delta F}$[1]. Even in the most precise measurements the rounding of results also becomes essential, including one resulting from the precision of processing in digital circuits [2–4].

The accuracy of each range of multivariable instrumental measurement system is up to now described by the maximal value (worth case) of limited absolute error $|\Delta_{y_i}|_{max}$. The absolute error of any output signal y_i has two components, i.e. $\Delta_{y_i} = \Delta_{y_{i0}} + \Delta_{(y_i - y_{i0})}$ and then the absolute error is

$$|\Delta_{y_i}| \leq |\Delta_{y_{i0}}| + y_i|\varepsilon_{S_i}(y_i)| \qquad \text{for } i = 1, \ldots m \tag{4}$$

where: $\Delta_{y_{i0}}$- absolute error of initial value y_{i0} of the range, $\varepsilon_{S_i} \equiv \Delta(y_i - y_{i0})/y_i$- relative error of the difference $(y_i - y_{i0})$.

If $|\Delta_{y_{i0}}| \ll |\Delta_{(y_i - y_{i0})}|$, then the relative limited error of the element y_i of Y is

$$|\Delta_{y_i}/y_i|_{max} \cong |\varepsilon_{S_i}|_{max} \tag{5}$$

3   What Is Needed

The description of the metrology data of measurement systems needs the accuracy for the whole values of measurement ranges of input and output signals. The probability of existence the maximal limited error in each instrument and in each its range is very low. Then the randomized description by uncertainties seems to be more valuable. It may be made also in the similar two component form as limited errors in the Eq. (4). Then it should contain the expanded absolute uncertainty U_0 of the initial value of each

---

[1] Such problems of measurement technology are included in the **measurement science,** discipline wider then metrology. This concept was proposed by prof. Ludvik Finkelstein from the City University of London during his IMEKO activity from 1970-s years.

range and expanded relative uncertainty U_r of its increase for all values of the range. Both these uncertainties should be given for defined P probability of the confidence level [1], e.g. for $P = 0.95$ is $U \approx 0.95\,u$ and $U_r \approx 0.95u_r$ (u, u_r - standard uncertainties marked as in GUM). In the most cases these components of uncertainty are non-correlated, and very often U_r is constant for the whole range or its function or maximal value can be given. Moreover, the type B components of standard uncertainties are significantly smaller then maximal limited errors because they are estimated as the square root of possible interactions, and not as their sum. The example of the output value standard uncertainty y_i is

$$u_{y_i} = \sqrt{u_{y_i0}^2 + (y_i - y_{i0})^2 u_{ry_i}^2} \tag{6}$$

If $u_{y_i0}^2 \ll (y_i - y_{i0})^2 u_{ry_i}^2$, then the accuracy of y_i is described by the single value of relative uncertainty $\delta_{y_i} \equiv u_{y_i} \cong u_{ry_i}$ unchanged in measuring range, or $\delta_{y_i} \leq u_{ry_i max}$.

Up to date there are no internationally accepted regulations how to statistically describe the accuracy of different kind of instrumental systems for indirect multivariable measurements. It will be clearer on analysis of two simply examples of indirect multivariate measurements given in this paper, i.e. measurement of star circuit resistances from its terminals and indirect measurements of three resistances by three different configurations of Wheatstone bridges made from these resistances and multi-decade laboratory resistor. For the second measurements we proposed to use directly the vector formula (3) between covariance matrixes $U_{\delta X}$ and $U_{\delta Y}$ of relative standard uncertainties for the multiplicative type of measurement equations. Some other conclusions regarding the description of the accuracy of multivariate measuring systems will be also discussed.

Authors developed the vector method for the description of accuracy of multivariate measurements systems with considering uncertainties $u_{\delta F}$ of the functional F parameters. This method is wider then recommendations given in GUM Supplement 2 [1], which do not consider inaccuracy of functional F. Some example is provided below in point 6.

4 Example of the Additive Multivariate Measurement Equations

In many applications of the three-terminal resistance circuits (Fig. 2) the direct connection to any their internal point including the common point 0, is impossible or the star structure is only the equivalent circuit. So, three resistances of this circuit must be determined indirectly from results of the input resistance measurements between pairs of its terminals A, B, C.

Next the values coming from measurements can be transferred to the module of performance **E,** in which finally the values of resistances and their uncertainties are calculated.

In the first step we assume, that the values of resistance of star are determined precisely without any disturbances and modifications by A/D converters and

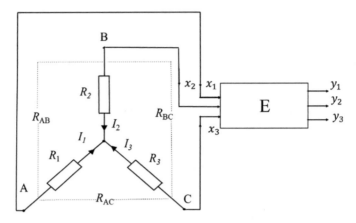

Fig. 2. The diagram of the star circuits with module E of performance measurements.

arithmetical modules located in **E**. In the second step we take into consideration the disturbances of functional **F** in the form of its attenuation during processing signal in channels. The main measurement equations are:

$$R_{AB} = R_1 + R_2, \quad R_{BC} = R_2 + R_3, \quad R_{AC} = R_1 + R_3 \tag{7}$$

or in the matrix form:

$$\begin{bmatrix} R_{AB} \\ R_{BC} \\ R_{AC} \end{bmatrix} = \begin{bmatrix} 1 & 1 & 0 \\ 0 & 1 & 1 \\ 1 & 0 & 1 \end{bmatrix} \begin{bmatrix} R_1 \\ R_2 \\ R_3 \end{bmatrix} \tag{7a}$$

To obtain solutions, the both sides of Eq. (7a) are left side multiplicated by the inverse to above matrix and the main formula (1) is here the matrix form

$$Y = F \cdot X \tag{8}$$

where:

$$X = \begin{bmatrix} x_1 \\ x_2 \\ x_3 \end{bmatrix} = \begin{bmatrix} R_{AB} \\ R_{BC} \\ R_{AC} \end{bmatrix}, \ Y = \begin{bmatrix} y_1 \\ y_2 \\ y_3 \end{bmatrix} = \begin{bmatrix} R_1 \\ R_2 \\ R_3 \end{bmatrix}, \ F = \frac{1}{2} \begin{bmatrix} 1 & -1 & 1 \\ 1 & 1 & -1 \\ -1 & 1 & 1 \end{bmatrix} \tag{8a, b, c}$$

The star circuit resistances are

$$R_1 = \frac{R_{AB}}{2} - \frac{R_{BC}}{2} + \frac{R_{AC}}{2}; \ R_2 = \frac{R_{AB}}{2} + \frac{R_{BC}}{2} - \frac{R_{AC}}{2}; \ R_3 = -\frac{R_{AB}}{2} + \frac{R_{BC}}{2} + \frac{R_{AC}}{2} \tag{9}$$

Then corrections are implemented for known systematic errors. Unknown systematic errors are randomized and estimated as components of the type B uncertainty u_B. Next the results of absolute standard uncertainties σ_{AB}, σ_{BC}, σ_{AC} are find as a square of quadratic values of uncertainties u_A and u_B (type A and B), and relative uncertainties δ_{AB}, δ_{BC}, δ_{AC} should be calculated.

To find the absolute uncertainties and correlation coefficients of output quantities, the vector method given in Supplement 2 to GUM is used [1]. Covariance matrices are related by formula (2), i.e. $U_Y = S \cdot U_X \cdot S^T$. In which: U_Y, U_Y – covariance matrixes of output vector Y and input vector X, S - the Jacobian matrix of uncertainty sensitive coefficients. For the resistances of stair circuit:

$$S = F = \begin{bmatrix} \partial R_1/\partial R_{AB} & \partial R_1/\partial R_{BC} & \partial R_1/\partial R_{AC} \\ \partial R_2/\partial R_{AB} & \partial R_2/\partial R_{BC} & \partial R_2/\partial R_{AC} \\ \partial R_3/\partial R_{AB} & \partial R_3/\partial R_{BC} & \partial R_3/\partial R_{AC} \end{bmatrix} = \frac{1}{2} \begin{bmatrix} 1 & -1 & 1 \\ 1 & 1 & -1 \\ -1 & 1 & 1 \end{bmatrix} \quad (10)$$

- **correlated variables in the input**

Let us consider the general case when absolute uncertainties of input quantities σ_{AB}, σ_{BC}, σ_{AC} are correlated. Then in the covariance matrix U_X of input quantities in such case the non-zero elements in non-diagonal positions are appearing. They are defined with correlation coefficients ρ_{AB}, ρ_{BC}, ρ_{AC}

$$U_X = \begin{bmatrix} \sigma_{AB}^2 & \rho_{AB}\sigma_{AB}\sigma_{BC} & \rho_{BC}\sigma_{AB}\sigma_{AC} \\ \rho_{AB}\sigma_{AB}\sigma_{BC} & \sigma_{BC}^2 & \rho_{AC}\sigma_{BC}\sigma_{AC} \\ \rho_{BC}\sigma_{AB}\sigma_{AC} & \rho_{AC}\sigma_{BC}\sigma_{AC} & \sigma_{AC}^2 \end{bmatrix},$$

$$U_Y = \begin{bmatrix} \sigma_{R1}^2 & \rho_{R1,2}\sigma_1\sigma_2 & \rho_{R1,3}\sigma_1\sigma_3 \\ \rho_{R1,2}\sigma_1\sigma_2 & \sigma_{R2}^2 & \rho_{R2,3}\sigma_2\sigma_3 \\ \rho_{R1,3}\sigma_1\sigma_3 & \rho_{R2,3}\sigma_2\sigma_3 & \sigma_{R3}^2 \end{bmatrix}$$

$$(11a, b)$$

If the relative uncertainties of measured resistances on stair circuit terminals are the same, i.e.: $\delta_{AB} = \delta_{BC} = \delta_{AC} = \delta$, then absolute uncertainties of input quantities are: $\sigma_{AB} = \delta \cdot R_{AB}$, $\sigma_{AC} = \delta \cdot R_{AC}$ and $\sigma_{BC} = \delta \cdot R_{BC}$. Then output absolute uncertainties are:

$$\sigma_{R1} = \frac{\delta}{2} \sqrt{R_{AB}^2 + R_{BC}^2 + R_{AC}^2 + 2(\rho_{BC}R_{AB}R_{AC} - \rho_{AB}R_{AB}R_{BC} - \rho_{AC}R_{BC}R_{AC})}$$

$$\sigma_{R2} = \frac{\delta}{2} \sqrt{R_{AB}^2 + R_{BC}^2 + R_{AC}^2 + 2(\rho_{AB}R_{AB}R_{BC} - \rho_{BC}R_{AB}R_{AC} - \rho_{AC}R_{BC}R_{AC})}$$

$$\sigma_{R3} = \frac{\delta}{2} \sqrt{R_{AB}^2 + R_{BC}^2 + R_{AC}^2 + 2(\rho_{AC}R_{BC}R_{AC} - \rho_{BC}R_{AB}R_{AC} - \rho_{AB}R_{AB}R_{BC})}$$

$$(12a, b, c)$$

and the output relative uncertainties:

$$\delta_{R1} = \frac{\delta\sqrt{1+\beta^2+\gamma^2+2(\rho_{BC}\gamma-\rho_{AB}\beta-\rho_{AC}\beta\gamma)}}{1-\beta+\gamma}$$

$$\delta_{R2} = \frac{\delta\sqrt{1+\beta^2+\gamma^2+2(\rho_{AB}\beta-\rho_{BC}\gamma-\rho_{AC}\beta\gamma)}}{1+\beta-\gamma} \qquad (13a, b, c)$$

$$\delta_{R3} = \frac{\delta\sqrt{1+\beta^2+\gamma^2+2(\rho_{AC}\beta\gamma-\rho_{BC}\gamma-\rho_{AB}\beta)}}{\beta+\gamma-1}$$

where: $\beta = \frac{R_{BC}}{R_{AB}}$ and $\gamma = \frac{R_{AC}}{R_{AB}}$.

The correlations coefficients of output quantities are defined as follows

$$\rho_{R1,2} = \frac{\delta^2}{4}\frac{R_{AB}^2 - R_{BC}^2 - R_{AC}^2 + 2\rho_{AC}R_{BC}R_{AC}}{\sigma_1\sigma_2},$$

$$\rho_{R1,3} = \frac{\delta^2}{4}\frac{R_{AC}^2 - R_{BC}^2 - R_{AB}^2 + 2\rho_{AB}R_{BC}R_{AC}}{\sigma_1\sigma_3}, \qquad (14a, b, c)$$

$$\rho_{R2,3} = \frac{\delta^2}{4}\frac{R_{BC}^2 - R_{AC}^2 - R_{AB}^2 + 2\rho_{BC}R_{AB}R_{AC}}{\sigma_2\sigma_3}.$$

If $R_{AB} = R_{BC} = R_{AC} = R$, the uncertainties are

$$\sigma_{R1} = \frac{\delta R}{2}\sqrt{3+2(\rho_{BC}-\rho_{AB}-\rho_{AC})}, \qquad \sigma_{R2} = \frac{\delta R}{2}\sqrt{3+2(\rho_{AB}-\rho_{BC}-\rho_{AC})}$$

$$\sigma_{R3} = \frac{\delta R}{2}\sqrt{3+2(\rho_{AC}-\rho_{AB}-\rho_{BC})}$$

$$(15a, b, c)$$

- **non-correlated variables in the input**

For non-correlated variables $\rho_{AB} = \rho_{BC} = \rho_{AC} = 0$, and from (12a, b, c) the absolute uncertainties are

$$\sigma_{R_1} = \sigma_{R_2} = \sigma_{R_3} = \frac{1}{2}\sqrt{\sigma_{R_{AB}}^2 + \sigma_{R_{AC}}^2 + \sigma_{R_{BC}}^2} \qquad (16)$$

And from (13a, b, c) correlation coefficients

$$\rho_{R1,2} = \frac{\sigma_{AB}^2-\sigma_{BC}^2-\sigma_{AC}^2}{\sigma_{AB}^2+\sigma_{BC}^2+\sigma_{AC}^2}, \qquad \rho_{R1,3} = \frac{-\sigma_{AB}^2-\sigma_{BC}^2+\sigma_{AC}^2}{\sigma_{AB}^2+\sigma_{BC}^2+\sigma_{AC}^2},$$

$$\rho_{R2,3} = \frac{-\sigma_{AB}^2+\sigma_{BC}^2-\sigma_{AC}^2}{\sigma_{AB}^2+\sigma_{BC}^2+\sigma_{AC}^2} \qquad (17a, b, c)$$

If $\quad \sigma_{AB} = \sigma_{BC} = \sigma_{AC} = \sigma$ then $\quad \sigma_{R1} = \sigma_{R2} = \sigma_{R3} = \frac{\sqrt{3}}{2}\sigma, \quad \rho_{R1,2} = \rho_{R1,3} = \rho_{R2,3} = -\frac{1}{3}.$

The determinant of matrix U_y is

$$w = 1 - \rho_{R1,2}^2 - \rho_{R1,3}^2 - \rho_{R2,3}^2 + 2 \cdot \rho_{R1,2} \cdot \rho_{R1,3} \cdot \rho_{R2,3} > 0 \qquad (18)$$

and this parameter has always the positive sign [9].

Defining $\varkappa = \frac{\sigma_{BC}^2 + \sigma_{AC}^2}{\sigma_{AB}^2} > 0$, $v = \frac{\sigma_{BC}^2 - \sigma_{AC}^2}{\sigma_{AB}^2}$ we express correlation coefficients by

$$\rho_{R1,2} = \frac{1 - \varkappa}{1 + \varkappa}; \qquad \rho_{R1,3} = -\frac{1 + v}{1 + \varkappa}; \qquad \rho_{R2,3} = \frac{v - 1}{1 + \varkappa} \qquad (19a, b, c)$$

so, the parameter $w = 4\frac{\varkappa^2 - v^2}{(1+\varkappa)^3}$ must be $\varkappa^2 > v^2$, what is always fulfilled. That is why the characteristic equation of inverse matrix has three positive roots.

The border of cover region for values of results with given probability $P \leq 0.95$ is ellipsoid, closed in solid cubic, and contiguous in six points the wall of cubic with edge distance

$$k_p \sqrt{\sigma_{AB}^2 + \sigma_{BC}^2 + \sigma_{AC}^2} \qquad (20)$$

$k_p = 2.8$ – cover factor/extension coefficients.

5   Summary of Solutions of Some Cases

- if $\sigma_{AB} = \sigma_{BC} = \sigma_{AC} = \sigma_{in}$, $\rho_{AB} = \rho_{BC} = \rho_{AC} = \rho_{in}$;
  $\sigma_{out} = \sigma_{R1} = \sigma_{R2} == \sigma_{R3} = \frac{1}{2}\sigma_{in}\sqrt{3 - 2\rho_{in}}$;
  $\rho_{R1,2} = \rho_{R1,3} = \rho_{R2,3} = \frac{2\rho_{in} - 1}{3 - 2\rho_{in}} \leq 0$ for $\rho_{in} \leq 1/2$.
- if $\rho_{in} = 0$: $\sigma_{out} = \sqrt{3}/2\sigma_{in}$, $\rho_{R1,2} = \rho_{R1,3} = \rho_{R2,3} = -\frac{1}{3}$;
  half axes: $1,4\,\sigma_{in}, 2,8\sigma_{in}, 2,8\sigma_{in}$
- if $\rho_{in} = 1$: min $\sigma_{out} = \frac{1}{2}\sigma_{in}$, $\rho_{R1,2} = \rho_{R1,3} = \rho_{R2,3} = 0$; radius $1,4\,\sigma_{in}$
- if $\rho_{in} = -1$: max $\sigma_{out} = \frac{\sqrt{5}}{2}\sigma_{in}$, $\rho_{R1,2} = \rho_{R1,3} = \rho_{R2,3} = -\frac{3}{5}$; $w < 0$.

6   Influence of Uncertainties u_F of Performance Matrix F

In the instrumental system for measurements the stair circuit resistances processing of output values and their uncertainties is made in digital unit **E**. The main matrix equation $Y = F \cdot X$ was given in (8) and (8a, b,c). Solution of vector Y elements is given in (9). Let us now consider uncertainties of amplification/attenuation of signals in measurement channels. The realization of signals processing has linear disturbances in channels changing levels of signals, e.g.:

$$X_S = \begin{bmatrix} k_1 x_1 \\ k_2 x_2 \\ k_3 x_3 \end{bmatrix} \tag{21}$$

where k_1, k_2, k_3 is amplifying coefficients.

The analog/digital processing input signals has uncertainty u_F. Therefore, the functional matrix F is must be modified, and a new matrix is defined as follows:

$$F_S = \frac{1}{2} \begin{bmatrix} k_1(1+\delta_1) & -k_2(1+\delta_1) & k_3(1+\delta_1) \\ k_1(1+\delta_2) & k_2(1+\delta_2) & -k_3(1+\delta_2) \\ -k_1(1+\delta_3) & k_2(1+\delta_3) & k_3(1+\delta_3) \end{bmatrix} \tag{22}$$

where: $\delta_1, \delta_2, \delta_3$ - coefficients dedicated to the components of output quantities.

The vector function of output quantities has additional uncertainties associated with zero set errors $\left(\frac{\Delta_{10}}{1+\delta_1} ; \frac{\Delta_{20}}{1+\delta_2} ; \frac{\Delta_{30}}{1+\delta_3} \right)$ and output components now are:

$$\begin{aligned} y_1 &= (1+\delta_1)\left(\frac{k_1 x_1 - k_2 x_2 + k_3 x_3}{2} + \frac{\Delta_{10}}{1+\delta_1} \right) \\ y_2 &= (1+\delta_2)\left(\frac{k_1 x_1 + k_2 x_2 - k_3 x_3}{2} + \frac{\Delta_{20}}{1+\delta_2} \right) \\ y_3 &= (1+\delta_3)\left(\frac{-k_1 x_1 + k_2 x_2 + k_3 x_3}{2} + \frac{\Delta_{30}}{1+\delta_3} \right) \end{aligned} \tag{23a, b, c}$$

Using formulas which are derived for absolute uncertainties of star circuit and modifying it, the absolute uncertainties after modification with zero set errors are more complicated as:

$$\sigma_{R_1} = (1+\delta_1)\sqrt{k_1^2\sigma_{x_1}^2 + k_2^2\sigma_{x_2}^2 + k_3^2\sigma_{x_3}^2 + 2\left(k_1 k_3 \rho_{x_1 x_3}\sigma_{x_1}\sigma_{x_3} - k_1 k_2 \rho_{x_1 x_2}\sigma_{x_1}\sigma_{x_2} - k_2 k_3 \rho_{x_2 x_3}\sigma_{x_2}\sigma_{x_3}\right) + \sigma^2\left(\frac{\Delta_{10}}{1+\delta_1}\right)}$$

$$\sigma_{R_2} = (1+\delta_2)\sqrt{k_1^2\sigma_{x_1}^2 + k_2^2\sigma_{x_2}^2 + k_3^2\sigma_{x_3}^2 + 2\left(k_1 k_2 \rho_{x_1 x_2}\sigma_{x_1}\sigma_{x_2} - k_1 k_3 \rho_{x_1 x_3}\sigma_{x_1}\sigma_{x_3} - k_2 k_3 \rho_{x_2 x_3}\sigma_{x_2}\sigma_{x_3}\right) + \sigma^2\left(\frac{\Delta_{20}}{1+\delta_2}\right)}$$

$$\sigma_{R_3} = (1+\delta_3)\sqrt{k_1^2\sigma_{x_1}^2 + k_2^2\sigma_{x_2}^2 + k_3^2\sigma_{x_3}^2 + 2\left(k_2 k_3 \rho_{x_2 x_3}\sigma_{x_2}\sigma_{x_3} - k_1 k_2 \rho_{x_1 x_2}\sigma_{x_1}\sigma_{x_2} - k_1 k_3 \rho_{x_1 x_3}\sigma_{x_1}\sigma_{x_3}\right) + \sigma^2\left(\frac{\Delta_{30}}{1+\delta_3}\right)}$$

$$\tag{24a, b, c}$$

7 Case of Multiplicative Measurement Equations

Values of resistances R_2, R_3, R_4 of three resistors can be determined in indirect measurements with required accuracy without use the professional DC or AC bridge or digital ohmmeter. For such measurements the own circuit of Wheatstone bridge from the common regulated multi-decade resistor R_1 and three measured resistors can be three times built and used. Measured resistances can be connected in three different orders in arms of the bridge circuit loop, i.e.: R_2, R_3, R_4 (Fig. 3), R_2, R_4, R_3 and R_3, R_2, R_4.

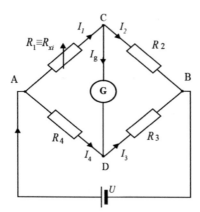

Fig. 3. The first structure from three variants of Wheatstone DC bridge connected from measured resistances R_2, R_3, R_4 and regulated multidecade resistance R_1.

Three settings R_{x1}, R_{x2}, R_{x3} of R_1, which satisfied balances of these bridge variants, are:

$$R_{x1} = R_2 \frac{R_4}{R_3}, \qquad R_{x2} = R_2 \frac{R_3}{R_4}, \qquad R_{x3} = R_3 \frac{R_4}{R_2} \qquad (25a,b,c)$$

From above relations the unknown resistances as elements of the output vector Y can be calculated

$$R_2 = \sqrt{R_{x1}R_{x2}}, \; R_3 = \sqrt{R_{x2}R_{x3}}, \; R_4 = \sqrt{R_{x1}R_{x3}} \qquad (26a,b,c)$$

As solutions (26a, b, c) are of the multiplicative type the Eq. (3) for direct calculating the relative uncertainties can be used. The measurement sensitivity function $\mathbf{S}_\delta$ for relative uncertainties is

$$\mathbf{S}_\delta = \begin{bmatrix} \frac{R_{x1}}{R_2}\frac{\partial R_2}{\partial R_{x1}} & \frac{R_{x2}}{R_2}\frac{\partial R_2}{\partial R_{x2}} & \frac{R_{x3}}{R_2}\frac{\partial R_2}{\partial R_{x3}} \\ \frac{R_{x1}}{R_3}\frac{\partial R_3}{\partial R_{x1}} & \frac{R_{x2}}{R_3}\frac{\partial R_3}{\partial R_{x2}} & \frac{R_{x3}}{R_3}\frac{\partial R_3}{\partial R_{x3}} \\ \frac{R_{x1}}{R_4}\frac{\partial R_4}{\partial R_{x1}} & \frac{R_{x2}}{R_4}\frac{\partial R_4}{\partial R_{x2}} & \frac{R_{x3}}{R_4}\frac{\partial R_4}{\partial R_{x3}} \end{bmatrix} = \begin{bmatrix} 1/2 & 1/2 & 0 \\ 0 & 1/2 & 1/2 \\ 1/2 & 0 & 1/2 \end{bmatrix} = \frac{1}{2}\begin{bmatrix} 1 & 1 & 0 \\ 0 & 1 & 1 \\ 1 & 0 & 1 \end{bmatrix} \qquad (27)$$

For estimation uncertainty of measured resistances, we assumed firstly that input variables R_{x1}, R_{x2}, R_{x3} are not correlated. and using formula (3) relative uncertainties, i.e. $U_{\delta y} = S_\delta U_{\delta x} S_\delta^T$ is:

$$
\begin{aligned}
\mathbf{u}_{\delta Y} &= \frac{1}{2}
\begin{bmatrix} 1 & 1 & 0 \\ 0 & 1 & 1 \\ 1 & 0 & 1 \end{bmatrix}
\begin{bmatrix} \delta_{R_{x1}}^2 & 0 & 0 \\ 0 & \delta_{R_{x1}}^2 & 0 \\ 0 & 0 & \delta_{R_{x1}}^2 \end{bmatrix}
\frac{1}{2}
\begin{bmatrix} 1 & 0 & 1 \\ 1 & 1 & 0 \\ 0 & 1 & 1 \end{bmatrix} \\
&= \frac{1}{4}
\begin{bmatrix}
\delta_{R_{x1}}^2 + \delta_{R_{x2}}^2 & \delta_{R_{x2}}^2 & \delta_{R_{x1}}^2 \\
\delta_{R_{x2}}^2 & \delta_{R_{x2}}^2 + \delta_{R_{x3}}^2 & \delta_{R_{x3}}^2 \\
\delta_{R_{x1}}^2 & \delta_{R_{x3}}^2 & \delta_{R_{x1}}^2 + \delta_{R_{x3}}^2
\end{bmatrix}
\end{aligned}
\tag{28}
$$

So, the standard relative uncertainties of output quantities are defined:

$$
\delta_{R_2} = \frac{1}{2}\sqrt{\delta_{R_{x1}}^2 + \delta_{R_{x2}}^2}, \ \delta_{R_3} = \frac{1}{2}\sqrt{\delta_{R_{x2}}^2 + \delta_{R_{x3}}^2}, \ \delta_{R_3} = \frac{1}{2}\sqrt{\delta_{R_{x1}}^2 + \delta_{R_{x3}}^2}
\tag{29a, b, c}
$$

and correlations coefficients:

$$
\rho_{R_2 R_3} = \frac{\frac{1}{4}\delta_{R_{x2}}^2}{\frac{1}{2}\sqrt{\delta_{R_{x1}}^2 + \delta_{R_{x2}}^2}\,\frac{1}{2}\sqrt{\delta_{R_{x2}}^2 + \delta_{R_{x3}}^2}} = \frac{1}{\sqrt{1 + \left(\frac{\delta_{R_{x1}}}{\delta_{R_{x2}}}\right)^2}\sqrt{1 + \left(\frac{\delta_{R_{x3}}}{\delta_{R_{x2}}}\right)^2}} > 0
$$

$$
\rho_{R_2 R_4} = \frac{\frac{1}{4}\delta_{R_{x1}}^2}{\frac{1}{2}\sqrt{\delta_{R_{x1}}^2 + \delta_{R_{x2}}^2}\,\frac{1}{2}\sqrt{\delta_{R_{x1}}^2 + \delta_{R_{x3}}^2}} = \frac{1}{\sqrt{1 + \left(\frac{\delta_{R_{x2}}}{\delta_{R_{x1}}}\right)^2}\sqrt{1 + \left(\frac{\delta_{R_{x3}}}{\delta_{R_{x1}}}\right)^2}} > 0
$$

$$
\rho_{R_3 R_4} = \frac{\frac{1}{4}\delta_{R_{x3}}^2}{\frac{1}{2}\sqrt{\delta_{R_{x2}}^2 + \delta_{R_{x3}}^2}\,\frac{1}{2}\sqrt{\delta_{R_{x1}}^2 + \delta_{R_{x3}}^2}} = \frac{1}{\sqrt{1 + \left(\frac{\delta_{R_{x2}}}{\delta_{R_{x3}}}\right)^2}\sqrt{1 + \left(\frac{\delta_{R_{x1}}}{\delta_{R_{x3}}}\right)^2}} > 0
$$

$$
\tag{30a, b, c}
$$

All above correlations coefficients are positive.

If $\delta_{R_{x1}} = \delta_{R_{x2}} = \delta_{R_{x3}} = \delta$, $\delta_{R_2} = \delta_{R_3} = \delta_{R_4} = \frac{\sqrt{2}}{2}\delta$, $\rho_{R_2 R_3} = \rho_{R_2 R_4} = \rho_{R_3 R_4} = \frac{1}{2}$, then the coverage region is ellipsoid of parameter $w = 1 - \frac{1}{2} - \frac{1}{2} - \frac{1}{2} + 2 \cdot \frac{1}{2} \cdot \frac{1}{2} \cdot \frac{1}{2} = \frac{1}{2} > 0$.

In this case the ellipsoidal coverage region for relative uncertainties of Y with probability 0.95 determines the ellipsoid with half axis $a = 2.8\delta$, $b = 1.4\delta$, $c = 1.4\delta$. In formulas it is used that the coverage factor/extension/coefficient for 95% of coverage region in 3D (three-dimensional) Gauss distribution is equal $k_p = 2.8$. This ellipsoid is contiguous in six points to the cube with edges $d = 2 \cdot 2.8 \frac{\sqrt{2}}{2}\delta = 3.96\,\delta$. The relations between capacity of ellipsoid and cube is $4\pi\,abc/(3d^3) = 37\%$.

8 Summary and Conclusion

Two examples of determining uncertainties of indirect multi-parameter measurements in case of linear and nonlinear multiplicative formulas have been presented, i.e. for measurements of three resistances connected for three variants of Wheatstone bridge, and measurements of the resistance star circuit. Proposed is also using the covariance matrix of relative uncertainties and corresponding formulas dedicated for relative uncertainties.

It is shown that in the case when two or more parameters/for example of elements of electrical circuits/must be measured together, i.e. without disconnection from the circuit, then the uncertainties of above parameters are correlated. So, if the above correlated elements are used in the circuit of other devices then in the determination of uncertainties of such new device, we should consider the corresponding correlations coefficients obtained from first measurements.

More of general information on uncertainty estimations of indirect multivariable measurements is given in [1–5] and about some electrical multivariate systems - in other of authors papers [6–9].

References

1. JCGM 102: 2011 Evaluation of measurement data – Supplement 2 to the Guide to the expression of uncertainty in measurement– Extension to any number of output quantities
2. Warsza, Z.L.: Evaluation and numerical presentation of the results of indirect multivariate measurements. In: Pavese, F., et al. (ed.) Advanced Mathematical and Computational Tools in Metrology and Testing IX. Advances in Mathematics for Applied Sciences, vol. 84, pp. 418–425. World Scientific Books (2012)
3. Warsza, Z.L.: About evaluation of multivariate measurements results. J. Autom. Mobile Robot. Intell. Syst. JAMRIS 6(4), 27–32 (2012)
4. Ramos, P.M., Janeiro, F.M., Girao, P.S.: Uncertainty evaluation of multivariate quantities: a case study on electrical impedance. Measurement 78, 397–411 (2016)
5. Hall, B.D.: Evaluating the measurement uncertainty of complex quantities: a selective review. Metrologia 53, S25–S31 (2016)
6. Warsza, Z.L., Puchalski, J.: Estimation of uncertainty of indirect measurement in multi-parametric systems with few examples. In: CD Proceedings of conference Problems and Progress of Metrology, PPM 2018, Szczyrk 04–06 June 2018, Series: Conferences No. 22. Katowice Branch of Polish Academy of Science (2018)
7. Warsza, Z.L., Puchalski, J.: Matrix estimation of uncertainty of indirect multi-parameter measurements with examples. Pomiary Automatyka Robotyka (PAR) 2, 31–40 (2018). (in Polish)
8. Warsza, Z.L., Puchalski, J.: Evaluation of the impedance uncertainties determined indirectly in polar or rectangular components. Pomiary Automatyka Robotyka (PAR) 3, 61–67 (2018). (in Polish)
9. Puchalski, J., Fotowicz, P.: Uncertainty propagation in determining of coverage region for three-dimension vector measurand. In: CD Proceedings of Symposium PPM 2018, Szczyrk, 04–06 June 2018, Series: Conference No. 22. Katowice Branch of Polish Academy of Science (2018). ISBN 978-83-7880-541-0 (in Polish)

Estimation of Linear Regression Parameters of Symmetric Non-Gaussian Errors by Polynomial Maximization Method

Serhii W. Zabolotnii[1], Zygmunt L. Warsza[2(✉)], and Oleksandr Tkachenko[1]

[1] Cherkasy State Technological University, Cherkasy, Ukraine
s.zabolotnii@chdtu.edu.ua
[2] Industrial Research Institute for Automation and Measurements PIAP,
Al. Jerozolimskie 202, 02-486 Warsaw, Poland
zlw1936@gmail.com

Abstract. In this paper, a new way of estimation of single-factor linear regression parameters of symmetrically distributed non-Gaussian errors is proposed. This new approach is based on the Polynomial Maximization Method (PMM) and uses the description of random variables by higher order statistics (moments and cumulants). Analytic expressions that allow to find estimates and analyze their asymptotic accuracy are obtained for the degree of polynomial $S = 3$. It is shown that the variance of polynomial estimates can be less than the variance of estimates of the ordinary least squares' method. The increase of accuracy depends on the values of cumulant coefficients of higher order of the random regression errors. The statistical modeling of the Monte Carlo method has been performed. The results confirm the effectiveness of the proposed approach.

Keywords: Linear regression · Symmetrical distribution of errors · Stochastic polynomial · Variance · High order statistics · Cumulants

1 Introduction

A linear one-factor relationship between two variables is very simple and is the most common type of regression models. In addition, many non-linear dependencies (power, exponential, logarithmic, etc.) can also be reduced to a linear model by corresponding transformations. The consequence of this is the widespread use of linear regression in technical, ecological, medical economic, and other applications.

The most common way of finding estimates of the parameters of linear regression is to use the method OLS (Ordinary Least Squares) [1, 12]. The OLS-estimations are linear, unmixed and have minimal variance, provided that the errors are homogeneous, uncorrelated and normally distributed. However, many researchers note that quite often the normal (Gaussian) law is not the only possible probabilistic model of regression errors. In earlier publications it was shown that regression errors can have larger tails [1, 2] in comparison with the Gaussian distribution or may be limited [3].

© Springer Nature Switzerland AG 2020
R. Szewczyk et al. (Eds.): AUTOMATION 2019, AISC 920, pp. 636–649, 2020.
https://doi.org/10.1007/978-3-030-13273-6_59

For the non-Gaussian nature of statistical data, an increase in accuracy can be achieved using a parametric approach based on the Maximum Likelihood Estimator (MLE) method. The use of MLE requires an adequate description of error models based on probability density (PDF). To solve the problems of regression analysis, many different types of symmetric PDFs are used: elliptic laws (Logistic, Cauchy, Student t) [4–6], the family of Exponential Power Family (EPF) exponential laws [7] and mixtures of Gaussian distributions [8, 9]. In addition, models of symmetric distributions, that allow changing the magnitude of the kurtosis coefficient and the severity of the tails of regression errors are specially developed [10–12].

From a computational point of view, the parametric approach is characterized by a significantly greater complexity (relative to the least-squares method), as well as a significant increase in the volume of a priori information. This is due to the necessity of preliminary specification (selection) of the probability distribution law for the error model and evaluation of non-informative parameters. Therefore, in practice, non-parametric methods as simpler from the implementation point of view are often used. These may be robust versions of the least-squares method [13], whose application is primarily aimed at ensuring stability against the influence of extreme deviations (emissions) [2], estimating the Least Absolute Deviation (LAD) method [14], quantile [15] and signed [16] methods of estimation.

A compromise from the point of view of the complexity and completeness of the probabilistic description of non-Gaussian random variables is the approach based on Higher-Order Statistics (HOS). Examples of its use for solving regression analysis problems are given in the papers [17–19].

2   The Aim of Research

This paper is a direct continuation of the paper [20], where the application of the Polynomial Maximization Method (PMM) [21] for the solution of the regression analysis (estimation of linear regression parameters for asymmetrically distributed errors) was firstly considered. Conceptually PMM is nearly like MLE, since it also uses the principle of maximizing statistics from sample data near the true value of the estimated parameter. However, for the formation of such statistics, descriptions of random variables by PDFs are not used, but by higher-order statisticians, for example, by moments or cumulants. We note the functionality of PMM, which can be used not only to find estimates of the scalar parameter [22–24], but also for finding the change-point problem of the properties of a random sequence in a posteriori formulation of the problem [25], as well as used in joint detection signals against the background of non-Gaussian noise [26]. The totality of the results obtained in these papers shows that for non-Gaussian statistics the PMM-estimations can be much more effective (have a smaller variance) compared with linear estimates.

The purpose of this study is to synthesize algorithms for finding adaptive PMM-estimators, and to estimate their accuracy by statistical Monte Carlo simulation using the model of a single-factor linear regression with symmetrically distributed non-Gaussian errors.

3 Mathematical Formulation of the Problem

Let's have a one-factor regression model of observations describing the values dependence of the target variable y on its predictor x:

$$y_v = f(x_v) + \xi_v, v = \overrightarrow{1, N}, \tag{1}$$

where $f(x_v) = a_0 + a_1 x_v$ - determined linear dependence component and ξ_v - random component (error) of the model ($E\{\xi\} = 0$).

Regression errors are a sequence of independent and identically distributed random variables that have a symmetric distribution, which is different from the Gaussian law. Probabilistic properties of errors can be described with the help of higher-order statistics (there are moments μ_r and cumulants κ_r, $r = \overrightarrow{2, 6}$), the values of which are a priori unknown. The problem consists in finding estimates of the vector parameter $\theta = \{a_0, a_1\}$ on the basis of a statistical analysis of the set of points (x_v, y_v), $v = \overrightarrow{1, N}$.

4 Estimation of the Vector Parameter by the Polynomial Maximization Method

To solve the problem, we use the variant of the polynomial maximization method (PMM), which is developed for the case of estimating the vector parameter for unequally distributed statistical data [20, 21]. According to the PMM, the estimate of a vector parameter θ of dimension Q can be found as the solution of Q stochastic power equation system:

$$\sum_{v=1}^{N} \sum_{i=1}^{S} k_{i,v}^{(p)} \left[(y_v)^i - \alpha_{i,v} \right] \Bigg|_{\theta_p = \hat{\theta}_p} = 0, p = \overrightarrow{0, Q - 1} \tag{2}$$

where: S – is the order of the polynomial used for parameter estimation, $\alpha_{i,v} = E\{(y_v)^i\}$ – are theoretical initial moments of the i-th order from a sequence y_v.

Coefficients $k_{i,v}^{(p)}$ (for each component of vector parameter $p = \overrightarrow{0, Q - 1}$) can be found by solving the system of S linear algebraic equations, given by conditions of minimization of variance (with the appropriate order S) of the estimate of the parameter θ, namely:

$$\sum_{i=1}^{S} k_{i,v}^{(p)} F_{(i,j)v} = \frac{\partial}{\partial \theta_p} \alpha_{j,v}, j = \overrightarrow{1, S}, p = \overrightarrow{0, Q - 1}, \tag{3}$$

where $F_{(i,j)v} = \alpha_{(i+j),v} - \alpha_{i,v} \alpha_{j,v}$.

Systems of Eq. (3) can be solved analytically using the Kramer method.

It was shown that PMM-estimations, which are the solutions of system of stochastic equations of the form (2), are consistent and asymptotically unbiased. To calculate the variance of parameter estimates is necessary to find the volume of extracted information on the estimated parameters θ, which generally are described by the equation:

$$J_{SN}^{(p,q)} = \sum_{v=1}^{N}\sum_{i=1}^{S}\sum_{j=1}^{S} k_{i,v}^{(p)} k_{j,v}^{(q)} F_{(i,j)v} = \sum_{v=1}^{N}\sum_{i=1}^{S} k_{i,v}^{(p)} \frac{\partial}{\partial\theta_q}\alpha_{i,v}, \quad p,q = \overrightarrow{0,Q-1}. \quad (4)$$

The statistical sense of function $J_{SN}^{(p,q)}$ is like the classical Fisher concept of information quantity. The asymptotic values (for $N \to \infty$) of the variances $\sigma_{(\theta_p)S}^2$ of PMM-estimates - components the vector parameter θ lie on the main diagonal of the variational matrix of estimates:

$$V_{(S)} = \left[J_{(S)} \right]^{-1}. \quad (5)$$

Obtained by inversion of the quadratic (dimension Q) matrix of the amount of information retrieved $J_{(S)}$, consisting of elements $J_{SN}^{(p,q)}$ like in (4).

5 Polynomial Estimation of Linear Regression Parameters

It was shown in [20] that when the degree of the polynomial of the PMM-estimator of the vector parameter $\theta = \{a_0, a_1\}$ of the linear regression model (1) is used, they completely coincide with linear OLS estimates. Such estimates are optimal (by the criterion of minimum variance) in a situation where the errors of the regression model have a Gaussian distribution. In addition, it was shown in [20–23] that when the statistical data are symmetrically distributed (which corresponds to zero values of unpaired order cumulant coefficients), PMM-estimator with the degree of polynomial $S = 2$ degenerate into linear estimates obtained for $S = 1$. Therefore, the case of finding PMM-estimators, which is based on the use of polynomials of degree, it is considered below. When PMM-estimates of two components from the sought vector parameter are found from the solution of a system of two equations formed on the general formula (2):

$$\sum_{v=1}^{N} \left\{ k_{1,v}^{(p)}[y_v - f(x_v)] + k_{2,v}^{(p)}\left[(y_v)^2 - \left([f(x_v)]^2 + \mu_2 \right) \right] \right.$$
$$\left. + k_{3,v}^{(p)}\left[(y_v)^3 - \left([f(x_v)]^3 + 3f(x_v)\mu_2 \right) \right] \right\} = 0, \quad p = \overrightarrow{0,1} \quad (6)$$

where the optimal coefficients $k_{i,v}^{(p)}$, $i = \overrightarrow{1,3}$ ensure the minimization of the variance of the estimates (for the corresponding degree of the polynomial). They can be found by solving systems of linear equations of the form (3) described by expressions:

$$k_{1,v}^{(p)} = \frac{1}{\Delta_3}\left[3[f(x_v)]^2(\mu_4 - 3\mu_2^2) + 3\mu_4\mu_2 - \mu_6 \right]\frac{\partial}{\partial a_p}f(x_v),$$
$$k_{2,v}^{(p)} = \frac{-3}{\Delta_3}f(x_v)[\mu_4 - 3\mu_2]\frac{\partial}{\partial a_p}f(x_v), \quad k_{3,v}^{(p)} = \frac{1}{\Delta_3}[\mu_4 - 3\mu_2]\frac{\partial}{\partial a_p}f(x_v), \quad p = \overrightarrow{0,1}, \quad (7)$$

where $\Delta_3 = \mu_2^{-2}(\mu_4^2 - \mu_2\mu_6)$.

Substituting the coefficients (7) in (6), after some transformations, the system of equations for finding the estimates of the components of the parameter can be represented in the form:

$$\sum_{v=1}^{N} (x_v)^{p-1} \left[A(a_0 + a_1 x_v)^3 + B(a_0 + a_1 x_v)^2 + C(a_0 + a_1 x_v) + D \right] = 0, \, p = \overrightarrow{0,1} \quad (8)$$

where $A = 1$, $B = -3\hat{\alpha}_1$, $C = 3\hat{\alpha}_2 - \frac{\mu_6 - 3\mu_4\mu_2}{\mu_4 - 3\mu_2^2}$, $D = \hat{\alpha}_1 \frac{\mu_6 - 3\mu_4\mu_2}{\mu_4 - 3\mu_2^2} - \hat{\alpha}_3$.

Note that in (8) the statistics $\hat{\alpha}_i = \frac{1}{N} \sum_{v=1}^{N} (y_v)^i$, $i = \overrightarrow{1,3}$ are sample initial moments, and μ_2, μ_4 and μ_6 - theoretical central moments of the regression errors.

An analysis of expression (8) shows that, using polynomials of degree, PMM estimates can only be found by means of a numerical solution of systems of nonlinear equations. As a first approximation, using the appropriate iterative procedures (for example, based on the Newton-Raphson method) it is logical to use OLS estimators to the required parameters.

6   Analysis the Accuracy of Polynomial Estimates

To quantify the magnitude of changes in the accuracy of estimates, we use the notion of a coefficient for reducing the variance of estimates of linear regression parameters:

$$g_{(\theta_p)s} = \frac{\sigma^2_{(\theta_p)s}}{\sigma^2_{(\theta_p)OLS}} = \frac{\sigma^2_{(\theta_p)s}}{\sigma^2_{(\theta_p)1}}, \, p = \overrightarrow{0,1}. \quad (9)$$

These coefficients are formed as the variance ratios of PMM parameter θ, estimates found using the polynomial of the first order to the variances of the OLS estimates of the corresponding components of this vector parameter. And since OLS estimates are equivalent to PMM estimates obtained for a $S = 1°$, their variances will also coincide, i.e. $\sigma^2_{(\theta_p)OLS} = \sigma^2_{(\theta_p)1}$ [20].

It is shown [20] that using the apparatus of the amount of extracted information, the variance of PMM-estimates (for $S = 1$) linear regression parameters can be found as elements of the main diagonal of the variance estimate matrix of the form:

$$V_{(1)} = \mu_2 \left[BB^T \right]^{-1}, \quad (10)$$

where B - matrix of $2 \times N$ with elements $b_{v,p} = (x_v)^{p-1}$, $p = \overrightarrow{0,1}$, $v = \overrightarrow{1,N}$.

Using expressions (7) describing the optimal coefficients $k_{i,v}^{(p)}$, $i = \overrightarrow{1, 3}$ based on (4) and (5) we can form the variational matrix of PMM-estimators for

$$V_{(3)} = \frac{\mu_2\mu_6 - \mu_4^2}{9\mu_2^3 - 6\mu_2\mu_4 + \mu_6} \left[BB^T\right]^{-1} = \kappa_2 \left(1 - \frac{\gamma_4^2}{6 + 9\gamma_4 + \gamma_6}\right) \left[BB^T\right]^{-1} \quad (11)$$

where $\gamma_4 = \frac{\mu_4}{\mu_2^2} - 3$, $\gamma_6 = \frac{\mu_6}{\mu_2^3} - 15\frac{\mu_4}{\mu_2^2} + 30$ – dimensionless cumulant coefficients.

The transition in the expressions (11) from the moment description to the cumulant one is caused not only by the greater compactness of the latter, but also by the fact that the deviation of the values of higher-order cumulant coefficients $\gamma_r = \kappa_r / \kappa_2^{r/2}$ from zero describes the degree of difference from the Gaussian model. This allows us to interpret the results more clearly. It is obvious from (11) that the values of the variance reduction coefficients $g_{(\theta_p)3}$ do not depend on the index p of the vector parameter component θ, and the potential decrease in the estimates variance is determined precisely by the degree of non-gaussness, expressed numerically by the cumulant coefficients of the 4th and 6th orders, e.g.:

$$g_{(\theta_p)3} = 1 - \frac{\gamma_4^2}{6 + 9\gamma_4 + \gamma_6}, \quad p = \overrightarrow{0, 1}. \quad (12)$$

It is necessary to mention the universality of formula (12), since it also describes the coefficient of decrease in the variances of PMM-estimators (at the degree $S = 3$) of the scalar parameter under the conditions of symmetric measurement errors [22, 23]. It should be noted that the coefficients of higher order cumulants of random variables are not arbitrary values, because their combination has the domain of admissible values [27]. For example, for symmetrically distributed random variables, probability properties of which are given by cumulant coefficients of the 4th and 6th order, the domain of admissible values of these parameters are limited to two inequalities: $\gamma_4 > -2$ and $\gamma_6 + 9\gamma_4 + 6 > \gamma_4^2$. Considered the last inequalities, from the analysis of (12), we can conclude that the coefficient $g_{(\theta_p)3}$ of variance reduction has the range $(0; 1]$ and is dimensionless. Figure 1 shows the graphs which are built with these limitations.

Since the reduction of variance coefficient $g_{(\theta_p)3}$ is a function of two variables γ_4 and γ_6, the set of their values will be a surface (Fig. 1a). For greater clarity, in addition to the 3D-graphic, projections of the contour plots on the plane are also presented. The darker regions in Fig. 1b correspond to the projections of large values of the function (12).

From these graphs, you can see that the variance of PMM-estimates greatly increases and tends asymptotically to zero when value of cumulant coefficients approaching to the border region bounded by a parabola $\gamma_6 = \gamma_4^2 - 9\gamma_4 - 6$. The reduction of estimate variances is not observed only in the case when the kurtosis coefficient is zero. In other cases (with $\gamma_4 \neq 0$) the PMM-estimates has the less variance than the OLS-estimates.

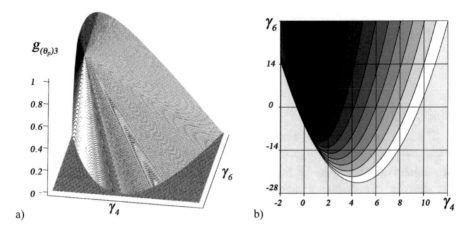

Fig. 1. Coefficient of reduction variance $g_{(\theta_p)3}$ dependency on cumulative coefficients γ_4, γ_6: (a) 3D- graphic; (b) Contour plots.

7   Features of the Algorithm for Adaptive Polynomial Estimates

The results of the theoretical analysis of the effectiveness of polynomial estimates presented in the previous section, as well as the corresponding results of [20], testify to the advisability of using PMM only if the distribution of regression errors is not of the Gaussian law. It was also noted that the inadequacy of the Gaussian model of regression errors is not critical from the point of view that OLS estimates remain unbiased and consistent, although they cease to be optimal. In this case, the OLS is inherently linear, and, consequently, the probabilistic properties of the regression residues after its application do not actually differ from the properties of the original random component of the regression model [28]. Such a factor is often used to obtain adaptive estimates using the maximum likelihood method, which additionally requires solving the problems of estimating the error distribution density [7, 29].

Once again, we note that it is important from the practical point of view that to obtain PMM-estimates, information is used not about the distribution of errors, but their moment-cumulant description. Thus, a fairly simple way of overcoming a priori uncertainty about the probabilistic properties of the error model arises by finding estimates of a limited number of their parameters: moments or cumulants. Moreover, the obtained estimates of the cumulative coefficients of the 3rd (asymmetry) and the 4th (kurtosis) orders can also be used to test the hypothesis of Gaussianity and the symmetry of regression errors [30, 31]. On the above, we modify the approach proposed in [20], which allows us to find adaptive PMM estimates of the regression parameters in accordance with the following algorithm:

Step 1 – finding OLS estimates of regression parameters;
Step 2 – formation of regression residues and finding estimates of their moments and cumulants up to the 4th order;

Step 3 – testing the Gaussian regression residues distribution hypothesis (in the case of its non-refutation, the algorithm ends);

Step 4 – testing hypothesis about the regression residues distribution symmetry (in case of its non-refutation, the transition to step 6);

Step 5 – finding PMM estimates using the polynomial of degree $S = 2$ ((according to [20]) and completing the algorithm;

Step 6 – estimates moments of 6th orders of regression residuals of OLS;

Step 7 – finding PMM-estimates using the degree polynomial $S = 3$ (by a numerical solution of the system (8)) and completing the algorithm.

8 Monte-Carlo Simulation

Based on the results obtained, the set program (for MATLAB/OCTAVE), used in [20, 22–24], was modernized. This set of m-scripts and m-functions based on the Monte Carlo statistical simulation compare the accuracy of OLS and PMM-estimates at different models of the regression of non-Gaussian errors.

Table 1 presents the Monte Carlo simulation results.

Table 1. The results of Monte-Carlo parameters estimation simulation.

Distribution		Theoretical values			Monte-Carlo simulation results					
		γ_4	γ_6	$g_{(\theta_p)3}$	$\hat{g}_{(\theta_p)3}$					
					$N = 20$		$N = 50$		$N = 200$	
					a_0	a_1	a_0	a_1	a_0	a_1
Arcsines		−1.3	8.2	0.2	0.51	0.53	0.31	0.32	0.22	0.23
Uniform		−1.2	6.9	0.3	0.67	0.67	0.44	0.45	0.33	0.33
Trapezoidal	$\beta = 0.75$	−1.1	6.4	0.36	0.70	0.70	0.50	0.50	0.40	0.40
	$\beta = 0.5$	−1	5	0.55	0.85	0.84	0.68	0.69	0.57	0.58
	$\beta = 0.25$	−0.7	2.9	0.76	**1.01**	**1.00**	0.89	0.90	0.80	0.81
Triangular		−0.6	1.7	0.84	**1.08**	**1.09**	0.97	0.97	0.88	0.89
Laplace		3	30	0.86	**1.47**	**1.50**	0.86	0.87	0.85	0.85

As a comparative criterion of effectiveness, estimates of the magnitude of variance reduction coefficients $\hat{g}_{(\theta_p)3}$ are used, whose are determined according to (9). Analysis of the results of Table 1 shows that experimental values $\hat{g}_{(\theta_p)3}$ differ from theoretical values, since the analytical formula (12) was obtained for the asymptotic case (for $N \to \infty$). In addition, the uncertainty of a posteriori estimates of regression error parameters (paired moments up to the 6th order), which also depends significantly on the sample size N, affects the accuracy of obtaining adaptive PMM-estimates of the desired components of the parameter θ. But to a much greater extent, the relative effectiveness of PMM depends on the probabilistic properties of regression errors and is more pronounced for distributions having a flat-topped character and negative values of the kurtosis coefficient. For example, for error models in which $\gamma_4 < -1$ the variance decreases from 15% (for small samples $N = 20$) to several times (with growth N or decrease γ_4).

In addition, in Figs. 2 and 3 examples of simulation results are given showing distributions of the experimental values of OLS and PMM-estimates (at $S = 3$) of components of vector parameters $\theta = \{a_0, a_1\}$.

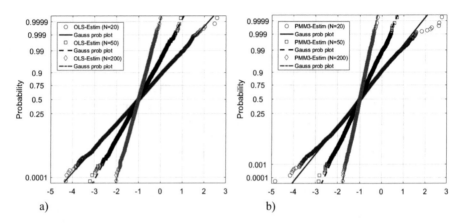

Fig. 2. Gaussian probability graphs approximating experimental values of the linear regression parameter $a_0 = -1$ of trapezoid error distribution: (a) OLS estimates; (b) PMM estimates.

In these examples the input data have the $M = 10^4$ samples and size ($N = 20, 50, 200$), containing the results estimation of the parameter $a_0 = -1$ and $a_1 = 2$ for model of errors based the random variable with Trapeze PDF ($\beta = 0.5$).

Analysis of these data and other results of experimental research shows that for the regression error, the distribution of which is a significantly different from the Gaussian model, the normalization of the distribution of estimates is observed only for a sufficiently large size of initial sample elements. This is explained by the presence of additional nonlinear transformations (calculation of quadratic and cubic statistics when PMM estimates are found for $S = 3$), which slows down the dynamics of the normalization of the empirical distribution of PMM-estimations relatively in OLS estimates.

9   Real Data Experiment

Let's test the proposed algorithm for finding the adaptive PMM estimations of linear regression parameters using a set of real data. This data set is atmospheric concentrations of CO_2 expressed in parts per million (ppm) reported in the preliminary 1997 SIO manometric mole derived from in situ air samples collected at Mauna Loa Observatory, Hawaii. [32]. For the experiment, part of this data was used by volume

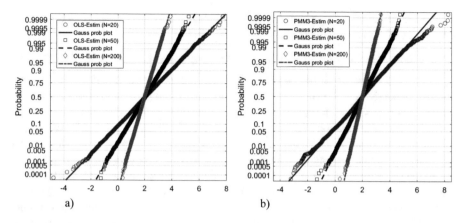

Fig. 3. Gaussian probability graphs approximating the experimental values of the linear regression parameter $a_1 = 2$ for the trapezoid error distribution: (a) OLS estimates; (b) PMM estimates.

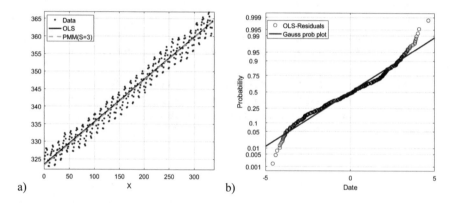

Fig. 4. Linear regression model (a) Experimental data and regression estimates; (b) Q-Q plot OLS-residuals.

$N = 336$ (monthly measurements from January 1970 to September 1997), which are described adequately (see Fig. 4a) by a linear regression model of the form (1).

The least-squares estimations of the parameters of such model are the values: $\hat{a}_0^{(1)} = 323.411$; $\hat{a}_1^{(1)} = 0.121$. OLS residuals is refuted by the Yarki-Ber test embedded in the MATLAB (*JBSTAT* $= 15.6$ at the threshold value $CV = 5.8$ at a fixed significance degree 0.05) [31]. The symmetry of the error distribution of the model is visually visible in Fig. 4(b), where the probabilistic graph of the Gaussian approximation (Q-Q plot) of the regression OLS residuals is presented and is confirmed by the small absolute value of the estimation of their asymmetry coefficient $\hat{\gamma}_3 = -0.15$.

Estimated values of even cumulant coefficients: $\hat{\gamma}_4 = -1$ and $\hat{\gamma}_6 = 5.2$ regression OLS residuals make it possible to determine (12) sufficiently accurately (taking into account a sufficiently large volume of initial data N) to determine the value of the coefficient of variance reduction $\hat{g}_{(\theta_p)3} = 0.5$.

We note that the refined values of PMM estimates themselves ($\hat{a}_0^{(3)} = 323.156$; $\hat{a}_1^{(3)} = 0.122$) differ from the values of OLS estimators not significantly. However, a significant decrease in the variance allows us to build more narrow confidence integrals, which are usually used in regression analysis. Considering the observance of the condition for the normalization of the distribution of OLS estimates (for large N), we get that at a confidence level of 0.95, the interval (322.936; 323.886) covers the value of the regression parameter a_0, and the interval (0.119; 0.124) – covers the value parameter a_1. At the same time, taking into account the value of the estimation of the coefficient of variance reduction, as well as the obtained point values of PMM- estimations, allows for a given level of confidence probability, it is justified to correct and reduce $(1 - \sqrt{0.5})\,100 \approx 30\,\%$ the width of the confidence interval, having obtained its limits (322.82; 323.491) for the parameter a_0, and (0.12; 0.123) for the parameter a_1.

To verify the correctness of the results, a statistical experiment based on the bootstrapping method. With the help of built-in MATLAB bootstrap-resampling tools the original sample was multiplied by 10^4 bootstrap samples with the return. For each of them, we found OLS and PMM estimates for linear regression parameters. The empirical distribution of these estimates is presented in Fig. 5 in the form of boxplot-graphs, the upper and lower bounds of which are respectively 2.5% and 97.5% percentile.

The visual analysis of Fig. 5 shows that boundaries of the 95% confidence intervals obtained as the result of statistical bootstrap modeling, and correlate with the calculated values of the interval estimates obtained above indicate their significant closeness. This generally confirms the reliability of the analytical calculations.

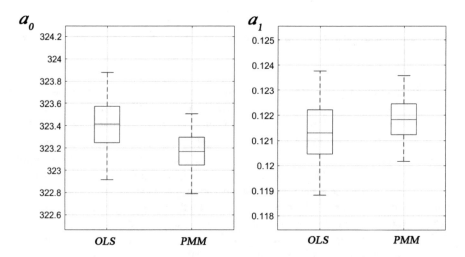

Fig. 5. Empirical (bootstrap) distribution of linear regression parameter estimates.

10   Conclusions

An analysis of the set of obtained results confirms the possibility and expediency of applying the method of polynomial maximization for finding estimates of single-factor linear regression parameters under the condition of a non-zero-symmetric distribution of errors, which are described using higher-order statistics.

Theoretical studies have shown that nonlinear PMM-estimates synthesized at a polynomial degree $S = 3$ are characterized by the greater accuracy than OLS-estimates. The coefficient of variance decrease is determined by the degree of non-Gaussianity of the random component of the regression model, expressed numerically by the absolute values of the cumulative coefficients of the 4th and 6th orders.

The results obtained in statistical modeling by the Monte Carlo method confirm the effectiveness of PMM (for $S = 3$) for situations where errors of the regression model have a symmetric distribution. The greatest efficiency (relative to OLS) of PMM is appears in situations where the kurtosis of regression errors has negative values, which is observed, for example in flat-top and double-modal distributions. For example, for model experimental data with the arcsine distribution, the variance of PMM estimates compared to OLS estimates is 5 times less. And for the considered example of real experimental data (CO_2 concentration in the atmosphere), the variance of estimates of the regression model parameters decreases by about 2 times. This made possible to reduce the width of their confidence intervals by 30%.

The asymptotic nature of PMM is confirmed in [20], since with the increase in the sample data the experimental values of the variance reduction coefficients tend to theoretically calculated values, and the distribution of the estimates is normalized.

In general, the proposed approach has less analytical and computational complexity than the parametric MLE and provides a reduction in uncertainty in comparison with OLS, which does not consider the difference in the probability distribution of statistical data from the Gaussian law.

Among many possible directions of further research, one should mention the following:

- carry out a comparative analysis of the efficiency of PMM with estimates of alternative nonparametric methods (least absolute deviation, sign, etc.);
- extend the proposed approach and explore the features of its application for non-linear single-factor regression models, and consider a more general multifactor case;
- investigate the possibility of approximating the distribution of PMM estimates, models based on Johnson curves for constructing confidence intervals for small sample sizes.

References

1. Anscombe, F.J.: Topics in the investigation of linear relations fitted by the method of least squares. J. R. Stat. Soc. Ser. B (Methodological) **29**, 1–52 (1967)
2. Cox, D.R., Hinkley, D.V.: A note on the efficiency of least-squares estimates. J. R. Stat. Soc. Ser. B (Methodological) **30**, 284–289 (1968)

3. Schechtman, E., Schechtman, G.: Estimating the parameters in regression with uniformly distributed errors. J. Stat. Comput. Simul. **26**(3–4), 269–281 (1986). https://doi.org/10.1080/00949658608810965

4. Galea, M., Paula, G.A., Bolfarine, H.: Local influence in elliptical linear regression models. J. R. Stat. Soc. Ser. D: Stat. **46**(1), 71–79 (1997)

5. Liu, S.: Local influence in multivariate elliptical linear regression models. Linear Algebra Appl. **354**(1–3), 159–174 (2002). https://doi.org/10.1016/S0024-3795(01)00585-7

6. Ganguly, S.S.: Robust regression analysis for non-normal situations under symmetric distributions arising in medical research. J. Modern Appl. Stat. Meth. **13**(1), 446–462 (2014). https://doi.org/10.22237/jmasm/1398918480

7. Zeckhauser, R., Thompson, M.: Linear regression with non-normal error terms. Rev. Econ. Stat. **52**(3), 280–286 (1970)

8. Bartolucci, F., Scaccia, L.: The use of mixtures for dealing with non-normal regression errors. Comput. Stat. Data Anal. **48**(4), 821–834 (2005). https://doi.org/10.1016/j.csda.2004.04.005

9. Seo, B., Noh, J., Lee, T., Yoon, Y.J.: Adaptive robust regression with continuous Gaussian scale mixture errors. J. Korean Stat. Soc. **46**(1), 113–125 (2017). https://doi.org/10.1016/j.jkss.2016.08.002

10. Tiku, M.L., Islam, M.Q., Selçuk, A.S.: Non-normal regression II. Symmetric distributions. Commun. Stat. Theory Meth. **30**(6), 1021–1045 (2001). https://doi.org/10.1081/STA-100104348

11. Andargie, A.A., Rao, K.S.: Estimation of a linear model with two-parameter symmetric platykurtic distributed errors. J. Uncertain. Anal. Appl. **1**(1), 1–19 (2013)

12. Atsedeweyn, A.A., Srinivasa Rao, K.: Linear regression model with generalized new symmetric error distribution. Math. Theory Model. **4**(2), 48–73 (2014). https://doi.org/10.1080/02664763.2013.839638

13. Huber, P.J., Ronchetti, E.M.: Robust Statistics. Wiley, Hoboken (2009). https://doi.org/10.1002/9780470434697

14. Narula, S.C., Wellington, J.F.: The minimum sum of absolute errors regression: a state of the art survey. Int. Stat. Rev. **50**(3), 317–326 (1982)

15. Koenker, R., Hallock, K.: Quantile regression: an introduction. J. Economic. Perspect. **15**(4), 43–56 (2001)

16. Tarassenko, P.F., Tarima, S.S., Zhuravlev, A.V., Singh, S.: On sign-based regression quantiles. J. Stat. Comput. Simul. **85**(7), 1420–1441 (2015). https://doi.org/10.1080/00949655.2013.875176

17. Dagenais, M.G., Dagenais, D.L.: Higher moment estimators for linear regression models with errors in the variables. J. Econom. **76**(1–2), 193–221 (1997). https://doi.org/10.1016/0304-4076(95)01789-5

18. Cragg, J.G.: Using higher moments to estimate the simple errors-in-variables model. RAND J. Econ. **28**, S71 (1997). https://doi.org/10.2307/3087456

19. Gillard, J.: Method of moments estimation in linear regression with errors in both variables. Commun. Stat. Theory Meth. **43**(15), 3208–3222 (2014)

20. Zabolotnii, S., Warsza, Z., Tkachenko, O.: Polynomial estimation of linear regression parameters for the asymmetric pdf of errors. In: Advances in Intelligent Systems and Computing. vol. 743, pp. 758–772. Springer (2018). https://doi.org/10.1007/978-3-319-77179-3_75

21. Kunchenko, Y.: Polynomial Parameter Estimations of Close to Gaussian Random variables. Shaker Verlag, Aachen (2002)

22. Warsza, Z.L., Zabolotnii, S.W.: A polynomial estimation of measurand parameters for samples of non-Gaussian symmetrically distributed data. In: Advances in Intelligent Systems and Computing, vol. 550, pp. 468–480. Springer (2017). http://doi.org/10.1007/978-3-319-54042-9_45

23. Warsza, Z.L., Zabolotnii, S.W.: Uncertainty of measuring data with trapeze distribution evaluated by the polynomial maximization method. Przemysł Chemiczny 1(12), 68–71 (2017). https://doi.org/10.15199/62.2017.12.6. (in Polish)

24. Warsza, Z., Zabolotnii, S.: Estimation of measurand parameters for data from asymmetric distributions by polynomial maximization method. In: Advances in Intelligent Systems and Computing, vol. 743, pp. 746–757. Springer (2018). https://doi.org/10.1007/978-3-319-77179-3_74

25. Zabolotnii, S.W., Warszam, Z.L.: Semi-parametric estimation of the change-point of parameters of non-Gaussian sequences by polynomial maximization method. In: Advances in Intelligent Systems and Computing, vol. 440, pp. 903–919. Springer (2016). http://doi.org/10.1007/978-3-319-29357-8_80

26. Palahin, V., Juh, J.: Joint signal parameter estimation in non–Gaussian noise by the method of polynomial maximization. J. Electr. Eng. 67, 217–221 (2016). https://doi.org/10.1515/jee-2016-0031

27. Cramér, H.: Mathematical Methods of Statistics, vol. 9. Princeton University Press, Princeton (2016)

28. Cook, R.D., Weisberg, S.: Residuals and Influence in Regression. Monographs on Statistics and Applied Probability. Chapman and Hall, New York (1982). https://doi.org/10.2307/1269506

29. Stone, C.J.: Adaptive maximum likelihood estimators of a location parameter. Annal. Stat. 3(2), 267–284 (1975). https://doi.org/10.1214/aos/1176343056

30. Boos, D.D.: Detecting skewed errors from regression residuals. Technometrics 29(1), 83–90 (1987). https://doi.org/10.1080/00401706.1987.10488185

31. Jarque, C.M., Bera, A.K.: A test for normality of observations and regression residuals. Int. Stat. Rev. 55(2), 163–172 (2012)

32. Keeling, C.D., Whorf, T.P.: Scripps Institution of Oceanography (SIO). University of California, La Jolla, California USA 92093-0220. ftp://cdiac.esd.ornl.gov/pub/maunaloa-co2/maunaloa.co2

Static Field Magnetic Flux Thickness Gauge

Maciej Szudarek[1]($\boxtimes$), Michał Nowicki[1], Filip Wierzbicki[1],
and Marcin Safinowski[2]

[1] Institute of Metrology and Biomedical Engineering,
Warsaw University of Technology,
sw. Andrzeja Boboli 8, 02-525 Warsaw, Poland
szudarek@mchtr.pw.edu.pl
[2] Industrial Research Institute for Automation and Measurements,
al. Jerozolimskie 202, 02-486 Warsaw, Poland

Abstract. Non-destructive testing of coating thickness has long been standard practice. Depending on material and geometric properties of base and coating, different measurement methods are applied. The article presents a simple device to measure coating thickness working on a principle of static field magnetic flux gauge. The device has been tested in the range of (2–6000) μm, which is between the usual range of typical thickness gauges operating on ultrasonic or eddy-currents principles. Measurement results are shown. The expanded uncertainty of thickness measurement ($k = 2$) does not exceed 5% in the range of (2–4) mm. Construction of a working test stand constitutes the first step in further study on possible solutions of reducing the measurement uncertainty.

Keywords: Non-destructive testing · Thickness measurement ·
Static field magnetic flux method

1 Introduction

Non-destructive testing of coating thickness is a highly valuable technique that has long been standard practice [1]. The most commonly applied methods include eddy-current, ultrasonic and magnetic methods. While ultrasonic methods can be used to measure thickness of coating on base of any kind [2], eddy-current methods require base and coating to have widely differing electrical conductivities [3]. Magnetic methods can be applied in two cases: on a condition that base is magnetizable and coating is non-magnetizable and also in the reverse case, when coatings are magnetizable and the base is not [4]. Thickness of non-magnetic materials such as plastic containers may also be measured with this method, in this case reference balls with a specially coated finish are used as a substitute for base.

In the presented article static field magnetic flux method has been used to construct a thickness gauge which covers measurement range between the usual range of eddy-current (0–2 mm) and ultrasonic (5–5000 mm) methods.

© Springer Nature Switzerland AG 2020
R. Szewczyk et al. (Eds.): AUTOMATION 2019, AISC 920, pp. 650–656, 2020.
https://doi.org/10.1007/978-3-030-13273-6_60

2   Theoretical Background

2.1   Magnetic Methods

The principle of operation of all the magnetic methods can be explained with the example of a ferrous ring with an air gap and excitation winding. Magnetic flux density B is a function of the air gap length x:

$$B = \frac{n \cdot I \cdot \mu_0 \cdot \mu_r}{2\pi r - x + \mu_r \cdot x} \tag{1}$$

where n is the number of windings, I is the electric current through the coil, μ_r is the relative magnetic permeability of the ferrous material, μ_o is the magnetic constant and r is the radius of the ferrous ring.

Similarly to air, a nonmagnetic coating can be treated as a gap between a magnetic field source and magnetic base material. Certain minimal thickness of base metal is required so that it has influence on magnetic flux density. This value depends on the strength of the magnet, as well as geometry and magnetic permeability of the base material and should be determined experimentally.

One of the most widely used magnetic thickness measurement methods is magnetic pull-off method [5]. It is based on measurement of attractive force between the base material and magnetic field source, which can be either a permanent magnet or an electromagnet. Magnetic flux density decreases with increasing thickness of nonmagnetic coating, so in this case measurement of coating thickness comes down to measurement of force. To reduce the impact of probe position on the measurement, precise instruments compensate the influence of gravity.

Magnetic inductive method bases on the fact that electrical inductivity of a coil changes when an iron core is inserted or approaches the coil. In this way electrical inductivity is a measure of distance between the coil and a ferromagnetic base. Usually two coils are used, where the first generates low frequency magnetic field and the second measures induced voltage. To reduce the influence of component geometry, a highly magnetizable core is used to concentrate the magnetic field. Depending on the application, coils can either be employed together and measure thickness locally or can be separated to form two pole probe with integrating capabilities [6].

Finally, magnetic flux gauges (Fig. 1) employ magnetic flux detector that measures the magnetic flux density close to magnetic field source, which is either constant or of alternating frequency [7]. Hall sensors or magneto resistive sensors are usually used as magnetic flux detectors. The device that is described in this article works on the principle of static magnetic flux gauge.

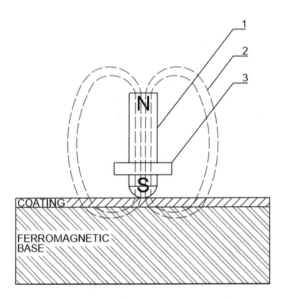

Fig. 1. Magnetic flux gauge using a Hall sensor, 1 – magnetic field source, 2 – lines of static magnetic field 3 – Hall sensor. Proximity of ferromagnetic base increases flux density at the sensor location.

2.2   Uncertainty Sources in Static Field Magnetic Flux Gauges

Crucial factors that influence measurement uncertainty are quality of surfaces, their curvature and homogeneity of magnetic properties [6]. In order to minimize the influence of surface curvature or other geometric features such as surface roughness, any adjustments to the instrument's calibration curve should take place using the same geometry with coating of a known thickness. Another option is to reduce the measurement area covered by magnetic flux sensor. Small graphene hall-effect sensors [8] or Matteucci effect sensors [9] are promising for this application.

Additional errors may appear in case of external magnetic fields, since this method is susceptible to such disturbances. Measurement probe should be placed in an adequate distance from any edges or drills to avoid edge effects on the measurement signal. To study the influence of geometry on measurement, magnetostatic modelling methods may be applied [10, 11]. In case of numerical modelling, simulation validation criteria need to be formulated, e.g. using methodology described in [12].

Finally, large errors can be introduced if remanence of the base material is present. It is worth noting that the base material may be magnetized through repeated measurements.

To improve repeatability of readings, commercially available devices are equipped with springs systems to reduce the effect of probe pressure.

3   Measurement Stand

Device components are shown in Fig. 2. Thickness standards were used to determine calibration curves of the constructed device in the range of (20–500) μm. For thicknesses up to 6000 μm non-magnetizable coatings were measured with micrometer screw as a reference. St3 steel plates of various thicknesses were employed as a ferromagnetic base.

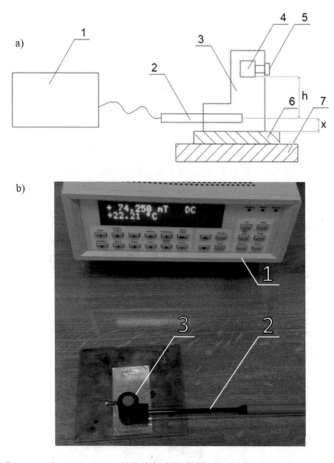

Fig. 2. a. Test stand components. Model 455 DSP magnetometer (1) was connected to Lakeshore HMNT-4E04-VR Hall sensor (2), which was placed in a 3D-printed mounting tube (3) at height $x = 2$ mm. Permanent magnet (4) was mounted by screw (5) at height $x + h = 10$ mm above the measured coating. (6) The coating was placed on a ferromagnetic base (7). b. Assembled test stand.

4   Discussion of Results

Magnetic flux gauge method is sensitive to base thickness (Fig. 3), thus the device should be calibrated using reference sample with known base thickness. What is more, as the magnetic flux density depends on magnetic permeabilities μ_r of the substances in the magnetic field, magnetic flux gauges require calibration on the same material.

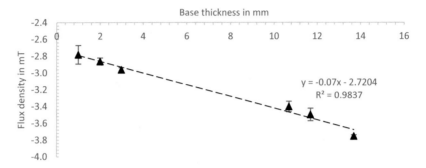

Fig. 3. Influence of base thickness on measured magnetic flux density

Regardless of magnet material, shape of calibration curves (Figs. 4 and 5) does not change. Strong magnets are preferable to achieve better sensitivity. The function between the coating thickness and Hall sensor output is nonlinear, which is in agreement with theoretical considerations. In the range of thin coating layers a small change of thickness causes proportionally larger change of magnetic reluctance than in the range of thick layers. Therefore slope of calibration curve decreases with increasing thickness of coating.

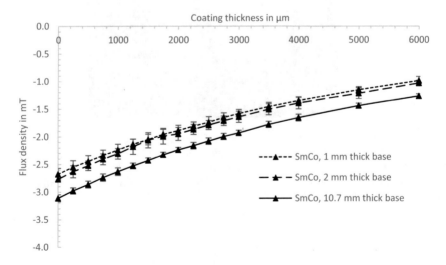

Fig. 4. Calibrations curves for SmCo magnet and various base thicknesses

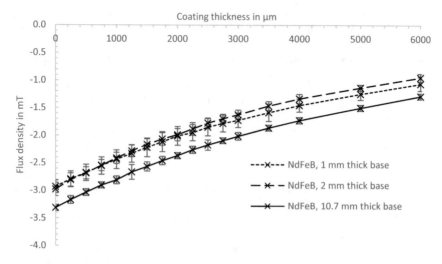

Fig. 5. Calibrations curves for neodymium magnet and various base thicknesses

To estimate uncertainty of the measurement, guidelines presented in [4] were used. Error bars in Figs. 3–5 present expanded uncertainty for $k = 2$.

5 Summary

The article presents a simple device to measure coating thickness working on a principle of static field magnetic flux gauge. The device has been tested in the range of (2–6000) μm, which is between the usual range of typical thickness gauges operating on ultrasonic (5–5000 mm) or eddy-currents (0–2 mm) principles. For neodymium magnet the expanded uncertainty of the measurement of flux density ($k = 2$) was in the range of (2–5)%. The constructed device can be used to measure thickness in the range of (2–4) mm with the expanded uncertainty not exceeding 5%.

Static field magnetic flux gauge principle poses many advantages over other types of thickness measurement methods and should be an object of further study. Construction of a working test stand constitutes the first step in research on possible solutions for reducing the measurement uncertainty.

References

1. Brenner, A.: Magnetic method of measuring the thickness of nonmagnetic coatings on iron and steel. Bur. Stand. J. Res. **20**, 357–368 (1938)
2. Lavrentyev, A.I., Rohlin, S.: An ultrasonic method for determination of elastic moduli, density, attenuation and thickness of a polymer coating on a stiff plate. Ultrasonic **39**(3), 211–221 (2001)

3. Wang, Z., Yu, Y.: Thickness and conductivity measurement of multilayered electricity-conducting coating by pulsed eddy current technique: experimental investigation. IEEE Trans. Instrum. Meas. (2018)
4. International Organization for Standardization. Electrodeposited nickel coatings on magnetic and non-magnetic substrates. ISO Standard No. 2361 (1982)
5. Beamish, D.: Coating thickness measurement. Met. Finish. **108**, 379–385 (2010)
6. International Organization for Standardization. Non-magnetic coatings on magnetic substrates. ISO Standard No. 2178 (2016)
7. Lukhvich, A., et al.: The magneto-dynamic method and gauges for two-layer coatings testing. In: 10th European Conference on Non-destructive Testing, Moscow, pp. 73–75 (2010)
8. Petruk, O., et al.: Sensitivity and offset voltage testing in the hall-effect sensors made of graphene. In: Recent Advances in Automation, Robotics and Measuring Techniques, pp. 631–640. Springer (2014)
9. Charubin, T., et al.: Spectral analysis of Matteucci effect based magnetic field sensor. In: AIP Conference Proceedings, vol. 1996, no. 1 (2018)
10. Jin, J.-M.: The Finite Element Method in Electromagnetics. Wiley, London (2015)
11. Szewczyk, R.: Magnetostatic Modelling of Thin Layers Using the Method of Moments and Its Implementation in Octave/Matlab, vol. 491. Springer, Heidelberg (2018)
12. Turkowski, M., Szufleński, P.: New criteria for the experimental validation of CFD simulations. Flow Meas. Instrum. **34**, 1–10 (2013)

Method for Living Cell Mechanical Properties Evaluation from Force-Indentation Curves

Inga Morkvenaite-Vilkonciene[1,2](✉) [ID], Raimundas Vilkoncius[1] [ID],
Juste Rozene[1] [ID], Antanas Zinovicius[2] [ID], Oleksii Balitskyi[3],
Almira Ramanaviciene[2] [ID], Arunas Ramanavicius[2] [ID],
Andrius Dzedzickis[1] [ID], and Vytautas Bučinskas[1] [ID]

[1] Vilnius Gediminas Technical University,
J. Basanaviciaus 28, 03224 Vilnius, Lithuania
{inga.morkvenaite-vilkonciene,
vytautas.bucinskas}@vgtu.lt
[2] Vilnius University, Naugarduko 24, 03225 Vilnius, Lithuania
[3] Center for Physical Sciences and Technology, Sauletekio 3, Vilnius, Lithuania

Abstract. Living cells mechanical properties establishment from Atomic force microscopy (AFM) force-separation curves is a challenge because the calculated Young's modulus depends on the applied mathematical model. The more reliable results can be obtained using finite element models. In this work, yeast cells with different mechanical properties were measured by AFM. To change cells mechanical properties, yeasts were immersed in 9, 10-phenanthrenequinone, which changed cells' membranes elasticity. 3D finite element model of the whole cell was created to calculate reacting force when AFM tip indents the cell in the same way as in the real experiment. It was found that our model is capable to draw the information about cells mechanical properties and visco-elastic behavior of cells membranes.

Keywords: AFM · Mechanical properties · Living cells · Finite element model

1 Introduction

Atomic force microscopy (AFM) is a valuable tool in living cells research [1]. Evaluation of cells' elastic properties, adhesion, and hydrophobicity of membrane, as well as morphological examination could be performed by AFM [2, 3].

The studies on mechanical properties of single cells show that cancerous cells are 'softer', highly deformable, and Young's modulus is lower compared to healthy ones [4]. Measured elastic properties of cells' membranes depend on the conditions of the experiment, such as cell fixation, ambient environment (solution or the air), temperature, measuring time, scanning speed, probe geometry, applied force, mode of measurement, the quality of tip and sample preparation procedures [5]. Therefore, usually, AFM results are evaluated by comparing; for example, the healthy cells to cancer cells, cancer cells to that treated by drugs and so on.

The Young's modulus for the cell at different experimental conditions can be calculated by performing the curve fitting to some mathematical models. Several

© Springer Nature Switzerland AG 2020
R. Szewczyk et al. (Eds.): AUTOMATION 2019, AISC 920, pp. 657–663, 2020.
https://doi.org/10.1007/978-3-030-13273-6_61

theories describe the elastic deformation of the sample. These theories have been developed by Hertz [6], Johnson-Kendall-Roberts (JKR) [7] and Derjaguin-Muller-Toporov (DMT) [8]. Sneddon contact behavior model [9, 10] is applied in order to gain a perfectly elastic indentation of a half-space material, in this case, force measurements are performed with an axisymmetric indenter and elimination of adhesion forces. JKR model can be applied to evaluate mechanical contacts of soft and elastic spherical bodies interacting by short-range adhesion forces. All these equations are valid for spherical tips. Each of models has shortcomings; therefore, to fix them the finite element models are applied [11]. The significant difference in hepatocellular carcinoma cells measuring results with spherical and conical tips was determined; it was found that Young's modulus is 15.0 ± 3.2 kPa and 52.7 ± 5.3 kPa, respectively [12]. Authors concluded that Hertz-Sneddon model cannot explain these results; finite element model results; solution of finite element model gives stresses in 8.52 kPa and 7.44 kPa for the conical tip and sphere-shaped tip, respectively.

Thus, the finite element models are highly reliable in living cells mechanical properties establishment from force-indentation curves. The main aim of our research was to compare two different approaches in living cells mechanical properties determination: one of them usual Hertz-Sneddon indentation model, and another – our created finite element model.

2   Materials and Methods

2.1   Materials

Chemicals were purchased from Merck (Sigma Aldrich), Carl Roth, Scharlau and Fluka companies and were of reagent grade. Phosphate buffer solution was prepared from 0.05 M CH_3COONa; 0.05 M NaH_2PO_4; 0.05 M Na_2HPO_4. Dried yeast was purchased from food supplier "Dr. Oetker Lietuva" (Vilnius, Lithuania).

2.2   Yeast Strain and Growth Conditions

1 g of YPD-broth was mixed with 20 ml of distilled water to get medium with YPD-broth of 50 g/l concentration. 100 mg of dried yeast was introduced to prepared suspension.

A further culture was grown in shaking incubator at 200 rpm till yeast reaches logarithmic phase of growth.

After 20–24 h of incubation medium with yeast cells was centrifuged in 2 ml "Eppendorf" at 3G for 3 min. The supernatant was carefully removed and the pellet was resuspended in distilled water and again centrifuged. After removing of the supernatant the pellet was resuspended in PBS, the final concentration of yeast cells was 0.5 mg/ml.

2.3   Yeast Cells Immobilization

Silicon wafer with square holes (size of hole 6 µm × 6 µm), obtained from Panevezys mechatronics center was used for cells separation and entrapment. Silicon wafer was rinsed with distilled water. The wafer was immersed in a in Piranha solution (3:1 - H_2SO_4: 30% H_2O_2) for 15 min to remove any residues from it. It was also washed with distilled water. For cells immobilization, the bottom of holes was covered by poly-L-lysine (0.1%). One drop of 0.5 µl Poly-L-lysine was added on the selected area and left for 2–3 min. One drop (0.5 µL) of earlier prepared yeast suspension was put on the same area and left to dry. To change membranes' elasticity, cells were filled by different concentration of 9,10-phenanthrenequinone (PQ) for 10 min.

2.4   Evaluation of Cells Mechanical Properties by Atomic Force Microscopy

Fundamental operation of AFM system is shown schematically in Fig. 1a. The probe is the tip fixed on a flexible cantilever end. Cantilever deflects dependently on applied and/or acting forces. This deflection is determined by the optical system: laser beam reflected from the cantilever falls on the photodetector, usually consisting of four light emitting diodes and differential signal between the detector components was recorded. Cells' elastic properties are calculated from the force-distance curves that are usually measured on stiff and compliant surfaces (Fig. 1b). This is represented by a straight–sloped line and it is usually applied as a reference line required for the force calibration.

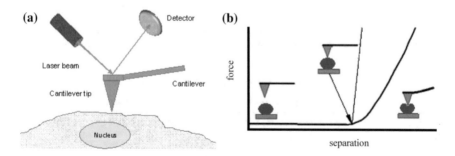

Fig. 1. (a) Atomic force microscopy scheme; (b) force-separation curves for hard and soft (thick line) surfaces. The arrow shows the point of tip contact with the surface.

For compliant samples like cells, cantilever deflection is much smaller and the resulting force curve has a non–linear character [13].

3   Results and Discussion

3.1   Force Measurements

The cells, affected by different concentrations of PQ, were entrapped in the holes and scanned horizontally (Fig. 2a). Then the force-separation curves were recorded approaching the silicon wafer and the cell (Fig. 2b). The force-separation curve, recorded approaching silicon's surface, show that surface is hard (non-deformable). The curve, recorded approaching the non-affected cell, shows that the surface is deformable. Force-separation curves for cells affected with PQ were recorded, and Young's modulus calculated by applying Hertz-Sneddon model for each measurement (Table 1). Young's modulus without PQ was lower if compared with this parameter obtained for cells affected by 15 μM PQ concentration of PQ solution: unaffected cells have Young's modulus of 1.25 ± 0.2 MPa; while cells affected by 15 μM PQ and 45 μM PQ have Young's modulus of 1.66 ± 0.33 MPa, and 1.47 ± 0.28 MPa, respectively. However, cells, affected by 90 μM PQ solution and 375 μM PQ, solution, had Young's modulus of 0.387 ± 0.073 MPa, and 0.149 ± 0.012 MPa, respectively.

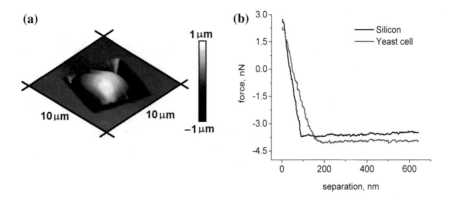

Fig. 2. AFM measurements of yeast cell trapped in the hole of silicon membrane: (a) horizontal scan; (b) repelling force vs. separation of the AFM cantilever from cell membrane.

Table 1. Young's modulus of the cells

	Un-affected	15 μM PQ	45 μM PQ	90 μM PQ	375 μM PQ
Young's modulus, MPa	1.25 ± 0.2	1.66 ± 0.33	1.47 ± 0.28	0.387 ± 0.073	0.149 ± 0.012

3.2 Finite Element Model

Cell finite element (FE) model contains membrane of the cell, cytoplasm, and nucleus (Fig. 4). The cell is fixed on the support. The places of contact are: tip-membrane, without friction; membrane-cytoplasm; cytoplasm-nucleus; membrane-poly-L-lysine; poly-L-lysine-support. Mesh density was variable in the model. Tip-membrane contact place meshing element was lower 0.01 μm. The theoretical experiment was performed by giving the displacement to the tip and measuring the reaction force from the cell. Start point is at the tip-membrane contact, endpoint is 1 μm going vertically. 3D model of the whole cell was calculated without dividing into parts. The experiment, performed at unaffected cells, was compared with solved FE model results (Fig. 3).

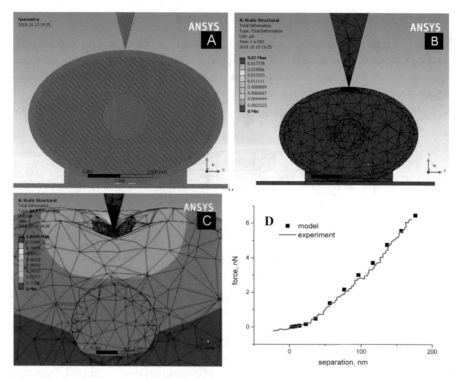

Fig. 3. Finite element model: A. The cross-section view; B. Meshing; C. Deformation of the cell during indentation; D. Force-separation curves, obtained by FE model and experimentally.

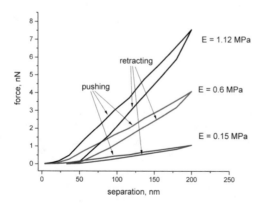

Fig. 4. Force-separation curves, obtained from FE model solution when cell was pushed by the tip and retracted from cells surface.

3.3   Results from Finite Element Model Solution

The results from the model were obtained by calculating approaching and retracting curves at different membrane stiffness (Fig. 4). It was observed that the difference between approaching and retracting curves is higher if the membrane's stiffness is higher (Table 2).

Table 2. The properties of materials, used in the model

	Cytoplasm	Nucleus	Membrane	Ref.
Elastic modulus, kPa	10	0.4	1	[11]
Poisson's ratio	0.37	0.37	0.3	

4   Conclusions

By comparing two different mathematical models – Hertz-Sneddon and our created finite element model, it could be concluded, that our model simulates cells mechanical properties in a range of 1.12–0.15 MPa. This result is comparable with that, usually calculated by other models, when living cells are evaluated by force-separation curves. Our model is promising in living cells mechanical properties determination. In further research, we are planning to study mechanical properties of living cells involving various phenomenas, such as properties of layers of membrane and cell wall, fixation of the cell, and size of pores in the membrane.

Acknowledgement. This research was funded by the European Social Fund according to the activity "Development of Competences of Scientists, other Researchers and Students through Practical Research Activities" of Measure No. 09.3.3-LMT-K-712. Project No. 09.3.3-LMT-K-712-02-0137.

References

1. Morkvenaite-Vilkonciene, I., Ramanavicius, A., Ramanaviciene, A.: Atomic force microscopy as a tool for the investigation of living cells. Medicina **49**(4), 155–164 (2013)
2. Roduit, C., Sekatski, S., Dietler, G., Catsicas, S., Lafont, F., Kasas, S.: Stiffness tomography by atomic force microscopy. Biophys. J. **97**(2), 674–677 (2009)
3. Guo, Q., Xia, Y., Sandig, M., Yang, J.: Characterization of cell elasticity correlated with cell morphology by atomic force microscope. J. Biomech. **45**(2), 304–309 (2012)
4. Lieber, S.C., Aubry, N., Pain, J., Diaz, G., Kim, S.-J., Vatner, S.F.: Aging increases stiffness of cardiac myocytes measured by atomic force microscopy nanoindentation. Am. J. Physiol. Heart Circ. Physiol. **287**(2), H645–H651 (2004)
5. Cross, S.E., Jin, Y.-S., Rao, J., Gimzewski, J.K.: Nanomechanical analysis of cells from cancer patients. Nat. Nanotechnol. **2**(12), 780 (2007)
6. Hertz, H.: On the contact of elastic bodies. Hertz's Miscellaneous Papers, pp. 146–162 (1881)

7. Dulińska, I., Targosz, M., Strojny, W., Lekka, M., Czuba, Z., Balwierz, W., Szymoński, M.: Stiffness of normal and pathological erythrocytes studied by means of atomic force microscopy. J. Biochem. Biophys. Methods **66**(1–3), 1–11 (2006)
8. Lekka, M., Laidler, P., Ignacak, J., Łabędź, M., Lekki, J., Struszczyk, H., Stachura, Z., Hrynkiewicz, A.: The effect of chitosan on stiffness and glycolytic activity of human bladder cells. Biochim. Biophys. Acta (BBA)-Mol. Cell Res. **1540**(2), 127–136 (2001)
9. Sneddon, I.N.: The relation between load and penetration in the axisymmetric boussinesq problem for a punch of arbitrary profile. Int. J. Eng. Sci. **3**, 47–57 (1965)
10. Lekka, M., Laidler, P., Gil, D., Lekki, J., Stachura, Z., Hrynkiewicz, A.: Elasticity of normal and cancerous human bladder cells studied by scanning force microscopy. Eur. Biophys. J. **28**(4), 312–316 (1999)
11. McGarry, J., Prendergast, P.: A three-dimensional finite element model of an adherent eukaryotic cell. Eur. Cell Mater. **7**, 27–33 (2004)
12. Cappella, B., Dietler, G.: Force-distance curves by atomic force microscopy. Surf. Sci. Rep. **34**(1), 1–104 (1999)
13. Matzke, R., Jacobson, K., Radmacher, M.: Direct, high-resolution measurement of furrow stiffening during division of adherent cells. Nat. Cell Biol. **3**(6), 607 (2001)

Explicitness of Parameters Identification in Anhysteretic Curve of Magnetic Materials with Strong Perpendicular Anisotropy

Roman Szewczyk[(⊠)]

Industrial Research Institute for Automation and Measurements,
Al. Jerozolimskie 202, 02-486 Warsaw, Poland
rszewczyk@onet.pl

Abstract. Paper presents the results of verification of the explicitness of determination of parameters for anhysteretic loop in Jiles-Atherton model of magnetic characteristics of strongly anisotropic material. For the experiments the VitroVac 6150F, cobalt-based amorphous alloy with strong perpendicular anisotropy was used. The results indicated, that anhysteretic curve very well represents the shape of magnetic hysteresis loop of such alloys. Moreover, results confirmed the explicitness of model parameters. As a result, anhysteretic curve may represent the shape of magnetic hysteresis loop during the development and optimisation of current transformers based on VitroVac 6150F amorphous alloy.

Keywords: Anhysteretic curve · Jiles-Atherton model · Differential evolution

1 Introduction

Jiles-Atherton model [1, 2] is one of the most popular models of magnetic hysteresis [3]. This model is especially useful for modelling the characteristics of amorphous alloys with strong perpendicular hysteresis. In the case of these alloys, hysteresis loop is negligible. As a result, the magnetization characteristics can be approximated by the anhysteretic magnetization curve [4, 5]. Such approximation is especially useful for modelling the functional characteristics of current transformers [6] developed on the base of amorphous alloys with strong perpendicular anisotropy.

In spite of the fact, that the anhysteretic curve well represents the shape of magnetic hysteresis of amorphous alloys with strong perpendicular anisotropy, the method of determination of model parameters requires application of stochastic optimisation methods [7, 8], such as differential evolution [9]. Due to stochastic character of such method, the explicitness of parameters of anhysteretic curve is not obvious. This problem is also very important due to the fact, that recently developed method of two-step identification [10] of Jiles-Atherton model parameters utilizes the identification of parameters of anhysteretic curve.

As a result the explicitness of determination of parameters of anhysteretic curve is significant problem both from theoretical and application point of view. Such explicitness was verified for simple case of isotropic material (Mn-Zn ferrite) [11], however,

© Springer Nature Switzerland AG 2020
R. Szewczyk et al. (Eds.): AUTOMATION 2019, AISC 920, pp. 664–671, 2020.
https://doi.org/10.1007/978-3-030-13273-6_62

the results of such analyse for more sophisticated case of anisotropic materials with strong uniaxial anisotropy seems to be still not presented.

2   The Model of Anhysteretic Curve in Jiles-Atherton Model of Magnetic Material with Uniaxial Anisotropy

The concept of anhysteretic magnetization curve is the key element of Jiles-Atherton model [1, 12, 13]. Anhysteretic curve can be measured experimentally by demagnetization with given offset constant value of magnetizing field [2]. Such measurements are commonly considered as extremely difficult from technical point of view, however, it can be carried out efficiently with digitally controlled hysteresisgraph [14].

In the Jiles-Atherton model, the concept of the model of anhysteretic curve is based on the model of paramagnetic material. In paramagnetic material, the magnetization M_{para} can be calculated from the Boltzmann statistical distribution of magnetic domain directions [1, 5], given by following equation:

$$M_{para} = M_s \frac{\int\limits_{0}^{\pi} e^{\frac{-E_m(\theta)}{k_B \cdot T}} \sin\theta \cdot \cos\theta \cdot d\theta}{\int\limits_{0}^{\pi} e^{\frac{-E_m(\theta)}{k_B \cdot T}} \sin\theta \cdot d\theta} \tag{1}$$

where M_s is the saturation magnetization of a paramagnetic material, θ is the angle between the atomic magnetic moment m_{at} and direction of the magnetizing field H, T is the temperature and k_B is the Boltzmann constant. In such a case, the energy of the magnetic moment $E_m(\theta)$ can be calculated from the following equation:

$$E_m(\theta) = -\mu_0 \cdot m_{at} \cdot H \cdot \cos\theta \tag{2}$$

In Jiles-Atherton model of the anhysteretic magnetization of isotropic ferromagnetic materials, the atomic magnetic moment m_{at} is substituted by the average magnetization of domain m_d, given as [1]:

$$m_d = \frac{M_s}{N} \tag{3}$$

where N is the average domain density in the material. Moreover, the effective magnetizing field H_e is introduced [1]:

$$H_e = H + \alpha \cdot M \tag{4}$$

where α is the interdomain coupling according to the Bloch model.

For anisotropic materials with uniaxial anisotropy, on the base of the Eq. (1), anisotropic anhysteretic magnetization M_{ah_aniso} is given as [11–13]:

$$M_{ah\_aniso} = M_s \left[\frac{\int_0^\pi e^{E(1)+E(2)} \sin\theta \cdot \cos\theta \cdot d\theta}{\int_0^\pi e^{E(1)+E(2)} \sin\theta \cdot d\theta} \right] \tag{5}$$

where:

$$E(1) = \frac{H_e}{a}\cos\theta - \frac{K_{an}}{M_s \cdot \mu_0 \cdot a}\sin^2(\psi - \theta) \tag{6}$$

$$E(2) = \frac{H_e}{a}\cos\theta - \frac{K_{an}}{M_s \cdot \mu_0 \cdot a}\sin^2(\psi + \theta) \tag{7}$$

$$a = \frac{N \cdot k_B \cdot T}{\mu_0 \cdot M_s} \tag{8}$$

In this model K_{an} is macroscopic, average energy density connected with uniaxial anisotropy in a magnetic material, and ψ is the angle between direction of the magnetizing field H and the easy axis of magnetization due to the uniaxial anisotropy.

It should be highlighted, that for isotropic materials, where $K_{an} = 0$, Eq. (5) reduces to the commonly known form of Langevin equation [1] describing isotropic anhysteretic magnetization M_{ah_iso}:

$$M_{ah\_iso} = M_s \left[\coth\left(\frac{H_e}{a}\right) - \left(\frac{a}{H_e}\right) \right] \tag{9}$$

Anhysteretic flux density B_{ah} in the core can be simply calculated as:

$$B_{ah} = \mu_0(M_{ah} + H) \tag{10}$$

3   Tested Material and the Method of Measurements

The magnetic hysteresis loops were experimentally measured on the ring-shaped samples made of VITROVAC 6125F. VITROVAC 6125F is cobalt-based amorphous alloy produced by Vacuumshmelze GmbH (Germany). Due to annealing in strong magnetic field, perpendicular anisotropy is generated in VITROVAC 6125F cores [15]. Such cores are used in current transformers as well as in other devices, where flat and narrow magnetic hysteresis loop is required.

Quasi-static magnetic hysteresis loops were measured by digitally controlled hysteresisgraph utilizing type 480 fluxmeter produced by Lakeshore. VITROVAC 6125F ring-shaped core with was magnetized by BOP36-6 high power voltage-current converter produced by Kepco. Measuring system was controlled by computer with data acquisition card produced by National Instruments and LabVIEW software.

It should be highlighted, that amorphous and nanocrystalline alloys with perpendicular anisotropy exhibit extremely narrow magnetic hysteresis loop, as it can be observed in Fig. 1. As a result, in the case of these alloys, the anhysteretic magnetic curve can be easily estimated [5] due to the fact, that it is inside very narrow magnetic hysteresis loop. This effect is very useful for validation of the models of anhysteretic magnetization of soft magnetic materials.

4 Method of Identification and Verification of Explicitness of Parameters for Anhysteretic Curve

One of the most sophisticated problems connected with the Jiles-Atherton model is the method of identification of model's parameters. As it was previously indicated, the most efficient method of parameters identification is optimisation utilizing differential evolution algorithm [9]. Previously published results confirmed [15–18] that such method enables to avoid local minima in the target function F given by the following equation:

$$F = \sum_{i=1}^{n} \left(B_{meas}(H_i) - B_{sym}(H_i)\right)^2 \tag{11}$$

where B_{meas} are the results of experimental measurements and B_{sym} are the results of calculations, both for the value H_i of magnetizing field.

Figure 1 presents the results of fitting the anhysteretic loop into the measured magnetic hysteresis loop of VITROVAC 6125F core, whereas Jiles-Atherton model's parameters are presented in Table 1.

Table 1. The result of identification of parameters anhysteretic curve of VITROVAC 6125F core.

Parameter	Units	Description	Jiles-Atherton
M_s	A/m	Saturation magnetization of the material	$8.28 \cdot 10^5$
a	A/m	Quantifies domain wall density	1.40
α	–	Bloch interdomain coupling	$7.95 \cdot 10^{-6}$
K_{an}	J/m^3	Average energy density connected with uniaxial anisotropy	288
ψ	–	The angle between direction of the magnetizing field	$\pi/2$

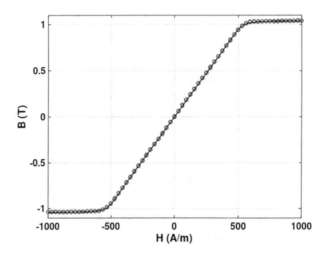

Fig. 1. The results of fitting the model of anhysteretic loop (black line) into the results of measurement of magnetic hysteresis loop of VITROVAC 6125F core (red circles).

As it can be seen from Eq. 5, four physical parameters describing the anhysteretic loop in Jiles-Atherton model can be identified: M_s, a, α and K_{an}. Parameter ψ can be estimated in advance due to the knowledge about the amorphous alloy production process.

Figure 2 presents the results of tests of the dependence of target function F on the changes of all pairs of model's parameters: M_s, a, α and K_{an}. For four parameters, six pairs can be analysed. The analyses were carried out in the neighbourhood of results of identification be differential evolution. To clarify the results, target function F and parameter a were presented in logarithmic scale.

As it can be seen in Fig. 2, dependencies $F(a,\alpha)$, $F(a, K_{an})$, $F(a, K_{an})$, $F(M_s,\alpha)$ and $F(M_s, K_{an})$ exhibit clear and explicit minima. However, the results of identification of function $F(K_{an}, \alpha)$ may lead to unexpected increase of parameter K_{an}. For this reason, results of identification of this parameter of anhysteretic curve of magnetic materials made of material with strong, perpendicular anisotropy should be validated on the base of following physical dependency presented previously in the literature [19]:

$$K_{an} = \frac{B_s^2}{2\mu_0\mu_i} \tag{12}$$

where μ_i is initial relative permeability of magnetic material and B_s is saturation flux density.

To enable validation and verification of results, the software for modelling is available at: http://www.github.com/romanszewczyk/JAmodel in the subdirectory *13_anhysteretic_6125F*.

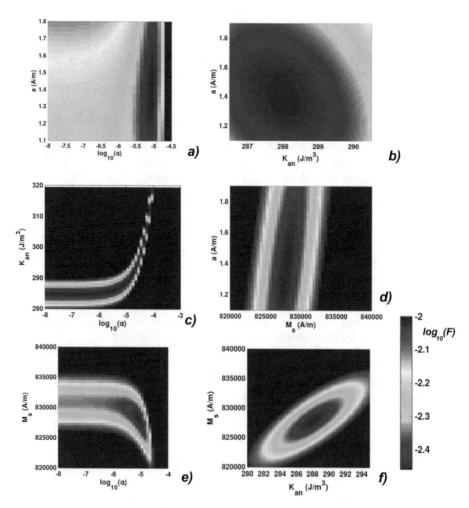

Fig. 2. The results of tests of the dependence of target function F on the changes of all pairs of model's parameters: (a) $F(a,\alpha)$, (b) $F(a, K_{an})$, (c) $F(K_{an}, \alpha)$, (d) $F(a, K_{an})$, (e) $F(M_s,\alpha)$, (f) $F(M_s, K_{an})$.

5   Conclusions

Presented results indicate, that in the case of magnetic materials with strong perpendicular anisotropy (such as VITROVAC 6125F), magnetic hysteresis loop can be efficiently described by anhysteretic magnetization curve used in the Jiles-Atherton model. Moreover, explicit parameters M_s, a, α and K_{an} of anhysteretic curve can be efficiently estimated during the differential evolution based optimisation process.

However, due to the fact, that function $F(K_{an}, \alpha)$ doesn't exhibit clear minima, final value of parameter K_{an} should be verified on the base of physical dependencies connected with hysteresis loop of magnetic materials with strong perpendicular anisotropy.

References

1. Jiles, D.C., Atherton, D.L.: Theory of ferromagnetic hysteresis. J. Mag. Magn. Mater. **61**, 48–60 (1986). https://doi.org/10.1016/0304-8853(86)90066-1
2. Jiles, D.C., Atherton, D.: Theory of ferromagnetic hysteresis. J. Appl. Phys. **55**, 2115 (1984). https://doi.org/10.1063/1.333582
3. Liorzou, F., Phelps, B., Atherton, D.L.: Macroscopic models of magnetization. IEEE Trans. Magn. **36**, 418 (2000). https://doi.org/10.1109/20.825802
4. Steentjes, S., Petrun, M., Glehn, G., Dolinar, D., Hameyer, K.: Suitability of the double Langevin function for description of anhysteretic magnetization curves in NO and GO electrical steel grades. AIP Advances **7**, 056013 (2017). https://doi.org/10.1063/1.4975135
5. Szewczyk, R.: Validation of the anhysteretic magnetization model for soft magnetic materials with perpendicular anisotropy. Materials **7**, 5109–5116 (2014). https://doi.org/10.3390/ma7075109
6. Soliman, E., Hoffmann, K., Reeg, H.: Sensor studies for DC current transformer application. In: Proceedings of IBIC2014, Monterey, CA, USA, pp. 624–628 (2014)
7. Wilson, P., Neil, R., Brown, A.: Optimizing the Jiles-Atherton model of hysteresis by a genetic algorithm. IEEE Trans. Magn. **37**, 989–993 (2001). https://doi.org/10.1109/20.917182
8. Bai, B., Wang, J., Zhu, K.: Identification of the Jiles-Atherton model parameters using simulated annealing method. In: International Conference on Electrical Machines and Systems, Beijing (2011). https://doi.org/10.1109/icems.2011.6073612
9. Biedrzycki, R., Jackiewicz, D., Szewczyk, R.: Reliability and efficiency of differential evolution based method of determination of Jiles-Atherton model parameters for X30Cr13 corrosion resisting martensitic steel. J. Autom. Mobile Robot. Intell. Syst. **8**, 63 (2014). https://doi.org/10.14313/JAMRIS_4-2014/39
10. Szewczyk, R.: Two step, differential evolution-based identification of parameters of Jiles-Atherton model of magnetic hysteresis loops. Adv. Intell. Syst. Comput. **743**, 795 (2018). https://doi.org/10.1007/978-3-319-77179-3
11. Szewczyk, R., Nowicki, M.: Sensitivity of Jiles-Atherton model parameters identified during the optimization process. AIP Conf. Proc. **1996**, 02004 (2018). https://doi.org/10.1063/1.5048898
12. Ramesh, A., Jiles, D.C., Roderik, J.: A model of anisotropic anhysteretic magnetization. IEEE Trans. Magn. **32**, 4234–4236 (1999). https://doi.org/10.1109/20.539344
13. Ramesh, A., Jiles, D.C., Bi, Y.: Generalization of hysteresis modeling to anisotropic materials. J. Appl. Phys. **81**, 5585 (1997). https://doi.org/10.1063/1.364843
14. Nowicki, M.: Anhysteretic magnetization measurement methods for soft magnetic materials. Materials **11**, 2021 (2018). https://doi.org/10.3390/ma11102021
15. https://www.vacuumschmelze.com/en/the-company/quality/information-sheets-msds/vitrovac.html
16. Chwastek, K., Szczygłowski, J.: Identification of a hysteresis model parameters with genetic algorithms. Math. Comput. Simul. **71**, 206–211 (2006). https://doi.org/10.1016/j.matcom.2006.01.002

17. Jiles, D.C., Thoelke, J.B., Devine, M.K.: Numerical determination of hysteresis parameters for the modeling of magnetic properties using the theory of ferromagnetic hysteresis. IEEE Trans. Magn. **28**, 27–35 (1992). https://doi.org/10.1109/20.119813
18. Pop, N.C., Caltun, O.F.: Jiles–atherton magnetic hysteresis parameters identification. Acta Phys. Pol. **A 120**, 491 (2011)
19. Buttino, G., Poppi, M.: Dependence on the temperature of magnetic anisotropies in Fe-based alloys of Finemet. J. Magn. Magn. Mater. **170**, 211 (1997). https://doi.org/10.1016/S0304-8853(97)00005-X

Uncertainty of Measurement and Reliability of the Decision Making on Compliance

Eugenij Volodarsky[1], Zygmunt L. Warsza[2(✉)], Larysa A. Kosheva[3], and Maryna A. Klevtsova[4]

[1] Department of Automation of Experimental Studies,
National Technical University of Ukraine "KPI", Kiev, Ukraine
vet-l@ukr.net

[2] Industrial Research Institute of Automation and Measurement (PIAP),
Warsaw, Poland
zlw@op.pl

[3] Department of Biocybernetics and Aerospace Medicine,
National Aviation University of Ukraine, Kiev, Ukraine
l.kosh@ukr.net

[4] LTD «Svityaz Factory», Kiev, Ukraine
switmak@gmail.com

Abstract. It is shown that the probability of making the right decision on the compliance of the object with the norms depends on the measurement uncertainty associated with the result of the measurement control of the object parameters and the length of the tolerance interval. The decisive rule on the suitability (unfitness) of the object of control should take into account both the possible uncertainty of the result and the requirements for the cost and complexity of the control procedure, as well as to the specified reliability. To improve the reliability of this procedure, a control method based on a sequential adaptive decision-making procedure is proposed, which takes into account the relationship of the parameters of the distribution law of the possible values of the controlled quantity, random effects in the measurement and the length of the tolerance interval. The decision-making procedure assumes consistently, depending on the result at the current stage, to introduce additional tolerance intervals, the length of which is determined by the parameters of the distribution function of random variables accompanying the measurements of the controlled parameters. The result of measurement of the controlled parameter is compared with these intervals, and decisions on continuation or completion of control with introduction of additional limits are made. This provides a reduction in the probability of erroneous decisions at each additional stage of control, since the time of the initial hit of the controlled value in a consistently calculated controlled interval is determined step by step. The use of a sequential adaptive procedure provides a given reliability of control and does not allow to carry out the control procedure in full for all possible values of the controlled value.

Keywords: Quality control · Technological process ·
Measurement inspection · Conformity assessment ·
Uncertainty of measurement · Adaptive successive algorithm ·
Probability of decision making

© Springer Nature Switzerland AG 2020
R. Szewczyk et al. (Eds.): AUTOMATION 2019, AISC 920, pp. 672–683, 2020.
https://doi.org/10.1007/978-3-030-13273-6_63

1   Introduction

Inspection as a procedure of conformity assessment of products to specified requirements is based, as a rule, on the results of measurement of a certain characteristic
property of products. For the measured value, the requirements are set by the tolerance
or compliance limits, which separate the intervals of permissible values of the measured value from their critical (invalid) values. The object meets the specified
requirements when the value of its property (parameter) is within the tolerance [1].

The readings of the measuring system [2, 3] reflect the information about the value
of the controlled value using the measurement model, including the effects of both
systematic and random effects (or their combination). If the influence of systematic
effects can be taken into account (corrected) during the calibration of the measuring
system, the effect of random effects remains and is estimated as uncertainty of the
measurement result. Because of the measurement uncertainty, there is always a risk
(probability) of making an erroneous decision about the compliance or non-compliance
of the object (its parameter) with the established requirements on the basis of the
measured value of the object property (its parameter). Thus, the assessment of compliance with the specified requirements is a probabilistic task based on measuring
information. This paper is a continuation of authors' works [4–7] in this field.

2   Stated Problem

Consider the case when the result of measurement z_1 of the controlled parameter is in
the zone of correspondence, for example, does not exceed the upper limit of the norm
x_L ($z_1 > x_L$). Naturally, the result of this measurement is decided that the product
corresponds to the norm. But this result is the sum of the possible values of x^0 of the
controlled quantity X, and the values of the influencing random variable y^0, i.e. $z_1 =
x^0 + y^0$. Therefore, the result z_1 can correspond to the combinations of realizations of
random variables (y_1, y_2) in the measurement of the relevant parameters x_1 and x_2.

Thus, there are many combinations: $z_1 = (x_1 + y_1)$, $z_1 = (x_2 + y_2)$, etc. However,
the same result of z_1 can be obtained as $z_1 = x_3 + y_3$, or $z_1 = x_4 + y_4$, (y_3 and y_4 are the
realizations of the random variable when measured, respectively, x_3 and x_4, which are
less than the limit value). In this case, an erroneous decision is made that the product
corresponds to the norm, although in reality it is not, since both x_3 and x_4 are less than
the limit value of x_L. Such an erroneous solution is called an indefinite nonconformity
of the product [8]. The probability of such an event depends on the possible value of
the controlled quantity, as well as the type and parameters of the distribution function
of possible values of the influencing random variable y^0. We accept, as it is done in
most practical cases, that the distribution of the value y^0 is normal. From the above we
can conclude that the farther from the limit value x_L is the value of the controlled value
of the property of the object X, not corresponding to the norm, the less likely an
erroneous decision on compliance.

3   Disadvantages of the Used Methods

The measurement accuracy, which depends on the accuracy of the measuring system used during the control procedure, is estimated by expanded uncertainty $U = 2u_m$ [7, 9] taking into account the standard uncertainty of measurement and u_m. In this regard, starting with the value of the controlled value, which is more than the limit value of the expanded uncertainty U, erroneous decisions can occur with a probability of 0.05, which is a satisfactory result in practice.

In the case where the interests of the customer have priority, in order to obtain the probability of making a decision is not less than the specified (permissible), in the documents [10–12], it is proposed to introduce a guard band w. To form a guard band allows the introduction of the acceptance interval, the acceptance limits of which are spaced from the tolerance limits (shifted to the middle of the tolerance) on the guard band

$$w = U = 2u_m.$$

Figure 1 shows as the upper acceptance limit $x_{U'}$ located on the inner side relative to the upper tolerance limit x_U, determines an acceptance interval that reduces the probability of mistakenly accepting non-conforming item (consumer's risk). By convention, the length parameter w associated with the guard band is a positive value:

$$w = x_U - x_{U'} > 0.$$

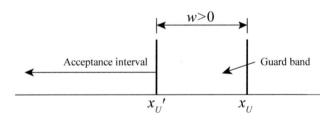

Fig. 1. Location guard band.

The decision rule based on the guard band is called guarded acceptance or guar banding [13]. In the literature, this concept is known as a specification zone in acceptance [14], in [15] to demonstrate compliance with technical requirements, default setting guarded acceptance rule, and an acceptance interval is called a conformance zone. The width of this band should be equal to the standard uncertainty, beyond which the residual effect of random variables does not exceed 5%. This is exactly what is recommended in the document [10].

The method of decision-making on compliance with the imposition of guard bands can reduce the probability of first-rank error by increasing the probability of second-rank error. This is achieved by a corresponding offset of the acceptance limits relative to the tolerance limits. Thus, the introduction of guard bands between the limits of the tolerance interval and the corresponding acceptance limit may lead to losses of the manufacturer, which, under certain conditions, may be significant.

The direct way to reduce the impact of measurement uncertainty on the correctness of the decision on compliance is to increase the requirements for the accuracy of the measuring system. But this leads to an increase in its complexity, and therefore to an increase in its cost, size and weight. This increases the measurement time, decreases the speed of the measuring system, which is inversely proportional to the complexity of the measuring system. It is known that the effectiveness of the decision-making procedure E on compliance with the control is a functional control reliability B, performance P and cost C

$$E = F(P \cdot B \cdot C).$$

Thus, solving the problem of improving the efficiency of control, the gain is achieved on one indicator-the probability of making a decision on compliance, and the loss on the other two – productivity and cost [16].

Another approach to reduce the influence of measurement uncertainty is to conduct multiple observations of the controlled value. As a result of measuring the characteristic is the arithmetic mean value. This approach also has significant drawbacks.

First, although the standard uncertainty u decreases by a factor $\sqrt{n}$, the volume of control and measurement operations increases by Wn times (W is the number of controlled objects, and n is the number of parallel observations). But the further "is" controlled value from the limiting value, the lower the probability of the situation when it can be taken wrong decision about compliance. Moreover, there is a certain range of possible values of the controlled value for which erroneous decisions on compliance cannot be made. In this case, you do not need to conduct additional experiments. Thus, the direct carrying out of additional observations with their subsequent averaging has redundancy.

Direct access of the situations considered above, is the use of positive properties of each approach: inserting the protective strip, that is, acceptance of the border $x_{L'}$, $x_{U'}$, which shifted inwards tolerance interval for the value of extended uncertainty, and the presence of the primary measurements n additional observations. This provides a reduction in material costs that arise when deciding on compliance. For objects, the results of the primary transformations which are in the interval $(x_{U'}, x_U)$ or $(x_{L'}, x_L)$, a decision is made under, and for objects, the results of which are not located in these intervals is carried out n parallel observations.

We will analyze how this will increase the volume of control and measurement operations. Without breaking the generality of the obtained conclusions, we will consider the situations that arise in the area of the lower limit value of x_L in the evaluation of the conformity of the object.

The number of objects s_0, which after the primary control and measurement operation requires additional observations, is a discrete random variable that can take the values $l = 0, 1, \ldots, W$, that is, the case when the interval $(x_L, x_{L'})$ gets the results of observations of all objects.

We denote by p the probability that a decision is made on the compliance of the object with the norms, and this control ends. Then $q = 1 - p$ is the probability that the result of the initial (primary) measurement was in the zone of uncertainty, and there is a need to move to a sequential procedure. The probability that the discrete random number of hits of the results of the primary results will take the value of l from a possible number W is equal to

$$P\{s_0 = l\} = C_W^l p^{W-1} q^l. \tag{1}$$

The expectation of a random variable s_0, which can take values $l = \overline{0, W}$, is defined as

$$M\{s_0\} = \sum_{l=0}^{W} l \cdot C_W^l p^{W-l} q = qW \tag{2}$$

Then the mathematical expectation of full volume of control and measuring operations taking into account the main primary and additional control and measuring operations is:

$$\tilde{V} = W + M\{s_0\}(n - 1) = W + Wq(n - 1)$$

or

$$\tilde{V} = W[1 + q(n - 1)]. \tag{3}$$

The increase in the volume of control and measurement operations in comparison with the initial (primary) volume $V_0 = W$ is determined as

$$\underset{W \to \infty}{Lim} \frac{\tilde{V}}{W} = 1 + q(n - 1). \tag{4}$$

However, the distance at which the value of the controlled quantity is relative to the limit value was not taken into account. This distance depends on the parameters of the distribution function of the possible values of the controlled value and may affect the adoption of an erroneous decision on the discrepancy.

4   Sequential Procedure of Control

To take into account the law of distribution of possible values of the controlled value, it is advisable to organize the control procedure so that, based on a preliminary assessment of the value of the controlled value, a decision on the need for additional observations.

In this case, the decision to extend or complete the control procedure is made at each current stage and depends on the influence of random variables in the previous stages. This procedure is called sequential in statistics [17]. A feature of the control with the introduction of a sequential procedure is a differentiated approach, which consists in the evaluation of each object entering the control, and individual clarification of its condition. In the traditional integral approach, a fixed number of multiple measurements for all controlled objects is carried out, followed by averaging the results.

The total number of measurements (volume), including the steps of the sequential procedure is

$$V_{W(r)} = W + s_0 + s_1 + \ldots + s_r,$$

where r – is the number of stages; $s_0, s_1 \ldots$ – numbers of objects in primary measurements of differently classified results.

$$MV_{Wp}(r) = W + Wq \frac{1 - q^{r-1}}{1 - q} = W(1 + q \frac{1 - q^{r-1}}{1 - q}).$$

The number of objects s_0 with results fell into the interval $(x_L, x_{L'})$ and for which it is decided to carry out an additional control and measurement operation; s_1 – the number of objects, the results of which after the first additional stage remained in the interval $(x_L, x_{L'})$, and it is decided to conduct a second additional control and measurement operation, etc.

Since the number r of additional stages of the sequential control and measurement procedure for each specific set (batch) of objects is a random value associated with the distribution function of the controlled quantity $f(x)$ and the possible values of the influencing quantities $f_1(y)$, then $s_0 + s_1 + \ldots + s_r$ are random values. Then the mathematical expectation of the total volume of control and measuring operations is

$$MV_{W(r)} = W + Ms_0 + Ms_1 + \ldots + Ms_r. \tag{5}$$

Thus, after controlling for s_0 object in the first additional step, the measurement results which fall into the interval $(x_L, x_{L'})$, part of the results will be more $x_{L'}$, i.e. $z_i > x_{L'}$ and the decision on the match object standards. There is a part of objects from the subset s_0, the results of which parameters measurement will be less than x_L, i.e. $z_i < x_L$, and the decision on discrepancy is made.

But there is still a part of objects s_1 – which after the first additional stage of n observations remained in the interval $(x_L, x_{L'})$. For them, it is decided to hold the second additional stage.

Since s_0 and s_1 are dependent quantities, it is necessary to determine the expectation conditions

$$Ms_1 = M(Ms_1|s_0).$$

Since the expectation of a discrete random variable $s_1|s_0$, which takes values from 0 to l, is equal to the sum of the products of the values of this value and the probability of occurrence of such an event, we obtain

$$M(s_1|s_0) = \sum_{j=0}^{l} j \cdot C_l^j \cdot p^{l-j} \cdot q^j = ql.$$

Thus, the number of control and measuring operations at the first additional stage is determined by the expression

$$M(s_1) = M(M(s_1|s_0)) = \sum_{l=0}^{W} q \cdot l \cdot p\{s_0 = l\} = q\sum_{l=0}^{W} l \cdot C_W^l p^{W-l} q^l = ql.$$

$$MV_{Wp}(r) = W + W \sum_{j=1}^{r} q^j \tag{6}$$

After similar mathematical transformations, we obtain an expression for the mathematical expectation of the total (taking into account r additional stages) control volume when applying a sequential procedure

Using the formula for the sum of the geometric progression, the expression (6) is written as

$$MV_{W\ \text{посл}}(r) = W + Wq\frac{1-q^{r-1}}{1-q} = W(1 + q\frac{1-q^{r-1}}{1-q}).$$

Thus, the increase in the volume of control and measurement operations is

$$\lim_{W \to \infty} \frac{M(V_p)}{W} = 1 + q\frac{1 - q^{r-1}}{1 - q}. \tag{7}$$

If we compare the expressions (7) and (4), characterizing the increase in the volume of control and measurement operations in relation to the volume with simple acceptance, we can conclude that the method of sequential decision-making has a much smaller increase in the volume of control and measurement operations.

Thus, the method of sequential decision-making with fixed limit values of the controlled value can improve the reliability of control at a lower time cost, and therefore reduces the cost of production without loss of quality.

5   Relationship of Measurement Uncertainty and Probability for a Decision of the Compliance of Controlled Objects

There are such enterprises in which the varieties of one type of product can change quite often. This is due to the change in the length of the tolerance interval, and the uncertainty of the measurement procedure, which precedes the control, remains unchanged.

We define the relationship between the probability of erroneous decisions and the standard uncertainty of measurements. To do this, we use the ratio of the length of the tolerance interval and the standard uncertainty u_m. In the document [10] the parameter characterizing quality of measurement concerning requirements to object of control which are set in the form of the admission is entered. This parameter is called the measurement capability index and is defined as

$$C_m = \frac{x_U - x_L}{4u_m} = \frac{T}{2U}, \tag{8}$$

where x_U, x_L – upper and lower tolerance limits, $U = 2u_m$ – expanded uncertainty with coverage factor $k = 2$; T – tolerance.

It follows from equality (8) that $C_m > 4$ at $u_m \leq T/16$. The coefficient 4 is selected using the coverage interval $[z - 2u_m; z + 2u_m]$. Based on the value of the measurement capability index C_m, it is possible to calculate the a priori probability of compliance of the object with the specified norms for the tolerance limits (x_U, x_L).

Consider the case of a two-sided tolerance interval, since it is known that a one-sided tolerance interval is a special case of a two-sided one, where one limit is explicitly set (its value is recorded in the relevant regulations), the other limit is set implicitly, based on physical or theoretical reasons.

According to the normal distribution function of possible deviations of the measurement results z from the true value of the measured value x, the conformance probability is calculated as

$$p_c = \Phi\left(\frac{x_e - x}{u_c}\right) - \Phi\left(\frac{x_u - x}{u_c}\right), \tag{9}$$

where $\Phi(a) = \int_{-\infty}^{a} e^{-t^2} dt$ – Laplace function.

The measured value which is within the tolerance interval $T = x_U - x_L$ is entered. For values in the tolerances area the relative value is

$$\tilde{x} = \frac{x - x_L}{T}, \tag{10}$$

which takes values:

$$\tilde{x} = \begin{cases} 0 & \text{when } x = x_L \\ 1 & \text{when } x = x_U \end{cases}.$$

Substituting the expression (10) in the expression (9), after some transformations, we obtain a dependence that relates the conformance probability and the measurement capability index, which considers the relative value of the controlled quantity

$$p_c = \Phi[4C_m(1 - \tilde{x})] - \Phi(-4C_m\tilde{x}) = p_c(\tilde{x}, C_m). \tag{11}$$

In Fig. 2 is diagram which shows at what ratios C_m and $\tilde{x}$, value of conformance probability p_c remains constant and equals 95% for values of the size lying in the range is given within the range $0 \leq \tilde{x} \leq 1$.

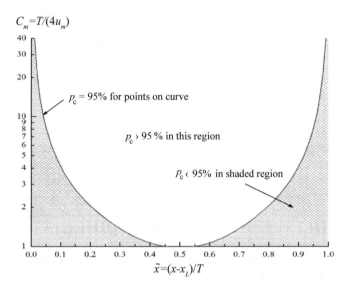

Fig. 2. Ratios C_m and $\tilde{x}$ of the conformance probability $p_c = 95\%$.

The curve separates the matching (shaded) and nonconformity (unshaded) areas. As you can see in Fig. 2, if $C_m = 1$ ($u_m = T/4$), the conformance probability $p_c \geq 95\%$ is valid only for the relative controlled values in the range $0.45 \leq \tilde{x} \leq 0.55$. To expand the range of possible controlled values, it is necessary to increase the value of C_m. A direct way to achieve this is to reduce the uncertainty of measurement u_m.

Since the measuring instruments on the production line remain the same, the instrumental component of measurement uncertainty remains the same. This leads to a change in C_m, which in turn affects the conformance probability, and ultimately, the probability of making the right decision on the results of inspection.

6   Implementation the Method of Adaptive Control Values

To avoid possible disadvantages associated with the introduction of guard bands, conducting multiple measurements and increase in the accuracy of devices, it is proposed a serial adaptive method of decision-making on conformity (adaptive control of limit values).

At the beginning of the decision-making procedure, based on the real ratio of the length of the tolerance interval and measurement uncertainty (Fig. 2), the output relative control limits $\tilde{x}_{1L}$ and $\tilde{x}_{1U}$ are defined, which correspond to the probability of making a decision on the conformity of the object $p_c = 95\%$. Based on the expression (10) the relative values of these control limits are

$$\tilde{x}_{1L} = \frac{x_{1L} - x_L}{T}, \tilde{x}_{1U} = \frac{x_{1U} - x_U}{T},$$

The control limits for the primary control measurement are

$$x_{1L} = \tilde{x}_{1L}T + x_L, \tag{12}$$

$$x_{1U} = \tilde{x}_{1U}T + x_U. \tag{13}$$

The primary measurement result z_1 is compared with these reference limit values. If z_1 is within range

$$x_{1H} \le z_1 \le x_{1B} \tag{14}$$

then with probability $p_c \ge 95\%$ the decision on compliance of object (parameter) to the set norms is made and the control procedure comes to the end.

If the inequality (14) is not satisfied then proceed to the procedure of adaptive determination of control limits and comparison with them of the calculated average value of the results of two parallel observations:

$$\bar{z}_2 = \frac{z_1 + z_2}{2},$$

It is assumed that the systematic effects were taken into account in the calibration of the measuring system.

The value of the relative control limits is found on the same chart (Fig. 2) for the adjusted measurement capability indicator

$$C_1 = T/4u_1,$$

Where $u_1 = u_c/\sqrt{2}$.

The absolute values of the control limits in the second stage x_{2L} and x_{2U} are calculated by formulas (12) and (13), where the output values are $\tilde{x}_{2L}$ and $\tilde{x}_{2U}$, which are found on the chart (Fig. 2) for C_1. Again, it is checked if the following inequality is satisfied but with different control limits

$$x_{2L} \le \bar{z}_2 \le x_{2U}.$$

If the inequality is satisfied, the decision on compliance is made. Otherwise, the third measurement is carried out, and the average value of the three measurements is compared with the new calculated limits x_{3L} and x_{3U}, etc.

The number of additional measurements for each controlled object depends on the probability of hitting $i + 1$ average value between the control limits of this interval, provided that in the previous stage the average value was between the control limits of the i-th control interval, however, the probability of compliance was less than 0.95.

7 Conclusions

Measurement uncertainty leads to erroneous decisions about product compliance with established standards. Introducing a guard band practically eliminates the influence of measurement uncertainty on the decision made. However, this leads to first-rank error – loss of the manufacturer.

Conducting a fixed number of n additional measurements of a controlled quantity for results that were in the guard band, increases the volume of control and measurement operations by qn. So, with the probability of a measurement result falling into the guard band $q = 0.1 - 0.15$ and $n = 2$, the relative increase in volume of measurements will be from 10% to 20%.

In the event of a measurement result falling into the guard band the use of a sequential procedure of control for detecting conformity of an object allows with a slight increase in the volume of control and measurement operations, to reduce the probability of first-rank error. So, for the same q values, the relative increase in the measurement volume will be from 1.1% to 2.5%. Such an algorithm is recommended for use in serial production, when it is possible to choose measuring equipment, and thereby provide the required measurement capability index.

With a small production, the product range can change even within one day, and the measuring equipment remains the same, which changes the measurement capability index, and hence the probability of compliance. In this case, it is effective to apply a sequential procedure with an adaptive change of control values.

References

1. ISO/IES 17000:2004 (rev. 2014). Conformity assessment. Vocabulary and general principles
2. International vocabulary of metrology - Basic and general concepts and associated terms (VIM), JCGM 200:2008, Joint Committee for Guides in Metrology (JCGM) (2008)
3. Tirkel, I., Rabinowitz, G.: The relationship between yield and flow time in a production system under inspection. Int. J. Prod. Res. **50**, 3686–3697 (2012)
4. Volodarsky, E., Warsza, Z., Kosheva L.A., Idzkowski, A.: Method of upgrading the reliability of measurement inspection. Advances in Intelligent Systems and Computing, vol. 393, pp. 431–438. Springer, Cham (2016)
5. Volodarsky, E., Warsza, Z., Kosheva, L.A., Idzkowski, A.: Precautionary statistical criteria in the monitoring quality of technological process. Advances in Intelligent Systems and Computing, vol. 543, pp. 740–750. Springer, Cham (2017)
6. Warsza, Z., Idzkowski, A.: About accuracy of the simultaneous measurement of temperature difference and average using transducers with Pt sensors. Przemys. Chem. **96**(2), 380–385 (2017). (in Polish)
7. Idzkowski, A., Walendziuk, W., Swietochowski, P., Warsza, Z.L.: Metrological properties of a two-output transducer for measuring sum and difference of small resistances. Elektron. Elektrotech. **23**(5), 41–45 (2017)
8. ISO 10576-1:2003. Statistical methods. Guidelines for the evaluation of conformity with specified requirements. Part 1: General principles
9. ISO/IEC Guide 98-3:2008 Uncertainty of measurement – Part 3: Guide to the expression of uncertainty in measurement (GUM:1995). Geneva: ISO/IEC (2008)

10. JCGM 106:2012. Evaluation of measurement data – The role of measurement uncertainty in conformity assessment
11. IEC Guide 115:2007. Application of uncertainty of measurement to conformity assessment activities in the electrotechnical sector
12. OIML G 19:2017. The role of measurement uncertainty in conformity assessment decisions in legal metrology
13. Deaver, D.: Guard banding with condense. In: 1994 NCSL Workshop and Symposium, pp. 383–394 (1994)
14. American Society of Mechanical Engineers. ASME B89.7.3.1:2001 Guidelines for decision rules: Considering measurement uncertainty in determining conformance to specifications, New York (2001)
15. ISO 14253-1:1998 Geometrical Product Specifications GPS - Inspection by measurement of workpieces and measuring equipment - Part 1: Decision rules for proving conformance or non-conformance with specifications
16. Sheu, S.-H., et al.: Economic optimization of off-line inspection with inspection errors. J. Oper. Res. Soc. **54**, 888–895 (2003)
17. Jonhson, N., Leone, F.: Statistical and Experimental Design in Engineering and the Physical Science, vol. 2, 2nd edn. Wiley and Sons, New York (1977)

The Effect of Ground Changes and the Setting of External Magnetic Field on Electroplating FeNi Layers

Anna Maria Białostocka$^{(\boxtimes)}$ [ID] and Adam Idźkowski [ID]

Faculty of Electrical Engineering, Bialystok University of Technology,
Wiejska 45D, 15-351 Bialystok, Poland
a.bialostocka@pb.edu.pl

Abstract. The electrodeposition of iron-nickel alloys has been studied with different conditions of substrate and magnetic field arrangement. All experimental conditions influence the resultant film quality, the alloy composition and morphology. Studies show differences in before mentioned features of ferromagnetic metals (Fe, Ni) electrodeposited on non-ferromagnetic substrates – paramagnetic (Cu) and diamagnetic (Ag). The presence and absence of the external magnetic field influences FeNi alloys too. The obtained layers were investigated by Scanning Electron Microscopy (SEM), X-ray Diffraction (XRD) and X-ray Fluorescence (XRF).

Keywords: Neodymium magnet · Electrocrystallization · Alloy · Morphology

1 Introduction

FeNi alloys have been the matters of interest for a long time (nearly 150 years). An intermetallic compounds Fe_3Ni, Ni_3Fe, Ni_2Fe were found in meteorites and FeNi is a native alloy of meteorites too. Researchers were motivated to carry out detailed studies on the FeNi alloy. They wanted to find the roots of the most interesting properties (strong directional bonding and ferromagnetism) of iron-nickel layers, which make them one of the toughest materials. The systems are manufactured with varied techniques, for example vacuum evaporation, cold rolling, single-roll rapid quenching and sputtering electrodeposition. As a result, alloys, multilayers and nanowires can be obtained. This is determined by the place where the FeNi systems will be used, such as high-performance transformer cores, read-write heads, magnetic actuators, magnetic shielding, thermostatic bimetals, glass sealing, integrated circuit packing, cathode ray tube shadow masks, composites molds/tooling, the membranes for liquid natural gas tankers etc. [1–4]. Depending on the thickness, the material can take form of coatings, a continuous thin foil (10–250 µm) and a thick plate (> than 5 mm). Magnetic, mechanical and corrosion properties of FeNi alloys have been studied by several researchers [5–7]. These properties are affected by different plating parameters, for instance electric, chemical and physical. The electric variables are represented by applied current density and voltage conditions [8], and the chemical variables are represented by plating bath chemistry, pH [9–11]. The physical parameters include,

© Springer Nature Switzerland AG 2020
R. Szewczyk et al. (Eds.): AUTOMATION 2019, AISC 920, pp. 684–696, 2020.
https://doi.org/10.1007/978-3-030-13273-6_64

among others, temperature, electrolyte mixing conditions [4, 12]. Hitherto, the FeNi structures were obtained as a result of using a direct current (DC) or pulse current (PC) power supply [8]. It was confirmed that the above mentioned deposition parameters can control primarily composition and structure of the deposited FeNi alloys.

Plating iron group metals (Fe, Co, Ni) is called an anomalous type and Brenner [13] was the first who implemented a definition of particular co-deposition. Then a lot of successors have continued his work (Dahms and Croll, Horkans, Lieder and Biallozor, Grimmett, Schwartz, and Nobe, Zech, Podhala, and Landolt) [14–19]. The anomalous codeposition (ACD) results in higher concentration of a less noble metal in the deposit in comparison with this metal's presence in the solution. The preferred bath (for iron-nickel electrodeposition) is sulfate solution of pH 1–3. The electrolyte solution could also contain chloride and fluoroborate compounds, citric acid, glycolic acid, boric acid and other organic additives which work as a moisturising agent (lauryl sulfate) or stress reducer (saccharin) [20]. A common condition to conduct the process is galvanostatic deposition [14, 21–23]. Firstly, the anomalous deposition was attributed to pH rise as a result of the hydrogen evolution [24]. Now, based on this knowledge, some scientists suggested that formation of $Fe(OH)_2$ inhibits nickel deposition [9, 23]. Hessami and Tobias (1989) attributed inhibition of Ni to $Ni(OH)^+$ and $Fe(OH)^+$ formation. The high concentration of the last compound near the electrode results in the appearance of $Fe(OH)^+$ over the entire surface and has a very negative impact on the Ni ions deposition. For years, scientists have proposed various models of the electrodeposition process and have tried to find correlations between modelling and the experiments [25].

These days some scientists reported that the morphology, a change in structure and composition of the obtained layers is influenced by both settings (perpendicular, parallel) of the external magnetic field (EMF), mainly Fahidy (1983) and Aogaki [26–29]. Magnetic induction affects the surface of the deposited layer by a magnetohydrodynamic (MHD) effect. Magnetic field application is the right approach to create additional convection (in an electrochemical cell) which is induced by the Lorentz force

$$F_L = j \times B, \qquad (1)$$

where: j – current density, B – magnetic flux density.

When B and j are orthogonal, the force from Eq. (1) reaches largest value. Unevenness of j (at the edge of the cathode, in particular) is the reason for the existence of the Lorentz force only in the places where the current density has a non-normal component. This causes uneven growth of the layer and occurrence of micro-MHD vortices. In the majority of cases the external magnetic field influences the hydrogen co-reduction as well. An increase in free hydrogen ions reduction was observed in the presence of convection (influenced by a magnetically-induced flow). Hydrogen bubbles sweep out of electrode surface and thereby facilitate the growth of the deposit. Thus, the magnetic field modifies a bubbling regime. The EMF interacts with local current density. The MHD-induced flow reduces thickness of a diffusion layer and enhances mass transport. The results from morphology, texture, roughness and dendritic electrodeposits depend on orientation of magnetic induction in relation to the cathode surface. Aogaki ascertained the facts that parallel (II) arrangement of field and electrode surface activates two-dimensional growth and smoother deposit. Rough precipitates are the result of the perpendicular (I_) relationship between EMF and a working electrode. The field configuration (II or I_) induces the preferred crystallographic texture of the FeNi deposits. The forced form of this texture is obtained by the modifications of the grain shape (rounded, needle-like). The EMF affects both parts of the electrodeposition process, which is growth and nucleation of the obtained alloys. Some investigators [28, 30, 31] have found that the nucleation process perfectly describes two parameters: A – nucleation rate constant and N_0 – nucleation site density (nuclei at saturation). Instantaneous nucleation ($A \gg 1$) is characterized by high growth rate of a new phase and a small number of formed active nucleation sites. Progressive nucleation ($A \ll 1$) implies slow growth rate and a large number of a new nuclei formed respectively. An increase of nucleation site density results in decrease of the growth rate. We can find in literature different values of the mentioned parameters. Researchers carried out the experiments under varied conditions of potential's value, the presence of additives, values of the current density etc. The mentioned requirements affect A and N_0 parameters. A magnetic field application retards N_0 but A is not altered by the magnetic field. Magnetic flux density and delaying AN_0 is proportional. Explorations have given response to the question if the action of EMF modulates early stages of the layer growth. Only the growth process is affected by the EMF. Nucleation process is unaffected [28, 32, 33]. Besides, the mentioned action of the Lorentz force and magnetic gradient force is being strengthened in the case of parallel configuration of both fields in particular

$$F_{\mathrm{B}} = \frac{\chi_m C}{\mu_0} (\boldsymbol{B} \nabla \boldsymbol{B}), \tag{2}$$

where: χ_m – molar magnetic susceptibility, μ_0 – magnetic permeability, C – concentration of metal ions.

Natural convection is mainly apparent the cathode surface is positioned vertically. In this case interaction between the Lorentz force and driving is force very important. Involvement of applied forces varies by type of substrates and embedded metal (ferromagnetic or non-ferromagnetic) [35–37]. Therefore, the substrate and its asperities influence an electroplating process too. Material of the substrate and finish of its surface affects particularly initial stages of deposition process. Nucleation on the cleaned substrate is significantly different from the same process taking place on a sub-coated substrate. It is particularly visible when a crystal structure of the substrate is similar to the deposit structure. The substrate was found to have an influence on the structure of the deposited layer up to 10 μm in depth. There are procedures which result in depletion of the substrate's influence. A cleaned, degreased and polished surface is necessary to carry out the process of depositing a continuous, high quality film with good adhesion. There is another interesting aspect. During the electroplating with the external magnetic field assistance, EMF can be deformed by the quality of the substrate and influence lines the electroplated film as a final outcome. Then a gradient effect starts playing a greater role. Current density-potential curves obtained in different substrate conditions exhibit varies. For example, hydrogen reduction rate (HER) is different for a various substrates. The hydrogen evolves on the substrate during initial stages of deposition (time up to several minutes). Hydrogen evolution reaction depends on coefficients a and b in the Tafel equation

$$E = a + b \, \ln i, \tag{3}$$

which are characteristic of the substrate.

Low overpotential (low coefficient a) on the substrate results in predominant hydrogen production. Appearance of a thin deposit causes an increase in the current efficiency of the electrodeposited metal. During these initial stages hydrogen concentration can achieve saturation on the thin film of the embedded metal and then diffuse into the substrate. It leads to damaging side effects (pores of the thin film or diffusion through the coating) [38]. The dominant importance of the hydrogen reduction current onto the paramagnetic substrate (Cu) in comparison to the diamagnetic substrate (Sn) was shown by Fritoceaux and Russo. It was noticed that kinetics transformations were more rapid during the electrodeposition on Cu. In this case, came round to modification of adsorption sites and changes in mechanism responsible for efficiency current. The efficiency of the current value decreased after using the copper substrate. Directions of changes in the hydrogen evolution rate affected the composition of a layer. Therefore, different substrates influenced iron adsorption and nickel percentage content [20, 34, 39, 40]. The topic of the article includes three scientific fields: electrochemistry, hydrodynamics and magnetism. In the present study, iron-nickel alloys were produced using direct current power supply in the presence or absence of EMF. Layers have been deposited on the substrates with different magnetic properties: silver (diamagnetic) and copper (paramagnetic) [41].

2 Materials and Methods

For the deposition of FeNi layers on Cu and Ag vertical plates (working electrode: width 10 mm × height 20 mm × thickness 0.25 mm), the following chemicals originating from POCH were used: 15% $FeSO_4 \cdot 7H_2O$, 30% $NiSO_4 \cdot 7H_2O$, 0.4% H_3BO_3. The cylindrical glass cell (in 45 mm diameter) was used in all experiments (20 ml of the electrolyte volume). The platinum vertical plate (width 6 mm × height 5 mm × thickness 0.5 mm) was used as a counter electrode at a 20 mm distance to the working electrode [42, 43]. Both electrodes were centered in the major axis of the magnetic field source (symmetrically arranged between two permanent magnets – 1T, NdFeB, IBS Magnet). The used magnets (width 75 mm × height 50 mm × thickness 10 mm) were distanced 55 mm from each other [44]. The permanent magnets are magnetized in direction normal to the largest magnet face (75 mm × 50 mm) and B is oriented horizontally in both cases (II and I_). The orientation of the magnetic field direction was out or in plane of the deposition plate. The magnetization vectors of both magnets were oriented anti-parallel (II and I_). The magnetic field distribution was measured along the largest magnet face with the gauge FH51 (Magnet-Physik). The magnetic field strength varied according to the distance from the magnet and ranged from 80 mT to 400 mT (with accuracy of ±2%). The distribution of the magnetic field around the electrodes was uniform. All experiments were conducted at the temperature of 21 ± 1 °C. The electrodeposition was performed galvanostatically using a potentiostat/galvanostat instrument (Matrix MPS-7163). The surface morphology of samples was investigated using a Scanning Electron Microscope (SEM), which provided information on the composition of the alloy and layer's thickness. X-Ray Diffractometer with Mo Kα radiation ($\chi = 0.713067$ Å) probe was used for crystal structure characterization.

3   Preparation of FeNi Layers

Each copper plate was electrochemically polished and activated by immersion into a mixture of perchloric acid and ethanol in a 25:75 ratio. Galvanostatic electrodeposition of the FeNi layers was carried out with two values of electric current density (50 mA/cm² and 100 mA/cm²) and time (60, 45, 30, 15, 7.5 min) conditions. The FeNi films were prepared from a plating solution with a mixture of Ni and Fe ions, with a composition 1:2 for Fe:Ni.

Experiments of the deposition of iron-nickel layers in the presence of the external magnetic field were done with a specially designed apparatus, where a set of two permanent magnets was rigidly placed in parallel (II, Fig. 1B) or perpendicularly (⊥, Fig. 1C) to the cathode surface (Cu or Ag). Figure 1 shows the setup [45].

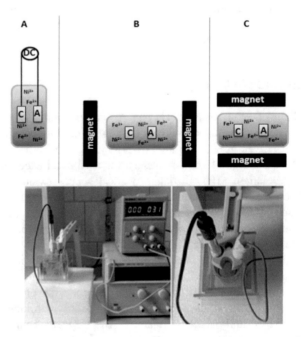

Fig. 1. A view (A – front view, B, C – top view) and photo of the experimental setup: C – cathode, A – anode, DC – power supply, magnet – neodymium magnet, Fe^{2+} – iron ions, Ni^{2+} – nickel ions, green color – electrolyte.

4 Results and Discussion

The authors found the strong effect of B on nucleation and growth mechanism. It could be classified as progressive due to different time of nuclei arrival and different rates of growing process. The opposite is the mechanism where the mentioned nuclei are formed in instantaneous nucleation. They are generated at the same time and grow at the same rates. This causes a high nucleus growth rate and a low number of sites active for the nucleation (lower than in the case without the EMF) [32]. The competition between the nucleation and the growth mechanisms seems to be the key to adjust the morphology of the deposited layers to individual needs. These facts allowed to accomplish an instantaneous nucleation during the electrodeposition with a perpendicular EMF setting and a progressive one without using the EMF.

In fact, the influence of the perpendicular magnetic field on the uniformity of the sediment decreases. This results from a decrease in the electrode's surface active for nucleation which is much lower compared to the case without EMF. After application of the parallel placed EMF, the number of grains is more or less the same as in the case without EMF assistance. Additionally, the noticeable effect is the formation of the specific morphology caused by the appearance of micro-turbulences. Based on the final effect, one can comment on the sensitivity of the process to the number of nucleation sites as a result of changes in process conditions. Additional convection (as result of MHD) is a way of environment mixing which is strongly different from mechanical stirring and results in pH stabilization. Changes in the EMF setting (parallel and perpendicular) lead to modifications of the nucleation and growth mechanism.

From the SEM image (Fig. 2), a greater uniformity of surfaces electrodeposited in EMF presence can be observed in comparison to surfaces obtained in EMF absence. This confirmed our earlier results [43]. Figure 2 shows an effect of applying a different EMF arrangement on morphologies of deposited FeNi layers. Deposition time was 60 min and various substrates (silver and copper) were employed. In the case of a silver substrate (without EMF assistance), the processes of nucleation and growth occurred simultaneously. This situation is characterized by nuclei at different ages and sizes. The nucleation occurs at many active places. The application of an external magnetic field starts with the excitation of the MHD effect. This leads to transporting of buoyant mass towards already existing nuclei. This phenomenon resembles behavior during electrodeposition with the parallel field assistance. The appearance of micro-vortices and an increase in the effect of the micro-MHD was the consequence of the magnetic field application. An evident change occurs when a copper substrate is used. The layers obtained with presence of the EMF are characterized by nodular particles with some flattened areas distributed all over the surface. Morphology of iron-nickel alloys deposited without the EMF evolves in clusters of fine particles which are connected with an increase of the iron content. As a result, the embedded layer exhibits 3D growth centres distributed all over the electrode's surface [28, 32, 40, 46].

In addition, the experiments were carried out on a copper substrate using successively increasing current density values (25, 35, 50, 75 mA/cm^2). Differences between embedded structures were observed. To a certain value of the current, the obtained layers were characterized by an increased volume of formed nuclei (Fig. 3). At

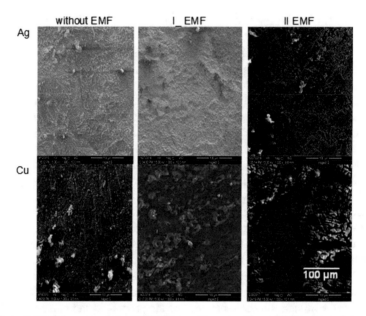

Fig. 2. Set of SEM images of the FeNi films with and without EMF; time of deposition – 60 min, current density – 50 mA/cm², magnification – 1000.

50 mA/cm², the layer became smoother and more even. Further increase of the current value caused the reappearance of ovules of various sizes. It was found that the thickness of electrodeposited FeNi layer was smaller during the process carried out with EMF assistance than without. It was proofed of XRF measurements given in Table 1. Outcomes confirmed the increase of alloy's mass electrodeposited in presence of the external magnetic field. To sum up, the iron-nickel surface became denser in a process carried out with the EMF assistance [47, 48].

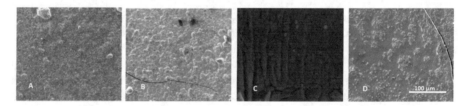

Fig. 3. Set of SEM images of the FeNi films without EMF; copper substrate, time of deposition – 60 min, current density: 25 mA/cm² (A), 35 mA/cm² (B), 50 mA/cm² (C), 75 mA/cm² (D), magnification – 1000.

Composition (EDX). Changes in the FeNi alloys composition during the deposition process were registered [49]. EDX analysis of conducted experiments was helpful to give an explanation of differences in the film's elemental composition. In the case of electrodeposition on Cu substrate, changes in the elemental composition were

Table 1. FeNi film thickness.

	Ag substrate [μm]	Cu substrate [μm]
Without EMF	0.91	1.76
I_ EMF	0.59	0.50
II EMF	0.66	0.59

influenced by a setting of the EMF. An increase of Fe content without magnetic field assistance and a decrease of iron with the magnetic field assistance were obtained. No effect of the magnetic field on Fe content was noticed during FeNi electrodeposition on a silver substrate (Fig. 4). The direction of changes of FeNi layer's elemental composition was the same in both cases (with and without the EMF). This is clearly shown in Fig. 5. It also shows an influence of the substrate on the Fe content in FeNi films. In our study found a clear correlation between the type of used bases and composition of the FeNi films. The external magnetic field affects the electrodeposition process. It was reported when the paramagnetic substrate (copper) was being tested. The shape of EMF lines was influenced by bases [50].

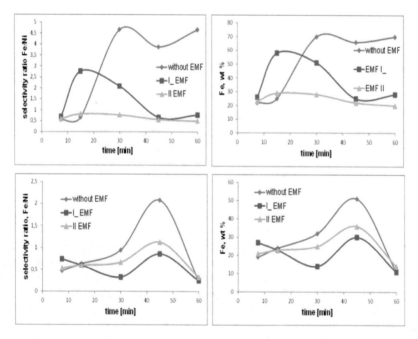

Fig. 4. Effect of time deposition on selectivity ratio Fe/Ni (without EMF – rhombs, I_ EMF – squares, II EMF – triangles) and on Fe content in FeNi films, current density – 50 mA/cm², top row – copper substrate, bottom row – silver substrate.

Crystal structure (XRD). Figure 6 shows XRD patterns of the FeNi alloys deposited on different conditions of a substrate and external magnetic field presence. The

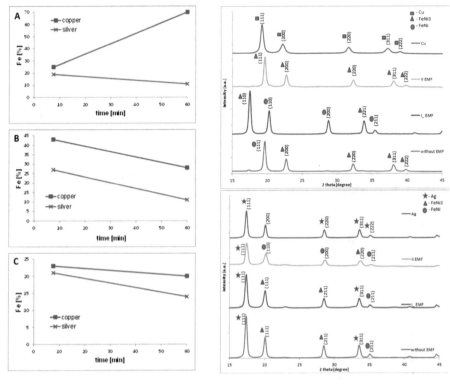

Fig. 5. Effect of time deposition on the composition of the Fe/Ni alloys (A – without EMF, B – I⊥ EMF, C – II EMF), current density – 50 mA/cm² (copper substrate – squares, silver substrate – crosses).

Fig. 6. Diffraction XRD pattern of FeNi electrodeposited films: top – copper substrate, bottom – silver substrate, current density – 50 mA/cm², time of deposition – 60 min.

crystalline structure of the film was analyzed. Diffractograms changed as a result of applying various kinds of the substrate. Different crystal structures were obtained and Miller indices were assigned. The copper substrate did not affect deposited films but the silver substrate did. It was pointed out that evident peaks came from the paramagnetic basis. The intensity of a set of patterns assigned to (111), (200), (220), (311) and (222) was dependent on a diamagnetic substrate (II EMF, without EMF). This illustrates the increase in FeNi₃ (fcc) structure. The case of the perpendicular EMF was characterized by a mixed state of fcc and bcc (FeNi) phases, just as layers on a Ag substrate. Stability of the fcc phase is associated with a rise in the iron content. Mixed structure of fcc and bcc is related to the nature of alloy composition changes (percentage content).

5 Conclusions

Iron-nickel electrocrystallization was studied in the presence and absence of the EMF. The electrochemical cell had the ability to change the orientation relative to the external magnetic field. It caused various correlations between the direction of the EMF and the electric current. A clear effect of applying the magnetic field on the course of the electrodeposition process and, hence, its result was noted. The number of grains deposited in both EMF orientations is much higher in comparison to a non-magnetic situation. The origin of this effect lays in MHD convection. This retards a steady state nucleation rate and increases the transport of electroactive species to existing nuclei. The mass of the alloy increases but its thickness declines. This shows that the layer gets denser. The influence of the EMF on the direction of the iron content in the alloy is also significant. This is visible especially during deposition on the copper surface.

The term introduced by Brenner has been verified here. Less noble metal (Fe) is preferentially deposited (without the EMF). Situation changes after EMF application. It has been noticed that the magnetic field does not affect Fe content during the deposition on the silver substrate. Structures of the obtained layers were additionally investigated to follow the relationship between the composition and the substrate. Outcomes allowed us to clarify effects of the diamagnetic basis on the crystal composition. In the case of paramagnetic basis application, the crystal composition was affected by the EMF. Better knowledge of this kind of relationship will be crucial to tailor material properties to the needs of industries and economy.

Acknowledgement. The work was partially financed by EU fund as part of the projects: POPW.01.03.00-20.034/09 and POPW.01.03.00-20-004/11 and supported by PSC within the project: S/WE/2/2018.

References

1. Rousse, C., Fricoteaux, P.: Electrodeposition of thin films and nanowires Ni-Fe alloys, study of their magnetic susceptibility. J. Mater. Sci. **18**(46), 6046–6053 (2011)
2. Cao, D., Wang, Z., Feng, E., Liu, Q.: Magnetic properties and microstructure investigation of electrodeposited FeNi/ITO films with different thickness. J. Alloys Compd. **581**, 66–70 (2013)
3. Hou, X.W., Liu, S.B., Yang, S.L., Li, J.P., Gou, B.: Electrical and magnetic properties of electrodeposited Fe-based alloys used for thin transformer. Sci. China Technol. Sci. **56**(1), 84–88 (2013)
4. Dang, M.Z., Rancourt, D.G.: Simultaneous magnetic and chemical order-disorder phenomena in Fe_3Ni, FeNi, $FeNi_3$. Phys. Rev. B **53**(5), 2291–2302 (1996)
5. Huang, Q.: Electrodeposition of FeCoNiCu quaternary system. Ph.D. thesis. Faculty of the Louisiana State University and Agricultural and Mechanical College, Louisiana (2004)
6. Shumskaya, A.E., Kaniukov, E.Y., Kozlovskiy, A.L., Shlimas, D.I., Zdorovets, M.V., Ibragimova, M.A., Rusakov, V.S., Kadyrzhanov, K.K.: Template synthesis and magnetic characterization of FeNi nanotubes. Prog. Electromagn. Res. C **75**, 23–30 (2017)
7. Dimitrievich, T.R.: Normal electrochemical deposition of FeNi films. J. Adv. Res. **11**(2), 1–10 (2017)

8. Sasaki, K.Y., Talbot, J.B.: Electrodeposition of iron-group metals and binary alloys from sulfate baths. J. Electrochem. Soc. **147**(1), 189–197 (2000)
9. Neuróhr, K., Csik, A., Vad, K., Molnár, G., Bakonyi, I., Péter, L.: Near-substrate composition depth profile of direct current-plated and pulse-plated Fe-Ni alloys. Electrochim. Acta **103**, 179–187 (2013)
10. Uhlemann, M., Tschulik, K., Gebert, A., Mutschke, G., Fröhlich, J., Bund, A., Yang, X., Eckert, K.: Structured electrodeposition in magnetic gradient fields. Eur. Phys. J. Spec. Top. **220**(1), 287–302 (2013)
11. Tabakovic, I., Gong, J., Riemer, S., Kautzky, M.: Influence of surface roughness and current efficiency on composition gradients of thin NiFe films obtained by electrodeposition. J. Electrochem. Soc. **162**(3), 102–108 (2015)
12. Tabakovic, I., Inturi, V., Thurn, J., Kief, M.: Properties of $Ni_{1-x}Fe_x$ ($0.1 < x < 0.9$) and Invar ($x = 0.64$) alloys obtained by electrodeposition. Electrochim. Acta **55**(22), 6749–6754 (2010)
13. Brenner, A.: Electrodeposition of Alloys. Academic Press, New York (1963)
14. Dragos, O., Chiriac, H., Lupu, N., Grigoras, M., Tabakovic, I.: Anomalous codeposition of *fcc* NiFe nanowires with 5–55% Fe and their morphology, crystal structure and magnetic properties. J. Electrochem. Soc. **163**(3), 83–94 (2016)
15. Dahms, H., Croll, I.M.: The anomalous codeposition of iron-nickel alloys. J. Electrochem. Soc. **112**(8), 771 (1965)
16. Horkans, J.: Effect of plating parameters on electrodeposited NiFe. J. Electrochem. Soc. **128** (1), 45–49 (1981)
17. Lieder, M., Biallozor, S.: Study of the electrodeposition process of Ni-Fe alloys from chloride electrolytes. Surf. Technol. **21**(1), 1–10 (1984)
18. Grimmett, D.L., Schwartz, M., Nobe, K.: A comparison of DC and pulsed Fe-Ni alloy deposits. J. Electrochem. Soc. **140**(4), 1136–1340 (1993)
19. Zech, N., Podhala, E.J., Landolt, D.: Anomalous codeposition of iron group metals. J. Electrochem. Soc. **146**(8), 2886–2891 (1999)
20. Torabinejad, V., Aliofkhazraei, M., Assreh, S., Allahyarzadeh, M.H., Rouhaghdam, A.S.: Electrodeposition of Ni-Fe alloys, composites, and nano coatings – A review. J. Alloys Compd. **691**, 841–859 (2017)
21. Sknar, Y., Sknar, I., Yermolenko, I., Karakurkchi, A., Mizin, V., Proskurina, V., Sachanova, Y.: Research into composition and properties of the Ni-Fe electrolytic alloy. Mater. Sci. **12** (88), 4–10 (2017)
22. Gong, J., Riemer, S., Morrone, A., Venkatasamy, V., Kautzky, M., Tabakovic, I.: Composition gradients and magnetic properties of 5–100 nm thin NiFe films obtained by electrodeposition. J. Magn. Magn. Mater. **398**, 64–69 (2016)
23. Su, X., Qiang, Ch.: Influence of pH and bath composition on properties of Ni-Fe alloy films synthesized by electrodeposition. Bull. Mater. Sci. **35**(2), 183–189 (2012)
24. Ispas, A., Matsushima, H., Plieth, W., Bund, A.: Influence of a magnetic field on the electrodeposition of nickel-iron alloys. Electrochim. Acta **52**(8), 2785–2795 (2007)
25. Trzaska, M., Białostocka, A.M.: Electrodes modelling in the copper electrocrystallization process. Przegląd Elektrotechniczny **2**, 95–98 (2007)
26. Aogaki, R.: Magnetic field effects in electrochemistry. Magnetohydrodynamics **37**(1/2), 143–150 (2001)
27. Rousse, C., Msellak, K., Fricoteaux, P., Merienne, E., Chopart, J.-P.: Magnetic and electrochemical studies on electrodeposited Ni-Fe alloys. Magnetohydrodynamics **42**(4), 371–378 (2006)

28. Ispas, A., Matsushima, H., Bund, A., Bozzini, B.: Nucleation and growth of thin nickel layers under the influence of a magnetic field. J. Electroanal. Chem. **626**(1–2), 174–182 (2009)

29. Białostocka, A.M., Żabiński, P.: Modification of FeNi alloys electrodeposited by applying external magnetic fields. Key Eng. Mater. **641**, 157–163 (2015)

30. Mogi, I., Marimoto, R., Aogaki, R., Watanabe, K.: Surface chirality induced by rotational electrodeposition in magnetic fields. Scientific Reports 3 (2013)

31. Matsushima, H., Fukunaka, Y., Kikuchi, S., Ispas, A., Bund, A.: Iron electrodeposition in a magnetic field. Int. J. Electrochem. Sci. **7**, 9345–9353 (2012)

32. Koza, A.J., Mogi, I., Tschulik, K., Uhlemann, M., Mickel, Ch., Gebert, A., Schultz, L.: Electrocrystallisation of metallic films under the influence of an external homogeneous magnetic field-early stages of the layer growth. Electrochim. Acta **55**, 6533–6541 (2010)

33. Monzon, L.M.A., Coey, J.M.D.: Magnetic fields in electrochemistry: the Lorentz force. A Mini-Rev. Electrochem. Commun. **42**, 38–41 (2014)

34. Mutschke, G., Bund, A.: On the 3D character of the magnetohydrodynamic effect during metal electrodeposition in cuboid cells. Electrochem. Commun. **10**(4), 597–601 (2008)

35. Mutschke, G., Tschulik, K., Weier, T., Uhlemann, M., Bund, A., Fröhlich, J.: On the action of magnetic gradient forces in micro-structured copper deposition. Electrochim. Acta **55**, 9060–9066 (2010)

36. Zhou, P., Zhong, Y., Wang, H., Fan, L., Dong, L., Li, F., Long, Q., Zheng, T.: Behavior of Fe/nano-Si particles composite electrodeposition with a vertical electrode system in a static parallel magnetic field. Electrochim. Acta **111**, 126–135 (2013)

37. Gamburg, Y.D., Zangari, G.: Theory and practice of metal electrodeposition. Springer Science, New York (2011)

38. Gong, J., Riemer, S., Kautzky, M., Tabakovic, I.: Composition gradient, structure, stress, roughness and magnetic properties of 5–500 nm thin NiFe films obtained by electrodeposition. J. Magn. Magn. Mater. **398**, 64–69 (2016)

39. Fricoteaux, P., Rousse, C.: Influence of substrate and magnetic field onto composition and current efficiency of electrodeposited Ni-Fe alloys. J. Electroanal. Chem. **612**(1), 9–14 (2008)

40. Abdel-Karim, R., Reda, Y., Muhammed, M., El-Raghy, S., Shoeib, M., Ahmed, H.: Electrodeposition and characterization of nanocrystalline Ni-Fe alloys. J. Nanomater. **7**, 1–8 (2011)

41. Zhao, Z.: Electrocrystallization nucleation and growth of NiFe alloys on brass substrate. Trans. Nonferrous Met. Soc. China **15**(4), 1–5 (2005)

42. Kalska-Szostko, B., Wykowska, U., Piekut, K., Zambrzycka, E.: Stability of iron (Fe) nanowires. Colloids Surf. **416**, 66–72 (2013)

43. Białostocka, A.M., Klekotka, U., Żabiński, P., Kalska-Szostko, B.: Microstructure evolution of Fe/Ni layers deposited by electroplating under an applied magnetic field. Magnetohydrodynamics **53**(2), 309–320 (2017)

44. Kalska-Szostko, B., Brancewicz, E., Mazalski, P., Szeklo, J., Olszewski, W., Szymanski, K., Sidor, A.: Electrochemical deposition of nanowires in porous alumina. Acta Phys. Pol. A **115**(2), 542–544 (2009)

45. Kalska-Szostko, B., Orzechowska, E.: Preparation of magnetic nanowires modified with functional groups. Curr. Appl. Phys. **11**(5), 103–108 (2011)

46. Edabi, M., Basirun, W.J., Alias, Y., Mahmoudian, M.: Electrodeposition of quaternary alloys in the presence of magnetic field. Chem. Cent. J. **4**, 1–14 (2010)

47. Mehdi, M., Basirun, W.J., Alias, Y.: Morphology and mass changes with magnetic field during the electrodeposition of Ni-Co. Adv. Mater. Res. **264–265**, 1389–1394 (2011)

48. Edabi, M., Basirun, W.J., Alias, Y.: Influence of magnetic field on corrosion resistance and microstructure of electrodeposited Ni coatings. Adv. Mater. Res. **264–265**, 1383–1388 (2011)
49. Solmaz, R., Kardas, G.: Electrochemical deposition and characterization of NiFe coatings as electrocatalytic materials for alkaline water electrolysis. Electrochim. Acta **54**(14), 3726–3734 (2009)
50. Ohno, I., Mukai, M.: The effect of a magnetic field on the electrodeposition of iron-nickel alloy. Electrodeposition Surf. Treat. **3**(3), 213–218 (1975)

Accelerating Image Fusion Algorithms Using CUDA on Embedded Industrial Platforms Dedicated to UAV and UGV

Artur Kaczmarczyk[✉] and Weronika Zatorska[✉]

WZE S.A., 1 Maja 1, 05-220 Zielonka, Poland
{artur.kaczmarczyk, weronika.zatorska}@wze.com.pl

Abstract. The objective of image fusion is to combine information from multiple images of the same scene, where the image result of fusion is more suitable for human and machine perception. This process is usually time consuming on most standard computing platforms that can be used in autonomous and semi-autonomous platforms, such as an unmanned ground vehicle (UGV) and an unmanned aerial vehicle (UAV). This creates a conflict between satisfying the hardware requirements and the platform's current demand. In this article, we propose a solution to this problem, being an image fusion algorithm utilizing CUDA technology, dedicated for an embedded platform (~ 5 W), allowing for image processing in 25 fps, which should be satisfactory for the needs of the mentioned platforms. During the development, we researched multimodal image fusion algorithms and implemented the chosen methods. The chosen testing environment and chosen measures of the quality of the fusion are presented as well.

Keywords: CUDA · GPU · Image fusion · Laplacian pyramid · Laplacian · Thermal image

1 Introduction

Image fusion is a synergistic combination of information coming from sensors with different characteristics, into a coherent whole. This process results in an improvement of image's quality. Due to the high number of imaging technologies and modalities, image fusion allows for collecting information from different sources and combining them in order to perform a comparative analysis, as well as to enrich the image and perform a quantitative assessment of the images, resulting in an increase of the system's efficiency. Image fusion, in particular, can involve infrared images, visible light and thermovision. Data cumulated this way allows for the emphasis of additional information in the images, not directly visible in the component images.

The effect of the integration of the images is a new, synthetic dataset, characterized by increased quality and reduction of redundant information [1]. Image fusion usually consists of two stages: fitting and aggregation of the images. As a part of our research we reviewed adequate literature and basing on our findings, chose image aggregation algorithms to be implemented and tested theirs performance in context of resulting

© Springer Nature Switzerland AG 2020
R. Szewczyk et al. (Eds.): AUTOMATION 2019, AISC 920, pp. 697–706, 2020.
https://doi.org/10.1007/978-3-030-13273-6_65

quality and efficiency on an embedded platform, aiming to lower the execution time for the image fusion below 0.05 s, to ensure smoothness of the operation, making it fitting for an online system.

Image aggregation algorithms can be divided into three categories: pixel-based, feature-based [2] and symbol-based [1]. From the listed groups, the group that has the widest range of use is the group of lowest level algorithms - the ones that operate directly on the pixels. After analyzing the literature, we decided to test methods of aggregation based on operating directly on the pixels, which use the laplacian pyramid. This choice is dictated by the fact that the target platform allows for parallelization and scaling of said algorithms and implementing them on GPU. The results were compared with image fusion methods, which use the wavelet transform [3] in order to demonstrate the validity of our choice of method.

2   Image Registration and Fusion Methods

The process of image registration in multisensor systems is based on fitting the obtained images together in the context of the same scene, observed from different points of data collection and with a different time characteristic. In result, when an image is registered from many sensors, a transformation model of said images as against the reference image, sourced from an arbitrarily marked reference camera, needs to be calculated. Owing to the different characteristics of used sensors, the image gained from the fusion should hold more data than any of the input images separately. The process of registration from multiple sensors can be treated as a composition of two or more input images into one, synthetic output image. Many algorithmic solutions to this problem exist in the literature, however when it comes to combining images coming from different sensors, the number of said algorithms is limited, due to issues such as different operating ranges of sensors on different wavelengths and geometrical deformations. The established cameras (in PAL mode) layout used for testing in the trials described, used for the collection of the data can be seen in Fig. 1. The camera layout consists of:

1. Day/night camera (CP) [4]:
   COPST Citadel Panoramic 170° FOV (two images with each 85.5);
   sensor element: (2x) high sensitivity 1/3" colour CCD, Focal length 2.6 mm, f/1.6;
   light sensitivity: 0.007 lx;
2. Thermal imaging camera (CT) [5]:
   COPST Citadel Thermal 90° FOV;
   sensor: uncooled VOx microbolometer, detector pitch 17 μm, spectral response 8–14 μm
   thermal sensitivity: NETD <50 mK;
3. Visible light camera (CD) [6]:
   sensor element: ¼" CMOS digital image sensor;
   light sensitivity: <0.05 lx;

The software used to collect data was implemented with the use of C++ and CUDA in a Linux environment, on the Nvidia Jetson TK1 platform [7].

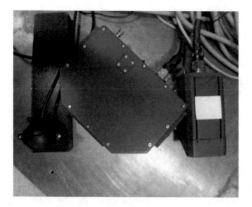

Fig. 1. The camera layout.

The camera layout alongside the acquisition block allowed for the generation of a set of images: a multispectral image from the CP camera and a panchromatic image from the CT camera and the CD camera. Assembling the color data from said images with the use of methods based on the IHS transform [8] is the most popular solution, which allows for increasing the amount of data in an image resulting from the image fusion. The general schematic of the process is presented on Fig. 2, where the panchromatic image is marked as PAN, whereas the following multispectral images are marked as MS1..3.

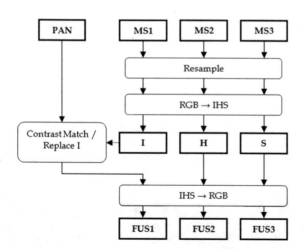

Fig. 2. The data fusion schematic from panchromatic and multispectral images.

The algorithm based on Laplacian pyramid [9] for the image fusion tests has been implemented.

The main advantage of this approach is operating on pixels in different scales of the examined image, which leads to the tested discontinuity including a smaller range than it would in case of the wavelet transform method, which was chosen as a point of reference. In the Laplacian pyramid subsequent images are formed from the difference between the image from the higher level of the pyramid and the image from the lower level of the Gaussian pyramid [9]. Every image in the pyramid is formed by reducing the number of pixels, where every new pixel is the mean of neighboring pixels. The output image is resized to fit the size of the image from a higher level of the pyramid by interpolation. Following steps of the algorithm are shown below:

- Step 1: Two input images from IR (gc) and CCD (gIR) are converted into 8-bit images
- Step 2: Input images are reduced by the Gaussian (R(gm,n)) in order to obtain k + 1 levels of Gaussian pyramids
- Step 3: Obtained Gaussian pyramid is used in the generation of images of the Laplacian pyramid L(m,n), by the difference of the image and the Gaussian.

$$L_{fk} = g_{fk} - E(g_{fk-1})$$ (1)

- Step 4: The higher value of analyzed pixels is chosen on the given level of the pyramid

$$L_{fk}(i,j) = sign_k(i,j) \times max(|L_{ck}(i,j), |L_{irk}(i,j)||)$$ (2)

- Step 5: The process is repeated on the subsequent levels of the pyramid
- Step 6: On the last level k + 1 of the pyramid, instead of the highest value of the analyzed pixels, the mean of the pixels is chosen
- Step 7: The reverse process of image generation based on assembling the Laplacians

$$g_{fk} = L_{fk} + E(g_{fk+1})$$ (3)

- Step 8: Output: a synthetic image from the fusion of the input images

The images shown below are the input images from a reference set of images TNO Image Fusion Dataset [10]: in infrared (Fig. 3a), in visible light (Fig. 3b) and fusion result (Fig. 3c). Because the dataset tested is normalized, image fitting was simplified.

a   b   c

Fig. 3. Image in infrared (4a); Image in visible light (4b); Fusion (4c).

The aforementioned set of images was used to demonstrate the difference in the fit of the images between the Laplacian pyramid method and the wavelet transform method, in favor of the Laplacian pyramid method [compared results in Table 1]. The method was then used in the test environment shown in Fig. 1. The analysis of the quality of the image fusion was based on chosen measures [11–13], specifically the comparison of the methods was based on the ERGAS factor [13], which clearly points to the Laplacian pyramid method as the more sufficient one [Table 1].

Table 1. ERGAS factor for the image fusion algorithms executed on TNO image dataset.

obrazy TNO Image Fusion Dataset	laplacian	wavelet
Balls nir-photo	56.5188	63.8235
barbed_wire_1 lwir-rgb	130.999	111.763
barbe_wire_2 b_nir-a_Vis	38.532	33.1301
Bosnia nir-daylight	82.122	49.969
Farm ir-photo	99.1719	91.9965
House nir-photo	94.5298	81.8391
houses_with_3_men lwir-vis	142.793	141.289
jeep_in_smoke nir-vis	21.4913	15.4235
Kaptein_01 IR-Vis	82.0483	76.2396
Kaptein_1123 IR-Vis	90.8836	83.9267
Kaptein_1654 IR-Vis	84.0691	82.7521
Kaptein_19 IR-Vis	94.1022	96.3425
Mame_01 IR-Vis	171.16	156.934
Mame_02 IR-Vis	147.042	143.953
Mame_03 IR-Vis	159.926	126.935
Mame_04 IR-Vis	148.434	106.123
Mame_06 IR-Vis	141.447	113.046
pancake_house NIR-photo	68.1395	62.7042
Reek IR-Vis	99.1891	98.4076
siokdier_behind_smoke lwir-rgb	62.8509	56.7777
soldiers_with_jeep IR-Vis	166.511	130.618
square_with_houses NIR-Vis	20.5441	15.6692
Veluwe NIR-photo	73.3774	58.4686
Vlasakkers NIR-Vis	28.1934	28.7055

The schematic of the image fusion with the use of the wavelet transform is shown in Fig. 4, with all its basic steps.

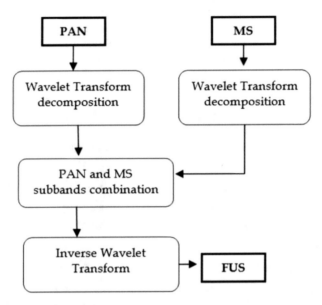

Fig. 4. The schematic of the IHS transform with the use of wavelet transform.

3   Conducted Tests and Results

The computing platform used is based on the Nvidia Jetson TK1 (Table 2). Because of the platform's small current demand (~ 5 W), this solution is best suited for embedded solutions, especially autonomous vehicles, drones included.

Table 2. Nvidia Jetson TK1 specification [7]

Tegra K1 SOC (CPU+GPU+ISP in a single chip, with typical power consumption between 1 to 5 W)	
GPU	NVIDIA Kepler "GK20a" GPU with 192 SM3.2 CUDA cores (up to 326 GFLOPS)
CPU	NVIDIA "4-Plus-1" 2.32 GHz ARM quad-core Cortex-A15 CPU with Cortex-A15 battery-saving shadow-core
RAM	2 GB DDR3L 933 MHz EMC $\times$16 using 64-bit data width
Storage	16 GB fast eMMC 4.51 (routed to SDMMC4)

Thanks to the use of the CUDA library, the implemented algorithm of image fusion based on the Laplacian pyramid is characterized by the short computation time, which would be acceptable in aforementioned solutions. The measured computation times shown in the Table 3 are indicating the possibility of a practical use of the implemented algorithm in autonomous and semi-autonomous systems for ground and aerial platforms, where a 25 fps framerate is satisfactory.

Table 3. Comparison between the algorithms implemented and executed on a CPU and, with the use of CUDA, on GPU.

Platform	RGB to HSV conversion [s]	Laplacian pyramid transform [s]	HSV to RGB conversion [s]
CPU: Intel Core i7-4770 3.4 GHz	0.39	0.36	0.57
GPU: Nvidia Jetson TK1	0.013	0.008	0.013

The tests were conducted in a closed climatic chamber, where the level of light intensity was measured to be 8 lx (lights on – variant A) and 0.5 lx (lights off – variant B). Captured images from the cameras presented, among others (like in Table 4 below):

Table 4. Conducted tests.

No	Test environment	Light intensity [Lux]	ERGAS for fusion image	Distance from object to the cameras [m]
1	Bright colored object	8	91.816	3
2	A cold and a hot object	8	82.939	3
3	A bright colored object hidden behind glass	8	86.462	3
4	A thermal blanket heated up with a stream of hot air	8	93.628	3
5	Glass heated up with a stream of hot air	8	87.402	3
6	A bright colored object hidden behind glass in a fog	8	49.176	3
7	A bright colored object hidden behind glass in a fog	8	79.590	1
8	A bright colored object hidden behind glass	0.5	52.683	3
9	A bright colored object hidden behind glass	0.5	41.728	1

The image fusion products shown below are the result of our tests conducted in laboratory conditions with the use of cameras described in Sect. 2. The ERGAS factor is also presented, as a measure of the fit of the images.

Fig. 5. Test no. 3 from Table 4 (a) CP camera (b) CT camera (c) fusion of CP and CT images.

Fig. 6. Test no. 9 from Table 4 (a) CP camera (b) CT camera (c) fusion of CP and CT images.

The test results demonstrate an increase in the output image's entropy. It's especially significant in the tests including a glass with an object hidden behind it in difficult lighting conditions, as seen in Figs. 5 and 6. As a result of the fusion, the object's color becomes visible, despite being hidden behind glass, in front of the thermal vision camera, which could prove to be useful in difficult lighting conditions while detecting glazed objects. It should be noted, that in case of the CD camera, the images produced during test no. 9 the image was inutile, whereas by means of the CP camera, the color of the object hidden behind the glass was visible. Fog in the climate chamber created a similar outcome, where the CP and CT camera's images fusion allowed for a successful read of the object's color, and the CD camera's image was inutile once again.

4 Conclusion

The prepared test environment, described in Sect. 2 and the conducted tests proved the possibility of the implementation of an image fusion algorithm, using a day/night and a thermal imaging camera, on an embedded platform with a low current demand (~ 5 W), performing at a speed of 25 fps. The chosen solution shows a considerable promise of being useful in autonomous and semi-autonomous vehicles like UGV and UAV, where the reduction of the current demand is pivotal. In comparison, a solution based on an Intel i7 processor, providing similar efficiency, would consume ~ 80 W, instead of ~ 5 W.

The combination of Laplacian pyramid image fusion of a thermal image with a day/night camera image allowed for a significant increase in the quality of the output image, in comparison with a visible light camera and the wavelet transform image fusion. Conducted tests, both in the variant A and B, shown an improved quality of the image fusion for the day/night camera in comparison to the visible light camera, which in the variant B was inutile and didn't bring on any new information to the output image.

The use of the day/night camera allowed for, most of all, the enrichment of the thermal picture with color and the visibility of glazed objects and objects hidden behind glass, which makes this solution promising in terms of object detection in difficult lighting conditions. The tests demonstrated the superiority of the Laplacian pyramid to the wavelet transform in this case.

Next potential steps in the development of this algorithm would concern procedures of initial image processing after the fusion, which should include: histogram alignment, normalization and filtration. Those operations would better the readability of the output image in difficult lighting conditions. Simultaneously, those operations are easily scalable and shouldn't overload the chosen platform, which is the Nvidia Jetson TK1/TX1 and so, the framerate should stay above 20 fps.

References

1. Goshtasby, A., Nikolov, S.: Image fusion: advances in the state of the art. Inf. Fusion **8**(2), 114–118 (2007). Special Issue on Image Fusion: Advances in the State of the Art
2. Zitova, B., Flusser, J.: Image registration methods: a survey. Image Vis. Comput. **21**(11), 977–1000 (2003)
3. Arthur, J.J., Kramer, L.J., Bailey, R.E.: Flight test comparison between enhanced vision (FLIR) and synthetic vision systems, SPIE-INT SOC Optical Engineering, BELLINGHAM, 25–36 (2005)
4. Products citadel. http://www.copst.com/productscitadel. Accessed 28 Oct 2018
5. Citadel thermal. http://www.copst.com/citadel-thermal. Accessed 28 Oct 2018
6. Orlaco famos camera. https://www.orlaco.com/product/orlaco-famos-camera. Accessed 28 Oct 2018
7. Jetson. https://elinux.org/Jetson_TK1. Accessed 28 Oct 2018
8. Choi, M.: A new intensity-hue-saturation fusion approach to image fusion with a tradeoff parameter. IEEE Trans. Geosci. Remote Sens. **44**(6), 1672–1682 (2006). Print ISSN 0196-2892
9. Burt, P.J., Adelson, E.H.: The Laplacian pyramid as a compact image code. IEEE Trans. Commun. **4**(31), 532–540 (1983)
10. TNO Image Fusion Dataset. https://figshare.com/articles/TNO_Image_Fusion_Dataset/1008029. Accessed 28 Oct 2018
11. Maruthi, R., Suresh, R.M.: Metrics for measuring the quality of fused images. In: International Conference on Computational Intelligence and Multimedia Applications (2007)
12. Xydeas, C., Petrovic, V.: Objective image fusion performance measure. Electron. Lett. **36**, 308–309 (2000)
13. Osińska-Skotak, K.: Ocena przydatności różnych metod integracji obrazów panchromaty-cznych i wielospektralnych w odniesieniu o zobrazowań Worldview-2. Arch. Fotogram. Kartogr. Teledetekcji **24**, 231–244 (2012)

Geophysical Measurements in the Aspect of Recognition of Discontinuous Deformation Processes – A Case Study

Rafał Jendruś[(✉)]

Faculty of Mining and Geology, Silesian University of Technology,
Akademicka 2A Street, 44-100 Gliwice, Poland
Rafal.Jendrus@polsl.pl

Abstract. Article discusses one of the geophysical prospecting methods (Electrical Resistivity Survey – Method) used for solving environmental geo-engineering problems. Studying bedding (rock mass) using non-invasive geophysical methods is beneficial particularly because of the ability to obtain complete and very accurate information, depending on adopted research and prospecting approach. Interpretation of geophysical data, should always include available geological, empirical and environmental information, while all results should be presented on integrated profiles and maps. The paper presents analysis of geophysical tests conducted in Triassic strata and assessment of their results, for purposes of an investment located in Tarnowskie Góry, on the post – ore mining area. Electrical resistance survey (ERS) demonstrated presence of numerous low – and high resistive anomalies located both in Triassic strata. ERS data, analysis of geological profiles, fill operations, post-completion ERS, and other observations, allowed evaluation of the ground surface safety in terms of mine subsidence appearance and formulation of recommendations for building designers. The purpose of the conducted geophysical research was to determine the occurrence and location of underground cavities in Triassic formations that could pose a threat of discontinuous deformations on the surface of the considered area.

Keywords: Shallow ore mining operations ·
Non-invasive geophysical measurements · Electrical resistivity survey ·
Geological setting

1 Introduction

Geophysical methods can be successfully used to solve multiple problems of environmental geoengineering. Their significant advantage is non-invasive measurement technique and the ability to conduct continuous studies on large areas (compared to "in situ" single-point measurements such as measurement holes – boreholes). The drawback of geophysical prospecting is their indirect presentation of information about structure and environmental pollution using studied physical field units. Scaling of physical field units to obtain information about geological structure or values, which characterize environmental pollution is an important process and geophysicist who

© Springer Nature Switzerland AG 2020
R. Szewczyk et al. (Eds.): AUTOMATION 2019, AISC 920, pp. 707–718, 2020.
https://doi.org/10.1007/978-3-030-13273-6_66

interprets field images is required to have significant knowledge in that field. Results of geophysical measurements must refer to geological information related to studied area. On the other hand, it is necessary for most of the research to adopt at least two independent geophysical methods. Only then, information obtained during the research can be considered reliable enough. Unfortunately, using several different methods raises research costs and eliminates many potential contractors who have only one type of geophysical equipment available (most often, electrical resistivity equipment). These factors successfully inhibit wider development of geophysical methods in Poland, which could otherwise be used for environmental protection and broadly understood geoengineering.

The article discusses the issue of application of chosen geophysical method, i.e. electrical resistivity, which is in many cases optimal for solving different problems of environmental geoengineering. In particular, the article presents environmental hazards in areas with past mining operations, in regards to future utilization of these areas and their role in spatial planning [1–4].

Mining and post mining areas are characterized by occurrence of hazards that are not present in any other part of the country. As a result of that, geophysicists, geologists and geotechnicians working in such areas are required to inform about the scale of these specific problems, danger that these problems can be for future generations, price of these research, cost of securing the areas, potential economic loss as a result of insufficient recognition of hazards and their detailed evaluation in areas of future investments. Many geologists, builders and geotechnicians who do not work in mining and post-mining areas might be unaware of the scale of certain geodynamic phenomena in these areas. Local authorities are quite aware of the hazard caused by shallow, old mining operation and possibility of formation of cavities, depressions and discontinuities. In some areas with investment plans, geological and geophysical surveys are conducted to detect and possibly fill underground cavities or check the efficiency of potential repairs of damaged land. Examination cavity hazards is most often performed via electrical resistivity method. Usually geophysical studies are designed with consideration for available geological and hydrogeological data and information from old mining maps (based on geological-mining expertise), which allows to choose optimal measurement techniques. Cavities may cause or are still causing, significant hazard to investment areas where mining operation took place in the past, which may generation of discontinuities mentioned above [5, 6].

It is also located in the post-mining area, which is under the direct influence of completed intensive exploitation of galena ores and the presence of galleries made in this deposit.

As part of exploratory research, a geophysical identification of the investment site was carried out, as a result of which the location of zones with increased resistance, which means potential places where the so-called non-continuous deformations may activate. Obtained results of geophysical surveys are an example of the analysis of the surface area threat from intermittent deformations in the rock mass, which are caused by underground ore mining in Triassic deposits lying under the layer of soluble rocks (carbonate dolomites) and disintegrating thin layer overlay composed of loose Quaternary formations in the form especially of sands and clays. The application of the electrical resistance survey method enabled an accurate understanding of the structure of the rock mass near the border of the Triassic and Quaternary layers, also in terms of the water conditions prevailing in it.

2 Geological Setting

In terms of geology, the discussed area lies within the Tarnogórska trough, built by the Middle Triassic rocks. This trough is part of the geological unit of the Silesian Triassic, which borders the north and north-east of the Upper Silesia Coal Basin.

In the geological structure occur the Quaternary and Triassic forms. The geological structure was characterized on the basis of a geological map and data contained in the ore mining map (see Figs. 1 and Fig. 2).

Fig. 1. Fragment of the geological map of the area around Tarnowskie Góry, Tarnowitz-Brinitz Berlin 1913). The scale of the original 1:25 000.

Fig. 2. Fragment of an appendix to the Prus geological map, the royal geological office in Berlin, 1913. The scale of the original 1:25 000.

The total thickness of Quaternary strata ranges from 12 m to 16 m. They are lithologically developed as sands of different grain-size and loamy sands as well as clays.

In the lower part of the geological profile, among interbeddings of clays and slits, limonite iron ores of a nesting nature are present. According to the available materials, discussed area was not subject to mining operations.

Under the Quaternary strata, down to a depth of 16 m below the surface ore bearing dolomites occur. Their lithological form is variable, dolomites occur as compacted rock, finely crystalline, cavernous and fractured. In the floor part of the dolomites, galena lead ore is found, which the subject of exploitation was. They are underlain by a complex of limestone rocks belonging to the Gogolińskie strata.

3 History of Ore Mining in the Region of Tarnowskie Góry

Mining in the wider region of Tarnowskie Góry (including the area of Bytom and Piekary Śląskie), dates back to the early Middle Ages making this area one of the oldest centres for the extraction of metal ores in Europe.

The oldest recorded mention of ore mining in this area comes from 1136 and concerns the silver mining in the Surroundings of Bytom. The document issued by Prince Władysław Opolski in 1247 already mentions ore mining in the region of Repty Śląskie, which is near the area being the subject of this paper.

Mining works in the region of Tarnowskie Góry were carried out virtually continuously from the early Middle Ages until 1912, when the region of the Tarnogórskie Góry deposits was finally recognized as exhausted.

The contemporary area of historical metal ore mining in the region of Tarnowskie Góry covers 185 km of preserved mine workings in the form of galleries, rooms and adits, as well as a huge amount of remnants of historical metal ore mining, mainly in the form of shafts and pits in a very differentiated state of preservation or decommissioning, whose total number is estimated at around 20,000. Geotechnically significant post-mining remains also include extensive areas with a surface formed of anthropogenic embanked land extracted from waste rock and waste from metal ore processing. Nowadays, the underground museum of the Historic Silver Mine operates within the best preserved part of underground workings. The section of one of the drainage drifts under the name of the Black Trout Drift is also available for tourists. Adaptation of selected fragments of historical mine workings for museum and tourist purposes are the only underground mining works carried out in the region of Tarnowskie Góry [7, 8].

4 Materials and Methods

4.1 Analysis of Ore Mining on the Basis of Historical Maps from the Tarnowskie Góry Region

Due to the fact that mining works in the region of Tarnowskie Góry have been completed in 1912, the richest in details are the geological and mining maps from this period, containing the most complete and up-to-date picture of the scope of mining works.

It should be noted that the accuracy of determining the location of underground workings is vitiated by an error related to the vectorization as well as the transformation of historical materials to the coordinate system being currently in use. The error of location of mining elements on the map in the 1:10 000 scale is estimated at MX = MY = ± 15 m and on maps of the "Fryderyk" mine in the scale 1:1000 on MX = MY = ± 10 m.

According to the map of ore mining in Upper Silesia of the former Central Mining Authority in Wrocław prepared in the scale of 1:10 000, sheet No. 11 (see Fig. 3) and maps of the "Fryderyk" mine in scale 1:1000 (see Fig. 4), the area of the parcel under consideration was under direct impact of intensive exploitation of galena ores and the occurrence of galleries made in this deposit.

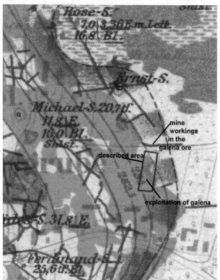 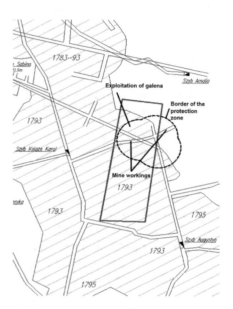

Fig. 3. Fragment of an ore mining map in Upper Silesia of the former Central Mining Office in Wrocław, Berlin 1911. (erzkarte) number 11. The scale of the original 1:10000.

Fig. 4. Fragment of the mining map of the ore Fryderyk mine from 1790–1860. The scale of the original 1:1000

Exploitation of the Galena deposits in the discussed region has been carried out since 1793 using a room mining system with caving. Old mining maps do not document the depth of mining, as well as the thickness of the extracted ore. From the experience of the author of the elaboration of the "Fryderyk" mine underground workings exploration, and also from the data from the shafts included in the ore maps, it was found that the probable thickness of ore mining was about 2 m, occasionally it reached 3 m. The analysis of the shaft descriptions and contours of the floor of the dolomite rocks shows that mining of galena ores in the discussed area was carried out at a depth of 15 m to 18 m below the ground surface.

According to the maps of the "Fryderyk" mine, these galleries in parallel in vertical plane – one below the other, being able to be in direct contact with each other (the central part of the plot).

The nearest documented shafts, according to the map of the "Fryderyk" mine, are: "Książe Karol" shaft located 18 m west and "Augustin" shaft, located 18 m south-east of the parcel boundaries - not posing a threat to the planned investment.

On the basis of the above analysis of archival materials, it was decided to carry out non-invasive geophysical surveys to detect anomalies in the rock mass, which might indicate existence of voids in the Triassic or Quaternary rock mass, posing a potential risk of discontinuous deformations in these areas. The above studies will enable the Investor to continue planning the design and construction works. It was decided to conduct research using electro-resistance geophysics.

4.2   Characteristics of an Electro-Resistance Method for Testing the Occurrence of Discontinuities in the Rock Mass

Geophysical measurements including geoelectrical methods are based on variation of electrical parameters of studied medium. To have a full understanding of phenomena taking place when electrical current flows through soil-rock environment, one has to be aware that such medium shows resistance to current flow. According to Ohm's law, this resistance is proportional to the voltage used in relation to the current. The most basic, analyzed parameter is the electrical resistance (resistivity). That is why geo-electrical methods are equated with electrical resistivity methods. Resistivity is a characteristic of each specific material and a basis for identification of studied medium.

The essence of electrical resistivity studies is the flow of electrical current of known intensity between current electrodes (powered). Electrical potential difference generated this way, i.e. voltage measured between other two electrodes (measuring), allows for designation of resistivity of specific medium. However, only so called apparent resistivity can be measured in the field and it is reliant on true resistivity of each component of the medium.

Depending on research, following electrical resistivity surveys can be carried out:

– vertical electrical soundings 1D (PSE, SGE, VES),
– electrical resistivity profiling 1D (PE),
– electrical resistivity tomography 2D.

The essence of electrical resistivity method (geoelectrical) is the flow of electric current of known intensity between current (supply) electrodes. The potential difference created in this way, i.e. voltage measured with potential (measuring) electrodes, allows determining the resistance of a given medium. The determined resistance in the field is, however, the so-called apparent resistance, which depends on the real (specific) resistance of the individual components of the medium (see Fig. 5).

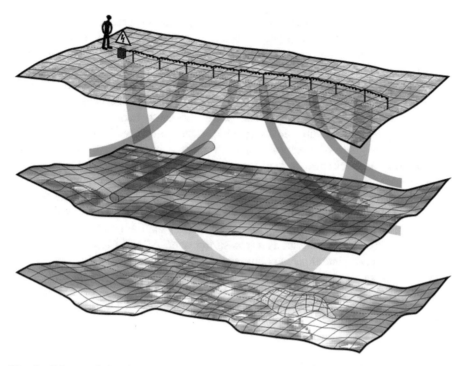

Fig. 5. Scheme of electrical resistivity method (geoelectrical) research (www.geospectrum.pl).

Choosing the method and specific measurement system relies on type and shape of surveyed structure, expected resistivity contrast, amount of noise and capabilities of equipment owned. Additionally, different measurement systems are characterized by various degree of profile covering, vertical and horizontal resolution and depth range. Surveys result in vertical profiles, cross sections and maps of resistivity variation along with their geological and engineering interpretation. Electrical resistivity method (geoelectrical) with utilization of techniques mentioned above is widely used for: identifying ground-water conditions and geological structure, localization of mining cavities (especially as a result of shallow mining operation), control of the efficiency of cavity filling and loosening the rock mass in shallow mining operation areas, localization of ground water level, water layers and underground water deposits or localization of faults [9–12].

4.3 Methodology of Field Surveys

Based on the above, it has been proposed to perform non-invasive geophysical surveys. Geophysical surveying for possible discontinuities, as well as detecting cavities in Triassic formations and potential loosening of Quarternary formations after finished exploitation of coal deposits, was performed using medium gradient electrical resistivity profiling. Electrical resistivity method is based on variation of resistivity between anomalies such as cavities, caverns, fracture zones etc. and surrounding rock mass [13, 14].

Measuring system generates artificial electrical field in the medium and allows for mearing of its voltage in specific segment of surface. Results are basis for evaluation of apparent resistivity of the ground. The choice of depth range (choosing specific measuring system) is conditioned by specific, often complicated, geological-mining situation. Depth range of electrical resistivity method is dependent on the distance between powered electrodes stuck in the ground (in geoelectrical methods marked as current electrodes A and B) and lithology of given area [3, 6, 9]. Potential variation is measured between measuring electrodes (marked as M and N). To perform the task that is a subject of the article, area was measured using two setups, AB = 60 m and AB = 90 m in a 5 m grid and measuring intervals of 5 m. The grid has been adjusted to fit measuring field dimensions. First measuring setup AB = 60 m has been done to check the rock mass conditions in depth range of the first shallow mining operation of iron ore. Its depth range, based on long, experimental research has been determined to be around 12–15 m. Second measuring setup AB = 150 m has been chosen based on deeper ore exploitation. Depth range of AB = 90 m electrode setup is around 18–22.5 m [14].

5   Results and Discussion

5.1   Geophysical Survey Interpretation

Electrical resistivity method has been chosen to determine condition of rock mass, which helps to evaluate potential threats to a surface. Basic notion of resistivity profiling interpretation is that, no matter the method, the results have common characteristic, which is the possibility to view resistivity as a sum of "background" and anomalies. Background can be understood as a value of resistivity obtained in case of uniform rock mass or a rock mass built of uniform horizontal layers. When we talk about physically diverse system (cavities, fissures, fractures, water), then the resistivity curve of such rock mass has variable course. Any distortions, exceeding the value of the background are called anomalies. Their intensity is dependent on dimensions and intensity of distortion. Absolute values of resistivity in anomalous zones also rely on resistivity of surrounding rocks, fracturing intensity caused by mining activity, water levels etc. Results of study, based on field survey interpretation, allowed for creation of apparent resistivity map and anomalous zone map for each measuring setup used (see Figs. 6 and 7). The main criteria for designating anomalous zones on measuring field was average value of apparent resistivity (Table 1). Any deviation from average resistivity value in studied rock medium, in form of significant growth or reduction is considered the anomalous zone (place). The following Table 1, shows compilation of resistivity values obtained as a result of electrical resistivity surveys [15–18].

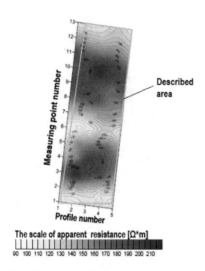

Fig. 6. Map of the apparent resistance, spacing AB = 60 m.

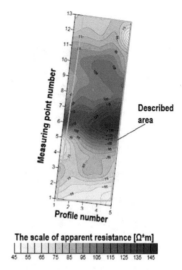

Fig. 7. Map of the apparent resistance, spacing AB = 90 m.

Table 1. Compilation of electrical resistivity values.

Area	Measuring field surface and dim. 1200 m²; 60 m × 20 m	Measuring field surface and dim. 1200 m²; 60 m × 20 m
Measuring setup	AB = 60 m Profiles direction E-W	AB = 90 m Profiles direction E-W
Minimum apparent resistivity [Ω m]	95.0	49.0
Maximum apparent resistivity [Ω m]	214.0	141.0
Average apparent resistivity [Ω m]	149.0	91.0

Due to the fact that in electro-resistance profiling the parameter being measured is apparent resistance, not the absolute one, it is not possible to determine the types of rocks, which form particular strata in the rock mass. For the measurement field in the light of obtained measurement results from the spacing AB = 60 m and AB = 90 m, high-resistance anomalous zones have been distinguished. (see Figs. 8, 9 and 10). Places characterized by high resistance are not saturated with water, dry or filled with air such as loosenings, cracks or empty voids in the maxima of anomaly. As previously discussed, discontinuous surface deformations are associated with the occurrence of physical discontinuities in the rock mass and their forms depend on the physic mechanical properties of the overburden.

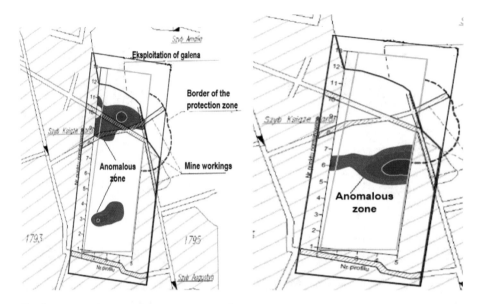

Fig. 8. Map of anomalous zones spacing AB = 60 m.

Fig. 9. Map of anomalous zones spacing AB = 90 m.

In loose and free flowing overburden conical sinkholes are usually formed, - irregular hollows in brittle rocks, whereas in plastic layers (cohesive soils) mainly regular troughs occur. On the basis of the analysis of archival mining maps, it is stated that practically all of the investment area has been undermined in the past. Particular attention should be paid to the intersection of galleries, where a high-resistance anomaly has been observed at the spacing AB = 60 m, with the maximum resistance of such a height that it may indicate a threat due to inappropriate liquidation of workings at this point.

Assuming that there was a slight mistake of transferring the parcel's location in relation to mine workings, if the mining situation had been moved 5 m towards NW, then the registered anomalies would have linear continuity, indicating the occurrence of poorly liquidated mining cavings or the effect of ground thinning.

In the light of the conducted research, it is stated that the indicated anomalous zones, and in particular their maximum values marked with a yellow border depicted in the form of maps of the resistance distribution of anomalous zones, indicate areas potentially endangered by possible occurrence of surface area deformation.

Increased activity of processes taking place deep into the rock mass over the years, both during active mining operations and after their completion, have undoubtedly disturbed the natural system of overburden strata, which in turn increased the destructive impact of processes related to the suffosion as well as chemical erosion of the rock mass.

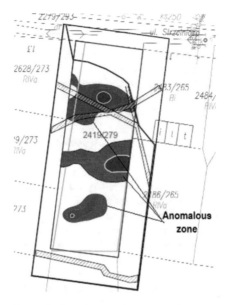

Fig. 10. Map with marked total anomalous zones from both depth spacing.

6   Conclusions

From the geomechanical point of view, on the post-mining and contaminated areas, a detailed geological and mining diagnosis is important, in relation to the possibility of surface deformation and the creation of loosening in the building ground foundation adversely affecting its bearing capacity.

The analysis of completed mining exploitation and the results of geophysical surveys showed that only the marginal southern part of the studied area of the future investment is not/should be exposed to the risk of discontinuous deformations of a sinkhole type, associated with the roof rocks collapsing and the self-filling of large cavities (caverns in carbonate rocks and relatively loose rock zones in Quaternary formations). In the central and northern parts of the research area may be present locally voids of larger sizes.

The main cause of unfavourable impacts on the surface and the ground foundation should be seen in the suffosive migration of fine granular material of loose rocks of the Quaternary overburden into the cracked-cavernous voids in the Triassic rock mass. The formation of these voids is, undoubtedly, mainly associated with the previously carried out room and pillar mining of galena. But the basic mechanism of the formation and development of the system of fissures and voids is associated primarily with the flow of waters from precipitation and snowmelt, through thin layers of Quaternary sediments into soluble carbonate rocks.

It should be mentioned here that the area in question was previously used for agriculture, and therefore possible ground discontinuities could persist in a relatively stable balance, which can be disturbed under the influence of increased earthworks or other "ground loads".

References

1. Zakolski, R.: Określenie nieciągłości górotworu metodami geofizycznymi na obszarze GZW. Praca GIG, Komunikat nr 662 (1974)
2. Dubiński, J., Mutke, G.: Geophysics application in environmental research. In: Workshops. Natural hazards in mining industry, Material Symposium, pp. 435–443 (2004)
3. Written, A.J.: Handbook of Geophysics and Archeology, pp. 1–329. Equinox Publishing Ltd., London (2006)
4. Kowalski, A., Jędrzejec, E., Gruchlik, P.: Linear discontinuous deformations of the surface in the upper silesian coal basin. Arch. Min. Sci. **55**(2), 331–346 (2010)
5. Zuberek, W.M., Żogała, B., Rusin, M., Pierwoła, J., Wzientek, K.: Geoelectrical and magnetic surveys in areas degraded by military activity, vol. 27, no. 352, pp. 195–208. Publications of Institute of Geophysics Polish Academy of Sciences (2002)
6. Zuberek, W.M.: Prediction of geological hazards and natural disasters – restrictions and possibilities. Min. Resour. Manage. **24**(2/3), 123–134 (2008)
7. Fajklewicz, Z.: Geneza anomalii siły ciężkości i pionowego gradientu nad pustkami powstającymi w skałach kruchych. Ochrona terenów górniczych 60, rok XIX, 3–13 (1985)
8. Paul, J., Moj, H.: Expertize of construction works impact on underground mining workings behavior beneath the surface of subjected area and usefulness evaluation of the 5 ha area lots located in Tarnowskie Góry-Bobrowniki Śląskie for single-family flat construction (2007)
9. Pilecki, Z., Ziętek, J., Karczewski, J., Pilecka, E., Kłosiński, J.: The effectiveness of recognizing of failure surface of the Carpathian flysch landslide using wave methods. In: Proceedings of 13th European Meeting of Environmental and Engineering Geophysics, Istambul, 3–5 Sept 2007 (2007)
10. Jendruś, R., Kłosiński, J.: Detection of underground cavities beneath buildings based on geophysical seismic method MASW. Scientific Books of Higher Technical School – Construction and Environmental Engineering, Katowice, pp. 49–56 (2011). Book no. 3
11. Tomecka-Suchoń, S., Marcak, H.: Interpretation of ground penetrating radar attributes in identifying the risk of mining subsidence. Arch. Min. Sci. **60**(2), 645–656 (2015)
12. Jendruś, R.: Geological and mining expertize for newly designed buildings on Skarbek street in Tarnowskie Góry, 690/44 and 691/44 lots, GeoRock company (2016, not published)
13. Marcak, H.: Powstanie zapadlisk i innych form deformacji nieciągłych powierzchni spowodowanych wystąpieniem pustek. Mat. Sym. Warsztaty **99**, 71–84 (1999)
14. Szymczyk, P., Marcak, H., Tomecka-Suchoń, S., Szymczyk, M., Gajer, M., Gołębiowski, T.: Zaawansowane metody przetwarzania danych georadarowych oraz automatyczne rozpoznawanie anomalii w strukturach geologicznych. Elektronika **12**, 56–61 (2014)
15. Sowiński, D., Bukowy-Olejnik, H.: Documentation of electrical resistivity geophysical survey results, conducted for the area of planned construction project in Tarnowskie Góry on Strzelnicza street, Geosolum company (2017, not published)
16. Dobecki, T.L., Upchurch, S.B.: Geophysical applications to detect sinkholes and ground subsidence. The Leading Edge **25**, 336–341 (2006)
17. Carcione, J.M.: Wave Fields in Real Media. Theory and Numerical Simulation of Wave Propagation in Anisotropic, Anelastic, Porous and Electromagnetic Media. Elsevier, Amsterdam (2007)
18. Coşkun, N.: The effectiveness of electrical resistivity imaging in sinkhole investigations. Int. J. Phys. Sci. **7**(15), 2398–2405 (2012)

Simplified Modelling the Demagnetization of H-Bar with Method of Moments

Roman Szewczyk[(✉)]

Industrial Research Institute for Automation and Measurements PIAP,
Al. Jerozolimskie 202, 02-486 Warsaw, Poland
rszewczyk@onet.pl

Abstract. Calculation of distribution of flux density in the constructional elements is required in non-destructive testing. Paper presents the simplified method of calculation of flux density distribution in H-bar based on the generalization of the method of moments. In opposite to finite elements method, the method of moments doesn't require to solve ill-posed differential equations. As a result, the solution together with software presented in the paper can be helpful in the process of non-destructive evaluation of the mechanical stress distribution in ferromagnetic construction elements.

Keywords: Demagnetization · Non-destructive testing ·
Magnetostatic method of moments

1 Introduction

Testing the magnetic and magnetoelastic properties of ferromagnetic construction elements (e.g. made of steel) is one of most promising area of development of non-destructive methods [1]. In specific group of such methods, magnetoelastic properties of the elements are tested in the Earth's magnetic field [2], what enable contactless and efficient stress estimation. However, analyse of the results of such methods of measurements requires the model of magnetic flux density distribution in the tested construction element. In the case of open magnetic circuit of the element (what is obvious in the case of bars) calculation of reduction of magnetizing field due to demagnetization, is not the obvious task [3]. It should be stressed, that constant value of demagnetization factor at all length of the bar occurs only in the case of ellipsoid-shaped elements [4]. However such elements are rarely used for the technical purposes. Schematic view of different shapes of constructional bars is presented in Fig. 1.

To calculate magnetic field distribution in the bar with given cross section, the finite elements method FEM [5, 6] methods may be used. Such methods are very efficient in the case of e.g. flow modelling [7, 8], however, magnetostatic modelling leads to the necessity of solving the ill-posed system of Maxwell equations [9]. Moreover, in the case of bars with thin walls, the number of tetrahedral elements of the mesh increases rapidly [10]. This is the significant barrier in practical application of finite elements methods in non-destructive evaluation of construction components on the base of magnetoelastic analyses.

R. Szewczyk et al. (Eds.): AUTOMATION 2019, AISC 920, pp. 719–724, 2020.
https://doi.org/10.1007/978-3-030-13273-6_67

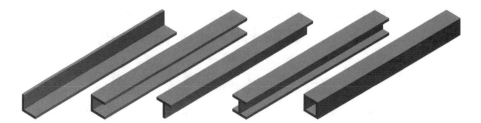

Fig. 1. Schematic view of different shapes of constructional bars available in the market.

To overcome this barrier the generalization of the magnetostatic method of the moments was proposed [11]. On the base of this generalization, the flux density distribution construction bar can be assessed by solving the set of linear equations. Moreover, method of moments uses uniform cuboid meshes, what prevent the uncontrolled increase of the number of elements for thin layer bars.

In the paper the analyses of demagnetization in the H-shaped bar are presented. However, the same method may be used for other shapes of bar's cress sections presented in the Fig. 1. To enable validation and practical application of solutions presented in the paper, the software used for calculation is available on the open-source MIT licence at: www.github.com/romanszewczyk/MoM in the directory **Hbar**.

2   Principles of Magnetostatic Modelling the Different Ferromagnetic Bars with the Method of Moments

Magnetostatic method of moments is edge element type method [12] which can be, in some specific applications [13], the important alternative to finite elements method. The method of moment is based on the assumption, that to each differential part ds_i of the

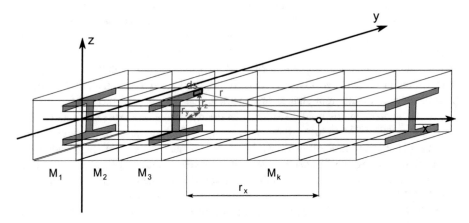

Fig. 2. Schematic view of H-bar meshed with cuboidal elements. The cross-section k of the H-bar exhibits given magnetization M_k.

wall of cuboid mesh, the differential magnetic moment dm_x in the x direction can be assigned accordingly to the following equation [14]:

$$dm_x = \sigma \cdot ds_i = \Delta M \cdot n \cdot \Delta L \cdot ds_i \tag{1}$$

where s is the surface charge density, n is normal unity vector to the cells border, ΔM is the difference between the cells magnetization, and ΔL is the cell's length in the x direction.

Considering the assumption, that cuboidal meshing cell is uniformly magnetized by the magnetization M_k, demagnetization is caused only due to the magnetic moments on the cell's walls. Value of differential demagnetizing field $dH_d(i,k)$, caused by the magnetic moments assigned to the border between i and $i + 1$ elements influencing on the barycentre of k element may be calculated as the magnetic field from the differential moment dm assigned to differential wall part ds [14]:

$$dH_d(i,k) = \frac{(M_i - M_{i+1}) \cdot \Delta L \cdot sign\left(i - k + \frac{1}{2}\right)}{4\pi} \left(\frac{3 \cdot r_x^2 - |r|^2}{|r|^5}\right) \cdot ds_i \tag{2}$$

where r is the vector connecting the barycentre of k element and differential element ds, as it is presented in the Fig. 2.

Total demagnetizing field $H_d(i,k)$ is the result of integration over the all cross section of the bar:

$$H_d(i,k) = \frac{(M_i - M_{i+1}) \cdot \Delta L \cdot sign\left(i - k + \frac{1}{2}\right)}{4\pi} \int_{-y_{min}}^{y_{max}} \int_{-z_{min}}^{z_{max}} \left(\frac{3 \cdot r_x^2 - |r|^2}{|r|^5}\right) \cdot p(y,z) \cdot dr_y dr_z \tag{3}$$

To simplify the calculation, function $p(y, z)$ is equal 1 for the presence of material ad 0 in the other case.

Finally, the set of i linear equations may be stated, describing the magnetization in the bar's elements and enabling assessment of the magnetization M_k of each cross-section [14]:

$$M_k + M_1 \cdot \frac{(\mu - 1) \cdot \Delta L}{4\pi} \cdot (g(1,k) - g(0,k)) + \ldots + M_n \cdot \frac{(\mu - 1) \cdot \Delta L}{4\pi} \cdot (g(n,k) - g(n-1,k))$$
$$= (\mu - 1) \cdot H_{ext} \tag{4}$$

where μ is the relative permeability of material and H_{ext} is the external magnetizing field in the x direction.

3 Magnetization Distribution in H-Bar

Schematic diagram of H-bar is presented in the Fig. 3. From the point of view of magnetostatic simulations, the H-bar is determined by its length L, cross section a and wall thickness k.

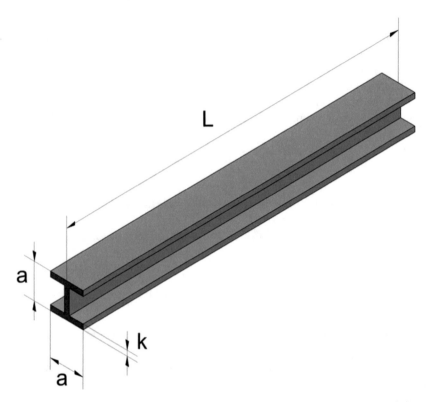

Fig. 3. Schematic view of H-bar meshed with cuboidal elements. The cross-section k of the H-bar exhibits given magnetization M_k.

Modelling was carried out for the assumptions, that the relative permeability of H-bar material μ is equal 1000, whereas external magnetizing field H_{ext} (acting in the x direction) is equal 100 A/m. The length L of H-bar is equal 2 m, whereas it's cross section a is equal 20 cm.

The modelling of flux density B distribution in the H-bar was performed for different values of wall thickness k. The results of modelling are presented in Fig. 4.

Presented results clearly indicate, that demagnetization significantly reduces the flux density B distribution in the H-bar. Moreover, maximal value of flux density B depends on H-bar wall thickness k.

The results of modelling confirm, that demagnetization of H-bars on have to be considered in analyses of magnetoelastic effects in non-destructive testing and

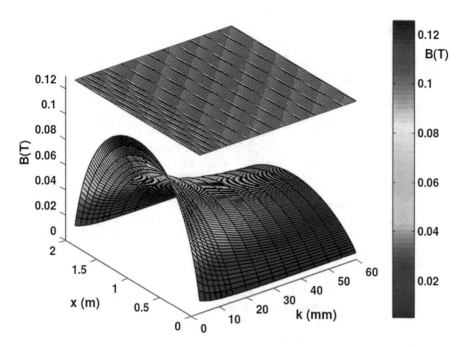

Fig. 4. The results modelling of flux density B distribution in the x axis of the H-bar. Modelling was carried out for different values of wall's thickness k. Red surface at the top indicates the flux density level without demagnetization ($\mu\mu_0 H_{ext}$).

magnetoelastic stress estimation. Moreover, it is not justified to develop specialized tables and nomograms to calculate flux density distributions for different shapes of cross sections of construction elements. More efficient solution is to use presented software for calculation of flux density B distribution in specific case required for non-destructive testing.

4 Conclusions

Presented results confirm that the use of method of moments can be efficient way to model magnetostatic characteristics of constructional bars with different cross sections. With the use of presented open-source software, the flux density B distribution can be estimated at the all length of construction element. Such approach is very useful for non-destructive testing techniques based on Earth's magnetic field magnetization and observation of magnetoelastic phenomena.

Proposed solution based on the method of moments allows magnetostatic calculations on the base of solving the set of linear equations. Such solution is much less computer resources consuming, than finite elements method – which requires solving sophisticated sets of ill-posed differential equations. Moreover, in the case of method of moments, there is no radical increase of the number of elements in the case of thin walled bars, what may happen in the case of tetrahedral meshing for the finite elements method.

References

1. Zhang, H., Liao, L., Zhao, R., Zhou, J., Yang, M., Xia, R.: The non-destructive test of steel corrosion in reinforced concrete bridges using a micro-magnetic sensor. Sensors **16**, 1439 (2016). https://doi.org/10.3390/s16091439
2. Gontarz, S., Radkowski, S.: Impact of various factors on relationships between stress and eigen magnetic field in a steel specimen. IEEE Trans. Magn. **48**, 1143–1154 (2012). https://doi.org/10.1109/TMAG.2011.2170845
3. Pardo, E., Chen, D.-X., Sanchez, A.: Demagnetizing factors for square bars. IEEE Trans. Magn. **40**, 1491–1498 (2004). https://doi.org/10.1109/tmag.2004.827186
4. Chen, D., Pardo, E., Sanchez, A.: Demagnetizing factors for rectangular prisms. IEEE Trans. Magn. **41**, 2077–2088 (2005). https://doi.org/10.1109/TMAG.2005.847634
5. Jin, J.-M.: The Finite Element Method in Electromagnetic. Wiley, Hoboken (1993)
6. Zlámal, M.: On the finite element method. Numer. Math. **12**, 394–409 (1968)
7. Turkowski, M., Szufleński, P.: New criteria for the experimental validation of CFD simulations. Flow Meas. Instrum. **34**, 1–10 (2013)
8. Turkowski, M.: Modeling of two-phase gas-liquid flow in laboratory conditions. Mach. Dyn. Probl. **28**, 159–164 (2004)
9. Logg, A., Mardal, K.A., Wells, G.: Automated Solution of Differential Equations by the Finite Element Method. Springer, Heidelberg (2012)
10. Szewczyk, R.: Generalization of magnetostatic method of moments for thin layers with regular rectangular grids. Acta Phys. Pol. A **131**, 845 (2017)
11. Szewczyk, R.: The method of moments in Jiles-Atherton model based magnetostatic modelling of thin layers. Arch. Electr. Eng. **6**, 27–35 (2018). https://doi.org/10.24425/118989
12. Harrington, R.F.: Field Computation by Moment Methods. Wiley, Hoboken (1968)
13. Chadebec, O., Coulomb, J.L., Bongiraud, J.P., Cauffet, G., Le Thiec, P.: Recent improvements for solving inverse magnetostatic problem applied to thin shells. IEEE Trans. Magn. **38**, 1005–1008 (2002)
14. Szewczyk, R.: Magnetostatic Modelling of Thin Layers Using the Method of Moments and Its Implementation in OCTAVE/MATLAB. Lecture Notes in Electrical Engineering, vol. 491. Springer, Heidelberg (2018). https://doi.org/10.1007/978-3-319-77985-0

Author Index

Printed in the United States
By Bookmasters